CHEMICAL RESISTANCE GUIDE FOR ELASTOMERS

Copyright © 1988 and 1994

Manufactured in the United States of America

CHEMICAL RESISTANCE GUIDE FOR ELASTOMERS II
Copyright 1988 and 1994

This guide contains recommendations for 48800 combinations of elastomeric materials in contact with various corrodents and other environmental conditions. All results are believed to be based on valid laboratory or field tests or experience; however, no guarantee is expressed or implied as to results that will be obtained by the user. It is recommended that the user test the anticipated combinations in his own laboratory before committing his product to any libelous application. The publisher is not responsible for any damage that might result from the use of this guide.

This guide is not to be used as an inducement to violate any patent, or any federal, state, local, dietary or insurance laws.

Written and researched by Kenneth M. Pruett
Published by Compass Publications
P.O. Box 1275
La Jolla, California USA 92038-1275
TEL (619) 551-9240   FAX (619) 551-9340

ISBN 1-889712-02-7

INTERPRETATION OF DATA

Symbolic Rating	VOLUMETRIC SWELLING	LOSS OF TENSILE STRENGTH	DESCRIPTION OF ATTACK
A	≤ 15% in 30 days to 1 year	≤ 15% in 30 days to 1 year	Excellent, little or no swelling or softening or surface deterioration.
B	≤ 30% in 30 days to 1 year	≤ 30% in 30 days to 1 year	Good chemical resistance. Minor chemical attack, swelling, softening or surface deterioration.
C	≤ 50% in 30 days to 1 year	≤ 60% in 30 days to 1 year	Limited chemical resistance. Moderate chemical attack. Conditional service.
NR	≥ 50% immediately to 1 year	> 60% immediately to 1 year	Severe attack, swelling, softening, or dissolved within minutes to years. Not recommended.
Q			Questionable resistance. Doubtful usage.
TEST			Test before using.

EXAMPLES OF CALLOUTS

A to 70°F	≤ 15% volume increase or ≤ 15% loss of tensile strength at 100% concentration or concentrated or saturated solution, to 70°F. Little or no chemical attack.
A to 100% to 70°F	≤ 15% volume increase or ≤ 15% loss of tensile strength in any concentration from 0 to 100% at 70°F. Little or no chemical attack.
A to 20% to 70°F	≤ 15% volume increase or ≤ 15% loss of tensile strength in any concentration from 0 to 20% at 70°F. Little or no chemical attack.
B 20-50% 70-140°F	≤ 15% to 30% increase or 15 to 30% loss of tensile strength, from 20% to 50% concentration of solution, between 70°F and 140°f.
B to conc to 212°F	≤ 15% to 30% increase or 15 to 30% loss of tensile strength, in any concentration to 100% at any temperature from 70°F to 212°F.
AB at 70°F	0% to 30% volume increase, or 0% to 30% loss of tensile strength at 100% concentration or concentrated, or saturated solution, at 70°F. Reported data varies widely possibly due to compounding diffrences within the same generic family.
NR at 70°F	> 50% volume increase or > 60% loss of tensile strength at 100% concentration, or concentrated, or saturated solution at 70°F. Severe attack. Not recommended.
A/NR at 70°F	≤ 15% to > 50% volume increase or ≤ 15% to > 60% loss of tensile strength at 100% concentration, or concentrated, or saturated solution at 70°F. Varying or variable rates reported by multiple sources. Use only after specific testing of all parameters to be encountered.

EXAMPLES OF SPECIFIC DATA CALLOUTS

+55% vol 3 days 70°F	55% volume increase during 3 day period at 100% concentration, at 70°F.
+1-5% vol 3 days 70°F	+1% to +5% volume increase during 3 day period at 100% concentration at 70°F.
-1 to +5% vol 3 days 70°F	1% decrease to +5% increase in volume during 3 day period at 100% concentration at 70°F.
+1.8% vol 28% conc 7 days 70°F	+1.8% volume increase at 28% concentration during 7 day period at 70°F.
-2 to -4% vol 3 days 70°F	2% to 4% decrease in volume during 4 day period at 100% concentration at 70°F.
-41% T.S. 50% conc 7 days 77°F	41% loss of tensile strength in 50% concentration solution during 7 day period at 77°F.

Maximum use temperatures listed are not necessarily upper limits of usability, but are limits of data available at time of printing.

Entries marked with an asterisk (*) are not specifically tested. They are either manufacturer's estimates based on performance of a particular elastomers compatibility with similar classes of chemicals, or they are performances available for some but not all compounds of a particular elastomer.

Note that when boiling temperature is indicated, the boiling temperature varies with the concentration of the media.

	RECOMMENDATIONS	NITRILE NBR	ETHYLENE PROPYLENE EP, EPDM	FLUOROCARBON FKM	CHLOROPRENE CR	HYDROGENATED NITRILE HNBR
ABSOLUT ALCOHOL		A to 70°F	A to 70°F	B at 70°F	A to 70°F	
ACETALDEHYDE	EP,SBR,VMQ	NR any conc at 70°F	A to 104°F AB to 200°F	NR 100% at 70°F B 40% to 104°F	NR any conc at 70°F	
ACETALDEHYDE				C 40% at 140°F NR 40% at 176°F		
ACETAMIDE (ACRYIMIDE)	NBR,EP,IIR,CR	A to 70°F	A to 140°F	AB to 140°F	AB to 70°F	A to 70°F
ACETATE OF LIME		B at 70°F	A to 70°F	NR at 70°F	B at 70°F	
ACETATE SOLVENTS	EP,IIR	NR at 70°F crude or pure	A to 200°F	NR at 70°F crude or pure	NR at 70°F crude B/NR at 70°F pure	
ACETIC ACID, GLACIAL	EP,IIR	+55% vol 3 days 70°F NR glacial at 70°F	+16% vol 3 days 158°F AB glacial to 200°F	+112% vol 3 days 70°F NR at 70°F	+16% vol 3 days 158°F	B at 70°F
ACETIC ACID	EP,IIR 100% conc	NR 100% at 70°F C 80% at 70°F	NR clacial boiling A to 100% to 70°F	NR 90-100% at 70°F B/NR 80% at 70°F	B 100% to 200°F BC 50% at 70°F	
ACETIC ACID	EP,IIR 30% conc	B to 30% at 70°F B to 20% to 185°F		B 50% to 140°F C 50% at 176°F	A to 30% to 70°F AB 10% at 100°F	
ACETIC ACID	EP,FKM,CR,IIR,5% conc			NR 50% at 212°F B/NR 30% at 70°F	BC 10% at 150°F	
ACETIC ACID				B 10-25% to 100°F C 25% at 140°F		
ACETIC ACID				C 20% at 176°F NR 20% at 212°F		
ACETIC ACID				B 10% at 176°F A to 10% to 70°F		
ACETIC ACID VAPORS	EP,IIR	BC to 150°F	A to 70°F	NR at 70°F	C/NR 70°F to hot	
ACETIC ALDEHYDE	EP,SBR,VMQ	NR any conc at 70°F	A to 104°F AB to 200°F	NR at 70°F	NR any conc at 70°F	
ACETIC ANHYDRIDE	CR,EP,IIR	B/NR at 70°F	B to 200°F	NR 100% at 70°F B 50% to 70°F	AB 100% to 70°F NR 100% at 140°F	NR at 70°F
ACETIC ANHYDRIDE				NR 50% at 100°F	B 50% at 70°F C 50% at 100°F	
ACETIC ANHYDRIDE					NR 50% at 140°F	
ACETIC ESTER	EP,IIR,T	NR any conc at 70°F	A to 130°F BC at 158°F	NR at 70°F	NR any conc at 70°F	NR at 70°F
ACETIC ETHER		NR any conc at 70°F	B at 70°F	NR at 70°F	NR any conc at 70°F	
ACETIC OXIDE		C 100% at 70°F NR 25-50% at 70°F	B at 70°F	NR at 70°F	B 100% at 70°F NR 25-50% at 70°F	
ACETIDIN		NR at 70°F	B at 70°F	NR at 70°F	NR at 70°F	
ACETOACETIC ACID		NR at 70°F			NR at 70°F	
ACETOACETIC ESTER		NR any conc at 70°F	B at 70°F		NR any conc at 70°F	
ACETONE	EP,IIR,NR	+125% vol 3 days 70°F NR any conc at 70°F	+2% vol 3 days 70°F A any conc to 200°F	+200% vol 7 days 70°F NR 100% at 70°F	+31% vol 3 days 70°F C/NR 100% at 70°F	NR at 70°F
ACETONE				C 10% to 104°F NR 10% at 140°F	NR 100% at 140°F C 10% to 104°F	
ACETONE CYANOHYDRIN	EP	NR at 70°F	A to 200°F	NR at 70°F	B at 70°F	
ACETONIC		NR at 70°F	A to 70°F	NR at 70°F	B at 70°F	
ACETONITRILE	CR,NR	C at 70°F	A to 70°F	C/NR at 70°F	A to 70°F	
ACETOPHENONE	EP,IIR	NR any conc at 70°F	A to 140°F B at 176°F	C/NR at 70°F NR at 104°F	NR any conc at 70°F	
ACETYL ACETONE	EP,IIR	NR at 70°F	A to 70°F	NR at 70°F	NR at 70°F	NR at 70°F
ACETYL ACETONIC		NR at 70°F	A to 70°F	NR at 70°F	NR at 70°F	
ACETYL BENZENE	EP,IIR	NR any conc at 70°F	A to 140°F B at 176°F	C/NR at 70°F NR at 104°F	NR any conc at 70°F	
ACETYL BROMIDE		NR at 70°F	NR at 70°F		NR at 70°F	
ACETYL CHLORIDE	FKM,FVMQ	NR at 70°F	C at 70°F NR at 70°F dry	B at 70°F A to 70°F dry	NR at 70°F	NR at 70°F
ACETYL OXIDE		C 100% at 70°F NR 25-50% at 70°F	B at 70°F	NR at 70°F	B 100% at 70°F NR 25-50% at 70°F	
ACETYL PROPANE		NR at 70°F	B at 70°F	NR at 70°F	NR at 70°F	
ACETYLACETIC ACID		NR at 70°F			NR at 70°F	
ACETYLENE GAS	EP,NBR,IIR,NR	A to 200°F	A to 200°F AB to 250°F	A to 200°F B at 212°F	A to 104°F B to 200°F	
ACETYLENE GAS + H₂O	NBR	A to 200°F	A to 200°F	A to 200°F	B to 200°F	
ACETYLENE DICHLORIDE	FKM	NR at 70°F	B/NR at 70°F	AB to 160°F	NR at 70°F	
ACETYLENE TETRABROMIDE	FKM,EP,IIR	NR at 70°F	A to 200°F	A to 200°F	B to 200°F	NR at 70°F
ACETYLENE TETRACHLORIDE	FKM,FVMQ	NR at 70°F	NR at 70°F	A to 200°F	NR at 70°F	NR at 70°F
ACROLEIN	EP	BC at 70°F	A to 120°F	A/NR at 70°F	C at 70°F	

	STYRENE BUTADIENE SBR	POLYACRYLATE ACM	POLYURETHANE AU, EU	ISOBUTYLENE ISOPRENE IIR	POLYBUTADIENE BR	® AEROQUIP AQP
ABSOLUT ALCOHOL			NR at 70°F	A to 70°F		
ACETALDEHYDE	NR at 70°F	NR at 70°F	C /NR at 70°F	A to 70°F	B at 70°F	
ACETALDEHYDE						
ACETAMIDE (ACRYIMIDE)	NR at 70°F	NR at 70°F	NR at 70°F	AB to 70°F	NR at 70°F	
ACETATE OF LIME				A to 70°F		
ACETATE SOLVENTS	NR at 70°F	NR at 70°F	NR at 70°F	AB to 70°F crude or pure		NR at 70°F crude or pure
ACETIC ACID, GLACIAL	+19% vol 3 days 158°F BC at 70°F	+154% vol 3 days 158°F NR at 70°F	+172% vol 3 days 158°F NR at 70°F	+0.5% vol 3 days 70°F B to 125°F	B at 70°F NR hot & high pressure	NR at 70°F
ACETIC ACID	NR hot & high pressure B 5-100% at 70°F	NR hot & high pressure NR 5-100% at 70°F	NR hot & high pressure NR 30-60% at 70°F	NR hot & high pressure AB to 50% to 80°F	B to 100% at 70°F	NR 30-100% at 70°F
ACETIC ACID			B /NR 5-20% at 70°F BC dilute at 70°F	AB to 25% to 150°F A 10% to 70°F		
ACETIC ACID						
ACETIC ACID						
ACETIC ACID						
ACETIC ACID						
ACETIC ACID VAPORS	C at 70°F	NR at 70°F	NR at 70°F	AB to 70°F NR hot & high pressure		NR at 70°F
ACETIC ALDEHYDE	NR at 70°F	NR at 70°F	C /NR at 70°F	A any conc to 80°F	B at 70°F	
ACETIC ANHYDRIDE	C /NR at 70°F	NR at 70°F	NR at 70°F	B 100% at 70°F AB 50% to 80°F		
ACETIC ANHYDRIDE				AB 25% to 150°F		
ACETIC ANHYDRIDE						
ACETIC ESTER	NR at 70°F	NR at 70°F	NR at 70°F	B any conc to 100°F	NR at 70°F	AB to 70°F
ACETIC ETHER			NR at 70°F	B any conc to 100°F		
ACETIC OXIDE			B at 70°F	B 100% at 70°F AB to 50% to 80°F		
ACETIDIN			NR at 70°F	B at 70°F		AB to 70°F
ACETOACETIC ACID				B at 70°F		
ACETOACETIC ESTER				AB any conc to 80°F		
ACETONE	+18% vol 3 days 70°F BC at 70°F	+201% vol 3 days 70°F NR at 70°F	+87% vol 3 days 70°F NR at 70°F	-2% vol 3 days 70°F A to 70°F	NR at 70°F	AB to 70°F
ACETONE			AB fumes to 70°F	AB any conc to 150°F		
ACETONE CYANOHYDRIN			NR at 70°F	A to 70°F		
ACETONIC			NR at 70°F	A to 70°F		
ACETONITRILE				A to 70°F		
ACETOPHENONE	NR at 70°F	NR at 70°F	NR at 70°F	AB any conc to 80°F	NR at 70°F	
ACETYL ACETONE	NR at 70°F	NR at 70°F	NR at 70°F	A to 70°F	NR at 70°F	
ACETYL ACETONIC			B at 70°F	B at 70°F		
ACETYL BENZENE			NR at 70°F	AB any conc to 80°F		
ACETYL BROMIDE			NR at 70°F			
ACETYL CHLORIDE	NR at 70°F	NR at 70°F	NR at 70°F	NR at 70°F	NR at 70°F	
ACETYL OXIDE				A 100% to 80°F AB 25% to 150°F		
ACETYL PROPANE				B at 70°F		
ACETYLACETIC ACID				B at 70°F		
ACETYLENE GAS	BC at 70°F		AB to 70°F	A any conc to 80°F	B at 70°F	AB to 70°F
ACETYLENE GAS + H₂O	C at 70°F		A to 70°F			
ACETYLENE DICHLORIDE				C at 70°F		
ACETYLENE TETRABROMIDE	NR at 70°F		NR at 70°F	A /NR at 70°F		
ACETYLENE TETRACHLORIDE				NR at 70°F		
ACROLEIN	C at 70°F	NR at 70°F	NR at 70°F	A to 70°F		

C 3

	SYNTHETIC ISOPRENE IR	NATURAL ISOPRENE NR	CHLOROSULFONATED POLYETHYLENE CSM	FLUOROSILICONE FVMQ	SILICONE VMQ	CHEMRAZ FFKM
ABSOLUT ALCOHOL	A to 70°F	A to 70°F	A to 70°F		A to 70°F	
ACETALDEHYDE	BC at 70°F	B/NR at 70°F	C/NR at 70°F	NR at 70°F	AB to 70°F	B/NR at 70°F
ACETALDEHYDE						
ACETAMIDE (ACRYIMIDE)	C/NR at 70°F	C/NR at 70°F	B at 70°F	A to 70°F	BC at 70°F	A to 70°F
ACETATE OF LIME	B at 70°F	B at 70°F	A to 70°F		A to 70°F	
ACETATE SOLVENTS	C/NR at 70°F crude or pure	C at 70°F	NR at 70°F	NR at 70°F	B/NR at 70°F	
ACETIC ACID, GLACIAL	BC at 70°F NR hot & high pressure	B/NR at 70°F NR hot & high pressure	C at 70°F C hot & high pressure	+12% vol 3 days 158°F NR at 70°F	+3% vol 7 days 70°F +3% vol 3 days 158°F	AB to 70°F
ACETIC ACID	C 50% at 70°F BC 30% at 70°F	B 80% at 70°F C/NR 50% at 70°F	A to 80% to 70°F AB to 20% to 200°F	NR hot & high pressure B to 100% at 70°F	B at 70°F NR hot & high pressure	A to 30% to 70°F
ACETIC ACID	B to 20% at 70°F	B 5-30% at 70°F B to 10% to 150°F				
ACETIC ACID						
ACETIC ACID						
ACETIC ACID						
ACETIC ACID						
ACETIC ACID VAPORS	NR at 70°F	C/NR at 70°F		C at 70°F	B at 70°F crude or pure	
ACETIC ALDEHYDE	BC at 70°F	B/NR at 70°F	C/NR at 70°F	NR at 70°F	AB to 70°F	B/NR at 70°F
ACETIC ANHYDRIDE	BC at 70°F	BC 100% at 70°F NR 25-50% at 70°F	A to 70°F AB to 200°F	NR at 70°F	BC at 70°F	AB to 70°F
ACETIC ANHYDRIDE						
ACETIC ANHYDRIDE						
ACETIC ESTER	NR at 70°F	NR any conc at 70°F	NR at 70°F	NR at 70°F	B at 70°F	A to 70°F
ACETIC ETHER	NR at 70°F	NR any conc at 70°F	NR at 70°F		B at 70°F	
ACETIC OXIDE	C at 70°F	C 100% at 70°F NR 25-50% at 70°F	A to 70°F		C at 70°F	
ACETIDIN	NR at 70°F	NR at 70°F	NR at 70°F		B at 70°F	
ACETOACETIC ACID	B at 70°F	B at 70°F				A to 70°F
ACETOACETIC ESTER	B at 70°F	B any conc at 70°F	NR at 70°F			
ACETONE	NR at 70°F	B/NR at 70°F	NR at 70°F	+205% vol 3 days 70°F NR at 70°F	+18% vol 7 days 70°F BC at 70°F	A to 70°F
ACETONE						
ACETONE CYANOHYDRIN	C at 70°F	C at 70°F	C at 70°F		A to 70°F	A to 70°F
ACETONIC	B at 70°F	B at 70°F	B at 70°F		A to 70°F	
ACETONITRILE	B at 70°F	B at 70°F	B at 70°F		TEST	A to 70°F
ACETOPHENONE	C/NR at 70°F	C/NR any conc at 70°F	NR at 70°F	NR at 70°F	NR at 70°F	A to 70°F
ACETYL ACETONE	NR at 70°F	NR at 70°F	NR at 70°F	NR at 70°F	NR at 70°F	A to 70°F
ACETYL ACETONIC	C at 70°F	C at 70°F	NR at 70°F			
ACETYL BENZENE	C/NR at 70°F	C/NR any conc at 70°F	NR at 70°F	NR at 70°F	NR at 70°F	A to 70°F
ACETYL BROMIDE						
ACETYL CHLORIDE	NR at 70°F	NR at 70°F	B/NR at 70°F	A to 70°F	C at 70°F	A to 70°F
ACETYL OXIDE	C at 70°F	C 100% at 70°F NR 25-50% at 70°F	NR at 70°F			
ACETYL PROPANE	NR at 70°F	NR at 70°F	NR at 70°F			
ACETYLACETIC ACID						
ACETYLENE GAS	B at 70°F	AB any conc at 70°F	BC at 70°F	BC at 70°F	BC at 70°F	A to 70°F liquid or gas
ACETYLENE GAS + H₂O					C at 70°F	
ACETYLENE DICHLORIDE	NR at 70°F	NR at 70°F	NR at 70°F	NR at 70°F	C/NR at 70°F	A to 70°F
ACETYLENE TETRABROMIDE	NR at 70°F	NR at 70°F				A to 70°F
ACETYLENE TETRACHLORIDE	NR at 70°F	NR at 70°F	NR at 70°F	B at 70°F	C/NR at 70°F	A to 70°F
ACROLEIN	B at 70°F	B at 70°F	B at 70°F			

	POLYSULFIDE T	CHLORINATED POLYETHYLENE CM	EPICHLOROHYDRIN CO, ECO	® HYTREL COPOLYESTER TPE	® SANTOPRENE COPOLYMER TPO	® C-FLEX STYRENIC TPE
ABSOLUT ALCOHOL						
ACETALDEHYDE	BC at 70°F			AB to 70°F		AC at 70°F
ACETALDEHYDE						
ACETAMIDE (ACRYIMIDE)	NR at 70°F					
ACETATE OF LIME						
ACETATE SOLVENTS	TEST	NR at 70°F crude or pure		NR at 70°F crude or pure		AC at 70°F
ACETIC ACID, GLACIAL	+22% vol 3 DAYS 70°F BC at 70°F	A to 70°F A hot & high press.	NR at 70°F	+16% to 39% vol 100°F AC at 70°F	NR at 70°F	
ACETIC ACID	NR hot & high press.	A to 100% to 70°F	B 30% at 70°F	C/NR at 100°F	C 50% at 70°F A to 30% to 70°F	
ACETIC ACID	B 5-30% at 70°F					
ACETIC ACID						
ACETIC ACID						
ACETIC ACID						
ACETIC ACID VAPORS				C/NR at 70°F		
ACETIC ALDEHYDE	BC at 70°F			AB to 70°F		AC at 70°F
ACETIC ANHYDRIDE	BC at 70°F	A to 70°F	NR at 70°F	C at 70°F	NR at 70°F	TEST
ACETIC ANHYDRIDE						
ACETIC ANHYDRIDE						
ACETIC ESTER	B at 70°F	AB to 70°F	NR at 70°F	BC at 70°F	*A to 70°F	A to 70°F
ACETIC ETHER						
ACETIC OXIDE						
ACETIDIN						
ACETOACETIC ACID						
ACETOACETIC ESTER						
ACETONE	+7.3% vol 3 days 70°F AC at 70°F	AB to 70°F	NR at 70°F	BC at 70°F	-11% vol 7 days 73°F A to 70°F	AB at 70°F
ACETONE						
ACETONE CYANOHYDRIN						
ACETONIC						
ACETONITRILE					*A to 70°F	AC to 70°F
ACETOPHENONE	NR at 70°F		NR at 70°F			
ACETYL ACETONE	B at 70°F	B at 70°F	NR at 70°F			
ACETYL ACETONIC						
ACETYL BENZENE						
ACETYL BROMIDE				NR at 70°F		AC to 70°F
ACETYL CHLORIDE	NR at 70°F	A to 70°F	NR at 70°F	NR at 70°F	*A to 70°F	AC at 70°F
ACETYL OXIDE						
ACETYL PROPANE						
ACETYLACETIC ACID						
ACETYLENE GAS	C at 70°F	AB to 70°F	B at 70°F *B at 140°F	AB to 70°F		
ACETYLENE GAS + H_2O						
ACETYLENE DICHLORIDE						
ACETYLENE TETRABROMIDE						
ACETYLENE TETRACHLORIDE						
ACROLEIN						

C 5

	ETHYLENE ACRYLIC EA	POLYALLOMER LINEAR COPOLYMER	NYLON 11 POLYAMIDE	NYLON 12 POLYAMIDE	ETHYLENE VINYL ACETATE EVA	POLYVINYLCHLORIDE FLEXIBLE PVC
ABSOLUT ALCOHOL						
ACETALDEHYDE	BC at 70°F	AB to 70°F NR at 122°F	AB to 70°F C at 105°F	AB 40% to 70°F	B/NR at 70°F NR at 140°F	NR any conc at 70°F
ACETALDEHYDE			NR at 140°F			
ACETAMIDE (ACRYIMIDE)	A to 122°F			AB 50% to 70°F		NR at 70°F
ACETATE OF LIME						
ACETATE SOLVENTS		A to 70°F crude or pure	A to 70°F crude or pure		A to 70°F crude or pure	NR at 70°F crude or pure
ACETIC ACID, GLACIAL	NR at 70°F	A to 70°F	C/NR at 70°F NR at 105°F	NR at 70°F	B/NR to 140°F	C/NR at 70°F NR at 140°F
ACETIC ACID	A to 30% to 70°F	A to 100% to 70°F A to 50% to 122°F	NR 98% at 70°F AC 50% to 70°F	NR 98% at 70°F BC 10-50% at 70°F	B/NR 50% at 70°F B 30% at 70°F	C/NR 85% at 70°F NR 85% at 140°F
ACETIC ACID			NR 50% at 140°F AB 20% at 70°F		AB 20% to 70°F	AB 50% to 70°F B 50% at 122°F
ACETIC ACID			A 10% to 70°F C/NR 10% at 150°F			NR 50% at 140°F B 10% at 70°F
ACETIC ACID						NR 10% at 140°F A to 5% at 122°F
ACETIC ACID						+0.1% vol 5% conc 14 days 70°F
ACETIC ACID VAPORS						AC to 70°F crude or pure
ACETIC ALDEHYDE	BC at 70°F	AB to 70°F NR at 122°F	AB to 70°F C at 105°F		B/NR at 70°F NR at 140°F	NR any conc at 70°F
ACETIC ANHYDRIDE			C/NR at 70°F NR at 105°F	C at 70°F		NR any conc at 70°F
ACETIC ANHYDRIDE						
ACETIC ANHYDRIDE						
ACETIC ESTER	NR at 70°F	A to 122°F	A to 70°F AB to 140°F		AC to 70°F NR at 140°F	NR any conc at 70°F
ACETIC ETHER						NR any conc at 70°F
ACETIC OXIDE						NR 25-50% at 70°F
ACETIDIN						
ACETOACETIC ACID						
ACETOACETIC ESTER						NR any conc at 70°F
ACETONE	NR at 70°F	A to 122°F	A to 70°F AB to 105°F (yellows)	AB to 70°F	C/NR at 70°F	NR any conc at 70°F
ACETONE			C/NR at 140°F NR at 194°F			
ACETONE CYANOHYDRIN						
ACETONIC						
ACETONITRILE		C at 70°F NR at 122°F				NR at 70°F
ACETOPHENONE		A to 70°F			BC at 70°F	NR any conc at 70°F
ACETYL ACETONE	NR at 70°F					
ACETYL ACETONIC						
ACETYL BENZENE						NR any conc at 70°F
ACETYL BROMIDE			NR at 70°F			NR any conc at 70°F
ACETYL CHLORIDE			NR at 70°F			NR at 70°F
ACETYL OXIDE						NR 25-50% at 70°F
ACETYL PROPANE						
ACETYLACETIC ACID						
ACETYLENE GAS		A to 70°F	A to 70°F AB to 140°F		A/NR at 70°F	NR at 70°F
ACETYLENE GAS + H₂O						NR at 70°F
ACETYLENE DICHLORIDE						
ACETYLENE TETRABROMIDE						
ACETYLENE TETRACHLORIDE						NR at 70°F
ACROLEIN						

	POLYETHYLENE LOW DENSITY LDPE	®TEFLON FEP	®KALREZ PERFLUORINATED ELASTOMER	®FLUORAZ FLUORINATED COPOLYMER	®AFLAS FLUORINATED COPOLYMER	®NORPRENE COPOLYMER TPO
ABSOLUT ALCOHOL		A to 70°F				
ACETALDEHYDE	C/NR 40-100% at 70°F NR 40-100% at 122°F	A to 100% to 200°F	A to 70°F *A to 212°F	C/NR at 70°F	+18% vol 7 days 158°F	A to 70°F
ACETALDEHYDE						
ACETAMIDE (ACRYIMIDE)	A to 122°F	A to 300°F	A to 70°F *A to 400°F	B at 70°F	A to 70°F	
ACETATE OF LIME		A to 70°F				
ACETATE SOLVENTS	AB to 70°F crude or pure	A to 300°F crude or pure		NR at 70°F		BC at 70°F
ACETIC ACID, GLACIAL	AB to 122°F NR at 140°F	A to boiling	A to 75°F B at 203°F	C/NR hot or cold	+71% vol 7 days 73°F +25% vol 7 days 73°F	A to 70°F
ACETIC ACID	AB 60% to 140°F A to 50% to 122°F	A to 100% boiling	BC glacial at 428°F A to 70% at 428°F	C/NR 100% to hot B 30% at 70°F	C/NR at 70°F A 34% to 203°F	A to 100% to 70°F
ACETIC ACID				NR 15% hot A 5% to 70°F		
ACETIC ACID						
ACETIC ACID						
ACETIC ACID						
ACETIC ACID						
ACETIC ACID VAPORS		A to 200°F				
ACETIC ALDEHYDE	C/NR 40-100% at 70°F NR 40-100% at 122°F	A to 100% to 200°F	A to 70°F *A to 212°F	C/NR at 70°F	+18% vol 7 days 158°F A to 70°F	A to 70°F
ACETIC ANHYDRIDE	A to 70°F NR at 140°F	A to 300°F	+2% vol 7 days 70°F A to 113°F	BC at 70°F	B at 70°F	A to 70°F
ACETIC ANHYDRIDE			*A to 212°F			
ACETIC ANHYDRIDE						
ACETIC ESTER	AB to 122°F NR at 140°F	A to 310°F	A to 212°F	NR at 70°F	NR at 70°F	
ACETIC ETHER		A to 70°F				
ACETIC OXIDE		A to 70°F				
ACETIDIN		A to 70°F				BC at 70°F
ACETOACETIC ACID		A to 70°F	*A to 212°F			
ACETOACETIC ESTER		A to 70°F				
ACETONE	C/NR at 70°F NR at 122°F	A to 200°F	+2% vol 7 days 70°F A to 113°F	NR at 70°F	+50% vol 7 days 70°F NR at 70°F	BC at 70°F
ACETONE			*A to 212°F			
ACETONE CYANOHYDRIN		A to 70°F	A to 212°F			
ACETONIC		A to 70°F				
ACETONITRILE		A to 200°F	A to 212°F	A to 70°F		
ACETOPHENONE	B at 70°F	+0.7% wt gain 394°F A to boiling	A to 212°F	NR at 70°F		
ACETYL ACETONE		A to 70°F	A to 212°F	NR at 70°F	+54% vol 3 days 73°F NR at 70°F	
ACETYL ACETONIC		A to 70°F				
ACETYL BENZENE	B at 70°F	A to boiling	A to 212°F	NR at 70°F		
ACETYL BROMIDE	NR at 70°F	A to 70°F	A to 212°F			
ACETYL CHLORIDE	NR at 70°F	A to 200°F	A to 212°F	A to 70°F		
ACETYL OXIDE		A to 70°F				
ACETYL PROPANE		A to 70°F				
ACETYLACETIC ACID		A to 70°F				
ACETYLENE GAS	AB to 70°F	A to 200°F	A to 212°F	A to 70°F		
ACETYLENE GAS + H₂O		A to 70°F				
ACETYLENE DICHLORIDE	NR at 70°F	A to 200°F	A to 212°F	A to 70°F		
ACETYLENE TETRABROMIDE		A to 70°F	A to 212°F	A to 70°F		
ACETYLENE TETRACHLORIDE	B at 70°F	A to 125°F	A to 212°F	NR at 70°F		
ACROLEIN		A to 70°F	A to 212°F			

C7

	RECOMMENDATIONS	NITRILE NBR	ETHYLENE PROPYLENE EP, EPDM	FLUOROCARBON FKM	CHLOROPRENE CR	HYDROGENATED NITRILE HNBR
ACRYLIC ALDEHYDE		B at 70°F		A to 70°F		
ACRYLONITRILE	NR,CR	NR any conc at 70°F	NR at 70°F	+115% vol 7 days 122°F C/NR at 70°F	C/NR any conc at 70°F	NR at 70°F
ADIPIC ACID	NBR	A any conc to 70°F	A to 140°F B at 212°F	A to 176°F B at 212°F	A to 140°F	A to 70°F
ADIPIC KETONE		NR at 70°F		NR at 70°F		
AERO LUBRIPLATE	NBR	A to 70°F	NR at 70°F	A to 70°F	A to 70°F	A to 70°F
AEROSAFE 2300	EP	NR at 70°F	A to 70°F	NR at 70°F	NR at 70°F	
AEROSAFE 2300W	EP	NR at 70°F	A to 70°F	NR at 70°F	NR at 70°F	NR at 70°F
AEROSHELL 1AC GREASE	NBR,FKM,ACM,AU,EU	A to 70°F	NR at 70°F	A to 70°F	B at 70°F	A to 70°F
AEROSHELL 7A GREASE	NBR,FKM,ACM,AU,EU	A to 70°F	NR at 70°F	A to 70°F	B at 70°F	A to 70°F
AEROSHELL 17 GREASE	NBR,FKM,ACM,AU,EU	A to 70°F	NR at 70°F	A to 70°F	B at 70°F	A to 70°F
AEROSHELL 750	FKM	B at 70°F	NR at 70°F	A to 70°F	NR at 70°F	B at 70°F
AEROZENE 50 (50% Hydrazine, 50% UDMH)	EP,IIR,VMQ,NBR	C at 70°F	A to 70°F	NR at 70°F	NR at 70°F	B at 70°F
AGAR		B any conc to 180°F			B any conc to 180°F	
AGRICULTURAL LIME		A to 70°F AB any conc to 200°F	A to 70°F	A to 70°F	A to 70°F AB any conc to 200°F	
AIR	EP,NBR,CR,FKM 200°F EP,NBR,IIR,VMQ 300°F	A to 200°F AB to 225°F	A to 200°F B at 300°F	A to 400°F C/NR at 500°F	A to 200°F B at 300°F	A to 70°F
AIR	VMQ,FKM,FVMQ 400°F VMQ at 500°F	B at 300°F NR at 400°F	NR at 400°F		NR at 400°F	
AIR GAS		A to 70°F	C at 70°F	A to 70°F	B at 70°F	
AIR SHOW		AB to 70°F				
AIR SLAKED LIME		A to 70°F AB any conc to 200°F			A to 70°F AB any conc to 200°F	
ALCOHOLS GENERAL		AB any conc to 105°F	AC to 70°F	AB to 70°F	AB any conc to 120°F	
ALDEHYDE		NR any conc at 70°F	A to 70°F	NR at 70°F	C/NR any conc at 70°F	
ALDEHYDE ACETALDEHYDE	EP	NR at 70°F	A to 120°F	NR at 70°F	C at 70°F	
ALICYCLIC HYDROCARBONS		B at 70°F	B/NR at 70°F	A to 70°F	NR at 70°F	
ALIPHATIC ALCOHOL		A to 70°F	A to 70°F	A to 70°F	A to 70°F	
ALIPHATIC HYDROCARBONS		A to 70°F	NR at 70°F	A to 70°F	B at 70°F	
ALIPHATIC NAPHTHA		C at 70°F		A to 70°F	NR at 70°F	
ALK-TRI		NR at 70°F		A to 70°F		
ALKAZENE	FKM,FVMQ,T	NR at 70°F	NR at 70°F	AB to 70°F	NR at 70°F	
ALKYL ARYL SULPHONIC ACID	EP	C at 70°F	A to 120°F	NR at 70°F	C at 70°F	
ALKYL BENZENE	FKM	NR at 70°F	NR at 70°F	A to 120°F	NR at 70°F	
ALKYLTATE	NBR	A to 200°F	NR at 70°F	A to 70°F	NR at 70°F	
ALLOMALEIC ACID		C at 70°F		A to 70°F	B at 70°F	
ALLYL ALCOHOL	NBR	A to 70°F	A to 70°F	AB to 140°F B at 176°F	A to 70°F	
ALLYL ALDEHYDE		B at 70°F		A to 70°F		
ALLYL AMINE	EP	A to 70°F	A to 250°F	NR at 70°F	A to 70°F	
ALLYL BROMIDE		NR at 70°F		B at 70°F	NR at 70°F	
ALLYL CHLORIDE	FKM	AB to 70°F NR at 140°F	NR at 70°F	AB to 120°F C at 140°F	NR at 70°F	A to 70°F
ALLYL KETONE	EP	NR at 70°F	A to 120°F	NR at 70°F	C at 70°F	
ALLYL PHENYL METHYL ETHER		NR at 70°F		B at 70°F		
ALMOND OIL, ARTIFICIAL		NR at 70°F	B at 70°F	NR at 70°F	NR at 70°F	
ALPHA CHLOROPROPYLENE		NR at 70°F			NR at 70°F	
ALPHA CHLOROTOLUENE		NR at 70°F			NR at 70°F	
ALPHA HYDROXY PROPIONIC ACID		B 100% at 70°F AB 50% to 80°F			B 100% at 70°F AB 50% to 80°F	
ALPHA HYDROXTOLUENE		NR any conc at 70°F			NR any conc at 70°F	

	STYRENE BUTADIENE SBR	POLYACRYLATE ACM	POLYURETHANE AU, EU	ISOBUTYLENE ISOPRENE IIR	POLYBUTADIENE BR	® AEROQUIP AQP
ACRYLIC ALDEHYDE				A to 70°F		
ACRYLONITRILE	C at 70°F	NR at 70°F	NR at 70°F	NR any conc at 70°F	C at 70°F	
ADIPIC ACID			A to 70°F	NR any conc at 70°F		
ADIPIC KETONE				NR at 70°F		
AERO LUBRIPLATE	B at 70°F	A to 70°F	A to 70°F	NR at 70°F	NR at 70°F	
AEROSAFE 2300	NR at 70°F	NR at 70°F	NR at 70°F	B at 70°F	NR at 70°F	AB to 200°F
AEROSAFE 2300W	NR at 70°F	NR at 70°F	NR at 70°F	B at 70°F	NR at 70°F	AB to 200°F
AEROSHELL 1AC GREASE	NR at 70°F	A to 70°F	A to 70°F	NR at 70°F	NR at 70°F	
AEROSHELL 7A GREASE	NR at 70°F	A to 70°F	A to 70°F	NR at 70°F	NR at 70°F	
AEROSHELL 17 GREASE	NR at 70°F	A to 70°F	A to 70°F	NR at 70°F	NR at 70°F	
AEROSHELL 750	NR at 70°F	B at 70°F	NR at 70°F	NR at 70°F	NR at 70°F	
AEROZENE 50 (50% Hydrazine, 50% UDMH)	NR at 70°F		NR at 70°F	A to 70°F	NR at 70°F	
AGAR				A to 70°F AB any conc to 185°F		
AGRICULTURAL LIME			A to 70°F	A to 70°F AB any conc to 185°F		
AIR	AB to 200°F NR at 300°F	A to 200°F B at 300°F	A to 70°F BC at 200°F	A to 70°F AB at 200°F	B to 200°F NR at 300°F	A to 200°F
AIR		NR at 400°F	C at 300°F NR at 400°F	B at 300°F NR at 400°F		
AIR GAS			A to 70°F	B at 70°F		
AIR SHOW				NR any conc at 70°F		
AIR SLAKED LIME				A to 70°F AB any conc to 185°F		
ALCOHOLS GENERAL	AB to 70°F		A/NR at 70°F	AB any conc to 185°F		
ALDEHYDE			C at 70°F	A any conc to 80°F		
ALDEHYDE ACETALDEHYDE	NR at 70°F	NR at 70°F	NR at 70°F			
ALICYCLIC HYDROCARBONS			B at 70°F	NR at 70°F		
ALIPHATIC ALCOHOL						
ALIPHATIC HYDROCARBONS			B at 70°F	NR at 70°F		
ALIPHATIC NAPHTHA				NR at 70°F		
ALK-TRI				NR at 70°F		
ALKAZENE	NR at 70°F	NR at 70°F	B /NR at 70°F	NR at 70°F	NR at 70°F	
ALKYL ARYL SULPHONIC ACID	NR at 70°F	NR at 70°F	NR at 70°F			
ALKYL BENZENE	NR at 70°F	NR at 70°F	NR at 70°F			
ALKYLTATE	NR at 70°F	A to 70°F	A to 70°F			
ALLOMALEIC ACID				NR at 70°F		
ALLYL ALCOHOL				A to 70°F		
ALLYL ALDEHYDE				A to 70°F		
ALLYL AMINE	NR at 70°F	NR at 70°F	NR at 70°F			
ALLYL BROMIDE				NR at 70°F		
ALLYL CHLORIDE	A to 70°F	NR at 70°F		C /NR at 70°F		
ALLYL KETONE	A to 70°F	NR at 70°F	NR at 70°F			
ALLYL PHENYL METHYL ETHER				NR at 70°F		
ALMOND OIL, ARTIFICIAL			NR at 70°F	B at 70°F		
ALPHA CHLOROPROPYLENE				NR at 70°F		
ALPHA CHLOROTOLUENE				NR at 70°F		
ALPHA HYDROXY PROPIONIC ACID				A 100% to 70°F AB 50% to 150°F		
ALPHA HYDROXTOLUENE				A to 70°F AB any conc to 185°F		

C 9

	SYNTHETIC ISOPRENE IR	NATURAL ISOPRENE NR	CHLOROSULFONATED POLYETHYLENE CSM	FLUOROSILICONE FVMQ	SILICONE VMQ	CHEMRAZ FFKM
ACRYLIC ALDEHYDE	B at 70°F	B at 70°F	B at 70°F	A to 70°F		
ACRYLONITRILE	BC at 70°F	BC at 70°F	BC at 70°F	NR at 70°F	NR at 70°F	A to 70°F
ADIPIC ACID	A any conc to 80°F	A to 70°F				A to 70°F
ADIPIC KETONE	NR at 70°F	NR at 70°F	NR at 70°F			
AERO LUBRIPLATE	NR at 70°F	NR at 70°F	A to 70°F	A to 70°F	B at 70°F	A to 70°F
AEROSAFE 2300	NR at 70°F	NR at 70°F	NR at 70°F	C at 70°F	C at 70°F	A to 70°F
AEROSAFE 2300W	NR at 70°F	NR at 70°F	NR at 70°F	C at 70°F	C at 70°F	A to 70°F
AEROSHELL 1AC GREASE	NR at 70°F	NR at 70°F	A to 70°F	A to 70°F	B at 70°F	A to 70°F
AEROSHELL 7A GREASE	NR at 70°F	NR at 70°F	A to 70°F	A to 70°F	B at 70°F	A to 70°F
AEROSHELL 17 GREASE	NR at 70°F	NR at 70°F	A to 70°F	A to 70°F	B at 70°F	A to 70°F
AEROSHELL 750	NR at 70°F	NR at 70°F	B at 70°F	NR at 70°F		A to 70°F
AEROZENE 50 (50% Hydrazine, 50% UDMH)	NR at 70°F	NR at 70°F	NR at 70°F	NR at 70°F	NR at 70°F	B at 70°F
AGAR	A to 70°F	A to 70°F AB at 150°F				
AGRICULTURAL LIME	A to 70°F	A to 70°F AB at 150°F	B at 70°F		A to 70°F	
AIR	AB to 70°F BC at 200°F	AB to 150°F BC at 200°F	A to 200°F B at 300°F	A to 300°F B/NR at 400°F	A to 300°F 1.5 yrs A to 400°F 1 yr	A to 500°F
AIR	NR at 300°F	NR at 300°F	NR at 400°F	NR at 500°F	-17% T.S. 1 day 437°F AB at 500°F	
AIR GAS	C at 70°F	C at 70°F	B at 70°F		B at 70°F	
AIR SHOW						
AIR SLAKED LIME	A to 70°F	A to 70°F AB any conc to 150°F				
ALCOHOLS GENERAL	AB to 70°F	AB any conc to 80°F	AB to 70°F		A/NR at 70°F	
ALDEHYDE	C at 70°F	C/NR any conc at 70°F	C at 70°F		A to 70°F	
ALDEHYDE ACETALDEHYDE				NR at 70°F	A to 70°F	
ALICYCLIC HYDROCARBONS	NR at 70°F	NR at 70°F	NR at 70°F		NR at 70°F	
ALIPHATIC ALCOHOL	A to 70°F	A to 70°F	A to 70°F			
ALIPHATIC HYDROCARBONS	NR at 70°F	NR at 70°F	B at 70°F		NR at 70°F	
ALIPHATIC NAPHTHA	NR at 70°F	NR at 70°F	B at 70°F			
ALK-TRI	NR at 70°F	NR at 70°F	NR at 70°F			
ALKAZENE	NR at 70°F	NR at 70°F	NR at 70°F	B at 70°F	NR at 70°F	A to 70°F
ALKYL ARYL SULPHONIC ACID				NR at 70°F	NR at 70°F	A to 70°F
ALKYL BENZENE				A to 70°F	NR at 70°F	A to 70°F
ALKYLTATE				A to 70°F	NR at 70°F	
ALLOMALEIC ACID	B at 70°F	AB any conc to 80°F	B at 70°F		B at 70°F	
ALLYL ALCOHOL	A to 70°F	A to 70°F	A to 70°F			
ALLYL ALDEHYDE	B at 70°F	B at 70°F	B at 70°F			
ALLYL AMINE				NR at 70°F	C at 70°F	
ALLYL BROMIDE	NR at 70°F	NR at 70°F	NR at 70°F			
ALLYL CHLORIDE	NR at 70°F	NR at 70°F	NR at 70°F	NR at 70°F	A to 70°F	A to 70°F
ALLYL KETONE				NR at 70°F	B at 70°F	
ALLYL PHENYL METHYL ETHER	NR at 70°F	NR at 70°F	NR at 70°F			
ALMOND OIL, ARTIFICIAL	NR at 70°F	NR at 70°F	NR at 70°F		C at 70°F	
ALPHA CHLOROPROPYLENE	NR at 70°F	NR at 70°F				
ALPHA CHLOROTOLUENE	NR at 70°F	NR at 70°F				
ALPHA HYDROXY PROPIONIC ACID	A to 70°F	A to 70°F AB 50% at 120°F				
ALPHA HYDROXTOLUENE	NR at 70°F	NR any conc at 70°F				

	POLYSULFIDE T	CHLORINATED POLYETHYLENE CM	EPICHLOROHYDRIN CO, ECO	®HYTREL COPOLYESTER TPE	®SANTOPRENE COPOLYMER TPO	®C-FLEX STYRENIC TPE
ACRYLIC ALDEHYDE						
ACRYLONITRILE	NR at 70°F	A to 70°F			*NR at 70°F	
ADIPIC ACID						
ADIPIC KETONE						
AERO LUBRIPLATE	A to 70°F					
AEROSAFE 2300	NR at 70°F			AB to 200°F		
AEROSAFE 2300W	NR at 70°F			AB to 200°F		
AEROSHELL 1AC GREASE	A to 70°F					
AEROSHELL 7A GREASE	A to 70°F					
AEROSHELL 17 GREASE	A to 70°F					
AEROSHELL 750	B at 70°F					
AEROZENE 50 (50% Hydrazine, 50% UDMH)	NR at 70°F					
AGAR						
AGRICULTURAL LIME						
AIR	A to 70°F B at 200°F	A to 70°F		A to 200°F A to 70°F wet	A to 300°F	A to 70°F
AIR	C/NR at 300°F NR at 400°F					
AIR GAS						
AIR SHOW						
AIR SLAKED LIME						
ALCOHOLS GENERAL	AB to 70°F	AB to 70°F		AB to 70°F		AB to 70°F
ALDEHYDE						
ALDEHYDE ACETALDEHYDE						
ALICYCLIC HYDROCARBONS						
ALIPHATIC ALCOHOL						
ALIPHATIC HYDROCARBONS						NR at 70°F
ALIPHATIC NAPHTHA						
ALK-TRI						
ALKAZENE	AB to 70°F					
ALKYL ARYL SULPHONIC ACID						
ALKYL BENZENE						
ALKYLTATE						
ALLOMALEIC ACID						
ALLYL ALCOHOL					*A to 70°F	
ALLYL ALDEHYDE						
ALLYL AMINE						
ALLYL BROMIDE						
ALLYL CHLORIDE						
ALLYL KETONE						
ALLYL PHENYL METHYL ETHER						
ALMOND OIL, ARTIFICIAL						
ALPHA CHLOROPROPYLENE						
ALPHA CHLOROTOLUENE						
ALPHA HYDROXY PROPIONIC ACID						
ALPHA HYDROXTOLUENE						

	ETHYLENE ACRYLIC EA	POLYALLOMER LINEAR COPOLYMER	NYLON 11 POLYAMIDE	NYLON 12 POLYAMIDE	ETHYLENE VINYL ACETATE EVA	POLYVINYLCHLORIDE FLEXIBLE PVC
ACRYLIC ALDEHYDE						
ACRYLONITRILE		C at 70°F NR at 122°F	A to 70°F			NR at 70°F
ADIPIC ACID		A to 122°F				A to 70°F B at 122°F
ADIPIC KETONE						
AERO LUBRIPLATE						
AEROSAFE 2300			AB to 200°F			
AEROSAFE 2300W			AB to 200°F			
AEROSHELL 1AC GREASE						
AEROSHELL 7A GREASE						
AEROSHELL 17 GREASE						
AEROSHELL 750						
AEROZENE 50 (50% Hydrazine, 50% UDMH)						
AGAR						AB any conc to 150°F
AGRICULTURAL LIME						AB any conc to 150°F
AIR	A to 300°F NR at 400°F	A to 70°F	A to 70°F AB at 200°F		A to 70°F	A to 70°F AB at 150°F
AIR						
AIR GAS						
AIR SHOW						
AIR SLAKED LIME						AB to 150°F
ALCOHOLS GENERAL		A to 70°F	A to 70°F	AB to 70°F	A to 70°F	AC to 70°F
ALDEHYDE						NR any conc at 70°F
ALDEHYDE ACETALDEHYDE						
ALICYCLIC HYDROCARBONS						
ALIPHATIC ALCOHOL						
ALIPHATIC HYDROCARBONS						C at 70°F
ALIPHATIC NAPHTHA						
ALK-TRI						
ALKAZENE						
ALKYL ARYL SULPHONIC ACID						
ALKYL BENZENE						
ALKYLTATE						
ALLOMALEIC ACID						
ALLYL ALCOHOL		A to 122°F		C at 70°F		B at 70°F C at 122°F
ALLYL ALDEHYDE						
ALLYL AMINE						
ALLYL BROMIDE						
ALLYL CHLORIDE						
ALLYL KETONE						
ALLYL PHENYL METHYL ETHER						
ALMOND OIL, ARTIFICIAL						
ALPHA CHLOROPROPYLENE						
ALPHA CHLOROTOLUENE						
ALPHA HYDROXY PROPIONIC ACID						
ALPHA HYDROXTOLUENE						

	POLYETHYLENE LOW DENSITY LDPE	®TEFLON FEP	®KALREZ PERFLUORINATED ELASTOMER	®FLUORAZ FLUORINATED COPOLYMER	®AFLAS FLUORINATED COPOLYMER	®NORPRENE COPOLYMER TPO
ACRYLIC ALDEHYDE		A to 70°F				
ACRYLONITRILE		A to 200°F	A to 100°F *A to 212°F	BC at 70°F		
ADIPIC ACID	A to 70°F AB to 140°F	A to 200°F	*A to 212°F	B at 70°F		
ADIPIC KETONE		A to 70°F				
AERO LUBRIPLATE				A to 70°F		
AEROSAFE 2300		AB to 400°F		A to 70°F		
AEROSAFE 2300W		AB to 400°F		B at 70°F		
AEROSHELL 1AC GREASE				A to 70°F		
AEROSHELL 7A GREASE				A to 70°F	A to 176°F	
AEROSHELL 17 GREASE				A to 70°F		
AEROSHELL 750				A to 70°F		
AEROZENE 50 (50% Hydrazine, 50% UDMH)				B at 70°F		
AGAR		A to 70°F				
AGRICULTURAL LIME		A to 70°F				
AIR	A to 70°F	A to 200°F	A to 500°F −10% tensile strength	A to 400°F	A to 390°F C at 446°F	A to 275°F
AIR			B at 550°F −39% tensile strength	B at 528°F −42% tensile strength	NR at 500°F	
AIR GAS		A to 70°F			NR at 70°F	
AIR SHOW		A to 70°F				
AIR SLAKED LIME		A to 70°F				
ALCOHOLS GENERAL	A to 70°F AB to 140°F	A to 300°F				
ALDEHYDE				NR at 70°F		
ALDEHYDE ACETALDEHYDE						
ALICYCLIC HYDROCARBONS		A to 70°F				
ALIPHATIC ALCOHOL		A to 70°F				
ALIPHATIC HYDROCARBONS		A to 70°F				BC at 70°F
ALIPHATIC NAPHTHA		A to 70°F				
ALK-TRI		A to 70°F				
ALKAZENE		A to 70°F	A to 70°F	B at 70°F		
ALKYL ARYL SULPHONIC ACID			A to 212°F	NR at 160°F		
ALKYL BENZENE			A to 212°F	NR at 70°F		
ALKYLTATE				NR at 70°F		
ALLOMALEIC ACID		A to 70°F				
ALLYL ALCOHOL	A/NR to 122°F	A to 300°F				
ALLYL ALDEHYDE		A to 200°F				
ALLYL AMINE						
ALLYL BROMIDE		A to 70°F				
ALLYL CHLORIDE		A to 70°F				
ALLYL KETONE		A to 70°F				
ALLYL PHENYL METHYL ETHER						
ALMOND OIL, ARTIFICIAL		A to 70°F				
ALPHA CHLOROPROPYLENE		A to 70°F				
ALPHA CHLOROTOLUENE		A to 70°F				
ALPHA HYDROXY PROPIONIC ACID		A to 70°F				
ALPHA HYDROXTOLUENE		A to 70°F				

	RECOMMENDATIONS	NITRILE NBR	ETHYLENE PROPYLENE EP, EPDM	FLUOROCARBON FKM	CHLOROPRENE CR	HYDROGENATED NITRILE HNBR
ALUM (SEE ALSO SPECIFIC TYPES)	NBR,EP,CR	A to 140°F B at 212°F	A to 140°F B at 212°F	A to 212°F	A to 70°F AB to 200°F	
ALUMS (NH3-Cr-K)	NBR,EP,CR,IIR	A to 160°F	A to 70°F	A/NR at 70°F	A to 70°F	A to 70°F
ALUMINUM ACETATE	EP,IIR,NBR,CR	B any conc to 120°F	A to 176°F	A/NR to 176°F	B any conc to 120°F	
ALUMINUM AMMONIUM SULPHATE		AB any conc to 140°F B at 176°F	A to 176°F	A to 212°F	B any conc to 120°F	A to 70°F
ALUMINUM BROMIDE	NBR,EP,CR,FKM	A to 140°F AB any conc to 180°F	A to 140°F	A to 176°F	A to 70°F AB any conc to 180°F	A to 70°F
ALUMINUM CHLORIDE	NBR,EP,CR,FKM	A to 160°F AB any conc to 200°F	A to 176°F B at 212°F	A to 212°F	A any conc to 140°F AB any conc to 200°F	A to 70°F
ALUMINUM FLUORIDE	NBR,EP,CR,FKM	A any conc to 176°F AB any conc to 200°F	A to 176°F	A any conc to 212°F	A any conc to 140°F AB any conc to 200°F	A to 70°F
ALUMINUM FORMATE		NR at 70°F		NR at 70°F		
ALUMINUM HYDROXIDE	NBR	A to 160°F B at 212°F	A to 140°F B at 212°F	A to 176°F B at 212°F	A to 70°F	
ALUMINUM NITRATE	NBR,EP,CR,FKM	A to 140°F B any conc at 200°F	A to 176°F B at 212°F	A to 212°F	A to 70°F AB any conc to 200°F	A to 70°F
ALUMINUM PHOSPHATE		A to 70°F	A to 70°F	A to 70°F	A to 70°F	A to 70°F
ALUMINUM POTASSIUM SULPHATE	NBR,EP,CR,IIR	A to 140°F AB to 176°F	A to 176°F B at 212°F	A to 176°F B at 212°F	A to 140°F AB any conc to 200°F	A to 70°F
ALUMINUM SALTS	NBR,EP,FKM,CR	A to 70°F AB any conc to 180°F	A to 70°F	A to 70°F	A to 70°F AB any conc to 200°F	A to 70°F
ALUMINUM SODIUM SULPHATE		A to 70°F AB any conc to 200°F	A to 70°F	A to 70°F	A to 70°F AB any conc to 200°F	A to 70°F
ALUMINUM SULPHATE	NBR,EP,CR,FKM	A to 160°F AB any conc to 200°F	A to 140°F	A to 140°F A 10% to boiling	A to 158°F AB any conc to 200°F	
AMBREX 33, MOBILE	NBR,FKM,ACM	A to 70°F	NR at 70°F	A to 70°F	B at 70°F	A to 70°F
AMBREX 830, MOBILE	NBR,FKM,ACM,AU	A to 70°F	C at 70°F	A to 70°F	B at 70°F	A to 70°F
AMERICAN ASHES		A to 70°F AB any conc to 180°F	A to 70°F	A to 70°F	A to 70°F AB any conc to 200°F	
AMINES	EP	C/NR at 70°F	AC to 70°F	NR at 70°F	NR at 70°F	
AMINES MIXED (EG: ALLYL, ETHYL, ETC.)	EP,CR,IIR	NR at 70°F	AB to 250°F	NR at 70°F	B/NR at 70°F	NR at 70°F
AMINOBENZENE	EP,IIR	+210% vol 3 days 70°F NR at 70°F	+1% vol 3 days 70°F AB to 200°F	A to 70°F	+60% vol 3 days 70°F NR at 70°F	
AMINO DIMETHYLBENZENE		C/NR at 70°F	C/NR at 70°F	C/NR at 70°F	NR at 70°F	
AMINO ETHANOL	EP,CR,NBR,IIR	B any conc to 80°F	AB to 200°F	NR at 70°F	B any conc to 80°F	
AMINO ETHYLETHANOLAMINE		B at 70°F				
AMINO HEXANE		C at 70°F		NR at 70°F		
AMINO PENTANE		C at 70°F	NR at 70°F	NR at 70°F	NR at 70°F	
AMINO XYLENE		NR at 70°F	C at 70°F	C at 70°F	NR at 70°F	
AMMONIA ALUM	EP	AB any conc to 120°F	A to 70°F	A to 70°F	AB any conc to 120°F	
AMMONIA CUPRIC SULFATE		A to 70°F		A to 70°F		
AMMONIA GAS, COLD	NBR,EP,CR,IIR SBR	A to 104°F	A to 140°F	NR at 70°F	A to 70°F	A to 70°F
AMMONIA GAS, HOT	VMQ,CR,EP,IIR	B to 140°F NR at 200°F	B at 176°F	NR	B hot NR at 200°F	NR
AMMONIA, ANHYDROUS LIQUID	EP,IIR,CR,NBR	AB to 70°F C 140-240°F	A to 140°F	NR at 70°F	A to 240°F	B at 70°F
AMMONIA, AQUEOUS LIQUID	EP,IIR,NBR	+1.8% vol 28% conc at 70°F	+0.01% vol 28% conc at 70°F	+5.3% vol 28% conc at 70°F	+9.5% vol 28% conc at 70°F	
AMMONIA, AQUEOUS LIQUID		A 100% to 70°F B 100% at 104°F	A 100% to 212°F	NR 100% at 70°F AB 30% to 70°F	A conc to 140°F AB conc to 160°F	
AMMONIA, AQUEOUS LIQUID		AB 38% to 200°F A 10% to 70°F		AB 10% to 70°F C 10% at 104°F	AB 38% to 200°F	
AMMONIA, AQUEOUS LIQUID		B 10% at 140°F		NR 10% at 140°F A ammonia H$_2$O to 212°F		
AMMONIA LIQUORS (NH$_4$OH·H$_2$O)		AB to 70°F		NR at 70°F	AB to 70°F	
AMMONIA + LITHIUM METAL SOLUTION	NBR,EP	B at 70°F	B at 70°F	NR at 70°F	NR at 70°F	B at 70°F
AMMONIA NITRATE		A to 70°F C at 140°F	A to 70°F	A to 70°F	BC at 70°F C at 140°F	
AMMONIAK		A to 70°F	A to 70°F	A to 70°F	A to 70°F	
AMMONIATED CITRIC ACID		B any conc to 110°F			B any conc to 150°F	
AMMONIATED LATEX		B at 70°F	B at 70°F	A to 70°F	A to 70°F	
AMMONIUM ACETATE		A to 140°F B at 176°F	A to 140°F B at 212°F	A to 140°F B at 212°F	A to 70°F	
AMMONIUM ALUM		B at 70°F			B at 70°F	

	STYRENE BUTADIENE SBR	POLYACRYLATE ACM	POLYURETHANE AU, EU	ISOBUTYLENE ISOPRENE IIR	POLYBUTADIENE BR	® AEROQUIP AQP
ALUM (SEE ALSO SPECIFIC TYPES)			C/NR at 70°F	A to 70°F AB any conc to 185°F		NR at 70°F
ALUMS (NH3-Cr-K)	A to 70°F	NR at 70°F	AB to 70°F	A to 70°F	A to 70°F	
ALUMINUM ACETATE	B/NR at 70°F	NR at 70°F	NR at 70°F	B any conc to 120°F	NR at 70°F	
ALUMINUM AMMONIUM SULPHATE				A to 70°F AB any conc to 150°F		
ALUMINUM BROMIDE	A to 70°F	A to 70°F	C at 70°F	A to 70°F AB any conc to 120°F	A to 70°F	
ALUMINUM CHLORIDE	A to 70°F	A to 70°F	B conc at 70°F A sol'n to 70°F	A to 70°F AB any conc to 150°F	A to 70°F	
ALUMINUM FLUORIDE	A to 70°F		BC at 70°F	A to 70°F AB any conc to 185°F	A to 70°F	
ALUMINUM FORMATE				B at 70°F		
ALUMINUM HYDROXIDE	B at 70°F	NR at 70°F	AB to 70°F	A to 70°F		
ALUMINUM NITRATE	A to 70°F		C at 70°F	A to 70°F AB any conc to 185°F	A to 70°F	
ALUMINUM PHOSPHATE	A to 70°F			A to 70°F	A to 70°F	
ALUMINUM POTASSIUM SULPHATE	AB to 70°F			A to 70°F AB any conc to 185°F		
ALUMINUM SALTS	A to 70°F	A to 70°F	C at 70°F	A to 70°F AB any conc to 150°F	A to 70°F	
ALUMINUM SODIUM SULPHATE			A to 70°F	A to 70°F AB any conc to 185°F		
ALUMINUM SULPHATE	B at 70°F	NR at 70°F	B at 70°F	A to 70°F AB any conc to 185°F	AB to 70°F	NR at 70°F
AMBREX 33, MOBILE	NR at 70°F	A to 70°F	B at 70°F	NR at 70°F	NR at 70°F	NR at 70°F
AMBREX 830, MOBILE	NR at 70°F	A to 70°F	A to 70°F	C/NR at 70°F	NR at 70°F	AB to 300°F
AMERICAN ASHES				A to 70°F AB any conc to 180°F		
AMINES						
AMINES MIXED (EG: ALLYL, ETHYL, ETC.)	B at 70°F	NR at 70°F	NR at 70°F	B at 70°F	B at 70°F	
AMINOBENZENE	+24% vol 3 days 70°F NR at 70°F	+272% vol 3 days 70°F NR at 70°F	+261% vol 3 days 70°F NR at 70°F	+5% vol 3 days 70°F A to 70°F	NR at 70°F	AB to 70°F
AMINO DIMETHYLBENZENE				B to 100°F		
AMINO ETHANOL			C at 70°F	A to 70°F AB any conc to 140°F		
AMINO ETHYLETHANOLAMINE				A to 70°F		
AMINO HEXANE				B at 70°F		
AMINO PENTANE				B at 70°F		
AMINO XYLENE				B to 100°F		
AMMONIA ALUM				B any conc to 120°F		
AMMONIA CUPRIC SULFATE				A to 70°F		
AMMONIA GAS, COLD	A to 70°F	NR at 70°F	B at 70°F	A to 70°F	A to 70°F	NR at 70°F
AMMONIA GAS, HOT	NR	NR	NR	B hot NR at 200°F	NR	NR
AMMONIA, ANHYDROUS LIQUID	B/NR to 140°F C/NR at 240°F	NR at 70°F	NR at 70°F	A to 70°F	NR at 70°F	NR at 70°F
AMMONIA, AQUEOUS LIQUID	C/NR conc to 140°F	NR at 70°F	A/NR at 70°F	+4.0% vol at 28% conc at 70°F		NR at 70°F
AMMONIA, AQUEOUS LIQUID				A 100% at 70°F C 100% at 150°F		
AMMONIA, AQUEOUS LIQUID				AB 38% to 185°F		
AMMONIA, AQUEOUS LIQUID						
AMMONIA LIQUORS (NH$_4$OH•H$_2$O)						
AMMONIA + LITHIUM METAL SOLUTION	NR at 70°F	NR at 70°F	NR at 70°F	B at 70°F	NR at 70°F	
AMMONIA NITRATE						
AMMONIAK			A to 70°F	A to 70°F		
AMMONIATED CITRIC ACID				A to 70°F AB any conc to 185°F		
AMMONIATED LATEX						
AMMONIUM ACETATE			C/NR at 70°F			
AMMONIUM ALUM				A to 70°F		

	SYNTHETIC ISOPRENE IR	NATURAL ISOPRENE NR	CHLOROSULFONATED POLYETHYLENE CSM	FLUOROSILICONE FVMQ	SILICONE VMQ	CHEMRAZ FFKM
ALUM (SEE ALSO SPECIFIC TYPES)	A to 70°F	A to 70°F AB any conc to 150°F	A to 70°F AB to 150°F	C at 70°F	AC to 70°F	A to 70°F
ALUMS (NH3-Cr-K)	A to 70°F	A to 70°F	A to 70°F	NR at 70°F	A to 70°F	A to 70°F
ALUMINUM ACETATE	A to 70°F	A to 70°F AB any conc to 150°F	B/NR at 70°F	NR at 70°F	NR at 70°F	A to 70°F
ALUMINUM AMMONIUM SULPHATE	A to 70°F	A to 70°F AB any conc to 150°F				A to 70°F
ALUMINUM BROMIDE	A to 70°F	A to 70°F AB any conc to 150°F	A to 70°F	A to 70°F	A to 70°F	A to 70°F
ALUMINUM CHLORIDE	A to 70°F	A any conc to 70°F AB any conc to 150°F	A to 70°F AB to 150°F	A to 70°F	B any conc at 70°F	A to 70°F
ALUMINUM FLUORIDE	A to 70°F	AB any conc to 150°F	A to 70°F AB to 150°F	A to 70°F	B at 70°F	A to 70°F
ALUMINUM FORMATE	NR at 70°F	NR at 70°F	NR at 70°F			A to 70°F
ALUMINUM HYDROXIDE	A to 70°F	A to 70°F	AB to 70°F	A sol'n to 70°F	A to 70°F	A to 70°F B at 70°F white
ALUMINUM NITRATE	A to 70°F	A to 70°F AB any conc to 150°F	A to 70°F		B at 70°F	A to 70°F
ALUMINUM PHOSPHATE	A to 70°F	A to 70°F	A to 70°F		A to 70°F	A to 70°F
ALUMINUM POTASSIUM SULPHATE	A to 70°F	A to 70°F AB any conc to 150°F	A to 70°F		A any conc to 70°F	A to 70°F
ALUMINUM SALTS	A to 70°F	A to 70°F AB any conc to 150°F	A to 70°F	A to 70°F	A to 70°F	A to 70°F
ALUMINUM SODIUM SULPHATE	A to 70°F	A to 70°F AB any conc to 150°F	A to 70°F		A to 70°F	A to 70°F
ALUMINUM SULPHATE	A to 70°F	A to 70°F AB any conc to 150°F	A to 250°F	A to 70°F	A to 70°F	A to 70°F
AMBREX 33, MOBILE	NR at 70°F	NR at 70°F	C at 70°F	C at 70°F	NR at 70°F	A to 70°F
AMBREX 830, MOBILE	NR at 70°F	NR at 70°F	B at 70°F	A to 70°F	B at 70°F	A to 70°F
AMERICAN ASHES	A to 70°F	A to 70°F AB any conc to 150°F	A to 70°F			
AMINES		C/NR at 70°F	NR at 70°F		BC at 70°F	AB to 70°F
AMINES MIXED (EG: ALLYL, ETHYL, ETC.)	AB to 70°F	AB to 70°F	NR at 70°F	NR at 70°F	AB to 70°F	AB to 70°F B at 70°F white
AMINOBENZENE	NR at 70°F	NR at 70°F	C/NR at 70°F NR at 100°F	−4% vol 3 days 70°F BC at 70°F	+3% vol 7 days 70°F AB to 70°F	A to 70°F
AMINO DIMETHYLBENZENE	NR at 70°F	NR at 70°F	NR at 70°F	NR at 70°F	NR at 70°F	
AMINO ETHANOL	B at 70°F	B any conc at 70°F	B at 70°F	NR at 70°F	B at 70°F	A to 70°F
AMINO ETHYLETHANOLAMINE	B at 70°F	B at 70°F	B at 70°F			A to 70°F
AMINO HEXANE	C at 70°F	C at 70°F	C at 70°F			
AMINO PENTANE	C at 70°F	C at 70°F	C at 70°F			
AMINO XYLENE	NR at 70°F	NR at 70°F	NR at 70°F		NR at 70°F	
AMMONIA ALUM	A to 70°F	A to 70°F AB any conc to 150°F				
AMMONIA CUPRIC SULFATE	C at 70°F	C at 70°F	A to 70°F			
AMMONIA GAS, COLD	A to 70°F	A to 70°F	A to 70°F	A/NR at 70°F	A to 70°F	A to 70°F
AMMONIA GAS, HOT	C/NR	C/NR	B	NR at 140°F	A	A
AMMONIA, ANHYDROUS LIQUID	NR at 70°F	NR at 70°F	B/NR at 70°F	NR at 70°F	BC to 140°F C at 240°F	AB to 70°F B at 70°F white
AMMONIA, AQUEOUS LIQUID	A sol'n to 70°F	A to 70°F AB any conc to 150°F	+18% vol 28% conc 70°F	NR 100% at 70°F B 30% at 70°F	−2.8% vol 28% conc 70°F	
AMMONIA, AQUEOUS LIQUID			B 28% at 70°F		AB to 70°F C at 200°F	
AMMONIA, AQUEOUS LIQUID						
AMMONIA, AQUEOUS LIQUID						
AMMONIA LIQUORS (NH$_4$OH·H$_2$O)		A/NR at 70°F	AB to 70°F			
AMMONIA + LITHIUM METAL SOLUTION	NR at 70°F	NR at 70°F	NR at 70°F	NR at 70°F	NR at 70°F	A/NR at 70°F
AMMONIA NITRATE		BC at 70°F				
AMMONIAK	A to 70°F	A to 70°F	A to 70°F		B at 70°F	
AMMONIATED CITRIC ACID	A to 70°F	A to 70°F AB any conc to 150°F				
AMMONIATED LATEX	C at 70°F	C at 70°F	C at 70°F			
AMMONIUM ACETATE	A to 70°F	AC to 70°F	A to 70°F			A to 70°F
AMMONIUM ALUM	A to 70°F	A to 70°F				

	POLYSULFIDE T	CHLORINATED POLYETHYLENE CM	EPICHLOROHYDRIN CO, ECO	® HYTREL COPOLYESTER TPE	® SANTOPRENE COPOLYMER TPO	® C-FLEX STYRENIC TPE
ALUM (SEE ALSO SPECIFIC TYPES)	AB to 150°F	A to 150°F		NR at 70°F		AB to 70°F
ALUMS (NH3-Cr-K)	NR at 70°F	A/NR at 70°F				
ALUMINUM ACETATE	NR at 70°F	A to 70°F	B at 70°F			
ALUMINUM AMMONIUM SULPHATE						
ALUMINUM BROMIDE	A to 70°F					
ALUMINUM CHLORIDE	NR at 70°F	A/NR at 70°F	A to 70°F	AB to 70°F *AB to 140°F		AB to 70°F
ALUMINUM FLUORIDE	NR at 70°F	NR at 70°F	A to 70°F			
ALUMINUM FORMATE						
ALUMINUM HYDROXIDE	AB to 70°F					
ALUMINUM NITRATE	B to 70°F		A to 70°F			
ALUMINUM PHOSPHATE			A to 70°F			
ALUMINUM POTASSIUM SULPHATE						
ALUMINUM SALTS	A to 70°F					
ALUMINUM SODIUM SULPHATE						
ALUMINUM SULPHATE	B/NR to 150°F	A to 158°F		AB to 70°F *AB to 150°F	A to 70°F	AB to 70°F
AMBREX 33, MOBILE	C at 70°F			AB to 200°F		
AMBREX 830, MOBILE	B at 70°F			AB to 200°F		
AMERICAN ASHES						
AMINES				A to 70°F NR at 158°F		
AMINES MIXED (EG: ALLYL, ETHYL, ETC.)	NR at 70°F			AB to 70°F		
AMINOBENZENE	+325% vol 3 days 70°F NR at 70°F	AB to 70°F	NR at 70°F	+31-93% vol 100% conc 7 days 70°F	*A to 70°F	AB to 70°F
AMINO DIMETHYLBENZENE						
AMINO ETHANOL						
AMINO ETHYLETHANOLAMINE						
AMINO HEXANE						
AMINO PENTANE						
AMINO XYLENE						
AMMONIA ALUM						
AMMONIA CUPRIC SULFATE						
AMMONIA GAS, COLD	A to 70°F	NR at 70°F		A/NR at 70°F		AB to 70°F
AMMONIA GAS, HOT	NR	NR				
AMMONIA, ANHYDROUS LIQUID	NR at 70°F	A to 70°F		NR at 70°F		AB to 70°F
AMMONIA, AQUEOUS LIQUID						AB to 70°F
AMMONIA, AQUEOUS LIQUID						
AMMONIA, AQUEOUS LIQUID						
AMMONIA, AQUEOUS LIQUID						
AMMONIA LIQUORS (NH$_4$OH·H$_2$O)						
AMMONIA + LITHIUM METAL SOLUTION	NR solution at 70°F					
AMMONIA NITRATE						
AMMONIAK						
AMMONIATED CITRIC ACID						
AMMONIATED LATEX						
AMMONIUM ACETATE						AB to 70°F
AMMONIUM ALUM						

	ETHYLENE ACRYLIC EA	POLYALLOMER LINEAR COPOLYMER	NYLON 11 POLYAMIDE	NYLON 12 POLYAMIDE	ETHYLENE VINYL ACETATE EVA	POLYVINYLCHLORIDE FLEXIBLE PVC
ALUM (SEE ALSO SPECIFIC TYPES)		A to 70°F	AB to 70°F *AB to 140°F		A to 70°F *AB to 140°F	A to 150°F
ALUMS (NH3-Cr-K)			C at 70°F	AB to 70°F	B at 70°F	A to 70°F
ALUMINUM ACETATE						
ALUMINUM AMMONIUM SULPHATE						
ALUMINUM BROMIDE						
ALUMINUM CHLORIDE	A to 70°F	A to 70°F	NR any conc at 70°F		B to 140°F	A to 70°F AB to 150°F
ALUMINUM FLUORIDE			AB conc to 70°F NR 20% at 70°F		C at 70°F	B at 70°F
ALUMINUM FORMATE						
ALUMINUM HYDROXIDE		A to 70°F B at 122°F	AB to 70°F			A to 70°F B at 122°F
ALUMINUM NITRATE						
ALUMINUM PHOSPHATE						
ALUMINUM POTASSIUM SULPHATE			AB to 70°F			AB to 100% to 70°F
ALUMINUM SALTS		A to 122°F	AB to 70°F	AB to 70°F		A to 122°F
ALUMINUM SODIUM SULPHATE						
ALUMINUM SULPHATE	A to 70°F	A to 70°F	A to 70°F AB to 194°F		A to 70°F AB to 140°F	A to 150°F
AMBREX 33, MOBILE			AB to 200°F			
AMBREX 830, MOBILE			AB to 200°F			
AMERICAN ASHES						
AMINES						A to 70°F
AMINES MIXED (EG: ALLYL, ETHYL, ETC.)	NR at 70°F					
AMINOBENZENE	*NR at 70°F	B at 70°F C at 122°F	C at 70°F NR at 105°F		C/NR at 70°F NR at 140°F	C/NR at 70°F NR (food grades)
AMINO DIMETHYLBENZENE						NR at 70°F
AMINO ETHANOL						
AMINO ETHYLETHANOLAMINE						
AMINO HEXANE						
AMINO PENTANE						
AMINO XYLENE						NR at 70°F
AMMONIA ALUM						
AMMONIA CUPRIC SULFATE						
AMMONIA GAS, COLD	NR at 70°F	A to 122°F	AB to 140°F	AB any conc to 70°F	AB to 140°F	BC to 122°F
AMMONIA GAS, HOT	NR		AB to 140°F		AB to 140°F	*C wet at 140°F
AMMONIA, ANHYDROUS LIQUID	NR at 70°F	A to 70°F	A to 70°F AB to 105°F		A to 70°F	AB to 150°F
AMMONIA, AQUEOUS LIQUID		A 100% to 70°F A 30% to 70°F	A to 70°F AB to 140°F	AB 10% to 70°F	AB to 140°F	AB to 70°F B at 122°F
AMMONIA, AQUEOUS LIQUID		B 30% at 122°F A 5% to 122°F				NR at 150°F A to 30% at 70°F
AMMONIA, AQUEOUS LIQUID						B 30% at 122°F C 20% at 150°F
AMMONIA, AQUEOUS LIQUID						AB 10% to 140°F
AMMONIA LIQUORS (NH4OH·H2O)						
AMMONIA + LITHIUM METAL SOLUTION	NR at 70°F					
AMMONIA NITRATE						B at 70°F
AMMONIAK						
AMMONIATED CITRIC ACID						
AMMONIATED LATEX						AB to 70°F NR at 150°F
AMMONIUM ACETATE		A to 122°F	A to 70°F		A to 70°F	A to 122°F
AMMONIUM ALUM						

	POLYETHYLENE LOW DENSITY LDPE	®TEFLON FEP	®KALREZ PERFLUORINATED ELASTOMER	®FLUORAZ FLUORINATED COPOLYMER	®AFLAS FLUORINATED COPOLYMER	®NORPRENE COPOLYMER TPO
ALUM (SEE ALSO SPECIFIC TYPES)	A to 100% to 70°F AB to 100% at 175°F	A to 150°F	A to 212°F	A to 70°F	A to 212°F	A to 70°F
ALUMS (NH3-Cr-K)	A to 70°F AB to 140°F	A to 300°F	A to 212°F	A to 70°F		
ALUMINUM ACETATE		A to 70°F	A to 212°F	A to 70°F		
ALUMINUM AMMONIUM SULPHATE		A to 70°F				
ALUMINUM BROMIDE		A to 70°F	A to 212°F	A to 70°F		
ALUMINUM CHLORIDE	AB to 100% to 150°F	A to boiling	A to 212°F	A to 100% to 70°F		A to 70°F
ALUMINUM FLUORIDE	A to 70°F	A to 200°F	A to 212°F	A to 70°F		
ALUMINUM FORMATE		A to 70°F	A to 212°F			
ALUMINUM HYDROXIDE	A to 70°F B at 122°F	A to 200°F	A to 212°F	A to 70°F		
ALUMINUM NITRATE		A to 200°F	A to 212°F	A to 70°F		
ALUMINUM PHOSPHATE		A to 70°F	A to 212°F	A to 70°F		
ALUMINUM POTASSIUM SULPHATE	B 100% to 175°F A 10% to 70°F	A to 300°F	A to 212°F	A to 70°F		
ALUMINUM SALTS	A to 122°F AB to 140°F	A to 122°F	A to 212°F	A to 70°F		
ALUMINUM SODIUM SULPHATE		A to 70°F	A to 212°F	A to 70°F		
ALUMINUM SULPHATE	AB to 140°F	A to 100% to 200°F	A to 212°F	A to 70°F		A to 70°F
AMBREX 33, MOBILE		AB to 450°F		A to 70°F		
AMBREX 830, MOBILE		AB to 450°F		A to 70°F		
AMERICAN ASHES		A to 70°F				
AMINES		A to 200°F	*A to 212°F	A to 70°F	AB to 70°F	
AMINES MIXED (EG: ALLYL, ETHYL, ETC.)				AB to 70°F	A to 70°F	
AMINOBENZENE	A to 70°F B at 122°F	+0.3% weight gain 7 days 365°F	A to 300°F	A to 70°F	+0.7% vol 7 days 70°F A to 70°F	
AMINO DIMETHYLBENZENE		A to 70°F	A to 70°F			
AMINO ETHANOL	AB to 70°F	A to 300°F	A to 113°F	A to 70°F	+2% vol 7 days 158°F A to 70°F	
AMINO ETHYLETHANOLAMINE			*A to 212°F			
AMINO HEXANE		A to 70°F				
AMINO PENTANE		A to 70°F				
AMINO XYLENE						
AMMONIA ALUM		A to 70°F				
AMMONIA CUPRIC SULFATE		A to 70°F				
AMMONIA GAS, COLD	A to 122°F AB to 140°F	A to 122°F	A to 70°F	A to 70°F		
AMMONIA GAS, HOT	AB	A to 122°F	*A to 212°F	AB		
AMMONIA, ANHYDROUS LIQUID	AB to 122°F	A to boiling	A to 75°F *A to 212°F	A to 120°F	A to 70°F	A to 70°F
AMMONIA, AQUEOUS LIQUID	AB to 100% to 150°F	A to 300°F		+3.2% vol 28% conc 70°F	A 28% to 158°F	
AMMONIA, AQUEOUS LIQUID		A to 70°F ammonia H2O		A to 70°F		
AMMONIA, AQUEOUS LIQUID						
AMMONIA, AQUEOUS LIQUID						
AMMONIA LIQUORS (NH4OH·H2O)		A to 70°F				
AMMONIA + LITHIUM METAL SOLUTION						
AMMONIA NITRATE						
AMMONIAK		A to 70°F				
AMMONIATED CITRIC ACID		A to 70°F				
AMMONIATED LATEX		A to 70°F				
AMMONIUM ACETATE	A to 122°F	A to 122°F	A to 212°F			
AMMONIUM ALUM		A to 70°F				

	RECOMMENDATIONS	NITRILE NBR	ETHYLENE PROPYLENE EP, EPDM	FLUOROCARBON FKM	CHLOROPRENE CR	HYDROGENATED NITRILE HNBR
AMMONIUM BICARBONATE	NBR,EP,CR	A to 140°F	AB to 70°F	NR at 70°F	A to 70°F	
AMMONIUM BICHROMATE		A to 70°F	A to 70°F		A to 70°F	
AMMONIUM BIFLUORIDE		A to 140°F B at 212°F	A to 140°F B at 212°F	A to 140°F B at 212°F	A 100% to 140°F NR 10% at 70°F	
AMMONIUM CARBONATE	EP,CR,NBR,IIR	AB any conc to 200°F	A to 176°F B at 212°F	A to 212°F	A to 140°F AB any conc to 200°F	NR at 70°F
AMMONIUM CHLORIDE	EP,NBR,FKM,CR,IIR	AB any conc to 200°F NR 10% at 212°F	A to 176°F B at 212°F	A to 100% to 212°F	A to 100% to 140°F B 176-212°F	A to 70°F
AMMONIUM CHROMIC SULFATE		A to 70°F			A to 70°F	
AMMONIUM DICHROMATE		A to 70°F	A to 70°F		A to 70°F	
AMMONIUM DIPHOSPHATE		A to 70°F	A to 70°F	A to 70°F	A to 70°F	
AMMONIUM FLUORIDE		AB any conc to 104°F	A to 100% to 140°F	A to 100% to 140°F	AB to 100% to 100°F A 10% to 70°F	
AMMONIUM FLUORIDE ACID		B at 70°F			NR at 70°F	
AMMONIUM HYDRATE		A 38% to 200°F			A 38% to 200°F	
AMMONIUM HYDROXIDE	EP,IIR,CR,NBR	A/NR conc to 140°F A to 38% to 200°F	A to 100% to 160°F AB to 200°F	+7.5% vol 28 days 75°F AB to 70°F	A to conc to 158°F BC 176-200°F	
AMMONIUM HYDROXIDE			C at 212°F	B 104-140°F C at 140°F	A to 38% to 200°F	
AMMONIUM HYDROXIDE				A to 46% to 70°F		
AMMONIUM HYPOSULPHITE		A to 70°F	A to 70°F	A to 70°F	A to 70°F	
AMMONIUM METAPHOSPHATE		A any conc to 200°F	A to 176°F	A to 176°F	A any conc to 200°F	
AMMONIUM MURIATE		A any conc to 200°F	A to 70°F	A to 70°F	A any conc to 200°F	
AMMONIUM NITRATE	NBR,EP,IIR	A any conc to 200°F	A to 176°F AB to 200°F	A to 100% to 176°F	A any conc to 200°F	A to 70°F
AMMONIUM NITRITE	NBR,EP,IIR,CR,SBR	A to 70°F	A to 70°F	A to 70°F	A to 70°F	A to 70°F
AMMONIUM OXALATE		A/NR to 140°F	A to 100% to 140°F		A any conc to 200°F	
AMMONIUM PERSULFATE	EP,IIR	A any conc to 200°F	B at 70°F A 10% to 70°F	A to 100% to 140°F	A/NR conc at 70°F A to 10% to 70°F	NR at 70°F
AMMONIUM PHOSPHATE	NBR,EP,IIR,CR,SBR	A any conc to 200°F	A to 176°F AB to 200°F	A to 176°F	A to 140°F AB to 200°F	
AMMONIUM PHOSPHATE DIBASIC	NBR,EP,CR,SBR	A to 160°F	A to 140°F	A to 140°F	A to 140°F	
AMMONIUM PHOSPHATE MONOBASIC	NBR,EP,CR,SBR	A to 160°F	A to 140°F	A to 140°F	A to 140°F	
AMMONIUM PHOSPHATE TRIBASIC	NBR,EP,CR,SBR	A to 160°F	A to 140°F	A to 140°F	A to 70°F C at 140°F	
AMMONIUM RHODANATE		A to 70°F	A to 70°F	A to 70°F	A to 70°F	
AMMONIUM SALTS	NBR,EP,CR,SBR	A any conc to 200°F	A to 70°F	C at 70°F	A any conc to 200°F	A to 70°F
AMMONIUM SULFATE	NBR,EP,CR,IIR	A any conc to 200°F	A to conc to 176°F AB to 200°F	A to 100% to 176°F A 10% to boiling	A any conc to 200°F	
AMMONIUM SULFIDE	NBR,EP,CR,IIR	A to 70°F	A to 70°F	NR at 70°F	A to 70°F	A to 70°F
AMMONIUM SULFITE	NBR,EP	A to 70°F B at 150°F	A to 70°F	A to boiling	A to 70°F	
AMMONIUM SULPHATE	NBR,EP	A to 70°F	A to 120°F	A/NR at 70°F A to 5% to 70°F	A to 158°F	A to 70°F
AMMONIUM THIOCYANATE		A to 70°F	A to 70°F	A to 70°F	A to 70°F	
AMMONIUM THIOCYANIDE		A to 70°F	A to 70°F	A to 70°F	A to 70°F	
AMMONIUM THIOSULFATE	NBR	A to 140°F	A to 70°F	A to 70°F	A to 140°F	
AMOIL		NR at 70°F		C at 70°F		
AMYL ACETATE	EP,IIR	NR at 70°F	A to 200°F	+410% vol 7 days 70°F NR at 70°F	NR at 70°F	NR at 70°F
AMYL ACETIC ESTER		NR at 70°F	B at 70°F	NR at 70°F	NR at 70°F	
AMYL ACETONE		NR at 70°F		NR at 70°F		
AMYL ACID PHOSPHATE					NR at 70°F	
AMYL ALCOHOL	EP,IIR,NBR	AB any conc to 180°F	A to 200°F	A to 212°F	A to 158°F AB any conc to 180°F	B at 70°F
AMYL AMINE		C at 70°F	NR at 70°F	NR at 70°F	NR at 70°F	
AMYL BORATE	NBR,CR,T	AB any conc to 100°F	NR at 70°F	A to 70°F	A to 70°F	A to 70°F
AMYL BROMIDE		NR at 70°F	NR at 70°F	AB to 70°F	NR at 70°F	
AMYL CARBINOL		A to 70°F AB any conc to 120°F	C at 70°F	B at 70°F	B/NR at 70°F	

	STYRENE BUTADIENE SBR	POLYACRYLATE ACM	POLYURETHANE AU, EU	ISOBUTYLENE ISOPRENE IIR	POLYBUTADIENE BR	® AEROQUIP AQP
AMMONIUM BICARBONATE			C at 70°F	A to 70°F		
AMMONIUM BICHROMATE				A to 70°F		
AMMONIUM BIFLUORIDE				NR 10% at 70°F		
AMMONIUM CARBONATE	A to 70°F		A to 70°F	A to 70°F AB any conc to 185°F	A to 70°F	
AMMONIUM CHLORIDE	A to 70°F	A to 70°F	AB to 70°F *C at 140°F	A to 70°F AB any conc to 185°F	A to 70°F	AB to 70°F
AMMONIUM CHROMIC SULFATE				A to 70°F		
AMMONIUM DICHROMATE				A to 70°F		
AMMONIUM DIPHOSPHATE				A to 70°F		
AMMONIUM FLUORIDE				A to 70°F AB any conc to 150°F		
AMMONIUM FLUORIDE ACID				NR at 70°F		
AMMONIUM HYDRATE				A 38% to 185°F		
AMMONIUM HYDROXIDE	NR conc at 70°F AB 3 molar at 70°F	NR conc at 70°F NR 3 molar at 70°F	NR conc at 70°F NR 10-50% at 70°F	A to conc to 70°F A 38% to 185°F	C /NR conc at 70°F B 3 molar at 70°F	TEST
AMMONIUM HYDROXIDE			AB dilute to 70°F AB fumes to 50% to 70°F			
AMMONIUM HYDROXIDE						
AMMONIUM HYPOSULPHITE				A to 70°F		
AMMONIUM METAPHOSPHATE				AB any conc to 185°F		
AMMONIUM MURIATE			A to 70°F	A any conc to 185°F		
AMMONIUM NITRATE	A to 100% to 70°F	AB conc to 70°F NR 2N at 70°F	B /NR at 70°F	A any conc to 185°F	A to 70°F	AB to 70°F
AMMONIUM NITRITE	A to 70°F	C at 70°F	A to 70°F	A to 70°F	A to 70°F	
AMMONIUM OXALATE						
AMMONIUM PERSULFATE	NR 10-100% at 70°F	NR 10-100% at 70°F	NR 5-100% at 70°F	A any conc to 185°F	NR at 70°F	
AMMONIUM PHOSPHATE	A to 70°F		B at 70°F	A any conc to 185°F	A to 70°F	NR at 70°F
AMMONIUM PHOSPHATE DIBASIC	A to 70°F			A to 70°F		
AMMONIUM PHOSPHATE MONOBASIC	A to 70°F		B at 70°F	A to 70°F	A to 70°F	NR at 70°F
AMMONIUM PHOSPHATE TRIBASIC	A to 70°F			A to 70°F		
AMMONIUM RHODANATE				A to 70°F		
AMMONIUM SALTS	A to 70°F	C at 70°F		A any conc to 185°F		
AMMONIUM SULFATE	B at 70°F	NR at 70°F	A to 100% to 70°F	A any conc to 185°F	A to 70°F	AB to 70°F
AMMONIUM SULFIDE	B at 70°F	NR at 70°F	AB to 70°F	A to 70°F	A to 70°F	
AMMONIUM SULFITE	A to 70°F	C at 70°F		A to 70°F		
AMMONIUM SULPHATE	AB to 70°F	NR at 70°F	AB to 70°F	A to 70°F	AC to 70°F	AB to 70°F
AMMONIUM THIOCYANATE			AB to 70°F			
AMMONIUM THIOCYANIDE			B at 70°F	A to 70°F		
AMMONIUM THIOSULFATE				A to 70°F		
AMOIL				A to 70°F		
AMYL ACETATE	C /NR at 70°F	NR at 70°F	NR at 70°F	AB to 70°F	C /NR at 70°F	NR at 70°F
AMYL ACETIC ESTER			NR at 70°F	B at 70°F		
AMYL ACETONE				B at 70°F		
AMYL ACID PHOSPHATE				NR at 70°F		
AMYL ALCOHOL	B at 70°F	NR at 70°F	C /NR at 70°F	A to 70°F AB any conc to 180°F	B at 70°F	AB to 70°F
AMYL AMINE				B at 70°F		
AMYL BORATE	NR at 70°F			NR any conc at 70°F		
AMYL BROMIDE				NR at 70°F		
AMYL CARBINOL			NR at 70°F	B at 70°F		

C 21

	SYNTHETIC ISOPRENE IR	NATURAL ISOPRENE NR	CHLOROSULFONATED POLYETHYLENE CSM	FLUOROSILICONE FVMQ	SILICONE VMQ	CHEMRAZ FFKM
AMMONIUM BICARBONATE	A to 70°F	AB to 70°F	A to 70°F			A to 100% to 70°F
AMMONIUM BICHROMATE	A to 70°F	A to 70°F	A to 70°F			
AMMONIUM BIFLUORIDE	NR 10% at 70°F	AB 100% to 70°F NR 10% at 70°F				A to 70°F
AMMONIUM CARBONATE	A to 70°F	A to 70°F AB any conc to 150°F	AB to 70°F		BC at 70°F	A to 70°F
AMMONIUM CHLORIDE	A to 70°F	A to 70°F AB any conc to 150°F	A to 70°F	AC to 70°F	BC at 70°F	A 2N to 70°F
AMMONIUM CHROMIC SULFATE	A to 70°F	A to 70°F				
AMMONIUM DICHROMATE	A to 70°F	A to 70°F	A to 70°F			A to 70°F
AMMONIUM DIPHOSPHATE	A to 70°F	A to 70°F	A to 70°F		A to 70°F	A to 70°F
AMMONIUM FLUORIDE	B at 70°F	B any conc to 80°F				
AMMONIUM FLUORIDE ACID	NR at 70°F	NR at 70°F				
AMMONIUM HYDRATE	A 38% to 70°F	A 38% to 150°F				
AMMONIUM HYDROXIDE	C/NR at 70°F B 3 molar at 70°F	C/NR at 70°F B 38% to 150°F	A to conc to 200°F	+3% vol conc sol'n 7 days 70°F	+7% vol conc sol'n 7 days 70°F	A conc to 70°F
AMMONIUM HYDROXIDE		B 3 molar at 70°F		A to conc to 70°F	A to conc to 70°F	
AMMONIUM HYDROXIDE						
AMMONIUM HYPOSULPHITE	A to 70°F	A to 70°F	A to 70°F			
AMMONIUM METAPHOSPHATE	A to 70°F	A any conc to 150°F	AB to 70°F			A to 70°F
AMMONIUM MURIATE	A to 70°F	A any conc to 150°F	A to conc to 70°F		B at 70°F	
AMMONIUM NITRATE	A to conc to 70°F	A any conc to 150°F	A to 70°F AB to 200°F	BC at 70°F	BC at 70°F	A to 70°F
AMMONIUM NITRITE	AB to 70°F	AB to 70°F	A to 70°F		B at 70°F	A to 70°F
AMMONIUM OXALATE						A to 70°F
AMMONIUM PERSULFATE	A to conc to 70°F	A any conc to 150°F	A any conc to 70°F		NR 10% at 70°F	A to 100% to 70°F
AMMONIUM PHOSPHATE	A to 70°F	A any conc to 150°F	A to 70°F AB to 200°F	NR at 70°F	A to 70°F	A to 70°F
AMMONIUM PHOSPHATE DIBASIC	A to 70°F	A to 70°F	A to 70°F		A to 70°F	A to 70°F
AMMONIUM PHOSPHATE MONOBASIC	AC to 70°F	A to 70°F	A to 70°F	NR at 70°F	A to 70°F	A to 70°F
AMMONIUM PHOSPHATE TRIBASIC	AC to 70°F	AB to 70°F	A to 70°F		A to 70°F	A to 70°F
AMMONIUM RHODANATE	A to 70°F	A to 70°F	A to 70°F			
AMMONIUM SALTS	A to 70°F	A any conc to 150°F	A to 70°F	C at 70°F	A to 70°F	A to 70°F
AMMONIUM SULFATE	A to 70°F	A any conc to 150°F	A to 200°F	A to 70°F	A to 70°F	A to 70°F
AMMONIUM SULFIDE	A to 70°F	A to 70°F	A to 70°F			A to 100% to 70°F
AMMONIUM SULFITE	A to 70°F	A to 70°F	A to 70°F			A to 70°F
AMMONIUM SULPHATE	A to 70°F	A to 70°F	A to 200°F	A to 70°F	A to 70°F	
AMMONIUM THIOCYANATE		A to 70°F	A to 70°F			A to 70°F
AMMONIUM THIOCYANIDE	A to 70°F	A to 70°F	A to 70°F			
AMMONIUM THIOSULFATE	A to 70°F	A to 70°F	A to 70°F			A to 70°F
AMOIL	NR at 70°F	NR at 70°F	NR at 70°F			
AMYL ACETATE	NR at 70°F	C/NR at 70°F	NR at 70°F	NR at 70°F	NR at 70°F	A to 70°F
AMYL ACETIC ESTER	C at 70°F	C at 70°F	C at 70°F		NR at 70°F	
AMYL ACETONE	NR at 70°F	NR at 70°F	NR at 70°F			
AMYL ACID PHOSPHATE	NR at 70°F	NR at 70°F				
AMYL ALCOHOL	AB to 70°F	A to 70°F AB any conc to 150°F	A to 200°F	A to 70°F	NR at 70°F	A to 70°F
AMYL AMINE	C at 70°F	C at 70°F	C at 70°F			
AMYL BORATE	NR at 70°F	NR any conc at 70°F	A to 70°F		NR at 70°F	A to 70°F
AMYL BROMIDE	NR at 70°F	NR at 70°F	NR at 70°F			
AMYL CARBINOL	B at 70°F	B/NR at 70°F	A to 70°F		B at 70°F	

	POLYSULFIDE T	CHLORINATED POLYETHYLENE CM	EPICHLOROHYDRIN CO, ECO	® HYTREL COPOLYESTER TPE	® SANTOPRENE COPOLYMER TPO	® C-FLEX STYRENIC TPE
AMMONIUM BICARBONATE						
AMMONIUM BICHROMATE						
AMMONIUM BIFLUORIDE						
AMMONIUM CARBONATE			B at 70°F			AB to 70°F
AMMONIUM CHLORIDE	A to 70°F	AB to 70°F	A to 70°F	A to 70°F *AB to 140°F		AB to 70°F
AMMONIUM CHROMIC SULFATE						
AMMONIUM DICHROMATE						
AMMONIUM DIPHOSPHATE						
AMMONIUM FLUORIDE						
AMMONIUM FLUORIDE ACID						
AMMONIUM HYDRATE						
AMMONIUM HYDROXIDE	NR at 70°F NR 3M at 70°F	A to 70°F	B at 70°F	A/NR to 140°F		AB to 140°F
AMMONIUM HYDROXIDE						
AMMONIUM HYDROXIDE						
AMMONIUM HYPOSULPHITE						
AMMONIUM METAPHOSPHATE						
AMMONIUM MURIATE						
AMMONIUM NITRATE	AB to 70°F	AB to 70°F	A to conc to 70°F	AB to 70°F		AB to 70°F
AMMONIUM NITRITE						
AMMONIUM OXALATE						
AMMONIUM PERSULFATE						
AMMONIUM PHOSPHATE	A to 70°F	AB to 70°F	A to 70°F	AB to 70°F		AB to 70°F
AMMONIUM PHOSPHATE DIBASIC	A to 70°F	A to 70°F				
AMMONIUM PHOSPHATE MONOBASIC	A to 70°F	AB to 70°F	A to 70°F	AB to 70°F		AB to 70°F
AMMONIUM PHOSPHATE TRIBASIC	A to 70°F					
AMMONIUM RHODANATE						
AMMONIUM SALTS						
AMMONIUM SULFATE	NR at 70°F	A to 70°F		AB to 70°F *C at 140°F		AB to 70°F
AMMONIUM SULFIDE	NR at 70°F	A to 70°F				
AMMONIUM SULFITE						
AMMONIUM SULPHATE	NR at 70°F	A to 70°F		AC at 70°F		AB to 70°F
AMMONIUM THIOCYANATE						
AMMONIUM THIOCYANIDE						
AMMONIUM THIOSULFATE						
AMOIL						
AMYL ACETATE	NR at 70°F	C/NR at 70°F	NR at 70°F	BC at 70°F	NR at 70°F	NR at 70°F
AMYL ACETIC ESTER						
AMYL ACETONE						
AMYL ACID PHOSPHATE						
AMYL ALCOHOL	B at 70°F	A to 70°F	A to 70°F	A to 70°F	A to 70°F	NR at 70°F
AMYL AMINE						
AMYL BORATE	A to 70°F					
AMYL BROMIDE						
AMYL CARBINOL						

	ETHYLENE ACRYLIC EA	POLYALLOMER LINEAR COPOLYMER	NYLON 11 POLYAMIDE	NYLON 12 POLYAMIDE	ETHYLENE VINYL ACETATE EVA	POLYVINYLCHLORIDE FLEXIBLE PVC
AMMONIUM BICARBONATE			A to 70°F		B at 70°F	AB to 70°F
AMMONIUM BICHROMATE						
AMMONIUM BIFLUORIDE						
AMMONIUM CARBONATE		A to 70°F	AB to 70°F		A to 70°F	AB to 70°F *AB to 140°F
AMMONIUM CHLORIDE	A to 70°F	A to 70°F	A/NR at 70°F	AB 10% to 70°F	AB to 140°F	AB to 150°F
AMMONIUM CHROMIC SULFATE						
AMMONIUM DICHROMATE						
AMMONIUM DIPHOSPHATE			AB to 70°F			A to 70°F
AMMONIUM FLUORIDE						
AMMONIUM FLUORIDE ACID						
AMMONIUM HYDRATE						
AMMONIUM HYDROXIDE	NR at 70°F A 3 molar at 70°F	A to 100% to 70°F B 30% at 122°F	+2.5% vol 132 days 70°F A to 70°F		AB to 70°F *AB to 140°F	+0.5% vol 10% conc 14 days 70°F
AMMONIUM HYDROXIDE		A 5% to 122°F	AB to 194°F			AB to 70°F B to 122°F
AMMONIUM HYDROXIDE						A 5% to 122°F
AMMONIUM HYPOSULPHITE						
AMMONIUM METAPHOSPHATE						
AMMONIUM MURIATE						
AMMONIUM NITRATE		A to 70°F	AB to 70°F		AB to 70°F	AB to 70°F *AB to 140°F
AMMONIUM NITRITE						AB to 70°F
AMMONIUM OXALATE		A to 70°F B at 122°F				A to 122°F
AMMONIUM PERSULFATE						*AB to 140°F
AMMONIUM PHOSPHATE		A to 70°F	A to 70°F		AB to 70°F	AB to 70°F *AB to 140°F
AMMONIUM PHOSPHATE DIBASIC						B at 70°F
AMMONIUM PHOSPHATE MONOBASIC		A to 70°F	A to 70°F		AB to 70°F	AB to 70°F
AMMONIUM PHOSPHATE TRIBASIC						
AMMONIUM RHODANATE						
AMMONIUM SALTS		A to 122°F	AB to 70°F	AB to 100% to 70°F		A to 70°F B at 122°F
AMMONIUM SULFATE	A to 70°F	A to 70°F	AB to 70°F C at 140°F		AB to 140°F	AB to 140°F
AMMONIUM SULFIDE						AB to 70°F NR at 140°F
AMMONIUM SULFITE						
AMMONIUM SULPHATE		A to 70°F	AB to 70°F		AB to 70°F	AB to 70°F
AMMONIUM THIOCYANATE						*AB to 140°F
AMMONIUM THIOCYANIDE						
AMMONIUM THIOSULFATE						
AMOIL						
AMYL ACETATE	NR at 70°F	B at 70°F C at 122°F	AB to 70°F BC at 140°F	BC at 70°F	B/NR at 70°F	NR at 70°F
AMYL ACETIC ESTER						
AMYL ACETONE						
AMYL ACID PHOSPHATE						
AMYL ALCOHOL	B/NR at 70°F	B at 70°F	A to 70°F *BC at 140°F	AB to 70°F	B at 70°F *NR at 140°F	AB to 70°F C at 70°F (food grades)
AMYL AMINE						
AMYL BORATE						
AMYL BROMIDE						
AMYL CARBINOL						

	POLYETHYLENE LOW DENSITY LDPE	®TEFLON FEP	® KALREZ PERFLUORINATED ELASTOMER	® FLUORAZ FLUORINATED COPOLYMER	® AFLAS FLUORINATED COPOLYMER	®NORPRENE COPOLYMER TPO
AMMONIUM BICARBONATE	A to 70°F	A to 70°F	A to 212°F	NR at 70°F A aqueous to 70°F		*A to 70°F
AMMONIUM BICHROMATE		A to 70°F	A to 212°F			*A to 70°F
AMMONIUM BIFLUORIDE		A to 100% to 200°F	A to 122°F *A to 212°F	AB to 70°F		
AMMONIUM CARBONATE	A to 70°F	A to 300°F	A to 212°F	NR at 70°F	+4.1% vol 7 days 212°F A to 212°F	*A to 70°F
AMMONIUM CHLORIDE	AB to 140°F	A to 100% to 250°F	A to 212°F	A/NR at 70°F A 2N to 70°F		*A to 70°F
AMMONIUM CHROMIC SULFATE		A to 70°F				*A to 70°F
AMMONIUM DICHROMATE		A to 70°F	A to 212°F			*A to 70°F
AMMONIUM DIPHOSPHATE		A to 70°F	A to 212°F	NR at 70°F		
AMMONIUM FLUORIDE	A 10% to 70°F	A to 100% to 70°F A to 25% to 200°F	*A to 212°F			
AMMONIUM FLUORIDE ACID		A to 70°F				
AMMONIUM HYDRATE		A 38% to 70°F				
AMMONIUM HYDROXIDE	AB to 140°F A 5% to 122°F	A to 100% to 300°F	A to 70°F A 35% to 113°F	A 3M to conc to 70°F	A 28% to 158°F	A to 70°F
AMMONIUM HYDROXIDE			*A conc to 212°F			
AMMONIUM HYDROXIDE						
AMMONIUM HYPOSULPHITE		A to 70°F				
AMMONIUM METAPHOSPHATE	A to 70°F	A to 70°F	A to 212°F			
AMMONIUM MURIATE		A to 70°F				
AMMONIUM NITRATE	A to 70°F	A to 200°F	A to 212°F	A/NR at 70°F A 2N to 70°F	+0% vol 7 days 212°F A to 212°F	
AMMONIUM NITRITE		A to 70°F	A to 300°F	A to 70°F		
AMMONIUM OXALATE	A to 70°F B at 122°F	A to 122°F	A to 212°F			
AMMONIUM PERSULFATE		A to 70°F A 25% to 200°F	A to 212°F	A to 100% to 70°F		
AMMONIUM PHOSPHATE	AB to 70°F	A to 70°F	A to 212°F	A/NR at 70°F		
AMMONIUM PHOSPHATE DIBASIC	B at 70°F	A to 200°F	A to 212°F	B/NR at 70°F		
AMMONIUM PHOSPHATE MONOBASIC	AB to 70°F	A to 70°F	A to 212°F	A/NR at 70°F		
AMMONIUM PHOSPHATE TRIBASIC	AB to 70°F	A to 70°F	A to 212°F	B/NR at 70°F		
AMMONIUM RHODANATE		A to 70°F				
AMMONIUM SALTS	A to 122°F AB at 140°F	A to 122°F	A to 212°F	A to 70°F		
AMMONIUM SULFATE	A to 70°F AB at 140°F	A to 100% to 200°F	A to 212°F	A/NR at 70°F		
AMMONIUM SULFIDE	A to 70°F	A to 70°F	A to 212°F	A to 70°F		
AMMONIUM SULFITE		A to 150°F NR boiling	A to 212°F			
AMMONIUM SULPHATE	A to 70°F	A to 70°F	A to 212°F			
AMMONIUM THIOCYANATE	A to 70°F	A to 70°F	A to 212°F	B at 70°F		
AMMONIUM THIOCYANIDE		A to 70°F		C at 70°F		
AMMONIUM THIOSULFATE		A to 70°F	A to 212°F			
AMOIL		A to 70°F				
AMYL ACETATE	A/NR to 140°F	A to 300°F	A to 212°F	NR at 70°F		A to 70°F
AMYL ACETIC ESTER		A to 70°F				
AMYL ACETONE		A to 70°F				
AMYL ACID PHOSPHATE		A to 70°F				
AMYL ALCOHOL	AB to 70°F B/NR at 140°F	A to 70°F	A to 70°F *A to 212°F	A to 70°F	C at 70°F	BC at 70°F
AMYL AMINE		A to 70°F				
AMYL BORATE		A to 70°F	A to 212°F	A to 70°F		
AMYL BROMIDE		A to 70°F				
AMYL CARBINOL		A to 70°F				

C 25

	RECOMMENDATIONS	NITRILE NBR	ETHYLENE PROPYLENE EP, EPDM	FLUOROCARBON FKM	CHLOROPRENE CR	HYDROGENATED NITRILE HNBR
AMYL CHLORIDE	FKM, FVMQ	NR at 70°F	NR at 70°F	A to 200°F	NR at 70°F	A to 70°F
AMYL CHLORONAPHTHALENE	FKM, T	B/NR to 80°F	NR at 70°F	A to 140°F	NR any conc at 70°F	NR at 70°F
AMYL ETHER		C at 70°F	NR at 70°F		NR at 70°F	
AMYL HYDRATE		AB any conc to 180°F			AB any conc to 180°F	
AMYL HYDRIDE	NBR, FKM	A to 250°F	NR at 70°F	A to 140°F	A/NR at 70°F	
AMYL IODIDE		NR at 70°F	NR at 70°F		NR at 70°F	
AMYL NAPHTHALENE	FKM, FVMQ, T	NR at 70°F	NR at 70°F	A to 70°F	NR at 70°F	NR at 70°F
AMYL NITRATE	EP	A to 70°F	AB to 200°F			
AMYL OLEATE		B at 70°F				
AMYL PHENOL		NR at 70°F		A to 70°F		
AMYL PHTHALATE		NR at 70°F		C at 70°F		
AMYLENE		B at 70°F	NR at 70°F	A to 70°F	A to 70°F	
ANDEROL L-774 (DIESTER)	NBR, FKM, FVMQ	A to 275°F	NR at 70°F	+8% vol 7 days 400°F A to 400°F	NR at 70°F	
ANDEROL L-826 (DIESTER)	FKM	B at 70°F	NR at 70°F	+11% vol 28 days 300°F A to 70°F	NR at 70°F	B at 70°F
ANDEROL L-829 (DIESTER	FKM	B at 70°F	NR at 70°F	+5% vol 28 days 300°F A to 300°F	NR at 70°F	B at 70°F
ANEHOLE (ANETHOL)		NR at 70°F		B at 70°F		
ANG-25 (GLYCERAL ESTER)	EP, FKM	B at 70°F	A to 70°F	A to 70°F	B at 70°F	B at 70°F
ANG-25 (DIESTER BASE)	FKM	B at 70°F	NR at 70°F	A to 70°F	NR at 70°F	B at 70°F
ANHYDROUS AMMONIA	EP, CR, IIR	AB to 70°F C 140–170°F	A to 140°F	NR at 70°F	A to 240°F	B at 70°F
ANHYDROUS HYDRAZINE	EP, SBR	NR at 70°F	AB to 70°F	NR at 70°F	B at 70°F	
ANHYDROUS HYDROGEN FLUORIDE	EP, IIR	NR at 70°F	AB to 70°F	NR at 70°F	NR at 70°F	
ANILINE	EP, IIR	+210% vol 3 days 70°F NR at 70°F	+1% vol 3 days 70°F AB to 200°F	+3% vol 3 days 70°F +25% vol 28 days 158°F	+60% vol 3 days 70°F NR at 70°F	
ANILINE				dissolved 7 days 300°F A to 70°F		
ANILINE				C at 158°F NR at 300°F		
ANILINE CHLORIDE		NR any conc at 70°F	B at 70°F	B at 70°F	NR at 70°F	
ANILINE DYES	EP, IIR	NR at 70°F	AB to 200°F	B at 70°F	BC at 70°F	NR at 70°F
ANILINE HYDROCHLORIDE	EP, IIR	NR any conc at 70°F	B to 200°F	AB to 140°F	NR at 70°F	
ANILINE OILS	EP	NR at 70°F	B at 70°F	C at 70°F	NR at 70°F	NR at 70°F
ANILINE SALTS		NR any conc at 70°F	B at 70°F	B at 70°F	NR at 70°F	
ANIMAL FATS	NBR, ACM, FKM, AU, EP	A to 200°F	A to 140°F	A to 140°F	B at 70°F C at 140°F	A to 70°F
ANIMAL GELATIN		A to 70°F	A to 70°F	A to 70°F	A to 70°F	
ANIMAL GLUE		A to 70°F	B at 70°F	A to 70°F	A to 70°F	
ANIMAL GLYCERIN		A to 70°F	B at 70°F	A to 70°F		
ANIMAL GREASE		A to 70°F	A to 70°F	A to 70°F	B at 70°F	
ANIMAL OIL	NBR, FKM, FVMQ, EP ACM	A to 140°F	A to 140°F	A to 70°F	AB to 70°F	A to 70°F
ANISE CAMPHOR		NR at 70°F		B at 70°F		
ANSUL ETHER 161, 181	T	C at 70°F	C at 70°F	NR at 70°F	NR at 70°F	C at 70°F
ANT OIL	EP, IIR	NR at 70°F	AB to 160°F NR at 212°F	NR at 70°F	BC at 70°F C/NR at 140°F	NR at 70°F
ANTICHLOR		A to 70°F AB any conc to 160°F	A to 70°F	A to 70°F	A to 70°F AB any conc to 160°F	
ANTIFREEZE		A to 140°F	A to 70°F	A to 70°F	A to 70°F C at 140°F	
ANTIMONIC CHLORIDE		NR at 70°F				
ANTIMONOUS CHLORIDE		B at 70°F	A to 70°F	A to 70°F		
ANTIMONY CHLORIDES	NBR	AB to 70°F	A to 70°F	AB to 70°F	NR at 70°F	
ANTIMONY PENTAFLUORIDES		NR at 70°F			NR at 70°F	

	STYRENE BUTADIENE SBR	POLYACRYLATE ACM	POLYURETHANE AU, EU	ISOBUTYLENE ISOPRENE IIR	POLYBUTADIENE BR	® AEROQUIP AQP
AMYL CHLORIDE	NR at 70°F	NR at 70°F	C at 70°F	NR at 70°F	NR at 70°F	
AMYL CHLORONAPHTHALENE	NR at 70°F	NR at 70°F	NR at 70°F	NR any conc at 70°F	NR at 70°F	
AMYL ETHER				NR at 70°F		
AMYL HYDRATE						
AMYL HYDRIDE	C /NR at 70°F	A to 70°F	B /NR at 70°F	C /NR at 70°F		
AMYL IODIDE				NR at 70°F		
AMYL NAPHTHALENE	NR at 70°F	B at 70°F	NR at 70°F	NR at 70°F		
AMYL NITRATE						
AMYL OLEATE				NR at 70°F		
AMYL PHENOL				NR at 70°F		
AMYL PHTHALATE				A to 70°F		
AMYLENE				NR at 70°F		
ANDEROL L-774 (DIESTER)	NR at 70°F	AB to 400°F	NR at 70°F	NR at 70°F	NR at 70°F	
ANDEROL L-826 (DIESTER)	NR at 70°F	B at 70°F	NR at 70°F	NR at 70°F	NR at 70°F	
ANDEROL L-829 (DIESTER	NR at 70°F	B at 70°F	NR at 70°F	NR at 70°F	NR at 70°F	
ANEHOLE (ANETHOL)				NR at 70°F		
ANG-25 (GLYCERAL ESTER)	B at 70°F	NR at 70°F	NR at 70°F	B at 70°F	B at 70°F	
ANG-25 (DIESTER BASE)	NR at 70°F	B at 70°F	NR at 70°F	NR at 70°F	NR at 70°F	
ANHYDROUS AMMONIA			NR at 70°F	NR at 70°F		
ANHYDROUS HYDRAZINE	A to 70°F	NR at 70°F	NR at 70°F	B at 70°F	NR at 70°F	
ANHYDROUS HYDROGEN FLUORIDE	NR at 70°F	NR at 70°F	NR at 70°F	A to 70°F	NR at 70°F	
ANILINE	+24% vol 3 days 70°F NR at 70°F	+272% vol 3 days 70°F NR at 70°F	+261% vol 3 days 70°F NR at 70°F	+5% vol 3 days 70°F A to 70°F	NR at 70°F	AB to 70°F
ANILINE						
ANILINE						
ANILINE CHLORIDE			NR at 70°F	NR any conc at 70°F		
ANILINE DYES	B at 70°F	NR at 70°F	NR at 70°F	AB to 80°F	B at 70°F	NR at 70°F
ANILINE HYDROCHLORIDE	C at 70°F	NR at 70°F	NR at 70°F	B /NR at 70°F	NR at 70°F	
ANILINE OILS	NR at 70°F	NR at 70°F	NR at 70°F	AB to 75°F	NR at 70°F	AB to 70°F
ANILINE SALTS			NR at 70°F	C /NR any conc at 70°F		
ANIMAL FATS	NR at 70°F	A to 70°F	AC to 70°F	BC at 70°F	NR at 70°F	
ANIMAL GELATIN			A to 70°F	A to 70°F		
ANIMAL GLUE			A to 70°F	B at 70°F		
ANIMAL GLYCERIN				B at 70°F		
ANIMAL GREASE			A to 70°F	C at 70°F		
ANIMAL OIL	NR at 70°F	A to 70°F	BC at 70°F	B at 70°F	NR at 70°F	
ANISE CAMPHOR				NR at 70°F		
ANSUL ETHER 161, 181	NR at 70°F	NR at 70°F	AB to 70°F	C at 70°F	NR at 70°F	
ANT OIL	NR at 70°F	NR at 70°F	C /NR at 70°F	A to 70°F B to 185°F	C /NR at 70°F	AB to 70°F
ANTICHLOR			A to 70°F	A to 70°F AB to 185°F		
ANTIFREEZE	AB to 70°F	NR at 70°F	NR at 70°F			
ANTIMONIC CHLORIDE				NR at 70°F		
ANTIMONOUS CHLORIDE				A to 70°F		
ANTIMONY CHLORIDES				A to 70°F		
ANTIMONY PENTAFLUORIDES				NR at 70°F		

	SYNTHETIC ISOPRENE IR	NATURAL ISOPRENE NR	CHLOROSULFONATED POLYETHYLENE CSM	FLUOROSILICONE FVMQ	SILICONE VMQ	CHEMRAZ FFKM
AMYL CHLORIDE	NR at 70°F	NR at 70°F	NR at 70°F	B at 70°F	NR at 70°F	A to 70°F
AMYL CHLORONAPHTHALENE	NR at 70°F	NR any conc at 70°F	NR at 70°F	B at 70°F	NR at 70°F	A to 70°F
AMYL ETHER	NR at 70°F	NR at 70°F	C at 70°F			
AMYL HYDRATE	A to 70°F	A to 70°F AB any conc to 150°F				
AMYL HYDRIDE	NR at 70°F	NR at 70°F	BC at 70°F	A to 70°F	NR at 70°F	
AMYL IODIDE	NR at 70°F	NR at 70°F	NR at 70°F			
AMYL NAPHTHALENE	NR at 70°F	NR at 70°F	NR at 70°F	A to 70°F	NR at 70°F	A to 70°F
AMYL NITRATE						A to 70°F
AMYL OLEATE	NR at 70°F	NR at 70°F	NR at 70°F		NR at 70°F	
AMYL PHENOL	NR at 70°F	NR at 70°F	NR at 70°F			A to 70°F
AMYL PHTHALATE	NR at 70°F	NR at 70°F	NR at 70°F			
AMYLENE	NR at 70°F	NR at 70°F	NR at 70°F			
ANDEROL L-774 (DIESTER)	NR at 70°F	NR at 70°F	NR at 70°F	B to 400°F	NR at 70°F	A to 70°F
ANDEROL L-826 (DIESTER)	NR at 70°F	NR at 70°F	NR at 70°F	B at 70°F	NR at 70°F	A to 70°F
ANDEROL L-829 (DIESTER	NR at 70°F	NR at 70°F	NR at 70°F	B at 70°F	NR at 70°F	A to 70°F
ANEHOLE (ANETHOL)	NR at 70°F	NR at 70°F	NR at 70°F			
ANG-25 (GLYCERAL ESTER)	B at 70°F	B at 70°F	B at 70°F	B at 70°F	B at 70°F	A to 70°F
ANG-25 (DIESTER BASE)	NR at 70°F	NR at 70°F	NR at 70°F	B at 70°F	B at 70°F	A to 70°F
ANHYDROUS AMMONIA	NR at 70°F	NR at 70°F	B/NR at 70°F	NR at 70°F	BC to 140°F C at 240°F	AB to 70°F B at 70°F white
ANHYDROUS HYDRAZINE	NR at 70°F	NR at 70°F	B at 70°F	NR at 70°F	B/NR at 70°F	B at 70°F
ANHYDROUS HYDROGEN FLUORIDE	NR at 70°F	NR at 70°F		NR at 70°F	NR at 70°F	
ANILINE	NR at 70°F	NR at 70°F	+35% vol 7 days 70°F C/NR at 70°F	-4% vol 3 days 70°F BC at 70°F	+2% vol 3 days 70°F +3% vol 7 days 70°F	A to 70°F
ANILINE			NR at 100°F		AB to 70°F	
ANILINE						
ANILINE CHLORIDE	B at 70°F	B any conc to 80°F	NR at 70°F		NR at 70°F	
ANILINE DYES	B at 70°F	B to 80°F	B at 70°F	B at 70°F	A/NR at 70°F	A to 70°F
ANILINE HYDROCHLORIDE	AB to 70°F	AB any conc to 80°F	NR at 70°F	B at 70°F	C/NR at 70°F	A to 70°F
ANILINE OILS	NR at 70°F	NR at 70°F	C/NR at 70°F	C at 70°F	NR at 70°F	A to 70°F
ANILINE SALTS	B at 70°F	B any conc to 80°F	NR at 70°F		NR at 70°F	
ANIMAL FATS	NR at 70°F	B/NR at 70°F	BC at 70°F	A to 140°F	B at 70°F NR at 140°F	A to 70°F
ANIMAL GELATIN	A to 70°F	A to 70°F	A to 70°F		A to 70°F	
ANIMAL GLUE	B at 70°F	B at 70°F	A to 70°F		A to 70°F	
ANIMAL GLYCERIN	NR at 70°F	NR at 70°F			B at 70°F	
ANIMAL GREASE	NR at 70°F	NR at 70°F	B/NR at 70°F	A to 70°F	B at 70°F	
ANIMAL OIL	NR at 70°F	NR at 70°F	B/NR at 70°F	A to 70°F	B/NR at 70°F	A to 70°F
ANISE CAMPHOR	NR at 70°F	NR at 70°F	NR at 70°F			
ANSUL ETHER 161, 181	NR at 70°F	NR at 70°F	NR at 70°F	C at 70°F	NR at 70°F	A to 70°F
ANT OIL	NR at 70°F	NR at 70°F	B/NR at 70°F	NR at 70°F	C/NR at 70°F	B at 70°F
ANTICHLOR	A to 70°F	A to 70°F AB any conc to 80°F	A to 70°F		A to 70°F	
ANTIFREEZE		A to 70°F	AB to 70°F		AC to 70°F	
ANTIMONIC CHLORIDE	NR at 70°F	NR at 70°F	NR at 70°F			
ANTIMONOUS CHLORIDE						
ANTIMONY CHLORIDES			AB 50% to 70°F			
ANTIMONY PENTAFLUORIDES	NR at 70°F	NR at 70°F	NR at 70°F			

	POLYSULFIDE T	CHLORINATED POLYETHYLENE CM	EPICHLOROHYDRIN CO, ECO	® HYTREL COPOLYESTER TPE	® SANTOPRENE COPOLYMER TPO	® C-FLEX STYRENIC TPE
AMYL CHLORIDE		C at 70°F				NR at 70°F
AMYL CHLORONAPHTHALENE	AC at 70°F					
AMYL ETHER						
AMYL HYDRATE						
AMYL HYDRIDE				AB to 70°F		
AMYL IODIDE						
AMYL NAPHTHALENE	BC at 70°F					
AMYL NITRATE						
AMYL OLEATE						
AMYL PHENOL						
AMYL PHTHALATE						
AMYLENE				NR at 70°F		
ANDEROL L-774 (DIESTER)	B at 70°F					
ANDEROL L-826 (DIESTER)	B at 70°F					
ANDEROL L-829 (DIESTER	B at 70°F					
ANEHOLE (ANETHOL)						
ANG-25 (GLYCERAL ESTER)	NR at 70°F					
ANG-25 (DIESTER BASE)	B at 70°F					
ANHYDROUS AMMONIA				NR at 70°F		
ANHYDROUS HYDRAZINE	NR at 70°F					
ANHYDROUS HYDROGEN FLUORIDE						
ANILINE	+325% vol 3 days 70°F NR at 70°F	AB to 70°F	NR at 70°F	+31-93% vol 7 days 70°F	*A to 70°F	A/NR at 70°F
ANILINE				NR at 70°F		
ANILINE						
ANILINE CHLORIDE						
ANILINE DYES	B/NR at 70°F	NR at 70°F	NR at 70°F	NR at 70°F		
ANILINE HYDROCHLORIDE	NR at 70°F		NR at 70°F			A/NR at 70°F
ANILINE OILS	B at 70°F	AB to 70°F		NR at 70°F		
ANILINE SALTS						
ANIMAL FATS	NR at 70°F		A to 70°F	AB to 70°F	A to 212°F	
ANIMAL GELATIN						
ANIMAL GLUE						
ANIMAL GLYCERIN						
ANIMAL GREASE						
ANIMAL OIL	NR at 70°F		A to 70°F	A to 70°F	A to 212°F	
ANISE CAMPHOR						
ANSUL ETHER 161, 181	A to 70°F					
ANT OIL	B/NR at 70°F	A to 70°F	NR at 70°F	AB to 70°F	A to 70°F	
ANTICHLOR						
ANTIFREEZE	AB to 70°F				A 50% to 212°F	
ANTIMONIC CHLORIDE						
ANTIMONOUS CHLORIDE						
ANTIMONY CHLORIDES						
ANTIMONY PENTAFLUORIDES						

	ETHYLENE ACRYLIC EA	POLYALLOMER LINEAR COPOLYMER	NYLON 11 POLYAMIDE	NYLON 12 POLYAMIDE	ETHYLENE VINYL ACETATE EVA	POLYVINYLCHLORIDE FLEXIBLE PVC
AMYL CHLORIDE		NR at 70°F	A to 70°F		C/NR at 70°F	C/NR at 70°F
AMYL CHLORONAPHTHALENE						NR any conc at 70°F
AMYL ETHER						
AMYL HYDRATE						
AMYL HYDRIDE			AB to 70°F			AC to 70°F NR at 150°F
AMYL IODIDE						
AMYL NAPHTHALENE						NR at 70°F
AMYL NITRATE						
AMYL OLEATE						
AMYL PHENOL						
AMYL PHTHALATE						
AMYLENE						
ANDEROL L-774 (DIESTER)						NR at 70°F
ANDEROL L-826 (DIESTER)						
ANDEROL L-829 (DIESTER						
ANEHOLE (ANETHOL)			AB to 70°F			
ANG-25 (GLYCERAL ESTER)						
ANG-25 (DIESTER BASE)						
ANHYDROUS AMMONIA			AB to 70°F			C/NR at 70°F
ANHYDROUS HYDRAZINE						
ANHYDROUS HYDROGEN FLUORIDE	NR at 70°F					
ANILINE	*NR at 70°F	B at 70°F C at 122°F	C at 70°F NR at 105°F	C/NR at 70°F	C/NR at 70°F NR at 140°F	C/NR at 70°F NR at 70°F (food grade
ANILINE						
ANILINE						
ANILINE CHLORIDE						NR any conc at 70°
ANILINE DYES			NR at 70°F		NR at 70°F	NR at 70°F
ANILINE HYDROCHLORIDE						NR any conc at 70° all grades
ANILINE OILS			NR at 70°F			C at 70°F NR at 150°F
ANILINE SALTS						NR any conc at 70°
ANIMAL FATS	A to 70°F	A to 70°F	A to 70°F		B at 70°F	AB to 70°F C at 150°F
ANIMAL GELATIN						
ANIMAL GLUE						
ANIMAL GLYCERIN						
ANIMAL GREASE						
ANIMAL OIL	A to 70°F		A to 70°F			
ANISE CAMPHOR						
ANSUL ETHER 161, 181						
ANT OIL	NR at 70°F	NR at 70°F	BC at 70°F NR at 194°F		B/NR at 70°F	A/NR at 70°F NR at 150°F
ANTICHLOR						
ANTIFREEZE				AB to 70°F		B at 70°F
ANTIMONIC CHLORIDE						
ANTIMONOUS CHLORIDE						
ANTIMONY CHLORIDES						
ANTIMONY PENTAFLUORIDES						

	POLYETHYLENE LOW DENSITY LDPE	® TEFLON FEP	® KALREZ PERFLUORINATED ELASTOMER	® FLUORAZ FLUORINATED COPOLYMER	® AFLAS FLUORINATED COPOLYMER	® NORPRENE COPOLYMER TPO
AMYL CHLORIDE	NR at 70°F	A to 200°F	A to 212°F	A to 70°F		
AMYL CHLORONAPHTHALENE		A to 70°F	A to 212°F	B at 70°F		
AMYL ETHER		A to 70°F				
AMYL HYDRATE		A to 70°F				
AMYL HYDRIDE	NR at 70°F	A to 70°F		A to 70°F		
AMYL IODIDE		A to 70°F				
AMYL NAPHTHALENE		A to 70°F	A to 212°F	B at 70°F		
AMYL NITRATE			A to 212°F		NR at 70°F	
AMYL OLEATE		A to 70°F				
AMYL PHENOL		A to 70°F	A to 212°F			
AMYL PHTHALATE		A to 70°F				
AMYLENE		A to 70°F				
ANDEROL L-774 (DIESTER)		A to 70°F	A to 400°F	A to 70°F	+6% vol 3 days 212°F A to 70°F	
ANDEROL L-826 (DIESTER)						
ANDEROL L-829 (DIESTER			A to 400°F	B at 70°F		
ANEHOLE (ANETHOL)	NR at 70°F	A to 70°F				
ANG-25 (GLYCERAL ESTER)				B at 70°F		
ANG-25 (DIESTER BASE)				B at 70°F		
ANHYDROUS AMMONIA	AB to 122°F	A to boiling	A to 75°F	AC to 120°F	A to 70°F	A to 70°F
ANHYDROUS HYDRAZINE						
ANHYDROUS HYDROGEN FLUORIDE	AB to 70°F	AB to 70°F	B at 70°F			
ANILINE	A to 70°F B at 122°F	+0.3% weight gain 7 days 365°F	A to 300°F	A to 70°F	+0.7% vol 7 days 70°F A to 70°F	
ANILINE		A to 365°F				
ANILINE						
ANILINE CHLORIDE		A to 70°F				
ANILINE DYES	B/NR at 70°F	A to 70°F	A to 212°F	A to 70°F		
ANILINE HYDROCHLORIDE		A to 212°F	A to 212°F	A to 70°F	A to 70°F	
ANILINE OILS		A to 70°F		B at 70°F		
ANILINE SALTS		A to 70°F				
ANIMAL FATS	AC to 70°F NR at 140°F	A to 300°F	A to 212°F	A to 70°F		
ANIMAL GELATIN		A to 70°F				
ANIMAL GLUE		A to 70°F				
ANIMAL GLYCERIN		A to 70°F				
ANIMAL GREASE		A to 70°F				
ANIMAL OIL	B/NR at 70°F	A to 356°F	A to 212°F	A to 70°F	A to 70°F	
ANISE CAMPHOR		A to 70°F				
ANSUL ETHER 161, 181		A to 70°F	A to 212°F	A to 70°F		
ANT OIL	A/NR at 70°F	A to 300°F	A to 70°F *A to 158°F	A to 70°F		BC at 70°F
ANTICHLOR		A to 70°F	A to 70°F *A to 158°F	A to 70°F		
ANTIFREEZE	B at 70°F	A to 70°F	*A to 212°F	AC at 70°F water, alcohol or glycol		
ANTIMONIC CHLORIDE		A to 70°F				
ANTIMONOUS CHLORIDE		A to 70°F				
ANTIMONY CHLORIDES	A to 70°F AB to 140°F	A to 70°F	A to 212°F			
ANTIMONY PENTAFLUORIDES		A to 70°F	B at 70°F			

	RECOMMENDATIONS	NITRILE NBR	ETHYLENE PROPYLENE EP, EPDM	FLUOROCARBON FKM	CHLOROPRENE CR	HYDROGENATED NITRILE HNBR
ANTIMONY SALTS	NBR	A to 250°F	A to 70°F	A to 70°F	A to 70°F	
ANTIMONY TRICHLORIDE		B at 70°F	AB to 70°F	A to 140°F B at 176°F	BC at 70°F	
AN-0-3 GRADE M	NBR, FKM, ACM, AU	A to 70°F	NR at 70°F	A to 70°F	B at 70°F	A to 70°F
AN-0-6	NBR, FKM, ACM, AU	A to 70°F	NR at 70°F	A to 70°F	B at 70°F	A to 70°F
AN-0-366	NBR, FKM, ACM, AU	A to 70°F	NR at 70°F	A to 70°F	B at 70°F	A to 70°F
AN-VV-0-366b HYDRAULIC FLUID	NBR, FKM	A to 70°F	NR at 70°F	A to 70°F	B at 70°F	A to 70°F
APPLE ACID	NBR, FKM, FVMQ	A to 70°F	NR at 70°F	A to 70°F	AB to 140°F	A to 70°F
AQUA AMMONIA		B at 70°F AB 30% to 100°F		C at 70°F	B at 70°F AB 30% to 150°F	
AQUA REGIA	FKM, EP, FVMQ	NR at 70°F	B to 104°F NR at 140°F	A to 70°F B to 185°F	NR at 70°F	C/NR at 70°F
ARACHIDONIC ACID		B to 80°F			NR at 70°F	
ARGON GAS	EP, FKM, ACM, AU	A to 250°F	A to 70°F AB 95% to 400°F	A to 70°F AB 95% to 500°F	NR at 70°F	A to 70°F
AROCHLORS	FKM	C at 70°F	BC at 70°F	A to 200°F	NR at 70°F	C/NR at 70°F
ARACHLOR 1248	FKM	C at 70°F NR at 140°F	B to 140°F	A to 250°F	NR at 70°F	C at 70°F
AROCHLOR 1254	FKM	NR at 70°F	B at 70°F	A to 300°F	NR at 70°F	NR at 70°F
AROCHLOR 1260	NBR, FKM, CR, SBR	A to 70°F		A to 70°F	A to 70°F	A to 70°F
AROMATIC ALCOHOL		C at 70°F	C at 70°F	A to 70°F	C at 70°F	
AROMATIC FUEL 50%	FKM, NBR 50% NBR, FVMQ, 30%	AB to 250°F	NR at 70°F	A to 70°F AB to 400°F	NR at 70°F	B at 70°F
AROMATIC HYDROCARBONS	FKM	NR at 70°F	NR at 70°F	A to 300°F	NR at 70°F	
AROMATIC SPIRITS		C at 70°F		A to 70°F		
AROMATIC TAR		C at 70°F		A to 70°F		
AROMATIC VINEGAR		C at 70°F	A to 70°F	A to 70°F	B at 70°F	
ARO-TOX SPRAY		C at 70°F			C at 70°F	
ARQUADS		A to 70°F		A to 70°F		
ARSENIC ACID	EP, IIR, NBR, FVMQ FKM	A to 140°F AB any conc to 200°F	A to 140°F B at 176°F	A to 140°F B at 176°F	A to 140°F AB any conc to 150°F	A to 70°F
ARSENIC ACID SOLUTION		A to 160°F	A to 140°F	A to 140°F	A to 140°F	
ARSENIC BUTTER		C at 70°F	NR at 70°F	NR at 70°F	A to 70°F	
ARSENIC CHLORIDE		C at 70°F	NR at 70°F	NR at 70°F	A to 70°F	
ARSENIC SALTS				AB to 70°F		
ARSENIC TRICHLORIDE		A to 70°F	NR at 70°F	NR at 70°F	A to 70°F	A to 70°F
ARSENOUS CHLORIDE		C at 70°F	NR at 70°F	NR at 70°F	A to 70°F	
ARSENOUS TRICHLORIDE	NBR, CR	C at 70°F	NR at 70°F	NR at 70°F	A to 70°F	
ARTIFICIAL VINEGAR		C at 70°F	A to 70°F	A to 70°F	B at 70°F	
ASKAREL TRANSFORMER OIL	FKM, NBR, FVMQ	B at 70°F	NR at 70°F	A to 70°F	NR at 70°F	B at 70°F
ASPHALT EMULSION	FKM, T, NBR	B at 70°F	NR at 70°F	A to 400°F	BC at 70°F	
ASPHALT TOPPING	FKM, T	B to 150°F	NR at 70°F	A to 212°F	AB to 80°F BC 104-140°F	
ASTM-REFERENCE FUEL A	NBR, FKM, ACM, FVMQ	+2% vol 3 days 70°F A to 250°F	+103% vol 3 days 70°F NR at 70°F	+0% vol 3 days 75°F A to 75°F	+14% vol 3 days 70°F B to 80°F	A to 70°F
ASTM-REFERENCE FUEL B	FKM	+31% vol 3 days 70°F B/NR at 70°F	+128% vol 3 days 70°F NR at 70°F	+2.5% vol 3 days 70°F A to 70°F	+65% vol 3 days 70°F NR at 70°F	A to 70°F
ASTM-REFERENCE FUEL C	FKM	B/NR at 70°F	NR at 70°F	+4.8% vol 7 days 75°F +18% vol 7 days 104°F	NR at 70°F	B at 70°F
ASTM-REFERENCE FUEL D	FKM	C at 70°F	NR at 70°F	A to 70°F	NR at 70°F	B at 70°F
ASTM-REFERENCE NO. 1 OIL	NBR, FKM, CR, ACM	-2% vol 3 days 212°F A to 180°F	+83% vol 3 days 212°F NR at 70°F	+0% vol 8 days 212°F A to 300°F	+8% vol 3 days 212°F A to 150°F	A to 70°F
ASTM-REFERENCE NO. 2 OIL	NBR, FKM, ACM	A to 70°F AB to 180°F	NR at 70°F	+1.3% vol 3 days 300°F A to 300°F	B to 150°F	A to 70°F
ASTM-REFERENCE NO. 3 OIL	NBR, FKM, ACM, FVMQ	+15% vol 3 days 212°F A to 180°F	+168% vol 3 days 212°F NR at 70°F	+5% vol 21 days 350°F A to 350°F	+68% vol 3 days 212°F BC to 158°F	A to 70°F
ASTM-REFERENCE NO. 4 OIL	FKM	B at 70°F	NR at 70°F	+6% vol 7 days 300°F A to 70°F	NR at 70°F	B at 70°F
ASTRAL OIL		A to 70°F	NR at 70°F	A to 70°F	B at 70°F	

	STYRENE BUTADIENE SBR	POLYACRYLATE ACM	POLYURETHANE AU, EU	ISOBUTYLENE ISOPRENE IIR	POLYBUTADIENE BR	® AEROQUIP AQP
ANTIMONY SALTS	A to 70°F	NR at 70°F	AB to 70°F			
ANTIMONY TRICHLORIDE			AB to 70°F	A to 70°F		
AN-0-3 GRADE M	NR at 70°F	A to 70°F	A to 70°F	NR at 70°F	NR at 70°F	
AN-0-6	NR at 70°F	A to 70°F	A to 70°F	NR at 70°F	NR at 70°F	
AN-0-366	NR at 70°F	A to 70°F	A to 70°F	NR at 70°F	NR at 70°F	
AN-VV-0-366b HYDRAULIC FLUID	NR at 70°F	B at 70°F	BC at 70°F	NR at 70°F	NR at 70°F	
APPLE ACID	B at 70°F	NR at 70°F		NR any conc at 70°F	B at 70°F	
AQUA AMMONIA				A to 70°F AB 30% to 185°F		
AQUA REGIA	NR at 70°F	NR at 70°F	NR at 70°F	C/NR at 70°F	NR at 70°F	
ARACHIDONIC ACID				A to 70°F AB to 150°F		
ARGON GAS	A/NR at 70°F	AB to 70°F	A to 70°F	B at 70°F	NR at 70°F	
AROCHLORS	NR at 70°F	NR at 70°F	C at 70°F	B/NR at 70°F	NR at 70°F	
ARACHLOR 1248	NR at 70°F	NR at 70°F	B at 70°F	B at 70°F	NR at 70°F	
AROCHLOR 1254	NR at 70°F	NR at 70°F	NR at 70°F	NR at 70°F	NR at 70°F	
AROCHLOR 1260	A to 70°F	NR at 70°F	A to 70°F	A to 70°F	A to 70°F	
AROMATIC ALCOHOL						
AROMATIC FUEL 50%	NR at 70°F	B/NR at 70°F	NR at 70°F	NR at 70°F	NR at 70°F	
AROMATIC HYDROCARBONS	NR at 70°F	NR at 70°F	C/NR at 70°F	NR at 70°F		
AROMATIC SPIRITS				NR at 70°F		
AROMATIC TAR				NR at 70°F		
AROMATIC VINEGAR			B at 70°F	A to 70°F		
ARO-TOX SPRAY						
ARQUADS				A to 70°F		
ARSENIC ACID	A to 70°F	C at 70°F	C at 70°F	A to 70°F AB any conc to 150°F	A to 70°F	
ARSENIC ACID SOLUTION	A to 70°F	C at 70°F	C at 70°F	A to 70°F AB to 150°F		
ARSENIC BUTTER				NR at 70°F		
ARSENIC CHLORIDE				NR at 70°F		
ARSENIC SALTS			AB to 70°F			
ARSENIC TRICHLORIDE				NR at 70°F		
ARSENOUS CHLORIDE				NR at 70°F		
ARSENOUS TRICHLORIDE				NR at 70°F		
ARTIFICIAL VINEGAR			B at 70°F	A to 70°F		
ASKAREL TRANSFORMER OIL	NR at 70°F	NR at 70°F	NR at 70°F	NR at 70°F	NR at 70°F	
ASPHALT EMULSION	NR at 70°F	B at 70°F	B at 70°F	NR at 70°F	NR at 70°F	AB to 180°F
ASPHALT TOPPING			AB to 180°F	NR at 70°F		
ASTM-REFERENCE FUEL A	+112% vol 3 days 70°F NR at 70°F	-2% vol 3 days 70°F A to 70°F	+1% vol 28 days 70°F B to 158°F	+111% vol 3 days 70°F NR at 70°F	NR at 70°F	
ASTM-REFERENCE FUEL B	NR at 70°F	NR at 70°F	+16% vol 28 days 70°F B at 70°F	NR at 70°F	NR at 70°F	
ASTM-REFERENCE FUEL C	NR at 70°F	NR at 70°F	+25% vol 28 days 70°F C/NR at 70°F	NR at 70°F	NR at 70°F	
ASTM-REFERENCE FUEL D	NR at 70°F	NR at 70°F	AB to 70°F			
ASTM-REFERENCE NO. 1 OIL	+44% vol 3 days 212°F NR at 70°F	+2% vol 3 days 212°F A to 212°F	+0% vol 28 days 212°F A to 212°F	NR at 70°F	NR at 70°F	
ASTM-REFERENCE NO. 2 OIL	NR at 70°F	A to 70°F	B at 70°F	NR at 70°F	NR at 70°F	
ASTM-REFERENCE NO. 3 OIL	+105% vol 3 days 212°F NR at 70°F	+16% vol 3 days 212°F A to 70°F	+0% vol 28 days 70°F +5% vol 3 days 212°F	+60% vol 3 days 212°F NR at 70°F	NR at 70°F	
ASTM-REFERENCE NO. 4 OIL	NR at 70°F	B at 70°F	NR at 70°F	NR at 70°F	NR at 70°F	
ASTRAL OIL			C at 70°F	NR at 70°F		AB to 200°F

	SYNTHETIC ISOPRENE IR	NATURAL ISOPRENE NR	CHLOROSULFONATED POLYETHYLENE CSM	FLUOROSILICONE FVMQ	SILICONE VMQ	CHEMRAZ FFKM
ANTIMONY SALTS				A to 70°F	A to 70°F	A to 70°F
ANTIMONY TRICHLORIDE		A to 70°F				A to 70°F
AN-0-3 GRADE M	NR at 70°F	NR at 70°F	B at 70°F	A to 70°F	B at 70°F	A to 70°F
AN-0-6	NR at 70°F	NR at 70°F	B at 70°F	A to 70°F	NR at 70°F	A to 70°F
AN-0-366	NR at 70°F	NR at 70°F	B at 70°F	A to 70°F	NR at 70°F	A to 70°F
AN-VV-0-366b HYDRAULIC FLUID	NR at 70°F	NR at 70°F	B at 70°F	A to 70°F	NR at 70°F	A to 70°F
APPLE ACID	A to 70°F	A any conc to 80°F	B at 70°F	A to 70°F	B at 70°F	A to 70°F
AQUA AMMONIA	A to 70°F	A to 70°F AB to 30% to 150°F	B at 70°F			
AQUA REGIA	NR at 70°F	NR at 70°F	B /NR at 70°F	C at 70°F	C /NR at 70°F	AB to 70°F NR at 70°F white
ARACHIDONIC ACID	NR at 70°F	NR at 70°F				
ARGON GAS	NR at 70°F	NR at 70°F	NR at 70°F	B at 70°F	B at 70°F	A to 70°F
AROCHLORS	NR at 70°F	NR at 70°F	NR at 70°F	B at 70°F	BC at 70°F	A to 70°F
ARACHLOR 1248	NR at 70°F	NR at 70°F	NR at 70°F	B at 70°F	B at 70°F	A to 70°F
AROCHLOR 1254	NR at 70°F	NR at 70°F	NR at 70°F	B at 70°F	C at 70°F	A to 70°F
AROCHLOR 1260	A to 70°F	A to 70°F	A to 70°F	A to 70°F	A to 70°F	A to 70°F
AROMATIC ALCOHOL	C at 70°F	C at 70°F				
AROMATIC FUEL 50%	NR at 70°F	NR at 70°F	NR at 70°F	B at 70°F	NR at 70°F	A to 70°F
AROMATIC HYDROCARBONS	NR at 70°F	NR at 70°F	NR at 70°F	A to 70°F	C /NR at 70°F	
AROMATIC SPIRITS	NR at 70°F	NR at 70°F	NR at 70°F			
AROMATIC TAR	NR at 70°F	NR at 70°F	NR at 70°F			
AROMATIC VINEGAR	A to 70°F	A to 70°F	B at 70°F		A to 70°F	
ARO-TOX SPRAY						
ARQUADS	A to 70°F	A to 70°F	A to 70°F			
ARSENIC ACID	A to 70°F	AB any conc to 150°F	A to 70°F	A to 70°F	A to 70°F	A to 70°F
ARSENIC ACID SOLUTION				A to 70°F	A to 70°F	
ARSENIC BUTTER	NR at 70°F	NR at 70°F	NR at 70°F			
ARSENIC CHLORIDE	NR at 70°F	NR at 70°F	NR at 70°F			
ARSENIC SALTS						
ARSENIC TRICHLORIDE	NR at 70°F	NR at 70°F	NR at 70°F			A to 70°F
ARSENOUS CHLORIDE	NR at 70°F	NR at 70°F	NR at 70°F			
ARSENOUS TRICHLORIDE	NR at 70°F	NR at 70°F	NR at 70°F			
ARTIFICIAL VINEGAR	B at 70°F	B at 70°F	B at 70°F		A to 70°F	
ASKAREL TRANSFORMER OIL	NR at 70°F	NR at 70°F	NR at 70°F	B at 70°F	NR at 70°F	A to 70°F
ASPHALT EMULSION	NR at 70°F	NR at 70°F	B /NR at 70°F	B at 70°F	NR at 70°F	A to 70°F
ASPHALT TOPPING		A to 70°F	C at 70°F	NR at 70°F	C /NR at 70°F	A to 70°F
ASTM-REFERENCE FUEL A	NR at 70°F	NR at 70°F	AB to 70°F	+24% vol 3 days 70°F AB to 70°F	+109% vol 3 days 70°F NR at 70°F	A to 70°F
ASTM-REFERENCE FUEL B	NR at 70°F	NR at 70°F	NR at 70°F	+22% vol 7 days 70°F +20% vol 3 days 150°F	+150% vol 7 days 70°F NR at 70°F	A to 70°F
ASTM-REFERENCE FUEL C	NR at 70°F	NR at 70°F	NR at 70°F	B at 70°F	NR at 70°F	A to 70°F
ASTM-REFERENCE FUEL D					NR at 70°F	A to 70°F
ASTM-REFERENCE NO. 1 OIL	NR at 70°F	NR at 70°F	AB to 70°F	no swell 3 days 212-300°F	+7-10% vol 3 days 212-300°F	A to 70°F
ASTM-REFERENCE NO. 2 OIL	NR at 70°F	NR at 70°F	NR at 70°F	A to 70°F	+19% vol 3 days 300°F A /NR at 70°F	A to 70°F
ASTM-REFERENCE NO. 3 OIL	NR at 70°F	NR at 70°F	+40% vol 3 days 212°F B to 158°F	+4% vol 3 days 300°F A to 300°F	+23% vol 3 days 212°F +30-60% vol 3 days 300°F	A to 70°F
ASTM-REFERENCE NO. 4 OIL	NR at 70°F	NR at 70°F	NR at 70°F	B at 70°F	NR at 70°F	A to 70°F
ASTRAL OIL	NR at 70°F	NR at 70°F	C at 70°F		NR at 70°F	

	POLYSULFIDE T	CHLORINATED POLYETHYLENE CM	EPICHLOROHYDRIN CO, ECO	® HYTREL COPOLYESTER TPE	® SANTOPRENE COPOLYMER TPO	® C-FLEX STYRENIC TPE
ANTIMONY SALTS				AB to 70°F		AB to 70°F
ANTIMONY TRICHLORIDE						
AN-0-3 GRADE M	A to 70°F					
AN-0-6	A to 70°F					
AN-0-366	A to 70°F					
AN-VV-0-366b HYDRAULIC FLUID	A to 70°F					
APPLE ACID						AB to 70°F
AQUA AMMONIA						
AQUA REGIA	NR at 70°F					AB to 70°F
ARACHIDONIC ACID						
ARGON GAS	NR at 70°F			A to 70°F		
AROCHLORS	NR at 70°F			C at 70°F		
ARACHLOR 1248	NR at 70°F					
AROCHLOR 1254	NR at 70°F					
AROCHLOR 1260	A to 70°F					
AROMATIC ALCOHOL						
AROMATIC FUEL 50%	B at 70°F					
AROMATIC HYDROCARBONS	TEST			C at 70°F		NR at 70°F
AROMATIC SPIRITS						
AROMATIC TAR						
AROMATIC VINEGAR						
ARO-TOX SPRAY						
ARQUADS						
ARSENIC ACID	A to 70°F	A to 70°F	A to 70°F			
ARSENIC ACID SOLUTION						
ARSENIC BUTTER						
ARSENIC CHLORIDE						
ARSENIC SALTS				AB to 70°F		AB to 70°F
ARSENIC TRICHLORIDE						
ARSENOUS CHLORIDE						
ARSENOUS TRICHLORIDE						
ARTIFICIAL VINEGAR						
ASKAREL TRANSFORMER OIL	NR at 70°F					
ASPHALT EMULSION	A to 70°F	AB to 70°F	A to 70°F	A to 180°F		
ASPHALT TOPPING	A to 185°F			AB to 185°F		
ASTM-REFERENCE FUEL A	+0.2% vol 3 days 70°F A to 70°F	A to 70°F	A to 70°F	A to 158°F	NR at 70°F	
ASTM-REFERENCE FUEL B	A to 70°F	B at 70°F	A to 70°F	+1-8% vol 3 days 70°F A to 158°F	NR at 70°F	
ASTM-REFERENCE FUEL C	B at 70°F	C at 70°F		A to 70°F AB at 158°F	NR at 70°F	
ASTM-REFERENCE FUEL D						
ASTM-REFERENCE NO. 1 OIL	-2% vol 3 days 212°F A to 70°F	A to 70°F		A to 300°F	NR at 70°F	
ASTM-REFERENCE NO. 2 OIL	BC at 70°F			A to 300°F	NR at 70°F	
ASTM-REFERENCE NO. 3 OIL	+5.4% vol 3 days 212°F AC at 70°F			A to 300°F	NR at 70°F	
ASTM-REFERENCE NO. 4 OIL	C at 70°F					
ASTRAL OIL				AB to 150°F		

	ETHYLENE ACRYLIC EA	POLYALLOMER LINEAR COPOLYMER	NYLON 11 POLYAMIDE	NYLON 12 POLYAMIDE	ETHYLENE VINYL ACETATE EVA	POLYVINYLCHLORIDE FLEXIBLE PVC
ANTIMONY SALTS			AB to 70°F			A to 70°F *AB to 140°F
ANTIMONY TRICHLORIDE						
AN-0-3 GRADE M						
AN-0-6						
AN-0-366						
AN-VV-0-366b HYDRAULIC FLUID						
APPLE ACID		B at 70°F	A to 70°F		C at 70°F	A to 70°F
AQUA AMMONIA						
AQUA REGIA	NR at 70°F			NR at 70°F		NR at 70°F
ARACHIDONIC ACID						
ARGON GAS			A to 70°F			A to 70°F
AROCHLORS			AB to 70°F			C at 70°F
ARACHLOR 1248						
AROCHLOR 1254						
AROCHLOR 1260						
AROMATIC ALCOHOL						
AROMATIC FUEL 50%	NR at 70°F					
AROMATIC HYDROCARBONS			AB to 70°F			NR at 70°F
AROMATIC SPIRITS						
AROMATIC TAR						
AROMATIC VINEGAR						
ARO-TOX SPRAY						
ARQUADS						
ARSENIC ACID			AB to 70°F		NR at 70°F	BC at 70°F
ARSENIC ACID SOLUTION						
ARSENIC BUTTER						
ARSENIC CHLORIDE						
ARSENIC SALTS			AB to 70°F			AB to 70°F
ARSENIC TRICHLORIDE						
ARSENOUS CHLORIDE						
ARSENOUS TRICHLORIDE						
ARTIFICIAL VINEGAR						
ASKAREL TRANSFORMER OIL						NR at 70°F
ASPHALT EMULSION	A to 70°F	A to 70°F	AB to 180°F		A to 70°F	NR at 70°F
ASPHALT TOPPING			AB to 180°F			AB to 70°F
ASTM-REFERENCE FUEL A	A to 70°F		A to 70°F	AB to 70°F	NR at 70°F	AB to 70°F C at 150°F
ASTM-REFERENCE FUEL B	B/NR at 70°F		A to 70°F			NR at 70°F
ASTM-REFERENCE FUEL C	NR at 70°F		A to 70°F			
ASTM-REFERENCE FUEL D						
ASTM-REFERENCE NO. 1 OIL	A to 350°F		A to 70°F		C/NR at 70°F	B to 150°F
ASTM-REFERENCE NO. 2 OIL	A to 70°F		A to 70°F			
ASTM-REFERENCE NO. 3 OIL	AB to 300°F		A to 70°F		C/NR at 70°F	C at 70°F NR at 150°F
ASTM-REFERENCE NO. 4 OIL						
ASTRAL OIL			AB to 180°F			

	POLYETHYLENE LOW DENSITY LDPE	®TEFLON FEP	®KALREZ PERFLUORINATED ELASTOMER	®FLUORAZ FLUORINATED COPOLYMER	®AFLAS FLUORINATED COPOLYMER	®NORPRENE COPOLYMER TPO
ANTIMONY SALTS	AB to 70°F		A to 212°F			A to 70°F
ANTIMONY TRICHLORIDE	A to 70°F	A to 200°F	A to 212°F			
AN-0-3 GRADE M				A to 70°F		
AN-0-6				A to 70°F		
AN-0-366				A to 70°F		
AN-VV-0-366b HYDRAULIC FLUID				A to 70°F		
APPLE ACID	B/NR at 70°F	A to 300°F	A to 70°F	A to 70°F		TEST
AQUA AMMONIA		A to 70°F				
AQUA REGIA	AB to 70°F BC at 104°F	A to 248°F	A to 212°F	C at 70°F		BC at 70°F
ARACHIDONIC ACID		A to 70°F				
ARGON GAS	A to 70°F	A to 70°F	A to 212°F	A to 70°F		
AROCHLORS	B at 70°F	A to 70°F	A to 212°F	A/NR at 70°F		
ARACHLOR 1248	B at 70°F		A to 212°F	A/NR at 70°F		
AROCHLOR 1254			A to 212°F	A to 70°F		
AROCHLOR 1260			A to 212°F	A to 70°F		
AROMATIC ALCOHOL		A to 70°F				
AROMATIC FUEL 50%			A to 212°F	A to 70°F		
AROMATIC HYDROCARBONS	C/NR at 70°F	A to 70°F	A to 70°F	NR at 70°F	B/NR at 70°F	NR at 70°F
AROMATIC SPIRITS		A to 70°F				
AROMATIC TAR		A to 70°F				
AROMATIC VINEGAR		A to 70°F				
ARO-TOX SPRAY		A to 70°F				
ARQUADS		A to 70°F				
ARSENIC ACID	AB to 70°F	A to 200°F	A to 212°F	A to 70°F		
ARSENIC ACID SOLUTION						
ARSENIC BUTTER		A to 70°F				
ARSENIC CHLORIDE		A to 70°F				
ARSENIC SALTS	AB to 70°F					A to 70°F
ARSENIC TRICHLORIDE		A to 70°F	A to 212°F			
ARSENOUS CHLORIDE		A to 70°F				
ARSENOUS TRICHLORIDE		A to 70°F				
ARTIFICIAL VINEGAR		A to 70°F				
ASKAREL TRANSFORMER OIL		A to 70°F	A to 212°F	A to 70°F		
ASPHALT EMULSION	A to 70°F	A to 180°F	A to 560°F	A to 70°F		
ASPHALT TOPPING	A/NR at 70°F	A to 180°F				
ASTM-REFERENCE FUEL A	B at 70°F	+1% wt gain 3 days 212°F A to 300°F	A to 212°F	BC at 70°F	+19% vol 7 days 70°F BC at 70°F	
ASTM-REFERENCE FUEL B		A to 70°F	A to 212°F	NR at 70°F	NR at 70°F	
ASTM-REFERENCE FUEL C		A to 70°F	A to 212°F	NR at 70°F	NR at 70°F	
ASTM-REFERENCE FUEL D				NR at 70°F		
ASTM-REFERENCE NO. 1 OIL	C at 70°F	A to 70°F	A to 212°F	A to 70°F	+1.8% vol 3 days 212°F +4.1% vol 3 days 350°F	A to 242°F
ASTM-REFERENCE NO. 2 OIL		A to 70°F	A to 212°F	B at 70°F		
ASTM-REFERENCE NO. 3 OIL	C at 70°F	A to 70°F	A to 212°F	C at 70°F	+6-8% vol 3 days 212°F +13% vol 3 days 350°F	
ASTM-REFERENCE NO. 4 OIL			A to 212°F	B at 70°F		
ASTRAL OIL		A to 70°F AB to 250°F				

	RECOMMENDATIONS	NITRILE NBR	ETHYLENE PROPYLENE EP, EPDM	FLUOROCARBON FKM	CHLOROPRENE CR	HYDROGENATED NITRILE HNBR
ATL-857	FKM	B at 70°F	NR at 70°F	A to 70°F	NR at 70°F	B at 70°F
ATLANTIC DOMINION F	NBR,FKM,ACM	A to 70°F	NR at 70°F	A to 70°F	B at 70°F	A to 70°F
ATLANTIC UTRO GEAR-EP LUBE		A to 70°F	NR at 70°F	A to 70°F	B at 70°F	A to 70°F
AUREX 903R, MOBIL	NBR,FKM,ACM,AU	A to 70°F	NR at 70°F	A to 70°F	B at 70°F	
AUTOMATIC TRANSMISSION FLUID	NBR,FKM,ACM	A to 70°F	NR at 70°F	A to 70°F	B at 70°F	A to 70°F
AUTOMOTIVE BRAKE FLUID	EP,SBR,NR	+90% vol 5 days 212°F C at 70°F	A to 70°F	NR at 70°F dissolved	B at 70°F	
AUTOMOTIVE GASOLINE	NBR,FKM	+6% vol 7 days 70°F A to 250°F	NR at 70°F	+3% vol 7 days 70°F A to 70°F	+55% vol 7 days 70°F NR at 70°F	
AVIATION GASOLINE	NBR,FVMQ,FKM	A to 200°F airborn A to 250°F ground	NR at 70°F ground or airborn	A to 70°F ground NR at 70°F airborn	NR at 70°F ground or airborn	
AZOLE'S		NR at 70°F	C at 70°F	C at 70°F	NR at 70°F	
BAKING SODA	NBR,EP,FKM,CR,IIR	A to 70°F AB any conc to 200°F	A to 100% to 176°F B at 212°F	A to 100% to 212°F	A to 70°F AB any conc to 200°F	A to 70°F
BANANA OIL		NR at 70°F	B at 70°F	NR at 70°F	NR at 70°F	
BARDOL B	FKM	NR at 70°F	NR at 70°F	A to 70°F	NR at 70°F	NR at 70°F
BARIUM CARBONATE		A to 140°F B at 176°F	A to 176°F	A to 248°F	A to 70°F	
BARIUM CHLORIDE	NBR,EP,CR,IIR	A to 140°F AB any conc to 200°F	A to 176°F AB to 200°F	A to 248°F	A to 140°F AB any conc to 200°F	A to 70°F
BARIUM CYANIDE		C to 140°F	A to 140°F	A to 140°F	A to 140°F	
BARIUM FLUORIDE		NR at 70°F	NR at 70°F	C at 70°F	B at 70°F	
BARIUM HYDRATE		A to 70°F AB any conc to 200°F	A to 70°F	A to 70°F	A to 70°F AB any conc to 200°F	
BARIUM HYDROXIDE	NBR,EP,CR,FKM	A to 140°F AB to 200°F	A to 176°F B at 212°F	A to 248°F	A to 158°F AB to 200°F	A to 70°F
BARIUM MONOHYDRATE		A to 70°F AB any conc to 200°F			A to 70°F AB any conc to 200°F	
BARIUM MONOSULFIDE		A to 70°F AB any conc to 200°F	A to 70°F	A to 70°F	A to 70°F AB any conc to 200°F	
BARIUM NITRATE	NBR	A to 140°F B at 176°F	A to 176°F B at 212°F	A to 248°F	A to 140°F	
BARIUM OCTAHYDRATE		A to 70°F AB any conc to 200°F			A to 70°F AB any conc to 200°F	
BARIUM SALTS	NBR,EP,FKM,CR	A to 70°F	A to 70°F	A to 70°F	A to 70°F	A to 70°F
BARIUM SULFATE	NBR,CR,EP	A to 140°F AB to 200°F	A to 176°F AB to 200°F	A to 248°F	A to 70°F AB to 200°F	A to 70°F
BARIUM SULFIDE	NBR,EP,CR,FKM	A to 140°F AB any conc to 200°F	A to 140°F	A to 248°F	A to 70°F AB any conc to 200°F	A to 70°F
BASIC COPPER ARSENATE						
BASIC IRON SULFATE		A to 70°F AB any conc to 200°F			A to 70°F AB any conc to 200°F	
BAY OIL				A to 70°F	C at 70°F	
BAYOL D	NBR,FKM,ACM	A to 70°F	NR at 70°F	A to 70°F	B at 70°F	A to 70°F
BAYOL 35	NBR,FKM,ACM	A to 70°F	NR at 70°F	A to 70°F	B at 70°F	A to 70°F
BEEF EXTRACT	NBR	A to 70°F		A to 70°F	A to 70°F	
BEER	NBR,EP,CR,FKM	A to 140°F AB to 250°F	A to 140°F AB to 176°F	A to 176°F	AB to 140°F	
BEER WORT	NBR	A to 250°F	A to 70°F	A to 70°F	AB to 70°F	
BEET JUICE/PULP	NBR	A to 200°F	A to 70°F	A to 70°F	B at 70°F	
BEET SUGAR LIQUID		A to 70°F	A to 70°F	A to 70°F	A to 70°F	A to 70°F
BEET SUGAR LIQUORS	NBR,EP,FKN,CR	A to 200°F	A to 140°F	A to 176°F	AB to 140°F	A to 70°F
BENGAL GELATIN		AB any conc to 180°F			AB any conc to 180°F	
BENZAL ALCOHOL		NR at 70°F	C at 70°F	A to 70°F	C at 70°F	
BENZAL CHLORIDE		NR at 70°F				
BENZALDEHYDE	EP,IIR,T	NR any conc at 70°F	A to 200°F	NR 10% to conc at 70°F	NR any conc at 70°F	NR at 70°F
BENZENE (BENZOL)	FKM,FVMQ,T	+155% vol 3 days 70°F NR at 70°F	+82% vol 3 days 70°F NR at 70°F	+20% vol 7 days 70°F B to 158°F	+170% vol 7 days 70°F NR at 70°F	NR at 70°F
BENZENE (PETROLEUM ETHER)	T	NR at 70°F	NR at 70°F	NR at 70°F	NR at 70°F	
BENZENE CARBINAL		NR any conc at 70°F	B at 70°F	NR at 70°F	NR any conc at 70°F	
BENENE CARBOXYLIC ACID	FKM,FVMQ,T	NR at 70°F	A/NR at 70°F	A to 176°F B at 212°F	A to 70°F AB any conc to 150°F	

	STYRENE BUTADIENE SBR	POLYACRYLATE ACM	POLYURETHANE AU, EU	ISOBUTYLENE ISOPRENE IIR	POLYBUTADIENE BR	® AEROQUIP AQP
ATL-857	NR at 70°F	B at 70°F	NR at 70°F	NR at 70°F	NR at 70°F	
ATLANTIC DOMINION F	NR at 70°F	A to 70°F	B at 70°F	NR at 70°F	NR at 70°F	
ATLANTIC UTRO GEAR-EP LUBE	NR at 70°F	A to 70°F	A to 70°F	NR at 70°F	NR at 70°F	
AUREX 903R, MOBIL	NR at 70°F	A to 70°F	A to 70°F	NR at 70°F	NR at 70°F	
AUTOMATIC TRANSMISSION FLUID	NR at 70°F	A to 70°F	B at 70°F	NR at 70°F	NR at 70°F	AB to 300°F
AUTOMOTIVE BRAKE FLUID	A to 70°F		NR at 70°F	B at 70°F		
AUTOMOTIVE GASOLINE	NR at 70°F	B at 70°F	B at 70°F	+240% vol 7 days 70°F NR at 70°F		
AVIATION GASOLINE	NR at 70°F ground or airborne	NR at 70°F ground or airborne	B at 70°F ground or airborne	NR at 70°F		
AZOLE'S				B at 70°F		
BAKING SODA	A to 100% to 70°F	NR sol'n at 70°F	A to 70°F	A to 70°F AB any conc to 185°F	A to 70°F	
BANANA OIL			NR at 70°F	B/NR at 70°F		
BARDOL B	NR at 70°F	NR at 70°F	NR at 70°F	NR at 70°F	NR at 70°F	
BARIUM CARBONATE			AB to 70°F	A to 70°F		
BARIUM CHLORIDE	A to 70°F	A to 70°F	A to 70°F	A to 70°F AB any conc to 185°F	A to 70°F	AB to 70°F
BARIUM CYANIDE						
BARIUM FLUORIDE			NR at 70°F	C at 70°F		
BARIUM HYDRATE			A to 70°F	A to 70°F AB any conc to 185°F		
BARIUM HYDROXIDE	A to 70°F	NR any conc at 70°F	A/NR at 70°F	A to 70°F AB any conc to 185°F	A to 70°F	AB to 70°F
BARIUM MONOHYDRATE				A to 70°F AB any conc to 185°F		
BARIUM MONOSULFIDE			A to 70°F	A to 70°F AB any conc to 185°F		
BARIUM NITRATE	A sol'n to 70°F	NR sol'n at 70°F	AB to 70°F	B at 70°F		
BARIUM OCTAHYDRATE				A to 70°F AB any conc to 185°F		
BARIUM SALTS	A to 70°F	A to 70°F	A to 70°F	A to 70°F	A to 70°F	
BARIUM SULFATE	A to 70°F		A to 70°F	A to 70°F	A to 70°F	
BARIUM SULFIDE	B conc at 70°F A sol'n at 70°F	NR any conc at 70°F	A to 70°F	A to 70°F AB any conc to 185°F	B at 70°F	NR at 70°F
BASIC COPPER ARSENATE			AB to 70°F			
BASIC IRON SULFATE				A to 70°F AB any conc to 185°F		
BAY OIL						
BAYOL D	NR at 70°F	A to 70°F	NR at 70°F	NR at 70°F	NR at 70°F	
BAYOL 35	NR at 70°F	A to 70°F	BC at 70°F	NR at 70°F	NR at 70°F	
BEEF EXTRACT					A to 70°F	
BEER	A to 70°F	NR at 70°F	A/NR at 70°F	A to 70°F	A to 70°F	
BEER WORT	A to 70°F	NR at 70°F	B at 70°F			
BEET JUICE/PULP	A to 70°F	NR at 70°F	B/NR at 70°F			
BEET SUGAR LIQUID						
BEET SUGAR LIQUORS	A to 70°F	NR at 70°F	NR at 70°F	A to 70°F AB to 150°F	A to 70°F	AB to 70°F
BENGAL GELATIN				A to 70°F AB any conc to 185°F		
BENZAL ALCOHOL				A to 70°F		
BENZAL CHLORIDE				B at 70°F		
BENZALDEHYDE	NR at 70°F	NR at 70°F	NR at 70°F	AB any conc to 75°F	NR at 70°F	
BENZENE (BENZOL)	+180% vol 3 days 70°F NR at 70°F	+214% vol 3 days 70°F NR at 70°F	+109% vol 3 days 70°F NR at 70°F	+75% vol 3 days 70°F NR any conc at 70°F	NR at 70°F	TEST
BENZENE (PETROLEUM ETHER)			C at 70°F	NR at 70°F		
BENZENE CARBINAL			NR at 70°F	B any conc to 75°F		
BENENE CARBOXYLIC ACID				A to 70°F		

	SYNTHETIC ISOPRENE IR	NATURAL ISOPRENE NR	CHLOROSULFONATED POLYETHYLENE CSM	FLUOROSILICONE FVMQ	SILICONE VMQ	CHEMRAZ FFKM
ATL-857	NR at 70°F	NR at 70°F	NR at 70°F	B at 70°F	NR at 70°F	A to 70°F
ATLANTIC DOMINION F	NR at 70°F	NR at 70°F	NR at 70°F	A to 70°F	NR at 70°F	A to 70°F
ATLANTIC UTRO GEAR-EP LUBE	NR at 70°F	NR at 70°F	NR at 70°F		NR at 70°F	A to 70°F
AUREX 903R, MOBIL	NR at 70°F	B at 70°F	NR at 70°F	NR at 70°F	NR at 70°F	A to 70°F
AUTOMATIC TRANSMISSION FLUID	NR at 70°F	NR at 70°F	C at 70°F		NR at 70°F	A to 70°F
AUTOMOTIVE BRAKE FLUID			B at 70°F	NR at 70°F	+2% vol 5 days 212°F A to 70°F	A to 70°F
AUTOMOTIVE GASOLINE			+85% vol 7 days 70°F NR at 70°F	A to 70°F	+260% vol 7 days 70°F NR at 70°F	
AVIATION GASOLINE	NR at 70°F	NR at 70°F	NR at 70°F	A to 70°F ground or airborne	NR at 70°F ground or airborne	
AZOLE'S	C at 70°F	C at 70°F			B at 70°F	
BAKING SODA	A to 70°F	A to 70°F AB any conc to 185°F	A to 70°F AB to 200°F	A to 100% to 70°F	A to 70°F	A to 70°F
BANANA OIL	NR at 70°F	NR at 70°F	C at 70°F		NR at 70°F	
BARDOL B	NR at 70°F	NR at 70°F	NR at 70°F	B at 70°F	NR at 70°F	A to 70°F
BARIUM CARBONATE	A to 70°F	A to 70°F	A to 70°F			A to 70°F
BARIUM CHLORIDE	A to 70°F	A to 70°F AB any conc to 185°F	A to 70°F AB to 200°F	A to 70°F	A to 70°F	A to 70°F
BARIUM CYANIDE						A to 70°F
BARIUM FLUORIDE		NR at 70°F	C at 70°F		NR at 70°F	
BARIUM HYDRATE	A to 70°F	A to 70°F AB any conc to 150°F	A to 70°F		A to 70°F	
BARIUM HYDROXIDE	A to 70°F	A to 70°F AB any conc to 185°F	A to 158°F AB to 200°F	A to 70°F	A to 70°F	A to 70°F B at 70°F white
BARIUM MONOHYDRATE	A to 70°F	A to 70°F AB any conc to 150°F				
BARIUM MONOSULFIDE	A to 70°F	A to 70°F AB any conc to 150°F	A to 70°F		B at 70°F	
BARIUM NITRATE		NR at 70°F	B at 70°F	A sol'n to 70°F	B at 70°F A sol'n to 70°F	A to 70°F
BARIUM OCTAHYDRATE	A to 70°F	A to 70°F AB any conc to 150°F				
BARIUM SALTS	A to 70°F	A to 70°F	A to 70°F	A to 70°F	A to 70°F	A to 70°F
BARIUM SULFATE	A to 70°F	A to 70°F AB any conc 185°F	A to 70°F AB to 200°F	A to 70°F	AB to 70°F	A to 70°F
BARIUM SULFIDE	A to 70°F	A to 70°F AB any conc to 150°F	A to 70°F AB to 150°F	A to 70°F	A to 70°F	A to 70°F
BASIC COPPER ARSENATE						
BASIC IRON SULFATE	A to 70°F	A to 70°F AB any conc to 150°F				
BAY OIL		C at 70°F				
BAYOL D	NR at 70°F	NR at 70°F	NR at 70°F	A to 70°F	NR at 70°F	A to 70°F
BAYOL 35	NR at 70°F	NR at 70°F	NR at 70°F	A to 70°F	NR at 70°F	A to 70°F
BEEF EXTRACT		A to 70°F				
BEER	A to 70°F	A to 70°F	A to 70°F	A to 70°F	A to 70°F	
BEER WORT				A to 70°F	AB to 70°F	
BEET JUICE/PULP				A to 70°F	A to 70°F	
BEET SUGAR LIQUID		A to 70°F	A to 70°F		A to 70°F	A to 70°F
BEET SUGAR LIQUORS	A to 70°F	A to 70°F	A to 70°F	A to 70°F	A to 70°F	A to 70°F
BENGAL GELATIN	A to 70°F	A to 70°F AB any conc to 150°F				
BENZAL ALCOHOL	C at 70°F	C/NR at 70°F	B at 70°F			
BENZAL CHLORIDE						A to 70°F
BENZALDEHYDE	NR at 70°F	NR any conc at 70°F	NR at 70°F	NR at 70°F	B/NR at 70°F	B at 70°F
BENZENE (BENZOL)	NR at 70°F	NR any conc at 70°F	+225% vol 7 days 70°F NR at 70°F	+23% vol 3 days 70°F AB at 70°F	+140% vol 7 days 70°F NR at 70°F	A to 70°F
BENZENE (PETROLEUM ETHER)			NR at 70°F			
BENZENE CARBINAL	NR at 70°F	NR any conc at 70°F	NR at 70°F		C at 70°F	
BENENE CARBOXYLIC ACID	B/NR at 70°F	B/NR to 150°F	B/NR at 70°F	B at 70°F	B/NR at 70°F	A to 70°F

	POLYSULFIDE T	CHLORINATED POLYETHYLENE CM	EPICHLOROHYDRIN CO, ECO	® HYTREL COPOLYESTER TPE	® SANTOPRENE COPOLYMER TPO	® C-FLEX STYRENIC TPE
ATL-857	B at 70°F					
ATLANTIC DOMINION F	C at 70°F					
ATLANTIC UTRO GEAR-EP LUBE	A to 70°F					
AUREX 903R, MOBIL	A to 70°F					
AUTOMATIC TRANSMISSION FLUID				A to 300°F	NR at 212°F	
AUTOMOTIVE BRAKE FLUID	NR at 70°F				BC at 212°F	
AUTOMOTIVE GASOLINE						
AVIATION GASOLINE						
AZOLE'S						
BAKING SODA	C at 70°F		A to 70°F	AB to 70°F		A to 70°F
BANANA OIL						
BARDOL B	B at 70°F					
BARIUM CARBONATE						
BARIUM CHLORIDE	A to 70°F	AB to 70°F	A to 70°F	AB to 70°F *AB to 140°F		
BARIUM CYANIDE						
BARIUM FLUORIDE						
BARIUM HYDRATE						
BARIUM HYDROXIDE	A to 70°F AB to 150°F	AB to 70°F	A to 70°F	AB to 70°F		
BARIUM MONOHYDRATE						
BARIUM MONOSULFIDE						
BARIUM NITRATE						
BARIUM OCTAHYDRATE						
BARIUM SALTS	A to 70°F			AB to 70°F		AB to 70°F
BARIUM SULFATE	A to 70°F	NR at 70°F	A to 70°F	NR at 70°F		
BARIUM SULFIDE	B to 150°F					
BASIC COPPER ARSENATE				AB to 70°F		
BASIC IRON SULFATE						
BAY OIL						
BAYOL D	B at 70°F					
BAYOL 35	C at 70°F			AB to 70°F		
BEEF EXTRACT						
BEER	NR at 70°F		A to 70°F	A to 70°F AB at 140°F		
BEER WORT						
BEET JUICE/PULP						
BEET SUGAR LIQUID						
BEET SUGAR LIQUORS	NR at 70°F	AB to 70°F		AB to 70°F		
BENGAL GELATIN						
BENZAL ALCOHOL						
BENZAL CHLORIDE						
BENZALDEHYDE	NR at 70°F		NR at 70°F	AB to 70°F	NR at 70°F	NR at 70°F
BENZENE (BENZOL)	BC at 70°F	C at 70°F	NR at 70°F	BC at 70°F *NR at 140°F	NR at 70°F	NR at 70°F
BENZENE (PETROLEUM ETHER)	AB to 70°F	AB to 70°F		AC at 70°F		
BENZENE CARBINAL						
BENENE CARBOXYLIC ACID						

	ETHYLENE ACRYLIC EA	POLYALLOMER LINEAR COPOLYMER	NYLON 11 POLYAMIDE	NYLON 12 POLYAMIDE	ETHYLENE VINYL ACETATE EVA	POLYVINYLCHLORIDE FLEXIBLE PVC
ATL-857						
ATLANTIC DOMINION F						
ATLANTIC UTRO GEAR-EP LUBE	A to 70°F					
AUREX 903R, MOBIL						
AUTOMATIC TRANSMISSION FLUID	A to 70°F		AB to 200°F			
AUTOMOTIVE BRAKE FLUID	NR at 70°F		AB to 70°F			
AUTOMOTIVE GASOLINE			AB to 70°F			
AVIATION GASOLINE						
AZOLE'S						
BAKING SODA	A to 70°F	A to 70°F	A to 70°F AB to 140°F	AB to 70°F aqueous	A to 70°F AB to 140°F	A to 70°F *AB to 130°F
BANANA OIL						NR at 70°F
BARDOL B						NR at 70°F
BARIUM CARBONATE						A to 70°F *AB to 140°F
BARIUM CHLORIDE			AB to 194°F		B to 140°F	AB to 150°F
BARIUM CYANIDE						
BARIUM FLUORIDE					NR at 70°F	NR at 70°F
BARIUM HYDRATE						
BARIUM HYDROXIDE	*AB to 70°F		AB to 70°F *C at 140°F		AB to 140°F	A to 150°F
BARIUM MONOHYDRATE						
BARIUM MONOSULFIDE						
BARIUM NITRATE					B at 70°F	B/NR at 70°F
BARIUM OCTAHYDRATE						
BARIUM SALTS	*AB to 70°F	A to 70°F	A to 70°F	AB to 100% to 70°F	A to 70°F	A to 70°F
BARIUM SULFATE					A to 70°F	A to 70°F *AB to 140°F
BARIUM SULFIDE			B/NR at 70°F			A to 150°F
BASIC COPPER ARSENATE			AB to 70°F			A to 70°F
BASIC IRON SULFATE						
BAY OIL						
BAYOL D						
BAYOL 35						
BEEF EXTRACT						
BEER	*AB to 70°F	A to 70°F	AB to 140°F	AB to 70°F	A to 70°F AB to 140°F	A to 150°F
BEER WORT					AB to 70°F	AB to 70°F
BEET JUICE/PULP					A to 70°F	A to 150°F
BEET SUGAR LIQUID						
BEET SUGAR LIQUORS		A to 70°F	A to 70°F		A to 70°F	
BENGAL GELATIN						
BENZAL ALCOHOL						
BENZAL CHLORIDE						
BENZALDEHYDE	NR at 70°F	A to 70°F B at 122°F	A to 70°F NR at 140°F	C 100% at 70°F AB 0.3% to 70°F	C/NR at 70°F	NR at 70°F
BENZENE (BENZOL)	NR at 70°F	B at 70°F C at 122°F	AB to 105°F C at 140°F	AB to 70°F	NR at 70°F	C/NR at 70°F
BENZENE (PETROLEUM ETHER)						
BENZENE CARBINAL						
BENENE CARBOXYLIC ACID						

	POLYETHYLENE LOW DENSITY LDPE	®TEFLON FEP	®KALREZ PERFLUORINATED ELASTOMER	®FLUORAZ FLUORINATED COPOLYMER	®AFLAS FLUORINATED COPOLYMER	®NORPRENE COPOLYMER TPO
ATL-857				A to 70°F		
ATLANTIC DOMINION F				A to 70°F		
ATLANTIC UTRO GEAR-EP LUBE				A to 70°F		
AUREX 903R, MOBIL			A to 212°F	A to 70°F		
AUTOMATIC TRANSMISSION FLUID		A to 450°F	A to 302°F	+7% vol 3 days 275°F A to 275°F	A to 325°F	
AUTOMOTIVE BRAKE FLUID		A to 70°F	A to 212°F	A to 70°F	+9% vol 3 days 300°F B at 300°F	A to 70°F
AUTOMOTIVE GASOLINE			*A to 70°F		+25% vol 7 days 70°F BC at 70°F	BC at 70°F
AVIATION GASOLINE		A to 70°F 115 to 145 octane	A to 113°F		BC at 70°F	
AZOLE'S		A to 70°F				
BAKING SODA	A to 70°F AB to 140°F	A to 300°F	A to 70°F	A to 70°F aqueous		A to 70°F
BANANA OIL		A to 70°F				
BARDOL B		A to 70°F		B at 70°F		
BARIUM CARBONATE	AB to 70°F	A to 200°F	A to 212°F			
BARIUM CHLORIDE	AB to 140°F	A to 100% to 200°F	A to 70°F A aqueous to 212°F	A 20-100% to 70°F		
BARIUM CYANIDE	B at 70°F	*A to 70°F	A to 212°F			
BARIUM FLUORIDE	NR at 70°F					
BARIUM HYDRATE		A to 70°F				
BARIUM HYDROXIDE	AB to 140°F	A to 300°F	A to 212°F	A to 70°F		
BARIUM MONOHYDRATE		A to 70°F				
BARIUM MONOSULFIDE		A to 70°F				
BARIUM NITRATE	AB to 70°F	A to 300°F	A to 212°F	A to 70°F		
BARIUM OCTAHYDRATE		A to 70°F				
BARIUM SALTS	A to 70°F AB to 140°F	A to 300°F	A to 212°F	A to 70°F		A to 70°F
BARIUM SULFATE	A to 70°F	A to 300°F	A to 70°F A aqueous to 212°F	A to 70°F		
BARIUM SULFIDE	AB to 70°F	A to 300°F	A to 70°F A aqueous to 212°F	A to 70°F	A sat'd to 70°F	
BASIC COPPER ARSENATE	AB to 70°F					
BASIC IRON SULFATE		A to 70°F				
BAY OIL						
BAYOL D				A to 70°F		
BAYOL 35				A to 70°F		
BEEF EXTRACT						
BEER	A to 70°F AB to 140°F	A to 200°F	A to 70°F	A to 70°F		
BEER WORT	AB to 70°F			A to 70°F		
BEET JUICE/PULP	A to 70°F			A to 70°F		
BEET SUGAR LIQUID	A to 70°F	A to 70°F		A to 70°F		
BEET SUGAR LIQUORS	A to 70°F	A to 200°F	A to 212°F	A to 70°F		
BENGAL GELATIN		A to 70°F				
BENZAL ALCOHOL		A to 70°F				
BENZAL CHLORIDE		A to 70°F				
BENZALDEHYDE	NR 10% to conc at 70°F	+0.5% wt 7 days 354°F A to 354°F	A to 212°F	B at 70°F	B at 70°F	BC at 70°F
BENZENE (BENZOL)	C/NR to 104°F NR at 122°F	A to 200°F	+3% vol 7 days 70°F A to 212°F	C/NR at 70°F	+21% vol 50% conc 7 days NR 50-100% at 70°F	NR at 70°F
BENZENE (PETROLEUM ETHER)		A to 300°F				
BENZENE CARBINAL		A to 70°F				
BENENE CARBOXYLIC ACID	A to 122°F AB any conc to 140°F	A to 300°F	A to 316°F	A to 70°F		A to 70°F

	RECOMMENDATIONS	NITRILE NBR	ETHYLENE PROPYLENE EP, EPDM	FLUOROCARBON FKM	CHLOROPRENE CR	HYDROGENATED NITRILE HNBR
BENZENE METHYLAL		NR at 70°F	B at 70°F	NR at 70°F	NR at 70°F	
BENZENE SULFONIC ACID	FKM,FVMQ,CR	NR 10% to conc at 70°F	NR 10% to conc at 70°F	AB conc to 185°F A 10% to 140°F	AB to conc to 70°F	
BENZIDAM		NR at 70°F	B at 70°F	A to 70°F	C at 70°F	
BENZIN		A to 70°F	NR at 70°F	A to 70°F	B at 70°F	
BENZINE (GASOLINE)	NBR,FKM,FVMQ,T	A to 250°F	NR at 70°F	+1.5% vol 28 days 100°F +29% vol 28 days 158°F	B at 70°F	
BENZINE (PETROLEUM NAPHTHA)		AC at 70°F	NR at 70°F	A to 70°F	B/NR at 70°F	
BENZINE SOLVENT		A to 70°F	NR at 70°F	A to 70°F	B at 70°F	
BENZOCHLORIDE	FKM,FVMQ,EP,T	NR at 70°F	A to 70°F	A to 70°F	NR at 70°F	
BENZOIC ACID	FKM,FVMQ,T	NR at 70°F	A/NR to 176°F	A to 176°F	A to 70°F AB 100% to 150°F	
BENZOIC ALDEHYDE		NR any conc at 70°F	B at 70°F	NR at 70°F	NR to 100% at 70°F	
BENZOL (BENZENE)	FKM,FVMQ,T	+155% vol 3 days 70°F NR at 70°F	+82% vol 3 days 70°F NR at 70°F	+20% vol 7 days 70°F B to 158°F	+170% vol 7 days 70°F NR any conc at 70°F	NR at 70°F
BENZOL HYDRIDE		NR any conc at 70°F	B at 70°F	NR at 70°F	NR any conc at 70°F	
BENZOLINE		A to 70°F	NR at 70°F	A to 70°F	B at 70°F	
BENZOPHENOL		NR any conc at 70°F			NR any conc at 70°F	
BENZOPHENONE	FKM,FVMQ,EP,T	A/NR at 70°F	B at 70°F	A to 212°F		
BENZOTRICHLORIDE		NR at 70°F				NR at 70°F
BENZOYLE CHLORIDE		NR at 70°F	NR at 70°F	BC at 70°F	NR at 70°F	
BENZYL ACETATE		NR at 70°F		NR at 70°F		
BENZYL ALCOHOL	FKM,FVMQ,EP,T	NR any conc at 70°F	B to 104°F C at 140°F	A to 140°F AB to 250°F	B/NR at 70°F	
BENZYL BENZOATE	FKM,FVMQ,EP,T	NR at 70°F	B at 70°F	A to 70°F	NR at 70°F	
BENZYL CHLORIDE	FKM,FVMQ	NR at 70°F	NR at 70°F	A to 200°F	NR at 70°F	
BETA CAROTENE		A to 70°F		A to 70°F	NR at 70°F	
BETULA OIL		NR any conc at 70°F	C at 70°F	B at 70°F	NR at 70°F	
BIBORATE OF SODA		B at 70°F	A to 70°F	A to 70°F	A to 70°F	
BICARBONATE OF SODA	NBR,EP,FKM,CR,IIR	A to 70°F AB to 200°F	A to 100% to 176°F B at 212°F	A to 100% to 212°F	A to 70°F AB to 200°F	
BICHLORIDE OF MERCURY	NBR,CR,EP	A to 70°F	A to 70°F	A to 200°F	AB to 70°F	
BICHROMATE OF SODA						
BIPHENYL	FKM,T,FVMQ	NR at 70°F	NR at 70°F	A to 300°F	NR at 70°F	NR at 70°F
BIPHENYL OXIDES		NR at 70°F	NR at 70°F	A to 70°F	NR at 70°F	
BIPHENYL PHTHALATE		NR at 70°F		C at 70°F		
BIRCH OIL		NR at 70°F	C at 70°F	B at 70°F	NR at 70°F	
BISMUTH CARBONATE		A to 70°F	A to 70°F	A to 70°F	A to 70°F	
BISMUTH SUBCARBONATE		A to 70°F	A to 70°F	A to 70°F	A to 70°F	
BISMUTHYL CARBONATE		A to 70°F	A to 70°F	A to 70°F	A to 70°F	
BLACK ASH		A to 70°F AB any conc to 200°F	A to 70°F	A to 70°F	A to 70°F AB any conc to 200°F	
BLACK LIQUOR		A to 140°F B at 176°F	A to 176°F	A to 212°F	A to 70°F	
BLACK LIQUOR WASTE		A to 70°F	NR at 70°F	A to 70°F	B at 70°F	
BLACK POINT 77	NBR,EP,FKM,IIR	A to 70°F	A to 70°F	A to 70°F	C at 70°F	A to 70°F
BLACK SULFATE LIQUOR	EP,NBR,CR,FKM	AB any conc to 200°F	AB to 250°F	A to 176°F B at 212°F	AB any conc to 100°F	
BLANC FIXE, SYNTHETIC	NBR,CR,EP	A to 140°F AB to 200°F	A to 176°F AB to 200°F	A to 248°F	A to 70°F AB to 200°F	A to 70°F
BLAST FURNACE GAS	VMQ,FKM,NBR	NR at 70°F	NR at 70°F	A to 70°F	NR at 70°F	NR at 70°F
BLEACH LIQUOR	EP,FKM,IIR	B/NR at 70°F	A to 70°F	A to 70°F	B/NR at 70°F	B at 70°F
BLEACH SOLUTIONS	EP,IIR	NR conc at 70°F C 5-12% at 70°F	A conc to 70°F A 5% to 104°F	A conc to 70°F A 5% to 104°F	NR 10-100% at 70°F	
BLEACHING POWDER, SOLUTIONS	EP	NR 35% at 70°F C sol'n at 70°F	A conc to 140°F A sol'n to 200°F	A to 140°F	A sol'n to 70°F B sol'n at 140°F	B conc at 70°F

	STYRENE BUTADIENE SBR	POLYACRYLATE ACM	POLYURETHANE AU, EU	ISOBUTYLENE ISOPRENE IIR	POLYBUTADIENE BR	® AEROQUIP AQP
BENZENE METHYLAL			NR at 70°F	B at 70°F		
BENZENE SULFONIC ACID	NR 10% to conc at 70°F	NR 10% at 70°F	NR 10% to conc at 70°F	NR 10% to conc at 70°F	NR 10% at 70°F	
BENZIDAM			C at 70°F	A at 70°F		
BENZIN				NR at 70°F		
BENZINE (GASOLINE)	NR at 70°F	A to 70°F	BC at 70°F	NR at 70°F	NR at 70°F	AB to 70°F
BENZINE (PETROLEUM NAPHTHA)				NR at 70°F		AB to 70°F
BENZINE SOLVENT				NR at 70°F		
BENZOCHLORIDE	NR at 70°F	NR at 70°F		B at 70°F	NR at 70°F	
BENZOIC ACID	NR at 70°F	NR at 70°F	NR at 70°F	A/NR at 70°F	NR at 70°F	
BENZOIC ALDEHYDE			NR at 70°F	B any conc at 75°F		
BENZOL (BENZENE)	+180% vol 3 days 70°F NR at 70°F	+214% vol 3 days 70°F NR at 70°F	+109% vol 3 days 70°F NR at 70°F	+75% vol 3 days 70°F NR any conc at 70°F	NR at 70°F	TEST
BENZOL HYDRIDE			NR at 70°F	B any conc at 70°F		
BENZOLINE				NR at 70°F		
BENZOPHENOL				B any conc to 100°F		
BENZOPHENONE	NR at 70°F	NR at 70°F	NR at 70°F	B at 70°F	NR at 70°F	
BENZOTRICHLORIDE						
BENZOYLE CHLORIDE			NR at 70°F	NR at 70°F		
BENZYL ACETATE				A to 70°F		
BENZYL ALCOHOL	NR at 70°F	NR at 70°F	NR at 70°F	AB any conc to 185°F	NR at 70°F	
BENZYL BENZOATE	NR at 70°F	NR at 70°F		B to 80°F	NR at 70°F	
BENZYL CHLORIDE	NR at 70°F	NR at 70°F	NR at 70°F	NR at 70°F	NR at 70°F	
BETA CAROTENE			A to 70°F			
BETULA OIL				B any conc to 80°F		
BIBORATE OF SODA			A to 70°F	A to 70°F		
BICARBONATE OF SODA	A to 100% to 70°F	NR sol'n at 70°F	A to 70°F	A to 70°F AB to 185°F	A to 70°F	
BICHLORIDE OF MERCURY	A to 70°F	NR at 70°F	A to 70°F	A to 70°F		
BICHROMATE OF SODA				A to 70°F		
BIPHENYL				NR at 70°F		
BIPHENYL OXIDES				NR at 70°F		
BIPHENYL PHTHALATE				A to 70°F		
BIRCH OIL				B at 70°F		
BISMUTH CARBONATE				A to 70°F		
BISMUTH SUBCARBONATE				A to 70°F		
BISMUTHYL CARBONATE				A to 70°F		
BLACK ASH				A to 70°F AB any conc to 185°F		
BLACK LIQUOR			C/NR at 70°F	A to 70°F		
BLACK LIQUOR WASTE				NR at 70°F		AB to 70°F
BLACK POINT 77	C at 70°F	C at 70°F	C at 70°F	A to 70°F	C at 70°F	
BLACK SULFATE LIQUOR	B at 70°F	NR at 70°F	NR at 70°F	AB any conc to 150°F	B at 70°F	
BLANC FIXE, SYNTHETIC			A to 70°F	A to 70°F		
BLAST FURNACE GAS	NR at 70°F	NR at 70°F	NR at 70°F	C/NR at 70°F	NR at 70°F	AB to 70°F
BLEACH LIQUOR	B/NR at 70°F	NR at 70°F	NR at 70°F	A to 70°F	B/NR at 70°F	
BLEACH SOLUTIONS	B/NR at 70°F	NR at 70°F	NR 7–100% at 70°F AB dilute to 70°F	AB to 70°F	NR at 70°F	
BLEACHING POWDER, SOLUTIONS	NR at 70°F	NR at 70°F	NR at 70°F	B at 70°F AB 35% to 150°F		

	SYNTHETIC ISOPRENE IR	NATURAL ISOPRENE NR	CHLOROSULFONATED POLYETHYLENE CSM	FLUOROSILICONE FVMQ	SILICONE VMQ	CHEMRAZ FFKM
BENZENE METHYLAL	NR at 70°F	NR at 70°F	NR at 70°F		C at 70°F	
BENZENE SULFONIC ACID	NR 10% to conc at 70°F	NR 10% to conc at 70°F	AB to 70°F A 10% to 70°F	B 10% to conc at 70°F	NR 10% to conc at 70°F	A 10% to 70°F
BENZIDAM	NR at 70°F	NR at 70°F	C at 70°F		B at 70°F	
BENZIN	NR at 70°F	NR at 70°F	NR at 70°F		A to 70°F	
BENZINE (GASOLINE)	NR at 70°F	NR at 70°F	C/NR at 70°F	A to 70°F	NR at 70°F	A to 70°F
BENZINE (PETROLEUM NAPHTHA)	NR at 70°F	NR at 70°F	NR at 70°F		A to 70°F	
BENZINE SOLVENT	NR at 70°F	NR at 70°F	NR at 70°F		A to 70°F	
BENZOCHLORIDE	NR at 70°F	NR at 70°F	NR at 70°F	A to 70°F		
BENZOIC ACID	B/NR at 70°F	B/NR to 150°F	B/NR at 70°F	B at 70°F	B/NR at 70°F	A to 70°F
BENZOIC ALDEHYDE	NR at 70°F	NR any conc at 70°F	NR at 70°F		C at 70°F	
BENZOL (BENZENE)	NR at 70°F	NR any conc at 70°F	+225% vol 7 days 70°F NR at 70°F	+23% vol 3 days 70°F AB to 70°F	+140% vol 7 days 70°F NR at 70°F	A to 70°F
BENZOL HYDRIDE	NR at 70°F	NR any conc at 70°F	NR at 70°F		C at 70°F	
BENZOLINE	NR at 70°F	NR at 70°F	NR at 70°F		A to 70°F	
BENZOPHENOL	NR at 70°F	NR any conc at 70°F				
BENZOPHENONE	NR at 70°F			A to 70°F	NR at 70°F	A to 70°F
BENZOTRICHLORIDE						A to 70°F
BENZOYLE CHLORIDE	NR at 70°F	NR at 70°F	NR at 70°F		NR at 70°F	A to 70°F
BENZYL ACETATE	NR at 70°F	NR at 70°F	B at 70°F			A to 70°F
BENZYL ALCOHOL	C/NR at 70°F	NR any conc at 70°F	B at 70°F	B at 70°F		A to 70°F
BENZYL BENZOATE	NR at 70°F	NR any conc at 70°F	NR at 70°F	A to 70°F		A to 70°F
BENZYL CHLORIDE	NR at 70°F	NR any conc at 70°F	NR at 70°F	AB to 70°F	NR at 70°F	A to 70°F
BETA CAROTENE						
BETULA OIL	NR at 70°F	NR any conc at 70°F				
BIBORATE OF SODA	A to 70°F	A to 70°F	A to 70°F		B at 70°F	
BICARBONATE OF SODA	A to 70°F	A to 70°F AB to 150°F	A to 70°F AB to 200°F	A to 100% to 70°F	A to 70°F	
BICHLORIDE OF MERCURY	B at 70°F	B at 70°F	A to 70°F	A to 70°F	A to 70°F	A to 70°F
BICHROMATE OF SODA			C at 70°F			
BIPHENYL	NR at 70°F	NR at 70°F	NR at 70°F	B at 70°F	NR at 70°F	A to 70°F
BIPHENYL OXIDES	NR at 70°F	NR at 70°F	C at 70°F		C at 70°F	
BIPHENYL PHTHALATE	NR at 70°F	NR at 70°F	NR at 70°F			
BIRCH OIL	NR at 70°F	NR at 70°F				
BISMUTH CARBONATE	A to 70°F	A to 70°F	A to 70°F			A to 70°F
BISMUTH SUBCARBONATE	A to 70°F	A to 70°F	A to 70°F			
BISMUTHYL CARBONATE	A to 70°F	A to 70°F	A to 70°F			
BLACK ASH	A to 70°F	A to 70°F AB any conc to 150°F	A to 70°F		A to 70°F	
BLACK LIQUOR	B at 70°F	B at 70°F	A to 70°F		A to 70°F	BC at 70°F NR at 70°F white
BLACK LIQUOR WASTE	NR at 70°F	NR at 70°F	C at 70°F			
BLACK POINT 77	C at 70°F	C at 70°F	C at 70°F	C at 70°F	C at 70°F	A to 70°F
BLACK SULFATE LIQUOR	AB to 70°F	AB any conc to 75°F	AB to 70°F	B at 70°F	AB to 70°F	
BLANC FIXE, SYNTHETIC	A to 70°F	A to 70°F AB any conc to 185°F	A to 70°F AB to 300°F	A to 70°F	AB to 70°F	A to 70°F
BLAST FURNACE GAS	NR at 70°F	NR at 70°F	NR at 70°F	B at 70°F	A to 250°F	A to 70°F
BLEACH LIQUOR	B/NR at 70°F	B/NR at 70°F	A to 70°F	B at 70°F	B at 70°F	A to 70°F
BLEACH SOLUTIONS	NR at 70°F	NR at 70°F	A to 70°F	B at 70°F	B at 70°F	
BLEACHING POWDER, SOLUTIONS	C at 70°F	C conc at 70°F AB 35% to 70°F	A to 70°F	B at 70°F	B at 70°F	A conc to 70°F

	POLYSULFIDE T	CHLORINATED POLYETHYLENE CM	EPICHLOROHYDRIN CO, ECO	® HYTREL COPOLYESTER TPE	® SANTOPRENE COPOLYMER TPO	® C-FLEX STYRENIC TPE
BENZENE METHYLAL						
BENZENE SULFONIC ACID	NR 10% to conc at 70°F					A to 70°F
BENZIDAM						
BENZIN						
BENZINE (GASOLINE)	A to 70°F					
BENZINE (PETROLEUM NAPHTHA)	AB to 70°F			AB to 70°F		
BENZINE SOLVENT				AB to 70°F		
BENZOCHLORIDE	NR at 70°F					
BENZOIC ACID	B at 70°F			NR at 70°F		C/NR at 70°F
BENZOIC ALDEHYDE						
BENZOL (BENZENE)	BC at 70°F	C at 70°F	NR at 70°F	BC at 70°F *NR at 140°F	NR at 70°F	NR at 70°F
BENZOL HYDRIDE						
BENZOLINE						
BENZOPHENOL						
BENZOPHENONE	B at 70°F					
BENZOTRICHLORIDE						
BENZOYLE CHLORIDE						
BENZYL ACETATE						
BENZYL ALCOHOL	NR at 70°F		NR at 70°F	C at 70°F	NR at 70°F	NR at 70°F
BENZYL BENZOATE	NR at 70°F					
BENZYL CHLORIDE	NR at 70°F					
BETA CAROTENE						
BETULA OIL						
BIBORATE OF SODA						
BICARBONATE OF SODA	C at 70°F		A to 70°F	AB to 70°F		A to 70°F
BICHLORIDE OF MERCURY						
BICHROMATE OF SODA						
BIPHENYL						
BIPHENYL OXIDES						
BIPHENYL PHTHALATE						
BIRCH OIL						
BISMUTH CARBONATE						
BISMUTH SUBCARBONATE						
BISMUTHYL CARBONATE						
BLACK ASH						
BLACK LIQUOR						
BLACK LIQUOR WASTE						
BLACK POINT 77	C at 70°F					
BLACK SULFATE LIQUOR	NR at 70°F	C at 70°F		AB to 70°F		
BLANC FIXE, SYNTHETIC						
BLAST FURNACE GAS	NR at 70°F	AB to 70°F		AB to 70°F		
BLEACH LIQUOR	NR at 70°F					AB to 70°F non aromatic
BLEACH SOLUTIONS	NR at 70°F			AB 5% to 70°F		
BLEACHING POWDER, SOLUTIONS						

	ETHYLENE ACRYLIC EA	POLYALLOMER LINEAR COPOLYMER	NYLON 11 POLYAMIDE	NYLON 12 POLYAMIDE	ETHYLENE VINYL ACETATE EVA	POLYVINYLCHLORIDE FLEXIBLE PVC
BENZENE METHYLAL						
BENZENE SULFONIC ACID						C at 70°F
BENZIDAM						
BENZIN						
BENZINE (GASOLINE)	NR at 70°F		AB to 70°F		NR at 70°F	AC to 70°F
BENZINE (PETROLEUM NAPHTHA)			AB to 70°F			
BENZINE SOLVENT						
BENZOCHLORIDE						
BENZOIC ACID		A to 70°F B at 122°F	A to 70°F	C at 70°F	AB to 70°F	A to 70°F B at 122°F
BENZOIC ALDEHYDE						
BENZOL (BENZENE)	NR at 70°F	B at 70°F C at 122°F	AB to 105°F yellows C at 140°F		NR at 70°F	C/NR at 70°F
BENZOL HYDRIDE						
BENZOLINE						
BENZOPHENOL						NR any conc at 70°F
BENZOPHENONE						
BENZOTRICHLORIDE						
BENZOYLE CHLORIDE	*NR at 70°F					
BENZYL ACETATE		A to 70°F B at 122°F				NR at 70°F
BENZYL ALCOHOL		NR at 70°F	C at 70°F NR at 105°F	NR at 70°F	*NR at 70°F	AB to 70°F C at 122°F
BENZYL BENZOATE						
BENZYL CHLORIDE						
BETA CAROTENE						
BETULA OIL						NR any conc at 70°F
BIBORATE OF SODA						
BICARBONATE OF SODA	A to 70°F	A to 70°F	A to 70°F AB to 140°F	AB to 70°F aqueous	A to 70°F AB to 140°F	A to 70°F *AB to 130°F
BICHLORIDE OF MERCURY						
BICHROMATE OF SODA						
BIPHENYL						
BIPHENYL OXIDES						
BIPHENYL PHTHALATE						
BIRCH OIL						
BISMUTH CARBONATE						
BISMUTH SUBCARBONATE						
BISMUTHYL CARBONATE						
BLACK ASH						*AB to 140°F
BLACK LIQUOR					C/NR at 70°F	A to 150°F
BLACK LIQUOR WASTE						
BLACK POINT 77						
BLACK SULFATE LIQUOR		A to 70°F	BC at 70°F		A to 70°F	
BLANC FIXE, SYNTHETIC						
BLAST FURNACE GAS			AB to 70°F			
BLEACH LIQUOR			TEST			AB to 70°F
BLEACH SOLUTIONS			C 7% to conc at 70°F NR at 194°F		AB to 140°F	A to 70°F C at 70°F (food grade
BLEACHING POWDER, SOLUTIONS						

	POLYETHYLENE LOW DENSITY LDPE	®TEFLON FEP	®KALREZ PERFLUORINATED ELASTOMER	®FLUORAZ FLUORINATED COPOLYMER	®AFLAS FLUORINATED COPOLYMER	®NORPRENE COPOLYMER TPO
BENZENE METHYLAL						
BENZENE SULFONIC ACID	A 100% at 70°F NR 10% at 70°F	A to 100% to 300°F	A to 212°F	A 10% to 70°F		
BENZIDAM		A to 70°F				
BENZIN		A to 70°F				
BENZINE (GASOLINE)	C/NR at 70°F NR at 122°F	A to 300°F	A to 70°F	BC at 70°F	B at 70°F	BC at 70°F
BENZINE (PETROLEUM NAPHTHA)		A to 70°F	A to 212°F			
BENZINE SOLVENT		A to 70°F				
BENZOCHLORIDE						
BENZOIC ACID	A to 122°F AB any conc to 140°F	A to 300°F	A to 316°F	A to 70°F		A to 70°F
BENZOIC ALDEHYDE		A to 70°F				
BENZOL (BENZENE)	C/NR to 104°F NR at 122°F	A to 122°F	+3% vol 7 days 70°F A to 113°F	C/NR at 70°F	+21% vol 50% conc 7 days +12% vol 30% conc 7 days	NR at 70°F
BENZOL HYDRIDE		A to 70°F				
BENZOLINE		A to 70°F				
BENZOPHENOL		A to 70°F				
BENZOPHENONE			A to 212°F	A to 70°F		
BENZOTRICHLORIDE		A to 70°F	A to 212°F	A to 70°F		
BENZOYLE CHLORIDE		A to 70°F	A to 212°F			
BENZYL ACETATE	A to 70°F B at 122°F	A to 122°F	A to 212°F			
BENZYL ALCOHOL	AB to 104°F C/NR at 140°F	+0.4% wt 7 days 400°F A to 400°F	A to 212°F	A to 70°F	+9% vol 7 days 250°F A to 250°F	BC at 70°F
BENZYL BENZOATE		A to 70°F	A to 212°F	A to 70°F		
BENZYL CHLORIDE		A to 387°F	A to 212°F	A to 70°F	+7% vol 7 days 70°F AB to 70°F	
BETA CAROTENE						
BETULA OIL		A to 70°F				
BIBORATE OF SODA		A to 70°F				
BICARBONATE OF SODA	A to 70°F AB to 140°F	A to 300°F	A to 70°F	A to 70°F aqueous		A to 70°F
BICHLORIDE OF MERCURY		A to 70°F		A to 70°F		
BICHROMATE OF SODA		A to 70°F				
BIPHENYL	C at 70°F	A to 300°F	A to 500°F	C/NR at 70°F		
BIPHENYL OXIDES		A to 70°F				
BIPHENYL PHTHALATE		A to 70°F				
BIRCH OIL		A to 70°F				
BISMUTH CARBONATE		A to 70°F	A to 212°F			
BISMUTH SUBCARBONATE		A to 70°F				
BISMUTHYL CARBONATE		A to 70°F				
BLACK ASH		A to 70°F				
BLACK LIQUOR	A to 70°F	A to 70°F		A to 70°F		
BLACK LIQUOR WASTE		A to 70°F				
BLACK POINT 77				A to 70°F		
BLACK SULFATE LIQUOR	A to 70°F	A to 70°F				
BLANC FIXE, SYNTHETIC	A to 70°F	A to 300°F	A to 70°F	A to 70°F		
BLAST FURNACE GAS		A to 70°F	A to 212°F	A to 70°F		
BLEACH LIQUOR	AB to 70°F	A to 70°F		A to 70°F		A to 70°F
BLEACH SOLUTIONS	A to 70°F AB 10% to 140°F	A to 70°F	A to 70°F *A to 212°F	A to 70°F		
BLEACHING POWDER, SOLUTIONS	A conc to 70°F AB conc to 140°F	A to conc to 300°F	A conc to 70°F	A conc to 122°F	+0% vol 7 days 212°F A to 212°F	

C 49

	RECOMMENDATIONS	NITRILE NBR	ETHYLENE PROPYLENE EP, EPDM	FLUOROCARBON FKM	CHLOROPRENE CR	HYDROGENATED NITRILE HNBR
BLOOD		C at 70°F	A to 70°F	AC to 70°F	A to 70°F	
BLUE COPPERAS	NBR,EP,CR,IIR	A to 70°F AB any conc to 200°F	A to 176°F	A to conc to 212°F	A to 70°F AB any conc to 200°F	A to 70°F
BLUE JACK		A to 70°F	A to 70°F	A to 70°F	A to 70°F	
BLUE SALTS		A to 70°F AB any conc to 200°F	A to 70°F	A to 70°F	A to 70°F AB any conc to 200°F	
BLUE STONE		A to 70°F	A to 70°F	A to 70°F	A to 70°F	
BLUE VITRIOL	NBR,EP,CR,IIR	A to 70°F AB any conc to 200°F	A to conc to 176°F AB to conc to 212°F	A to conc to 140°F A conc at 212°F	A to conc to 140°F AB to conc to 200°F	A to 70°F
BOGHEAD NAPHTHA		A to 70°F	A to 70°F	A to 70°F	C at 70°F	
BOILED LINSEED OIL		+5.4% vol 3 days 212°F A to 212°F	B at 70°F	A to 70°F	+40% vol 3 days 212°F AB to 80°F	
BOILER FEED WATER	EP	AB to 70°F	A to 250°F	B/NR at 70°F	AC to 70°F	
BOLETIC ACID		NR any conc at 70°F				
BONE OIL		A to 70°F		A to 70°F		
BONE TAR		A to 70°F		A to 70°F		
BORACIC ACID	NBR,EP,CR,FKM,IIR	A to 140°F AB any conc to 200°F	A to 140°F AB to 200°F	A to 176°F B at 212°F	A to 158°F AB any conc to 200°F	
BORAX SOLUTIONS	NBR,EP,IIR,CR,AU	AB any conc to 200°F	A to 200°F	A to 176°F	A to 158°F AB any conc to 200°F	A to 70°F
BORAX DEHYDRATE		A to 70°F AB any conc to 200°F			B any conc to 200°F	
BORDEAUX MIXTURE	NBR,FKM,EP,IIR	A to 160°F AB any conc to 200°F	A to 70°F	A to 70°F	AB any conc to 200°F	
BORIC ACID	NBR,EP,CR,FKM,IIR	A to 140°F AB any conc to 200°F	A to 140°F AB to 200°F	A to 176°F B at 212°F	A to 158°F AB any conc to 200°F	A to 70°F
BORIC COPPER SULFATE						
BORON FLUIDS (HIGH ENERGY FUELS)	FKM,VMQ,FVMQ	B at 70°F	NR at 70°F	A to 70°F	NR at 70°F	B at 70°F
BRAKE FLUID	EP,SBR	+90% vol 5 days 212°F NR at 70°F	A to 250°F	NR at 70°F	B at 70°F	
BRAKE FLUID - VEGETABLE	EP,SBR,IIR	BC to 150°F	A to 70°F	NR at 70°F	AB to 100°F	
BRANDY		A to 70°F	A to 70°F	B at 70°F	A to 70°F	
BRANOL		C at 70°F	B at 70°F	A to 70°F	C at 70°F	
BRAY GG-130	FKM	B at 70°F	NR at 70°F	A to 70°F	NR at 70°F	
BRAYCO 719-R (WH-910)	FVMQ,EP	C at 70°F	A to 70°F	NR at 70°F	B at 70°F	
BRAYCO 885 (MIL-L-6085A)	FKM,AU	B at 70°F	NR at 70°F	A to 70°F	NR at 70°F	B at 70°F
BRAYCO 910	EP,IIR	B at 70°F	A to 70°F	NR at 70°F	B at 70°F	B at 70°F
BRET 710	EP	B at 70°F	A to 70°F	NR at 70°F	B at 70°F	
BRINE	NBR,CR,EP	A to 176°F B at 212°F	A to 140°F B at 212°F	A to 248°F	AB to 140°F B at 176°F	
BRINE, CaCl, SOL'N & INHIBITED	NBR	A to 200°F	A to 70°F	A to 70°F	A to 70°F	
BRINE, CuCl	NBR	A to 200°F	A to 70°F	A to 70°F	B at 70°F	
BRINE, NaCl	NBR	A to 200°F	A to 70°F	A to 70°F	B at 70°F	
BRINE, SEA WATER	NBR	AB to 212°F NR at 248°F	AB to 212°F C at 220°F	A to 248°F	AB to 140°F C at 176°F	A to 70°F
BRINE ACID		A to 70°F	A to 70°F	A to 70°F		
BROM-113	NBR	C at 70°F	NR at 70°F		NR at 70°F	
BROM-114	NBR,FKM	B at 70°F	NR at 70°F	B at 70°F	B at 70°F	
BROMALLYENE		NR at 70°F		B at 70°F	NR at 70°F	
BROMINE ANHYDROUS LIQUID		NR at 70°F	NR at 70°F	A to 70°F AB to 212°F	NR at 70°F	
BROMINE GAS	FKM,FVMQ,T	NR any conc at 70°F wet or dry	NR any conc at 70°F wet or dry	A to 200°F wet or dry	NR any conc at 70°F wet or dry	
BROMINE PENTAFLUORIDE		NR at 70°F	NR at 70°F	NR at 70°F	NR at 70°F	NR at 70°F
BROMINE TRIFLUORIDE		NR at 70°F	NR at 70°F	NR at 70°F	NR at 70°F	NR at 70°F
BROMINE WATER		NR at 70°F	NR at 70°F	A to 104°F AB to 200°F	B/NR at 70°F	C at 70°F
BROMOBENZENE	FKM,FVMQ,T	NR at 70°F	NR at 70°F	A to 70°F	NR at 70°F	NR at 70°F
BROMOCHLOROMETHANE		NR at 70°F	B at 70°F	C at 70°F	NR at 70°F	

	STYRENE BUTADIENE SBR	POLYACRYLATE ACM	POLYURETHANE AU, EU	ISOBUTYLENE ISOPRENE IIR	POLYBUTADIENE BR	®AEROQUIP AQP
BLOOD			A/NR at 70°F			
BLUE COPPERAS			A to 70°F	A to 70°F AB any conc to 185°F		
BLUE JACK			A to 70°F	A to 70°F		
BLUE SALTS			A to 70°F	A to 70°F AB any conc to 185°F		
BLUE STONE			A to 70°F	A to 70°F		
BLUE VITRIOL		NR at 70°F	AB 10% & 100% to 70°F C 50% at 70°F	A to 70°F AB any conc to 185°F		NR at 70°F
BOGHEAD NAPHTHA			C at 70°F	NR at 70°F		
BOILED LINSEED OIL	+56% vol 3 days 212°F NR at 212°F	+4% vol 3 days 212°F A to 212°F	B at 70°F	+12% vol 3 days 212°F AB to 150°F		
BOILER FEED WATER	AC to 70°F	NR at 70°F	NR at 70°F			
BOLETIC ACID						
BONE OIL				B at 70°F		
BONE TAR				B at 70°F		
BORACIC ACID	A to 70°F	NR at 70°F	A to 70°F	A to 70°F AB any conc to 185°F	A to 70°F	NR at 70°F
BORAX SOLUTIONS	B at 70°F	B at 70°F	A to 70°F	A to 70°F AB any conc to 185°F	B at 70°F	AB to 70°F
BORAX DEHYDRATE				A to 70°F AB any conc to 185°F		
BORDEAUX MIXTURE	B at 70°F	NR at 70°F	NR at 70°F	A to 70°F AB any conc to 185°F	B at 70°F	
BORIC ACID	A to 70°F	NR at 70°F	A to 70°F	A to 70°F AB any conc to 185°F	A to 70°F	NR at 70°F
BORIC COPPER SULFATE			AB to 70°F			
BORON FLUIDS (HIGH ENERGY FUELS)	NR at 70°F	NR at 70°F	NR at 70°F	NR at 70°F	NR at 70°F	
BRAKE FLUID	A to 70°F	NR at 70°F	B/NR at 70°F AB fumes at 70°F	A to 70°F		
BRAKE FLUID - VEGETABLE	A to 70°F	NR at 70°F	NR at 70°F	AB to 150°F		
BRANDY			NR at 70°F	A to 70°F		
BRANOL			B at 70°F	C at 70°F		
BRAY GG-130	NR at 70°F	NR at 70°F	NR at 70°F	B/NR at 70°F	B at 70°F	
BRAYCO 719-R (WH-910)			NR at 70°F	B at 70°F		
BRAYCO 885 (MIL-L-6085A)	NR at 70°F	BC at 70°F	AC to 70°F	NR at 70°F	NR at 70°F	
BRAYCO 910	B at 70°F	C at 70°F	C at 70°F	A to 70°F	A to 70°F	
BRET 710	B at 70°F	C at 70°F	C at 70°F	A to 70°F	A to 70°F	
BRINE			A/NR at 70°F	A to 70°F		AB to 70°F
BRINE, CaCl, SOL'N & INHIBITED	A to 70°F	NR at 70°F	B at 70°F			
BRINE, CuCl	A to 70°F	A to 70°F	AB to 70°F			
BRINE, NaCl	A to 70°F	NR at 70°F	B at 70°F			
BRINE, SEA WATER	A to 70°F	NR at 70°F	B at 70°F			
BRINE ACID						
BROM-113	NR at 70°F			NR at 70°F	NR at 70°F	
BROM-114	NR at 70°F			NR at 70°F	NR at 70°F	
BROMALLYENE				NR at 70°F		
BROMINE ANHYDROUS LIQUID			C/NR at 70°F			
BROMINE GAS	NR at 70°F	NR at 70°F	NR at 70°F wet or dry	NR any conc at 70°F	NR at 70°F	
BROMINE PENTAFLUORIDE	NR at 70°F	NR at 70°F	NR at 70°F	NR at 70°F	NR at 70°F	
BROMINE TRIFLUORIDE	NR at 70°F	NR at 70°F	NR at 70°F	NR at 70°F	NR at 70°F	
BROMINE WATER	NR at 70°F	NR at 70°F	NR at 70°F	NR at 70°F	NR at 70°F	
BROMOBENZENE	NR at 70°F	NR at 70°F	NR at 70°F	NR at 70°F	NR at 70°F	
BROMOCHLOROMETHANE				B/NR at 70°F		

	SYNTHETIC ISOPRENE IR	NATURAL ISOPRENE NR	CHLOROSULFONATED POLYETHYLENE CSM	FLUOROSILICONE FVMQ	SILICONE VMQ	CHEMRAZ FFKM
BLOOD					AB to 70°F	
BLUE COPPERAS	BC at 70°F	B any conc 70–150°F / A 15–50% to 70°F	A to conc to 70°F / AB to conc to 200°F	A to 100% to 70°F	A to 100% to 70°F	A to 70°F
BLUE JACK	C at 70°F	C at 70°F	A to 70°F		A to 70°F	
BLUE SALTS	A to 70°F	A to 70°F / AB any conc to 150°F	A to 70°F		A to 70°F	
BLUE STONE	C at 70°F	C at 70°F	A to 70°F		A to 70°F	
BLUE VITRIOL	BC at 70°F	B 10& 100% to 150°F / A 15 to 50% to 70°F	A to conc to 70°F / AB to conc at 200°F	A to 100% to 70°F	A to 100% to 70°F	A to 70°F
BOGHEAD NAPHTHA	NR at 70°F	NR at 70°F	C at 70°F		C at 70°F	
BOILED LINSEED OIL	NR at 70°F	NR at 70°F	B at 70°F		+3% vol 3 days 212°F / A/NR at 70°F	
BOILER FEED WATER				B at 70°F	C at 70°F	
BOLETIC ACID	B at 70°F	B any conc to 80°F				
BONE OIL			NR at 70°F			
BONE TAR	NR at 70°F	NR at 70°F	NR at 70°F			
BORACIC ACID	A to 70°F		A to 200°F	A to 70°F	A to 70°F	
BORAX SOLUTIONS	AB to 70°F	AB any conc to 185°F	A to 200°F	B at 70°F	B at 70°F	A to 70°F
BORAX DEHYDRATE	A to 70°F	A to 70°F / AB any conc to 150°F				
BORDEAUX MIXTURE	AC to 70°F	AB any conc to 150°F	A to 70°F	B at 70°F	B at 70°F	B at 70°F
BORIC ACID	A to 70°F	A to 70°F / AB any conc to 185°F	A to 200°F	A to 70°F	A to 70°F	A to 70°F
BORIC COPPER SULFATE						
BORON FLUIDS (HIGH ENERGY FUELS)	NR at 70°F	NR at 70°F	NR at 70°F	B at 70°F	NR at 70°F	
BRAKE FLUID		NR at 70°F	B/NR at 70°F	NR at 70°F	+2.8% vol 5 days 212°F / C at 70°F	
BRAKE FLUID - VEGETABLE	NR at 70°F	NR at 70°F	B at 70°F	NR at 70°F	C/NR at 70°F	
BRANDY	A to 70°F	A to 70°F	A to 70°F		A to 70°F	
BRANOL	C at 70°F	C at 70°F	A to 70°F		NR at 70°F	
BRAY GG-130	NR at 70°F	NR at 70°F	NR at 70°F	B at 70°F	NR at 70°F	A to 70°F
BRAYCO 719-R (WH-910)	B at 70°F	B at 70°F	B at 70°F	B at 70°F	B at 70°F	A to 70°F
BRAYCO 885 (MIL-L-6085A)	NR at 70°F	NR at 70°F	NR at 70°F	B at 70°F	NR at 70°F	A to 70°F
BRAYCO 910	A to 70°F	A to 70°F	A to 70°F	NR at 70°F	NR at 70°F	A to 70°F
BRET 710	A to 70°F	A to 70°F	A to 70°F	NR at 70°F	NR at 70°F	A to 70°F
BRINE	A to 70°F	A to 70°F	A to 70°F	A to 70°F	A to 70°F	
BRINE, CaCl, SOL'N & INHIBITED				A to 70°F	A to 70°F	A to 70°F
BRINE, CuCl				A to 70°F	A to 70°F	
BRINE, NaCl				A to 70°F	A to 70°F	A to 70°F
BRINE, SEA WATER				A to 70°F	A to 70°F	A to 70°F
BRINE ACID						
BROM-113			NR at 70°F		NR at 70°F	
BROM-114	NR at 70°F	NR at 70°F	B at 70°F		NR at 70°F	
BROMALLYENE	NR at 70°F	NR at 70°F	NR at 70°F			
BROMINE ANHYDROUS LIQUID		NR at 70°F	BC at 70°F	B at 70°F	C at 70°F	
BROMINE GAS	NR at 70°F	NR any conc at 70°F	NR at 70°F	B at 70°F	NR at 70°F	A to 70°F wet or dry
BROMINE PENTAFLUORIDE	NR at 70°F	NR at 70°F	NR at 70°F	NR at 70°F	NR at 70°F	BC at 70°F B at 70°F white
BROMINE TRIFLUORIDE	NR at 70°F	NR at 70°F	NR at 70°F	NR at 70°F	NR at 70°F	BC at 70°F B at 70°F white
BROMINE WATER	NR at 70°F	NR at 70°F	A/NR at 70°F	B at 70°F	NR at 70°F	A to 70°F
BROMOBENZENE	NR at 70°F	NR at 70°F	NR at 70°F	AB to 70°F	NR at 70°F	A to 70°F
BROMOCHLOROMETHANE	NR at 70°F	NR at 70°F	NR at 70°F			

	POLYSULFIDE T	CHLORINATED POLYETHYLENE CM	EPICHLOROHYDRIN CO, ECO	® HYTREL COPOLYESTER TPE	® SANTOPRENE COPOLYMER TPO	® C-FLEX STYRENIC TPE
BLOOD						
BLUE COPPERAS						
BLUE JACK						
BLUE SALTS						
BLUE STONE						
BLUE VITRIOL	NR 10 & 100% at 70°F A 50% to 70°F	NR at 70°F		A to 70°F *AB to 140°F		
BOGHEAD NAPHTHA						
BOILED LINSEED OIL	+0.5% vol 3 days 212°F A to 212°F					
BOILER FEED WATER						
BOLETIC ACID						
BONE OIL						
BONE TAR						
BORACIC ACID	NR at 70°F	NR at 70°F	A to 70°F	A to 70°F *AB to 140°F	A to 70°F	AB to 70°F
BORAX SOLUTIONS	NR at 70°F	AB to 70°F		A to 70°F *AB to 140°F		
BORAX DEHYDRATE						
BORDEAUX MIXTURE	NR at 70°F			AB to 70°F		
BORIC ACID	NR at 70°F	NR at 70°F	A to 70°F	A to 70°F *AB to 140°F	A to 70°F	AB to 70°F
BORIC COPPER SULFATE				AB to 70°F		
BORON FLUIDS (HIGH ENERGY FUELS)	B at 70°F					
BRAKE FLUID					+9-18% vol 7 days 212°F AB to 212°F	
BRAKE FLUID - VEGETABLE	NR at 70°F					
BRANDY						
BRANOL						
BRAY GG-130	B at 70°F					
BRAYCO 719-R (WH-910)	NR at 70°F					
BRAYCO 885 (MIL-L-6085A)	B at 70°F		B at 70°F			
BRAYCO 910	NR at 70°F					
BRET 710	NR at 70°F					
BRINE		AB to 70°F		AB to 70°F *AB to 140°F		
BRINE, CaCl, SOL'N & INHIBITED						
BRINE, CuCl						
BRINE, NaCl						
BRINE, SEA WATER						
BRINE ACID						
BROM-113	B at 70°F					
BROM-114	A to 70°F					
BROMALLYENE						
BROMINE ANHYDROUS LIQUID	B at 70°F			*NR at 70°F		
BROMINE GAS	B at 70°F	NR at 70°F		NR at 70°F		AB to 70°F
BROMINE PENTAFLUORIDE	NR at 70°F					
BROMINE TRIFLUORIDE	NR at 70°F		NR at 70°F			
BROMINE WATER	B at 70°F					
BROMOBENZENE	AC at 70°F		NR at 70°F		NR at 70°F	
BROMOCHLOROMETHANE						

	ETHYLENE ACRYLIC EA	POLYALLOMER LINEAR COPOLYMER	NYLON 11 POLYAMIDE	NYLON 12 POLYAMIDE	ETHYLENE VINYL ACETATE EVA	POLYVINYLCHLORIDE FLEXIBLE PVC
BLOOD					A to 70°F	A/NR at 70°F
BLUE COPPERAS						
BLUE JACK						
BLUE SALTS						
BLUE STONE						
BLUE VITRIOL	*A to 70°F	A to 70°F	AB to 194°F		B to 140°F	AB to 140°F
BOGHEAD NAPHTHA						
BOILED LINSEED OIL						
BOILER FEED WATER						
BOLETIC ACID						
BONE OIL						
BONE TAR						
BORACIC ACID	A to 70°F	A to 70°F	AC to 70°F		AB to 140°F	A to 150°F
BORAX SOLUTIONS	A to 212°F	A to 70°F	AB to 70°F *AB to 140°F	AB to 100% to 70°F	A to 70°F AB to 140°F	A to 70°F AB to 140°F
BORAX DEHYDRATE						
BORDEAUX MIXTURE			AB to 70°F			A to 150°F
BORIC ACID	A to 70°F	A to 70°F	AC to 70°F	AB 10% to 70°F	AB to 140°F	A to 150°F
BORIC COPPER SULFATE			AB to 70°F			A to 70°F
BORON FLUIDS (HIGH ENERGY FUELS)						
BRAKE FLUID			AB to 70°F	AB to 70°F	NR at 70°F	NR at 70°F
BRAKE FLUID - VEGETABLE	NR at 70°F					
BRANDY		A to 70°F	B at 70°F	AB to 70°F	B at 70°F	
BRANOL						
BRAY GG-130						
BRAYCO 719-R (WH-910)						
BRAYCO 885 (MIL-L-6085A)						
BRAYCO 910						
BRET 710						
BRINE			AB to 140°F		A to 70°F AB to 140°F	A to 70°F *AB to 140°F
BRINE, CaCl, SOL'N & INHIBITED						
BRINE, CuCl						
BRINE, NaCl						
BRINE, SEA WATER						
BRINE ACID						
BROM-113						
BROM-114						
BROMALLYENE						
BROMINE ANHYDROUS LIQUID			NR at 70°F	NR at 70°F	*NR at 70°F	AB to 70°F may discolor
BROMINE GAS		NR at 70°F	NR 20-100% at 70°F	NR at 70°F	NR at 70°F	B at 70°F NR at 122°F
BROMINE PENTAFLUORIDE					NR at 70°F	NR at 70°F
BROMINE TRIFLUORIDE						
BROMINE WATER		NR at 70°F	NR at 70°F		NR at 70°F	NR at 70°F
BROMOBENZENE		NR at 70°F				NR at 70°F
BROMOCHLOROMETHANE						NR at 70°F

	POLYETHYLENE LOW DENSITY LDPE	®TEFLON FEP	®KALREZ PERFLUORINATED ELASTOMER	®FLUORAZ FLUORINATED COPOLYMER	®AFLAS FLUORINATED COPOLYMER	®NORPRENE COPOLYMER TPO
BLOOD	A to 70°F	*A to 70°F		C at 70°F		
BLUE COPPERAS	AB to conc to 140°F	A to conc to 200°F	A to 70°F	AB to 100% to 70°F		
BLUE JACK		A to 70°F				
BLUE SALTS		A to 70°F				
BLUE STONE		A to 70°F				
BLUE VITRIOL	AB to conc to 140°F	A to conc to 200°F	A to 70°F	A to 100% to 70°F		
BOGHEAD NAPHTHA		A to 70°F				
BOILED LINSEED OIL		A to 70°F				
BOILER FEED WATER				A to 70°F		
BOLETIC ACID		A to 70°F				
BONE OIL		A to 70°F		A to 70°F		
BONE TAR		A to 70°F				
BORACIC ACID	A to 70°F AB to 140°F	A to 100% to 300°F	A to 70°F A 10% to 212°F	A to 70°F A 50% to 200°F		A to 70°F
BORAX SOLUTIONS	A to 70°F AB to 140°F	A to 300°F	A to 212°F	A to 70°F		A to 70°F
BORAX DEHYDRATE		A to 70°F				
BORDEAUX MIXTURE	AB to 70°F	A to 70°F	A to 212°F			
BORIC ACID	A to 70°F AB to 140°F	A to 100% to 300°F	A to 100% to 212°F	A to 70°F A 50% to 200°F		A to 70°F
BORIC COPPER SULFATE	AB to 70°F					
BORON FLUIDS (HIGH ENERGY FUELS)				A to 70°F		
BRAKE FLUID	C/NR at 70°F	A to 70°F		A to 70°F	+8.8% vol 5 days 212°F AB to 212°F	
BRAKE FLUID - VEGETABLE		A to 70°F				
BRANDY	A to 70°F	A to 70°F				
BRANOL		A to 70°F				
BRAY GG-130				B at 70°F		
BRAYCO 719-R (WH-910)				C at 70°F		
BRAYCO 885 (MIL-L-6085A)				B at 70°F		
BRAYCO 910				C at 70°F		
BRET 710						
BRINE	A to 70°F AB to 140°F	A to 70°F	A to 70°F AB to 212°F			
BRINE, CaCl, SOL'N & INHIBITED				A to 70°F		
BRINE, CuCl						
BRINE, NaCl				A to 70°F		
BRINE, SEA WATER	AB to 104°F C at 140°F			A to 180°F		
BRINE ACID	A to 70°F	A to 70°F				
BROM-113						
BROM-114						
BROMALLYENE		A to 70°F				
BROMINE ANHYDROUS LIQUID	NR at 70°F wet or dry	A to 300°F	A to 212°F			
BROMINE GAS	NR any conc at 70°F	A to 200°F wet or dry	A to 138°F	AC to 70°F C wet at 70°F	+6% vol 7 days 212°F 32% to concentrated	A to 70°F
BROMINE PENTAFLUORIDE	C/NR at 70°F		*B at 70°F	NR at 70°F		
BROMINE TRIFLUORIDE		A to 70°F	BC at 70°F	NR at 70°F		
BROMINE WATER	NR at 70°F	A to 300°F	*A to 212°F	A to 70°F		
BROMOBENZENE	NR at 70°F	A to 122°F	A to 212°F	NR at 70°F		
BROMOCHLOROMETHANE		A to 70°F				

	RECOMMENDATIONS	NITRILE NBR	ETHYLENE PROPYLENE EP, EPDM	FLUOROCARBON FKM	CHLOROPRENE CR	HYDROGENATED NITRILE HNBR
BROMOCHLOROTRIFLUOROMETHANE	FKM	NR at 70°F	NR at 70°F	A to 70°F	NR at 70°F	NR at 70°F
BROMOETHYLENE		NR at 70°F	C at 70°F	B at 70°F	NR at 70°F	
BROMOMETHANE	FKM, FVMQ	B at 70°F NR at 140°F	A/NR at 70°F	A to 160°F	NR at 70°F	B at 70°F
BROMOPENTANE		NR at 70°F	NR at 70°F	B at 70°F	NR at 70°F	
BROMOTOLUENE		NR at 70°F		B at 70°F		
BRONZING LIQUID		A to 70°F	B at 70°F	NR at 70°F	NR at 70°F	
BROWN ACETATE		B any conc to 120°F	A to 70°F	NR at 70°F	B/NR at 70°F	
BROWN LINSEED OIL		A to 70°F AB to 180°F	B at 70°F	A to 70°F	B to 80°F	
BRUCEITE	EP, FKM, IIR, NBR	B at 70°F	A to 176°F AB to 212°F	A to 100% to 212°F	A to 100% to 158°F AB to 100% to 200°F	B at 70°F
BRUDIUM				A to 140°F		
BUNKER "C"	NBR	A to 250°F	NR at 70°F	A to 70°F	B/NR at 70°F	
BUNKER OIL	NBR, FKM, ACM, FVMQ	A to 70°F AB to 150°F	NR at 70°F	A to 70°F	B/NR at 70°F	A to 70°F
BURNT ALUM		A to 70°F AB any conc to 200°F	A to 70°F	A to 70°F	A to 70°F AB any conc to 200°F	
BURNT LIME		A to 70°F	A to 70°F		A to 70°F	
BURNT POTASH		B at 70°F			A to 70°F	
BUROWS SOLUTION		B at 70°F			B at 70°F	
BUTADIEN (MONOMER)	FKM, FVMQ	A/NR to 140°F	A/NR to 140°F	AB to 250°F	B/NR at 70°F	
BUTANAL (BUTAL)		NR at 70°F	B at 70°F	NR at 70°F	C at 70°F	
BUTANE	NBR, FKM, CR, ACM, T	A to 200°F AB to 250°F	NR at 70°F	A to 176°F	AC to 140°F	A to 70°F
BUTANE, 2, 2-DIMETHYL	NBR, FKM, ACM	A to 70°F	NR at 70°F	A to 70°F	B at 70°F	A to 70°F
BUTANE, 2, 3-DIMETHYL	NBR, FKM, ACM	A to 70°F	NR at 70°F	A to 70°F	B at 70°F	A to 70°F
BUTANEDIOL		AB to 70°F	NR at 70°F	A to 70°F		
BUTANOIC ACID	EP, FKM, IIR	NR at 70°F B dilute at 70°F	AB to 130°F B at 140°F	AB to 130°F	NR at 70°F	NR at 70°F
BUTANOL	NBR, FKM, EP, CR, SBR ACM	A to 70°F AB any conc to 150°F	AB to 250°F	A to 250°F	A to 140°F B at 176°F	A to 70°F
BUTANONE	EP, IIR	NR any conc at 70°F	A to 140°F AB to 200°F	+240% vol 7 days 70°F NR at 70°F	NR any conc at 70°F	
BUTAROL		B any conc to 150°F			B any conc to 150°F	
1-BUTENE		A to 70°F	NR at 70°F	A to 70°F	A to 70°F	
1-BUTENE, 2-ETHYL	NBR, FKM, ACM	A to 70°F	NR at 70°F	A to 70°F	B/NR at 70°F	
BUTOXYETHANOL	EP, IIR	C any conc to 150°F	A to 200°F	NR at 70°F	NR any conc at 70°F	C at 70°F
BUTTER	NBR, EP, FKM, ACM	A to 200°F	A to 140°F	A to 140°F	B at 70°F C at 140°F	A to 70°F
BUTTER OF ANTIMONY		B at 70°F	AB to 70°F	A to 140°F C at 176°F	BC at 70°F	
BUTTER OF TIN		A to 70°F AB any conc to 150°F	B at 70°F		A/NR at 70°F	
BUTTER OF ZINC		B any conc to 150°F	A to 70°F	A to 70°F	B any conc to 150°F	
BUTTERMILK	NBR	A to 180°F	A to 70°F	A to 140°F	A to 140°F	
BUTYL ACETATE	EP, IIR	+116% vol 3 days 70°F NR any conc at 70°F	+18% vol 3 days 70°F NR at 140°F	+295% vol 3 days 70°F NR at 70°F	+107% vol 3 days 70°F NR any conc at 70°F	
N-BUTYL ACETATE	EP	NR at 70°F	B at 70°F	NR at 70°F	NR at 70°F	
BUTYL ACETATE RICINOLEATE	EP, FKM, IIR	B/NR at 70°F	A to 70°F	A to 70°F	NR at 70°F	
BUTYL ACETO ACETATE				NR at 70°F		
BUTYL ACETYL RICINOLEATE		BC at 70°F	A to 70°F	A to 70°F	B at 70°F	B at 70°F
BUTYL ACRYLATE	T	NR any conc at 70°F	A/NR to 104°F	+188% vol 3 days 122°F NR at 70°F	NR any conc at 70°F	NR at 70°F
BUTYL ALCOHOL	NBR, CR, FKM, SBR	A to 70°F AB any conc to 150°F	AB to 250°F	A to 250°F	A to 140°F B at 176°F	A to 70°F
N-BUTYL ALCOHOL		B at 70°F	A to 70°F	A to 70°F	A to 70°F	
SEC BUTYL ALCOHOL		A to 70°F	A to 70°F	A to 70°F	A to 70°F	
BUTYL ALDEHYDE		NR at 70°F	B at 70°F	NR at 70°F	C at 70°F	

	STYRENE BUTADIENE SBR	POLYACRYLATE ACM	POLYURETHANE AU, EU	ISOBUTYLENE ISOPRENE IIR	POLYBUTADIENE BR	® AEROQUIP AQP
BROMOCHLOROTRIFLUOROMETHANE	NR at 70°F	NR at 70°F	NR at 70°F	NR at 70°F	NR at 70°F	
BROMOETHYLENE				NR at 70°F		
BROMOMETHANE	NR at 70°F	C at 70°F	NR at 70°F	B /NR at 70°F	NR at 70°F	
BROMOPENTANE				NR at 70°F		
BROMOTOLUENE				NR at 70°F		
BRONZING LIQUID			NR at 70°F	B at 70°F		
BROWN ACETATE				A to 70°F AB any conc to 180°F		
BROWN LINSEED OIL			B at 70°F	A to 70°F AB at 150°F		
BRUCEITE			A to 70°F	A to 70°F		
BRUDIUM						
BUNKER "C"	NR at 70°F	B at 70°F	B at 70°F	NR at 70°F		
BUNKER OIL	NR at 70°F	A to 70°F	B at 70°F	NR at 70°F	NR at 70°F	
BURNT ALUM				A to 70°F AB any conc to 185°F		
BURNT LIME				A to 70°F		
BURNT POTASH				A to 70°F		
BUROWS SOLUTION				B at 70°F		
BUTADIEN (MONOMER)	NR at 70°F	NR at 70°F	NR at 70°F	A /NR at 70°F	NR at 70°F	
BUTANAL (BUTAL)			C at 70°F	B at 70°F		
BUTANE	C /NR at 70°F	A to 70°F	A /NR at 70°F	NR at 70°F	NR at 70°F	
BUTANE, 2, 2-DIMETHYL	C at 70°F	A to 70°F	NR at 70°F	NR at 70°F	NR at 70°F	
BUTANE, 2, 3-DIMETHYL	C at 70°F	A to 70°F	NR at 70°F	NR at 70°F	NR at 70°F	
BUTANEDIOL			AB to 70°F			
BUTANOIC ACID				NR any conc at 70°F		
BUTANOL	A to 250°F	NR at 70°F	B /NR at 70°F	B any conc to 150°F	A to 70°F	AB to 70°F
BUTANONE	NR at 70°F	NR at 70°F	NR at 70°F	A any conc to 100°F	NR at 70°F	A to 70°F
BUTAROL				A to 70°F AB any conc to 180°F		
1-BUTENE				B /NR at 70°F		
1-BUTENE, 2-ETHYL	NR at 70°F	A to 70°F	NR at 70°F	NR at 70°F	NR at 70°F	
BUTOXYETHANOL				A to 70°F AB any conc to 150°F		
BUTTER	NR at 70°F	A to 70°F	AC to 70°F	BC at 70°F	NR at 70°F	
BUTTER OF ANTIMONY				A to 70°F		
BUTTER OF TIN			B at 70°F	B any conc to 150°F		
BUTTER OF ZINC			A to 70°F	A to 70°F AB any conc to 185°F		
BUTTERMILK	A to 70°F	NR at 70°F	A to 70°F			
BUTYL ACETATE	+99% vol 3 days 70°F NR at 70°F	+175% vol 3 days 70°F NR at 70°F	+100% vol 3 days 70°F NR at 70°F	+27% vol 3 days 70°F BC at 70°F		AB to 70°F
N-BUTYL ACETATE	NR at 70°F	NR at 70°F	NR at 70°F	BC at 70°F		
BUTYL ACETATE RICINOLEATE				NR at 70°F		
BUTYL ACETO ACETATE						
BUTYL ACETYL RICINOLEATE	NR at 70°F		NR at 70°F	A to 70°F	NR at 70°F	
BUTYL ACRYLATE	NR at 70°F	NR at 70°F		NR at 70°F	NR at 70°F	
BUTYL ALCOHOL	A to 250°F	NR at 70°F	B /NR at 70°F	B any conc to 150°F	A to 70°F	AB to 70°F
N-BUTYL ALCOHOL			NR at 70°F	A to 70°F		
SEC BUTYL ALCOHOL			NR at 70°F	A to 70°F		
BUTYL ALDEHYDE			C at 70°F	B at 70°F		

	SYNTHETIC ISOPRENE IR	NATURAL ISOPRENE NR	CHLOROSULFONATED POLYETHYLENE CSM	FLUOROSILICONE FVMQ	SILICONE VMQ	CHEMRAZ FFKM
BROMOCHLOROTRIFLUOROMETHANE	NR at 70°F	NR at 70°F	NR at 70°F	B at 70°F	NR at 70°F	A to 70°F
BROMOETHYLENE	NR at 70°F	NR at 70°F	NR at 70°F			
BROMOMETHANE	C/NR at 70°F	C/NR at 70°F	NR at 70°F	A to 70°F		A to 70°F
BROMOPENTANE	NR at 70°F	NR at 70°F	NR at 70°F			
BROMOTOLUENE	NR at 70°F	NR at 70°F	NR at 70°F			
BRONZING LIQUID	C at 70°F	C at 70°F	C at 70°F		NR at 70°F	
BROWN ACETATE	B at 70°F	B any conc to 80°F	A to 70°F			
BROWN LINSEED OIL	NR at 70°F	NR at 70°F	B at 70°F		NR at 70°F	
BRUCEITE	A to 70°F	A to 70°F AB any conc to 185°F	A to 100% to 158°F AB to 200°F		AB to 70°F	AB to 70°F C at 70°F white
BRUDIUM						
BUNKER "C"	NR at 70°F	NR at 70°F	NR at 70°F	A to 70°F	B/NR at 70°F	
BUNKER OIL	NR at 70°F	NR at 70°F	NR at 70°F	A to 70°F	B/NR at 70°F	A to 70°F
BURNT ALUM	A to 70°F	A to 70°F AB any conc to 150°F	A to 70°F		A to 70°F	
BURNT LIME	A to 70°F	A to 70°F	A to 70°F			
BURNT POTASH	A to 70°F	A to 70°F				
BUROWS SOLUTION	A to 70°F	A to 70°F				
BUTADIEN (MONOMER)	NR at 70°F	NR at 70°F	B/NR at 70°F	AB to 70°F	NR at 70°F	A to 70°F
BUTANAL (BUTAL)	NR at 70°F	NR at 70°F	C at 70°F		C at 70°F	
BUTANE	NR at 70°F gas or liquid	NR at 70°F gas or liquid	A/NR at 70°F	AB to 70°F	NR at 70°F	A to 70°F
BUTANE, 2, 2-DIMETHYL	NR at 70°F	NR at 70°F	B at 70°F	A to 70°F	NR at 70°F	A to 70°F
BUTANE, 2, 3-DIMETHYL	NR at 70°F	NR at 70°F	B at 70°F	A to 70°F	NR at 70°F	A to 70°F
BUTANEDIOL						A to 70°F
BUTANOIC ACID	C at 70°F	C/NR at 70°F	C at 70°F	B at 70°F	NR at 70°F	A to 70°F
BUTANOL	A to 70°F	A to 70°F AB any conc to 150°F	A to 70°F AB to 140°F	A to 70°F	B at 70°F	A to 70°F
BUTANONE	NR at 70°F	NR at 70°F	NR at 70°F	NR at 70°F	NR at 70°F	A to 70°F
BUTAROL	A to 70°F	A to 70°F AB any conc to 150°F				
1-BUTENE	NR at 70°F	NR at 70°F				
1-BUTENE, 2-ETHYL	NR at 70°F	NR at 70°F	NR at 70°F	C at 70°F	NR at 70°F	
BUTOXYETHANOL	NR at 70°F	NR at 70°F	B/NR at 70°F	NR at 70°F	NR at 70°F	A to 70°F
BUTTER	NR at 70°F	NR at 70°F	B at 70°F	A to 70°F	AB to 70°F	A to 70°F
BUTTER OF ANTIMONY		A to 70°F				A to 70°F
BUTTER OF TIN	A to 70°F	A to 70°F AB any conc to 150°F	A to 70°F		B at 70°F	
BUTTER OF ZINC	B at 70°F	B any conc to 150°F	A to 70°F		A to 70°F	
BUTTERMILK		NR at 70°F		A to 70°F	A to 70°F	
BUTYL ACETATE	NR at 70°F	NR at 70°F	NR at 70°F	+175% vol 3 days 70°F NR at 70°F	+78% vol 3 days 70°F NR at 70°F	A to 70°F
N-BUTYL ACETATE	C/NR at 70°F	NR at 70°F	NR at 70°F	NR at 70°F	NR at 70°F	
BUTYL ACETATE RICINOLEATE	NR at 70°F	NR at 70°F				
BUTYL ACETO ACETATE	NR at 70°F	NR at 70°F	NR at 70°F			
BUTYL ACETYL RICINOLEATE	NR at 70°F	NR at 70°F	B at 70°F	B at 70°F		A to 70°F
BUTYL ACRYLATE	NR at 70°F	NR at 70°F	NR at 70°F	NR at 70°F	A to 70°F	A to 70°F
BUTYL ALCOHOL	A to 70°F	A to 70°F AB any conc to 150°F	A to 70°F AB to 140°F	A to 70°F	B at 70°F	A to 70°F
N-BUTYL ALCOHOL	A to 70°F	A to 70°F	A to 70°F		B at 70°F	
SEC BUTYL ALCOHOL	A to 70°F	A to 70°F	A to 70°F		B at 70°F	
BUTYL ALDEHYDE	NR at 70°F	NR at 70°F	C at 70°F		C at 70°F	

	POLYSULFIDE T	CHLORINATED POLYETHYLENE CM	EPICHLOROHYDRIN CO, ECO	® HYTREL COPOLYESTER TPE	® SANTOPRENE COPOLYMER TPO	® C-FLEX STYRENIC TPE
BROMOCHLOROTRIFLUOROMETHANE	NR at 70°F					
BROMOETHYLENE						
BROMOMETHANE				NR at 70°F	NR at 70°F	
BROMOPENTANE						
BROMOTOLUENE						
BRONZING LIQUID						
BROWN ACETATE						
BROWN LINSEED OIL						
BRUCEITE						
BRUDIUM						
BUNKER "C"						
BUNKER OIL	A to 70°F					
BURNT ALUM						
BURNT LIME						
BURNT POTASH						
BUROWS SOLUTION						
BUTADIEN (MONOMER)	C at 70°F		NR at 70°F			
BUTANAL (BUTAL)						
BUTANE	A to 70°F		A to 70°F	A to 70°F *AB to 140°F		NR at 70°F
BUTANE, 2, 2-DIMETHYL	A to 70°F			AB to 70°F		
BUTANE, 2, 3-DIMETHYL	A to 70°F					
BUTANEDIOL						
BUTANOIC ACID						
BUTANOL	B at 70°F	AB to 70°F		+2–19% vol 7 days 70°F AB to 70°F	B at 70°F	AB to 70°F
BUTANONE	AB to 70°F	AB to 70°F	NR at 70°F	A to 70°F	-4 to -11% vol 3 days 70°F A/NR at 70°F	
BUTAROL						
1-BUTENE						
1-BUTENE, 2-ETHYL	A to 70°F					
BUTOXYETHANOL						
BUTTER	NR at 70°F		A to 70°F	AB to 70°F	NR at 70°F	
BUTTER OF ANTIMONY						
BUTTER OF TIN						
BUTTER OF ZINC						
BUTTERMILK						
BUTYL ACETATE	+33% vol 3 days 70°F C/NR at 70°F	C at 70°F	NR at 70°F	BC at 70°F		NR at 70°F
N-BUTYL ACETATE	NR at 70°F		NR at 70°F			
BUTYL ACETATE RICINOLEATE						
BUTYL ACETO ACETATE						
BUTYL ACETYL RICINOLEATE						
BUTYL ACRYLATE	B at 70°F					
BUTYL ALCOHOL	B at 70°F	AB to 70°F		+2–19% vol 7 days 70°F AB to 70°F	B at 70°F	AB to 70°F
N-BUTYL ALCOHOL						
SEC BUTYL ALCOHOL						
BUTYL ALDEHYDE						

	ETHYLENE ACRYLIC EA	POLYALLOMER LINEAR COPOLYMER	NYLON 11 POLYAMIDE	NYLON 12 POLYAMIDE	ETHYLENE VINYL ACETATE EVA	POLYVINYLCHLORIDE FLEXIBLE PVC
BROMOCHLOROTRIFLUOROMETHANE						
BROMOETHYLENE						
BROMOMETHANE			AB to 70°F NR at 105°F			C/NR at 70°F
BROMOPENTANE						
BROMOTOLUENE						
BRONZING LIQUID						
BROWN ACETATE						
BROWN LINSEED OIL						
BRUCEITE						
BRUDIUM						
BUNKER "C"						
BUNKER OIL					NR at 70°F	C at 70°F
BURNT ALUM						
BURNT LIME						
BURNT POTASH						
BUROWS SOLUTION						
BUTADIEN (MONOMER)		NR at 70°F				BC at 70°F NR at 122°F
BUTANAL (BUTAL)						
BUTANE	*AB to 105°F	B at 70°F	A to 70°F AB to 140°F	AB to 70°F	A/NR at 70°F	B/NR at 70°F
BUTANE, 2, 2-DIMETHYL						
BUTANE, 2, 3-DIMETHYL						
BUTANEDIOL			AB to 70°F			AB to 70°F
BUTANOIC ACID						
BUTANOL	A to 70°F		B at 70°F C at 104°F	C at 70°F	BC to 140°F	AC to 70°F C/NR at 140°F
BUTANONE	NR at 70°F	A to 70°F NR at 122°F	AB to 105°F NR at 194°F	AB to 70°F	*C at 70°F *NR at 140°F	NR at 70°F
BUTAROL						
1-BUTENE						AB to 70°F
1-BUTENE, 2-ETHYL						AB to 70°F
BUTOXYETHANOL						
BUTTER	A to 70°F	A to 70°F	A to 70°F	AB to 70°F	B at 70°F	AB to 70°F C at 150°F
BUTTER OF ANTIMONY						
BUTTER OF TIN						
BUTTER OF ZINC						
BUTTERMILK				AB to 70°F		B at 70°F
BUTYL ACETATE	*NR at 70°F	C at 70°F	AB to 140°F C at 194°F	AB to 70°F	B/NR at 70°F	NR at 70°F
N-BUTYL ACETATE	NR at 70°F	B at 70°F C at 122°F				NR at 70°F
BUTYL ACETATE RICINOLEATE						
BUTYL ACETO ACETATE						
BUTYL ACETYL RICINOLEATE						
BUTYL ACRYLATE						NR at 70°F
BUTYL ALCOHOL	A to 70°F		B at 70°F C at 104°F	C at 70°F	BC to 140°F	AC to 70°F C/NR at 150°F
N-BUTYL ALCOHOL		A to 122°F				B at 70°F C at 122°F
SEC BUTYL ALCOHOL		A to 70°F B at 122°F				B to 122°F
BUTYL ALDEHYDE						

	POLYETHYLENE LOW DENSITY LDPE	®TEFLON FEP	®KALREZ PERFLUORINATED ELASTOMER	®FLUORAZ FLUORINATED COPOLYMER	®AFLAS FLUORINATED COPOLYMER	®NORPRENE COPOLYMER TPO
BROMOCHLOROTRIFLUOROMETHANE				A to 70°F		
BROMOETHYLENE		A to 70°F				
BROMOMETHANE	NR at 70°F	A to 300°F	A to 212°F	AC at 70°F		
BROMOPENTANE		A to 70°F				
BROMOTOLUENE		A to 70°F				
BRONZING LIQUID		A to 70°F				
BROWN ACETATE		A to 70°F				
BROWN LINSEED OIL		A to 70°F				
BRUCEITE	A to 70°F AB to 140°F	A to 300°F	A to 70°F	A to 70°F		
BRUDIUM		A to 70°F				
BUNKER "C"		A to 70°F	A to 212°F	A to 70°F		
BUNKER OIL	AB to 70°F	A to 70°F	A to 70°F	A to 70°F		
BURNT ALUM		A to 70°F				
BURNT LIME		A to 70°F				
BURNT POTASH		A to 70°F				
BUROWS SOLUTION		A to 70°F				
BUTADIEN (MONOMER)	A/NR at 70°F	A to 300°F	A to 212°F	C/NR at 70°F		
BUTANAL (BUTAL)		A to 70°F				
BUTANE	AC to 70°F	A to 300°F liquid or gas	A to 212°F	AC to 70°F		BC at 70°F
BUTANE, 2, 2-DIMETHYL				B at 70°F		
BUTANE, 2, 3-DIMETHYL				B at 70°F		
BUTANEDIOL	A to 70°F		*A to 212°F			
BUTANOIC ACID	A/NR at 70°F	A to 100% to 200°F	*B at 70°F	A to 70°F		A to 70°F
BUTANOL	A to 70°F *AB to 140°F	A to 70°F	A to 241°F	AB to 70°F	+5% vol 7 days 250°F B at 250°F	BC at 70°F
BUTANONE	AB to 122°F C/NR at 140°F	A to 300°F	A to 113°F	NR at 70°F	+58% vol 7 days 70°F NR at 70°F	BC at 70°F
BUTAROL		A to 70°F				
1-BUTENE	AB to 70°F	A to 70°F				
1-BUTENE, 2-ETHYL	AB to 70°F	A to 70°F				
BUTOXYETHANOL	AB to 70°F	A to 70°F	A to 113°F	C at 70°F		
BUTTER	A to 70°F	A to 70°F	A to 70°F	A to 70°F		
BUTTER OF ANTIMONY	A to 70°F	A to 200°F				
BUTTER OF TIN		A to 70°F				
BUTTER OF ZINC		A to 70°F				
BUTTERMILK		A to 70°F				
BUTYL ACETATE	B/NR at 70°F	A to 300°F	A to 212°F	NR at 70°F	NR at 70°F	
N-BUTYL ACETATE	B at 70°F C at 122°F	A to 122°F			NR at 70°F	
BUTYL ACETATE RICINOLEATE		A to 70°F				
BUTYL ACETO ACETATE		A to 70°F				
BUTYL ACETYL RICINOLEATE		A to 70°F	A to 212°F	A to 70°F		
BUTYL ACRYLATE		A to 70°F	A to 212°F	NR at 70°F		
BUTYL ALCOHOL	A to 70°F *AB to 140°F	A to 70°F	A to 241°F	AB to 70°F	+5% vol 7 days 250°F B at 250°F	BC at 70°F
N-BUTYL ALCOHOL	A to 122°F	A to 122°F			A to 70°F	
SEC BUTYL ALCOHOL	A to 70°F B at 122°F	A to 122°F				
BUTYL ALDEHYDE		A to 70°F				

C61

	RECOMMENDATIONS	NITRILE NBR	ETHYLENE PROPYLENE EP, EPDM	FLUOROCARBON FKM	CHLOROPRENE CR	HYDROGENATED NITRILE HNBR
BUTYL AMINE	VMQ,EP,IIR,NBR	A/NR at 70°F	A/NR at 70°F	NR at 70°F disintegrated 3 days	NR at 70°F	C at 70°F
N-BUTYL AMINE	VMQ,EP	C/NR at 70°F	A/NR to 250°F	NR at 70°F	NR at 70°F	
BUTYL BENZENE		NR at 70°F		A to 70°F		
N-BUTYL BENZOATE	EP,FKM,IIR	NR at 70°F	A to 70°F	A to 70°F	NR at 70°F	
BUTYL BENZYL PHTHALATE		NR at 70°F		C at 70°F		
BUTYL BROMIDE		NR at 70°F		B at 70°F		
N-BUTYL BUTYRATE	EP,FKM,IIR	NR at 70°F	A to 70°F	A to 70°F	NR at 70°F	
N-BUTYL, N-BUTYRATE				A to 70°F	NR at 70°F	
BUTYL CARBITOL	EP,IIR	+40% vol 3 days 70°F NR at 70°F	+1% vol 3 days 70°F A to 200°F	+40% vol 3 days 70°F C at 70°F	+42% vol 3 days 70°F C at 70°F	NR at 70°F
BUTYL CELLOSOLVE	EP,IIR	+38% vol 3 days 70°F C 100% to 150°F	+2% vol 3 days 70°F A to 200°F	+6% vol 3 days 70°F NR at 70°F	+31% vol 3 days 70°F NR 100% at 70°F	C at 70°F
BUTYL CELLOSOLVE ADIPATE	EP	NR at 70°F	B at 70°F	B at 70°F	NR at 70°F	NR at 70°F
BUTYL CHLORIDE		NR at 70°F		A to 70°F		
N-BUTYL ETHER	T	B/NR to 150°F	C/NR at 70°F	C/NR at 70°F	B/NR at 70°F	NR at 70°F
BUTYL ETHER ACETALDEHYDE		NR at 70°F		NR at 70°F		
BUTYL ETHYL ETHER		B at 70°F				
BUTYL FORMATE		NR at 70°F			NR at 70°F	
BUTYL HYDRATE		A to 70°F	AB to 70°F	A to 70°F	A to 70°F	
BUTYL HYDRIDE		A/NR at 70°F	C/NR at 70°F	A to 70°F	A/NR at 70°F	
BUTYL HYDROXIDE		A to 70°F	AB to 70°F	A to 70°F	A to 70°F	
BUTYL IODIDE		NR at 70°F			NR at 70°F	
BUTYL METHYL KETONE						
BUTYL MERCAPTAN				A to 70°F		
BUTYL OLEATE	FKM,EP,IIR	NR at 70°F	B at 70°F	A to 70°F	NR at 70°F	NR at 70°F
BUTYL OXIDE		A to 70°F		C at 70°F	B at 70°F	
BUTYL PHTHALATE	EP,T,IIR	NR at 70°F	A to 250°F	BC to 104°F NR at 140°F	NR at 70°F	NR at 70°F
BUTYL STEARATE	FKM,NBR,FVMQ	B to 104°F C at 140°F	B/NR at 70°F	A to 104°F	NR at 70°F	B at 70°F
BUTYL TERTIARY ALCOHOL		A to 70°F	A to 70°F	B at 70°F	A to 70°F	
BUTYLENE	NBR,FKM,FVMQ,T,AU	A to 250°F gas or liquid	NR at 70°F gas or liquid	A to 140°F gas or liquid	BC to 120°F gas or liquid	NR at 70°F
BUTYRALDEHYDE	EP,IIR,T	NR at 70°F	BC at 70°F	NR at 70°F	B/NR at 70°F	
BUTYRIC ACID	EP,FKM,IIR	NR at 70°F B dilute at 70°F	AB to 130°F B at 140°F	+35% vol 28 days 212°F AB to 130°F	NR at 70°F	
BUTYRIC ALCOHOL		B any conc to 150°F			B any conc to 150°F	
BUTYRIC ANHYDRIDE		C at 70°F				
BURYRONE		NR at 70°F		NR at 70°F		
BURYRONITRILE		NR at 70°F	A to 70°F		NR at 70°F	
CADMIUM ACETATE		NR at 70°F		NR at 70°F		
CADMIUM CYANIDE		AB to 200°F	A to 70°F AB to 200°F	A to 70°F AB to 200°F	A to 70°F AB to 200°F	
CAJEPUTENE		C/NR at 70°F	NR at 70°F	A to 70°F	NR at 70°F	
CAKE ALUM		A to 70°F AB any conc to 150°F			A to 70°F AB any conc to 150°F	
CALAMINE		B any conc to 80°F		A to 70°F	B any conc to 80°F	
CALCINE LIQUORS	NBR,EP,IIR,FKM	A to 70°F AB any conc to 200°F	A to 70°F	A to 70°F		A to 70°F
CALCIUM ACETATE	EP,IIR,NBR	AB any conc to 140°F	A to 176°F	NR at 70°F	B at 70°F	B at 70°F
CALCIUM ALUMINATE		A to 70°F		A to 70°F		
CALCIUM ACID SULFATE	EP	C at 70°F	AB to 70°F	NR at 70°F	C at 70°F	
CALCIUM ARSENATE						

	STYRENE BUTADIENE SBR	POLYACRYLATE ACM	POLYURETHANE AU, EU	ISOBUTYLENE ISOPRENE IIR	POLYBUTADIENE BR	® AEROQUIP AQP
BUTYL AMINE	NR at 70°F	NR at 70°F	NR at 70°F	NR at 70°F	NR at 70°F	
N-BUTYL AMINE	NR at 70°F	NR at 70°F	NR at 70°F	NR at 70°F	NR at 70°F	
BUTYL BENZENE						
N-BUTYL BENZOATE	NR at 70°F	NR at 70°F		A to 70°F	NR at 70°F	
BUTYL BENZYL PHTHALATE				A to 70°F		
BUTYL BROMIDE				NR at 70°F		
N-BUTYL BUTYRATE	NR at 70°F	NR at 70°F		A to 70°F	NR at 70°F	
N-BUTYL, N-BUTYRATE						
BUTYL CARBITOL	+18% vol 3 days 70°F NR at 70°F	+84% vol 3 days 70°F NR at 70°F	+81% vol 3 days 70°F NR at 70°F	A to 70°F AB to 180°F	NR at 70°F	
BUTYL CELLOSOLVE	+23% vol 3 days 70°F NR at 70°F	+89% vol 3 days 70°F NR at 70°F	+72% vol 3 days 70°F NR at 70°F	A to 70°F AB any conc to 150°F	NR at 70°F	
BUTYL CELLOSOLVE ADIPATE	NR at 70°F	NR at 70°F	NR at 70°F	B at 70°F	NR at 70°F	
BUTYL CHLORIDE				C at 70°F		
N-BUTYL ETHER	NR at 70°F	NR at 70°F	B at 70°F	C/NR at 70°F	NR at 70°F	
BUTYL ETHER ACETALDEHYDE				B at 70°F		
BUTYL ETHYL ETHER				NR at 70°F		
BUTYL FORMATE						
BUTYL HYDRATE			NR at 70°F	A to 70°F		
BUTYL HYDRIDE			A to 70°F	NR at 70°F		
BUTYL HYDROXIDE			NR at 70°F	A to 70°F		
BUTYL IODIDE						
BUTYL METHYL KETONE						
BUTYL MERCAPTAN			C/NR at 70°F			
BUTYL OLEATE	NR at 70°F			B at 70°F	NR at 70°F	
BUTYL OXIDE				NR at 70°F		
BUTYL PHTHALATE	NR at 70°F	NR at 70°F	NR at 70°F	B at 70°F		
BUTYL STEARATE	NR at 70°F		AB to 70°F	B/NR at 70°F	NR at 70°F	
BUTYL TERTIARY ALCOHOL			NR at 70°F	A to 70°F		
BUTYLENE	NR at 70°F	A/NR at 70°F	AC to 70°F	NR at 70°F	NR at 70°F	
BUTYRALDEHYDE	NR at 70°F	NR at 70°F	NR at 70°F	B at 70°F	C/NR at 70°F	
BUTYRIC ACID	NR at 70°F	NR at 70°F	NR at 70°F	B at 70°F	NR at 70°F	
BUTYRIC ALCOHOL				A to 70°F AB any conc to 180°F		
BUTYRIC ANHYDRIDE				C at 70°F		
BURYRONE				B at 70°F		
BURYRONITRILE				A to 70°F		
CADMIUM ACETATE				A to 70°F		
CADMIUM CYANIDE						
CAJEPUTENE				NR at 70°F		
CAKE ALUM				A to 70°F AB any conc to 185°F		
CALAMINE				A to 70°F AB any conc to 140°F		
CALCINE LIQUORS		NR at 70°F	NR at 70°F	A to 70°F AB to 200°F		
CALCIUM ACETATE	NR at 70°F	NR at 70°F	NR at 70°F	A to 70°F AB any conc to 180°F	NR at 70°F	
CALCIUM ALUMINATE				A to 70°F		
CALCIUM ACID SULFATE						
CALCIUM ARSENATE			AB to 70°F			

	SYNTHETIC ISOPRENE IR	NATURAL ISOPRENE NR	CHLOROSULFONATED POLYETHYLENE CSM	FLUOROSILICONE FVMQ	SILICONE VMQ	CHEMRAZ FFKM
BUTYL AMINE	NR at 70°F	NR at 70°F	B/NR at 70°F	NR at 70°F	B at 70°F	AB to 70°F B at 70°F white
N-BUTYL AMINE	NR at 70°F	NR at 70°F	NR at 70°F	NR at 70°F	B at 70°F	
BUTYL BENZENE	NR at 70°F	NR at 70°F	NR at 70°F			
N-BUTYL BENZOATE	NR at 70°F	NR at 70°F	NR at 70°F	A to 70°F		A to 70°F
BUTYL BENZYL PHTHALATE	NR at 70°F	NR at 70°F	NR at 70°F			
BUTYL BROMIDE	NR at 70°F	NR at 70°F	NR at 70°F			
N-BUTYL BUTYRATE	NR at 70°F	NR at 70°F	NR at 70°F	A to 70°F		A to 70°F
N-BUTYL, N-BUTYRATE						
BUTYL CARBITOL	NR at 70°F	NR at 70°F	A/NR at 70°F	+5% vol 3 days 70°F NR at 70°F	+5% vol 3 days 70°F NR at 70°F	A to 70°F
BUTYL CELLOSOLVE	NR at 70°F	NR at 70°F	B/NR at 70°F	+7% vol 3 days 70°F NR at 70°F	+100% vol 3 days 70°F NR at 70°F	A to 70°F
BUTYL CELLOSOLVE ADIPATE	NR at 70°F	NR at 70°F	NR at 70°F	B at 70°F	B at 70°F	A to 70°F
BUTYL CHLORIDE	NR at 70°F	NR at 70°F	NR at 70°F			A to 70°F
N-BUTYL ETHER	NR at 70°F	NR at 70°F	C/NR at 70°F	C at 70°F	NR at 70°F	A to 70°F
BUTYL ETHER ACETALDEHYDE	NR at 70°F	NR at 70°F	NR at 70°F			
BUTYL ETHYL ETHER	NR at 70°F	NR at 70°F	B/NR at 70°F	C at 70°F	NR at 70°F	
BUTYL FORMATE	NR at 70°F	NR at 70°F				
BUTYL HYDRATE	A to 70°F	A to 70°F	A to 70°F		B at 70°F	
BUTYL HYDRIDE	NR at 70°F	NR at 70°F	AB to 70°F			
BUTYL HYDROXIDE	A to 70°F	A to 70°F	A to 70°F		B at 70°F	
BUTYL IODIDE	NR at 70°F	NR at 70°F				
BUTYL METHYL KETONE						
BUTYL MERCAPTAN						A to 70°F
BUTYL OLEATE	NR at 70°F	NR at 70°F	NR at 70°F	B at 70°F	B at 70°F	A to 70°F
BUTYL OXIDE	NR at 70°F	NR at 70°F	C at 70°F			
BUTYL PHTHALATE	NR at 70°F	NR at 70°F	NR at 70°F	A to 70°F	AC to 70°F	A to 70°F
BUTYL STEARATE	NR at 70°F	NR at 70°F	NR at 70°F	B at 70°F		A to 70°F
BUTYL TERTIARY ALCOHOL	A to 70°F	A to 70°F	A to 70°F		B at 70°F	
BUTYLENE	NR at 70°F gas or liquid	NR at 70°F gas or liquid	B/NR at 70°F	B at 70°F	NR at 70°F	A to 70°F
BUTYRALDEHYDE	NR at 70°F	NR at 70°F	B/NR at 70°F	NR at 70°F	NR at 70°F	B/NR at 70°F NR at 70°F white
BUTYRIC ACID	C at 70°F	C/NR at 70°F	C at 70°F	B at 70°F	NR at 70°F	A to 70°F
BUTYRIC ALCOHOL	A to 70°F	A to 70°F AB any conc to 150°F				
BUTYRIC ANHYDRIDE	C at 70°F	C at 70°F	B at 70°F			A to 70°F
BURYRONE	NR at 70°F	NR at 70°F	NR at 70°F			
BURYRONITRILE	NR at 70°F	NR at 70°F				
CADMIUM ACETATE	NR at 70°F	NR at 70°F	A to 70°F			
CADMIUM CYANIDE		NR at 70°F	A to 70°F			
CAJEPUTENE	NR at 70°F	NR at 70°F	NR at 70°F			
CAKE ALUM	A to 70°F	A to 70°F AB any conc to 150°F				
CALAMINE	B at 70°F	B any conc to 80°F	A to 70°F			
CALCINE LIQUORS				A to 70°F		AC to 70°F C at 70°F white
CALCIUM ACETATE	A to 70°F	A to 70°F AB any conc to 80°F	A/NR at 70°F	NR at 70°F	NR at 70°F	A to 70°F
CALCIUM ALUMINATE	A to 70°F	A to 70°F	A to 70°F			
CALCIUM ACID SULFATE						A to 70°F
CALCIUM ARSENATE						

	POLYSULFIDE T	CHLORINATED POLYETHYLENE CM	EPICHLOROHYDRIN CO, ECO	® HYTREL COPOLYESTER TPE	® SANTOPRENE COPOLYMER TPO	® C-FLEX STYRENIC TPE
BUTYL AMINE	NR at 70°F					
N-BUTYL AMINE	NR at 70°F					
BUTYL BENZENE						
N-BUTYL BENZOATE	NR at 70°F					
BUTYL BENZYL PHTHALATE						
BUTYL BROMIDE						
N-BUTYL BUTYRATE	NR at 70°F					
N-BUTYL, N-BUTYRATE						
BUTYL CARBITOL	NR at 70°F					
BUTYL CELLOSOLVE	AB to 70°F				*A to 70°F	
BUTYL CELLOSOLVE ADIPATE	B at 70°F					
BUTYL CHLORIDE						
N-BUTYL ETHER	A to 70°F					
BUTYL ETHER ACETALDEHYDE						
BUTYL ETHYL ETHER						
BUTYL FORMATE						
BUTYL HYDRATE						
BUTYL HYDRIDE						
BUTYL HYDROXIDE						
BUTYL IODIDE						
BUTYL METHYL KETONE						
BUTYL MERCAPTAN						
BUTYL OLEATE						
BUTYL OXIDE						
BUTYL PHTHALATE						
BUTYL STEARATE	A to 70°F					
BUTYL TERTIARY ALCOHOL						
BUTYLENE	A to 70°F gas or liquid		A to 70°F	AB to 70°F		NR at 70°F
BUTYRALDEHYDE	BC at 70°F				*A to 70°F	
BUTYRIC ACID				*AB to 70°F	*A to 70°F	AB to 70°F
BUTYRIC ALCOHOL						
BUTYRIC ANHYDRIDE						
BURYRONE						
BURYRONITRILE						
CADMIUM ACETATE						
CADMIUM CYANIDE						
CAJEPUTENE						
CAKE ALUM						
CALAMINE						
CALCINE LIQUORS						
CALCIUM ACETATE	NR at 70°F					
CALCIUM ALUMINATE						
CALCIUM ACID SULFATE						
CALCIUM ARSENATE				AB to 70°F		

	ETHYLENE ACRYLIC EA	POLYALLOMER LINEAR COPOLYMER	NYLON 11 POLYAMIDE	NYLON 12 POLYAMIDE	ETHYLENE VINYL ACETATE EVA	POLYVINYLCHLORIDE FLEXIBLE PVC
BUTYL AMINE						
N-BUTYL AMINE						
BUTYL BENZENE						
N-BUTYL BENZOATE						
BUTYL BENZYL PHTHALATE						
BUTYL BROMIDE						
N-BUTYL BUTYRATE						
N-BUTYL, N-BUTYRATE						
BUTYL CARBITOL					B at 70°F	NR at 70°F
BUTYL CELLOSOLVE					NR at 70°F	NR at 70°F
BUTYL CELLOSOLVE ADIPATE						
BUTYL CHLORIDE						
N-BUTYL ETHER						
BUTYL ETHER ACETALDEHYDE						
BUTYL ETHYL ETHER						
BUTYL FORMATE						
BUTYL HYDRATE						
BUTYL HYDRIDE						
BUTYL HYDROXIDE						
BUTYL IODIDE						
BUTYL METHYL KETONE						
BUTYL MERCAPTAN					B at 70°F	NR at 70°F
BUTYL OLEATE	NR at 70°F					
BUTYL OXIDE						
BUTYL PHTHALATE						
BUTYL STEARATE	A to 70°F					
BUTYL TERTIARY ALCOHOL						
BUTYLENE	A to 70°F	B/NR at 70°F	AB to 140°F	AB to 70°F	NR at 70°F	A to 70°F B at 122°F
BUTYRALDEHYDE	NR at 70°F					
BUTYRIC ACID	*NR at 70°F	NR at 70°F		AB to 70°F	C/NR at 70°F	B at 70°F NR at 122°F
BUTYRIC ALCOHOL						
BUTYRIC ANHYDRIDE						
BURYRONE						
BURYRONITRILE						
CADMIUM ACETATE						
CADMIUM CYANIDE						
CAJEPUTENE						NR at 70°F
CAKE ALUM						
CALAMINE						
CALCINE LIQUORS						
CALCIUM ACETATE						
CALCIUM ALUMINATE						
CALCIUM ACID SULFATE						
CALCIUM ARSENATE			AB to 140°F			AB to 70°F

	POLYETHYLENE LOW DENSITY LDPE	®TEFLON FEP	®KALREZ PERFLUORINATED ELASTOMER	®FLUORAZ FLUORINATED COPOLYMER	®AFLAS FLUORINATED COPOLYMER	®NORPRENE COPOLYMER TPO
BUTYL AMINE		+0.4% wt 7 days 172°F A to 300°F	A to 70°F *A to 212°F	BC at 70°F		
N-BUTYL AMINE				B at 70°F		
BUTYL BENZENE		A to 70°F				
N-BUTYL BENZOATE		A to 70°F	A to 212°F	A to 70°F		
BUTYL BENZYL PHTHALATE		A to 70°F				
BUTYL BROMIDE		A to 70°F				
N-BUTYL BUTYRATE		A to 70°F	A to 212°F	A to 70°F		
N-BUTYL, N-BUTYRATE		A to 70°F				
BUTYL CARBITOL		A to 70°F	A to 212°F *A to 250°F	B at 70°F		
BUTYL CELLOSOLVE	AB to 70°F	A to 70°F	A to 212°F	C at 70°F		
BUTYL CELLOSOLVE ADIPATE		A to 70°F		B at 70°F		
BUTYL CHLORIDE		A to 200°F	A to 212°F			
N-BUTYL ETHER		A to 300°F	A to 212°F	NR at 70°F		
BUTYL ETHER ACETALDEHYDE		A to 70°F				
BUTYL ETHYL ETHER		A to 70°F				
BUTYL FORMATE		A to 70°F				
BUTYL HYDRATE		A to 70°F				
BUTYL HYDRIDE		A to 70°F				
BUTYL HYDROXIDE		A to 70°F				
BUTYL IODIDE		A to 70°F				
BUTYL METHYL KETONE		A to 70°F				
BUTYL MERCAPTAN	AB to 70°F	A to 70°F	A to 212°F			
BUTYL OLEATE		A to 70°F	A to 212°F	A to 70°F		
BUTYL OXIDE		A to 70°F				
BUTYL PHTHALATE	AC at 70°F NR at 140°F	A to 300°F	A to 70°F	B/NR at 70°F		
BUTYL STEARATE		A to 70°F	A to 212°F	A to 70°F	A to 70°F	
BUTYL TERTIARY ALCOHOL		A to 70°F				
BUTYLENE	AB to 70°F	A to 200°F gas or liquid	A to 212°F	A to 70°F		
BUTYRALDEHYDE		A to 70°F	A to 104°F *A to 212°F	NR at 70°F		
BUTYRIC ACID	A/NR at 70°F	A to 100% to 200°F	A to 212°F	A to 70°F		A to 70°F
BUTYRIC ALCOHOL		A to 70°F				
BUTYRIC ANHYDRIDE		A to 70°F	A to 212°F			
BURYRONE		A to 70°F				
BURYRONITRILE		A to 70°F				
CADMIUM ACETATE		A to 70°F				
CADMIUM CYANIDE		A to 70°F	A to 212°F			
CAJEPUTENE		A to 70°F				
CAKE ALUM		A to 70°F				
CALAMINE		A to 70°F				
CALCINE LIQUORS		A to 70°F		A to 70°F		
CALCIUM ACETATE		A to 70°F	A to 212°F	A to 70°F	+0% vol 7 days 212°F A to 212°F	
CALCIUM ALUMINATE		A to 70°F				
CALCIUM ACID SULFATE						
CALCIUM ARSENATE	AB to 70°F		A to 212°F			

	RECOMMENDATIONS	NITRILE NBR	ETHYLENE PROPYLENE EP, EPDM	FLUOROCARBON FKM	CHLOROPRENE CR	HYDROGENATED NITRILE HNBR
CALCIUM BICHROMATE						
CALCIUM BISULFATE		A to 70°F	A to 70°F	A to 100% to 70°F	AC to 70°F	
CALCIUM BISULFIDE		A to 70°F AB to 140°F	A/NR to 200°F	A to 140°F B at 176°F	AC to 140°F	
CALCIUM BISULFITE	NBR,CR,FKM	A to 104°F AB any conc to 200°F	B/NR to 104°F NR at 130°F	A to 176°F AB to 200°F	A to 158°F AB any conc to 200°F	A to 70°F
CALCIUM CARBONATE	NBR,EP,CR,FKM	A to 200°F	A to 140°F	A to 248°F	A to 140°F	A to 70°F
CALCIUM DISULFATE		A to 70°F	A to 70°F	A to 70°F	AC to 70°F	
CALCIUM CHLORATE	NBR,CR,EP	A to 70°F AB any conc to 200°F	A to 140°F	A to 140°F	A to 70°F AB any conc to 200°F	
CALCIUM CHLORIDE	NBR,EP,FKM,CR	A to 70°F AB 140-200°F	A to 176°F B at 212°F	A to 140°F AB 176-212°F	A to 140°F AB any conc to 200°F	A to 70°F
CALCIUM CYANIDE	NBR,EP,SBR	A to 70°F	A to 70°F	A to 70°F	A to 70°F	A to 70°F
CALCIUM FLUORPHOSPHATE				A to 70°F	A to 70°F	
CALCIUM HYDRATE		A to 70°F AB any conc to 200°F	A to 70°F	A to 70°F	A to 70°F AB any conc to 200°F	
CALCIUM HYDROXIDE	NBR,EP,CR,FKM,SBR	A to 140°F AB any conc at 200°F	A to 176°F B at 212°F	A to 212°F	A to 158°F AB any conc to 200°F	A to 70°F
CALCIUM HYPOCHLORIDE		NR at 70°F	A to 70°F	A to 70°F	NR at 70°F	
CALCIUM HYPOCHLORITE	EP,FKM,IIR	B/NR 100% to 140°F AC 35-80% at 70°F	A to conc to 250°F	A to 70°F AB any conc 200°F	BC 5% & conc at 70°F NR at 140°F	B at 70°F
CALCIUM MONOXIDE	EP,IIR,NBR	A to 140°F AB any conc to 200°F	A to 140°F	A to conc to 140°F	A to 140°F AB any conc to 200°F	
CALCIUM NITRATE	NBR,EP,CR,FKM	A to 176°F AB any conc to 200°F	A to 176°F AB any conc to 200°F	A to 212°F	A to 140°F AB any conc to 200°F	A to 70°F
CALCIUM OXIDE	EP,IIR,NBR	A to 140°F AB any conc to 200°F	A to 140°F	A to conc to 140°F	A to 140°F AB any conc to 200°F	
CALCIUM OXYCHLORIDE		BC 15% to 80°F	A 15% to 70°F	A 15% to 70°F	NR 15% at 70°F	
CALCIUM PHOSPHATE	NBR,EP,FKM	A to 200°F	A to 70°F AB to 200°F	A to 70°F AB to 200°F	B to 200°F	A to 70°F
CALCIUM SALTS	NBR,EP,FKM,CR	A to 70°F	A to 70°F	A to 70°F	A to 70°F	A to 70°F
CALCIUM SILICATE	NBR,EP,FKM,CR	A to 70°F	A to 70°F	A to 70°F	A to 70°F	A to 70°F
CALCIUM SILICO-ALUMINATE		NR at 70°F			NR at 70°F	
CALCIUM SULFATE	NBR,EP	A to 140°F AB 176-200°F	A to 176°F	A to 140°F AB to 212°F	A/NR to 200°F	
CALCIUM SULFHYDRATE		A to 70°F		A to 70°F		
CALCIUM SULFIDE	NBR,EP,FKM,CR	A to 140°F AB any conc to 200°F	A to 176°F	A to 212°F	A to 70°F AB any conc to 150°F	A to 70°F
CALCIUM SULFITE	NBR,EP,FKM,CR	A to 70°F	A to 70°F	A to 70°F	A to 70°F	A to 70°F
CALCIUM THIOSULFATE	EP,FKM,CR	B at 70°F	A to 70°F	A to 70°F	A to 70°F	B at 70°F
CALGON	NBR,EP	A to 200°F	A to 70°F	A to 70°F	A any conc to 150°F	
CALICHE		C at 70°F	A to 70°F	A to 70°F	B at 70°F	
CALICHE LIQUORS	NBR,CR,EP,FKM	A any conc to 200°F	A to 176°F	A to 212°F	A to 70°F AB any conc to 200°F	A to 70°F
CALX		A to 70°F	A to 70°F		A to 70°F	
CANDOL		A to 70°F	NR at 70°F	A to 70°F	B at 70°F	
CANE SUGAR JUICE		A to 140°F			A to 140°F	
CANE SUGAR LIQUORS	NBR,EP,CR,FKM	A to 140°F AB to 200°F	A to 176°F	A to 212°F	A to 70°F AB to 200°F	
CAPRILIC ACID		C at 70°F				
CAPRILIC ALDEHYDE	EP	NR at 70°F	B at 70°F	NR at 70°F	BC to 80°F	
N-CAPRONITRILE					NR at 70°F	
CAPROXYL ALCOHOL		NR at 70°F			NR at 70°F	
CAPROYL ALCOHOL		A to 70°F AB to 120°F	A to 70°F	B at 70°F	B/NR at 70°F	
CAPROYL HYDRIDE		A to 70°F	NR at 70°F	A to 70°F	B at 70°F	
CAPRYL ACETATE		NR at 70°F		NR at 70°F		
CAPRYL ALCOHOL		A to 70°F AB to 100°F	C at 70°F	B at 70°F	B/NR at 70°F	
CAPRYLIC ACID		C at 70°F				
CAPRYLIC ALCOHOL	EP,NBR,FKM,CR	A to 70°F AB to 100°F	A to 70°F	AB to 70°F	AB to 70°F	

	STYRENE BUTADIENE SBR	POLYACRYLATE ACM	POLYURETHANE AU, EU	ISOBUTYLENE ISOPRENE IIR	POLYBUTADIENE BR	® AEROQUIP AQP
CALCIUM BICHROMATE				A to 70°F		
CALCIUM BISULFATE			A to 70°F	A/NR at 70°F		
CALCIUM BISULFIDE	B at 70°F	C at 70°F	AC to 70°F			
CALCIUM BISULFITE	NR at 70°F	NR at 70°F	A to conc to 70°F	B/NR to 120°F	NR at 70°F	NR at 70°F
CALCIUM CARBONATE	A to 70°F	C/NR at 70°F	C/NR at 70°F	A to 70°F	A to 70°F	
CALCIUM DISULFATE			A to 70°F	A/NR at 70°F		
CALCIUM CHLORATE			AB to 70°F	A to 70°F AB any conc to 185°F		
CALCIUM CHLORIDE	A to 70°F	A to 70°F	+1% vol 28 days 70°F A to conc to 70°F	A to 70°F AB any conc to 185°F	A to 70°F	AB to 70°F
CALCIUM CYANIDE	A to 70°F			A to 70°F	A to 70°F	
CALCIUM FLUORPHOSPHATE						
CALCIUM HYDRATE			A to 70°F	A to 70°F AB any conc to 185°F		
CALCIUM HYDROXIDE	A to 70°F	NR at 70°F	A/NR at 70°F NR 10% boiling	A to 70°F AB any conc to 185°F	A to 70°F	AB to 70°F
CALCIUM HYPOCHLORIDE	NR at 70°F	NR at 70°F	NR at 70°F	A to 70°F	NR at 70°F	
CALCIUM HYPOCHLORITE	B/NR at 70°F	NR at 70°F	NR 5-100% at 70°F AB dilute at 70°F	A to 70°F AB 15% to 150°F	B/NR at 70°F	AB to 70°F
CALCIUM MONOXIDE			B to 100% to 70°F	A to 70°F AB any conc to 185°F		
CALCIUM NITRATE	A to 70°F	A to 70°F	A/NR at 70°F	A to 70°F AB any conc to 185°F	A to 70°F	
CALCIUM OXIDE			B to 100% to 70°F	A to 70°F AB any conc to 185°F		
CALCIUM OXYCHLORIDE			NR 15% at 70°F	B 15% to 185°F		
CALCIUM PHOSPHATE	A to 70°F	A to 70°F	A to 70°F	A to 70°F	A to 70°F	
CALCIUM SALTS	A to 70°F	A to 70°F	A to 70°F	A to 70°F	A to 70°F	
CALCIUM SILICATE	A to 70°F			A to 70°F	A to 70°F	
CALCIUM SILICO-ALUMINATE						
CALCIUM SULFATE			B at 70°F	A to 70°F		
CALCIUM SULFHYDRATE				A to 70°F		
CALCIUM SULFIDE	B/NR at 70°F	NR at 70°F	A to 70°F	A to 70°F AB any conc to 185°F	B/NR at 70°F	
CALCIUM SULFITE	B at 70°F	NR at 70°F	A to 70°F	A to 70°F	B at 70°F	
CALCIUM THIOSULFATE	B at 70°F	NR at 70°F	A to 70°F	A to 70°F	B at 70°F	
CALGON	AB to 70°F	NR at 70°F	NR at 70°F	A to 70°F AB any conc to 185°F		
CALICHE				A to 70°F		
CALICHE LIQUORS	A to 70°F	A to 70°F	A to 70°F	A to 70°F AB any conc to 185°F	A to 70°F	AB to 70°F
CALX				A to 70°F		
CANDOL			B at 70°F	NR at 70°F		
CANE SUGAR JUICE			A to 70°F			
CANE SUGAR LIQUORS	A to 70°F	NR at 70°F	NR at 70°F	A to 70°F AB to 150°F	A to 70°F	AB to 70°F
CAPRILIC ACID				C at 70°F		
CAPRILIC ALDEHYDE		NR at 70°F	NR at 70°F	B to 150°F	B at 70°F	
N-CAPRONITRILE						
CAPROXYL ALCOHOL				NR at 70°F		
CAPROYL ALCOHOL			NR at 70°F	B/NR at 70°F		
CAPROYL HYDRIDE			B at 70°F	NR at 70°F		
CAPRYL ACETATE				A to 70°F		
CAPRYL ALCOHOL			NR at 70°F	B/NR at 70°F		
CAPRYLIC ACID				C at 70°F		
CAPRYLIC ALCOHOL			NR at 70°F	B at 70°F		

	SYNTHETIC ISOPRENE IR	NATURAL ISOPRENE NR	CHLOROSULFONATED POLYETHYLENE CSM	FLUOROSILICONE FVMQ	SILICONE VMQ	CHEMRAZ FFKM
CALCIUM BICHROMATE			C at 70°F			
CALCIUM BISULFATE	A to 70°F	A to 70°F	A to 70°F	A to 70°F	AC to 70°F	
CALCIUM BISULFIDE		NR at 70°F	B/NR at 70°F	C at 70°F	C at 70°F	A to 70°F
CALCIUM BISULFITE	C/NR at 70°F	A/NR at 70°F	A to 200°F	A to 70°F	A to 70°F	A to 70°F
CALCIUM CARBONATE	A to 70°F	A to 70°F	A to 70°F	A to 70°F	AC to 70°F	A to 70°F
CALCIUM DISULFATE	A to 70°F	A to 70°F	A to 70°F	A to 70°F	AC to 70°F	
CALCIUM CHLORATE	A to 70°F	A to 70°F AB any conc to 150°F	A to 70°F			A aqueous to 70°F
CALCIUM CHLORIDE	A to 70°F	A to 70°F AB any conc to 150°F	A 50-100% to 70°F	A to 70°F	A 50-100% to 70°F	A to 70°F
CALCIUM CYANIDE	A to 70°F	A to 70°F	A to 70°F		A to 70°F	A to 70°F
CALCIUM FLUORPHOSPHATE						
CALCIUM HYDRATE	A to 70°F	A to 70°F AB any conc to 150°F	B at 70°F		A to 70°F	
CALCIUM HYDROXIDE	A to 70°F	A to 70°F AB any conc to 150°F	AB to 200°F	A to 70°F	A to 70°F	A to 70°F B at 70°F white
CALCIUM HYPOCHLORIDE	NR at 70°F	NR at 70°F	A to 70°F	A to 70°F		A to 70°F
CALCIUM HYPOCHLORITE	B/NR at 70°F C 35% at 70°F	B/NR at 80°F BC 35% at 70°F	A to 70°F A to 30% to 200°F	AB to 70°F	B 35-100% at 70°F	A to 70°F
CALCIUM MONOXIDE	A to 70°F	A to 70°F AB any conc to 150°F	A to 100% to 70°F AB 30% to 125°F		C conc at 70°F A dilute to 70°F	
CALCIUM NITRATE	A to 70°F	A to 70°F AB any conc to 150°F	A to 70°F	A to 70°F	AB to 70°F	A to 70°F
CALCIUM OXIDE	A to 70°F	A to 70°F AB any conc to 150°F	A to 100% to 70°F AB 30% to 125°F		C conc at 70°F A dilute to 70°F	A to 70°F
CALCIUM OXYCHLORIDE	C 15% at 70°F	BC 15% to 150°F	A 15% to 70°F		B 15% at 70°F	
CALCIUM PHOSPHATE	A to 70°F	A to 70°F	A to 70°F	A to 70°F	A to 70°F	A to 70°F
CALCIUM SALTS	A to 70°F	A to 70°F	A to 70°F	A to 70°F	B at 70°F	A to 70°F
CALCIUM SILICATE	A to 70°F	A to 70°F	A to 70°F			A to 70°F
CALCIUM SILICO-ALUMINATE	B at 70°F	B any conc to 80°F				
CALCIUM SULFATE	A to 70°F	AC to 70°F	A to 70°F			A to 70°F
CALCIUM SULFHYDRATE	A to 70°F	A to 70°F	A to 70°F			
CALCIUM SULFIDE	B/NR at 70°F	B/NR any conc to 150°F	A to 70°F	A to 70°F	AB to 70°F	A to 70°F
CALCIUM SULFITE	AB to 70°F	AB to 70°F	A to 70°F	A to 70°F	A to 70°F	A to 70°F
CALCIUM THIOSULFATE	B at 70°F	B at 70°F	A to 70°F	A to 70°F	A to 70°F	A to 70°F
CALGON	A to 70°F	A to 70°F AB any conc to 150°F	AB to 70°F	A to 70°F	A to 70°F	
CALICHE	B at 70°F	B at 70°F	A to 70°F AB to 200°F	A to 70°F	A to 70°F	
CALICHE LIQUORS	A to 70°F	A to 70°F AB any conc to 150°F	A to 70°F	A to 70°F	B at 70°F	A to 70°F
CALX	A to 70°F	A to 70°F	A to 70°F			
CANDOL	NR at 70°F	NR at 70°F	NR at 70°F			
CANE SUGAR JUICE		A to 70°F			A to 70°F	
CANE SUGAR LIQUORS	A to 70°F	A to 70°F AB to 150°F	A to 70°F	A to 70°F	A to 70°F	
CAPRILIC ACID	C at 70°F	C at 70°F	B at 70°F			
CAPRILIC ALDEHYDE	B/NR at 70°F	B/NR at 70°F		NR at 70°F	B at 70°F	
N-CAPRONITRILE		NR at 70°F				
CAPROXYL ALCOHOL	NR at 70°F	NR at 70°F				
CAPROYL ALCOHOL	B at 70°F	B/NR at 70°F	A to 70°F		B at 70°F	
CAPROYL HYDRIDE	NR at 70°F	NR at 70°F	B at 70°F			
CAPRYL ACETATE	NR at 70°F	NR at 70°F	A to 70°F			
CAPRYL ALCOHOL	B at 70°F	B/NR at 70°F	A to 70°F		B at 70°F	
CAPRYLIC ACID	C at 70°F	C at 70°F	B at 70°F			
CAPRYLIC ALCOHOL	B at 70°F	B at 70°F	A to 70°F		B at 70°F	

	POLYSULFIDE T	CHLORINATED POLYETHYLENE CM	EPICHLOROHYDRIN CO, ECO	® HYTREL COPOLYESTER TPE	® SANTOPRENE COPOLYMER TPO	® C-FLEX STYRENIC TPE
CALCIUM BICHROMATE						
CALCIUM BISULFATE	AB to 70°F					
CALCIUM BISULFIDE				B at 70°F		
CALCIUM BISULFITE	NR at 70°F	NR at 70°F		B/NR at 70°F		
CALCIUM CARBONATE	NR at 70°F					
CALCIUM DISULFATE	AB to 70°F					
CALCIUM CHLORATE	AB to 70°F					
CALCIUM CHLORIDE	A to 70°F	AB to 70°F	A to 70°F	A to 70°F *AB to 140°F		
CALCIUM CYANIDE	A to 70°F			AB to 70°F		
CALCIUM FLUORPHOSPHATE						
CALCIUM HYDRATE						
CALCIUM HYDROXIDE	NR at 70°F	AB to 70°F	A to 70°F	AB to 70°F		
CALCIUM HYPOCHLORIDE	NR at 70°F		B at 70°F			
CALCIUM HYPOCHLORITE	NR at 70°F	AB to 70°F	B at 70°F	BC at 70°F A 5-6% at 70°F		
CALCIUM MONOXIDE				AB to 70°F		AB dilute to 70°F
CALCIUM NITRATE	A to 70°F		A to 70°F			
CALCIUM OXIDE				AB to 70°F		AB dilute to 70°F
CALCIUM OXYCHLORIDE						
CALCIUM PHOSPHATE						
CALCIUM SALTS	A to 70°F			AB to 70°F		
CALCIUM SILICATE						
CALCIUM SILICO-ALUMINATE						
CALCIUM SULFATE	AB to 70°F					
CALCIUM SULFHYDRATE						
CALCIUM SULFIDE	NR at 70°F		B at 70°F			
CALCIUM SULFITE	B at 70°F					
CALCIUM THIOSULFATE	B at 70°F					
CALGON	AB to 70°F					
CALICHE						
CALICHE LIQUORS	A to 70°F	AB to 70°F		AB to 70°F		
CALX						
CANDOL						
CANE SUGAR JUICE						
CANE SUGAR LIQUORS	NR at 70°F	AB to 70°F	A to 70°F	AB to 70°F		
CAPRILIC ACID						
CAPRILIC ALDEHYDE	B at 70°F					
N-CAPRONITRILE						
CAPROXYL ALCOHOL						
CAPROYL ALCOHOL						
CAPROYL HYDRIDE						
CAPRYL ACETATE						
CAPRYL ALCOHOL						
CAPRYLIC ACID						
CAPRYLIC ALCOHOL						

	ETHYLENE ACRYLIC EA	POLYALLOMER LINEAR COPOLYMER	NYLON 11 POLYAMIDE	NYLON 12 POLYAMIDE	ETHYLENE VINYL ACETATE EVA	POLYVINYLCHLORIDE FLEXIBLE PVC
CALCIUM BICHROMATE						
CALCIUM BISULFATE						A/NR at 70°F
CALCIUM BISULFIDE			AB to 70°F			B/NR at 70°F
CALCIUM BISULFITE	*AB to 70°F	A to 70°F	A/NR at 70°F		A to 70°F	A to 70°F *AB to 140°F
CALCIUM CARBONATE	A to 70°F					
CALCIUM DISULFATE						A/NR at 70°F
CALCIUM CHLORATE						A to 70°F *AB to 140°F
CALCIUM CHLORIDE	A to 70°F		A to 194°F	AB to 20% to 70°F	A to 70°F AB to 140°F	AC to 70°F
CALCIUM CYANIDE						
CALCIUM FLUORPHOSPHATE						
CALCIUM HYDRATE						
CALCIUM HYDROXIDE		A to 122°F	A to 70°F		A to 70°F AB to 140°F	A to 150°F
CALCIUM HYPOCHLORIDE						
CALCIUM HYPOCHLORITE	*AB to 20% to 70°F	A to 122°F	NR at 70°F A 6% to 70°F		A to 70°F AB to 140°F	AB to 70°F AC at 150°F
CALCIUM MONOXIDE			AB to 70°F		AB to 140°F	A to conc to 70°F AB at 130°F
CALCIUM NITRATE			A to 70°F			
CALCIUM OXIDE			AB to 70°F		AB to 140°F	A to conc to 70°F AB at 130°F
CALCIUM OXYCHLORIDE						
CALCIUM PHOSPHATE						
CALCIUM SALTS		A to 70°F	A to 70°F		A to 70°F	A to 70°F
CALCIUM SILICATE						
CALCIUM SILICO-ALUMINATE						
CALCIUM SULFATE			AB to 70°F			A to 70°F *AB to 140°F
CALCIUM SULFHYDRATE						
CALCIUM SULFIDE		A to 70°F	A to 70°F		A to 70°F	
CALCIUM SULFITE						
CALCIUM THIOSULFATE						
CALGON			AB to 70°F			AB to 70°F
CALICHE						
CALICHE LIQUORS			AB to 70°F		C/NR at 70°F	A to 70°F
CALX						
CANDOL						
CANE SUGAR JUICE						A to 70°F
CANE SUGAR LIQUORS	A to 70°F	A to 70°F	A to 70°F		A to 70°F	A to 70°F
CAPRILIC ACID						
CAPRILIC ALDEHYDE						NR at 70°F
N-CAPRONITRILE						
CAPROXYL ALCOHOL						
CAPROYL ALCOHOL						
CAPROYL HYDRIDE						
CAPRYL ACETATE						
CAPRYL ALCOHOL						NR at 70°F
CAPRYLIC ACID						
CAPRYLIC ALCOHOL						

C 72

	POLYETHYLENE LOW DENSITY LDPE	® TEFLON FEP	® KALREZ PERFLUORINATED ELASTOMER	® FLUORAZ FLUORINATED COPOLYMER	® AFLAS FLUORINATED COPOLYMER	® NORPRENE COPOLYMER TPO
CALCIUM BICHROMATE		A to 70°F	A to 212°F			
CALCIUM BISULFATE		A to 70°F				
CALCIUM BISULFIDE	B at 70°F	A to 70°F	A to 212°F	A to 70°F		
CALCIUM BISULFITE	A to 70°F	A to 200°F	A to 212°F	A to 70°F		
CALCIUM CARBONATE	AB to 70°F	A to 300°F	A to 212°F	A to 70°F		
CALCIUM DISULFATE		A to 70°F				
CALCIUM CHLORATE	A to 70°F	A to 70°F	A to 212°F	A to 70°F		
CALCIUM CHLORIDE	AB to 140°F	A to boiling	A to 212°F	A to 70°F	+0% vol 7 days 212°F A to 212°F	
CALCIUM CYANIDE		A to 100% to 300°F	A to 212°F	A to 70°F		
CALCIUM FLUORPHOSPHATE		A to 70°F				
CALCIUM HYDRATE		A to 70°F				
CALCIUM HYDROXIDE	A to 122°F AB to 140°F	A to 122°F	A to 212°F	A to 70°F	A sat'd to 70°F	
CALCIUM HYPOCHLORIDE		A to conc to 300°F		A to 70°F		
CALCIUM HYPOCHLORITE	A to 122°F AB to 140°F	A to conc to 300°F	A to 212°F	A to 122°F	A 10% to 212°F	
CALCIUM MONOXIDE	A to 70°F AB to 140°F	A to 70°F				A to 70°F dilute
CALCIUM NITRATE	A to 70°F	A to 200°F	A to 212°F	A to 70°F	A sat'd to 70°F	
CALCIUM OXIDE	A to 70°F AB to 140°F	A to 200°F	A to 212°F			A to 70°F dilute
CALCIUM OXYCHLORIDE		A to 70°F				
CALCIUM PHOSPHATE		A to 70°F	A to 212°F	A to 70°F		
CALCIUM SALTS	A to 70°F AB to 140°F	A to 70°F	A to 212°F	A to 70°F		A to 70°F
CALCIUM SILICATE		A to 70°F		A to 70°F		
CALCIUM SILICO-ALUMINATE		A to 70°F				
CALCIUM SULFATE	AB to 70°F	A to 300°F	A to 212°F			
CALCIUM SULFHYDRATE		A to 70°F				
CALCIUM SULFIDE	A to 70°F	A to 300°F	A to 212°F	A to 70°F	A sat'd to 70°F	
CALCIUM SULFITE		A to 70°F	A to 212°F	A to 70°F		
CALCIUM THIOSULFATE		A to 70°F		A to 70°F		
CALGON	NR at 70°F	A to 70°F				
CALICHE		A to 70°F	A to 212°F			
CALICHE LIQUORS	A to 70°F AB to 140°F	A to 300°F	A to 70°F	A to 70°F		AB to 70°F
CALX		A to 70°F				
CANDOL		A to 70°F				
CANE SUGAR JUICE					A to 70°F	
CANE SUGAR LIQUORS	A to 70°F	A to 70°F	A to 212°F			
CAPRILIC ACID		A to 70°F	A to 212°F		+16.8% vol 7 days 70°F AB to 70°F	
CAPRILIC ALDEHYDE		A to 70°F				
N-CAPRONITRILE			*A to 212°F			
CAPROXYL ALCOHOL		A to 70°F				
CAPROYL ALCOHOL		A to 70°F				
CAPROYL HYDRIDE		A to 70°F				
CAPRYL ACETATE		A to 70°F				
CAPRYL ALCOHOL		A to 70°F				
CAPRYLIC ACID		A to 70°F				
CAPRYLIC ALCOHOL		A to 70°F				

	RECOMMENDATIONS	NITRILE NBR	ETHYLENE PROPYLENE EP, EPDM	FLUOROCARBON FKM	CHLOROPRENE CR	HYDROGENATED NITRILE HNBR
CAPRYLIC ALDEHYDE		NR at 70°F		NR at 70°F		
CARBAMATE	FKM	C at 70°F	B at 70°F	A to 70°F	B at 70°F	
CARBAMIDE		B any conc to 150°F			B any conc to 150°F	
CARBAZOLE						
CARBAZOTIC ACID		C/NR at 70°F	B at 70°F	A to 70°F	C/NR at 70°F	
CARBINOL	NBR,EP,CR,IIR,SBR	A to 70°F AB any conc to 150°F	A to 160°F AB at 176°F	BC at 70°F NR at 140°F	A to 100% to 140°F NR at 212°F	
CARBITOL 2 (2-ETHOXYETHOXY-ETHANOL)	EP,NBR,CR,IIR	AB to 70°F BC 104–150°F	AB to 70°F	AB to 70°F	BC at 70°F	
CARBITOL ACETATE		C/NR at 70°F	NR at 70°F	AB to 70°F	NR at 70°F	
CARBOLIC ACID	FKM,FVMQ,IIR	NR 90–100% at 70°F	NR 5–100% at 70°F	A to 100% to 140°F AB to 200°F	NR 5–100% at 70°F	NR 90–100% at 70°F
CARBON BISULFIDE	FKM,FVMQ,T	C/NR at 70°F	NR at 70°F	A to 200°F	NR at 70°F	NR at 70°F
CARBON DIOXIDE, DRY	NBR,EP,AU,IIR	A to 200°F	B to 176°F	A to 212°F	AB to 200°F	A to 70°F
CARBON DIOXIDE, WET/DRY	NBR	A to 160°F B at 176°F	AB to 176°F B at 212°F	AB to 212°F	AB to 70°F	A to 70°F
CARBON DIOXIDE, WET	NBR,IIR,EP	A to 140°F B at 176°F	AB to 176°F B at 212°F	AB to 212°F	B at 70°F	A to 70°F
CARBON DISULFIDE	FKM,FVMQ,VMQ	+56% vol 3 days 70°F C/NR at 70°F	+135% vol 3 days 70°F NR at 70°F	no swell 3 days 70°F +2% vol 28 days 75°F	+163% vol 3 days 70°F NR at 70°F	NR at 70°F
CARBON DISULFIDE		NR at 104°F		A to 200°F		
CARBON MONOXIDE	NBR,EP,IIR,FKM,VMQ	A to 140°F B at 176°F	A to 176°F	A to 212°F	A to 140°F C/NR hot	A to 70°F
CARBON SULFIDE		AB to 70°F		NR at 70°F	AB to 70°F	
CARBON TETRACHLORIDE	FKM,FVMQ	+81% vol 3 days 70°F NR at 70°F	+152% vol 3 days 70°F NR at 70°F	+1.3% vol 7 days 75°F +11.8% vol 28 days 158°F	+142% vol 3 days 70°F +330% vol 7 days 70°F	B at 70°F
CARBON TETRACHLORIDE		C at 70°F dry	B at 70°F dry	A to 158°F A to 250°F dry	NR at 70°F	
CARBON TETRAFLUORIDE		NR at 70°F	B at 70°F		NR at 70°F	
CARBONATED BEVERAGES	EP,NBR,CR	A to 140°F	A to 250°F	A to 140°F	A to 140°F	
CARBONIC ACID	EP,CR,FKM,IIR,NBR	AB to 200°F NR at 212°F	A to 176°F AB to 185°F	A to 176°F AB 185–212°F	AB any conc to 200°F NR at 212°F	A to 70°F
CARBONIC ANHYDRIDE		A to 70°F	A to 70°F	A to 70°F	A to 70°F	
CARBOXY BENZENE					B any conc to 150°F	
CASEIN (CASYMEN)	NBR	A to 200°F	A to 176°F	A to 176°F	A to 140°F B at 176°F	
CASING HEAD GASOLINE	IIR,FKM	A to 70°F		A to 70°F		
CASTOR OIL	NBR,CR,FKM,IIR	+6% vol 3 days 212°F A to 212°F	−2% vol 3 days 212°F AB to 140°F	no swell 3 days 212°F A to 212°F	+10% vol 3 days 212°F A to 158°F	A to 70°F
CATSUP		A to 140°F		A to 140°F	B/NR to 140°F	
CAUSTIC-BARYTA	NBR,EP,CR,FKM	A to 70°F AB any conc to 200°F	A to 176°F B at 212°F	A to 248°F	A to 158°F AB any conc to 200°F	A to 70°F
CAUSTIC-LIME	NBR,EP,CR,FKM,SBR	A to 140°F AB any conc to 200°F	A to 176°F	A to 212°F	A to 158°F AB any conc to 200°F	A to 70°F
CAUSTIC-POTASH	EP,IIR,CR,NBR	AB any conc to 150°F NR 30% at 175°F	A to 100% to 200°F B 25–100% at 212°F	AB to 100% to 70°F NR conc at 140°F	AB conc to 70°F B/NR conc at 140°F	B at 70°F
CAUSTIC-POTASH		A to 5% to 150°F		AB to 70% to 140°F C 25–50% at 176°F	AB to 50% to boiling A to 25% to 140°F	
CAUSTIC-SODA	EP,IIR,CR,NBR	+1% vol 3 days 70°F	no swell 3 days 70°F	no swell 28 days 70°F	+1% vol 3 days 70°F	B at 70°F
CAUSTIC-SODA		−5% vol 20% conc 3 days 212°F	−1.2% vol 20% conc 3 days 212°F	+46% vol 20% conc 3 days 212°F	−1% vol 20% conc 3 days 212°F	
CAUSTIC-SODA		NR 80–100% at 70°F B/NR 50% at 175°F	A to 100% to 70°F AB 100% to 125°F	B 100% at 70°F NR 100% at 104°F	A to conc to 70°F A to 50% to 104°F	
CAUSTIC-SODA		A to 50% to 176°F C/NR 30% at 212°F	A to 50% to 176°F B 30–50% at 212°F	B 80% at 140°F C/NR 40–80% at 176°F	AB to 50% to 200°F NR 50% at 212°F	
CAUSTIC-SODA		A to 20% to 212°F NR molten	AB 20% at 200°F B 20% at 212°F	AC 15–50% at 140°F NR 15–30% at 150°F		
CAUSTIC-SODA			B 15% at 176°F A sol'n to 250°F	A sol'n to 70°F		
CELLOSIZE		NR at 70°F				NR at 70°F
CELLOSOLVE	EP,T,IIR	C/NR at 70°F NR at 104°F	B at 70°F	C/NR at 70°F NR at 104°F	NR at 70°F	
CELLOSOLVE ACETATE	EP,T,IIR	NR at 70°F	AB to 70°F	NR at 70°F	NR at 70°F	NR at 70°F
CELLOSOLVE, BUTYL	EP,IIR,T	NR at 70°F	AB to 70°F	NR at 70°F	NR at 70°F	
CELLUGUARD	NBR,EP,FKM,CR	A to 70°F	A to 70°F	A to 70°F	A to 70°F	A to 70°F
CELLULOSE	NBR	B to 200°F	B at 70°F	NR at 70°F	B at 70°F	

	STYRENE BUTADIENE SBR	POLYACRYLATE ACM	POLYURETHANE AU, EU	ISOBUTYLENE ISOPRENE IIR	POLYBUTADIENE BR	® AEROQUIP AQP
CAPRYLIC ALDEHYDE				C at 70°F		
CARBAMATE	NR at 70°F	NR at 70°F	NR at 70°F	B at 70°F	NR at 70°F	
CARBAMIDE				AB any conc to 150°F		
CARBAZOLE						
CARBAZOTIC ACID			B at 70°F	C at 70°F		
CARBINOL	A to 200°F	NR at 70°F	NR 50-100% at 70°F	A to 70°F AB any conc to 185°F	A to 70°F	A to 70°F
CARBITOL 2 (2-ETHOXYETHOXY-ETHANOL)	B at 70°F	NR at 70°F	NR at 70°F	A to 70°F AB to 150°F	B at 70°F	
CARBITOL ACETATE	NR at 70°F	NR at 70°F	NR at 70°F	B at 70°F		
CARBOLIC ACID	NR 70-100% at 70°F	NR 70-100% at 70°F	NR 70-100% at 70°F	B any conc to 100°F	NR 70-100% at 70°F	AB 70-100% to 70°F
CARBON BISULFIDE	NR at 70°F	C/NR at 70°F	C/NR at 70°F	NR at 70°F	NR at 70°F	NR at 70°F
CARBON DIOXIDE, DRY	B at 70°F C at 70°F at H.P.	B at 70°F NR at 70°F at H.P.	A to 70°F A to 70°F at H.P.	B to 185°F	B at 70°F	AB to 70°F
CARBON DIOXIDE, WET/DRY	B at 70°F	B at 70°F	A to 70°F *AB to 140°F	B at 70°F	B at 70°F	
CARBON DIOXIDE, WET	B at 70°F	B at 70°F	A/NR at 70°F	AB to 185°F	B at 70°F	
CARBON DISULFIDE	+191% vol 3 days 70°F NR at 70°F	+51% vol 3 days 70°F NR at 70°F	+42% vol 3 days 70°F C/NR at 70°F	NR at 70°F	NR at 70°F	NR at 70°F
CARBON DISULFIDE			NR at 140°F			
CARBON MONOXIDE	B at 70°F		A to 70°F AC hot	A to 70°F AC hot	B at 70°F	AB to 140°F
CARBON SULFIDE				AB to 70°F		
CARBON TETRACHLORIDE	+207% vol 3 days 70°F NR at 70°F	+214% vol 3 days 70°F NR at 70°F	+76% vol 3 days 70°F NR at 70°F	+173% vol 3 days 70°F +275% vol 7 days 70°F	NR at 70°F	C at 70°F
CARBON TETRACHLORIDE			AB fumes to 70°F	NR at 70°F		
CARBON TETRAFLUORIDE				B at 70°F		
CARBONATED BEVERAGES	B at 70°F		B at 70°F			
CARBONIC ACID	B at 70°F	A to 70°F	A to 70°F	A to 70°F AB any conc to 185°F	B at 70°F	NR at 70°F
CARBONIC ANHYDRIDE			A to 70°F	A to 70°F		
CARBOXY BENZENE						
CASEIN (CASYMEN)	A to 70°F					
CASING HEAD GASOLINE				NR at 70°F		
CASTOR OIL	+4% vol 3 days 212°F A to 70°F	+11% vol 3 days 212°F A to 70°F	+19% vol 3 days 212°F A to 70°F	B to 150°F	A to 70°F	AB to 70°F
CATSUP			NR at 70°F			
CAUSTIC-BARYTA			A to 70°F	A to 70°F AB any conc to 185°F		
CAUSTIC-LIME	A to 70°F	NR at 70°F	A/NR at 70°F NR 10% boiling	A to 70°F AB any conc to 185°F	A to 70°F	AB to 70°F
CAUSTIC-POTASH	B 50% to conc at 70°F	NR 50% to conc at 70°F	B/NR 50-100% at 70°F C 20% at 70°F	A 50% to conc to 70°F AB any conc to 185°F	B 50% to conc at 70°F	NR at 70°F
CAUSTIC-POTASH			AB 10% to 70°F			
CAUSTIC-SODA	-2% vol 3 days 70°F A to 70°F	no swell 3 days 70°F A/NR conc at 70°F	+2% vol 3 days 70°F no swell 10% conc 70°F	+0.6% vol 3 days 70°F	A 100% to 70°F	C 50% at 70°F
CAUSTIC-SODA	B 3 molar at 70°F B sol'n at 70°F	NR 3 molar at 70°F NR sol'n at 70°F	-35% T.S. 10%conc 70°F B 50% to conc at 70°F	-2.7% vol 20% conc 3 days 70°F		
CAUSTIC-SODA			AB to 50% to 70°F AB 50% fumes to 70°F	A to conc to 70°F AB any conc to 185°F		
CAUSTIC-SODA				A 40-80% to 175°F A 20% to 212°F		
CAUSTIC-SODA						
CAUSTIC-SODA						
CELLOSIZE				A to 70°F		
CELLOSOLVE	NR at 70°F	NR at 70°F	NR at 70°F	B any conc to 150°F	NR at 70°F	
CELLOSOLVE ACETATE	NR at 70°F	NR at 70°F	NR at 70°F AB fumes to 70°F	B any conc to 150°F	NR at 70°F	AB to 70°F
CELLOSOLVE, BUTYL	NR at 70°F	NR at 70°F	NR at 70°F	B any conc to 150°F	NR at 70°F	
CELLUGUARD	A to 70°F	C at 70°F	NR at 70°F	A to 70°F	A to 70°F	
CELLULOSE	B at 70°F	NR at 70°F	B at 70°F			

C 75

	SYNTHETIC ISOPRENE IR	NATURAL ISOPRENE NR	CHLOROSULFONATED POLYETHYLENE CSM	FLUOROSILICONE FVMQ	SILICONE VMQ	CHEMRAZ FFKM
CAPRYLIC ALDEHYDE	NR at 70°F	NR at 70°F	NR at 70°F			
CARBAMATE	NR at 70°F	NR at 70°F	B at 70°F	A to 70°F		A to 70°F
CARBAMIDE	A to 70°F	A to 70°F AB any conc to 150°F	A to 70°F			
CARBAZOLE						A to 70°F
CARBAZOTIC ACID	C at 70°F	C/NR at 70°F	A to 70°F		NR at 70°F	
CARBINOL	A to 70°F	A to 100% to 70°F AB any conc to 100°F	A to 100% to 70°F A 100% to 158°F	A to 158°F	A to 100% to 70°F A 100% to 158°F	A to 70°F
CARBITOL 2 (2-ETHOXYETHOXY-ETHANOL)	B/NR at 70°F	B/NR at 70°F	B at 70°F	B at 70°F	B at 70°F	A to 70°F
CARBITOL ACETATE	NR at 70°F	NR at 70°F	B at 70°F		NR at 70°F	
CARBOLIC ACID	NR 70-100% at 70°F	NR 70-100% at 70°F C/NR 10% at 70°F	C/NR at 70°F	AB 70-100% to 70°F	NR 10-100% at 70°F	A to 70°F
CARBON BISULFIDE	NR at 70°F	NR at 70°F	NR at 70°F	B at 70°F	NR at 70°F	A to 70°F
CARBON DIOXIDE, DRY	B at 70°F	B to 150°F	A to 200°F	B at 70°F C at 70°F (at H.P.)	B at 70°F NR at 70°F (at H.P.)	A to 70°F
CARBON DIOXIDE, WET/DRY	B at 70°F	AB to 70°F	AB to 70°F	B at 70°F B at 70°F (at H.P.)	AB to 70°F B at 70°F (at H.P.)	A to 70°F
CARBON DIOXIDE, WET	B at 70°F	B to 150°F	B at 70°F	B at 70°F	B at 70°F	A to 70°F
CARBON DISULFIDE	NR at 70°F	NR at 70°F	NR at 70°F	+20% vol 3 days 70°F B at 70°F	+48% vol 3 days 70°F NR at 70°F	A to 70°F
CARBON DISULFIDE						
CARBON MONOXIDE	B at 70°F C at 140°F	BC at 70°F C at 140°F	A to 200°F	B at 70°F	A to 140°F	B at 70°F
CARBON SULFIDE		AB to 70°F	TEST			
CARBON TETRACHLORIDE	NR at 70°F	NR at 70°F	+350% vol 7 days 70°F NR at 70°F	+12% vol 3 days 70°F +20% vol 7 days 70°F	+103% vol 3 days 70°F +150% vol 7 days 70°F	A to 70°F
CARBON TETRACHLORIDE				BC at 70°F	NR at 70°F	
CARBON TETRAFLUORIDE	NR at 70°F	NR at 70°F	NR at 70°F			
CARBONATED BEVERAGES				B at 70°F	C at 70°F	
CARBONIC ACID	A to 70°F	A to 70°F AB any conc to 150°F	A to 70°F AB to 200°F	A to 70°F	A to 70°F	
CARBONIC ANHYDRIDE	A to 70°F	A to 70°F	A to 70°F		A to 70°F	
CARBOXY BENZENE	A to 70°F	A to 70°F AB any conc to 150°F				
CASEIN (CASYMEN)	A to 70°F	A to 70°F AB to 160°F	A to 70°F	A to 70°F	A to 70°F	
CASING HEAD GASOLINE	NR at 70°F	NR at 70°F	NR at 70°F			
CASTOR OIL	A to 70°F	A to 80°F	A to 158°F	no swell 3 days 212°F A to 70°F	+2% vol 3 days 212°F A to 70°F	A to 70°F
CATSUP		A to 70°F				
CAUSTIC-BARYTA	A to 70°F	A to 70°F AB any conc to 150°F	A to 70°F	A to 70°F	A to 70°F	A to 70°F B at 70°F white
CAUSTIC-LIME	A to 70°F	A to 70°F AB any conc to 150°F	A to 200°F	A to 70°F	A to 70°F	A to 70°F B at 70°F white
CAUSTIC-POTASH	B 50-100% at 70°F	B any conc to 150°F	A to 100% to 70°F AB any conc to 200°F	C 50-100% at 70°F	C 50-100% at 70°F AB 1% to 70°F	A to 70°F
CAUSTIC-POTASH						
CAUSTIC-SODA	A to 70°F A 3 molar to 70°F	A to conc to 70°F AB any conc to 150°F	-2% vol 20% conc 3 days 212°F	no swell 10% conc 7 days 75°F	no swell 3 days 70°F	AB to 100% to 70°F C 100% at 70°F white
CAUSTIC-SODA		NR 20% at 160°F	A conc to 70°F A to 73% to 280°F	+1% vol 50% conc 7 days 75°F	+1-9% vol 50% conc 7 days 70°F	
CAUSTIC-SODA				+1% vol 20% conc 7 days 70°F	-2% vol 20% conc 7 days 70°F	
CAUSTIC-SODA				B conc at 70°F B 3 molar at 70°F	+1.2% vol 20% conc 3 days 212°F	
CAUSTIC-SODA					+3% vol 10% conc 7 days 70°F	
CAUSTIC-SODA					A to conc to 70°F A 20% to 212°F	
CELLOSIZE	NR at 70°F	NR at 70°F	NR at 70°F			
CELLOSOLVE	NR at 70°F	NR at 70°F	B/NR at 70°F	NR at 70°F	NR at 70°F	A to 70°F
CELLOSOLVE ACETATE	C/NR at 70°F	NR at 70°F	NR at 70°F	NR at 70°F	NR at 70°F	A to 70°F
CELLOSOLVE, BUTYL	NR at 70°F	NR at 70°F	B/NR at 70°F	NR at 70°F	NR at 70°F	A to 70°F
CELLUGUARD	A to 70°F	A to 70°F	A to 70°F	A to 70°F	A to 70°F	A to 70°F
CELLULOSE				B at 70°F	B at 70°F	

	POLYSULFIDE T	CHLORINATED POLYETHYLENE CM	EPICHLOROHYDRIN CO, ECO	® HYTREL COPOLYESTER TPE	® SANTOPRENE COPOLYMER TPO	® C-FLEX STYRENIC TPE
CAPRYLIC ALDEHYDE						
CARBAMATE	B at 70°F					
CARBAMIDE						
CARBAZOLE						
CARBAZOTIC ACID						
CARBINOL	B at 70°F	AB to 70°F	B at 70°F	A to 70°F	A to 70°F	
CARBITOL 2 (2-ETHOXYETHOXY-ETHANOL)	B at 70°F					
CARBITOL ACETATE						
CARBOLIC ACID	NR 70-100% at 70°F	AB to 70°F	C 70% at 70°F	NR at 70°F	NR at 70°F	NR at 70°F
CARBON BISULFIDE	C at 70°F		NR at 70°F	AC to 70°F	*B/NR at 70°F	NR at 70°F
CARBON DIOXIDE, DRY	B at 70°F	AB to 70°F	A to 70°F	A to 70°F *AB to 140°F		AB to 70°F
CARBON DIOXIDE, WET/DRY	AB to 70°F		A to 70°F			
CARBON DIOXIDE, WET	B at 70°F		A to 70°F			
CARBON DISULFIDE				C at 70°F NR at 140°F		NR at 70°F
CARBON DISULFIDE						
CARBON MONOXIDE	NR at 70°F	AB at 140°F	AB to 70°F	A to 70°F AB at 140°F		
CARBON SULFIDE						
CARBON TETRACHLORIDE	+52% vol 3 days 70°F NR at 70°F	C at 70°F	B at 70°F	B/NR at 70°F	B/NR at 70°F	A/NR at 70°F
CARBON TETRACHLORIDE						
CARBON TETRAFLUORIDE						
CARBONATED BEVERAGES						
CARBONIC ACID	A to 70°F	NR at 70°F	A to 70°F	C/NR at 70°F	NR at 70°F	
CARBONIC ANHYDRIDE						
CARBOXY BENZENE						
CASEIN (CASYMEN)						
CASING HEAD GASOLINE						
CASTOR OIL	-2.9% vol 3 days 212°F C at 70°F	AB to 70°F	A to 70°F	BC at 70°F NR at 140°F		
CATSUP						
CAUSTIC-BARYTA						
CAUSTIC-LIME	NR at 70°F	AB to 70°F	A to 70°F	AB to 70°F		
CAUSTIC-POTASH	B 50% to conc at 70°F	NR at 70°F	A 50% to conc to 70°F	C/NR 50-100% at 70°F C 20% at 70°F	AB to conc to 70°F	A to conc to 70°F
CAUSTIC-POTASH				A solution to 70°F		
CAUSTIC-SODA	+0.2% vol 3 days 70°F NR conc at 70°F	C at 70°F	B conc at 70°F B 3 molar at 70°F	+1% vol 20% conc 3 days 70°F	no swell 15-20% conc 7 days 70°F	A to conc to 70°F
CAUSTIC-SODA	NR 3 molar at 70°F AB 20-50% at 70°F			NR 73% at 70°F BC 46-50% at 70°F	A to 50% to 70°F	
CAUSTIC-SODA				AC to 20% to 70°F		
CAUSTIC-SODA						
CAUSTIC-SODA						
CAUSTIC-SODA						
CELLOSIZE						
CELLOSOLVE	B at 70°F			NR at 70°F		
CELLOSOLVE ACETATE	B at 70°F	NR at 70°F		NR at 70°F	*A to 70°F	
CELLOSOLVE, BUTYL	B at 70°F			NR at 70°F		
CELLUGUARD	A to 70°F					
CELLULOSE						

	ETHYLENE ACRYLIC EA	POLYALLOMER LINEAR COPOLYMER	NYLON 11 POLYAMIDE	NYLON 12 POLYAMIDE	ETHYLENE VINYL ACETATE EVA	POLYVINYLCHLORIDE FLEXIBLE PVC
CAPRYLIC ALDEHYDE						
CARBAMATE						
CARBAMIDE						
CARBAZOLE		A to 122°F				NR at 70°F
CARBAZOTIC ACID						NR at 70°F
CARBINOL	A to 70°F	A to 70°F	AB to 70°F NR at 140°F	C at 70°F	B at 70°F NR at 140°F	AB to 100% to 70°F C/NR 122–150°F
CARBITOL 2 (2-ETHOXYETHOXY-ETHANOL)			A to 70°F			
CARBITOL ACETATE						NR at 70°F
CARBOLIC ACID	*NR at 70°F C 70% at 70°F	A to 70°F NR crystals at 122°F	NR at 70°F	NR at 70°F	C/NR at 70°F	C at 70°F NR at 150°F
CARBON BISULFIDE	*NR at 70°F		AB to 70°F		NR at 70°F	NR at 70°F
CARBON DIOXIDE, DRY	A to 450°F	A to 70°F	AB to 70°F *AB to 140°F		A to 70°F AB to 140°F	A to 150°F
CARBON DIOXIDE, WET/DRY	A to 70°F	A to 70°F	A to 70°F		A to 70°F	A to 70°F
CARBON DIOXIDE, WET	A to 70°F	A to 70°F	AC to 70°F		A to 70°F	A to 70°F
CARBON DISULFIDE		NR at 70°F	BC at 70°F C at 104°F	AB to 70°F	C at 70°F	NR at 70°F
CARBON DISULFIDE			NR at 70°F			
CARBON MONOXIDE	A to 70°F		A to 70°F C at 140°F			AB to 70°F *AB to 140°F
CARBON SULFIDE			TEST			B/NR at 70°F
CARBON TETRACHLORIDE	NR at 70°F	BC at 70°F C at 122°F	B/NR at 70°F	BC at 70°F	C/NR at 70°F	NR at 70°F
CARBON TETRACHLORIDE						
CARBON TETRAFLUORIDE						
CARBONATED BEVERAGES						
CARBONIC ACID			AB to 70°F			A to 150°F
CARBONIC ANHYDRIDE						
CARBOXY BENZENE						
CASEIN (CASYMEN)						A to 70°F C at 150°F
CASING HEAD GASOLINE						
CASTOR OIL			AB to 70°F		A/NR at 70°F	A to 70°F
CATSUP					A to 70°F	AB to 70°F B at 150°F
CAUSTIC-BARYTA						
CAUSTIC-LIME		A to 122°F	A to 70°F		A to 70°F AB to 140°F	A to 150°F
CAUSTIC-POTASH		A to conc to 122°F	AB to conc to 70°F C 50% at 105°F	AB to 50% to 70°F	AB conc to 70°F AB to 20% to 140°F	AB to conc to 122°F NR conc at 130°F
CAUSTIC-POTASH			NR 50% at 140°F			C 50% at 130°F AB to 30% at 130°F
CAUSTIC-SODA	NR 73% at 70°F A 20–40% to 70°F	A to conc to 122°F	+0% vol 10% conc 14 days 70°F	AB 40% to 70°F	A to 70°F AB to 140°F	AB to conc to 70°F B conc at 122°F
CAUSTIC-SODA			+1.2% vol 10% conc 28 days 70°F			NR conc at 140°F B/NR 40–80% at 130°F
CAUSTIC-SODA			+2.5% vol 10% conc 180 days 70°F			BC 10% at 150°F A 1% to 122°F
CAUSTIC-SODA			AB to conc to 70°F NR 50% at 140°F			
CAUSTIC-SODA						
CAUSTIC-SODA						
CELLOSIZE						
CELLOSOLVE	NR at 70°F		AB to 70°F			NR at 70°F
CELLOSOLVE ACETATE	NR at 70°F	A to 70°F B at 122°F	AB to 70°F		B at 70°F	C/NR at 70°F NR at 122°F
CELLOSOLVE, BUTYL			AB to 70°F			NR at 70°F
CELLUGUARD			AB to 70°F			
CELLULOSE						

	POLYETHYLENE LOW DENSITY LDPE	®TEFLON FEP	®KALREZ PERFLUORINATED ELASTOMER	®FLUORAZ FLUORINATED COPOLYMER	®AFLAS FLUORINATED COPOLYMER	®NORPRENE COPOLYMER TPO
CAPRYLIC ALDEHYDE		A to 70°F				
CARBAMATE		A to 70°F	A to 212°F			
CARBAMIDE		A to 70°F				
CARBAZOLE	A to 122°F	A to 122°F	A to 212°F			
CARBAZOTIC ACID		A to 70°F				
CARBINOL	AB to 100% to 122°F B/NR at 140°F	A to boiling	A 50-100% to 113°F	A to 70°F	A to 70°F	A to 70°F
CARBITOL 2 (2-ETHOXYETHOXY-ETHANOL)		A to 70°F	A to 212°F			
CARBITOL ACETATE		A to 70°F				
CARBOLIC ACID	AB to 70°F NR crystals at 122°F	A to 300°F A crystals to 122°F	A to 212°F A 90% to 203°F	AB to 100% to 400°F		B/NR at 70°F
CARBON BISULFIDE	NR at 70°F	A to 300°F	A to 212°F	A/NR at 70°F	+10% vol 7 days 78°F B at 70°F	NR at 70°F
CARBON DIOXIDE, DRY	A to 70°F AB to 140°F	A to 450°F	A to 212°F	AC at 80°F		A to 70°F
CARBON DIOXIDE, WET/DRY	A to 70°F AB to 140°F	A to 300°F	A to 212°F	AC at 70°F		
CARBON DIOXIDE, WET	A to 70°F AB to 140°F	A to 300°F	A to 212°F	AC at 70°F		
CARBON DISULFIDE	NR at 70°F	A to 300°F	A to 212°F	A/NR at 70°F	+10% vol 7 days 78°F B at 70°F	NR at 70°F
CARBON DISULFIDE			A to 212°F			
CARBON MONOXIDE	AB to 140°F	A to 300°F	A to 212°F	A to 70°F		
CARBON SULFIDE	AB to 70°F	A to 70°F				
CARBON TETRACHLORIDE	C/NR at 70°F NR at 122°F	A to 200°F	+4% vol 7 days 70°F AB to 70°F	NR 10-100% at 70°F	+86% vol 7 days 70°F NR at 70°F	NR at 70°F
CARBON TETRACHLORIDE			A to 212°F			
CARBON TETRAFLUORIDE		A to 70°F	A to 212°F			
CARBONATED BEVERAGES						
CARBONIC ACID	AB to 140°F	A to 70°F	A to 212°F	A to 70°F		
CARBONIC ANHYDRIDE		A to 70°F				
CARBOXY BENZENE		A to 70°F				
CASEIN (CASYMEN)		A to 70°F	A to 212°F			
CASING HEAD GASOLINE		A to 70°F				
CASTOR OIL	A/NR at 70°F	A to 70°F	A to 212°F	A to 70°F	A to 70°F	
CATSUP	A to 70°F			A to 70°F		
CAUSTIC-BARYTA	AB to 140°F	A to 300°F	A to 70°F	A to 70°F		
CAUSTIC-LIME	A to 122°F AB to 140°F	A to 122°F	A to 212°F	A to 70°F		
CAUSTIC-POTASH	A to conc to 122°F AB to conc at 140°F	A to conc to 300°F	A conc to 212°F A to 42% to 250°F	A to conc to 70°F		A to 100% to 70°F
CAUSTIC-POTASH						
CAUSTIC-SODA	AB conc to 140°F C50-80% at 70°F	no swell to 50% conc 3 days 212°F	A to conc to 212°F A to 50% to 302°F	A any conc to 250°F	+2% vol 50% conc 3 days 212°F	AB 50% to conc to 70°F BC 46% at 70°F
CAUSTIC-SODA	AB 5-20% to 140°F	A to 100% to 300°F			+2% vol 20% conc 3 days 212°F	A to 40% to 70°F
CAUSTIC-SODA		NR molten			A to 50% to 212°F	
CAUSTIC-SODA						
CAUSTIC-SODA						
CAUSTIC-SODA						
CELLOSIZE		A to 70°F				
CELLOSOLVE	NR at 70°F	A to 300°F	A to 212°F			
CELLOSOLVE ACETATE	AC to 122°F	A to 122°F	A to 212°F			
CELLOSOLVE, BUTYL	NR at 70°F	A to 70°F				
CELLUGUARD		A to 70°F				
CELLULOSE						

	RECOMMENDATIONS	NITRILE NBR	ETHYLENE PROPYLENE EP, EPDM	FLUOROCARBON FKM	CHLOROPRENE CR	HYDROGENATED NITRILE HNBR
CELLULOSE ACETATE		B any conc to 150°F		C at 70°F	B/NR at 70°F	
CELLULUBE HYDRAULIC FLUIDS	EP, T	NR at 70°F C at 70°F (150 & 220)	A to 250°F	+4% vol 7 days 158°F A to 158°F	NR at 70°F C at 70°F (150 & 220)	
CELLULUBE A60	EP, IIR, T, AU, FKM	NR at 70°F	AB to 250°F	B/NR at 70°F	C/NR at 70°F	NR at 70°F
CELLUTHERM 2505A	FKM	B at 70°F	NR at 70°F	A at 70°F	NR at 70°F	B at 70°F
CETANE	NBR, FKM, ACM	A to 70°F	NR at 70°F	A to 70°F	B at 70°F	A to 70°F
CEYLON GELETIN		B any conc to 180°F			B any conc to 180°F	
CHASSIS GREASE		AB to 70°F	NR at 70°F	AB to 70°F	C at 70°F	
CHILE NITER		AB any conc to 150°F	A to 70°F	A to 70°F	AB any conc to 200°F	
CHILE NITRATE	EP, IIR, NBR, CR, NR	AB any conc to 150°F	A to 176°F	A to 212°F	AB any conc to 200°F	
CHILE SALTPETER	EP, IIR, NBR, CR, NR	AB any conc to 150°F B at 176°F	A to 100% to 176°F A sol'n to 200°F	A to 212°F	AB any conc to 200°F	
CHINA BEAN OIL		A to 70°F AB to 120°F	C at 70°F	B at 70°F	AB to 80°F	
CHINA WOOD OIL	NBR, FKM, FVMQ	A to 250°F	NR at 70°F	A to 70°F	A to 70°F AB to 120°F	A to 70°F
CHINESE BEAN OIL		A to 70°F AB to 120°F	C at 70°F	B at 70°F	AB to 80°F	
CHINESE GELATIN		AB any conc to 180°F			AB any conc to 180°F	
CHINESE WOOD OIL	NBR, FKM	A to 200°F	NR at 70°F	A to 70°F	A to 70°F AB to 120°F	A to 70°F
CHLORACETALDEHYDE	EP, IIR	NR at 70°F	A to 160°F	NR at 70°F	NR at 70°F	
CHLORACETIC ACID	EP, IIR	NR 10-100% at 70°F	AB to 160°F	NR at 70°F	A to 70°F C at 104°F	NR at 70°F
CHLORDANE	FKM	B at 70°F	NR at 70°F	A to 70°F	C at 70°F	B at 70°F
CHLOREXTOL	FKM	B at 70°F	NR at 70°F	A to 70°F	B at 70°F	B at 70°F
CHLORIDE OF LIME (CHLORATE OF LIME)	NBR, CR, EP	C at 70°F	A to 70°F	A to 70°F	NR at 70°F	B at 70°F
CHLORINATED BIPHENYL (DIPHENYL)	FKM	B/NR at 70°F	NR at 70°F	A to 160°F	C/NR at 70°F	NR at 70°F
CHLORINATED HYDROCARBON		NR at 70°F	NR at 70°F	A/NR at 70°F	NR at 70°F	
CHLORINATED LIME	EP, FKM, IIR	B/NR 100% to 140°F AC 35-80% to 70°F	A to conc to 250°F	A to 70°F AB any conc to 200°F	BC 5% & conc to 70°F NR 35% at 70°F	B at 70°F
CHLORINATED SALT BRINE	FKM	NR at 70°F	NR at 70°F	A to 70°F	NR at 70°F	NR at 70°F
CHLORINATED SOLVENTS, DRY	FKM, FVMQ	NR at 70°F	NR at 70°F	A to 200°F	NR at 70°F	NR at 70°F
CHLORINATED SOLVENTS, WET	FKM, FVMQ	NR at 70°F	NR at 70°F	A to 200°F	NR at 70°F	NR at 70°F
CHLORINATED TAR CAMPHOR		NR at 70°F			NR at 70°F	
CHLORINE, DRY	FKM, FVMQ	C/NR at 70°F	A/NR at 70°F	A to 70°F AB to 400°F	C/NR at 70°F	C at 70°F
CHLORINE, WET	FKM	C/NR at 70°F	B/NR at 70°F	A to 70°F BC to 300°F	C/NR at 70°F	C at 70°F
CHLORINE, DRY GAS		C/NR at 70°F	B/NR at 70°F	-20% T.S. 5 days 212°F A to 212°F	BC at 70°F	
CHLORINE, WET GAS		C/NR 97-100% at 70°F	B/NR at 70°F C/NR at 104°F	BC at 70°F NR at 140°F	NR 97-100% at 70°F	
CHLORINE, DRY LIQUID		C/NR at 70°F NR at 140°F	B/NR to 140°F	AC to 140°F	NR at 70°F	
CHLORINE AQUEON						
CHLORINE DIOXIDE	FKM, EP, IIR	NR 8-100% at 70°F	C 100% at 70°F NR 8% at 70°F	A to 100% to 70°F	NR 8-100% at 70°F	NR at 70°F
CHLORINE PEROXIDE		NR at 70°F	C at 70°F	A to 70°F	NR at 70°F	
CHLORINE TRIFLUORIDE		NR at 70°F	NR at 70°F	NR at 70°F	NR at 70°F	NR at 70°F
CHLORINE WATER		C sat'd at 70°F B/NR 3% at 70°F	BC sat'd to 104°F NR sat'd at 140°F	C/NR sat'd at 70°F NR sat'd at 140°F	NR sat'd at 70°F C/NR 3% at 70°F	C at 70°F
CHLORINE WATER		NR 400 ppm at 70°F	B 400 ppm at 70°F C 400 ppm at 104°F	C 400 ppm at 70°F NR 400 ppm at 104°F		
CHLOROACETIC ACID	EP, IIR	NR 10-100% at 70°F	AB to 160°F AC to 200°F	B 50-100% at 70°F C 100% at 104°F	NR at 70°F	NR at 70°F
CHLOROACETONE	EP, IIR	NR at 70°F	A to 70°F	B/NR at 70°F	C at 70°F	NR at 70°F
CHLOROACETONITRILE		C/NR at 70°F			C/NR at 70°F	
CHLOROALLYENE		NR at 70°F	NR at 70°F	B at 70°F	NR at 70°F	C at 70°F
CHLOROAZOTIC ACID	FKM, EP, FVMQ	NR at 70°F	C at 70°F NR at 104°F	A to 70°F B to 185°F	NR at 70°F	
CHLOROBENZAL		NR at 70°F				

	STYRENE BUTADIENE SBR	POLYACRYLATE ACM	POLYURETHANE AU, EU	ISOBUTYLENE ISOPRENE IIR	POLYBUTADIENE BR	® AEROQUIP AQP
CELLULOSE ACETATE			NR at 70°F			
CELLULUBE HYDRAULIC FLUIDS	NR at 70°F	NR at 70°F	NR at 70°F	A to 100°F	NR at 70°F	
CELLULUBE A60	B/NR at 70°F	NR at 70°F	NR at 70°F	B at 70°F	NR at 70°F	
CELLUTHERM 2505A	NR at 70°F	B at 70°F	NR at 70°F	NR at 70°F	NR at 70°F	
CETANE	NR at 70°F	A to 70°F	NR at 70°F	NR at 70°F	NR at 70°F	
CEYLON GELETIN				A to 70°F AB any conc to 185°F		
CHASSIS GREASE	NR at 70°F	AB to 70°F				
CHILE NITER				A to 70°F AB any conc to 150°F		
CHILE NITRATE				A to 70°F AB any conc to 150°F		
CHILE SALTPETER	B to 100% at 70°F		B at 70°F	A to 70°F AB any conc to 150°F	A to 70°F	AB to 70°F
CHINA BEAN OIL			B at 70°F	B/NR at 70°F		
CHINA WOOD OIL	NR at 70°F	A to 70°F	BC at 70°F	B/NR at 70°F	NR at 70°F	AB to 70°F
CHINESE BEAN OIL			B at 70°F	B/NR at 70°F		
CHINESE GELATIN				AB any conc to 185°F		
CHINESE WOOD OIL	NR at 70°F		BC at 70°F	BC at 70°F	NR at 70°F	AB to 70°F
CHLORACETALDEHYDE	NR at 70°F	NR at 70°F	NR at 70°F			
CHLORACETIC ACID	NR at 70°F	NR at 70°F	NR at 70°F	B at 70°F AB 10% to 150°F	NR at 70°F	NR at 70°F
CHLORDANE	NR at 70°F		B/NR at 70°F	NR at 70°F	NR at 70°F	
CHLOREXTOL	NR at 70°F	B at 70°F	NR at 70°F	NR at 70°F	NR at 70°F	
CHLORIDE OF LIME (CHLORATE OF LIME)			C/NR at 70°F	A to 70°F		
CHLORINATED BIPHENYL (DIPHENYL)	NR at 70°F	NR at 70°F	NR at 70°F	NR at 70°F		
CHLORINATED HYDROCARBON			C at 70°F			
CHLORINATED LIME	B/NR at 70°F	NR at 70°F	NR 5-100% at 70°F AB dilute to 70°F	A conc to 70°F AB 15-35% to 150°F	B/NR at 70°F	AB to 70°F
CHLORINATED SALT BRINE	NR at 70°F	NR at 70°F	NR at 70°F	NR at 70°F	NR at 70°F	
CHLORINATED SOLVENTS, DRY	NR at 70°F	NR at 70°F	NR at 70°F	NR at 70°F	NR at 70°F	TEST
CHLORINATED SOLVENTS, WET	NR at 70°F	NR at 70°F	NR at 70°F	NR at 70°F	NR at 70°F	TEST
CHLORINATED TAR CAMPHOR				NR at 70°F		
CHLORINE, DRY	NR at 70°F	NR at 70°F	NR at 70°F	C/NR at 70°F	NR at 70°F	NR at 70°F
CHLORINE, WET	C/NR at 70°F	NR at 70°F	NR at 70°F	C/NR at 70°F	NR at 70°F	NR at 70°F
CHLORINE, DRY GAS			B/NR at 70°F	C/NR at 70°F		
CHLORINE, WET GAS			*NR at 70°F	C/NR at 70°F		
CHLORINE, DRY LIQUID						
CHLORINE AQUEON				NR at 70°F		
CHLORINE DIOXIDE	NR 8-100% at 70°F	NR 8-100% at 70°F	NR 8-100% at 70°F	C/NR conc at 70°F NR 8% at 70°F	NR 8-100% at 70°F	
CHLORINE PEROXIDE				NR at 70°F		
CHLORINE TRIFLUORIDE	NR at 70°F	NR at 70°F	NR at 70°F	NR at 70°F	NR at 70°F	
CHLORINE WATER			NR at 70°F	NR 3-100% at 70°F		
CHLORINE WATER						
CHLOROACETIC ACID	NR at 70°F	NR at 70°F	NR at 70°F	B at 70°F AB 10% to 150°F	NR at 70°F	NR at 70°F
CHLOROACETONE	NR at 70°F	NR at 70°F	NR at 70°F	BC at 70°F	NR at 70°F	
CHLOROACETONITRILE				C/NR at 70°F		
CHLOROALLYENE				C/NR at 70°F		
CHLOROAZOTIC ACID	NR at 70°F	NR at 70°F	NR at 70°F	C/NR at 70°F	NR at 70°F	
CHLOROBENZAL				B at 70°F		

C81

	SYNTHETIC ISOPRENE IR	NATURAL ISOPRENE NR	CHLOROSULFONATED POLYETHYLENE CSM	FLUOROSILICONE FVMQ	SILICONE VMQ	CHEMRAZ FFKM
CELLULOSE ACETATE	B at 70°F	B at 70°F				A to 70°F
CELLULUBE HYDRAULIC FLUIDS	C/NR at 70°F	B/NR at 70°F	NR at 70°F	B at 70°F	A to 70°F	A to 70°F
CELLULUBE A60	NR at 70°F	NR at 70°F	NR at 70°F	C/NR at 70°F	C at 70°F	A to 70°F
CELLUTHERM 2505A	NR at 70°F	NR at 70°F	NR at 70°F	B at 70°F	NR at 70°F	A to 70°F
CETANE	NR at 70°F	NR at 70°F	B at 70°F	C at 70°F	NR at 70°F	A to 70°F
CEYLON GELETIN	A to 70°F	A to 70°F AB any conc to 150°F				
CHASSIS GREASE					NR at 70°F	
CHILE NITER	B at 70°F	B any conc to 150°F	A to 70°F		C at 70°F	
CHILE NITRATE	B at 70°F	B any conc to 150°F	A to 70°F AB to 200°F		C/NR at 70°F	A to 70°F
CHILE SALTPETER	B at 70°F	B any conc to 150°F	A to 70°F AB to 200°F		C/NR at 70°F NR sol'n at 70°F	A to 70°F
CHINA BEAN OIL	NR at 70°F	NR at 70°F	B at 70°F		C at 70°F	
CHINA WOOD OIL	NR at 70°F	NR at 70°F	A to 70°F	B at 70°F	NR at 70°F	
CHINESE BEAN OIL	NR at 70°F	NR at 70°F	B at 70°F		C at 70°F	A to 70°F
CHINESE GELATIN	A to 70°F	A to 70°F AB any conc to 150°F				
CHINESE WOOD OIL	NR at 70°F	NR at 70°F	A to 70°F	*B at 70°F	NR at 70°F	A to 70°F
CHLORACETALDEHYDE				C at 70°F	NR at 70°F	*B at 70°F
CHLORACETIC ACID	NR at 70°F	NR 10-100% at 70°F	AB to 200°F	NR at 70°F	C/NR at 70°F	A to 70°F
CHLORDANE	NR at 70°F	NR at 70°F	C at 70°F	B at 70°F	NR at 70°F	A to 70°F
CHLOREXTOL	NR at 70°F	NR at 70°F	NR at 70°F	B at 70°F	NR at 70°F	A to 70°F
CHLORIDE OF LIME (CHLORATE OF LIME)	C at 70°F	C at 70°F	A to 70°F AB to 35% to 200°F	AB to 70°F	B 35-100% at 70°F	A to 70°F
CHLORINATED BIPHENYL (DIPHENYL)	NR at 70°F	NR at 70°F	B at 70°F	NR at 70°F	C/NR at 70°F	
CHLORINATED HYDROCARBON	NR at 70°F	NR at 70°F	NR at 70°F			
CHLORINATED LIME	B/NR conc at 70°F C 35% at 70°F	B/NR conc at 70°F BC 35% at 70°F	A conc to 70°F A to 35% to 200°F	AB to 70°F	B 35-100% at 70°F	A to 70°F
CHLORINATED SALT BRINE	B at 70°F	B at 70°F	B at 70°F	A to 70°F	NR at 70°F	A to 70°F
CHLORINATED SOLVENTS, DRY	NR at 70°F	NR at 70°F	NR at 70°F	A to 70°F	NR at 70°F	A to 70°F
CHLORINATED SOLVENTS, WET	NR at 70°F	NR at 70°F	NR at 70°F	A to 70°F	NR at 70°F	A to 70°F
CHLORINATED TAR CAMPHOR	NR at 70°F	NR at 70°F				
CHLORINE, DRY	C/NR at 70°F	B/NR at 70°F	B/NR at 70°F	A to 70°F	NR at 70°F	AB to 70°F A to 70°F white
CHLORINE, WET	NR at 70°F	C/NR at 70°F	C/NR at 70°F	B at 70°F	NR at 70°F	AC to 70°F B at 70°F white
CHLORINE, DRY GAS	C/NR at 70°F	C/NR at 70°F	BC at 70°F	A to 70°F	NR at 70°F	A to 70°F
CHLORINE, WET GAS	NR at 70°F	NR at 70°F	BC at 70°F	B at 70°F		A to 70°F
CHLORINE, DRY LIQUID						
CHLORINE AQUEON	NR at 70°F	NR at 70°F				
CHLORINE DIOXIDE	NR 8-100% at 70°F	NR 8-100% at 70°F	C 100% at 70°F NR 8% at 70°F	B 8-100% at 70°F	C at 70°F	AC to 70°F B at 70°F white
CHLORINE PEROXIDE	NR at 70°F	NR at 70°F	B at 70°F		C at 70°F	
CHLORINE TRIFLUORIDE	NR at 70°F	NR at 70°F	NR at 70°F	B/NR at 70°F	NR at 70°F	BC at 70°F B at 70°F white
CHLORINE WATER	NR sat'd at 70°F B 3% at 70°F	NR sat'd at 70°F B 3% to 150°F	B 3% to sat'd at 70°F		C/NR at 70°F	A to 70°F
CHLORINE WATER						
CHLOROACETIC ACID	B/NR at 70°F	B/NR 10-100% at 70°F	AB to 200°F	NR at 70°F	B/NR at 70°F	A to 70°F
CHLOROACETONE	B/NR at 70°F	B/NR at 70°F	B/NR at 70°F	NR at 70°F	NR at 70°F	A to 70°F
CHLOROACETONITRILE	C at 70°F	C at 70°F				
CHLOROALLYENE	NR at 70°F	NR at 70°F	NR at 70°F			
CHLOROAZOTIC ACID	NR at 70°F	NR at 70°F	B/NR at 70°F	C at 70°F	C/NR at 70°F	AB to 70°F NR at 70°F white
CHLOROBENZAL						

	POLYSULFIDE T	CHLORINATED POLYETHYLENE CM	EPICHLOROHYDRIN CO, ECO	® HYTREL COPOLYESTER TPE	® SANTOPRENE COPOLYMER TPO	® C-FLEX STYRENIC TPE
CELLULOSE ACETATE						
CELLULUBE HYDRAULIC FLUIDS	NR at 70°F			C at 70°F		
CELLULUBE A60	B/NR at 70°F					
CELLUTHERM 2505A	A to 70°F					
CETANE	A to 70°F					
CEYLON GELETIN						
CHASSIS GREASE						
CHILE NITER						
CHILE NITRATE						
CHILE SALTPETER	AB to 70°F	AB to 70°F	A to 70°F	AB to 70°F		A to 70°F
CHINA BEAN OIL						
CHINA WOOD OIL	BC at 70°F	C at 70°F		AB to 70°F		
CHINESE BEAN OIL						
CHINESE GELATIN						
CHINESE WOOD OIL	BC at 70°F	C at 70°F		AB to 70°F		
CHLORACETALDEHYDE						
CHLORACETIC ACID	NR at 70°F	NR at 70°F		NR at 70°F	NR at 70°F	AB to 70°F
CHLORDANE				B/NR at 70°F		
CHLOREXTOL	NR at 70°F					
CHLORIDE OF LIME (CHLORATE OF LIME)						
CHLORINATED BIPHENYL (DIPHENYL)						
CHLORINATED HYDROCARBON				C at 70°F		
CHLORINATED LIME	NR at 70°F	AB to 70°F	B at 70°F	BC at 70°F A to 6% at 70°F		
CHLORINATED SALT BRINE	NR at 70°F					
CHLORINATED SOLVENTS, DRY	NR at 70°F	NR at 70°F		NR at 70°F		
CHLORINATED SOLVENTS, WET	NR at 70°F	NR at 70°F		NR at 70°F		
CHLORINATED TAR CAMPHOR						
CHLORINE, DRY	NR at 70°F	NR at 70°F	B at 70°F	NR at 70°F	NR at 70°F	AB to 70°F
CHLORINE, WET	C at 70°F	NR at 70°F	B at 70°F	NR at 70°F	NR at 70°F	AB to 70°F
CHLORINE, DRY GAS						
CHLORINE, WET GAS						
CHLORINE, DRY LIQUID						
CHLORINE AQUEON						
CHLORINE DIOXIDE	NR 8-100% at 70°F					
CHLORINE PEROXIDE						
CHLORINE TRIFLUORIDE	NR at 70°F		NR at 70°F			
CHLORINE WATER				NR at 70°F		
CHLORINE WATER						
CHLOROACETIC ACID	NR at 70°F	NR at 70°F		NR at 70°F	NR at 70°F	AB to 70°F
CHLOROACETONE	NR at 70°F					
CHLOROACETONITRILE						
CHLOROALLYENE						
CHLOROAZOTIC ACID	NR at 70°F					AB to 70°F
CHLOROBENZAL						

C83

	ETHYLENE ACRYLIC EA	POLYALLOMER LINEAR COPOLYMER	NYLON 11 POLYAMIDE	NYLON 12 POLYAMIDE	ETHYLENE VINYL ACETATE EVA	POLYVINYLCHLORIDE FLEXIBLE PVC
CELLULOSE ACETATE						
CELLULUBE HYDRAULIC FLUIDS			A to 70°F (150 & 220)			AB to 70°F (150 & 220)
CELLULUBE A60						
CELLUTHERM 2505A						
CETANE						
CEYLON GELETIN						
CHASSIS GREASE						
CHILE NITER						
CHILE NITRATE						
CHILE SALTPETER		A to 70°F	A to 70°F *AB to 140°F		A to 70°F AB to 140°F	A 30-100% to 70°F *AB 30-100% to 140°F
CHINA BEAN OIL						
CHINA WOOD OIL	*AB to 70°F		AB to 70°F			
CHINESE BEAN OIL						
CHINESE GELATIN						
CHINESE WOOD OIL			AB to 70°F			
CHLORACETALDEHYDE						
CHLORACETIC ACID		A to 70°F B at 122°F	NR at 70°F		C/NR at 70°F	NR 10-100% at 70°F
CHLORDANE			AB to 70°F		B at 70°F	B/NR at 70°F
CHLOREXTOL						
CHLORIDE OF LIME (CHLORATE OF LIME)					B at 70°F	C at 70°F
CHLORINATED BIPHENYL (DIPHENYL)						
CHLORINATED HYDROCARBON						NR at 70°F
CHLORINATED LIME	*AB to 20% to 70°F	A to 122°F	NR at 70°F A to 6% to 70°F	NR sol'n at 70°F	A to 70°F AB to 140°F	AB to 70°F AC at 150°F
CHLORINATED SALT BRINE						
CHLORINATED SOLVENTS, DRY			A/NR at 70°F		B at 70°F	C/NR at 70°F
CHLORINATED SOLVENTS, WET			A/NR at 70°F		B at 70°F	C/NR at 70°F
CHLORINATED TAR CAMPHOR						NR at 70°F
CHLORINE, DRY	NR at 70°F	NR at 70°F B 10% at 70°F	NR at 70°F	NR at 70°F	C/NR at 70°F	AC to 70°F
CHLORINE, WET	NR at 70°F	NR at 70°F B 10% at 70°F	NR at 70°F	NR conc at 70°F C to 5% at 70°F	C/NR at 70°F	AB to 70°F B/NR at 122°F
CHLORINE, DRY GAS			NR at 70°F	C to 5% at 70°F		
CHLORINE, WET GAS						
CHLORINE, DRY LIQUID						
CHLORINE AQUEON						
CHLORINE DIOXIDE						NR at 70°F
CHLORINE PEROXIDE						
CHLORINE TRIFLUORIDE					NR at 70°F	NR at 70°F
CHLORINE WATER			NR at 70°F		A sat'd to 70°F *NR sat'd at 140°F	AC sat'd to 70°F NR sat'd at 140°F
CHLORINE WATER				C to 5% at 70°F	*AB to 25% to 140°F	
CHLOROACETIC ACID		A to 70°F B at 122°F	NR at 70°F	NR 10% at 70°F	C/NR at 70°F	NR 10-100% at 70°F
CHLOROACETONE	NR at 70°F					NR at 70°F
CHLOROACETONITRILE						
CHLOROALLYENE						
CHLOROAZOTIC ACID	NR at 70°F	A to 70°F				NR at 70°F
CHLOROBENZAL						

	POLYETHYLENE LOW DENSITY LDPE	® TEFLON FEP	® KALREZ PERFLUORINATED ELASTOMER	® FLUORAZ FLUORINATED COPOLYMER	® AFLAS FLUORINATED COPOLYMER	® NORPRENE COPOLYMER TPO
CELLULOSE ACETATE		A to 70°F	A to 212°F			
CELLULUBE HYDRAULIC FLUIDS		A to 70°F	A to 212°F	AB to 70°F		
CELLULUBE A60				B at 70°F		
CELLUTHERM 2505A				A to 70°F		
CETANE		A to boiling	A to 212°F	A to 70°F		
CEYLON GELETIN		A to 70°F				
CHASSIS GREASE						
CHILE NITER		A to 70°F				
CHILE NITRATE	A to 70°F AB to 140°F	A to 300°F	A to 70°F	A/NR at 70°F		AB to 70°F
CHILE SALTPETER	A to 70°F AB to 140°F	A to 300°F	A to 70°F	A/NR at 70°F		AB to 70°F
CHINA BEAN OIL		A to 70°F				
CHINA WOOD OIL		A to 70°F	A to 212°F	A to 70°F		
CHINESE BEAN OIL		A to 70°F				
CHINESE GELATIN		A to 70°F				
CHINESE WOOD OIL		A to 70°F	A to 70°F	A to 70°F		
CHLORACETALDEHYDE						
CHLORACETIC ACID	A/NR at 122°F	A to 122°F	A to 70°F A 10% to 104°F	B/NR at 70°F		A to 70°F
CHLORDANE	AB to 70°F	A 5% to 70°F	A to 212°F	A to 70°F		
CHLOREXTOL				A to 70°F		
CHLORIDE OF LIME (CHLORATE OF LIME)	A to 122°F AB to 140°F	A to 300°F	A to 70°F	A to 122°F	A 10% to 70°F	
CHLORINATED BIPHENYL (DIPHENYL)		A to 70°F				
CHLORINATED HYDROCARBON		A to 70°F				
CHLORINATED LIME	A to 122°F AB to 140°F	A to conc to 300°F	A to 70°F	A to 122°F	A 10% to 212°F	
CHLORINATED SALT BRINE		A to 70°F		A to 70°F		
CHLORINATED SOLVENTS, DRY	AC to 70°F	A to 70°F	A to 212°F	C/NR at 70°F		
CHLORINATED SOLVENTS, WET	AC to 70°F	A to 70°F	A to 212°F	C/NR at 70°F		
CHLORINATED TAR CAMPHOR		A to 70°F				
CHLORINE, DRY	B/NR at 70°F B 10% at 70°F	A to 300°F	A to 212°F	NR at 70°F		A to 70°F
CHLORINE, WET	B/NR at 70°F B 10% at 70°F	A to 300°F	B at 70°F	NR at 70°F		A to 70°F
CHLORINE, DRY GAS	C/NR to 104°F NR at 140°F	A to 70°F	A to 212°F			
CHLORINE, WET GAS	NR 97% at 70°F	A to 70°F	B at 70°F			
CHLORINE, DRY LIQUID	C/NR at 70°F	C at 70°F	A to 212°F			
CHLORINE AQUEON		A to 70°F				
CHLORINE DIOXIDE		A to 70°F	AB to 70°F	C at 70°F		
CHLORINE PEROXIDE		A to 70°F				
CHLORINE TRIFLUORIDE	C/NR at 70°F	A to 70°F	B at 70°F	NR at 70°F		
CHLORINE WATER	AB conc to 70°F C conc at 104°F	A to 300°F		A to 70°F		
CHLORINE WATER	NR conc at 140°F AB 25% to 140°F					
CHLOROACETIC ACID	A/NR to 122°F	A to 200°F	A to 212°F A 10% to 104°F	B/NR at 70°F		A to 70°F
CHLOROACETONE		A to 70°F	A to 212°F	NR at 70°F	NR at 70°F	
CHLOROACETONITRILE		A to 70°F				
CHLOROALLYENE		A to 70°F				
CHLOROAZOTIC ACID	AB to 70°F BC at 104°F	A to 248°F	A to 70°F	C at 70°F		BC at 70°F
CHLOROBENZAL		A to 70°F				

	RECOMMENDATIONS	NITRILE NBR	ETHYLENE PROPYLENE EP, EPDM	FLUOROCARBON FKM	CHLOROPRENE CR	HYDROGENATED NITRILE HNBR
CHLOROBENZENE, MONO, DI, TRI	FKM,FVMQ,T	NR at 70°F	NR at 70°F	+10% vol 30 days 75°F A to 200°F	NR at 70°F	NR at 70°F (mono)
CHLOROBENZOL	FKM,FVMQ,T	NR at 70°F	NR at 70°F	A to 200°F	NR at 70°F	NR at 70°F
CHLOROBROMOMETHANE	FKM,FVMQ,EP,IIR	NR at 70°F	B at 70°F	+9% vol 7 days 75°F A to 200°F	NR at 70°F	NR at 70°F
CHLOROBUTADIENE	FKM,FVMQ,EP	NR at 70°F	NR at 70°F	+5% vol 2 days -4°F A to 70°F	NR at 70°F	NR at 70°F
CHLOROBUTANE		NR at 70°F		A to 70°F		
CHLORODODECANE	FKM,FVMQ,EP	NR at 70°F	NR at 70°F	A to 70°F	NR at 70°F	NR at 70°F
CHLOROETHANE	NBR,EP,IIR,FKM	A to 140°F	A to 140°F	A to 140°F AB to 200°F	NR at 70°F	A to 70°F
CHLOROETHANOIC ACID	EP,IIR	NR to 100% at 70°F	AB to 160°F AC to 200°F	B 50-100% at 70°F C 100% at 104°F	NR at 70°F	NR at 70°F
CHLOROETHANOL	FKM,EP,CR	NR at 70°F	AB to 70°F	A to 70°F	B at 70°F	NR at 70°F
CHLOROETHYL ALCOHOL	FKM,EP,CR	NR at 70°F	AB to 70°F	A to 70°F	B at 70°F	NR at 70°F
CHLOROETHYLBENZENE		NR at 70°F	NR at 70°F	A to 70°F	NR at 70°F	
CHLOROFORM	FKM,FVMQ	+279% vol 3 days 70°F NR at 70°F	+113% vol 3 days 70°F NR at 70°F	+11.4% vol 7 days 75°F AB to 200°F	+191% vol 3 days 70°F NR at 70°F	NR at 70°F
CHLOROMETHANE	FKM,FVMQ,T,IIR	C/NR at 70°F	C/NR at 70°F	AB to 140°F	NR at 70°F	
CHLOROMETHYL ETHER	T	NR at 70°F	C at 70°F	NR at 70°F	NR at 70°F	
O-CHLORONAPHTHALENE	FKM,FVMQ	NR at 70°F	NR at 70°F	A to 70°F	NR at 70°F	
1-CHLORO 1-NITROETHANE		NR at 70°F	NR at 70°F	C/NR at 70°F	NR at 70°F	
CHLOROPENTANE	FKM,FVMQ,EP	NR at 70°F	NR at 70°F	A to 200°F	NR at 70°F	A to 70°F
CHLOROPHENIC ACID		NR at 70°F	NR at 70°F	B at 70°F	NR at 70°F	
O-CHLOROPHENOL	FKM,FVMQ	NR at 70°F	NR at 70°F	AB to 70°F	NR at 70°F	
CHLOROPRENE	FKM,FVMQ	NR at 70°F	NR at 70°F	A to 70°F	NR at 70°F	NR at 70°F
CHLOROPROPANONE		NR at 70°F		NR at 70°F	NR at 70°F	
CHLOROPROPENE	FKM	AB to 140°F	NR at 70°F	AB to 120°F	NR at 70°F	A to 70°F
CHLOROPROPYLENE	FKM	AB to 140°F	NR at 70°F	AB to 120°F C at 140°F	NR at 70°F	A to 70°F
CHLOROPROPYLENE OXIDE		NR at 70°F	B at 70°F	NR at 70°F	NR at 70°F	
CHLOROSULFONIC ACID	EP	NR any conc at 70°F wet or dry	NR at 70°F wet C at 70°F dry	NR at 70°F wet C at 70°F dry	NR at 70°F wet or dry	
CHLOROTHENE	FKM,FVMQ	NR at 70°F	NR at 70°F	A to 140°F AB to 200°F	NR at 70°F	NR at 70°F
CHLOROTOLUENE	FKM,FVMQ	NR at 70°F	NR at 70°F	A to 200°F	NR at 70°F	NR at 70°F
CHLOROTRIFLUOROETHYLENE		NR at 70°F				
CHLOROX	FKM,EP,FVMQ,CR	B to 140°F	B to 140°F	A to 250°F	B to 140°F	B at 70°F
CHOCOLATE SYRUP		A to 140°F		A to 70°F	A to 140°F	
CHRESYLIC ACID		NR 50% at 70°F		A 50% to 70°F	NR 50% at 70°F	
CHROME ALUM	NBR,EP,FKM,CR	A to 140°F AB any conc to 200°F	A to conc to 140°F B at 212°F	A to conc to 212°F	A to 70°F AB any conc to 200°F	
CHROME AMMONIUM ALUM		A to 70°F AB any conc to 200°F			A to 70°F AB any conc to 200°F	
CHROME PLATING SOLUTIONS	FKM,IIR	NR at 70°F	NR at 70°F	A to 70°F	NR at 70°F	NR at 70°F
CHROMIC ACID	FKM,EP,IIR	disintegrated 10% conc 3 days 70°F	-3% vol 10% conc 3 days 70°F	no swell 10% conc 3 days 70°F	+3% vol 10% conc 3 days 70°F	NR at 70°F
CHROMIC ACID		NR 5-100% at 70°F	NR conc at 70°F AC to 50% to 140°F	A conc to 70°F	NR 5-100% at 70°F	
CHROMIC OXIDE		NR 0.9% at 70°F	B 0.9% at 70°F	A 0.8% at 70°F	NR 0.8% at 70°F	NR at 70°F
CHROMICOAT					NR at 70°F	
CHROMIUM AMMONIUM SULFATE		A to 70°F AB any conc to 200°F			A to 70°F AB any conc to 200°F	
CHROMIUM POTASSIUM SULFATE	NBR,EP,FKM,CR	A to 140°F AB any conc to 200°F	A to conc to 140°F B at 212°F	A to conc to 212°F	A to 70°F AB any conc to 200°F	
CHROMIUM SALTS						
CHRYSOLEPIC ACID		C at 70°F	B at 70°F	A to 70°F	C at 70°F	
CIDER	EP,NBR,CR	A to 140°F	A to 275°F	A to 140°F	A to 140°F	
CINENE		C/NR at 70°F	NR at 70°F	A to 70°F	NR at 70°F	

	STYRENE BUTADIENE SBR	POLYACRYLATE ACM	POLYURETHANE AU, EU	ISOBUTYLENE ISOPRENE IIR	POLYBUTADIENE BR	® AEROQUIP AQP
CHLOROBENZENE, MONO, DI, TRI	NR at 70°F	NR at 70°F	NR at 70°F	NR at 70°F	NR at 70°F	
CHLOROBENZOL	NR at 70°F	NR at 70°F	NR at 70°F	NR at 70°F		
CHLOROBROMOMETHANE	NR at 70°F	NR at 70°F	NR at 70°F	B at 70°F	NR at 70°F	
CHLOROBUTADIENE	NR at 70°F	NR at 70°F	NR at 70°F	NR at 70°F	NR at 70°F	
CHLOROBUTANE				C at 70°F		
CHLORODODECANE	NR at 70°F	NR at 70°F	NR at 70°F	NR at 70°F	NR at 70°F	
CHLOROETHANE			C at 70°F	A/NR at 70°F		
CHLOROETHANOIC ACID			NR at 70°F	C at 70°F		
CHLOROETHANOL				B at 70°F		
CHLOROETHYL ALCOHOL				B any conc to 80°F		
CHLOROETHYLBENZENE	NR at 70°F	NR at 70°F	B at 70°F	NR at 70°F	NR at 70°F	
CHLOROFORM	+217% vol 3 days 70°F NR at 70°F	+296% vol 3 days 70°F NR at 70°F	+263% vol 3 days 70°F NR at 70°F	NR at 70°F	NR at 70°F	NR at 70°F
CHLOROMETHANE	NR at 70°F	NR at 70°F	NR at 70°F	BC any conc to 80°F	NR at 70°F	TEST
CHLOROMETHYL ETHER	NR at 70°F	NR at 70°F	NR at 70°F			
O-CHLORONAPHTHALENE	NR at 70°F	NR at 70°F	NR at 70°F	NR at 70°F	NR at 70°F	
1-CHLORO 1-NITROETHANE	NR at 70°F	NR at 70°F	NR at 70°F	NR at 70°F	NR at 70°F	
CHLOROPENTANE				C at 70°F		
CHLOROPHENIC ACID				NR at 70°F		
O-CHLOROPHENOL	NR at 70°F	NR at 70°F	NR at 70°F	NR at 70°F	NR at 70°F	
CHLOROPRENE				NR at 70°F		
CHLOROPROPANONE				NR at 70°F		
CHLOROPROPENE				C/NR at 70°F		
CHLOROPROPYLENE				NR at 70°F		
CHLOROPROPYLENE OXIDE				C/NR at 70°F		
CHLOROSULFONIC ACID	NR at 70°F	NR at 70°F	NR at 70°F wet or dry	NR at 70°F	NR at 70°F	NR at 70°F
CHLOROTHENE			NR at 70°F	NR at 70°F		
CHLOROTOLUENE	NR at 70°F	NR at 70°F	NR at 70°F	NR at 70°F	NR at 70°F	
CHLOROTRIFLUOROETHYLENE				NR at 70°F		
CHLOROX	NR at 70°F	NR at 70°F	NR at 70°F	AB to 150°F	NR at 70°F	
CHOCOLATE SYRUP						
CHRESYLIC ACID						
CHROME ALUM	A to 70°F	NR at 70°F	B at 70°F	A to 70°F AB any conc to 185°F	A to 70°F	
CHROME AMMONIUM ALUM				A to 70°F AB any conc to 185°F		
CHROME PLATING SOLUTIONS	NR at 70°F	NR at 70°F	NR at 70°F	NR at 70°F	NR at 70°F	
CHROMIC ACID	dissolved 10% conc 3 days 70°F	+6% vol 10% conc 3 days 70°F	softened 10% conc 3 days 70°F	-0.1% vol 10% conc 3 days 70°F	NR 25-100% at 70°F	NR 30% at 70°F
CHROMIC ACID	NR 10-100% at 70°F	NR 25-100% at 70°F	NR 5-100% at 70°F AB 2% to 70°F	BC 100% to 100°F AB to 25% to 75°F		
CHROMIC OXIDE	NR 0.9% at 70°F	NR 0.9% at 70°F	NR 0.9% at 70°F	B 0.9% at 70°F	NR 0.9% at 70°F	
CHROMICOAT						
CHROMIUM AMMONIUM SULFATE				A to 70°F AB any conc to 185°F		
CHROMIUM POTASSIUM SULFATE	A to 70°F	NR at 70°F	B at 70°F	A to 70°F AB any conc to 185°F	A to 70°F	
CHROMIUM SALTS			AB to 70°F			A to 70°F
CHRYSOLEPIC ACID			B at 70°F	C at 70°F		
CIDER	B at 70°F	NR at 70°F	B/NR at 70°F AB dilute to 70°F			
CINENE				NR at 70°F		

C 87

	SYNTHETIC ISOPRENE IR	NATURAL ISOPRENE NR	CHLOROSULFONATED POLYETHYLENE CSM	FLUOROSILICONE FVMQ	SILICONE VMQ	CHEMRAZ FFKM
CHLOROBENZENE, MONO, DI, TRI	NR at 70°F	NR at 70°F	NR at 70°F	B at 70°F	NR at 70°F	A to 70°F (mono)
CHLOROBENZOL	NR at 70°F	NR at 70°F	NR at 70°F	B at 70°F	NR at 70°F	A to 70°F
CHLOROBROMOMETHANE	NR at 70°F	NR at 70°F	NR at 70°F	B at 70°F	NR at 70°F	A to 70°F
CHLOROBUTADIENE	NR at 70°F	NR at 70°F	NR at 70°F	B at 70°F	NR at 70°F	A to 70°F
CHLOROBUTANE	NR at 70°F	NR at 70°F	NR at 70°F			A to 70°F
CHLORODODECANE	NR at 70°F	NR at 70°F	NR at 70°F	A to 70°F	NR at 70°F	A to 70°F
CHLOROETHANE	BC at 70°F	B/NR at 70°F	B/NR at 70°F	A to 70°F	C/NR at 70°F	A to 70°F
CHLOROETHANOIC ACID	B/NR at 70°F	B/NR 10-100% at 70°F	AB to 200°F	NR at 70°F	B/NR at 70°F	A to 70°F
CHLOROETHANOL	BC at 70°F	BC at 70°F	BC at 70°F	B at 70°F	C at 70°F	A to 70°F
CHLOROETHYL ALCOHOL	BC at 70°F	BC at 70°F	BC at 70°F	B at 70°F	C at 70°F	A to 70°F
CHLOROETHYLBENZENE	NR at 70°F	NR at 70°F	NR at 70°F	B at 70°F	NR at 70°F	A to 70°F
CHLOROFORM	NR at 70°F	NR at 70°F	NR at 70°F	+36% vol 3 days 70°F B/NR at 70°F	+99% vol 3 days 70°F NR at 70°F	A to 70°F
CHLOROMETHANE	NR at 70°F	NR at 70°F	NR at 70°F	B at 70°F	NR at 70°F	A to 70°F
CHLOROMETHYL ETHER				NR at 70°F	NR at 70°F	
O-CHLORONAPHTHALENE	NR at 70°F	NR at 70°F	NR at 70°F	B at 70°F	NR at 70°F	A to 70°F
1-CHLORO 1-NITROETHANE	NR at 70°F	NR at 70°F	NR at 70°F	NR at 70°F	NR at 70°F	NR at 70°F
CHLOROPENTANE	NR at 70°F	NR at 70°F	NR at 70°F	B at 70°F	NR at 70°F	A to 70°F
CHLOROPHENIC ACID	NR at 70°F	NR at 70°F	NR at 70°F			
O-CHLOROPHENOL	NR at 70°F	NR at 70°F	NR at 70°F	B at 70°F	NR at 70°F	
CHLOROPRENE	NR at 70°F	NR at 70°F	NR at 70°F	B at 70°F	NR at 70°F	A to 70°F
CHLOROPROPANONE	NR at 70°F	NR at 70°F	NR at 70°F			
CHLOROPROPENE	NR at 70°F	NR at 70°F	NR at 70°F	NR at 70°F	A to 70°F	A to 70°F
CHLOROPROPYLENE	NR at 70°F	NR at 70°F	NR at 70°F	NR at 70°F	A to 70°F	A to 70°F
CHLOROPROPYLENE OXIDE	NR at 70°F	NR at 70°F	C at 70°F			
CHLOROSULFONIC ACID	NR at 70°F	NR at 70°F	NR at 70°F	NR at 70°F	NR at 70°F	A to 70°F
CHLOROTHENE	NR at 70°F	NR at 70°F	NR at 70°F	+14% vol 120 days 70°F B at 70°F	NR at 70°F	A to 70°F
CHLOROTOLUENE	NR at 70°F	NR at 70°F	NR at 70°F	AB to 70°F	NR at 70°F	A to 70°F
CHLOROTRIFLUOROETHYLENE	NR at 70°F	NR at 70°F				*B at 70°F
CHLOROX	NR at 70°F	NR at 70°F	B at 70°F	A to 70°F		A to 70°F
CHOCOLATE SYRUP						
CHRESYLIC ACID						
CHROME ALUM	A to 70°F	A to 70°F AB any conc to 150°F	A to 70°F		A to 70°F	A to 100% to 70°F
CHROME AMMONIUM ALUM	A to 70°F	A to 70°F AB any conc to 150°F				A to 70°F
CHROME PLATING SOLUTIONS	NR at 70°F	NR at 70°F	C/NR at 70°F	AB to 70°F	B at 70°F	A to 70°F
CHROMIC ACID	NR any conc at 70°F	NR 10-100% at 70°F B 5% at 70°F	A to 50% to 158°F AB to 30% at 185°F	-1% vol 10% conc 3 days 70°F	+0.2% vol 10% conc 3 days 70°F	A 50-100% to 70°F B at 70°F white
CHROMIC ACID				C 25-100% at 70°F A 10% at 70°F	C 5-50% at 70°F	
CHROMIC OXIDE	NR 0.9% at 70°F	NR 0.9% at 70°F	A 0.9% at 70°F	B 0.9% at 70°F	B 0.9% at 70°F	A to 70°F
CHROMICOAT	NR at 70°F	NR at 70°F				
CHROMIUM AMMONIUM SULFATE	A to 70°F	A to 70°F AB any conc to 150°F				A to 70°F
CHROMIUM POTASSIUM SULFATE	A to 70°F	A to 70°F AB any conc to 150°F	A to 70°F		A to 70°F	A to 100% to 70°F
CHROMIUM SALTS						A to 70°F
CHRYSOLEPIC ACID	C at 70°F	C at 70°F	A to 70°F		NR at 70°F	
CIDER		C at 70°F		A to 70°F	B at 70°F	
CINENE	NR at 70°F	NR at 70°F	NR at 70°F			

	POLYSULFIDE T	CHLORINATED POLYETHYLENE CM	EPICHLOROHYDRIN CO, ECO	® HYTREL COPOLYESTER TPE	® SANTOPRENE COPOLYMER TPO	® C-FLEX STYRENIC TPE
CHLOROBENZENE, MONO, DI, TRI	NR at 70°F (mono)		NR at 70°F (mono)	NR at 70°F (mono)	NR at 70°F (mono)	NR at 70°F (mono)
CHLOROBENZOL						
CHLOROBROMOMETHANE	NR at 70°F					NR at 70°F
CHLOROBUTADIENE	NR at 70°F					
CHLOROBUTANE						
CHLORODODECANE	NR at 70°F					
CHLOROETHANE						
CHLOROETHANOIC ACID						
CHLOROETHANOL						
CHLOROETHYL ALCOHOL						
CHLOROETHYLBENZENE	B /NR at 70°F					
CHLOROFORM	NR at 70°F	NR at 70°F		NR at 70°F	NR at 70°F	NR at 70°F
CHLOROMETHANE	C at 70°F	C at 70°F		NR at 70°F	NR at 70°F	B /NR at 70°F
CHLOROMETHYL ETHER	A to 160°F					
O-CHLORONAPHTHALENE	NR at 70°F					
1-CHLORO 1-NITROETHANE	NR at 70°F					
CHLOROPENTANE						
CHLOROPHENIC ACID						
O-CHLOROPHENOL	NR at 70°F					
CHLOROPRENE						
CHLOROPROPANONE						
CHLOROPROPENE						
CHLOROPROPYLENE						
CHLOROPROPYLENE OXIDE						
CHLOROSULFONIC ACID	NR at 70°F	NR at 70°F		NR at 70°F		AB to 70°F
CHLOROTHENE						
CHLOROTOLUENE	NR at 70°F					
CHLOROTRIFLUOROETHYLENE						
CHLOROX	NR at 70°F					
CHOCOLATE SYRUP						
CHRESYLIC ACID						
CHROME ALUM	NR at 70°F					
CHROME AMMONIUM ALUM						
CHROME PLATING SOLUTIONS	NR at 70°F					
CHROMIC ACID	disintegrated 10% conc 3 days 70°F	NR 25-100% at 70°F		NR 10-100% at 70°F	NR 10-50% at 70°F	AB to 30% to 70°F
CHROMIC ACID	NR 10-100% at 70°F					
CHROMIC OXIDE	NR 0.9% at 70°F					
CHROMICOAT						
CHROMIUM AMMONIUM SULFATE						
CHROMIUM POTASSIUM SULFATE	NR at 70°F					
CHROMIUM SALTS				AB to 70°F		AB to 70°F
CHRYSOLEPIC ACID						
CIDER				AB to 70°F		
CINENE						

C 89

	ETHYLENE ACRYLIC EA	POLYALLOMER LINEAR COPOLYMER	NYLON 11 POLYAMIDE	NYLON 12 POLYAMIDE	ETHYLENE VINYL ACETATE EVA	POLYVINYLCHLORIDE FLEXIBLE PVC
CHLOROBENZENE, MONO, DI, TRI	NR at 70°F (mono)		A to 70°F (mono)	C/NR at 70°F	NR at 70°F (mono)	NR at 70°F (mono)
CHLOROBENZOL						
CHLOROBROMOMETHANE				C at 70°F		NR at 70°F
CHLOROBUTADIENE						NR at 70°F
CHLOROBUTANE						
CHLORODODECANE						
CHLOROETHANE						
CHLOROETHANOIC ACID						
CHLOROETHANOL						
CHLOROETHYL ALCOHOL						NR at 70°F
CHLOROETHYLBENZENE						
CHLOROFORM	NR at 70°F	B at 70°F C at 122°F	B/NR at 70°F	NR at 70°F	NR at 70°F	NR at 70°F
CHLOROMETHANE	NR at 70°F		AB to 70°F NR at 105°F		NR at 70°F	NR at 70°F
CHLOROMETHYL ETHER						
O-CHLORONAPHTHALENE						
1-CHLORO 1-NITROETHANE						NR at 70°F
CHLOROPENTANE						NR at 70°F
CHLOROPHENIC ACID						
O-CHLOROPHENOL						
CHLOROPRENE						
CHLOROPROPANONE						NR at 70°F
CHLOROPROPENE						
CHLOROPROPYLENE						
CHLOROPROPYLENE OXIDE						NR at 70°F
CHLOROSULFONIC ACID	NR at 70°F		NR at 70°F		C/NR at 70°F	C/NR at 70°F
CHLOROTHENE						
CHLOROTOLUENE	NR at 70°F					
CHLOROTRIFLUOROETHYLENE						NR at 70°F
CHLOROX						B at 70°F
CHOCOLATE SYRUP		A to 70°F	A to 70°F		A to 70°F	AB to 70°F C at 150°F
CHRESYLIC ACID						
CHROME ALUM						*AB to 140°F
CHROME AMMONIUM ALUM						
CHROME PLATING SOLUTIONS						AB to 70°F
CHROMIC ACID	NR 50% at 70°F		B 50% at 70°F C 50% at 122°F	NR 10% at 70°F C 1% at 70°F	NR conc at 70°F BC 25% at 70°F	AC to conc to 70°F C 50% at 122°F
CHROMIC ACID			A 10% to 122°F			B 10% at 122°F C 10% at 150°F
CHROMIC OXIDE						
CHROMICOAT						
CHROMIUM AMMONIUM SULFATE						
CHROMIUM POTASSIUM SULFATE						*AB to 140°F
CHROMIUM SALTS			AB to 70°F	AB to 70°F		A to 70°F
CHRYSOLEPIC ACID						
CIDER			A to 70°F		*AB to 140°F	AB to 70°F
CINENE			AB to 70°F			NR at 70°F

C 90

	POLYETHYLENE LOW DENSITY LDPE	® TEFLON FEP	® KALREZ PERFLUORINATED ELASTOMER	® FLUORAZ FLUORINATED COPOLYMER	® AFLAS FLUORINATED COPOLYMER	® NORPRENE COPOLYMER TPO
CHLOROBENZENE, MONO, DI, TRI	NR at 70°F (mono)	A to 300°F (mono, di)	+1% vol 7 days 70°F A to 212°F	C/NR at 70°F (mono)		NR at 70°F (mono)
CHLOROBENZOL	NR at 70°F	A to 300°F	A to 212°F	C/NR at 70°F		NR at 70°F
CHLOROBROMOMETHANE		A to 70°F	A to 212°F	C at 70°F		
CHLOROBUTADIENE		A to 70°F	A to 212°F	C at 70°F	A to 70°F	
CHLOROBUTANE		A to 200°F	A to 212°F			
CHLORODODECANE		A to 70°F	A to 212°F	B at 70°F		
CHLOROETHANE	C/NR at 70°F NR at 140°F	A to 125°F dry A to 300°F wet	A to 212°F	AB to 70°F		NR at 70°F
CHLOROETHANOIC ACID	A/NR at 70°F	A to 200°F	A to 70°F A 10% to 104°F	B/NR at 70°F		A to 70°F
CHLOROETHANOL		A to 300°F	A to 70°F B to 113°F	A to 70°F		NR at 70°F
CHLOROETHYL ALCOHOL		A to 300°F	A to 70°F B to 113°F	A to 70°F		NR at 70°F
CHLOROETHYLBENZENE		A to 70°F	A to 212°F			
CHLOROFORM	NR at 70°F	A to 300°F	A to 212°F	NR at 70°F	+112% vol 7 days 70°F NR at 70°F	NR at 70°F
CHLOROMETHANE	NR at 70°F	A to 200°F wet or dry	A to 70°F	NR at 70°F		NR at 70°F
CHLOROMETHYL ETHER						
O-CHLORONAPHTHALENE		A to 70°F	A to 212°F	NR at 70°F		
1-CHLORO 1-NITROETHANE		A to 70°F	A to 212°F	NR at 70°F	A to 70°F	
CHLOROPENTANE	NR at 70°F	A to 200°F		A to 70°F		
CHLOROPHENIC ACID		A to 70°F				
O-CHLOROPHENOL		A to 70°F	A to 212°F	A to 70°F		
CHLOROPRENE		A to 70°F	A to 212°F	C at 70°F	A to 70°F	
CHLOROPROPANONE		A to 70°F				
CHLOROPROPENE		A to 70°F				
CHLOROPROPYLENE		A to 70°F				
CHLOROPROPYLENE OXIDE		A to 70°F				
CHLOROSULFONIC ACID	NR at 70°F	+0.8% wt 7 days 302°F A to boiling	A to 212°F	C to 50% at 70°F	*A to 70°F	
CHLOROTHENE	NR at 70°F	A to 200°F	A to 70°F *A to 212°F	NR at 70°F wet or dry		NR at 70°F
CHLOROTOLUENE		A to 387°F	A to 212°F	NR at 70°F		
CHLOROTRIFLUOROETHYLENE		A to 70°F	B at 70°F			
CHLOROX		A to 70°F	A to 212°F	A to 70°F		
CHOCOLATE SYRUP	A to 70°F					
CHRESYLIC ACID	NR 50% at 70°F					
CHROME ALUM	A to 70°F AB to 140°F	A to 70°F	A to 212°F	AB conc to 70°F C 10% at 70°F		
CHROME AMMONIUM ALUM		A to 70°F				
CHROME PLATING SOLUTIONS	AB to 140°F	A to 70°F	A to 212°F	A to 70°F		
CHROMIC ACID	B conc at 70°F AB to 50% to 70°F	+0.1% wt 50% conc 7 days 248°F	A to 212°F A 30% to 70°F	A to 100% to 70°F	+1.7% vol 62% conc 7 days 70°F	A to 70°F
CHROMIC ACID	C 50% at 104°F A to 10% to 122°F	A conc to 70°F A to 50% to 300°F			A 62% to 70°F	
CHROMIC OXIDE		A to 70°F	A to 212°F	A to 100% to 70°F		
CHROMICOAT		A to 70°F				
CHROMIUM AMMONIUM SULFATE		A to 70°F				
CHROMIUM POTASSIUM SULFATE	A to 70°F AB to 140°F	A to 70°F	A to 212°F	AB conc to 70°F C 10% at 70°F		
CHROMIUM SALTS	AB to 70°F		A to 212°F			A to 70°F
CHRYSOLEPIC ACID		A to 70°F				
CIDER	AB to 140°F	A to 122°F				
CINENE		A to 70°F				

	RECOMMENDATIONS	NITRILE NBR	ETHYLENE PROPYLENE EP, EPDM	FLUOROCARBON FKM	CHLOROPRENE CR	HYDROGENATED NITRILE HNBR
CINNAMON OIL					C at 70°F	
CIRCO LIGHT PROCESS OIL	NBR,FKM,ACM,AU	A to 70°F AB to 180°F	NR at 70°F	A to 70°F	B to 150°F	A to 70°F
CITRIC ACID	NBR,CR,EP,FKM	A to 200°F B at 212°F	A to conc to boiling	A to conc to boiling	A to conc to boiling	A to 70°F
CITRIC ACID, AMMONIATED		B at 70°F			B at 70°F	
CITROUS OILS		A to 140°F	B at 70°F	A to 140°F	NR at 70°F	
CITROUS PECTIN LIQUOR		A to 70°F		C at 70°F	A to 70°F	
CITY SERVICE KOOL MOTOR OIL NO. 140	NBR,FKM,ACM,AU	A to 70°F	NR at 70°F	A to 70°F	B at 70°F	A to 70°F
CITY SERVICE PACEMAKER NO. 2	NBR,FKM,ACM	A to 70°F (2,FR-15,20,25,FR)	NR at 70°F	A to 70°F	B at 70°F (NO.2) A to 70°F (FR 15,20,25)	A to 70°F
CITY SERVICE 65, 120, 250	NBR,FKM,ACM	A to 70°F	NR at 70°F	A to 70°F	B at 70°F	A to 70°F
CLAY SLURRY	NBR	A to 200°F	A to 70°F	B at 70°F	A to 70°F	
CLEANERS NAPHTHA		B to 80°F		A to 70°F	NR at 70°F	
CLEANING SOL'N. FOOD GRADE ACID		A to 70°F		A to 70°F	A to 70°F	
COAL GAS	IIR,FKM	NR at 70°F	A to 70°F	A to 70°F	A to 70°F	
COAL OIL		A to 70°F	NR at 70°F	A to 70°F	B at 70°F	
COAL TAR, BITUMINOUS	FKM,NBR	AB to 70°F	NR at 70°F	A to 212°F	B/NR at 70°F	
COAL TAR, CREOSOTE	NBR,FKM,FVMQ,ACM	AB to 70°F	NR at 70°F	A to 212°F	B/NR at 70°F	
COAL TAR NAPHTHA		NR at 70°F	NR at 70°F	A to 70°F	NR at 70°F	
COBALT CHLORIDE	NBR,EP,FKM,CR,IIR	A 2M to conc to 70°F	A 2M to conc to 70°F	A 2M to conc to 70°F	A 2M to conc to 70°F	A to 70°F
® COCA-COLA	NBR,CR,EP	A to 180°F	A to 70°F	B at 70°F	B at 70°F	
COCOA BUTTER	NBR	A to 180°F	NR at 70°F	A to 70°F	B at 70°F	
COCONUT FATTY ACID	NBR	A to 200°F	NR at 70°F	A to 70°F	B at 70°F	
COCONUT OIL (BUTTER)	NBR,FKM,ACM,EP,IIR	A to 250°F	B to 104°F	A to 212°F	AC to 140°F NR at 180°F	A to 70°F
COD LIVER OIL	NBR,EP,FKM,IIR	A to 70°F AB to 120°F	A to 70°F	+4% vol 7 days 75°F A to 75°F	B at 70°F	A to 70°F
COFFEE	NBR,EP,FKM,CR	A to 140°F	A to 70°F	A to 140°F	A to 140°F	
COFFEE EXTRACT	NBR	A to 250°F	A to 70°F	A to 70°F	A to 70°F	
COKE OVEN GAS	FKM,FVMQ,VMQ	B/NR at 70°F C/NR at 104°F	A/NR to 104°F	A to 250°F	C/NR at 70°F	NR at 70°F
COLICHE LIQUORS	CR	B at 70°F	B at 70°F		A to 70°F	
COLOGNE SPIRITS		A to 70°F AB to 150°F	A to 70°F	B at 70°F	A to 70°F AB to 150°F	
COLONIAL SPIRIT		A to 70°F	B at 70°F	C at 70°F	A to 70°F	
COLUMBIAN SPIRITS		A to 70°F	B at 70°F	C at 70°F	A to 70°F	
COLZA OIL		B at 70°F	A to 70°F	A to 70°F	B at 70°F	
COMMON ALUM		A to 70°F	A to 70°F	A to 70°F	A to 70°F	
CONVELEX 10	T,FKM	NR at 70°F		A to 70°F	NR at 70°F	NR at 70°F
COOLANOL (MONSANTO)	NBR,FKM,CR,FVMQ	A to 70°F	NR at 70°F	A to 70°F	AB to 70°F	A to 70°F
COOLANOL 25, 45	NBR,FKM	A to 70°F	C/NR at 70°F	A to 70°F	A to 70°F	A to 70°F
COPPER ACETATE (BLUE VERDIGRIS)	EP,NBR,IIR	B at 70°F	A to 200°F	NR at 70°F	B at 70°F	B at 70°F
COPPER AMMONIUM ACETATE	EP	A/NR at 70°F	A to 200°F	NR at 70°F	C at 70°F	
COPPER ARSENATE, BASIC		A to 70°F	A to 70°F	A to 70°F	A to 70°F	
COPPER BOROFLUORIDE		A to 70°F	A to 176°F B at 212°F	A to 70°F	A to 70°F	
COPPER CARBONATE		NR at 70°F	A to 70°F	A to 70°F		
COPPER CHLORIDE	NBR,EP,FKM,IIR,CR	A to 176°F B at 212°F	A to 176°F B at 212°F	A to 212°F	AB any conc to 200°F	A to 70°F
COPPER CYANIDE	NBR,EP,FKM,IIR,CR	A to boiling	A to 140°F AB to 200°F	A to boiling	A to boiling	A to 70°F
COPPER FLUOBORATE		B to 140°F		A to 140°F	A to 140°F	
COPPER FLUORIDE			A to 70°F	A to 70°F		

	STYRENE BUTADIENE SBR	POLYACRYLATE ACM	POLYURETHANE AU, EU	ISOBUTYLENE ISOPRENE IIR	POLYBUTADIENE BR	® AEROQUIP AQP
CINNAMON OIL						
CIRCO LIGHT PROCESS OIL	NR at 70°F	A to 70°F	A to 70°F	NR at 70°F	NR at 70°F	
CITRIC ACID	A to 70°F		A to conc to 70°F	A any conc to 185°F	A to 70°F	NR 10% at 70°F
CITRIC ACID, AMMONIATED				A to 70°F		
CITROUS OILS				B at 70°F		
CITROUS PECTIN LIQUOR			C at 70°F			
CITY SERVICE KOOL MOTOR OIL NO. 140	NR at 70°F	A to 70°F	A to 70°F	NR at 70°F	NR at 70°F	
CITY SERVICE PACEMAKER NO. 2	NR at 70°F	A to 70°F	B at 70°F	NR at 70°F	NR at 70°F	AB to 180°F
CITY SERVICE 65, 120, 250	NR at 70°F	A to 70°F	B at 70°F	NR at 70°F	NR at 70°F	
CLAY SLURRY	A to 70°F		A to 70°F			
CLEANERS NAPHTHA				NR at 70°F		
CLEANING SOL'N. FOOD GRADE ACID			A to 70°F			
COAL GAS			AB to 70°F			
COAL OIL			C at 70°F	NR at 70°F		
COAL TAR, BITUMINOUS			BC at 70°F	NR at 70°F		
COAL TAR, CREOSOTE	NR at 70°F	A to 70°F	B /NR at 70°F	NR at 70°F	NR at 70°F	AB to 70°F
COAL TAR NAPHTHA			NR at 70°F	NR at 70°F		
COBALT CHLORIDE	A 2M to conc to 70°F	A /NR conc at 70°F NR 2 molar at 70°F	A /NR conc at 70°F NR 2 molar at 70°F	A 2M to conc to 70°F	A 2M to conc to 70°F	
® COCA-COLA	B at 70°F	NR at 70°F	B at 70°F			
COCOA BUTTER	NR at 70°F		B at 70°F			
COCONUT FATTY ACID	NR at 70°F	A to 70°F	A to 70°F			
COCONUT OIL (BUTTER)	NR at 70°F	A to 70°F	AC to 70°F	AB to 150°F	NR at 70°F	
COD LIVER OIL	NR at 70°F	A to 70°F	A to 70°F	A to 70°F AB to 150°F	NR at 70°F	
COFFEE	A to 70°F	NR at 70°F	NR at 70°F	A to 70°F	A to 70°F	
COFFEE EXTRACT	A to 70°F	NR at 70°F	NR at 70°F			
COKE OVEN GAS	NR at 70°F	NR at 70°F	NR at 70°F	NR at 70°F	NR at 70°F	
COLICHE LIQUORS	B at 70°F			B at 70°F	A to 70°F	
COLOGNE SPIRITS			NR at 70°F	A to 70°F AB to 185°F		
COLONIAL SPIRIT			NR at 70°F	A to 70°F		
COLUMBIAN SPIRITS			NR at 70°F	A to 70°F		
COLZA OIL			B at 70°F	A to 70°F		
COMMON ALUM				A to 70°F		
CONVELEX 10	NR at 70°F		B at 70°F	NR at 70°F	NR at 70°F	
COOLANOL (MONSANTO)	NR at 70°F	NR at 70°F	NR at 70°F	NR at 70°F	NR at 70°F	
COOLANOL 25, 45	NR at 70°F	NR at 70°F	NR at 70°F	NR at 70°F	NR at 70°F	
COPPER ACETATE (BLUE VERDIGRIS)	NR at 70°F	NR at 70°F	NR at 70°F	A to 70°F	NR at 70°F	
COPPER AMMONIUM ACETATE	NR at 70°F	NR at 70°F	C /NR at 70°F			
COPPER ARSENATE, BASIC				A to 70°F		
COPPER BOROFLUORIDE						
COPPER CARBONATE						
COPPER CHLORIDE	A to 70°F	A to 70°F	AB to 70°F A to 5% to 70°F	A to 70°F AB any conc to 185°F	A to 70°F	
COPPER CYANIDE	A to 70°F	A to 70°F	A to 70°F	A to 70°F	A to 70°F	
COPPER FLUOBORATE						
COPPER FLUORIDE						

	SYNTHETIC ISOPRENE IR	NATURAL ISOPRENE NR	CHLOROSULFONATED POLYETHYLENE CSM	FLUOROSILICONE FVMQ	SILICONE VMQ	CHEMRAZ FFKM
CINNAMON OIL		C at 70°F				
CIRCO LIGHT PROCESS OIL	NR at 70°F	NR at 70°F	B at 70°F	A to 70°F	NR at 70°F	A to 70°F
CITRIC ACID	A conc to 70°F	A to 70°F	A to 70°F	A 10% to conc at 70°F	A to 70°F	A to 70°F
CITRIC ACID, AMMONIATED	A to 70°F	A to 70°F				
CITROUS OILS	NR at 70°F	NR at 70°F	NR at 70°F		C at 70°F	
CITROUS PECTIN LIQUOR						
CITY SERVICE KOOL MOTOR OIL NO. 140	NR at 70°F	NR at 70°F	B at 70°F	A to 70°F	NR at 70°F	A to 70°F
CITY SERVICE PACEMAKER NO. 2	NR at 70°F	NR at 70°F	NR at 70°F	A to 70°F	NR at 70°F	A to 70°F
CITY SERVICE 65, 120, 250	NR at 70°F	NR at 70°F	NR at 70°F	A to 70°F	NR at 70°F	A to 70°F
CLAY SLURRY				C at 70°F	C at 70°F	
CLEANERS NAPHTHA	NR at 70°F	NR at 70°F				
CLEANING SOL'N. FOOD GRADE ACID						
COAL GAS						
COAL OIL	NR at 70°F	NR at 70°F	C at 70°F		NR at 70°F	
COAL TAR, BITUMINOUS	NR at 70°F	NR at 70°F	C/NR at 70°F	A to 70°F	C/NR at 70°F	A to 70°F
COAL TAR, CREOSOTE	NR at 70°F	NR at 70°F	C/NR at 70°F	A to 70°F	C/NR at 70°F	A to 70°F
COAL TAR NAPHTHA	NR at 70°F	NR at 70°F	NR at 70°F		NR at 70°F	
COBALT CHLORIDE	A 2M to conc to 70°F	A 2M to conc to 70°F	A 2M to conc to 70°F	A 2M to conc to 70°F	B conc at 70°F / A 2M at 70°F	A to 70°F 2N to conc
® COCA-COLA				A to 70°F	A to 70°F	
COCOA BUTTER				A to 70°F	NR at 70°F	
COCONUT FATTY ACID				A to 70°F	A to 70°F	
COCONUT OIL (BUTTER)	NR at 70°F	NR at 70°F	BC at 70°F	A to 70°F	A to 70°F	A to 70°F
COD LIVER OIL	NR at 70°F	NR at 70°F	B at 70°F	A to 70°F	B at 70°F	A to 70°F
COFFEE	A to 70°F	A to 70°F	A to 70°F	A to 70°F	A to 70°F	
COFFEE EXTRACT				A to 70°F	A to 70°F	
COKE OVEN GAS	C/NR at 70°F	B/NR at 70°F	A/NR at 70°F	B at 70°F	B at 70°F	A to 70°F
COLICHE LIQUORS	A to 70°F	A to 70°F				A to 70°F
COLOGNE SPIRITS	A to 70°F	A to 70°F AB any conc to 150°F	A to 70°F		A to 70°F	
COLONIAL SPIRIT	A to 70°F	A to 70°F	A to 70°F		A to 70°F	
COLUMBIAN SPIRITS	A to 70°F	A to 70°F	A to 70°F		A to 70°F	
COLZA OIL	NR at 70°F	NR at 70°F	B at 70°F		NR at 70°F	
COMMON ALUM	A to 70°F	A to 70°F	A to 70°F		A to 70°F	
CONVELEX 10	NR at 70°F	NR at 70°F	NR at 70°F		NR at 70°F	A to 70°F
COOLANOL (MONSANTO)	NR at 70°F	NR at 70°F	B at 70°F	B at 70°F	NR at 70°F	A to 70°F
COOLANOL 25, 45	NR at 70°F	NR at 70°F	B at 70°F	B at 70°F	NR at 70°F	A to 70°F
COPPER ACETATE (BLUE VERDIGRIS)	A to 70°F	A to 70°F	B/NR at 70°F	NR at 70°F	NR at 70°F	A to 70°F
COPPER AMMONIUM ACETATE				NR at 70°F	NR at 70°F	A to 70°F
COPPER ARSENATE, BASIC	A to 70°F	A to 70°F	A to 70°F			
COPPER BOROFLUORIDE						
COPPER CARBONATE						
COPPER CHLORIDE	A to 70°F	AB any conc to 150°F	AB to 185°F	A to 70°F	A to 70°F	A to 70°F
COPPER CYANIDE	A to 70°F	A to 70°F	AB to 185°F	A to 70°F	A to 70°F	A to 70°F
COPPER FLUOBORATE		A 20% to 70°F				A to 70°F
COPPER FLUORIDE						

	POLYSULFIDE T	CHLORINATED POLYETHYLENE CM	EPICHLOROHYDRIN CO, ECO	® HYTREL COPOLYESTER TPE	® SANTOPRENE COPOLYMER TPO	® C-FLEX STYRENIC TPE
CINNAMON OIL						
CIRCO LIGHT PROCESS OIL	A to 70°F					
CITRIC ACID	NR at 70°F	NR at 70°F	A to 70°F	AB to conc to 70°F *AB to 140°F	A to 70°F	
CITRIC ACID, AMMONIATED						
CITROUS OILS						
CITROUS PECTIN LIQUOR						
CITY SERVICE KOOL MOTOR OIL NO. 140	A to 70°F					
CITY SERVICE PACEMAKER NO. 2				AB to 150°F		
CITY SERVICE 65, 120, 250						
CLAY SLURRY						
CLEANERS NAPHTHA						
CLEANING SOL'N. FOOD GRADE ACID						
COAL GAS				AB to 70°F		
COAL OIL					NR at 70°F	
COAL TAR, BITUMINOUS						
COAL TAR, CREOSOTE	C at 70°F	AB to 70°F	NR at 70°F	NR at 70°F	NR at 70°F	
COAL TAR NAPHTHA						
COBALT CHLORIDE	NR 2M to conc at 70°F					
® COCA-COLA						
COCOA BUTTER						
COCONUT FATTY ACID						
COCONUT OIL (BUTTER)	NR at 70°F					
COD LIVER OIL	NR at 70°F					
COFFEE	NR at 70°F					
COFFEE EXTRACT						
COKE OVEN GAS	NR at 70°F					
COLICHE LIQUORS	NR at 70°F					
COLOGNE SPIRITS						
COLONIAL SPIRIT						
COLUMBIAN SPIRITS						
COLZA OIL						
COMMON ALUM						
CONVELEX 10	A to 70°F					
COOLANOL (MONSANTO)	NR at 70°F					
COOLANOL 25, 45	NR at 70°F					
COPPER ACETATE (BLUE VERDIGRIS)	NR at 70°F					
COPPER AMMONIUM ACETATE						
COPPER ARSENATE, BASIC						
COPPER BOROFLUORIDE						
COPPER CARBONATE						
COPPER CHLORIDE	NR at 70°F	NR at 70°F		A to 70°F		
COPPER CYANIDE						
COPPER FLUOBORATE						
COPPER FLUORIDE						

	ETHYLENE ACRYLIC EA	POLYALLOMER LINEAR COPOLYMER	NYLON 11 POLYAMIDE	NYLON 12 POLYAMIDE	ETHYLENE VINYL ACETATE EVA	POLYVINYLCHLORIDE FLEXIBLE PVC
CINNAMON OIL		NR at 70°F				NR at 70°F
CIRCO LIGHT PROCESS OIL	A to 70°F					
CITRIC ACID	A to 70°F	A conc to 70°F A 10% to 122°F	AB to 105°F C/NR at 140°F	AB 50% to 70°F	A to 70°F AB to 140°F	AB to conc to 150°F
CITRIC ACID, AMMONIATED						
CITROUS OILS						
CITROUS PECTIN LIQUOR						
CITY SERVICE KOOL MOTOR OIL NO. 140	A to 70°F					
CITY SERVICE PACEMAKER NO. 2			AB to 160°F (no.2) A to 70°F (FR)			
CITY SERVICE 65, 120, 250						
CLAY SLURRY						
CLEANERS NAPHTHA						
CLEANING SOL'N. FOOD GRADE ACID						
COAL GAS			A to 105°F			A/NR at 70°F
COAL OIL			A to 70°F			
COAL TAR, BITUMINOUS						
COAL TAR, CREOSOTE	*NR at 70°F	A to 70°F	C/NR at 70°F		B/NR at 70°F	BC at 70°F
COAL TAR NAPHTHA						
COBALT CHLORIDE						
® COCA-COLA				AB to 70°F		A to 150°F
COCOA BUTTER						
COCONUT FATTY ACID						
COCONUT OIL (BUTTER)	A to 70°F			AB to 70°F	NR at 70°F	BC at 70°F NR at 150°F
COD LIVER OIL				AB to 70°F		
COFFEE	A to 70°F					AB to 70°F
COFFEE EXTRACT						
COKE OVEN GAS		A to 70°F	A to 70°F		BC at 70°F	B at 70°F
COLICHE LIQUORS						
COLOGNE SPIRITS						
COLONIAL SPIRIT						
COLUMBIAN SPIRITS						
COLZA OIL						
COMMON ALUM						
CONVELEX 10						
COOLANOL (MONSANTO)						
COOLANOL 25, 45						
COPPER ACETATE (BLUE VERDIGRIS)						A to 70°F
COPPER AMMONIUM ACETATE					B at 70°F	NR at 70°F
COPPER ARSENATE, BASIC					A to 70°F	A to 70°F
COPPER BOROFLUORIDE						
COPPER CARBONATE						
COPPER CHLORIDE	A to 70°F		B/NR at 70°F		B at 70°F	AB to 70°F *AB to 140°F
COPPER CYANIDE			AB to 70°F			AB to 140°F
COPPER FLUOBORATE						
COPPER FLUORIDE						

	POLYETHYLENE LOW DENSITY LDPE	® TEFLON FEP	® KALREZ PERFLUORINATED ELASTOMER	® FLUORAZ FLUORINATED COPOLYMER	® AFLAS FLUORINATED COPOLYMER	® NORPRENE COPOLYMER TPO
CINNAMON OIL	NR at 70°F	A to 122°F				
CIRCO LIGHT PROCESS OIL		A to 70°F		A to 70°F		
CITRIC ACID	AB to conc to 140°F A 10% to 122°F	A to 100% to 300°F	A to 212°F	A to 140°F	+0% vol 7 days 158°F A to 158°F	
CITRIC ACID, AMMONIATED		A to 70°F				
CITROUS OILS		A to 70°F				
CITROUS PECTIN LIQUOR						
CITY SERVICE KOOL MOTOR OIL NO. 140				A to 70°F		
CITY SERVICE PACEMAKER NO. 2		A to 70°F AB to 250°F		A to 70°F		
CITY SERVICE 65, 120, 250				A to 70°F		
CLAY SLURRY						
CLEANERS NAPHTHA		A to 70°F				
CLEANING SOL'N. FOOD GRADE ACID						
COAL GAS	AB to 70°F					
COAL OIL		A to 70°F		A to 70°F		
COAL TAR, BITUMINOUS	A/NR at 70°F	A to 70°F	A to 212°F	A to 70°F		
COAL TAR, CREOSOTE	A/NR at 70°F	A to 70°F	A to 212°F	A to 70°F		
COAL TAR NAPHTHA		A to 70°F				
COBALT CHLORIDE		A 2 molar to 70°F	A to 212°F	A to 70°F		
® COCA-COLA						
COCOA BUTTER				A to 70°F		
COCONUT FATTY ACID				A to 70°F		
COCONUT OIL (BUTTER)	AB to 140°F	A to 70°F	A to 212°F	A to 70°F		
COD LIVER OIL		A to 70°F	A to 212°F	A to 70°F		
COFFEE						
COFFEE EXTRACT						
COKE OVEN GAS	AB to 70°F	AC to 70°F	A to 212°F	A to 70°F		
COLICHE LIQUORS						
COLOGNE SPIRITS		A to 70°F				
COLONIAL SPIRIT		A to 70°F				
COLUMBIAN SPIRITS		A to 70°F				
COLZA OIL		A to 70°F				
COMMON ALUM		A to 70°F				
CONVELEX 10				B at 70°F		
COOLANOL (MONSANTO)				A to 70°F		
COOLANOL 25, 45				A to 70°F		
COPPER ACETATE (BLUE VERDIGRIS)		A to 70°F	A to 212°F	NR at 70°F		
COPPER AMMONIUM ACETATE	BC at 70°F		A to 212°F	NR at 70°F		
COPPER ARSENATE, BASIC		A to 70°F				
COPPER BOROFLUORIDE		A to 70°F				
COPPER CARBONATE		A to 70°F	A to 212°F			
COPPER CHLORIDE	B at 70°F	A to 150°F	A to 212°F	AB to 70°F		
COPPER CYANIDE	B at 70°F	A to 300°F	A to 212°F	BC at 70°F		
COPPER FLUOBORATE	A to 70°F	A to 70°F				
COPPER FLUORIDE		A to 200°F				

	RECOMMENDATIONS	NITRILE NBR	ETHYLENE PROPYLENE EP, EPDM	FLUOROCARBON FKM	CHLOROPRENE CR	HYDROGENATED NITRILE HNBR
COPPER HYDRATE		B at 70°F		C at 70°F		
COPPER HYDROXIDE		B at 70°F		C at 70°F		
COPPER LASUR		A to 70°F		A to 70°F		
COPPER NITRATE	NBR, CR, EP	A to 140°F AB any conc to 200°F	A to 176°F AB to 200°F	A to 212°F	A to conc to 140°F AB any conc to 200°F	
COPPER NITRITE		A to 70°F		A to 70°F		
COPPER PLATING SOLUTION	EP	A to 140°F	A to 200°F	A to 140°F	A to 140°F NR at 70°F electroless	
COPPER PLATING SOLUTION, ACID	EP	B at 70°F	A to 250°F	A to 140°F	A to 70°F	
COPPER SALTS	NBR, EP, FKM, CR	A to 70°F	A to 70°F	A to 70°F	A to 70°F	A to 70°F
COPPER SULFATE	NBR, EP, CR, IIR	A to conc to 176°F AB any conc to 212°F	A to conc to 176°F AB any conc to 212°F	A to conc to 140°F A conc to 212°F	A to conc to 140°F AB any conc to 200°F	A to 70°F
COPPER SULFIDE		A to 70°F		A to 70°F		
COPPER SULPHATE	NBR, EP, FKM, CR	A to conc to 70°F	A to conc to 70°F	A to conc to 70°F	A to conc to 70°F	
COPPERAS	NBR, CR, EP	A to 140°F AB any conc to 200°F	A to conc to 176°F B at 212°F	A to conc to 176°F B at 212°F	A to 140°F AB any conc to 200°F	A to 70°F
COPRA		B at 70°F	A to 70°F	A to 70°F	B at 70°F	
CORE OIL		A to 70°F			AB to 70°F	
CORN OIL	NBR, FKM, ACM, VMQ	A to 250°F	B/NR to 104°F	A to 140°F	AC to 70°F	A to 70°F
CORN STARCH SLURRY	NBR	A to 200°F	A to 70°F	A to 70°F	B at 70°F	
CORN SYRUP	NBR, FKM, EP	A to conc to 140°F AB conc to 200°F	A to conc to 176°F B at 212°F	A to 212°F	B any conc to 120°F	
CORROSIVE SUBLIMATE		A to 70°F AB any conc to 150°F	A to 70°F	A to conc to 70°F	B/NR at 70°F	
COTTONSEED OIL	NBR, FKM, ACM, VMQ CR	A to 200°F B at 212°F	A to 70°F C at 176°F	no swell 7 days 158°F A to 300°F	AC to 150°F	A to 70°F
CREOSOLS (2-METHYL - 4 METHYLPHENYL)	FKM, FVMQ	NR at 70°F	NR at 70°F	A to 70°F	NR at 70°F	
CREOSOTE	NBR, FKM, FVMQ	AB to 70°F NR at 140°F	NR at 70°F	+9.3% vol 7 days 212°F A to 212°F	NR at 70°F	
CREOSOTE-COAL TAR	NBR, FKM, FVMQ, ACM	AB to 70°F	NR at 70°F	A to 212°F	B/NR at 70°F	
CREOSOTE-WOOD TAR	NBR, FKM, FVMQ, ACM	A to 200°F	NR at 70°F	A to 200°F	BC at 70°F	A to 70°F
CRESOL (M, O or P)	FKM, FVMQ	C/NR at 70°F	NR any conc at 70°F	A to 104°F AB to 350°F	NR at 70°F any cresol	
CRESOLS (M, O and P)	FKM, FVMQ	C/NR at 70°F NR at 140°F	NR at 70°F	A to 350°F	C/NR at 70°F NR at 140°F	
CRESYL ALCOHOL		C/NR at 70°F	NR at 70°F	A to 70°F	NR at 70°F	
CRESYL HYDRATE		C at 70°F	NR at 70°F	A to 70°F	NR at 70°F	
CRESYLIC ACID	FKM, FVMQ	+200% vol 3 days 158°F C/NR at 70°F	+1% vol 3 days 158°F NR at 70°F	+1% vol 3 days 158°F +25% vol 28 days 300°F	+11% vol 3 days 158°F NR at 70°F	A to 70°F
CROTONALDEHYDE	EP	C/NR at 70°F	A to 200°F	A/NR at 70°F	NR at 70°F	
CRUDE OIL (ASPHALT BASE)	FKM	B at 70°F	NR at 70°F	A to 250°F	C at 70°F	A to 70°F
CRUDE OIL (EXCEPT ASPHALT BASE)	FKM, FVMQ, ACM, NBR	AB to 250°F	NR at 70°F	A to 300°F	C/NR at 70°F	
CRYOLITE		B to conc to 180°F	B/NR conc at 70°F A 10% to 70°F	A to conc to 70°F	AB to conc to 180°F	
CRYSCOAT F.H. RINSE					NR at 70°F	
CRYSCOAT H.C.					NR at 70°F	
CRYSCOAT L.T. & S.W.					NR at 70°F	
CRYSCOAT 47, 87, 89 & 89M					NR at 70°F	
CRYSTAL AMMONIA		A to 70°F AB any conc to 200°F			A to 70°F AB any conc to 200°F	
CUBIC NITER		C at 70°F	A to 70°F	A to 70°F	B at 70°F	
CUBIC SALTPETER		C at 70°F	A to 70°F	A to 70°F	B at 70°F	
CUBNIC		C at 70°F	A to 70°F	A to 70°F	B at 70°F	
CUMENE	FKM, FVMQ	NR at 70°F	A/NR at 70°F	A to 200°F	B/NR at 70°F	NR at 70°F
CUPRIC ACETATE		B at 70°F	A to 70°F		B at 70°F	
CUPRIC ARSENATE		A to 70°F	A to 70°F	A to 70°F	A to 70°F	
CUPRIC CARBONATE		A to 70°F		A to 70°F	A to 70°F	

	STYRENE BUTADIENE SBR	POLYACRYLATE ACM	POLYURETHANE AU, EU	ISOBUTYLENE ISOPRENE IIR	POLYBUTADIENE BR	® AEROQUIP AQP
COPPER HYDRATE				A to 70°F		
COPPER HYDROXIDE				A to 70°F		
COPPER LASUR				A to 70°F		
COPPER NITRATE			B conc at 70°F A to 50% to 140°F	A to 70°F AB any conc to 185°F		
COPPER NITRITE				A to 70°F		
COPPER PLATING SOLUTION	B at 70°F	NR at 70°F	NR at 70°F	A to 70°F AB to 180°F		
COPPER PLATING SOLUTION, ACID	B at 70°F	NR at 70°F	NR at 70°F			
COPPER SALTS	A to 70°F	A to 70°F	A to 70°F	A to 70°F	A to 70°F	
COPPER SULFATE	B to 100% at 70°F	NR 10% to conc at 70°F	AB 10% & conc to 70°F C 50% at 70°F	A to 70°F B any conc to 185°F	B to 100% at 70°F	NR at 70°F
COPPER SULFIDE				A to 70°F		
COPPER SULPHATE		NR at 70°F	AB 10% & conc to 70°F C 50% at 70°F	A to 70°F AB any conc to 185°F		NR at 70°F
COPPERAS	A sol'n to 70°F		A to conc to 70°F	A to 70°F AB any conc to 185°F		
COPRA			A to 70°F	A to 70°F		
CORE OIL						
CORN OIL	NR at 70°F	A to 70°F	A to 70°F	AB to 100°F	NR at 70°F	
CORN STARCH SLURRY	B at 70°F	NR at 70°F	B at 70°F			
CORN SYRUP	B at 70°F	NR at 70°F	A to 70°F	A to 70°F AB any conc to 185°F		
CORROSIVE SUBLIMATE				A to 70°F AB any conc to 150°F		
COTTONSEED OIL	NR at 70°F	A to 70°F	A to 70°F	AC to 100°F	NR at 70°F	AB to 70°F
CREOSOLS (2-METHYL - 4 METHYLPHENYL)	NR at 70°F	NR at 70°F	NR at 70°F	NR at 70°F	NR at 70°F	
CREOSOTE	NR at 70°F	A to 70°F	BC at 70°F	NR at 70°F	NR at 70°F	AB to 70°F
CREOSOTE-COAL TAR	NR at 70°F	A to 70°F	B /NR at 70°F	NR at 70°F	NR at 70°F	AB to 70°F
CREOSOTE-WOOD TAR	NR at 70°F	A to 70°F	BC at 70°F	NR at 70°F	NR at 70°F	
CRESOL (M, O or P)	NR at 70°F	NR at 70°F	NR at 70°F	NR at 70°F	NR at 70°F	
CRESOLS (M, O and P)	NR at 70°F	NR at 70°F	NR at 70°F	NR at 70°F	NR at 70°F	
CRESYL ALCOHOL			NR at 70°F	NR at 70°F		
CRESYL HYDRATE			NR at 70°F	NR at 70°F		
CRESYLIC ACID	+32% vol 3 days 70°F NR at 70°F	+266% vol 3 days 70°F NR at 70°F	+404% vol 3 days 158°F NR at 70°F	+0.3% vol 3 days 70°F NR at 70°F	NR at 70°F	
CROTONALDEHYDE	C at 70°F	NR at 70°F	NR at 70°F	A to 70°F		
CRUDE OIL (ASPHALT BASE)	NR at 70°F	A to 70°F	A to 70°F			
CRUDE OIL (EXCEPT ASPHALT BASE)	NR at 70°F	A to 70°F	B at 70°F *BC to 140°F	NR at 70°F	NR at 70°F	
CRYOLITE				A to conc to 70°F AB conc to 180°F		
CRYSCOAT F.H. RINSE						
CRYSCOAT H.C.						
CRYSCOAT L.T. & S.W.						
CRYSCOAT 47, 87, 89 & 89M						
CRYSTAL AMMONIA				A to 70°F AB any conc to 185°F		
CUBIC NITER				A to 70°F		
CUBIC SALTPETER				A to 70°F		
CUBNIC				A to 70°F		
CUMENE	NR at 70°F	NR at 70°F	NR at 70°F	NR at 70°F	NR at 70°F	
CUPRIC ACETATE				A to 70°F		
CUPRIC ARSENATE				A to 70°F		
CUPRIC CARBONATE				A to 70°F		

	SYNTHETIC ISOPRENE IR	NATURAL ISOPRENE NR	CHLOROSULFONATED POLYETHYLENE CSM	FLUOROSILICONE FVMQ	SILICONE VMQ	CHEMRAZ FFKM
COPPER HYDRATE	C at 70°F	C at 70°F	B at 70°F			
COPPER HYDROXIDE	C at 70°F	C at 70°F	B at 70°F			
COPPER LASUR	C at 70°F	C at 70°F	A to 70°F			
COPPER NITRATE	B at 70°F	B any conc to 150°F	A to 70°F AB to 200°F			A to 70°F
COPPER NITRITE	C at 70°F	C at 70°F	A to 70°F			
COPPER PLATING SOLUTION	B at 70°F	B to 100°F	A to 70°F	C at 70°F	C/NR at 70°F	
COPPER PLATING SOLUTION, ACID				C at 70°F	C at 70°F	
COPPER SALTS	A to 70°F	A to 70°F	A to 70°F	A to 70°F	A to 70°F	A to 70°F
COPPER SULFATE	BC at 70°F	B 10% & conc to 150°F A 15 & 50% to 70°F	A to conc to 70°F AB to conc at 200°F	A to 100% to 70°F	A to 100% to 70°F	A to 70°F
COPPER SULFIDE	C at 70°F	C at 70°F	A to 70°F			
COPPER SULPHATE	BC at 70°F	B 10% & conc to 150°F A 15 to 50% to 70°F	A to conc to 70°F AB to conc at 200°F	A to 100% to 70°F	A to 100% to 70°F	
COPPERAS	A to 70°F	A to 70°F AB any conc to 150°F	A to 70°F AB to 200°F	A to conc to 70°F	B to conc at 70°F BC conc at 70°F	A to 70°F
COPRA	NR at 70°F	NR at 70°F	B at 70°F			
CORE OIL						
CORN OIL	NR at 70°F	NR at 70°F	B at 70°F	A to 70°F	A to 70°F	A to 70°F
CORN STARCH SLURRY				A to 70°F	B at 70°F	
CORN SYRUP	B at 70°F	B any conc to 120°F	A to 70°F	A to 70°F	A to 70°F	
CORROSIVE SUBLIMATE	B at 70°F	B any conc to 150°F	A to 70°F		A to 70°F	
COTTONSEED OIL	NR at 70°F	NR at 70°F	AB to 70°F	A to 70°F	A to 70°F	A to 70°F
CREOSOLS (2-METHYL - 4 METHYLPHENYL)	NR at 70°F	NR at 70°F	NR at 70°F	B at 70°F	NR at 70°F	A to 70°F
CREOSOTE	NR at 70°F	NR at 70°F	C/NR at 70°F	A to 70°F	NR at 70°F	A to 70°F
CREOSOTE-COAL TAR	NR at 70°F	NR at 70°F	C/NR at 70°F	A to 70°F	C/NR at 70°F	A to 70°F
CREOSOTE-WOOD TAR	NR at 70°F	NR at 70°F	C/NR at 70°F	A to 70°F	C/NR at 70°F	A to 70°F
CRESOL (M, O or P)	NR at 70°F	NR at 70°F	B/NR at 70°F	AB to 70°F	B/NR at 70°F	A to 70°F (M)
CRESOLS (M, O and P)	NR at 70°F	NR at 70°F	B/NR at 70°F	AB to 70°F	B/NR at 70°F	
CRESYL ALCOHOL	NR at 70°F	NR at 70°F	B at 70°F		C at 70°F	
CRESYL HYDRATE	NR at 70°F	NR at 70°F	C at 70°F		C at 70°F	
CRESYLIC ACID	NR at 70°F	NR at 70°F	C/NR at 70°F			A to 70°F liquid or vapor
CROTONALDEHYDE	NR at 70°F	NR at 70°F	NR at 70°F	+2% vol 3 days 158°F AB to 70°F	+4% vol 3 days 70°F NR at 70°F	A to 70°F
CRUDE OIL (ASPHALT BASE)				B at 70°F	NR at 70°F	A to 70°F
CRUDE OIL (EXCEPT ASPHALT BASE)	NR at 70°F	NR at 70°F	NR at 70°F	AB at 70°F	NR at 70°F	
CRYOLITE	A to 70°F	A to 70°F AB to 150°F				
CRYSCOAT F.H. RINSE	NR at 70°F	NR at 70°F				
CRYSCOAT H.C.	NR at 70°F	NR at 70°F				
CRYSCOAT L.T. & S.W.	NR at 70°F	NR at 70°F				
CRYSCOAT 47, 87, 89 & 89M		NR at 70°F				
CRYSTAL AMMONIA						
CUBIC NITER	B at 70°F	B at 70°F	A to 70°F		C at 70°F	
CUBIC SALTPETER	B at 70°F	B at 70°F	A to 70°F		C at 70°F	
CUBNIC	B at 70°F	B at 70°F	A to 70°F		C at 70°F	
CUMENE	NR at 70°F	NR at 70°F	NR at 70°F	AB at 70°F	NR at 70°F	A to 70°F
CUPRIC ACETATE			B at 70°F			
CUPRIC ARSENATE	A to 70°F	A to 70°F	A to 70°F			
CUPRIC CARBONATE	C at 70°F	C at 70°F	A to 70°F			

	POLYSULFIDE T	CHLORINATED POLYETHYLENE CM	EPICHLOROHYDRIN CO, ECO	® HYTREL COPOLYESTER TPE	® SANTOPRENE COPOLYMER TPO	® C-FLEX STYRENIC TPE
COPPER HYDRATE						
COPPER HYDROXIDE						
COPPER LASUR						
COPPER NITRATE						
COPPER NITRITE						
COPPER PLATING SOLUTION						
COPPER PLATING SOLUTION, ACID						
COPPER SALTS	TEST					AB to 70°F
COPPER SULFATE	NR 10% & 100% at 70°F A 50% at 70°F	NR at 70°F		A to 70°F *AB to 140°F		
COPPER SULFIDE						
COPPER SULPHATE	NR 10% & 100% at 70°F A 50% at 70°F	NR at 70°F		A to 70°F *AB to 140°F		
COPPERAS	AB to 70°F			A to 70°F		A to 70°F
COPRA						
CORE OIL						
CORN OIL	NR at 70°F		A to 70°F	no swell at 212°F A to 212°F	+0.8% vol 7 days 212°F A to 212°F	
CORN STARCH SLURRY						
CORN SYRUP				*AB to 140°F		
CORROSIVE SUBLIMATE						
COTTONSEED OIL	NR at 70°F	AB to 70°F	A to 70°F	A to 70°F		
CREOSOLS (2-METHYL - 4 METHYLPHENYL)	NR at 70°F		NR at 70°F	NR at 70°F		
CREOSOTE	C at 70°F		NR at 70°F	NR at 70°F		
CREOSOTE-COAL TAR	C at 70°F	AB to 70°F	NR at 70°F	NR at 70°F	NR at 70°F	
CREOSOTE-WOOD TAR	C at 70°F	AB to 70°F	NR at 70°F	NR at 70°F	NR at 70°F	
CRESOL (M, O or P)	NR at 70°F			NR at 70°F	NR at 70°F	NR at 70°F
CRESOLS (M, O and P)	NR at 70°F			NR at 70°F	NR at 70°F	NR at 70°F
CRESYL ALCOHOL						
CRESYL HYDRATE						
CRESYLIC ACID	+260% vol 3 days 70°F NR at 70°F			NR at 70°F		
CROTONALDEHYDE						
CRUDE OIL (ASPHALT BASE)						
CRUDE OIL (EXCEPT ASPHALT BASE)	AB to 70°F			+21% vol 3 days 158°F AB to 70°F		
CRYOLITE						
CRYSCOAT F.H. RINSE						
CRYSCOAT H.C.						
CRYSCOAT L.T. & S.W.						
CRYSCOAT 47, 87, 89 & 89M						
CRYSTAL AMMONIA						
CUBIC NITER						
CUBIC SALTPETER						
CUBNIC						
CUMENE	B at 70°F					
CUPRIC ACETATE						
CUPRIC ARSENATE						
CUPRIC CARBONATE						

	ETHYLENE ACRYLIC EA	POLYALLOMER LINEAR COPOLYMER	NYLON 11 POLYAMIDE	NYLON 12 POLYAMIDE	ETHYLENE VINYL ACETATE EVA	POLYVINYLCHLORIDE FLEXIBLE PVC
COPPER HYDRATE						
COPPER HYDROXIDE						
COPPER LASUR						
COPPER NITRATE					B at 70°F	A to 70°F *AB to 140°F
COPPER NITRITE						
COPPER PLATING SOLUTION						NR at 70°F electroless
COPPER PLATING SOLUTION, ACID						
COPPER SALTS		A to 70°F	A to 70°F	AB 10% to 70°F	B at 70°F	A to 70°F
COPPER SULFATE	*A to 70°F	A to 70°F	AB to 194°F		B to 140°F	AB to 140°F
COPPER SULFIDE						
COPPER SULPHATE	*A to 70°F	A to 70°F	AB to 194°F		B to 140°F	AB to 140°F
COPPERAS		A to 70°F	A to 70°F		A to 70°F	A to 150°F
COPRA						
CORE OIL		A to 70°F	A to 70°F		C at 70°F	
CORN OIL	A to 70°F		AB to 70°F		B at 70°F	AB to 70°F B at 150°F
CORN STARCH SLURRY						
CORN SYRUP			*AB to 140°F		AB to 140°F	*AB to 140°F
CORROSIVE SUBLIMATE						
COTTONSEED OIL	A to 70°F	A to 70°F	A to 70°F		B at 70°F	AB to 70°F NR at 150°F
CREOSOLS (2-METHYL - 4 METHYLPHENYL)						
CREOSOTE	*NR at 70°F	A to 70°F	NR at 70°F		BC at 70°F	B/NR at 70°F
CREOSOTE-COAL TAR	*NR at 70°F	A to 70°F	C/NR at 70°F		B/NR at 70°F	BC at 70°F
CREOSOTE-WOOD TAR			NR at 70°F			C at 70°F
CRESOL (M, O or P)		B at 70°F C at 122°F	NR at 70°F	NR at 70°F	NR at 70°F	NR at 70°F
CRESOLS (M, O and P)		B at 70°F C at 122°F	NR at 70°F		NR at 70°F	NR at 70°F
CRESYL ALCOHOL						NR at 70°F
CRESYL HYDRATE						
CRESYLIC ACID			NR at 70°F		NR at 70°F	A/NR at 70°F
CROTONALDEHYDE						
CRUDE OIL (ASPHALT BASE)	A to 70°F		A to 70°F AB to 140°F (swells)		NR at 70°F	BC at 70°F NR at 140°F
CRUDE OIL (EXCEPT ASPHALT BASE)						
CRYOLITE						
CRYSCOAT F.H. RINSE						
CRYSCOAT H.C.						
CRYSCOAT L.T. & S.W.						
CRYSCOAT 47, 87, 89 & 89M						
CRYSTAL AMMONIA						
CUBIC NITER						
CUBIC SALTPETER						
CUBNIC						
CUMENE		C at 70°F NR at 122°F				NR at 70°F
CUPRIC ACETATE						
CUPRIC ARSENATE						
CUPRIC CARBONATE						

	POLYETHYLENE LOW DENSITY LDPE	® TEFLON FEP	® KALREZ PERFLUORINATED ELASTOMER	® FLUORAZ FLUORINATED COPOLYMER	® AFLAS FLUORINATED COPOLYMER	® NORPRENE COPOLYMER TPO
COPPER HYDRATE		A to 70°F				
COPPER HYDROXIDE		A to 70°F				
COPPER LASUR		A to 70°F				
COPPER NITRATE	AB to 70°F	A to 200°F	A to 212°F	AB to 70°F		
COPPER NITRITE		A to 70°F				
COPPER PLATING SOLUTION	A to 70°F	A to 70°F				
COPPER PLATING SOLUTION, ACID						
COPPER SALTS	B at 70°F	A to 70°F	A to 212°F	A to 70°F		A to 70°F
COPPER SULFATE	AB to conc to 140°F	A to conc to 200°F	A to 212°F	AB to 100% to 70°F		
COPPER SULFIDE		A to 70°F				
COPPER SULPHATE	AB to conc to 140°F	A to conc to 150°F	A to 212°F	AB to 50% to 70°F		
COPPERAS	A to 70°F	A to 300°F	A to 70°F	A to 70°F		
COPRA		A to 70°F				
CORE OIL	B at 70°F	A to 70°F				
CORN OIL	A to 70°F	A to 70°F	A to 212°F	A to 70°F		
CORN STARCH SLURRY						
CORN SYRUP	A to 70°F AB to 140°F	A to 70°F				
CORROSIVE SUBLIMATE		A to 70°F				
COTTONSEED OIL	A to 70°F	A to 300°F	A to 212°F	AB to 70°F	A to 70°F	
CREOSOLS (2-METHYL - 4 METHYLPHENYL)						
CREOSOTE	C/NR at 70°F	A to 70°F	A to 150°F	A to 70°F		
CREOSOTE-COAL TAR	A/NR at 70°F	A to 70°F	A to 212°F	A to 70°F		
CREOSOTE-WOOD TAR	C/NR at 70°F	A to 70°F	A to 70°F	A to 70°F		
CRESOL (M, O or P)	NR at 70°F	A to 376°F	A to 212°F		+9% vol 7 days 158°F A to 158°F	
CRESOLS (M, O and P)	NR at 70°F	A to 122°F	A to 212°F			
CRESYL ALCOHOL		A to 70°F				
CRESYL HYDRATE		A to 70°F		A to 70°F		
CRESYLIC ACID	AC to 70°F	A to 200°F	A to 212°F	AB to 70°F		
CROTONALDEHYDE		A to 70°F	A to 212°F			
CRUDE OIL (ASPHALT BASE)		A to 300°F	A to 212°F sweet or sour	A to 70°F		
CRUDE OIL (EXCEPT ASPHALT BASE)	NR at 70°F	A to 70°F	A to 212°F	AB to 70°F A to 70°F (sweet)	+8% vol 6 days 350°F AB to 350°F	
CRYOLITE		A to conc to 70°F				
CRYSCOAT F.H. RINSE		A to 70°F				
CRYSCOAT H.C.		A to 70°F				
CRYSCOAT L.T. & S.W.		A to 70°F				
CRYSCOAT 47, 87, 89 & 89M						
CRYSTAL AMMONIA		A to 70°F				
CUBIC NITER		A to 70°F				
CUBIC SALTPETER		A to 70°F				
CUBNIC		A to 70°F				
CUMENE	C at 70°F NR at 122°F	A to 122°F	A to 212°F	C at 70°F		
CUPRIC ACETATE		A to 70°F				
CUPRIC ARSENATE		A to 70°F				
CUPRIC CARBONATE		A to 70°F				

	RECOMMENDATIONS	NITRILE NBR	ETHYLENE PROPYLENE EP, EPDM	FLUOROCARBON FKM	CHLOROPRENE CR	HYDROGENATED NITRILE HNBR
CUPRIC CHLORIDE	FKM	AB any conc to 200°F	A to 176°F B at 212°F	A to conc to 212°F	A to 140°F AB any conc to 200°F	
CUPRIC CYANIDE		AB to 70°F	A to 70°F	A to 70°F	AB to 70°F	
CUPRIC FLUORIDE	FKM	A to 70°F	A to 140°F	A to 140°F	A to 70°F	
CUPRIC HYDROXIDE		B at 70°F		C at 70°F		
CUPRIC NITRATE	NBR, CR	A to 70°F AB any conc to 200°F	A to 70°F	A to 70°F	A to 70°F AB any conc to 200°F	
CUPRIC NITRITE	NBR	A to 70°F		A to 70°F		
CUPRIC SULFATE		A to 176°F AB any conc to 212°F	A to 176°F B to 212°F	A to 212°F	A to 70°F AB any conc to 200°F	
CUPRIC SULFIDE		A to 70°F		A to 70°F		
CUPRIC SULPHATE		A to 176°F AB any conc to 212°F	A to 176°F B to 212°F	A to 212°F	A to 70°F AB any conc to 200°F	
CUPROUS AMMONIA ACETATE	EP	A/NR at 70°F	A to 200°F	NR at 70°F	NR at 70°F	
CUPROUS AMMONIUM CHLORIDE			A to 70°F	A to 70°F	A to 70°F	
CUTTING OIL	NBR, FKM, FVMQ, ACM	A to 250°F	NR at 70°F	A to 70°F	B at 70°F	A to 70°F
CYANIC ACID SOULTIONS	EP	AC to 70°F	A to 200°F	NR at 70°F	NR at 70°F	
CYANIC COMPOUNDS	EP	A/NR at 70°F	A to 200°F	NR at 70°F	NR at 70°F	
CYANIDE						
CYANOGEN	EP	NR at 70°F	A to 200°F	NR at 70°F	NR at 70°F	
CYANOMETHANE		C at 70°F	A to 70°F	NR at 70°F	A to 70°F	
CYCLOHEXANE	NBR, FKM, FVMQ, T, AU	+11% vol 3 days 70°F AB to 250°F	+148% vol 3 days 70°F NR at 70°F	+4% vol 7 days 75°F A to 200°F	+73% vol 3 days 70°F NR at 70°F	A to 70°F
CYCLOHEXANOL	NBR, FKM	AB to 70°F	C/NR at 70°F	A to 104°F	AB to 70°F	A to 70°F
CYCLOHEXANONE	EP	NR at 70°F	BC at 70°F	NR at 70°F	NR at 70°F	NR at 70°F
CYCLOHEXATRIENE		NR at 70°F	NR at 70°F	A to 70°F	NR at 70°F	
CYCLOHEXYLAMINE		A/NR at 70°F	A to 200°F	NR at 70°F	A/NR at 70°F	
CYCLOPENTANE		B at 70°F	NR at 70°F	A to 70°F	A to 70°F	
CYCLOPENTANOL		B at 70°F		B at 70°F		
CYCLOPENTANONE		NR at 70°F		NR at 70°F		
CYMENE		C/NR at 70°F	NR at 70°F	A to 70°F	NR at 70°F	
P-CYMENE	FKM, FVMQ	NR at 70°F	NR at 70°F	A to 70°F	NR at 70°F	
D.B.T. (DIBUTYL PHTHALATE)	EP, T, IIR	NR at 70°F	A to 250°F	BC to 250°F	NR at 70°F	NR at 70°F
DC 200, 510, 550 & 710 SILICONE OILS	NBR, EP, FKM, CR, IIR	-4% vol 3 days 300°F A to 140°F	-5% vol 3 days 300°F A to 140°F	no swell 7 days 350°F A to 400°F	-2% vol 3 days 300°F A to 140°F	A to 70°F
DDT IN DEIONIZED KEROSENE	NBR	A to 200°F	NR at 70°F	A to 70°F	BC to 150°F	
DDT IN TOLUENE	FKM	A/NR at 70°F	NR at 70°F	A to 300°F	NR at 70°F	
DMF (DIMETHYL FORMAMIDE)	EP, NBR, VMQ	B/NR at 70°F	AB to 70°F B to 200°F	NR at 70°F	NR at 70°F	
DMP (DIMETHYL PHTHALATE)	EP, FKM, T, IIR	NR at 70°F	B to 200°F	A to 70°F B at 250°F	NR at 70°F	
DOP (DIOCTYL PHTHALATE)	EP, FKM, IIR, T	NR at 70°F	B to 200°F	B at 70°F	NR at 70°F	
DANFORTHS OIL		A to 70°F	NR at 70°F	A to 70°F	C at 70°F	
DEAD OIL	FKM	A to 70°F	NR at 70°F	A to 300°F	C at 70°F	
DECAHYDRONAPHTHALENE (DECALIN)	FKM, FVMQ, T	NR at 70°F	NR at 70°F	A to 200°F	NR at 70°F	
DECANAL		NR at 70°F	NR at 70°F	NR at 70°F		
DECANE	NBR, FKM, FVMQ, ACM	AB to 70°F	NR at 70°F	A to 70°F	C/NR at 70°F	A to 70°F
DECANOL (DECYL ALCOHOL)		A to 70°F AB to 100°F		B at 70°F	NR at 70°F	
DECYL ALDEHYDE		NR at 70°F		NR at 70°F		
DECYL BUTYL PHTHALATE		NR at 70°F		C at 70°F		
DEGREASING FLUID	FKM, FVMQ (chlorinated)	NR at 70°F	NR at 70°F	A to 70°F	NR at 70°F	
DEHYDRATED ALCOHOL		A to 70°F	A to 70°F	B at 70°F	A to 70°F	

	STYRENE BUTADIENE SBR	POLYACRYLATE ACM	POLYURETHANE AU, EU	ISOBUTYLENE ISOPRENE IIR	POLYBUTADIENE BR	® AEROQUIP AQP
CUPRIC CHLORIDE			A to 70°F	A to 70°F AB any conc to 185°F		
CUPRIC CYANIDE			A to 70°F	A to 70°F		
CUPRIC FLUORIDE						
CUPRIC HYDROXIDE				A to 70°F		
CUPRIC NITRATE				A to 70°F AB any conc to 185°F		
CUPRIC NITRITE				A to 70°F		
CUPRIC SULFATE			A/NR at 70°F	A to 70°F AB any conc to 185°F		
CUPRIC SULFIDE				A to 70°F		
CUPRIC SULPHATE			A/NR at 70°F	A to 70°F AB any conc to 185°F		
CUPROUS AMMONIA ACETATE	NR at 70°F	NR at 70°F	NR at 70°F			
CUPROUS AMMONIUM CHLORIDE						
CUTTING OIL	NR at 70°F	A to 70°F	A to 70°F	NR at 70°F	NR at 70°F	
CYANIC ACID SOULTIONS	NR at 70°F	NR at 70°F	NR at 70°F			
CYANIC COMPOUNDS	NR at 70°F	NR at 70°F	NR at 70°F			
CYANIDE			A to 70°F			
CYANOGEN	NR at 70°F	NR at 70°F	B/NR at 70°F			
CYANOMETHANE				A to 70°F		
CYCLOHEXANE	+143% vol 3 days 70°F NR at 70°F	+24% vol 3 days 70°F B at 70°F	+21% vol 3 days 70°F AB to 70°F	NR at 70°F	NR at 70°F	
CYCLOHEXANOL	NR at 70°F			NR at 70°F	NR at 70°F	
CYCLOHEXANONE	NR at 70°F	NR at 70°F	NR at 70°F	B/NR at 70°F	NR at 70°F	
CYCLOHEXATRIENE			NR at 70°F	NR at 70°F		
CYCLOHEXYLAMINE	NR at 70°F	NR at 70°F	NR at 70°F	A to 70°F		
CYCLOPENTANE				NR at 70°F		
CYCLOPENTANOL				NR at 70°F		
CYCLOPENTANONE				NR at 70°F		
CYMENE				NR at 70°F		
P-CYMENE	NR at 70°F	NR at 70°F	NR at 70°F	NR at 70°F	NR at 70°F	
D.B.T. (DIBUTYL PHTHALATE)	NR at 70°F	NR at 70°F	C/NR at 70°F NR at 158°F	BC any conc to 75°F	NR at 70°F	
DC 200, 510, 550 & 710 SILICONE OILS	-3% vol 3 days 300°F A to 70°F	-3% vol 3 days 300°F A to 70°F	-9% vol 3 days 300°F A to 70°F	-5.3% vol 3 days 300°F A to 70°F	A to 70°F	
DDT IN DEIONIZED KEROSENE	NR at 70°F	B at 70°F	B at 70°F	NR at 70°F		
DDT IN TOLUENE	NR at 70°F	NR at 70°F	NR at 70°F			
DMF (DIMETHYL FORMAMIDE)	NR at 70°F	NR at 70°F	NR at 70°F	B any conc to 100°F	NR at 70°F	
DMP (DIMETHYL PHTHALATE)	NR at 70°F	NR at 70°F	NR at 70°F	AB to 70°F	NR at 70°F	
DOP (DIOCTYL PHTHALATE)	NR at 70°F	NR at 70°F	NR at 70°F	B to 100°F	NR at 70°F	
DANFORTHS OIL			C at 70°F	NR at 70°F		
DEAD OIL	NR at 70°F	A to 70°F	B at 70°F			
DECAHYDRONAPHTHALENE (DECALIN)	NR at 70°F	NR at 70°F	NR at 70°F	NR at 70°F	NR at 70°F	
DECANAL				C at 70°F		
DECANE	NR at 70°F	A to 70°F	B at 70°F	NR at 70°F	NR at 70°F	
DECANOL (DECYL ALCOHOL)				B/NR at 70°F		
DECYL ALDEHYDE				C at 70°F		
DECYL BUTYL PHTHALATE				A to 70°F		
DEGREASING FLUID			A to 70°F	C at 70°F		
DEHYDRATED ALCOHOL			NR at 70°F	A to 70°F		

C105

	SYNTHETIC ISOPRENE IR	NATURAL ISOPRENE NR	CHLOROSULFONATED POLYETHYLENE CSM	FLUOROSILICONE FVMQ	SILICONE VMQ	CHEMRAZ FFKM
CUPRIC CHLORIDE	B at 70°F	B any conc to 150°F	A to 70°F		A to 70°F	
CUPRIC CYANIDE	A to 70°F	A to 70°F	B at 70°F		A to 70°F	
CUPRIC FLUORIDE						
CUPRIC HYDROXIDE	C at 70°F	C at 70°F	B at 70°F			
CUPRIC NITRATE	B at 70°F	B any conc to 150°F	A to 70°F			
CUPRIC NITRITE	C at 70°F	C at 70°F	A to 70°F			
CUPRIC SULFATE	C at 70°F	AB any conc to 150°F	A to 70°F		A to 70°F	A to 70°F
CUPRIC SULFIDE	C at 70°F	C at 70°F	A to 70°F			
CUPRIC SULPHATE	C at 70°F	AB any conc to 150°F	A to 70°F		A to 70°F	
CUPROUS AMMONIA ACETATE				NR at 70°F	NR at 70°F	A to 70°F
CUPROUS AMMONIUM CHLORIDE			A to 70°F			
CUTTING OIL	NR at 70°F	A to 70°F water base NR at 70°F sulfur base	B at 70°F	A to 70°F	NR at 70°F	A to 70°F
CYANIC ACID SOULTIONS		B/NR at 70°F	A to 70°F	NR at 70°F	NR at 70°F	
CYANIC COMPOUNDS				NR at 70°F	NR at 70°F	
CYANIDE						A to 70°F
CYANOGEN				NR at 70°F	NR at 70°F	A to 70°F
CYANOMETHANE	B at 70°F	B at 70°F	B at 70°F			
CYCLOHEXANE	NR at 70°F	NR at 70°F	B/NR at 70°F	+15% vol 3 days 70°F AB to 70°F	+102% vol 3 days 70°F NR at 70°F	A to 70°F
CYCLOHEXANOL	B/NR at 70°F	B/NR at 70°F	AB to 70°F	A to 70°F	NR at 70°F	A to 70°F
CYCLOHEXANONE	NR at 70°F	NR at 70°F	NR at 70°F	NR at 70°F	NR at 70°F	B at 70°F
CYCLOHEXATRIENE	NR at 70°F	NR at 70°F	NR at 70°F		NR at 70°F	
CYCLOHEXYLAMINE				NR at 70°F	NR at 70°F	A to 70°F
CYCLOPENTANE	NR at 70°F	NR at 70°F	NR at 70°F			A to 70°F
CYCLOPENTANOL	NR at 70°F	NR at 70°F	NR at 70°F			
CYCLOPENTANONE	NR at 70°F	NR at 70°F	NR at 70°F			
CYMENE	NR at 70°F	NR at 70°F	NR at 70°F		NR at 70°F	
P-CYMENE	NR at 70°F	NR at 70°F	NR at 70°F	B at 70°F	NR at 70°F	A to 70°F
D.B.T. (DIBUTYL PHTHALATE)	NR at 70°F	NR at 70°F	NR at 70°F	C at 70°F	B at 70°F	A to 70°F
DC 200, 510, 550 & 710 SILICONE OILS	AC to 70°F	AC to 70°F	A to 70°F	no swell 3 days 300°F A to 70°F	+10% vol 3 days 300°F C/NR at 70°F	A to 70°F
DDT IN DEIONIZED KEROSENE	NR at 70°F	NR at 70°F	NR at 70°F	A to 70°F	NR at 70°F	
DDT IN TOLUENE				A to 70°F	NR at 70°F	
DMF (DIMETHYL FORMAMIDE)	NR at 70°F	NR at 70°F	NR at 70°F	NR at 70°F	+20% vol 7 days 70°F BC at 70°F	*B at 70°F
DMP (DIMETHYL PHTHALATE)	NR at 70°F	NR at 70°F	NR at 70°F	B at 70°F	C at 70°F	A to 70°F
DOP (DIOCTYL PHTHALATE)	NR at 70°F	NR at 70°F	NR at 70°F	B at 70°F	C at 70°F	A to 70°F
DANFORTHS OIL	NR at 70°F	NR at 70°F	C at 70°F		C at 70°F	
DEAD OIL					NR at 70°F	
DECAHYDRONAPHTHALENE (DECALIN)	NR at 70°F	NR at 70°F	NR at 70°F	A to 70°F	NR at 70°F	
DECANAL	NR at 70°F	NR at 70°F	NR at 70°F			
DECANE	NR at 70°F	NR at 70°F	C/NR at 70°F	A to 70°F	B at 70°F	A to 70°F
DECANOL (DECYL ALCOHOL)	B at 70°F	B/NR at 70°F	A to 70°F			
DECYL ALDEHYDE	NR at 70°F	NR at 70°F	NR at 70°F			
DECYL BUTYL PHTHALATE	NR at 70°F	NR at 70°F	NR at 70°F			
DEGREASING FLUID	NR at 70°F	NR at 70°F	NR at 70°F			
DEHYDRATED ALCOHOL	A to 70°F	A to 70°F	A to 70°F		A to 70°F	

	POLYSULFIDE T	CHLORINATED POLYETHYLENE CM	EPICHLOROHYDRIN CO, ECO	® HYTREL COPOLYESTER TPE	® SANTOPRENE COPOLYMER TPO	® C-FLEX STYRENIC TPE
CUPRIC CHLORIDE						
CUPRIC CYANIDE						
CUPRIC FLUORIDE						
CUPRIC HYDROXIDE						
CUPRIC NITRATE						
CUPRIC NITRITE						
CUPRIC SULFATE				NR at 70°F		
CUPRIC SULFIDE						
CUPRIC SULPHATE				NR at 70°F		
CUPROUS AMMONIA ACETATE						
CUPROUS AMMONIUM CHLORIDE						
CUTTING OIL	A to 70°F					
CYANIC ACID SOULTIONS						
CYANIC COMPOUNDS						
CYANIDE						
CYANOGEN						
CYANOMETHANE						
CYCLOHEXANE	A to 70°F			A to 70°F	+20% to 62% vol 70°F C/NR at 70°F	NR at 70°F
CYCLOHEXANOL	B at 70°F				NR at 70°F	
CYCLOHEXANONE					NR at 70°F	NR at 70°F
CYCLOHEXATRIENE	B at 70°F		NR at 70°F	AB to 70°F		
CYCLOHEXYLAMINE						
CYCLOPENTANE						
CYCLOPENTANOL						
CYCLOPENTANONE						
CYMENE						
P-CYMENE	NR at 70°F					
D.B.T. (DIBUTYL PHTHALATE)	AB to 70°F		B at 70°F	+1-12% vol 7 days 70°F AB to 70°F	*A to 70°F	
DC 200, 510, 550 & 710 SILICONE OILS	softened 3 days 300°F A to 70°F		A to 70°F	AB to 70°F		A to 70°F
DDT IN DEIONIZED KEROSENE	AB to 70°F					
DDT IN TOLUENE						
DMF (DIMETHYL FORMAMIDE)	B at 70°F			+17% vol 7 days 70°F B at 70°F	+1% vol 3 days 70°F A to 70°F	AB to 70°F
DMP (DIMETHYL PHTHALATE)	B at 70°F				+10% vol 3 days 70°F AB to 70°F	
DOP (DIOCTYL PHTHALATE)	B at 70°F		B at 70°F	+3% vol 7 days 70°F A to 70°F	*A to 70°F	
DANFORTHS OIL						
DEAD OIL						
DECAHYDRONAPHTHALENE (DECALIN)	B at 70°F				+30-80% vol 70°F C/NR at 70°F	
DECANAL						
DECANE	A to 70°F				*AB to 70°F	
DECANOL (DECYL ALCOHOL)						
DECYL ALDEHYDE						
DECYL BUTYL PHTHALATE						
DEGREASING FLUID						
DEHYDRATED ALCOHOL						

	ETHYLENE ACRYLIC EA	POLYALLOMER LINEAR COPOLYMER	NYLON 11 POLYAMIDE	NYLON 12 POLYAMIDE	ETHYLENE VINYL ACETATE EVA	POLYVINYLCHLORIDE FLEXIBLE PVC
CUPRIC CHLORIDE						
CUPRIC CYANIDE						
CUPRIC FLUORIDE						
CUPRIC HYDROXIDE						
CUPRIC NITRATE						
CUPRIC NITRITE						
CUPRIC SULFATE			NR at 70°F			A to 70°F
CUPRIC SULFIDE						
CUPRIC SULPHATE			NR at 70°F			A to 70°F
CUPROUS AMMONIA ACETATE						
CUPROUS AMMONIUM CHLORIDE						
CUTTING OIL						
CYANIC ACID SOULTIONS						
CYANIC COMPOUNDS						
CYANIDE					A to 70°F *AB to 140°F	A to 70°F *AB to 140°F
CYANOGEN					A to 70°F	B at 70°F
CYANOMETHANE						
CYCLOHEXANE	BC at 70°F	C at 70°F NR at 122°F	AB to 105°F C at 140°F	AB to 70°F	NR at 70°F	C/NR at 70°F
CYCLOHEXANOL	A to 70°F	A to 70°F	AB to 70°F NR at 140°F	AB to 70°F	NR at 70°F	*NR at 70°F
CYCLOHEXANONE		B at 70°F	AB to 70°F NR at 140°F	AB to 70°F	NR at 70°F	NR at 70°F
CYCLOHEXATRIENE						
CYCLOHEXYLAMINE						
CYCLOPENTANE						
CYCLOPENTANOL						
CYCLOPENTANONE						
CYMENE						NR at 70°F
P-CYMENE						
D.B.T. (DIBUTYL PHTHALATE)	NR at 70°F	A to 70°F	A to 70°F		C/NR at 70°F NR at 140°F	NR at 70°F
DC 200, 510, 550 & 710 SILICONE OILS	A to 70°F		A to 70°F	AB to 70°F	B at 70°F	AB to 70°F
DDT IN DEIONIZED KEROSENE			A to 70°F			A to 70°F C at 150°F
DDT IN TOLUENE			A to 70°F			
DMF (DIMETHYL FORMAMIDE)		A to 122°F	AB to 70°F			C/NR at 70°F NR at 122°F
DMP (DIMETHYL PHTHALATE)	NR at 70°F					
DOP (DIOCTYL PHTHALATE)	NR at 70°F		A to 70°F AB to 140°F		*C at 70°F *NR at 140°F	C/NR at 70°F
DANFORTHS OIL						
DEAD OIL						
DECAHYDRONAPHTHALENE (DECALIN)		B at 70°F C at 122°F	AB to 140°F C at 194°F	AB to 70°F		A to 70°F B at 122°F
DECANAL						
DECANE						
DECANOL (DECYL ALCOHOL)						NR at 70°F
DECYL ALDEHYDE						
DECYL BUTYL PHTHALATE						
DEGREASING FLUID						
DEHYDRATED ALCOHOL						

	POLYETHYLENE LOW DENSITY LDPE	®TEFLON FEP	®KALREZ PERFLUORINATED ELASTOMER	®FLUORAZ FLUORINATED COPOLYMER	®AFLAS FLUORINATED COPOLYMER	®NORPRENE COPOLYMER TPO
CUPRIC CHLORIDE		A to saturated to 300°F				
CUPRIC CYANIDE	A to 70°F	A to 70°F				
CUPRIC FLUORIDE	A to 70°F	A to 70°F				
CUPRIC HYDROXIDE		A to 70°F				
CUPRIC NITRATE	A to 70°F	A to 70°F				
CUPRIC NITRITE		A to 70°F				
CUPRIC SULFATE	A to 70°F	A to 70°F		BC at 70°F B aqueous at 70°F	+0% vol 7 days 212°F A to 212°F	
CUPRIC SULFIDE		A to 70°F				
CUPRIC SULPHATE	A to 70°F	A to 70°F		BC at 70°F B aqueous at 70°F		
CUPROUS AMMONIA ACETATE						
CUPROUS AMMONIUM CHLORIDE						
CUTTING OIL		A to 70°F	A to 212°F	A to 70°F		
CYANIC ACID SOULTIONS		A to 70°F				
CYANIC COMPOUNDS						
CYANIDE	A to 70°F AB to 140°F		A to 212°F	NR at 70°F		
CYANOGEN	A to 70°F		A to 212°F	NR at 70°F in H$_2$O		
CYANOMETHANE	NR at 70°F	A to 70°F				
CYCLOHEXANE	C at 70°F NR at 122°F	A to 300°F	+1% vol 7 days 70°F A to 212°F	A to 70°F	+13% vol 7 days 70°F AB to 70°F	NR at 70°F
CYCLOHEXANOL	C at 70°F	A to 200°F	A to 212°F	A to 70°F		
CYCLOHEXANONE	C/NR at 70°F	A to 312°F	A to 212°F	C at 70°F	+22% vol 7 days 70°F B at 70°F	
CYCLOHEXATRIENE		A to 70°F				
CYCLOHEXYLAMINE			*A to 212°F			
CYCLOPENTANE		A to 70°F	A to 212°F			
CYCLOPENTANOL						
CYCLOPENTANONE		A to 70°F				
CYMENE		A to 70°F				
P-CYMENE			A to 212°F	C at 70°F		
D.B.T. (DIBUTYL PHTHALATE)	AC to 70°F NR at 140°F	A to 300°F	A to 212°F	BC at 70°F		
DC 200, 510, 550 & 710 SILICONE OILS	A to 70°F	A to 300°F	A to 70°F	A to 70°F		TEST
DDT IN DEIONIZED KEROSENE		A to 70°F	A to 200°F	NR at 70°F		
DDT IN TOLUENE		A to 70°F		NR at 70°F		
DMF (DIMETHYL FORMAMIDE)	A to 122°F	A to 309°F	A to 212°F	NR at 70°F		
DMP (DIMETHYL PHTHALATE)		A to 392°F	A to 212°F	C at 70°F		
DOP (DIOCTYL PHTHALATE)	C/NR at 70°F NR at 140°F	A to 300°F	A to 212°F	A to 70°F		
DANFORTHS OIL		A to 70°F				
DEAD OIL						
DECAHYDRONAPHTHALENE (DECALIN)	B at 70°F C at 122°F	A to 122°F	A to 212°F	B at 70°F		
DECANAL		A to 70°F				
DECANE		A to 70°F	A to 212°F	A to 70°F		
DECANOL (DECYL ALCOHOL)		A to 70°F				
DECYL ALDEHYDE		A to 70°F				
DECYL BUTYL PHTHALATE		A to 70°F				
DEGREASING FLUID		A to 70°F				
DEHYDRATED ALCOHOL		A to 70°F				

	RECOMMENDATIONS	NITRILE NBR	ETHYLENE PROPYLENE EP, EPDM	FLUOROCARBON FKM	CHLOROPRENE CR	HYDROGENATED NITRILE HNBR
DEIONIZED WATER	NBR,EP,CR,IIR	A to 70°F AB to 200°F	A to 70°F AB to 200°F	A to 70°F AB to 200°F	A to 70°F AB to 150°F	
DELCO BRAKE FLUID	EP,SBR	C at 70°F	A at 70°F	NR at 70°F	B at 70°F	C at 70°F
DENATURED ALCOHOL	NBR,EP,CR,FKM,IIR	A to 70°F AB to 180°F	A to 70°F	A to 70°F	A to 70°F AB to 150°F	A to 70°F
DETERGENT SOLUTIONS	NBR,EP,IIR,FKM	+3% vol 3 days 212°F A to 200°F	+22% vol 3 days 212°F A to 200°F	no swell 3 days 212°F A to 212°F	+7% vol 3 days 212°F AB to 212°F	A to 70°F
DEVELOPING FLUIDS	NBR,FKM,CR,EP,IIR	A to 70°F AB any conc to 200°F	AB to 104°F	A to 140°F	A to 140°F AB any conc to 200°F	A to 70°F
DEXTRIN		A to 140°F AB at 176°F	A to 176°F	A to 212°F	A to 70°F	
DEXTRON	NBR,FKM,ACM	A to 70°F	NR at 70°F	A to 70°F	B at 70°F	A to 70°F
DEXTRONIC ACID		C at 70°F				
DEXTROSE	FKM,NBR,EP	A to conc to 200°F	A to conc to 176°F B at 212°F	A to 212°F	AB any conc to 140°F	A to 70°F
DIACETIC ACID		NR at 70°F			NR at 70°F	
DIACETIC ESTER		NR at 70°F			NR at 70°F	
DIACETIC ETHER		NR at 70°F			NR at 70°F	
DIACETONE	EP,IIR	NR at 70°F	A to 70°F	NR at 70°F	NR at 70°F	
DIACETONE ALCOHOL (DIACETOL)	EP,IIR	+119% vol 3 days 158°F NR at 70°F	−1% vol 3 days 158°F A to 160°F	+178% vol 3 days 158°F NR at 70°F	+27% vol 3 days 158°F A/NR at 70°F	NR at 70°F
DIACETYLACETIC ACID		NR at 70°F			NR at 70°F	
DIAMINE (DIAMIDOGEN)		B at 70°F	A to 70°F	C at 70°F	C at 70°F	
DIAMINOETHANE	EP,NBR,CR,IIR	AB any conc to 80°F	A to 120°F	NR at 70°F	A any conc to 80°F	A to 80°F
DIAMMONIUM ORTHOPHOSPHATE		A to 70°F	A to 70°F	A to 70°F	A to 70°F	
DIAMYLAMINE (DIPENTYLAMINE)	EP	AB to 70°F	A to 200°F	NR at 70°F	A/NR at 70°F	
DIAMYL NAPHTHYLENE		NR at 70°F		C at 70°F		
DIAMYL PHENOL		NR at 70°F		A to 70°F		
DIAMYLENE		C at 70°F	NR at 70°F	A to 70°F	NR at 70°F	
DIATOL		NR at 70°F			NR at 70°F	
DIAZON (INSECTICIDE)	FKM	C at 70°F	NR at 70°F	B at 70°F	C at 70°F	
DIBENZYL ETHER	T,EP,IIR	NR at 70°F	BC at 70°F	NR at 70°F	NR at 70°F	NR at 70°F
DIBENZYL SEBACATE	FKM,EP,IIR	NR at 70°F	B at 70°F	B at 70°F	NR at 70°F	NR at 70°F
DIBROMOBENZENE		NR at 70°F		A to 70°F		
1, 2, DIBROMOMETHANE	FKM	NR at 70°F	C at 70°F	AB to 200°F	NR at 70°F	
DIBROMOETHYL BENZENE	FKM,FVMQ	NR at 70°F	NR at 70°F	A to 70°F	NR at 70°F	NR at 70°F
DIBUTYL		A to 70°F	NR at 70°F	A to 70°F		
DIBUTYL ACETATE					NR at 70°F	
DIBUTYLAMINE	EP,CR,IIR	NR at 70°F	NR at 70°F	NR at 70°F	C/NR at 70°F	
DIBUTYL CARBITOL				AB to 70°F		
DIBUTYL CELLOSOLVE ADIPATE	EP	NR at 70°F	A to 200°F	NR at 70°F	NR at 70°F	
DIBUTYL ETHER	T	B/NR to 150°F	C/NR at 70°F	C/NR at 70°F	B/NR at 70°F	NR at 70°F
DIBUTYL PHTHALATE	EP,T,IIR	NR at 70°F	A to 250°F	BC to 104°F NR at 140°F	NR at 70°F	NR at 70°F
DIBUTYL SEBACATE	EP,T,IIR	NR at 70°F	B at 70°F	A to 75°F B 158-250°F	NR at 70°F	NR at 70°F
DICALCIUM PHOSPHATE		A to 70°F		A to 70°F		
DICHLOROACETIC ACID		NR at 70°F		NR at 70°F		
DICHLOROANILINE		NR at 70°F	NR at 70°F	BC at 70°F	NR at 70°F	
O-DICHLOROBENZENE	FKM,FVMQ	NR at 70°F	NR at 70°F	+10.5% vol 28 days 158°F +25% vol 28 days 300°F	NR at 70°F	
P-DICHLOROBENZENE	FKM,T	NR at 70°F	NR at 70°F	A to 200°F	NR at 70°F	
DICHLOROBUTANE	FKM,NBR	B at 70°F	NR at 70°F	A to 70°F	NR at 70°F	B at 70°F
DICHLORO DIFLUORO METHANE (F-12)	CR,NBR,FKM,T,SBR	A to 250°F dry C at 70°F wet	BC at 70°F dry AB to 130°F wet	BC to 130°F dry B to 130°F wet	A to 250°F dry BC at 70°F wet	A to 70°F

	STYRENE BUTADIENE SBR	POLYACRYLATE ACM	POLYURETHANE AU, EU	ISOBUTYLENE ISOPRENE IIR	POLYBUTADIENE BR	®AEROQUIP AQP
DEIONIZED WATER				A to 70°F AB to 150°F		
DELCO BRAKE FLUID	A to 70°F			B at 70°F		
DENATURED ALCOHOL	A to 70°F	NR at 70°F	NR at 70°F	A to 70°F AB to 150°F	A to 70°F	
DETERGENT SOLUTIONS	+4% vol 3 days 212°F AB to 70°F	+31% vol 3 days 212°F NR at 70°F	+15% vol 3 days 212°F AB to 70°F	no swell 3 days 212°F A to 70°F	B at 70°F	
DEVELOPING FLUIDS	B at 70°F		A/NR at 70°F	B any conc to 185°F	B at 70°F	
DEXTRIN			A to 70°F			
DEXTRON	NR at 70°F	A to 70°F	B at 70°F	NR at 70°F	NR at 70°F	
DEXTRONIC ACID				C at 70°F		
DEXTROSE	B at 70°F	NR at 70°F	A to 70°F	A to 70°F AB any conc to 185°F		
DIACETIC ACID				B to 80°F		
DIACETIC ESTER				B to 80°F		
DIACETIC ETHER				B at 70°F		
DIACETONE	NR at 70°F	NR at 70°F	NR at 70°F	A to 70°F	NR at 70°F	
DIACETONE ALCOHOL (DIACETOL)	+11% vol 3 days 158°F C/NR at 70°F	+198% vol 3 days 158°F NR at 70°F	+131% vol 3 days 158°F NR at 70°F	+0.8% vol 3 days 158°F A to 70°F	NR at 70°F	
DIACETYLACETIC ACID				B at 70°F		
DIAMINE (DIAMIDOGEN)			NR at 70°F	A any conc to 100°F		
DIAMINOETHANE	B at 70°F	NR at 70°F	NR at 70°F	A any conc to 100°F	B at 70°F	
DIAMMONIUM ORTHOPHOSPHATE			NR at 70°F	A to 70°F		
DIAMYLAMINE (DIPENTYLAMINE)	NR at 70°F	NR at 70°F	NR at 70°F	A to 70°F		
DIAMYL NAPHTHYLENE				A to 70°F		
DIAMYL PHENOL				NR at 70°F		
DIAMYLENE				NR at 70°F		
DIATOL				NR at 70°F		
DIAZON (INSECTICIDE)	NR at 70°F		NR at 70°F	NR at 70°F	NR at 70°F	
DIBENZYL ETHER	NR at 70°F		B at 70°F	B to 75°F	NR at 70°F	
DIBENZYL SEBACATE	NR at 70°F	NR at 70°F	B at 70°F	B at 70°F	NR at 70°F	
DIBROMOBENZENE				NR at 70°F		
1, 2, DIBROMOMETHANE				NR at 70°F		
DIBROMOETHYL BENZENE	NR at 70°F	NR at 70°F	NR at 70°F	NR at 70°F	NR at 70°F	
DIBUTYL				NR at 70°F		
DIBUTYL ACETATE						
DIBUTYLAMINE	NR at 70°F	NR at 70°F	NR at 70°F	NR at 70°F	NR at 70°F	
DIBUTYL CARBITOL				NR at 70°F		
DIBUTYL CELLOSOLVE ADIPATE	NR at 70°F	NR at 70°F	NR at 70°F	A to 70°F		
DIBUTYL ETHER	NR at 70°F	C at 70°F	B/NR at 70°F	C/NR at 70°F	NR at 70°F	
DIBUTYL PHTHALATE	NR at 70°F	NR at 70°F	C/NR at 70°F NR at 158°F	BC any conc to 75°F	NR at 70°F	
DIBUTYL SEBACATE	NR at 70°F	NR at 70°F	NR at 70°F	B at 70°F	NR at 70°F	
DICALCIUM PHOSPHATE				A to 70°F		
DICHLOROACETIC ACID				C at 70°F		
DICHLOROANILINE	NR at 70°F	NR at 70°F	NR at 70°F	NR at 70°F	NR at 70°F	
O-DICHLOROBENZENE	NR at 70°F	NR at 70°F	NR at 70°F	NR at 70°F	NR at 70°F	
P-DICHLOROBENZENE	NR at 70°F	NR at 70°F	NR at 70°F	NR at 70°F	NR at 70°F	
DICHCLOROBUTANE	NR at 70°F	NR at 70°F	NR at 70°F	NR at 70°F	NR at 70°F	
DICHLORO DIFLUORO METHANE (F-12)	A to 70°F		A to 130°F	B/NR at 70°F	NR at 70°F	NR at 70°F

	SYNTHETIC ISOPRENE IR	NATURAL ISOPRENE NR	CHLOROSULFONATED POLYETHYLENE CSM	FLUOROSILICONE FVMQ	SILICONE VMQ	CHEMRAZ FFKM
DEIONIZED WATER	A to 70°F	A to 70°F AB to 150°F	A to 70°F AB to 200°F			A to 70°F NR at 70°F white
DELCO BRAKE FLUID			B at 70°F	NR at 70°F	C at 70°F	A to 70°F
DENATURED ALCOHOL	A to 70°F	A to 70°F AB to 150°F	A to 70°F	+3% vol 3 days 212°F A to 70°F	+2% vol 3 days 212°F A to 70°F	A to 70°F
DETERGENT SOLUTIONS						A to 70°F
DEVELOPING FLUIDS	A to 70°F	A to 70°F AB any conc to 150°F	A to 70°F	A to 70°F	AB to 70°F	A to 70°F
DEXTRIN		A to 70°F	A to 70°F			A to 70°F
DEXTRON	NR at 70°F	NR at 70°F	NR at 70°F	B at 70°F	NR at 70°F	A to 70°F
DEXTRONIC ACID	NR at 70°F	NR at 70°F	B at 70°F			
DEXTROSE	AB to 70°F	A to 70°F AB any conc to 120°F	A to 158°F	A to 70°F	A to 70°F	A to 70°F
DIACETIC ACID	B at 70°F	B any conc to 80°F				
DIACETIC ESTER	B at 70°F	B any conc to 80°F				
DIACETIC ETHER	B at 70°F	B at 70°F				
DIACETONE	C/NR at 70°F	C/NR at 70°F	NR at 70°F	NR at 70°F	NR at 70°F	A to 70°F
DIACETONE ALCOHOL (DIACETOL)	B/NR at 70°F	B/NR at 70°F	A/NR at 70°F	+49% vol 3 days 158°F NR at 70°F	+6% vol 3 days 158°F A/NR at 70°F	A to 70°F
DIACETYLACETIC ACID	B at 70°F	B at 70°F				
DIAMINE (DIAMIDOGEN)			B at 70°F		C at 70°F	
DIAMINOETHANE	AB to 70°F	AB any conc to 80°F	AB to 70°F	NR at 70°F	A to 70°F	B at 70°F
DIAMMONIUM ORTHOPHOSPHATE	A to 70°F	A to 70°F	A to 70°F		A to 70°F	
DIAMYLAMINE (DIPENTYLAMINE)	B at 70°F	B at 70°F	C at 70°F	NR at 70°F	NR at 70°F	A to 70°F
DIAMYL NAPHTHYLENE	NR at 70°F	NR at 70°F	NR at 70°F			
DIAMYL PHENOL	NR at 70°F	NR at 70°F	NR at 70°F			
DIAMYLENE	NR at 70°F	NR at 70°F	NR at 70°F			
DIATOL	NR at 70°F	NR at 70°F				
DIAZON (INSECTICIDE)	NR at 70°F	NR at 70°F	C at 70°F	B at 70°F	NR at 70°F	A to 70°F
DIBENZYL ETHER	NR at 70°F	NR at 70°F	NR at 70°F			A to 70°F
DIBENZYL SEBACATE	NR at 70°F	NR at 70°F	NR at 70°F	C at 70°F	C at 70°F	A to 70°F
DIBROMOBENZENE	NR at 70°F	NR at 70°F	NR at 70°F			
1, 2, DIBROMOMETHANE	NR at 70°F	NR at 70°F	NR at 70°F	C at 70°F	NR at 70°F	A to 70°F
DIBROMOETHYL BENZENE	NR at 70°F	NR at 70°F	NR at 70°F	B at 70°F	NR at 70°F	A to 70°F
DIBUTYL	NR at 70°F	NR at 70°F	NR at 70°F		NR at 70°F	
DIBUTYL ACETATE						
DIBUTYLAMINE	NR at 70°F	NR at 70°F	C/NR at 70°F	NR at 70°F	C at 70°F	A to 70°F
DIBUTYL CARBITOL					NR at 70°F	
DIBUTYL CELLOSOLVE ADIPATE				NR at 70°F	NR at 70°F	A to 70°F
DIBUTYL ETHER	NR at 70°F	NR at 70°F	C/NR at 70°F	C at 70°F	NR at 70°F	A to 70°F
DIBUTYL PHTHALATE	NR at 70°F	NR at 70°F	NR at 70°F	AC to 70°F	B at 70°F	A to 70°F
DIBUTYL SEBACATE	NR at 70°F	NR at 70°F	NR at 70°F	B at 70°F	B at 70°F	A to 70°F
DICALCIUM PHOSPHATE	A to 70°F	A to 70°F	A to 70°F			
DICHLOROACETIC ACID	B at 70°F	B at 70°F	NR at 70°F			A to 70°F
DICHLOROANILINE					NR at 70°F	A to 70°F
O-DICHLOROBENZENE	NR at 70°F	NR at 70°F	NR at 70°F	B at 70°F	NR at 70°F	A to 70°F
P-DICHLOROBENZENE	NR at 70°F	NR at 70°F	NR at 70°F	B at 70°F	NR at 70°F	A to 70°F
DICHLOROBUTANE	NR at 70°F	NR at 70°F	NR at 70°F	B at 70°F	NR at 70°F	A to 70°F
DICHLORO DIFLUORO METHANE (F-12)	NR at 70°F	B/NR at 70°F (dry) NR at 70°F (wet)	A to 130°F	C/NR at 70°F	NR at 70°F (wet or dry)	AB to 70°F

	POLYSULFIDE T	CHLORINATED POLYETHYLENE CM	EPICHLOROHYDRIN CO, ECO	® HYTREL COPOLYESTER TPE	® SANTOPRENE COPOLYMER TPO	® C-FLEX STYRENIC TPE
DEIONIZED WATER						
DELCO BRAKE FLUID	NR at 70°F					
DENATURED ALCOHOL	A to 70°F		A to 70°F			
DETERGENT SOLUTIONS	+11.4% vol 3 days 212°F A to 70°F		A to 70°F	*AB to 140°F		
DEVELOPING FLUIDS	A to 70°F			NR at 70°F		
DEXTRIN				*AB to 70°F		
DEXTRON						
DEXTRONIC ACID						
DEXTROSE				*AB to 140°F		
DIACETIC ACID						
DIACETIC ESTER						
DIACETIC ETHER						
DIACETONE	B at 70°F					
DIACETONE ALCOHOL (DIACETOL)	+34.7% vol 3 days 158°F B at 70°F		NR at 70°F	C at 70°F		AB to 70°F
DIACETYLACETIC ACID						
DIAMINE (DIAMIDOGEN)						
DIAMINOETHANE	NR at 70°F		A to 70°F			
DIAMMONIUM ORTHOPHOSPHATE				C at 70°F		
DIAMYLAMINE (DIPENTYLAMINE)						
DIAMYL NAPHTHYLENE						
DIAMYL PHENOL						
DIAMYLENE						
DIATOL						
DIAZON (INSECTICIDE)				NR at 70°F		
DIBENZYL ETHER	B at 70°F		NR at 70°F			
DIBENZYL SEBACATE	B at 70°F					
DIBROMOBENZENE						
1, 2, DIBROMOMETHANE					*B/NR at 70°F	
DIBROMOETHYL BENZENE	NR at 70°F					
DIBUTYL						
DIBUTYL ACETATE						
DIBUTYLAMINE	NR at 70°F					
DIBUTYL CARBITOL						
DIBUTYL CELLOSOLVE ADIPATE						
DIBUTYL ETHER	A to 120°F				*A to 70°F	
DIBUTYL PHTHALATE	AB to 70°F		B at 70°F	+1-12% vol 7 days 70°F AB to 70°F	*A to 70°F	
DIBUTYL SEBACATE	B at 70°F					
DICALCIUM PHOSPHATE						
DICHLOROACETIC ACID						
DICHLOROANILINE						
O-DICHLOROBENZENE	A/NR at 70°F			NR at 70°F	NR at 70°F	
P-DICHLOROBENZENE	A to 70°F			NR at 70°F	NR at 70°F	
DICHCLOROBUTANE	C at 70°F					
DICHLORO DIFLUORO METHANE (F-12)	A to 70°F	C at 70°F	A to 70°F	A to 70°F	NR at 70°F	

	ETHYLENE ACRYLIC EA	POLYALLOMER LINEAR COPOLYMER	NYLON 11 POLYAMIDE	NYLON 12 POLYAMIDE	ETHYLENE VINYL ACETATE EVA	POLYVINYLCHLORIDE FLEXIBLE PVC
DEIONIZED WATER						
DELCO BRAKE FLUID	NR at 70°F					
DENATURED ALCOHOL	A to 70°F					
DETERGENT SOLUTIONS	A to 70°F			AB to 70°F	A to 70°F *AB to 140°F	AB to 140°F
DEVELOPING FLUIDS					A to 70°F	*AB to 140°F
DEXTRIN			*AB to 140°F	AB to 70°F aqueous	AB to 140°F	AB to 70°F *AB to 140°F
DEXTRON	A to 70°F					
DEXTRONIC ACID						
DEXTROSE			*AB to 140°F		AB to 140°F	*AB to 140°F
DIACETIC ACID						NR at 70°F
DIACETIC ESTER						NR at 70°F
DIACETIC ETHER						
DIACETONE	NR at 70°F					
DIACETONE ALCOHOL (DIACETOL)			AB to 105°F NR at 194°F		C at 70°F	NR at 70°F
DIACETYLACETIC ACID						
DIAMINE (DIAMIDOGEN)						
DIAMINOETHANE					C at 70°F	C at 70°F
DIAMMONIUM ORTHOPHOSPHATE			AB to 105°F C at 140°F			AB to 70°F
DIAMYLAMINE (DIPENTYLAMINE)						
DIAMYL NAPHTHYLENE						
DIAMYL PHENOL						
DIAMYLENE			AB to 70°F			
DIATOL						
DIAZON (INSECTICIDE)						
DIBENZYL ETHER						NR at 70°F
DIBENZYL SEBACATE	NR at 70°F					
DIBROMOBENZENE						
1, 2, DIBROMOMETHANE						NR at 70°F
DIBROMOETHYL BENZENE						
DIBUTYL						
DIBUTYL ACETATE						
DIBUTYLAMINE						NR at 70°F
DIBUTYL CARBITOL						NR at 70°F
DIBUTYL CELLOSOLVE ADIPATE						
DIBUTYL ETHER						
DIBUTYL PHTHALATE	NR at 70°F	A to 70°F	A to 70°F	AB to 70°F	C/NR at 70°F NR at 140°F	NR at 70°F
DIBUTYL SEBACATE	NR at 70°F					
DICALCIUM PHOSPHATE						
DICHLOROACETIC ACID						
DICHLOROANILINE						
O-DICHLOROBENZENE		C at 70°F NR at 122°F	+0.7% vol 180 days 70°F A to 70°F			NR at 70°F
P-DICHLOROBENZENE		B at 70°F C at 122°F	A to 70°F			NR at 70°F
DICHCLOROBUTANE						
DICHLORO DIFLUORO METHANE (F-12)	AC to 70°F		AB to 70°F			B/NR at 70°F (dry) NR at 70°F (wet)

	POLYETHYLENE LOW DENSITY LDPE	® TEFLON FEP	® KALREZ PERFLUORINATED ELASTOMER	® FLUORAZ FLUORINATED COPOLYMER	® AFLAS FLUORINATED COPOLYMER	® NORPRENE COPOLYMER TPO
DEIONIZED WATER	A to 70°F	A to 212°F	*A to 212°F			
DELCO BRAKE FLUID				A to 70°F		
DENATURED ALCOHOL		A to 70°F	A to 212°F	A to 70°F		
DETERGENT SOLUTIONS	AB to 140°F	A to 300°F	A to 212°F	A to 70°F		
DEVELOPING FLUIDS	A to 104°F B at 140°F	A to 300°F	A to 212°F	AC to 70°F		
DEXTRIN	A to 140°F	A to 70°F	A to 212°F	A to 70°F		
DEXTRON				A to 70°F	+13% vol 3 days 350°F B at 350°F	
DEXTRONIC ACID		A to 70°F				
DEXTROSE	A to 70°F AB to 140°F	A to 300°F	A to 212°F	A to 70°F		
DIACETIC ACID		A to 70°F				
DIACETIC ESTER		A to 70°F				
DIACETIC ETHER		A to 70°F				
DIACETONE		A to 70°F	A to 212°F	NR at 70°F		
DIACETONE ALCOHOL (DIACETOL)	A to 70°F	A to 200°F	A to 212°F	NR at 70°F		
DIACETYLACETIC ACID		A to 70°F				
DIAMINE (DIAMIDOGEN)		A to 70°F				
DIAMINOETHANE	A to 70°F	A to 242°F	A to 104°F B at 203°F	AB to 70°F		
DIAMMONIUM ORTHOPHOSPHATE	AB to 70°F	A to 70°F				
DIAMYLAMINE (DIPENTYLAMINE)		A to 70°F	*A to 212°F			
DIAMYL NAPHTHYLENE		A to 70°F				
DIAMYL PHENOL		A to 70°F				
DIAMYLENE		A to 70°F				
DIATOL		A to 70°F				
DIAZON (INSECTICIDE)			A to 212°F	B at 70°F		
DIBENZYL ETHER		A to 70°F	A to 212°F	C at 70°F		
DIBENZYL SEBACATE		A to 70°F	A to 212°F	A to 70°F		
DIBROMOBENZENE		A to 70°F				
1, 2, DIBROMOMETHANE		A to 300°F		B at 70°F		
DIBROMOETHYL BENZENE			A to 212°F	NR at 70°F		
DIBUTYL		A to 70°F				
DIBUTYL ACETATE		A to 70°F				
DIBUTYLAMINE		A to 70°F	A to 70°F *A to 212°F	B/NR at 70°F		
DIBUTYL CARBITOL						
DIBUTYL CELLOSOLVE ADIPATE			A to 212°F	NR at 70°F		
DIBUTYL ETHER		A to 300°F	A to 212°F	NR at 70°F		
DIBUTYL PHTHALATE	AC to 70°F NR at 140°F	A to 300°F	A to 212°F	B/NR at 70°F	+33% vol 7 days 302°F BC at 302°F	
DIBUTYL SEBACATE		A to boiling	A to 212°F	B at 70°F		
DICALCIUM PHOSPHATE		A to 70°F				
DICHLOROACETIC ACID		A to 70°F	A to 212°F			
DICHLOROANILINE			A to 338°F			
O-DICHLOROBENZENE	C at 70°F NR at 140°F	A to 125°F	A to 338°F	NR at 70°F		
P-DICHLOROBENZENE	C at 70°F NR at 122°F	A to 122°F		NR at 70°F		
DICHCLOROBUTANE		A to 70°F	A to 70°F	A to 70°F	A to 70°F	
DICHLORO DIFLUORO METHANE (F-12)	AB to 70°F (dry) NR at 140°F (dry)	A to 70°F (wet or dry)	BC at 70°F	NR al 70°F		

C 115

	RECOMMENDATIONS	NITRILE NBR	ETHYLENE PROPYLENE EP, EPDM	FLUOROCARBON FKM	CHLOROPRENE CR	HYDROGENATED NITRILE HNBR
DICHLOROETHANE	FKM	NR at 70°F	C/NR at 70°F	+16% vol 7 days 75°F AB to 200°F	NR at 70°F	
DICHLOROETHANOIC ACID		NR at 70°F		NR at 70°F		
DICHLOROETHER		NR at 70°F				
DICHLOROETHYLENE	FKM	NR at 70°F	B/NR at 70°F	AB to 160°F	NR at 70°F	
DICHLOROETHYL ETHER	T	NR at 70°F				
DICHLOROFLUOROMETHANE		NR at 70°F	NR at 70°F	NR at 70°F	NR at 70°F	
DICHLOROHEXANE		NR at 70°F		A to 70°F		
DICHLORO-ISOPROPYL ETHER	T	NR at 70°F	C at 70°F	C at 70°F	NR at 70°F	NR at 70°F
DICHLOROMETHANE	FKM, FVMQ	NR at 70°F	BC to 130°F NR at 140°F	+29% vol 28 days 100°F B at 70°F	NR at 70°F	
DICHLOROPENTANE	FKM	NR at 70°F	NR at 70°F	A to 200°F	NR at 70°F	
DICHLOROPROPANE	FKM	NR at 70°F	NR at 70°F	A to 300°F	NR at 70°F	
DICHLOROTETRAFLUOROETHANE (F-114)	CR, NBR, FKM	A to 250°F	NR at 70°F	A to 104°F	A to 250°F	A to 70°F
DICYCLOHEXYLAMINE	NBR	C/NR at 70°F	NR at 70°F	NR at 70°F	NR at 70°F	C at 70°F
DIELDRIN IN XYLENE		NR at 70°F	NR at 70°F	B at 70°F	NR at 70°F	
DIELDRIN IN XYLENE & WATER SPRAY		B at 70°F	NR at 70°F	B at 70°F	B at 70°F	
DIELINE		NR at 70°F		A to 70°F	NR at 70°F	
DIESEL OIL	NBR, FKM, FVMQ, ACM	A to 250°F	NR at 70°F	A to 200°F	B/NR at 70°F	A to 70°F
DIESTER LUBRICANT (MIL-L-7808)	FKM, FVMQ, NBR, ACM	AB to 70°F	NR at 70°F	+7.5% vol 7 days 350°F +20.5% vol 28 days 400°F	NR at 70°F	B at 70°F
DIESTER SYNTHETIC LUBRICATING OILS	FKM, FVMQ	B/NR at 70°F	NR at 70°F	A to 70°F	NR at 70°F	B at 70°F
DIETHANOLAMINE	EP, IIR	A/NR at 70°F	A to 160°F	NR at 70°F	A/NR at 70°F	
DIETHYLAMINE	EP, IIR, CR	BC to 140°F	B to 140°F	B/NR to 140°F	B any conc to 140°F	
DIETHYLANILINE	EP, IIR	NR at 70°F	A to 160°F	NR at 70°F	NR at 70°F	
DIETHYLBENZENE	FKM, FVMQ	NR at 70°F	NR at 70°F	A to 200°F	NR at 70°F	
DIETHYLCARBINOL		A to 70°F	A to 70°F	AB to 70°F	A to 70°F	
DIETHYL CARBONATE	FKM	NR at 70°F	A/NR at 70°F	A to 200°F	NR at 70°F	
DIETHYL ETHER	T	NR at 70°F	NR at 70°F	NR at 70°F	NR at 70°F	NR at 70°F
DIETHYL FORMALDEHYDE	EP	NR at 70°F	A to 200°F	NR at 70°F	NR at 70°F	
DIETHYL HYDRAZINE	EP	C at 70°F	A to 160°F	NR at 70°F	C at 70°F	
DIETHYL KETONE		NR at 70°F	B at 70°F	NR at 70°F	NR at 70°F	
DIETHYL MALEATE	EP	C at 70°F	A to 160°F	NR at 70°F	C at 70°F	
DIETHYL OXALATE		NR at 70°F	NR at 70°F		NR at 70°F	
DIETHYL OXIDE		B/NR at 70°F	NR at 70°F	NR at 70°F	C/NR at 70°F	
DIETHYL PHTHALATE		NR at 70°F		C at 70°F		
DIETHYL SEBACATE	FKM, EP, IIR	NR at 70°F	B at 70°F	B at 70°F	NR at 70°F	C at 70°F
DIETHYL SULFATE	EP, CR	NR at 70°F	A to 200°F	NR at 70°F	A to 70°F	
DIETHYL TRIAMINE		B at 70°F				
DIETHYL DIOXIDE		NR at 70°F	A to 70°F	NR at 70°F	NR at 70°F	
DIETHYLENE ETHER		NR at 70°F			NR at 70°F	
DIETHYLENE GLYCOL	NBR, EP, CR, IIR	A to 200°F A any conc to 180°F	A to 140°F	A to 140°F	A to 140°F AB any conc to 180°F	
DIETHYLENE GLYCOL-DIALKYL ETHER		A to 70°F AB any conc to 150°F			NR at 70°F	
DIETHYLENE GLYCOL-MONOALKYL ETHER		A to 70°F AB any conc to 140°F			NR at 70°F	
DIETHYLENE GLYCOL-MONOBUTYL ETHER		A to 70°F AB to 150°F	A to 70°F	A to 70°F	B at 70°F	
DIETHYLENE GLYCOL-MONOETHYL ETHER		B at 70°F	B at 70°F	A to 70°F	B at 70°F	
DIETHYLENE OXIDE		NR at 70°F	A to 70°F	NR at 70°F	NR at 70°F	

	STYRENE BUTADIENE SBR	POLYACRYLATE ACM	POLYURETHANE AU, EU	ISOBUTYLENE ISOPRENE IIR	POLYBUTADIENE BR	® AEROQUIP AQP
DICHLOROETHANE	NR at 70°F	NR at 70°F	NR at 70°F	BC to 75°F	NR at 70°F	NR at 70°F
DICHLOROETHANOIC ACID				C at 70°F		
DICHLOROETHER				NR at 70°F		
DICHLOROETHYLENE			C at 70°F	C at 70°F		
DICHLOROETHYL ETHER				NR at 70°F		
DICHLOROFLUOROMETHANE				C at 70°F		
DICHLOROHEXANE				NR at 70°F		
DICHLORO-ISOPROPYL ETHER	NR at 70°F	BC at 70°F	B to 70°F	C/NR at 70°F	NR at 70°F	
DICHLOROMETHANE	NR at 70°F	NR at 70°F	NR at 70°F	NR at 70°F	NR at 70°F	
DICHLOROPENTANE	NR at 70°F	NR at 70°F	NR at 70°F	NR at 70°F		
DICHLOROPROPANE				NR at 70°F		
DICHLOROTETRAFLUOROETHANE (F-114)	A to 70°F		A to 70°F	A/NR at 70°F	A to 70°F	
DICYCLOHEXYLAMINE	NR at 70°F	NR at 70°F	NR at 70°F	NR at 70°F	NR at 70°F	
DIELDRIN IN XYLENE						
DIELDRIN IN XYLENE & WATER SPRAY						
DIELINE				C at 70°F		
DIESEL OIL	NR at 70°F	AB to 70°F	AC to 140°F	NR at 70°F	NR at 70°F	AB to 70°F
DIESTER LUBRICANT (MIL-L-7808)	NR at 70°F	B at 70°F	NR at 70°F	NR at 70°F	NR at 70°F	
DIESTER SYNTHETIC LUBRICATING OILS	NR at 70°F	B at 70°F	B/NR at 70°F	NR at 70°F	NR at 70°F	
DIETHANOLAMINE	NR at 70°F	NR at 70°F	NR at 70°F C 20% at 70°F	A to 70°F		
DIETHYLAMINE	B at 70°F	NR at 70°F	C at 70°F	B any conc to 140°F	B at 70°F	
DIETHYLANILINE	NR at 70°F	NR at 70°F	A/NR at 70°F	A to 70°F		
DIETHYLBENZENE	NR at 70°F	NR at 70°F	NR at 70°F	NR at 70°F	NR at 70°F	
DIETHYLCARBINOL			C at 70°F	A to 70°F		
DIETHYL CARBONATE	NR at 70°F	NR at 70°F	NR at 70°F	NR at 70°F		
DIETHYL ETHER	NR at 70°F	C at 70°F	A to 70°F	NR at 70°F	NR at 70°F	
DIETHYL FORMALDEHYDE	NR at 70°F	NR at 70°F	NR at 70°F			
DIETHYL HYDRAZINE	B at 70°F	NR at 70°F	NR at 70°F			
DIETHYL KETONE				B at 70°F		
DIETHYL MALEATE	C at 70°F	NR at 70°F	NR at 70°F			
DIETHYL OXALATE				C/NR at 70°F		
DIETHYL OXIDE			A to 70°F	C/NR at 70°F		
DIETHYL PHTHALATE				A to 70°F		
DIETHYL SEBACATE	NR at 70°F	NR at 70°F	NR at 70°F	B at 70°F	NR at 70°F	
DIETHYL SULFATE	A to 70°F	NR at 70°F	NR at 70°F	B at 70°F		
DIETHYL TRIAMINE				A to 70°F		
DIETHYL DIOXIDE				A/NR at 70°F		
DIETHYLENE ETHER				NR at 70°F		
DIETHYLENE GLYCOL	A to 70°F	NR at 70°F	NR at 70°F	A to 70°F AB any conc to 180°F	A to 70°F	
DIETHYLENE GLYCOL-DIALKYL ETHER				A to 70°F AB any conc to 150°F		
DIETHYLENE GLYCOL-MONOALKYL ETHER				A to 70°F AB any conc to 150°F		
DIETHYLENE GLYCOL-MONOBUTYL ETHER				A to 70°F AB any conc to 180°F		
DIETHYLENE GLYCOL-MONOETHYL ETHER			NR at 70°F	A to 70°F		
DIETHYLENE OXIDE				A/NR at 70°F		

	SYNTHETIC ISOPRENE IR	NATURAL ISOPRENE NR	CHLOROSULFONATED POLYETHYLENE CSM	FLUOROSILICONE FVMQ	SILICONE VMQ	CHEMRAZ FFKM
DICHLOROETHANE	NR at 70°F	NR at 70°F	NR at 70°F	C at 70°F	NR at 70°F	A to 70°F
DICHLOROETHANOIC ACID	B at 70°F	B at 70°F	NR at 70°F			
DICHLOROETHER	NR at 70°F	NR at 70°F	NR at 70°F			
DICHLOROETHYLENE	NR at 70°F	NR at 70°F	NR at 70°F	NR at 70°F	C/NR at 70°F	A to 70°F
DICHLOROETHYL ETHER	NR at 70°F	NR at 70°F	NR at 70°F		NR at 70°F	A to 70°F
DICHLOROFLUOROMETHANE	NR at 70°F	NR at 70°F	NR at 70°F			A to 70°F
DICHLOROHEXANE	NR at 70°F	NR at 70°F	NR at 70°F			
DICHLORO-ISOPROPYL ETHER	NR at 70°F	NR at 70°F	NR at 70°F	C at 70°F	NR at 70°F	A to 70°F
DICHLOROMETHANE	NR at 70°F	NR at 70°F	NR at 70°F	+14% vol 120 days 70°F B at 70°F	-4% vol 120 days 70°F NR at 70°F	A to 70°F
DICHLOROPENTANE	NR at 70°F	NR at 70°F	NR at 70°F	C at 70°F	NR at 70°F	
DICHLOROPROPANE	NR at 70°F	NR at 70°F	NR at 70°F	B at 70°F	NR at 70°F	A to 70°F
DICHLOROTETRAFLUOROETHANE (F-114)	A/NR at 70°F	A/NR at 70°F	A to 70°F *AB to 130°F	B at 70°F	NR at 70°F	C at 70°F
DICYCLOHEXYLAMINE	NR at 70°F	NR at 70°F	NR at 70°F	NR at 70°F	NR at 70°F	A to 70°F
DIELDRIN IN XYLENE	NR at 70°F	NR at 70°F	NR at 70°F			
DIELDRIN IN XYLENE & WATER SPRAY	NR at 70°F	NR at 70°F	NR at 70°F			A to 70°F
DIELINE	NR at 70°F	NR at 70°F	NR at 70°F			
DIESEL OIL	NR at 70°F	NR at 70°F	BC to 185°F	A to 70°F	NR at 70°F	A to 70°F
DIESTER LUBRICANT (MIL-L-7808)	NR at 70°F	NR at 70°F	NR at 70°F	A to 300°F	NR at 70°F	A to 70°F
DIESTER SYNTHETIC LUBRICATING OILS	NR at 70°F	NR at 70°F	NR at 70°F	B at 70°F	NR at 70°F	A to 70°F
DIETHANOLAMINE	B at 70°F	B at 70°F	C at 70°F	NR at 70°F	NR at 70°F	A to 70°F
DIETHYLAMINE	B at 70°F	B any conc to 100°F	C at 70°F	NR at 70°F	B at 70°F	A to 70°F
DIETHYLANILINE				NR at 70°F	NR at 70°F	A to 70°F
DIETHYLBENZENE	NR at 70°F	NR at 70°F	NR at 70°F	AC to 70°F	NR at 70°F	A to 70°F
DIETHYLCARBINOL	A to 70°F	A to 70°F	A to 70°F		NR at 70°F	
DIETHYL CARBONATE	NR at 70°F	NR at 70°F		B at 70°F	NR at 70°F	A to 70°F
DIETHYL ETHER	NR at 70°F	NR at 70°F	NR at 70°F	NR at 70°F	NR at 70°F	A to 70°F
DIETHYL FORMALDEHYDE				NR at 70°F	NR at 70°F	
DIETHYL HYDRAZINE				NR at 70°F	NR at 70°F	
DIETHYL KETONE	NR at 70°F	NR at 70°F	NR at 70°F			
DIETHYL MALEATE				NR at 70°F	NR at 70°F	
DIETHYL OXALATE	C at 70°F	C/NR at 70°F	NR at 70°F			
DIETHYL OXIDE	NR at 70°F	NR at 70°F	C at 70°F		NR at 70°F	
DIETHYL PHTHALATE	NR at 70°F	NR at 70°F	NR at 70°F			A to 70°F
DIETHYL SEBACATE	NR at 70°F	NR at 70°F	BC at 70°F	B at 70°F	B at 70°F	A to 70°F
DIETHYL SULFATE	NR at 70°F	NR at 70°F	NR at 70°F	C at 70°F	A to 70°F	A to 70°F
DIETHYL TRIAMINE	B at 70°F	B at 70°F	C at 70°F			
DIETHYL DIOXIDE	NR at 70°F	NR at 70°F	NR at 70°F		NR at 70°F	
DIETHYLENE ETHER	NR at 70°F	NR at 70°F				
DIETHYLENE GLYCOL	A to 70°F	A to 70°F AB any conc to 150°F	A to 70°F	A to 70°F	B at 70°F	A to 70°F
DIETHYLENE GLYCOL-DIALKYL ETHER	NR at 70°F	NR at 70°F				
DIETHYLENE GLYCOL-MONOALKYL ETHER	NR at 70°F	NR at 70°F				
DIETHYLENE GLYCOL-MONOBUTYL ETHER	NR at 70°F	NR at 70°F	A to 70°F		A to 70°F	
DIETHYLENE GLYCOL-MONOETHYL ETHER	NR at 70°F	NR at 70°F	B at 70°F		B at 70°F	
DIETHYLENE OXIDE	NR at 70°F	NR at 70°F	NR at 70°F		NR at 70°F	

	POLYSULFIDE T	CHLORINATED POLYETHYLENE CM	EPICHLOROHYDRIN CO, ECO	® HYTREL COPOLYESTER TPE	® SANTOPRENE COPOLYMER TPO	® C-FLEX STYRENIC TPE
DICHLOROETHANE	NR at 70°F	NR at 70°F	NR at 70°F	NR at 70°F	NR at 70°F	AB to 70°F
DICHLOROETHANOIC ACID						
DICHLOROETHER						
DICHLOROETHYLENE				NR at 70°F	NR at 70°F	
DICHLOROETHYL ETHER	AC to 70°F					
DICHLOROFLUOROMETHANE						
DICHLOROHEXANE						
DICHLORO-ISOPROPYL ETHER	A to 70°F					
DICHLOROMETHANE	NR at 70°F			NR at 70°F	NR at 70°F	
DICHLOROPENTANE						
DICHLOROPROPANE						
DICHLOROTETRAFLUOROETHANE (F-114)	A to 70°F		A to 70°F	A to 70°F		
DICYCLOHEXYLAMINE	NR at 70°F					
DIELDRIN IN XYLENE						
DIELDRIN IN XYLENE & WATER SPRAY						
DIELINE						
DIESEL OIL	A to 70°F	A to 70°F	A to 70°F	AB to 70°F *AB to 140°F	A/NR at 70°F	
DIESTER LUBRICANT (MIL-L-7808)	B at 70°F					
DIESTER SYNTHETIC LUBRICATING OILS	B at 70°F					
DIETHANOLAMINE				NR at 70°F C 20% at 70°F		
DIETHYLAMINE	B at 70°F					
DIETHYLANILINE						
DIETHYLBENZENE	B at 70°F					
DIETHYLCARBINOL						
DIETHYL CARBONATE						
DIETHYL ETHER	A to 70°F			C at 70°F	A to 70°F	
DIETHYL FORMALDEHYDE					*A to 70°F	
DIETHYL HYDRAZINE						
DIETHYL KETONE						
DIETHYL MALEATE						
DIETHYL OXALATE						
DIETHYL OXIDE						
DIETHYL PHTHALATE						
DIETHYL SEBACATE	B at 70°F			A to 70°F	A to 70°F	
DIETHYL SULFATE						
DIETHYL TRIAMINE						
DIETHYL DIOXIDE						
DIETHYLENE ETHER						
DIETHYLENE GLYCOL	A/NR at 70°F		A to 70°F	A to 70°F	*A to 70°F	
DIETHYLENE GLYCOL-DIALKYL ETHER						
DIETHYLENE GLYCOL-MONOALKYL ETHER						
DIETHYLENE GLYCOL-MONOBUTYL ETHER						
DIETHYLENE GLYCOL-MONOETHYL ETHER						
DIETHYLENE OXIDE						

	ETHYLENE ACRYLIC EA	POLYALLOMER LINEAR COPOLYMER	NYLON 11 POLYAMIDE	NYLON 12 POLYAMIDE	ETHYLENE VINYL ACETATE EVA	POLYVINYLCHLORIDE FLEXIBLE PVC
DICHLOROETHANE	*NR at 70°F		AC to 70°F		*NR at 70°F	NR at 70°F
DICHLOROETHANOIC ACID						
DICHLOROETHER						
DICHLOROETHYLENE		A to 70°F	BC at 70°F		NR at 70°F	
DICHLOROETHYL ETHER						
DICHLOROFLUOROMETHANE						
DICHLOROHEXANE						
DICHLORO-ISOPROPYL ETHER						
DICHLOROMETHANE	NR at 70°F	C at 70°F NR at 122°F	AC to 70°F NR at 140°F	NR at 70°F	NR at 70°F	NR at 70°F
DICHLOROPENTANE						
DICHLOROPROPANE						
DICHLOROTETRAFLUOROETHANE (F-114)						NR at 70°F
DICYCLOHEXYLAMINE						
DIELDRIN IN XYLENE						
DIELDRIN IN XYLENE & WATER SPRAY						
DIELINE						
DIESEL OIL	A to 70°F		A to 70°F BC at 140°F	AB to 70°F	*C at 70°F *NR at 140°F	C at 70°F NR at 140°F
DIESTER LUBRICANT (MIL-L-7808)						
DIESTER SYNTHETIC LUBRICATING OILS			AB to 70°F			NR at 70°F
DIETHANOLAMINE			AB to conc to 140°F C at 194°F			
DIETHYLAMINE						A/NR at 70°F
DIETHYLANILINE						
DIETHYLBENZENE		NR at 70°F				NR at 70°F
DIETHYLCARBINOL						
DIETHYL CARBONATE						
DIETHYL ETHER		NR at 70°F	A to 70°F	AB to 70°F	*NR at 70°F	C/NR at 70°F NR at 140°F
DIETHYL FORMALDEHYDE						
DIETHYL HYDRAZINE						
DIETHYL KETONE		B to 122°F				NR at 70°F
DIETHYL MALEATE		A to 122°F				B at 70°F NR at 122°F
DIETHYL OXALATE						NR at 70°F
DIETHYL OXIDE						NR at 70°F
DIETHYL PHTHALATE	NR at 70°F					NR at 70°F
DIETHYL SEBACATE						
DIETHYL SULFATE						
DIETHYL TRIAMINE						
DIETHYL DIOXIDE						
DIETHYLENE ETHER						
DIETHYLENE GLYCOL	A to 70°F	A to 122°F				AC to 70°F A/NR at 150°F
DIETHYLENE GLYCOL-DIALKYL ETHER						
DIETHYLENE GLYCOL-MONOALKYL ETHER						
DIETHYLENE GLYCOL-MONOBUTYL ETHER						
DIETHYLENE GLYCOL-MONOETHYL ETHER		A to 122°F				C at 70°F NR at 122°F
DIETHYLENE OXIDE						

	POLYETHYLENE LOW DENSITY LDPE	® TEFLON FEP	® KALREZ PERFLUORINATED ELASTOMER	® FLUORAZ FLUORINATED COPOLYMER	® AFLAS FLUORINATED COPOLYMER	® NORPRENE COPOLYMER TPO
DICHLOROETHANE	B/NR at 70°F NR at 120°F	A to 300°F	A to 212°F	A/NR at 70°F	no swell 7 days 70°F A to 70°F	
DICHLOROETHANOIC ACID		A to 70°F				
DICHLOROETHER	NR at 70°F	A to 70°F	A to 212°F	A to 70°F		
DICHLOROETHYLENE		A to 200°F				
DICHLOROETHYL ETHER		A to 70°F	A to 212°F			
DICHLOROFLUOROMETHANE		A to 70°F				
DICHLOROHEXANE		A to 70°F				
DICHLORO-ISOPROPYL ETHER		A to 70°F	A to 212°F			
DICHLOROMETHANE	BC at 70°F NR at 122°F	A to 300°F	A to 300°F	B/NR at 70°F		
DICHLOROPENTANE		A to 70°F				
DICHLOROPROPANE	C at 70°F	A to 70°F	A to 212°F			
DICHLOROTETRAFLUOROETHANE (F-114)		A to 70°F	BC at 70°F	NR at 70°F		
DICYCLOHEXYLAMINE		A to 70°F	*A to 212°F	C at 70°F		
DIELDRIN IN XYLENE		A to 70°F				
DIELDRIN IN XYLENE & WATER SPRAY		A to 70°F	A to 212°F			
DIELINE		A to 70°F				
DIESEL OIL	AC to 70°F C/NR at 140°F	A to 300°F	A to 212°F	A to 70°F	+9% vol 3 days 212°F +20% vol 3 days 302°F	
DIESTER LUBRICANT (MIL-L-7808)		A to 70°F	A to 212°F	A to 70°F	+6% vol 3 days 212°F AB to 212°F	
DIESTER SYNTHETIC LUBRICATING OILS	NR at 70°F	A to 70°F		A to 70°F		
DIETHANOLAMINE		A to 70°F	*A to 212°F	A to 70°F		
DIETHYLAMINE		A to 300°F	A to 75°F *A to 212°F	B at 70°F		
DIETHYLANILINE			A to 212°F			
DIETHYLBENZENE	NR at 70°F	A to 122°F	A to 212°F	NR at 70°F		
DIETHYLCARBINOL		A to 70°F				
DIETHYL CARBONATE		A to boiling	A to 212°F	A to 70°F		
DIETHYL ETHER	B/NR at 70°F	A to 300°F	A to 212°F	NR at 70°F	+42% vol 1 day 70°F NR at 70°F	
DIETHYL FORMALDEHYDE						
DIETHYL HYDRAZINE						
DIETHYL KETONE	B at 70°F C at 122°F	A to 122°F				
DIETHYL MALEATE	A to 122°F	A to 122°F				
DIETHYL OXALATE		A to 70°F				
DIETHYL OXIDE		A to 70°F				
DIETHYL PHTHALATE		A to 70°F	A to 212°F			
DIETHYL SEBACATE		A to 70°F	A to 212°F	B at 70°F		
DIETHYL SULFATE		A to 70°F	A to 212°F	AB to 70°F		
DIETHYL TRIAMINE		A to 70°F				
DIETHYL DIOXIDE		A to 70°F				
DIETHYLENE ETHER		A to 70°F				
DIETHYLENE GLYCOL	AB to 122°F	A to 125°F	A to 212°F	AB to 70°F		
DIETHYLENE GLYCOL-DIALKYL ETHER		A to 70°F				
DIETHYLENE GLYCOL-MONOALKYL ETHER						
DIETHYLENE GLYCOL-MONOBUTYL ETHER		A to 70°F				
DIETHYLENE GLYCOL-MONOETHYL ETHER	A to 122°F	A to 122°F				
DIETHYLENE OXIDE		A to 70°F				

	RECOMMENDATIONS	NITRILE NBR	ETHYLENE PROPYLENE EP, EPDM	FLUOROCARBON FKM	CHLOROPRENE CR	HYDROGENATED NITRILE HNBR
DIETHYLENE TRIAMINE	EP, IIR	B/NR at 70°F	A to 200°F	NR at 70°F	NR at 70°F	
DIFLUORODIBROMOMETHANE	EP, IIR	NR at 70°F	B at 70°F		NR at 70°F	
m-DIGALLIC ACID	NBR, EP, FKM, IIR, CR	A to 100% to 200°F	A to 100% to 140°F AB to 200°F	A to 100% to 140°F AB to 200°F	A to 100% to 70°F AB to 200°F	A to 70°F
DIHYDROXYDIETHYL AMINE	IIR	B at 70°F				
DIHYDROXYDIETHEL ETHER	NBR, EP, CR, IIR	A to 70°F AB any conc to 180°F	A to 140°F	A to 140°F	A to 140°F	
DIHYDROXYETHEL AMINE	IIR	B at 70°F				
DIHYDROXYPROPANE	NBR	A to 70°F	A to 70°F	A to 70°F		
DIHYDROXYSUCCINIC ACID	FKM	AB any conc to 150°F	B at 70°F	A to 70°F	BC any conc to 100°F	
DIISOBUTYL KETONE	EP, IIR	NR at 70°F	A to 200°F	NR at 70°F	NR at 70°F	
DIISOBUTYLENE	FKM, NBR, T	A to 250°F	NR at 70°F	A to 140°F	B/NR at 70°F	A to 70°F
DIISODECYL ADIPATE	IIR	NR at 70°F		C at 70°F		
DIISODECYL PHTHALATE	EP, IIR	NR at 70°F	A to 70°F	C at 70°F	NR at 70°F	
DIISOOCTYL ADIPATE	IIR	NR at 70°F		C at 70°F		
DIISOOCTYL PHTHALATE	IIR	NR at 70°F	B at 70°F	BC at 70°F		
DIISOOCTYL SEBACATE	FKM, EP, IIR	C at 70°F	C at 70°F	B at 70°F	NR at 70°F	NR at 70°F
DIISOPRENE	FKM	C at 70°F	NR at 70°F	A to 70°F	NR at 70°F	
DIISOPROPANOLAMINE	IIR	B at 70°F				
DIISOPROPYLAMINE	IIR	B at 70°F				
DIISOPROPYL BENZENE	FKM, FVMQ	NR at 70°F	NR at 70°F	A to 200°F	NR at 70°F	
DIISOPROPYL ETHER	NBR, T, FVMQ	AB any conc to 140°F	NR at 70°F	NR at 70°F	NR at 70°F	B at 70°F
DIISOPROPYL KETONE	EP, IIR	NR at 70°F	A to 70°F	NR at 70°F	NR at 70°F	
DILAUREL ETHER		B at 70°F				
DIMETHYLAMINE	IIR	B/NR at 70°F	A/NR at 70°F	NR at 70°F	B/NR at 70°F	
DIMETHYLANILINE	EP, IIR	NR at 70°F	AB to 160°F	NR at 70°F	NR at 70°F	
DIMETHYLBENZENE	FKM, FVMQ, T	NR at 70°F	NR at 70°F	A to 140°F BC at 158°F	NR at 70°F	NR at 70°F
DIMETHYLCARBINOL	FKM, CR, IIR	B any conc to 120°F	B at 70°F	A to 70°F	A to 70°F AB any conc to 120°F	
DIMETHYLETHER	EP, NBR, FKM, IIR	A to 70°F	AB to 70°F	A to 70°F	C at 70°F	A to 70°F
DIMETHYLFORMAMIDE	EP, NBR, VMQ	B/NR at 70°F	AB to 70°F	NR at 70°F	NR at 70°F	
DIMETHYLKETAL	EP, IIR	NR at 70°F	A to 70°F	NR at 70°F	C at 70°F	
DIMETHYLKETONE	EP, IIR, NR	+125% vol 3 days 70°F NR any conc at 70°F	+2% vol 3 days 70°F A any conc to 200°F	+200% vol 7 days 70°F NR at 70°F	C/NR 100% at 70°F C 10% to 104°F	NR at 70°F
DIMETHYLMETHANE	NBR, FKM, FVMQ, T	A to 250°F gas or liquid	NR at 70°F gas or liquid	A to 176°F gas or liquid	AC to 140°F gas or liquid	A to 70°F
DIMETHYLPHENOL	FKM	NR at 70°F		A to 70°F		
DIMETHYLPHTHALATE	EP, FKM, T, IIR	NR at 70°F	B to 200°F	A to 70°F B at 250°F	NR at 70°F	NR at 70°F
DIMETHYLSULFATE		NR at 70°F		NR at 70°F		
DIMETHYLSULFIDE		NR at 70°F				
DIMETHYLSULFOXIDE				NR at 70°F		
DIMETHYL TEREPHTHALATE	FKM	NR at 70°F	A/NR to 200°F	A to 75°F B at 250°F	NR at 70°F	
DINITROBENZENE	FKM	NR at 70°F		A to 70°F		
DINITROCHLOROBENZENE (DNCB)	FKM	NR at 70°F	NR at 70°F	A to 120°F	NR at 70°F	
DINITROTOLUENE		NR at 70°F	NR at 70°F	NR conc at 70°F BC 60% at 170°F	NR at 70°F	NR at 70°F
DIOCTYL ADIPATE	IIR	NR at 70°F	B at 70°F	C at 70°F	NR at 70°F	
DIOCTYL AMINE	EP	B/NR at 70°F	A to 120°F	NR at 70°F	NR at 70°F	
DIOCTYL PHOSPHATE						
DIOCTYL PHTHALATE (DOP)	EP, FKM, IIR, T	NR at 70°F	B to 200°F	B at 70°F	NR at 70°F	

	STYRENE BUTADIENE SBR	POLYACRYLATE ACM	POLYURETHANE AU, EU	ISOBUTYLENE ISOPRENE IIR	POLYBUTADIENE BR	®AEROQUIP AQP
DIETHYLENE TRIAMINE	NR at 70°F	NR at 70°F	NR at 70°F	A to 70°F		
DIFLUORODIBROMOMETHANE	NR at 70°F	NR at 70°F	NR at 70°F	B at 70°F	NR at 70°F	NR at 70°F
m-DIGALLIC ACID	B to 100% to 70°F	NR 10–100% at 70°F	A at 70°F	A to 100% to 70°F AB to 100% to 185°F	B 100% at 70°F A 10% to 70°F	NR at 70°F
DIHYDROXYDIETHYL AMINE				A to 70°F		
DIHYDROXYDIETHEL ETHER				A to 70°F		
DIHYDROXYETHEL AMINE				A to 70°F		
DIHYDROXYPROPANE				A to 70°F		
DIHYDROXYSUCCINIC ACID			A to 70°F	B at 70°F		
DIISOBUTYL KETONE	A to 70°F	NR at 70°F	NR at 70°F	A to 70°F		
DIISOBUTYLENE	NR at 70°F	NR at 70°F	NR at 70°F	NR at 70°F	NR at 70°F	
DIISODECYL ADIPATE				A to 70°F		
DIISODECYL PHTHALATE				A to 70°F		
DIISOOCTYL ADIPATE				A to 70°F		
DIISOOCTYL PHTHALATE				A to 70°F		
DIISOOCTYL SEBACATE	NR at 70°F	NR at 70°F	NR at 70°F	NR at 70°F	NR at 70°F	
DIISOPRENE				NR at 70°F		
DIISOPROPANOLAMINE				A to 70°F		
DIISOPROPYLAMINE				A to 70°F		
DIISOPROPYL BENZENE	NR at 70°F	NR at 70°F	NR at 70°F	NR at 70°F	NR at 70°F	
DIISOPROPYL ETHER				NR at 70°F		
DIISOPROPYL KETONE	NR at 70°F	NR at 70°F	NR at 70°F	A to 70°F	NR at 70°F	
DILAUREL ETHER	NR at 70°F	C at 70°F	AB to 70°F			
DIMETHYLAMINE				AB to 70°F		
DIMETHYLANILINE	NR at 70°F	NR at 70°F	NR at 70°F	A/NR to 100°F	NR at 70°F	
DIMETHYLBENZENE	NR at 70°F	NR at 70°F	C/NR at 70°F	NR at 70°F	NR at 70°F	TEST
DIMETHYLCARBINOL			NR at 70°F	A to 70°F AB any conc to 120°F		
DIMETHYLETHER	A to 70°F	NR at 70°F		AC to 70°F	A to 70°F	
DIMETHYLFORMAMIDE	NR at 70°F	NR at 70°F	NR at 70°F	B any conc to 100°F	NR at 70°F	
DIMETHYLKETAL			NR at 70°F	A to 70°F		
DIMETHYLKETONE	+18% vol 3 days 70°F BC at 70°F	+201% vol 3 days 70°F NR at 70°F	+87% vol 3 days 70°F NR at 70°F	A to 70°F AB any conc to 150°F	NR at 70°F	AB to 70°F
DIMETHYLMETHANE			B at 70°F	NR at 70°F		
DIMETHYLPHENOL				NR at 70°F		
DIMETHYLPHTHALATE	NR at 70°F	NR at 70°F	NR at 70°F	B to 70°F	NR at 70°F	
DIMETHYLSULFATE				B at 70°F		
DIMETHYLSULFIDE				C at 70°F		
DIMETHYLSULFOXIDE	NR at 70°F	NR at 70°F	NR at 70°F			
DIMETHYL TEREPHTHALATE	NR at 70°F	NR at 70°F	NR at 70°F			
DINITROBENZENE				C at 70°F		
DINITROCHLOROBENZENE (DNCB)	NR at 70°F	NR at 70°F	NR at 70°F			
DINITROTOLUENE	NR at 70°F	NR at 70°F	NR at 70°F	NR at 70°F	NR at 70°F	
DIOCTYL ADIPATE				A to 70°F		
DIOCTYL AMINE	NR at 70°F	NR at 70°F	NR at 70°F	A to 70°F		
DIOCTYL PHOSPHATE			C at 70°F			
DIOCTYL PHTHALATE (DOP)	NR at 70°F	NR at 70°F	NR at 70°F	B to 100°F	NR at 70°F	

C123

	SYNTHETIC ISOPRENE IR	NATURAL ISOPRENE NR	CHLOROSULFONATED POLYETHYLENE CSM	FLUOROSILICONE FVMQ	SILICONE VMQ	CHEMRAZ FFKM
DIETHYLENE TRIAMINE	B at 70°F	B at 70°F	C at 70°F	NR at 70°F	NR at 70°F	A to 70°F
DIFLUORODIBROMOMETHANE	NR at 70°F	NR at 70°F	NR at 70°F		NR at 70°F	A to 70°F
m-DIGALLIC ACID	A to 100% to 70°F	A to 100% to 70°F AB to 100% to 185°F	A to 100% to 140°F AB to 200°F	A 10% at 70°F	B to 100% at 70°F	A to 70°F
DIHYDROXYDIETHYL AMINE	B at 70°F	B at 70°F	C at 70°F			
DIHYDROXYDIETHEL ETHER	A to 70°F	A to 70°F AB any conc to 150°F	A to 70°F	A to 70°F	B at 70°F	A to 70°F
DIHYDROXYETHEL AMINE	B at 70°F	B at 70°F	C at 70°F			
DIHYDROXYPROPANE	A to 70°F	A to 70°F	A to 70°F		A to 70°F	
DIHYDROXYSUCCINIC ACID	A to 70°F	A to 70°F	A to 70°F		A to 70°F	
DIISOBUTYL KETONE	NR at 70°F	NR at 70°F	NR at 70°F	NR at 70°F	NR at 70°F	A to 70°F
DIISOBUTYLENE	NR at 70°F	NR at 70°F	C/NR at 70°F	C at 70°F	NR at 70°F	A to 70°F
DIISODECYL ADIPATE	NR at 70°F	NR at 70°F	NR at 70°F			
DIISODECYL PHTHALATE	NR at 70°F	NR at 70°F	NR at 70°F			
DIISOOCTYL ADIPATE	NR at 70°F	NR at 70°F	NR at 70°F			
DIISOOCTYL PHTHALATE	NR at 70°F	NR at 70°F	NR at 70°F			A to 70°F
DIISOOCTYL SEBACATE	NR at 70°F	NR at 70°F	NR at 70°F	C at 70°F	C at 70°F	A to 70°F
DIISOPRENE	NR at 70°F	NR at 70°F	NR at 70°F			
DIISOPROPANOLAMINE	B at 70°F	B at 70°F	C at 70°F			
DIISOPROPYLAMINE	B at 70°F	B at 70°F	C at 70°F			
DIISOPROPYL BENZENE	NR at 70°F	NR at 70°F	NR at 70°F	AB to 70°F	NR at 70°F	A to 70°F
DIISOPROPYL ETHER	NR at 70°F	AB any conc to 80°F	BC at 70°F	C at 70°F	NR at 70°F	A to 70°F
DIISOPROPYL KETONE	NR at 70°F	NR at 70°F	NR at 70°F	NR at 70°F	NR at 70°F	A to 70°F
DILAUREL ETHER	NR at 70°F	NR at 70°F	B at 70°F		NR at 70°F	
DIMETHYLAMINE	B at 70°F	B at 70°F	NR at 70°F			A to 70°F
DIMETHYLANILINE	NR at 70°F	NR at 70°F	NR at 70°F	NR at 70°F	NR at 70°F	
DIMETHYLBENZENE	NR at 70°F	NR at 70°F	NR at 70°F	+20% vol 7 days 70°F AC to 70°F	+150% vol 7 days 70°F NR at 70°F	A to 70°F
DIMETHYLCARBINOL	A to 70°F	A to 70°F AB any conc to 120°F	A to 70°F		A to 70°F	
DIMETHYLETHER	A/NR at 70°F	A/NR at 70°F	BC at 70°F	A to 70°F	A to 70°F	A to 70°F
DIMETHYLFORMAMIDE	NR at 70°F	NR at 70°F	NR at 70°F	NR at 70°F	+20% vol 7 days 70°F BC at 70°F	*B at 70°F
DIMETHYLKETAL	B at 70°F	B at 70°F	B at 70°F		A to 70°F	
DIMETHYLKETONE	NR at 70°F	NR at 70°F	NR at 70°F	+205% vol 3 days 70°F NR at 70°F	+18% vol 7 days 70°F BC at 70°F	A to 70°F
DIMETHYLMETHANE	NR at 70°F gas or liquid	NR at 70°F gas or liquid	A/NR at 70°F	AB to 70°F	NR at 70°F	A to 70°F
DIMETHYLPHENOL	NR at 70°F	NR at 70°F	NR at 70°F			
DIMETHYLPHTHALATE	NR at 70°F	NR at 70°F	NR at 70°F	B at 70°F		A to 70°F
DIMETHYLSULFATE	NR at 70°F	NR at 70°F	NR at 70°F			
DIMETHYLSULFIDE	NR at 70°F	NR at 70°F	NR at 70°F			
DIMETHYLSULFOXIDE						A to 70°F
DIMETHYL TEREPHTHALATE				B at 70°F	NR at 70°F	A to 70°F
DINITROBENZENE	NR at 70°F	NR at 70°F	NR at 70°F			
DINITROCHLOROBENZENE (DNCB)				B at 70°F	NR at 70°F	A to 70°F
DINITROTOLUENE	NR at 70°F	NR at 70°F	NR at 70°F	NR at 70°F	NR at 70°F	A to 70°F
DIOCTYL ADIPATE	NR at 70°F	NR at 70°F	NR at 70°F			
DIOCTYL AMINE	B at 70°F	B at 70°F	C at 70°F	NR at 70°F	NR at 70°F	A to 70°F
DIOCTYL PHOSPHATE						
DIOCTYL PHTHALATE (DOP)	NR at 70°F	NR at 70°F	NR at 70°F	B at 70°F	C at 70°F	A to 70°F

	POLYSULFIDE T	CHLORINATED POLYETHYLENE CM	EPICHLOROHYDRIN CO, ECO	® HYTREL COPOLYESTER TPE	® SANTOPRENE COPOLYMER TPO	® C-FLEX STYRENIC TPE
DIETHYLENE TRIAMINE						
DIFLUORODIBROMOMETHANE	NR at 70°F	NR at 70°F		B/NR conc at 70°F A to 10% to 70°F		AB to 70°F
m-DIGALLIC ACID	A to 70°F	NR at 70°F	B 10% at 70°F	AB 100% to 70°F A 10% to 70°F	A to 70°F	A to 70°F
DIHYDROXYDIETHYL AMINE						
DIHYDROXYDIETHEL ETHER						
DIHYDROXYETHEL AMINE						
DIHYDROXYPROPANE						
DIHYDROXYSUCCINIC ACID						
DIISOBUTYL KETONE						
DIISOBUTYLENE	A to 70°F					
DIISODECYL ADIPATE						
DIISODECYL PHTHALATE						
DIISOOCTYL ADIPATE						
DIISOOCTYL PHTHALATE						
DIISOOCTYL SEBACATE	C at 70°F					
DIISOPRENE						
DIISOPROPANOLAMINE						
DIISOPROPYLAMINE						
DIISOPROPYL BENZENE	B at 70°F					
DIISOPROPYL ETHER					*A to 70°F	
DIISOPROPYL KETONE	B at 70°F					
DILAUREL ETHER						
DIMETHYLAMINE						
DIMETHYLANILINE						
DIMETHYLBENZENE	B at 70°F	C at 70°F	NR at 70°F	+13% to + 88% vol 7 days 70°F	NR at 70°F	NR at 70°F
DIMETHYLCARBINOL						
DIMETHYLETHER	A to 70°F					
DIMETHYLFORMAMIDE	B at 70°F			+17% vol 7 days 70°F B at 70°F	+1% vol 3 days 70°F A to 70°F	AB to 70°F
DIMETHYLKETAL						
DIMETHYLKETONE	+7.3% vol 3 days 70°F AC to 70°F	AB to 70°F	NR at 70°F	BC at 70°F	-11% vol 168 days 70°F A to 70°F	AB to 70°F
DIMETHYLMETHANE						
DIMETHYLPHENOL						
DIMETHYLPHTHALATE	B at 70°F			A to 70°F	+10% vol 3 days 70°F AB to 70°F	
DIMETHYLSULFATE						
DIMETHYLSULFIDE						
DIMETHYLSULFOXIDE					*A to 70°F	
DIMETHYL TEREPHTHALATE						
DINITROBENZENE						
DINITROCHLOROBENZENE (DNCB)						
DINITROTOLUENE	NR at 70°F					
DIOCTYL ADIPATE					*A to 70°F	
DIOCTYL AMINE						
DIOCTYL PHOSPHATE				C at 70°F		
DIOCTYL PHTHALATE (DOP)	B at 70°F		B at 70°F	+3% vol 3 days 70°F A to 70°F	*A to 70°F	

	ETHYLENE ACRYLIC EA	POLYALLOMER LINEAR COPOLYMER	NYLON 11 POLYAMIDE	NYLON 12 POLYAMIDE	ETHYLENE VINYL ACETATE EVA	POLYVINYLCHLORIDE FLEXIBLE PVC
DIETHYLENE TRIAMINE						
DIFLUORODIBROMOMETHANE						
m-DIGALLIC ACID	*AB 10% to 70°F	A to 70°F	AB to 70°F		BC at 70°F	A to 150°F
DIHYDROXYDIETHYL AMINE						
DIHYDROXYDIETHEL ETHER						
DIHYDROXYETHEL AMINE						
DIHYDROXYPROPANE						
DIHYDROXYSUCCINIC ACID						
DIISOBUTYL KETONE						
DIISOBUTYLENE						
DIISODECYL ADIPATE						
DIISODECYL PHTHALATE						NR at 70°F
DIISOOCTYL ADIPATE						
DIISOOCTYL PHTHALATE						
DIISOOCTYL SEBACATE	NR at 70°F					
DIISOPRENE						
DIISOPROPANOLAMINE						
DIISOPROPYLAMINE						
DIISOPROPYL BENZENE						
DIISOPROPYL ETHER						NR at 70°F
DIISOPROPYL KETONE	NR at 70°F					NR at 70°F
DILAUREL ETHER						
DIMETHYLAMINE						
DIMETHYLANILINE						NR at 70°F
DIMETHYLBENZENE	NR at 70°F	C at 70°F NR at 122°F	AB to 105°F C 140–194°F	AB to 70°F	NR at 70°F	NR at 70°F
DIMETHYLCARBINOL						
DIMETHYLETHER						
DIMETHYLFORMAMIDE		A to 122°F	AB to 70°F	C at 70°F		C/NR at 70°F NR at 140°F
DIMETHYLKETAL						
DIMETHYLKETONE	NR at 70°F	A to 122°F	A to 70°F AB to 105°F		C/NR at 70°F	NR any conc at 70°F
DIMETHYLMETHANE						NR at 70°F
DIMETHYLPHENOL						
DIMETHYLPHTHALATE	NR at 70°F					
DIMETHYLSULFATE						
DIMETHYLSULFIDE						
DIMETHYLSULFOXIDE		A to 122°F	NR at 70°F			NR at 70°F
DIMETHYL TEREPHTHALATE						
DINITROBENZENE						
DINITROCHLOROBENZENE (DNCB)						
DINITROTOLUENE						
DIOCTYL ADIPATE						
DIOCTYL AMINE						
DIOCTYL PHOSPHATE			AB to 140°F C at 194°F			AB to 70°F
DIOCTYL PHTHALATE (DOP)	NR at 70°F		A to 70°F AB to 140°F	AB to 70°F	*C at 70°F *NR at 140°F	C/NR at 70°F

	POLYETHYLENE LOW DENSITY LDPE	® TEFLON FEP	® KALREZ PERFLUORINATED ELASTOMER	® FLUORAZ FLUORINATED COPOLYMER	® AFLAS FLUORINATED COPOLYMER	® NORPRENE COPOLYMER TPO
DIETHYLENE TRIAMINE	A to 70°F	A to 70°F	*A to 212°F	AB to 70°F		
DIFLUORODIBROMOMETHANE			A to 212°F	NR at 70°F		A to 70°F
m-DIGALLIC ACID	AB 100% to 70°F A 10% to 70°F	A to 100% to 300°F	A to 212°F	A to 150°F		A to 70°F
DIHYDROXYDIETHYL AMINE		A to 70°F				
DIHYDROXYDIETHEL ETHER	AB to 122°F	A to 125°F	A to 70°F	AB to 70°F		
DIHYDROXYETHEL AMINE		A to 70°F				
DIHYDROXYPROPANE		A to 70°F				
DIHYDROXYSUCCINIC ACID		A to 70°F				
DIISOBUTYL KETONE		A to 70°F	A to 212°F	NR at 70°F		
DIISOBUTYLENE	B at 70°F	A to 300°F	A to 212°F	BC at 70°F	BC at 70°F	
DIISODECYL ADIPATE		A to 70°F				
DIISODECYL PHTHALATE		A to 70°F				
DIISOOCTYL ADIPATE		A to 70°F				
DIISOOCTYL PHTHALATE		A to 70°F				
DIISOOCTYL SEBACATE				B at 70°F		
DIISOPRENE		A to 70°F				
DIISOPROPANOLAMINE		A to 70°F				
DIISOPROPYLAMINE		A to 70°F				
DIISOPROPYL BENZENE		A to 70°F	A to 212°F	C at 70°F		
DIISOPROPYL ETHER	B/NR at 70°F NR at 140°F	A to 300°F	A to 212°F	NR at 70°F		
DIISOPROPYL KETONE		A to 70°F	A to 212°F	NR at 70°F		
DILAUREL ETHER		A to 70°F				
DIMETHYLAMINE	AB to 70°F	A to 70°F	*A to 212°F			
DIMETHYLANILINE		A to 300°F	*A to 212°F			
DIMETHYLBENZENE	B/NR at 70°F NR at 122°F	A to 300°F	A to 450°F	NR at 70°F	+30% vol 7 days 70°F C/NR at 70°F	NR at 70°F
DIMETHYLCARBINOL		A to 70°F				
DIMETHYLETHER	AB to 70°F	A to 70°F	A to 212°F	NR at 70°F		
DIMETHYLFORMAMIDE	A to 122°F	A to 309°F	A to 212°F	B/NR at 70°F	A to 70°F	
DIMETHYLKETAL		A to 70°F				
DIMETHYLKETONE	C/NR at 70°F NR at 122°F	A to boiling	+2% vol 7 days 70°F A to 113°F	NR at 70°F	+50% vol 7 days 70°F NR at 70°F	BC at 70°F
DIMETHYLMETHANE	AC to 70°F	A to 300°F gas or liquid	A to 70°F	AC to 70°F		
DIMETHYLPHENOL		A to 70°F				
DIMETHYLPHTHALATE		A to 392°F	A to 212°F	BC at 70°F		
DIMETHYLSULFATE		A to 70°F				
DIMETHYLSULFIDE		A to 70°F				
DIMETHYLSULFOXIDE	A to 122°F	A to 372°F	A to 212°F			
DIMETHYL TEREPHTHALATE			A to 212°F	A to 70°F		
DINITROBENZENE		A to 70°F				
DINITROCHLOROBENZENE (DNCB)			A to 212°F	NR at 70°F	A to 70°F	
DINITROTOLUENE		A to 70°F	A to 212°F	NR at 70°F		
DIOCTYL ADIPATE		A to 70°F				
DIOCTYL AMINE		A to 70°F	A to 212°F	A to 70°F		
DIOCTYL PHOSPHATE	NR at 70°F					
DIOCTYL PHTHALATE (DOP)	C/NR at 70°F NR at 140°F	A to 300°F	A to 212°F	A to 70°F	+20% vol 7 days 302°F B at 302°F	

C 127

	RECOMMENDATIONS	NITRILE NBR	ETHYLENE PROPYLENE EP, EPDM	FLUOROCARBON FKM	CHLOROPRENE CR	HYDROGENATED NITRILE HNBR
DIOCTYL SEBACATE	FKM, EP, IIR, VMQ	NR at 70°F	B at 70°F	+94% vol 14 days 300°F B at 70°F	NR at 70°F	NR at 70°F
DIOFORM		NR at 70°F		A to 70°F	NR at 70°F	
DIOXANE	EP, IIR	NR at 70°F	AB to 160°F	+205% vol 7 days 77°F NR at 70°F	NR at 70°F	B at 70°F
DIOXETHYLENE ETHER		NR at 70°F			NR at 70°F	
DIOXOLANES (DIOXOLANS)	EP, IIR	NR at 70°F	B at 70°F	NR at 70°F	NR at 70°F	NR at 70°F
DIPENTENE	FKM, T, NBR	AB to 200°F	NR at 70°F	A to 70°F	NR at 70°F	B at 70°F
DIPHENYL	FKM, T, FVMQ	dissolved 3 days 300°F NR at 70°F	NR at 70°F	A to 300°F	dissolved 3 days 300°F NR at 70°F	NR at 70°F
DIPHENYL, CHLORINATED	FKM	NR at 70°F	NR at 70°F	A to 300°F	NR at 70°F	
DIPHENYL ETHER				AB to 70°F		
DIPHENYL OXIDES	FKM, FVMQ	NR at 70°F	NR at 70°F	A to 140°F	NR at 70°F	NR at 70°F
DIPHENYL PHTHALATE		NR at 70°F		C at 70°F		
DIPPEL'S OIL		A to 70°F		A to 70°F		
DIPROPYLAMINE		B at 70°F				
DIPROPYL KETONE		NR at 70°F		NR at 70°F		
DIPROPYL METHANE		A to 70°F	NR at 70°F	A to 70°F	A to 70°F	
DIPROPYLENE GLYCOL	NBR, CR, EP	A to 70°F		A to 70°F		
DISODIUM PHOSPHATE		A to 70°F	A to 70°F	A to 70°F		
DISPERSING OIL NO. 10		NR at 70°F	NR at 70°F	C at 70°F	NR at 70°F	
DISTILLED VINEGAR		C at 70°F	A to 70°F	A to 70°F	A to 70°F	
DIVINYL BENZENE	FKM	NR at 70°F	NR at 70°F	A to 300°F	NR at 70°F	
DIVINYL ETHER		B at 70°F		NR at 70°F		
DMP (DIMETHAL PHTHALATE)	EP, FKM, T, IIR	NR at 70°F	B to 200°F	A to 70°F B at 250°F	NR at 70°F	NR at 70°F
DMT (DIMETHYL TEREPHTHALATE)	FKM	NR at 70°F	A/NR to 200°F	A to 70°F B at 230°F	NR at 70°F	
DODECANOL (DODECYL ALCOHOL)		A to 70°F		B at 70°F		
DODECYL BENZENE		NR at 70°F		A to 70°F		
DODECYL TOLUENE		NR at 70°F		A to 70°F		
DOLOMITE		A to 70°F	B at 70°F	A to 70°F	A to 70°F	
DOUGLAS FIR OIL		AB to 150°F			NR at 70°F	
DOW CHEMICAL 50-4, ET588	EP, SBR	C at 70°F	A to 70°F	NR at 70°F	BC at 70°F	C at 70°F
DOW CHEMICAL ET378	T	NR at 70°F			NR at 70°F	NR at 70°F
DOW CORNING 3, 4, 11	EP, FKM, NBR	A to 70°F	A to 70°F	A to 70°F	A to 70°F	A to 70°F
DOW CORNING 5, 33, 44, 200, 220, 510, 550, 705, 710	EP, FKM, NBR	A to 70°F	A to 70°F	A to 70°F	A to 70°F	A to 70°F
DOW CORNING 55	EP, FKM, NBR	A to 70°F	A to 70°F	A to 70°F	A to 70°F	A to 70°F
DOW CORNING 1265 FLUOROSILICONE FLUIDS	EP, FKM, NBR	AB to 70°F	A to 70°F	A to 70°F	A to 70°F	
DOW CORNING 1208, 4050, 6620, F-60, XF-60		A to 70°F	A to 70°F	A to 70°F	A to 70°F	A to 70°F
DOW CORNING F-61		A to 70°F	A to 70°F	A to 70°F	A to 70°F	A to 70°F
DOW GAUGE FLUID R-200		A to 70°F	A to 70°F	A to 70°F	A to 70°F	
DOW GENERAL WEED KILLER (PHENOL)		NR at 70°F				
DOW GENERAL WEED KILLER (WATER)		B at 70°F				
DOW GUARD	NBR, EP, FKM, CR	A to 70°F	A to 70°F	A to 70°F	A to 70°F	A to 70°F
DOW PER		C at 70°F		A to 70°F		
DOW PURIFLOC C-31		A to 70°F AB to 150°F			A to 70°F AB to 150°F	
DOWANOLS	EP	AB to 70°F	A to 120°F	NR at 70°F	NR at 70°F	
DOWFUME W-40		NR at 70°F	C at 70°F	C at 70°F	NR at 70°F	

	STYRENE BUTADIENE SBR	POLYACRYLATE ACM	POLYURETHANE AU, EU	ISOBUTYLENE ISOPRENE IIR	POLYBUTADIENE BR	® AEROQUIP AQP
DIOCTYL SEBACATE	NR at 70°F	NR at 70°F	B at 70°F	AB to 100°F	NR at 70°F	
DIOFORM				C at 70°F		
DIOXANE	NR at 70°F	NR at 70°F	NR at 70°F	AB to 70°F	NR at 70°F	
DIOXETHYLENE ETHER				NR at 70°F		
DIOXOLANES (DIOXOLANS)	NR at 70°F	NR at 70°F	NR at 70°F	C/NR at 70°F	NR at 70°F	
DIPENTENE	NR at 70°F	NR at 70°F	NR at 70°F	NR at 70°F	NR at 70°F	
DIPHENYL	dissolved 3 days 300°F NR at 70°F	dissolved 3 days 300°F NR at 70°F	NR at 70°F	dissolved 3 days 300°F NR at 70°F	NR at 70°F	
DIPHENYL, CHLORINATED	NR at 70°F	NR at 70°F	NR at 70°F			
DIPHENYL ETHER				NR at 70°F		
DIPHENYL OXIDES	NR at 70°F	NR at 70°F	NR at 70°F	NR at 70°F	NR at 70°F	
DIPHENYL PHTHALATE				A to 70°F		
DIPPEL'S OIL				B at 70°F		
DIPROPYLAMINE				A to 70°F		
DIPROPYL KETONE				B at 70°F		
DIPROPYL METHANE				NR at 70°F		
DIPROPYLENE GLYCOL				A to 70°F		
DISODIUM PHOSPHATE				A to 70°F		
DISPERSING OIL NO. 10				NR at 70°F		
DISTILLED VINEGAR			B at 70°F	A to 70°F		
DIVINYL BENZENE	NR at 70°F	NR at 70°F		NR at 70°F		
DIVINYL ETHER				NR at 70°F		
DMP (DIMETHAL PHTHALATE)	NR at 70°F	NR at 70°F	NR at 70°F	B at 70°F	NR at 70°F	
DMT (DIMETHYL TEREPHTHALATE)	NR at 70°F	NR at 70°F	NR at 70°F			
DODECANOL (DODECYL ALCOHOL)				A to 70°F		
DODECYL BENZENE				NR at 70°F		
DODECYL TOLUENE				NR at 70°F		
DOLOMITE						
DOUGLAS FIR OIL						
DOW CHEMICAL 50-4, ET588	A to 70°F			B at 70°F		
DOW CHEMICAL ET378	NR at 70°F	C at 70°F	B at 70°F	NR at 70°F	NR at 70°F	
DOW CORNING 3, 4, 11	A to 70°F	A to 70°F	A to 70°F	A to 70°F	A to 70°F	
DOW CORNING 5, 33, 44, 200, 220, 510, 550, 705, 710	A to 70°F	A to 70°F	A to 70°F	A to 70°F	A to 70°F	AB to 70°F
DOW CORNING 55	A to 70°F	A to 70°F	A to 70°F	A to 70°F	A to 70°F	
DOW CORNING 1265 FLUOROSILICONE FLUIDS	A to 70°F	A to 70°F	A to 70°F	A to 70°F	A to 70°F	
DOW CORNING 1208, 4050, 6620, F-60, XF-60	A to 70°F	A to 70°F	A to 70°F	A to 70°F	A to 70°F	
DOW CORNING F-61	A to 70°F	A to 70°F	A to 70°F	A to 70°F	A to 70°F	
DOW GAUGE FLUID R-200			A to 70°F	A to 70°F		
DOW GENERAL WEED KILLER (PHENOL)						
DOW GENERAL WEED KILLER (WATER)						
DOW GUARD	A to 70°F	C at 70°F	C at 70°F	A to 70°F	A to 70°F	
DOW PER				NR at 70°F		
DOW PURIFLOC C-31						
DOWANOLS		NR at 70°F	NR at 70°F			
DOWFUME W-40				NR at 70°F		

	SYNTHETIC ISOPRENE IR	NATURAL ISOPRENE NR	CHLOROSULFONATED POLYETHYLENE CSM	FLUOROSILICONE FVMQ	SILICONE VMQ	CHEMRAZ FFKM
DIOCTYL SEBACATE	NR at 70°F	NR at 70°F	NR at 70°F	C at 70°F	C at 70°F	A to 70°F
DIOFORM	NR at 70°F	NR at 70°F	NR at 70°F			
DIOXANE	NR at 70°F	NR at 70°F	NR at 70°F	NR at 70°F	NR at 70°F	A to 70°F
DIOXETHYLENE ETHER	NR at 70°F	NR at 70°F				
DIOXOLANES (DIOXOLANS)	NR at 70°F	NR at 70°F	NR at 70°F	NR at 70°F	NR at 70°F	A to 70°F
DIPENTENE	NR at 70°F	NR at 70°F	NR at 70°F	C at 70°F	NR at 70°F	A to 70°F
DIPHENYL	NR at 70°F	NR at 70°F	NR at 70°F	B at 70°F	+15% vol 3 days 300°F NR at 70°F	A to 70°F
DIPHENYL, CHLORINATED				B at 70°F	NR at 70°F	
DIPHENYL ETHER					NR at 70°F	A to 70°F
DIPHENYL OXIDES	NR at 70°F	NR at 70°F	NR at 70°F	B at 70°F	C at 70°F	A to 70°F
DIPHENYL PHTHALATE	NR at 70°F	NR at 70°F	NR at 70°F			
DIPPEL'S OIL	NR at 70°F	NR at 70°F	NR at 70°F			
DIPROPYLAMINE	B at 70°F	B at 70°F	C at 70°F			
DIPROPYL KETONE	NR at 70°F	NR at 70°F	NR at 70°F			
DIPROPYL METHANE	NR at 70°F	NR at 70°F	B at 70°F			
DIPROPYLENE GLYCOL	A to 70°F	A to 70°F	A to 70°F			
DISODIUM PHOSPHATE	A to 70°F	A to 70°F	A to 70°F			
DISPERSING OIL NO. 10	NR at 70°F	NR at 70°F				
DISTILLED VINEGAR	A to 70°F	A to 70°F	B at 70°F		A to 70°F	
DIVINYL BENZENE	NR at 70°F	NR at 70°F	NR at 70°F	B at 70°F	NR at 70°F	
DIVINYL ETHER	NR at 70°F	NR at 70°F	B at 70°F			
DMP (DIMETHAL PHTHALATE)	NR at 70°F	NR at 70°F	NR at 70°F	B at 70°F		A to 70°F
DMT (DIMETHYL TEREPHTHALATE)				B at 70°F	NR at 70°F	A to 70°F
DODECANOL (DODECYL ALCOHOL)	A to 70°F	A to 70°F	A to 70°F			
DODECYL BENZENE	NR at 70°F	NR at 70°F	NR at 70°F			
DODECYL TOLUENE	NR at 70°F	NR at 70°F	NR at 70°F			A to 70°F
DOLOMITE			A to 70°F			
DOUGLAS FIR OIL	NR at 70°F	NR at 70°F				
DOW CHEMICAL 50-4, ET588			B at 70°F	NR at 70°F		B at 70°F
DOW CHEMICAL ET378	NR at 70°F	NR at 70°F	NR at 70°F		NR at 70°F	B at 70°F
DOW CORNING 3, 4, 11	A to 70°F	A to 70°F	A to 70°F	A to 70°F	C at 70°F	A to 70°F
DOW CORNING 5, 33, 44, 200, 220, 510, 550, 705, 710	A to 70°F	A to 70°F	A to 70°F	no swell 3 days 300°F AB to 70°F	+28-35% vol 3 days 300°F C at 70°F	A to 70°F
DOW CORNING 55	A to 70°F	A to 70°F	A to 70°F	A to 70°F	C at 70°F	A to 70°F
DOW CORNING 1265 FLUOROSILICONE FLUIDS	A to 70°F	A to 70°F	A to 70°F	C at 70°F	A to 70°F	
DOW CORNING 1208, 4050, 6620, F-60, XF-60	A to 70°F	A to 70°F	A to 70°F	A to 70°F	C at 70°F	A to 70°F
DOW CORNING F-61	A to 70°F	A to 70°F	A to 70°F	A to 70°F	B at 70°F	A to 70°F
DOW GAUGE FLUID R-200						
DOW GENERAL WEED KILLER (PHENOL)						
DOW GENERAL WEED KILLER (WATER)						
DOW GUARD	A to 70°F	A to 70°F	A to 70°F	A to 70°F	A to 70°F	A to 70°F
DOW PER	NR at 70°F	NR at 70°F	NR at 70°F			
DOW PURIFLOC C-31	A to 70°F	A to 70°F AB to 150°F				
DOWANOLS				NR at 70°F	NR at 70°F	
DOWFUME W-40	NR at 70°F	NR at 70°F				

	POLYSULFIDE T	CHLORINATED POLYETHYLENE CM	EPICHLOROHYDRIN CO, ECO	® HYTREL COPOLYESTER TPE	® SANTOPRENE COPOLYMER TPO	® C-FLEX STYRENIC TPE
DIOCTYL SEBACATE	C at 70°F		C at 70°F			
DIOFORM						
DIOXANE	NR at 70°F				*A to 70°F	
DIOXETHYLENE ETHER						
DIOXOLANES (DIOXOLANS)	NR at 70°F					
DIPENTENE	A to 70°F					
DIPHENYL	disint. 3 days 300°F B at 70°F					
DIPHENYL, CHLORINATED						
DIPHENYL ETHER						
DIPHENYL OXIDES	NR at 70°F					
DIPHENYL PHTHALATE						
DIPPEL'S OIL						
DIPROPYLAMINE						
DIPROPYL KETONE						
DIPROPYL METHANE						
DIPROPYLENE GLYCOL					*A to 70°F	
DISODIUM PHOSPHATE						
DISPERSING OIL NO. 10						
DISTILLED VINEGAR						
DIVINYL BENZENE						
DIVINYL ETHER						
DMP (DIMETHAL PHTHALATE)	B at 70°F			A to 70°F	+10% vol 3 days 70°F AB to 70°F	
DMT (DIMETHYL TEREPHTHALATE)						
DODECANOL (DODECYL ALCOHOL)						
DODECYL BENZENE						
DODECYL TOLUENE						
DOLOMITE	NR at 70°F					
DOUGLAS FIR OIL						
DOW CHEMICAL 50-4, ET588						
DOW CHEMICAL ET378	A/NR at 70°F					
DOW CORNING 3, 4, 11	A to 70°F					
DOW CORNING 5, 33, 44, 200, 220, 510, 550, 705, 710	A to 70°F			AB to 200°F		
DOW CORNING 55	A to 70°F					
DOW CORNING 1265 FLUOROSILICONE FLUIDS	A to 70°F					
DOW CORNING 1208, 4050, 6620, F-60, XF-60	A to 70°F					
DOW CORNING F-61	A to 70°F					
DOW GAUGE FLUID R-200						
DOW GENERAL WEED KILLER (PHENOL)						
DOW GENERAL WEED KILLER (WATER)						
DOW GUARD						
DOW PER						
DOW PURIFLOC C-31						
DOWANOLS						
DOWFUME W-40						

	ETHYLENE ACRYLIC EA	POLYALLOMER LINEAR COPOLYMER	NYLON 11 POLYAMIDE	NYLON 12 POLYAMIDE	ETHYLENE VINYL ACETATE EVA	POLYVINYLCHLORIDE FLEXIBLE PVC
DIOCTYL SEBACATE	NR at 70°F					NR at 70°F
DIOFORM						
DIOXANE		BC at 70°F C at 122°F	A to 70°F	AB to 70°F	NR at 70°F	C/NR at 70°F NR at 122°F
DIOXETHYLENE ETHER						NR at 70°F
DIOXOLANES (DIOXOLANS)						
DIPENTENE						NR at 70°F
DIPHENYL					NR at 70°F	NR 99-100% at 70°F
DIPHENYL, CHLORINATED						
DIPHENYL ETHER						
DIPHENYL OXIDES						A to 70°F
DIPHENYL PHTHALATE						
DIPPEL'S OIL						
DIPROPYLAMINE						
DIPROPYL KETONE						
DIPROPYL METHANE						
DIPROPYLENE GLYCOL		A to 122°F				B at 70°F C at 122°F
DISODIUM PHOSPHATE					AB to 140°F	AB to 140°F
DISPERSING OIL NO. 10						
DISTILLED VINEGAR						
DIVINYL BENZENE						
DIVINYL ETHER						
DMP (DIMETHAL PHTHALATE)	NR at 70°F					
DMT (DIMETHYL TEREPHTHALATE)						
DODECANOL (DODECYL ALCOHOL)						
DODECYL BENZENE						
DODECYL TOLUENE						
DOLOMITE						
DOUGLAS FIR OIL						
DOW CHEMICAL 50-4, ET588						
DOW CHEMICAL ET378						
DOW CORNING 3, 4, 11						
DOW CORNING 5, 33, 44, 200, 220, 510, 550, 705, 710	A to 70°F (200)		A to 70°F AB to 200°F			
DOW CORNING 55						
DOW CORNING 1265 FLUOROSILICONE FLUIDS			A to 70°F			
DOW CORNING 1208, 4050, 6620, F-60, XF-60						
DOW CORNING F-61						
DOW GAUGE FLUID R-200						
DOW GENERAL WEED KILLER (PHENOL)						
DOW GENERAL WEED KILLER (WATER)						
DOW GUARD						
DOW PER						
DOW PURIFLOC C-31						
DOWANOLS						
DOWFUME W-40						NR at 70°F

	POLYETHYLENE LOW DENSITY LDPE	®TEFLON FEP	®KALREZ PERFLUORINATED ELASTOMER	®FLUORAZ FLUORINATED COPOLYMER	®AFLAS FLUORINATED COPOLYMER	®NORPRENE COPOLYMER TPO
DIOCTYL SEBACATE		A to 300°F	A to 212°F	A to 70°F	+8.8% vol 3 days 212°F +20% vol 3 days 350°F	
DIOFORM		A to 70°F				
DIOXANE	B/NR at 70°F	A to 300°F	A to 212°F	NR at 70°F	+57% vol 3 days 70°F NR at 70°F	
DIOXETHYLENE ETHER		A to 70°F				
DIOXOLANES (DIOXOLANS)		A to 70°F	A to 212°F	NR at 70°F		
DIPENTENE		A to 70°F	A to 212°F	C at 70°F		
DIPHENYL	C at 70°F	A to 300°F	A to 500°F	C/NR at 70°F	B at 70°F	
DIPHENYL, CHLORINATED						
DIPHENYL ETHER		*A to 125°F	A to 212°F			
DIPHENYL OXIDES		A to 70°F A to 75% to 300°F	A to 212°F	B at 70°F	B at 70°F	
DIPHENYL PHTHALATE		A to 70°F				
DIPPEL'S OIL		A to 70°F				
DIPROPYLAMINE		A to 70°F				
DIPROPYL KETONE		A to 70°F				
DIPROPYL METHANE		A to 70°F				
DIPROPYLENE GLYCOL	A to 122°F	A to 125°F				
DISODIUM PHOSPHATE	A to 70°F AB to 140°F	A to 70°F				
DISPERSING OIL NO. 10		A to 70°F				
DISTILLED VINEGAR		A to 70°F				
DIVINYL BENZENE		A to 70°F				
DIVINYL ETHER		A to 70°F				
DMP (DIMETHAL PHTHALATE)		A to 392°F	A to 212°F	BC at 70°F		
DMT (DIMETHYL TEREPHTHALATE)			A to 212°F	A to 70°F		
DODECANOL (DODECYL ALCOHOL)		A to 70°F				
DODECYL BENZENE		A to 70°F	A to 212°F			
DODECYL TOLUENE		A to 70°F				
DOLOMITE						
DOUGLAS FIR OIL		A to 70°F				
DOW CHEMICAL 50-4, ET588		A to 70°F		NR at 70°F		
DOW CHEMICAL ET378				C at 70°F		
DOW CORNING 3, 4, 11				A to 70°F		
DOW CORNING 5, 33, 44, 200, 220, 510, 550, 705, 710		A to 70°F AB to 450°F		A to 70°F		
DOW CORNING 55				A to 70°F		
DOW CORNING 1265 FLUOROSILICONE FLUIDS		A to 70°F				
DOW CORNING 1208, 4050, 6620, F-60, XF-60				A to 70°F		
DOW CORNING F-61		A to 70°F		A to 70°F		
DOW GAUGE FLUID R-200						
DOW GENERAL WEED KILLER (PHENOL)		A to 70°F				
DOW GENERAL WEED KILLER (WATER)		A to 70°F				
DOW GUARD				A to 70°F		
DOW PER		A to 70°F				
DOW PURIFLOC C-31		A to 70°F				
DOWANOLS			A to 212°F			
DOWFUME W-40		A to 70°F				

C 133

	RECOMMENDATIONS	NITRILE NBR	ETHYLENE PROPYLENE EP, EPDM	FLUOROCARBON FKM	CHLOROPRENE CR	HYDROGENATED NITRILE HNBR
DOWTHERM A	FKM, FVMQ	+145% vol 3 days 70°F dissolved 3 days 300°F	+16% vol 3 days 70°F +77% vol 3 days 300°F	no swell 3 days 70°F +11% vol 3 days 300°F	+185% vol 3 days 70°F dissolved 3 days 300°F	NR at 70°F
DOWTHERM A		NR at 70°F	NR at 70°F	A to 300°F BC at 400°F	NR at 70°F	
DOWTHERM E	FKM, FVMQ	NR at 70°F	NR at 70°F	A to 300°F	NR at 70°F	NR at 70°F
DOWTHERM OIL			NR at 70°F	A to 70°F C 50% at 257°F	NR at 70°F	
DOWTHERM S.R.-1		A to 70°F		A to 70°F		
DOWTHERM 209	EP (50% conc)	C 50% at 70°F	A 50% to 70°F	NR 50% at 70°F	B 50% at 70°F	C at 70°F
DRINKING WATER	EP, NBR, FKM	+12% vol 3 days 212°F A to 70°F	no swell 3 days 212°F A to 275°F	+5.9% vol 3 days 212°F A to 212°F	+12% vol 3 days 212°F AB to 212°F	A to 70°F
DRINKING WATER		B 180-212°F NR at 275°F		B at 275°F	NR at 275°F	
DRINOX, MORTON CHEMICAL CO.		NR at 70°F		NR at 70°F	NR at 70°F	
DRYCID					BC to 85°F	
DRY CLEANING FLUIDS	FKM, FVMQ	C at 70°F	NR at 70°F	A to 70°F	NR at 70°F	C at 70°F
DRY CLEANING SOLVENT		A to 70°F	NR at 70°F	A to 70°F	C at 70°F	
DTE LIGHT OIL	NBR, FKM	A to 70°F	NR at 70°F	A to 70°F	B at 70°F	A to 70°F
DUTCH OIL		NR at 70°F	NR at 70°F		NR at 70°F	
DUTCH LIQUID		NR at 70°F	NR at 70°F		NR at 70°F	
DYES, ABRASIVE		NR at 70°F		A/NR at 70°F	C/NR at 70°F	
DYES, WATER BASE				A to 70°F	AC to 70°F	
EARTH PITCH		A to 70°F	NR at 70°F	A to 70°F	B at 70°F	
ELCO 28 EP LUBRICANT	NBR, FKM	A to 70°F	NR at 70°F	A to 70°F	C at 70°F	A to 70°F
EMBALMING FLUID	NBR	A to 160°F	NR at 70°F	NR at 70°F	NR at 70°F	
EMK	EP, IIR	NR any conc at 70°F	A to 140°F AB to 200°F	NR at 70°F	NR at 70°F	
ENAMEL		A to 70°F			A to 70°F	
EPI		NR at 70°F			NR at 70°F	
EPICHLOROHYDRIN	EP, IIR	NR at 70°F	BC at 70°F	+94% vol 7 days 122°F NR at 70°F	NR at 70°F	NR at 70°F
EPOXY RESINS	EP, CR		A to 70°F	NR at 70°F	A to 70°F	
ESAM-6 FLUID	EP, SBR		A to 70°F	NR at 70°F	B at 70°F	
ESSENCE OF MYRBANE		NR at 70°F	C at 70°F	A to 70°F	NR at 70°F	
ESSO FUEL 208	NBR, FKM	A to 70°F	NR at 70°F	A to 70°F	B at 70°F	A to 70°F
ESSO GOLDEN GASOLINE	FKM	B at 70°F	NR at 70°F	+4% vol 28 days 75°F A to 70°F	NR at 70°F	B at 70°F
ESSO MOTOR OIL	NBR, FKM	A to 70°F	NR at 70°F	A to 300°F	C at 70°F	A to 70°F
ESSO TRANSMISSION FLUID (TYPE A)	NBR, FKM	A to 70°F	NR at 70°F	A to 212°F B at 300°F	B at 70°F	A to 70°F
ESSO TURBO OIL	FKM, NBR	A to 250°F NR at 400°F	NR at 70°F	A to 400°F	NR at 70°F	
ESSO WS2812 (MIL-L-7808A)	FKM, FVMQ, NBR	AB to 70°F	NR at 70°F	A to 300°F	NR at 70°F	A to 70°F
ESSO XP90-EP LUBRICANT	NBR, FKM	A to 70°F	NR at 70°F	A to 70°F	B at 70°F	A to 70°F
ESSTIC 42, 43	NBR, FKM	A to 70°F	NR at 70°F	A to 70°F	B at 70°F	A to 70°F
ETHAMINE		NR at 70°F	A to 70°F	NR at 70°F	NR at 70°F	
ETHANAL	EP, SBR, VMQ	NR any conc at 70°F	A to 104°F AB to 200°F	NR 100% at 70°F B 40% to 104°F	NR any conc at 70°F	
ETHANALAMINE		AB to 70°F B to 140°F	A to 70°F	B/NR at 70°F NR at 140°F	AB to 200°F	
ETHANE	NBR, FKM	A to 140°F	NR at 70°F	A to 140°F	B at 70°F NR at 140°F	
ETHANE NITRILE		C at 70°F	A to 70°F	NR at 70°F	A to 70°F	
ETHANEDIOIC ACID		AB to 100% to 140°F NR 10% boiling	A to 100% to 200°F	A to 100% to 140°F A to 50% to 176°F	B to 100% to 140°F NR 20-50% at 176°F	B at 70°F
ETHANETHIOL		NR at 70°F	NR at 70°F	B at 70°F	NR at 70°F	
ETHANOIC ACID	EP, IIR	NR 100% at 70°F C 80% at 70°F	AB 100% to 200°F A to 100% to 70°F	NR 90-100% at 70°F B/NR 80% at 70°F	B to 200°F A to 30% to 70°F	B 100% at 70°F
ETHANOL	EP, CR, SBR, NBR	A to 140°F B to 185°F	A to 200°F	+23% vol 7 days 104°F +6% vol 21 days 70°F	A to 158°F B at 176°F	A to 70°F

	STYRENE BUTADIENE SBR	POLYACRYLATE ACM	POLYURETHANE AU, EU	ISOBUTYLENE ISOPRENE IIR	POLYBUTADIENE BR	® AEROQUIP AQP
DOWTHERM A	+174% vol 3 days 70°F dissolved 3 days 300°F	+375% vol 3 days 70°F dissolved 3 days 300°F	+110% vol 3 days 70°F dissolved 3 days 300°F	NR at 70°F	NR at 70°F	NR at 70°F
DOWTHERM A	NR at 70°F	NR at 70°F	NR at 70°F			
DOWTHERM E	NR at 70°F	NR at 70°F	NR at 70°F	NR at 70°F	NR at 70°F	NR at 70°F
DOWTHERM OIL	NR at 70°F		B at 70°F	NR at 70°F	NR at 70°F	
DOWTHERM S.R.-1				A to 70°F		
DOWTHERM 209				B 50% at 70°F		
DRINKING WATER	A to 70°F B at 180°F	NR at 70°F	A/NR at 70°F NR at 180°F	+0.1% vol 3 days 212°F A to 212°F	A to 70°F	AB to 160°F
DRINKING WATER	NR at 275°F					
DRINOX, MORTON CHEMICAL CO.			NR at 70°F			
DRYCID						
DRY CLEANING FLUIDS	NR at 70°F	NR at 70°F	NR at 70°F	NR at 70°F	NR at 70°F	
DRY CLEANING SOLVENT			A to 70°F	NR at 70°F		
DTE LIGHT OIL	NR at 70°F	A to 70°F	B at 70°F light, medium, heavy	NR at 70°F	NR at 70°F	AB to 300°F
DUTCH OIL				BC at 70°F		
DUTCH LIQUID				BC at 70°F		
DYES, ABRASIVE			NR at 70°F			
DYES, WATER BASE						
EARTH PITCH			B at 70°F	NR at 70°F		
ELCO 28 EP LUBRICANT	NR at 70°F	A to 70°F	A to 70°F	NR at 70°F	NR at 70°F	
EMBALMING FLUID						
EMK				NR at 70°F		
ENAMEL			AB to 70°F			
EPI				NR at 70°F		
EPICHLOROHYDRIN	NR at 70°F	NR at 70°F	NR at 70°F	B at 70°F	NR at 70°F	
EPOXY RESINS				A to 70°F		
ESAM-6 FLUID	A to 70°F			B at 70°F		
ESSENCE OF MYRBANE			NR at 70°F	BC any conc to 80°F		
ESSO FUEL 208	NR at 70°F	A to 70°F	NR at 70°F	NR at 70°F	NR at 70°F	
ESSO GOLDEN GASOLINE	NR at 70°F	NR at 70°F	NR at 70°F	NR at 70°F	NR at 70°F	
ESSO MOTOR OIL	NR at 70°F	A to 70°F	NR at 70°F	NR at 70°F	NR at 70°F	
ESSO TRANSMISSION FLUID (TYPE A)	NR at 70°F	A to 70°F	C at 70°F	NR at 70°F	NR at 70°F	
ESSO TURBO OIL	NR at 70°F	B to 250°F NR at 400°F	NR at 70°F			
ESSO WS2812 (MIL-L-7808A)	NR at 70°F	B at 70°F	NR at 70°F	NR at 70°F	NR at 70°F	
ESSO XP90-EP LUBRICANT	NR at 70°F	A to 70°F	A to 70°F	NR at 70°F	NR at 70°F	
ESSTIC 42, 43	NR at 70°F	A to 70°F	B at 70°F	NR at 70°F	NR at 70°F	
ETHAMINE				A to 70°F		
ETHANAL			C at 70°F	A any conc to 80°F		
ETHANALAMINE			NR at 70°F	A to 70°F		
ETHANE	NR at 70°F	A to 70°F	BC at 70°F	NR at 70°F	NR at 70°F	
ETHANE NITRILE				A to 70°F		
ETHANEDIOIC ACID				A to 70°F		
ETHANETHIOL				NR at 70°F		
ETHANOIC ACID	B 5-100% at 70°F NR hot	NR 5-100% at 70°F	NR 10-100% at 70°F BC dilute at 70°F	B 100% to 125°F AB to 50% to 80°F	B 100% to 70°F NR hot	NR 30-100% at 70°F
ETHANOL	A to 200°F	NR at 70°F	C/NR at 70°F NR at 140°F	A to 70°F AB to 185°F	A to 70°F	AB to 70°F

	SYNTHETIC ISOPRENE IR	NATURAL ISOPRENE NR	CHLOROSULFONATED POLYETHYLENE CSM	FLUOROSILICONE FVMQ	SILICONE VMQ	CHEMRAZ FFKM
DOWTHERM A	NR at 70°F	NR at 70°F	BC at 70°F	+2.5% vol 28 days 70°F +7.3% vol 28 days 212°F	no swell 3 days 70°F +15% vol 3 days 300°F	A to 70°F
DOWTHERM A				+22.6% vol 28 days 400°F A to 300°F	B/NR at 70°F	
DOWTHERM E	NR at 70°F	NR at 70°F	B/NR at 70°F	B at 70°F	A/NR at 70°F	A to 70°F
DOWTHERM OIL	NR at 70°F	NR at 70°F	NR at 70°F	A to 70°F	B at 70°F	
DOWTHERM S.R.-1	A to 70°F	A to 70°F	A to 70°F			
DOWTHERM 209				C 50% at 70°F	C 50% at 70°F	A 50-100% to 70°F
DRINKING WATER	A to 70°F	A to 70°F	+4% vol 3 days 212°F A to 212°F	A to 275°F	+2% vol 3 days 212°F A to 70°F	
DRINKING WATER					B at 180°F NR at 275°F	A to 70°F B at 70°F white
DRINOX, MORTON CHEMICAL CO.		NR at 70°F				
DRYCID	C at 70°F	BC at 70°F				
DRY CLEANING FLUIDS	NR at 70°F	NR at 70°F	NR at 70°F	B at 70°F	NR at 70°F	B at 70°F
DRY CLEANING SOLVENT	NR at 70°F	NR at 70°F	NR at 70°F		NR at 70°F	
DTE LIGHT OIL	NR at 70°F	NR at 70°F	NR at 70°F	A to 70°F	NR at 70°F	A to 70°F
DUTCH OIL	NR at 70°F	NR at 70°F				
DUTCH LIQUID	NR at 70°F	NR at 70°F				
DYES, ABRASIVE						
DYES, WATER BASE						
EARTH PITCH	NR at 70°F	NR at 70°F	B at 70°F		NR at 70°F	
ELCO 28 EP LUBRICANT	NR at 70°F	NR at 70°F	NR at 70°F	A to 70°F	B at 70°F	A to 70°F
EMBALMING FLUID						
EMK	NR at 70°F	NR at 70°F	NR at 70°F	NR at 70°F	NR at 70°F	A to 70°F
ENAMEL						
EPI		NR at 70°F				
EPICHLOROHYDRIN	NR at 70°F	NR at 70°F	NR at 70°F	NR at 70°F	NR at 70°F	B at 70°F
EPOXY RESINS						A to 70°F
ESAM-6 FLUID			B at 70°F	NR at 70°F		A to 70°F
ESSENCE OF MYRBANE	NR at 70°F	NR at 70°F	NR at 70°F		C at 70°F	
ESSO FUEL 208	NR at 70°F	NR at 70°F	C at 70°F	A to 70°F	NR at 70°F	A to 70°F
ESSO GOLDEN GASOLINE	NR at 70°F	NR at 70°F	NR at 70°F	A to 70°F	NR at 70°F	A to 70°F
ESSO MOTOR OIL	NR at 70°F	NR at 70°F	NR at 70°F	A to 70°F	NR at 70°F	A to 70°F
ESSO TRANSMISSION FLUID (TYPE A)	NR at 70°F	NR at 70°F	NR at 70°F	A to 70°F	NR at 70°F	A to 70°F
ESSO TURBO OIL				A to 400°F	NR at 70°F	
ESSO WS2812 (MIL-L-7808A)	NR at 70°F	NR at 70°F	NR at 70°F	A to 300°F	NR at 70°F	A to 70°F
ESSO XP90-EP LUBRICANT	NR at 70°F	NR at 70°F	B at 70°F	A to 70°F	NR at 70°F	A to 70°F
ESSTIC 42, 43	NR at 70°F	NR at 70°F	NR at 70°F	A to 70°F	NR at 70°F	A to 70°F
ETHAMINE	B at 70°F	B at 70°F	C at 70°F			
ETHANAL	BC at 70°F	B/NR at 70°F	C/NR at 70°F	NR at 70°F	AB to 70°F	B/NR at 70°F
ETHANALAMINE	C at 70°F	BC any conc to 80°F	B to 200°F		BC at 70°F	
ETHANE	NR at 70°F	NR at 70°F	B at 70°F	B at 70°F	NR at 70°F	A to 70°F
ETHANE NITRILE	B at 70°F	B at 70°F	B at 70°F			
ETHANEDIOIC ACID	B at 70°F	BC at 70°F AB 10% to 150°F	A to 70°F AB to 185°F	A to 70°F	B at 70°F	A to 70°F
ETHANETHIOL	NR at 70°F	NR at 70°F	NR at 70°F			
ETHANOIC ACID	BC 30-100% at 70°F NR hot	B/NR 80-100% at 70°F B to 30% at 70°F	C 100% at 70°F A to 80% to 70°F	NR 100% at 70°F B to 90% to 70°F	B to 70°F C/NR hot	AB 100% to 70°F A to 30% to 70°F
ETHANOL	A to 70°F	A to 70°F AB to 150°F	A to 200°F	+5% vol 7 days 70°F A to 70°F	+20% vol 7 days 70°F B at 70°F	A to 70°F

C 136

	POLYSULFIDE T	CHLORINATED POLYETHYLENE CM	EPICHLOROHYDRIN CO, ECO	® HYTREL COPOLYESTER TPE	® SANTOPRENE COPOLYMER TPO	® C-FLEX STYRENIC TPE
DOWTHERM A	NR at 70°F	NR at 70°F		AB to 70°F		
DOWTHERM A						
DOWTHERM E	NR at 70°F	NR at 70°F		AB to 70°F		
DOWTHERM OIL			NR at 70°F			
DOWTHERM S.R.-1						
DOWTHERM 209						
DRINKING WATER	NR at 70°F	AB to 70°F	B at 70°F	AC to 212°F	+4% vol 7 days 212°F A to 212°F	AB to 70°F
DRINKING WATER						
DRINOX, MORTON CHEMICAL CO.						
DRYCID						
DRY CLEANING FLUIDS	NR at 70°F					
DRY CLEANING SOLVENT						
DTE LIGHT OIL				AB to 200°F		
DUTCH OIL						
DUTCH LIQUID						
DYES, ABRASIVE						
DYES, WATER BASE						
EARTH PITCH						
ELCO 28 EP LUBRICANT	B at 70°F					
EMBALMING FLUID						
EMK						
ENAMEL				AB to 70°F		
EPI						
EPICHLOROHYDRIN	NR at 70°F			*NR at 70°F		
EPOXY RESINS						
ESAM-6 FLUID	NR at 70°F					
ESSENCE OF MYRBANE						
ESSO FUEL 208	A to 70°F					
ESSO GOLDEN GASOLINE	B at 70°F					
ESSO MOTOR OIL	A to 70°F					
ESSO TRANSMISSION FLUID (TYPE A)	A to 70°F					
ESSO TURBO OIL						
ESSO WS2812 (MIL-L-7808A)	B at 70°F					
ESSO XP90-EP LUBRICANT	A to 70°F					
ESSTIC 42, 43						
ETHAMINE						
ETHANAL						
ETHANALAMINE						
ETHANE	A to 70°F					
ETHANE NITRILE						
ETHANEDIOIC ACID						
ETHANETHIOL						
ETHANOIC ACID	BC at 70°F B 5-30% at 70°F	A to 100% to 70°F	NR at 70°F B 30% at 70°F	AC to 70°F C/NR at 100°F	NR at 70°F A to 30% to 70°F	
ETHANOL	A to 70°F	AB to 70°F	A to 70°F	A to 70°F	A to 70°F	AB to 70°F

	ETHYLENE ACRYLIC EA	POLYALLOMER LINEAR COPOLYMER	NYLON 11 POLYAMIDE	NYLON 12 POLYAMIDE	ETHYLENE VINYL ACETATE EVA	POLYVINYLCHLORIDE FLEXIBLE PVC
DOWTHERM A			NR at 70°F			NR at 70°F
DOWTHERM A						
DOWTHERM E			NR at 70°F			NR at 70°F
DOWTHERM OIL						
DOWTHERM S.R.-1						
DOWTHERM 209	A 50% to 70°F					
DRINKING WATER	A to 212°F	A to 70°F	A to 70°F C/NR 160-212°F		A to 70°F	A to 70°F
DRINKING WATER						
DRINOX, MORTON CHEMICAL CO.						
DRYCID						
DRY CLEANING FLUIDS						
DRY CLEANING SOLVENT						
DTE LIGHT OIL			AB to 70°F			
DUTCH OIL						NR at 70°F
DUTCH LIQUID						NR at 70°F
DYES, ABRASIVE						C at 70°F
DYES, WATER BASE						C at 70°F
EARTH PITCH						
ELCO 28 EP LUBRICANT	A to 70°F					
EMBALMING FLUID						
EMK						
ENAMEL			AB to 70°F		B at 70°F	C/NR at 70°F
EPI						
EPICHLOROHYDRIN						NR at 70°F
EPOXY RESINS						
ESAM-6 FLUID						
ESSENCE OF MYRBANE						NR at 70°F
ESSO FUEL 208						
ESSO GOLDEN GASOLINE	C at 70°F					
ESSO MOTOR OIL	A to 70°F			AB to 70°F		no swell 14 days 70°F A to 70°F
ESSO TRANSMISSION FLUID (TYPE A)	A to 70°F					
ESSO TURBO OIL						
ESSO WS2812 (MIL-L-7808A)						
ESSO XP90-EP LUBRICANT	A to 70°F					
ESSTIC 42, 43						
ETHAMINE						
ETHANAL						NR at 70°F
ETHANALAMINE						AB to 70°F
ETHANE					NR at 70°F	A/NR at 70°F
ETHANE NITRILE						
ETHANEDIOIC ACID						
ETHANETHIOL						NR at 70°F
ETHANOIC ACID	NR at 70°F A to 30% to 70°F	A to 100% to 70°F A to 50% to 122°F	NR 98-100% at 70°F AC 50% to 70°F	NR 40-100% at 70°F C 10% at 70°F	B/NR to 140°F B 30% at 70°F	C/NR 85-100% at 70°F AB 50% at 70°F
ETHANOL	AC to 70°F	A 40-100% to 70°F B 40-100% at 122°F	+8.8% vol 180 days 70°F NR at 140°F	C/NR 50-96% at 70°F AB 10% to 70°F	A to 70°F AB 35-100% to 140°F	A/NR at 70°F

	POLYETHYLENE LOW DENSITY LDPE	®TEFLON FEP	®KALREZ PERFLUORINATED ELASTOMER	®FLUORAZ FLUORINATED COPOLYMER	®AFLAS FLUORINATED COPOLYMER	®NORPRENE COPOLYMER TPO
DOWTHERM A	C at 70°F	A to 300°F	A to 500°F	AC at 70°F		
DOWTHERM A		A to 70°F	A to 212°F	A to 70°F		
DOWTHERM E			A to 212°F	A to 70°F		
DOWTHERM OIL		A to 70°F	A to 212°F			
DOWTHERM S.R.-1		A to 70°F	A to 212°F			
DOWTHERM 209			A to 212°F	A 50-100% to 70°F	+6.3% vol 3 days 212°F AB at 212°F	
DRINKING WATER	A to 70°F	A to 212°F	A to 212°F	A to 70°F	+1.1% vol 3 days 212°F A to 212°F	A to 70°F
DRINKING WATER				A to 70°F		
DRINOX, MORTON CHEMICAL CO.						
DRYCID		A to 70°F				
DRY CLEANING FLUIDS		A to 70°F	A to 212°F	C at 70°F		
DRY CLEANING SOLVENT		A to 70°F				
DTE LIGHT OIL		AB to 400°F		A to 70°F	A to 70°F	
DUTCH OIL		A to 70°F				
DUTCH LIQUID		A to 70°F				
DYES, ABRASIVE						
DYES, WATER BASE						
EARTH PITCH		A to 70°F				
ELCO 28 EP LUBRICANT				A to 70°F	A to 70°F	
EMBALMING FLUID						
EMK						
ENAMEL	AB to 70°F			A to 70°F		
EPI						
EPICHLOROHYDRIN		A to 300°F	*A to 212°F	NR at 70°F		
EPOXY RESINS				B at 70°F		
ESAM-6 FLUID				NR at 70°F		
ESSENCE OF MYRBANE		A to 70°F				
ESSO FUEL 208				A to 70°F		
ESSO GOLDEN GASOLINE				C at 70°F		
ESSO MOTOR OIL				A to 70°F		
ESSO TRANSMISSION FLUID (TYPE A)				A to 70°F		
ESSO TURBO OIL						
ESSO WS2812 (MIL-L-7808A)		A to 70°F		A to 70°F		
ESSO XP90-EP LUBRICANT				A to 70°F	A to 70°F	
ESSTIC 42, 43				A to 70°F		
ETHAMINE		A to 70°F				
ETHANAL	C/NR 40-100% at 70°F NR 40-100% at 122°F	A to 100% to 200°F	A to 70°F	C/NR at 70°F		A to 70°F
ETHANALAMINE		A to 70°F				
ETHANE	A/NR at 70°F	A to 70°F	A to 212°F	A to 70°F		
ETHANE NITRILE		A to 70°F				
ETHANEDIOIC ACID	A to 70°F AB any conc to 150°F	A to 200°F	AB to 70°F A dilute to 70°F	A to 70°F		A to 70°F
ETHANETHIOL		A to 70°F				
ETHANOIC ACID	AB to 122°F NR at 140°F	A to 100% to boiling	A to 75°F BC at 428°F	C/NR 100% at 70°F B 30% at 70°F	NR at 70°F	A to 100% to 70°F
ETHANOL	A/NR to 140°F AB to 40% to 70°F	A to 392°F	no swell 7 days 70°F A to 212°F	A to 70°F	A to 70°F AB to 158°F	BC at 70°F

	RECOMMENDATIONS	NITRILE NBR	ETHYLENE PROPYLENE EP, EPDM	FLUOROCARBON FKM	CHLOROPRENE CR	HYDROGENATED NITRILE HNBR
ETHANOLAMINE	EP,CR,NBR,IIR	B any conc to 80°F	AB to 200°F	NR at 70°F	B any conc to 80°F	
ETHANOLAMINE, MONO, DI, TRI	EP,IIR,CR	A/NR at 70°F	A to 120°F	NR at 70°F	A/NR at 70°F	C at 70°F(TRI)
ETHANOYL CHLORIDE	FKM,FVMQ	NR at 70°F	C at 70°F	B at 70°F	NR at 70°F	NR at 70°F
ETHER (ETHERS)	T	B/NR to 120°F NR at 140°F	C to 140°F	C/NR at 70°F	NR at 70°F	NR at 70°F
ETHINE		A any conc to 80°F	A to 70°F	A to 70°F	AB any conc to 80°F	
ETHOCEL		B at 70°F	B at 70°F	A to 70°F	B at 70°F	
ETHOXYETHANE	T	NR at 70°F	NR at 70°F	NR at 70°F	NR at 70°F	NR at 70°F
ETHOXYTHANOL	EP,NBR,IIR	AB to 70°F BC 104-150°F	AB to 70°F	B/NR at 70°F	C/NR at 70°F	
ETHYL ACETATE	EP,IIR,T	NR at 70°F	A to 130°F BC at 158°F	+425% vol 1 day 75°F NR at 70°F	NR at 70°F	NR at 70°F
ETHYL ACETIC ACID		C at 70°F	C at 70°F	C at 70°F	C/NR at 70°F	
ETHYL ACETOACETATE	EP,IIR,T	NR at 70°F	AB to 104°F	NR at 70°F	NR at 70°F	
ETHYL ACETYLENE					AB to 70°F	
ETHYL ACRYLATE	EP,IIR,T	NR at 70°F	B at 70°F	+234% vol 7 days 75°F NR at 70°F	NR at 70°F	NR at 70°F
ETHYL ACRYLIC ACID	EP	NR at 70°F	B at 70°F		B at 70°F	
ETHYL ALCOHOL	EP,CR,IIR,NBR	A to 140°F B to 185°F	A to 200°F	+23% vol 7 days 104°F A to 70°F	A to 158°F B at 176°F	A to 70°F
ETHYL ALDEHYDE		NR at 70°F	A to 70°F	NR at 70°F	C/NR at 70°F	
ETHYL ALUMINUM DICHLORIDE		NR at 70°F		B at 70°F		
ETHYL AMINE	EP	C/NR at 70°F	A to 250°F	NR at 70°F	NR at 70°F	
ETHYL BENZENE	FKM,FVMQ	NR at 70°F	NR at 70°F	A to 70°F	NR at 70°F	NR at 70°F
ETHYL BENZOATE	FKM,FVMQ,EP,T	NR at 70°F	B/NR at 70°F	A to 70°F	NR at 70°F	NR at 70°F
ETHYL BROMIDE	FKM	B at 70°F	NR at 70°F	A to 200°F	NR at 70°F	
ETHYL BUTANOATE		NR at 70°F			NR at 70°F	
ETHYL BUTANOL		A to 70°F		B at 70°F		
ETHYL BUTYL ACETATE		NR at 70°F		NR at 70°F		
ETHYL BUTYL ALCOHOL		A to 70°F		B at 70°F		
ETHYL BUTYL AMINE		B at 70°F				
ETHYL BUTYL KETONE		NR at 70°F		NR at 70°F		
ETHYL BUTYRALDEHYDE		NR at 70°F		NR at 70°F		
ETHYL BUTYRATE		NR at 70°F	NR at 70°F	C at 70°F	NR at 70°F	
ETHYL BUTYRIC ESTER		NR at 70°F	NR at 70°F	C at 70°F	NR at 70°F	
ETHYL CAPRYLATE		NR at 70°F	NR at 70°F		NR at 70°F	
ETHYL CAPRYLIC ESTER		NR at 70°F	NR at 70°F		NR at 70°F	
ETHYL CELLOSOLVE	EP,NBR,IIR,T	AB to 70°F BC 104-150°F	AB to 70°F	B/NR at 70°F	C/NR at 70°F	
ETHYL CELLULOSE	NBR,CR,EP	B to 200°F	B at 70°F	NR at 70°F	B at 70°F	
ETHYL CHLORIDE	NBR,EP,IIR,FKM	A to 140°F wet or dry	A to 140°F	A to 140°F AB to 200°F	NR at 70°F	A to 70°F
ETHYL CHLOROCARBONATE	FKM,FVMQ	NR at 70°F	NR at 70°F	A to 200°F	C/NR at 70°F	NR at 70°F
ETHYL CHLOROFORMATE	FKM,FVMQ	NR at 70°F	NR at 70°F	A to 70°F	C/NR at 70°F	NR at 70°F
ETHYL CYANIDE		NR at 70°F	A to 70°F	NR at 70°F	B at 70°F	
ETHYL CYCLOPENTANE	NBR,FKM,AU	A to 70°F	NR at 70°F	A to 70°F	C at 70°F	A to 70°F
ETHYL DIACETATE		NR at 70°F			NR at 70°F	
ETHYL DICHLORIDE		NR at 70°F	B/NR at 70°F	AB to 70°F	NR at 70°F	NR at 70°F
ETHYL DIMETHYL ACETATE		NR at 70°F	NR at 70°F		NR at 70°F	
ETHYL ETHER	T	+44% vol 3 days 70°F NR at 70°F	+58% vol 3 days 70°F NR at 70°F	+97% vol 7 days 70°F NR at 70°F	+64% vol 3 days 70°F NR at 70°F	NR at 70°F
ETHYL FORMATE	FKM,FVMQ,EP	NR at 70°F	B at 70°F	A to 70°F	B at 70°F	NR at 70°F

	STYRENE BUTADIENE SBR	POLYACRYLATE ACM	POLYURETHANE AU, EU	ISOBUTYLENE ISOPRENE IIR	POLYBUTADIENE BR	® AEROQUIP AQP
ETHANOLAMINE	B/NR at 70°F	NR at 70°F	C at 70°F	B any conc to 140°F	B at 70°F	
ETHANOLAMINE, MONO, DI, TRI	B/NR at 70°F	NR at 70°F	C/NR at 70°F	AB any conc to 150°F	B at 70°F	
ETHANOYL CHLORIDE				C at 70°F		
ETHER (ETHERS)	NR at 70°F	C/NR at 70°F	B/NR at 70°F	NR at 70°F	NR at 70°F	AB to 70°F
ETHINE				AB any conc to 80°F		
ETHOCEL			B at 70°F	A to 70°F		
ETHOXYETHANE			A to 70°F	C at 70°F		
ETHOXYTHANOL	B at 70°F	NR at 70°F	NR at 70°F	A to 70°F AB to 150°F	B at 70°F	
ETHYL ACETATE	NR at 70°F	NR at 70°F	NR at 70°F AB to 70°F fumes	B any conc to 100°F	NR at 70°F	AB to 70°F
ETHYL ACETIC ACID				C/NR at 70°F		
ETHYL ACETOACETATE	C at 70°F	NR at 70°F	NR at 70°F	B any conc to 80°F	C at 70°F	
ETHYL ACETYLENE				AB to 70°F		
ETHYL ACRYLATE	NR at 70°F	NR at 70°F	NR at 70°F	B at 70°F	NR at 70°F	
ETHYL ACRYLIC ACID	NR at 70°F	NR at 70°F	NR at 70°F	B at 70°F	NR at 70°F	
ETHYL ALCOHOL	A to 200°F	NR at 70°F	C/NR at 70°F NR at 140°F	A to 70°F AB to 185°F	A to 70°F	AB to 70°F
ETHYL ALDEHYDE			C at 70°F	AB any conc to 80°F		
ETHYL ALUMINUM DICHLORIDE				NR at 70°F		
ETHYL AMINE	B/NR at 70°F	NR at 70°F	C/NR at 70°F	B any conc to 140°F	B at 70°F	
ETHYL BENZENE	NR at 70°F	NR at 70°F	NR at 70°F	NR at 70°F	NR at 70°F	
ETHYL BENZOATE	NR at 70°F	NR at 70°F	NR at 70°F	NR at 70°F	NR at 70°F	
ETHYL BROMIDE	NR at 70°F	NR at 70°F	C/NR at 70°F	NR at 70°F	NR at 70°F	
ETHYL BUTANOATE				NR at 70°F		
ETHYL BUTANOL				A to 70°F		
ETHYL BUTYL ACETATE				A to 70°F		
ETHYL BUTYL ALCOHOL				A to 70°F		
ETHYL BUTYL AMINE				A to 70°F		
ETHYL BUTYL KETONE				B at 70°F		
ETHYL BUTYRALDEHYDE				B at 70°F		
ETHYL BUTYRATE				C/NR at 70°F		
ETHYL BUTYRIC ESTER				C at 70°F		
ETHYL CAPRYLATE				NR at 70°F		
ETHYL CAPRYLIC ESTER				NR at 70°F		
ETHYL CELLOSOLVE	NR at 70°F	NR at 70°F	NR at 70°F	AB to 70°F	NR at 70°F	
ETHYL CELLULOSE	B at 70°F	NR at 70°F	B at 70°F	B at 70°F	B at 70°F	AB to 70°F
ETHYL CHLORIDE	B at 70°F	C at 70°F	BC at 70°F	A to 70°F	B at 70°F	NR at 70°F
ETHYL CHLOROCARBONATE	NR at 70°F	NR at 70°F	NR at 70°F	NR at 70°F	NR at 70°F	
ETHYL CHLOROFORMATE	NR at 70°F	NR at 70°F	NR at 70°F	NR at 70°F	NR at 70°F	
ETHYL CYANIDE				A to 70°F		
ETHYL CYCLOPENTANE	NR at 70°F	B at 70°F	A to 70°F	NR at 70°F	NR at 70°F	
ETHYL DIACETATE				B at 70°F		
ETHYL DICHLORIDE	NR at 70°F	NR at 70°F	NR at 70°F	C at 70°F		
ETHYL DIMETHYL ACETATE				NR at 70°F		
ETHYL ETHER	+82% vol 3 days 70°F NR at 70°F	+99% vol 3 days 70°F NR at 70°F	+36% vol 3 days 70°F B/NR at 70°F	+51% vol 3 days 70°F C/NR at 70°F	NR at 70°F	
ETHYL FORMATE	NR at 70°F			B at 70°F	NR at 70°F	

	SYNTHETIC ISOPRENE IR	NATURAL ISOPRENE NR	CHLOROSULFONATED POLYETHYLENE CSM	FLUOROSILICONE FVMQ	SILICONE VMQ	CHEMRAZ FFKM
ETHANOLAMINE	B at 70°F	B at 70°F	B/NR at 70°F	NR at 70°F	B at 70°F	A to 70°F
ETHANOLAMINE, MONO, DI, TRI	B at 70°F	B at 70°F	AC to 70°F	NR at 70°F	B/NR at 70°F	A to 70°F
ETHANOYL CHLORIDE	NR at 70°F	NR at 70°F	B/NR at 70°F	A to 70°F	C at 70°F	A to 70°F
ETHER (ETHERS)	NR at 70°F	NR at 70°F	NR at 70°F	C at 70°F	NR at 70°F	A to 70°F
ETHINE	A to 70°F	A any conc to 80°F	B at 70°F		C at 70°F	
ETHOCEL	B at 70°F	B at 70°F	B at 70°F		C at 70°F	
ETHOXYETHANE	NR at 70°F	NR at 70°F	NR at 70°F	NR at 70°F	NR at 70°F	A to 70°F
ETHOXYTHANOL	B/NR at 70°F	B/NR at 70°F	B/NR at 70°F	B/NR at 70°F	B at 70°F	
ETHYL ACETATE	NR at 70°F	NR at 70°F	NR at 70°F	NR at 70°F	B at 70°F	A to 70°F
ETHYL ACETIC ACID	C at 70°F	C at 70°F	B at 70°F			
ETHYL ACETOACETATE	C at 70°F	BC to 80°F	NR at 70°F	NR at 70°F	B at 70°F	A to 70°F
ETHYL ACETYLENE						
ETHYL ACRYLATE	NR at 70°F	NR at 70°F	NR at 70°F	NR at 70°F	B at 70°F	A to 70°F
ETHYL ACRYLIC ACID	NR at 70°F	NR at 70°F	NR at 70°F	NR at 70°F	NR at 70°F	A to 70°F
ETHYL ALCOHOL	A to 70°F	A to 70°F AB to 150°F	A to 200°F	+5% vol 7 days 70°F A to 70°F	+20% vol 7 days 70°F B at 70°F	A to 70°F
ETHYL ALDEHYDE	C at 70°F	C/NR at 70°F	C at 70°F		A to 70°F	
ETHYL ALUMINUM DICHLORIDE	NR at 70°F	NR at 70°F	NR at 70°F			A to 70°F
ETHYL AMINE	BC at 70°F	BC at 70°F	BC at 70°F	NR at 70°F	BC at 70°F	
ETHYL BENZENE	NR at 70°F	NR at 70°F	NR at 70°F	A to 70°F	NR at 70°F	A to 70°F
ETHYL BENZOATE	NR at 70°F	NR at 70°F	NR at 70°F	A to 70°F	NR at 70°F	A to 70°F
ETHYL BROMIDE	B/NR at 70°F	B/NR at 70°F	NR at 70°F	A to 70°F	NR at 70°F	A to 70°F
ETHYL BUTANOATE	NR at 70°F	NR at 70°F	NR at 70°F			A to 70°F
ETHYL BUTANOL	A to 70°F	A to 70°F	A to 70°F			
ETHYL BUTYL ACETATE	NR at 70°F	NR at 70°F	B at 70°F			
ETHYL BUTYL ALCOHOL	A to 70°F	A to 70°F	A to 70°F			
ETHYL BUTYL AMINE	B at 70°F	B at 70°F	C at 70°F			
ETHYL BUTYL KETONE	NR at 70°F	NR at 70°F	NR at 70°F			
ETHYL BUTYRALDEHYDE	NR at 70°F	NR at 70°F	NR at 70°F			
ETHYL BUTYRATE	NR at 70°F	NR at 70°F	NR at 70°F			A to 70°F
ETHYL BUTYRIC ESTER	NR at 70°F	NR at 70°F	NR at 70°F			
ETHYL CAPRYLATE	NR at 70°F	NR at 70°F				
ETHYL CAPRYLIC ESTER	NR at 70°F	NR at 70°F				
ETHYL CELLOSOLVE	B/NR at 70°F	B/NR at 70°F	B/NR at 70°F	B/NR at 70°F	NR at 70°F	A to 70°F
ETHYL CELLULOSE	B at 70°F	B at 70°F	B at 70°F	NR at 70°F	BC at 70°F	A to 70°F
ETHYL CHLORIDE	BC at 70°F	B/NR at 70°F	B/NR at 70°F	A to 70°F	C/NR at 70°F	A to 70°F
ETHYL CHLOROCARBONATE	NR at 70°F	NR at 70°F	C/NR at 70°F	B at 70°F	NR at 70°F	A to 70°F
ETHYL CHLOROFORMATE	NR at 70°F	NR at 70°F	C/NR at 70°F	B at 70°F	NR at 70°F	
ETHYL CYANIDE	A to 70°F	A to 70°F				
ETHYL CYCLOPENTANE	NR at 70°F	NR at 70°F	NR at 70°F	A to 70°F	NR at 70°F	A to 70°F
ETHYL DIACETATE	B at 70°F	B at 70°F				
ETHYL DICHLORIDE	NR at 70°F	NR at 70°F	NR at 70°F	A to 70°F	C/NR at 70°F	A to 70°F
ETHYL DIMETHYL ACETATE	NR at 70°F	NR at 70°F				
ETHYL ETHER	NR at 70°F	NR at 70°F	NR at 70°F	+60% vol 3 days 70°F NR at 70°F	+103% vol 3 days 70°F NR at 70°F	A to 70°F
ETHYL FORMATE	NR at 70°F	NR at 70°F	B at 70°F	A to 70°F		BC at 70°F C at 70°F white

C 142

	POLYSULFIDE T	CHLORINATED POLYETHYLENE CM	EPICHLOROHYDRIN CO, ECO	® HYTREL COPOLYESTER TPE	® SANTOPRENE COPOLYMER TPO	® C-FLEX STYRENIC TPE
ETHANOLAMINE	B at 70°F		B at 70°F			
ETHANOLAMINE, MONO, DI, TRI	B/NR at 70°F		B/NR at 70°F	NR at 70°F		
ETHANOYL CHLORIDE						
ETHER (ETHERS)	A to 160°F	AB to 70°F		B/NR at 70°F		NR at 70°F
ETHINE						
ETHOCEL						
ETHOXYETHANE						
ETHOXYTHANOL	B at 70°F					
ETHYL ACETATE	B at 70°F	AB to 70°F	NR at 70°F	BC at 70°F	*A to 70°F	A to 70°F
ETHYL ACETIC ACID						
ETHYL ACETOACETATE	B at 70°F					
ETHYL ACETYLENE						
ETHYL ACRYLATE	B at 70°F		NR at 70°F			
ETHYL ACRYLIC ACID						
ETHYL ALCOHOL	A to 70°F	AB to 70°F	A to 70°F	A to 70°F	A to 70°F	AB to 70°F
ETHYL ALDEHYDE						
ETHYL ALUMINUM DICHLORIDE						
ETHYL AMINE	B at 70°F		B at 70°F			A to 70°F
ETHYL BENZENE	NR at 70°F		NR at 70°F		*B/NR at 70°F	
ETHYL BENZOATE	B at 70°F					
ETHYL BROMIDE	TEST					A to 70°F
ETHYL BUTANOATE						
ETHYL BUTANOL						
ETHYL BUTYL ACETATE						
ETHYL BUTYL ALCOHOL						
ETHYL BUTYL AMINE						
ETHYL BUTYL KETONE						
ETHYL BUTYRALDEHYDE						
ETHYL BUTYRATE						
ETHYL BUTYRIC ESTER						
ETHYL CAPRYLATE						
ETHYL CAPRYLIC ESTER						
ETHYL CELLOSOLVE	B at 70°F					
ETHYL CELLULOSE	NR at 70°F	AB to 70°F		AB to 70°F		
ETHYL CHLORIDE	NR at 70°F	NR at 70°F	B at 70°F	NR at 70°F	NR at 70°F	AB to 70°F
ETHYL CHLOROCARBONATE	NR at 70°F					
ETHYL CHLOROFORMATE	NR at 70°F					
ETHYL CYANIDE						
ETHYL CYCLOPENTANE	B at 70°F					
ETHYL DIACETATE						
ETHYL DICHLORIDE						
ETHYL DIMETHYL ACETATE						
ETHYL ETHER	+12% vol 3 days 70°F AB to 160°F	AB to 70°F	B at 70°F			NR at 70°F
ETHYL FORMATE	NR at 70°F		NR at 70°F			

	ETHYLENE ACRYLIC EA	POLYALLOMER LINEAR COPOLYMER	NYLON 11 POLYAMIDE	NYLON 12 POLYAMIDE	ETHYLENE VINYL ACETATE EVA	POLYVINYLCHLORIDE FLEXIBLE PVC
ETHANOLAMINE						NR at 70°F
ETHANOLAMINE, MONO, DI, TRI	A to 158°F		AB to 140°F			NR at 70°F
ETHANOYL CHLORIDE						
ETHER (ETHERS)		C/NR at 70°F	A to 70°F	AB to 70°F	C/NR at 70°F	C/NR at 70°F NR at 122°F
ETHINE						
ETHOCEL						
ETHOXYETHANE						
ETHOXYTHANOL			A to 70°F			
ETHYL ACETATE	NR at 70°F	A to 122°F	A to 70°F AB to 140°F	AB to 70°F	AC to 70°F NR at 140°F	+132% vol 7 days 70°F NR at 70°F
ETHYL ACETIC ACID						
ETHYL ACETOACETATE						NR at 70°F
ETHYL ACETYLENE						
ETHYL ACRYLATE	NR at 70°F					NR at 70°F
ETHYL ACRYLIC ACID						
ETHYL ALCOHOL	AC to 70°F	A 40-100% to 70°F B 40-100% at 122°F	+8.8% vol 180 days 70°F NR at 70°F	C/NR 50-96% at 70°F AB 10% to 70°F	A to 70°F AB 35-100% to 140°F	A/NR at 70°F
ETHYL ALDEHYDE						NR at 70°F
ETHYL ALUMINUM DICHLORIDE						
ETHYL AMINE						NR at 70°F
ETHYL BENZENE		C at 70°F NR at 122°F				NR at 70°F
ETHYL BENZOATE		B at 70°F C at 122°F				NR at 70°F
ETHYL BROMIDE			TEST			NR at 70°F
ETHYL BUTANOATE						
ETHYL BUTANOL						
ETHYL BUTYL ACETATE						
ETHYL BUTYL ALCOHOL						
ETHYL BUTYL AMINE						
ETHYL BUTYL KETONE						
ETHYL BUTYRALDEHYDE						
ETHYL BUTYRATE		B at 70°F NR at 122°F				NR at 70°F
ETHYL BUTYRIC ESTER						
ETHYL CAPRYLATE						
ETHYL CAPRYLIC ESTER						
ETHYL CELLOSOLVE	NR at 70°F					
ETHYL CELLULOSE			BC at 70°F		B at 70°F	NR at 70°F
ETHYL CHLORIDE	NR at 70°F	C at 70°F NR at 122°F	A to 70°F		NR at 70°F	NR at 70°F
ETHYL CHLOROCARBONATE						
ETHYL CHLOROFORMATE						
ETHYL CYANIDE						
ETHYL CYCLOPENTANE						
ETHYL DIACETATE						
ETHYL DICHLORIDE						
ETHYL DIMETHYL ACETATE						
ETHYL ETHER	BC at 70°F	C at 70°F	A to 70°F		C/NR at 70°F	C/NR at 70°F
ETHYL FORMATE						

	POLYETHYLENE LOW DENSITY LDPE	®TEFLON FEP	®KALREZ PERFLUORINATED ELASTOMER	®FLUORAZ FLUORINATED COPOLYMER	®AFLAS FLUORINATED COPOLYMER	®NORPRENE COPOLYMER TPO
ETHANOLAMINE	A to 70°F B at 122°F	A to 300°F	A to 150°F *A to 212°F	A to 70°F	+2% vol 7 days 158°F A to 70°F	TEST
ETHANOLAMINE, MONO, DI, TRI	AB to 70°F	A to 300°F	A to 113°F *A to 212°F	A to 70°F	A to 70°F (mono & tri)	
ETHANOYL CHLORIDE	NR at 70°F	A to 200°F	A to 70°F	A to 70°F		
ETHER (ETHERS)	B/NR at 70°F	A to 300°F	A to 212°F	NR at 70°F		BC at 70°F
ETHINE		A to 70°F				
ETHOCEL		A to 70°F				
ETHOXYETHANE	B/NR at 70°F	A to 300°F	A to 158°F	NR at 70°F	NR at 70°F	
ETHOXYTHANOL		A to 70°F	A to 212°F	A/NR at 70°F		
ETHYL ACETATE	AB to 122°F NR at 140°F	+1% wt 365 days 122°F A to 300°F	+3% vol 7 days 70°F A to 212°F	NR at 70°F	+88% vol 7 days 70°F NR at 70°F	
ETHYL ACETIC ACID		A to 70°F				
ETHYL ACETOACETATE		A to 70°F	A to 212°F			
ETHYL ACETYLENE	AB to 70°F	A to 70°F				
ETHYL ACRYLATE		A to 70°F	A to 212°F	C/NR at 70°F		
ETHYL ACRYLIC ACID				C at 70°F		
ETHYL ALCOHOL	A/NR to 140°F AB to 40% to 70°F	A to 392°F	no swell 7 days 70°F A to 212°F	A to 70°F	+9% vol 7 days 158°F A to 70°F	BC at 70°F
ETHYL ALDEHYDE		A to 70°F				
ETHYL ALUMINUM DICHLORIDE		A to 70°F	A to 212°F			
ETHYL AMINE	AB to 70°F	A to 70°F	A to 113°F	A to 70°F		
ETHYL BENZENE	C at 70°F NR at 122°F	A to 125°F	A to 212°F	B/NR at 70°F	+22% vol 7 days 70°F BC at 70°F	
ETHYL BENZOATE	C to 122°F	A to 125°F	A to 212°F	C at 70°F	+27% vol 7 days 70°F BC at 70°F	
ETHYL BROMIDE	NR at 70°F	A to 70°F	A to 212°F	B at 70°F		
ETHYL BUTANOATE	C/NR at 70°F	A to 125°F		C at 70°F		
ETHYL BUTANOL		A to 70°F				
ETHYL BUTYL ACETATE		A to 70°F				
ETHYL BUTYL ALCOHOL		A to 70°F				
ETHYL BUTYL AMINE		A to 70°F				
ETHYL BUTYL KETONE		A to 70°F				
ETHYL BUTYRALDEHYDE		A to 70°F				
ETHYL BUTYRATE	C/NR at 70°F	A to 125°F	*A to 212°F	C at 70°F		
ETHYL BUTYRIC ESTER		A to 70°F				
ETHYL CAPRYLATE		A to 70°F				
ETHYL CAPRYLIC ESTER		A to 70°F				
ETHYL CELLOSOLVE		A to 70°F	A to 212°F	A/NR at 70°F		
ETHYL CELLULOSE	AB to 70°F	A to 70°F	A to 212°F	A to 70°F		
ETHYL CHLORIDE	C/NR at 70°F NR at 140°F	A to 125°F (dry) A to 300°F (wet)	A to 212°F	AB to 70°F		NR at 70°F
ETHYL CHLOROCARBONATE		A to 70°F	A to 212°F	BC at 70°F		
ETHYL CHLOROFORMATE		A to 70°F	A to 212°F			
ETHYL CYANIDE		A to 70°F				
ETHYL CYCLOPENTANE				B at 70°F		
ETHYL DIACETATE		A to 70°F				
ETHYL DICHLORIDE		A to 70°F	A to 302°F	A to 70°F		
ETHYL DIMETHYL ACETATE		A to 70°F				
ETHYL ETHER	B/NR at 70°F	A to 300°F	A to 212°F	NR at 70°F	NR at 70°F	
ETHYL FORMATE		A to 70°F	A to 212°F	A to 70°F		

	RECOMMENDATIONS	NITRILE NBR	ETHYLENE PROPYLENE EP, EPDM	FLUOROCARBON FKM	CHLOROPRENE CR	HYDROGENATED NITRILE HNBR
ETHYL FORMIC ESTER		NR at 70°F	B at 70°F	C at 70°F	B at 70°F	
ETHYL HEXANOL	NBR,EP,FKM,IIR	A to 70°F	A to 70°F	A to 250°F	A to 70°F	
ETHYL HEXOIC ACID		C at 70°F				
ETHYL HEXYL ACETATE		NR at 70°F		NR at 70°F		
ETHYL HEXYL ALCOHOL	NBR,EP,FKM,IIR	A to 70°F	A to 70°F	A to 250°F	A to 70°F	
ETHYL HEXYL PHTHALATE		NR at 70°F			NR at 70°F	
ETHYL HYDRATE		A to 70°F	A to 70°F	B at 70°F	A to 70°F	
ETHYL HYDROXIDE		AB to 150°F			AB to 150°F	
ETHYL IODIDE		NR at 70°F	C at 70°F	B at 70°F	NR at 70°F	
ETHYL ISOBUTYL ETHER		NR at 70°F	NR at 70°F		NR at 70°F	
ETHYL ISOBUTYATE		NR at 70°F	NR at 70°F		NR at 70°F	
ETHYL LACTATE						
ETHYL MERCAPTAN	FKM,IIR	NR at 70°F	A/NR to 200°F	AB to 140°F	C/NR at 70°F	
ETHYL METHYL CARBINOL		A to 70°F		A to 70°F	A to 70°F	
ETHYL METHYL KETONE	EP,IIR	NR any conc at 70°F	A to 140°F AB to 200°F	+240% vol 7 days 70°F NR at 70°F	NR any conc at 70°F	
ETHYL MORPHOLENE, 50% / STANNOUS OCTOATE, 50% MIXTURE	EP	NR at 70°F	B at 70°F	NR at 70°F		
ETHYL ORTHOSILICATE		A to 70°F	A to 70°F	A to 70°F	A to 70°F	
ETHYL OXALATE	FKM,T,IIR,FVMQ	NR at 70°F	A/NR at 70°F	A to 70°F	NR at 70°F	
ETHYL OXIDE		B any conc to 80°F	NR at 70°F	NR at 70°F	NR at 70°F	
ETHYL PENTACHLOROBENZENE	FKM,FVMQ	NR at 70°F	NR at 70°F	A to 70°F	NR at 70°F	
ETHYL PHTHALATE		NR at 70°F		C at 70°F		
ETHYL PROPIONATE		NR at 70°F	NR at 70°F		NR at 70°F	
ETHYL PROPYL ETHER		AC to 70°F	NR at 70°F		NR at 70°F	
ETHYL PROPYL KETONE		NR at 70°F		NR at 70°F		
ETHYL PROPYL OXIDE		C at 70°F	NR at 70°F		NR at 70°F	
ETHYL PYRIDENE	EP	NR at 70°F	A to 200°F	NR at 70°F	A/NR at 70°F	
ETHYL SILICATE	EP,NBR,CR,FKM,IIR	A to 70°F AB to 180°F	A to 70°F	A to 70°F	A to 70°F AB to 180°F	
ETHYL SULFATE	EP,CR	NR at 70°F	A to 200°F	NR at 70°F	A to 70°F	
ETHYL SULFAHYDRATE		NR at 70°F	NR at 70°F	B at 70°F	NR at 70°F	
ETHYLENE	NBR	A to 200°F	C/NR at 70°F	A to 70°F	AC to 70°F	
ETHYLENE ALCOHOL		A to 70°F AB to 180°F	A to 70°F	A to 70°F	A to 70°F AB to 150°F	
ETHYLENE BROMIDE	FKM	NR at 70°F	BC at 70°F	AB to 200°F	NR at 70°F	
ETHYLENE CHLORIDE	FKM	NR at 70°F	C/NR at 140°F	AB to 200°F	NR at 70°F	
ETHYLENE CHLOROHYDRIN	FKM,EP,CR	NR at 70°F	AB to 70°F	A to 70°F	B at 70°F	NR at 70°F
ETHYLENE DIAMINE	EP,NBR,CR,IIR	AB any conc to 80°F	A to 120°F	NR at 70°F	A any conc to 80°F	A to 70°F
ETHYLENE DIBROMIDE	FKM	NR at 70°F	BC at 70°F	AB to 70°F	NR at 70°F	
ETHYLENE DICHLORIDE	FKM	NR at 70°F	C/NR to 140°F	+16% vol 7 days 75°F AB to 200°F	NR at 70°F	
ETHYLENE GLYCOL	EP,NBR,CR,FKM,IIR	+3% vol 3 days 212°F A to 212°F	+2% vol 3 days 212°F A to 212°F	+2% vol 14 days 212°F A to 250°F	+7% vol 3 days 212°F A to 212°F	A to 70°F
ETHYLENE GLYCOL MONOBUTYL ETHER	EP,IIR	+38% vol 3 days 70°F C at 70°F	+2% vol 3 days 70°F A to 200°F	+6% vol 3 days 70°F NR at 70°F	+31% vol 3 days 70°F C at 70°F	NR at 70°F
ETHYLENE GLYCOL MONOETHYL ETHER ACETATE		C at 70°F	A to 70°F	A to 70°F	NR at 70°F	
ETHYLENE GLYCOL MONOMETHYL ETHER		C at 70°F	B at 70°F	NR at 70°F	A to 70°F	
ETHYLENE MONOACETATE		C at 70°F	A to 70°F	A to 70°F	NR at 70°F	
ETHYLENE OXIDE	EP,IIR	NR 12-100% at 70°F	A/NR to 120°F B 12% at 70°F	NR 12-100% at 70°F	NR at 70°F	
ETHYLENE OXIDE, 12%/FREON 12, 80%	EP,IIR	C/NR at 70°F	B at 70°F	NR at 70°F	NR at 70°F	

	STYRENE BUTADIENE SBR	POLYACRYLATE ACM	POLYURETHANE AU, EU	ISOBUTYLENE ISOPRENE IIR	POLYBUTADIENE BR	® AEROQUIP AQP
ETHYL FORMIC ESTER				B at 70°F		
ETHYL HEXANOL	A to 70°F	NR at 70°F	NR at 70°F	A to 70°F	A to 70°F	
ETHYL HEXOIC ACID				C at 70°F		
ETHYL HEXYL ACETATE				A to 70°F		
ETHYL HEXYL ALCOHOL	A to 70°F	NR at 70°F	NR at 70°F	A to 70°F	A to 70°F	
ETHYL HEXYL PHTHALATE				B at 70°F		
ETHYL HYDRATE			NR at 70°F	A to 70°F		
ETHYL HYDROXIDE				A to 70°F AB to 185°F		
ETHYL IODIDE				C at 70°F		
ETHYL ISOBUTYL ETHER				NR at 70°F		
ETHYL ISOBUTYATE				NR at 70°F		
ETHYL LACTATE						
ETHYL MERCAPTAN	NR at 70°F		A to 70°F	NR at 70°F	NR at 70°F	
ETHYL METHYL CARBINOL				A to 70°F		
ETHYL METHYL KETONE	NR at 70°F	NR at 70°F	NR at 70°F	A any conc to 100°F	NR at 70°F	A to 70°F
ETHYL MORPHOLENE, 50% / STANNOUS OCTOATE, 50% MIXTURE	NR at 70°F			B at 70°F		
ETHYL ORTHOSILICATE				A to 70°F		
ETHYL OXALATE	NR at 70°F	NR at 70°F	A to 70°F	NR at 70°F	NR at 70°F	
ETHYL OXIDE			C at 70°F	C/NR at 70°F		
ETHYL PENTACHLOROBENZENE	NR at 70°F	NR at 70°F	C/NR at 70°F	NR at 70°F	NR at 70°F	
ETHYL PHTHALATE				A to 70°F		
ETHYL PROPIONATE				NR at 70°F		
ETHYL PROPYL ETHER				NR at 70°F		
ETHYL PROPYL KETONE				B at 70°F		
ETHYL PROPYL OXIDE				NR at 70°F		
ETHYL PYRIDENE	NR at 70°F	NR at 70°F	NR at 70°F			
ETHYL SILICATE	B at 70°F		C/NR at 70°F	A to 70°F AB to 180°F	B at 70°F	
ETHYL SULFATE	A to 70°F	NR at 70°F	NR at 70°F	B at 70°F		
ETHYL SULFAHYDRATE				NR at 70°F		
ETHYLENE	NR at 70°F	B at 70°F	B at 70°F	AB to 70°F		
ETHYLENE ALCOHOL			B at 70°F	A to 70°F AB to 180°F		
ETHYLENE BROMIDE	NR at 70°F	NR at 70°F	NR at 70°F	NR at 70°F		
ETHYLENE CHLORIDE	NR at 70°F	NR at 70°F	NR at 70°F	C/NR at 70°F	NR at 70°F	
ETHYLENE CHLOROHYDRIN	B at 70°F	NR at 70°F	NR at 70°F	AB any conc to 80°F	B at 70°F	
ETHYLENE DIAMINE	B at 70°F	NR at 70°F	NR at 70°F	A any conc to 100°F	B at 70°F	
ETHYLENE DIBROMIDE	NR at 70°F	NR at 70°F	NR at 70°F	C/NR at 70°F	NR at 70°F	
ETHYLENE DICHLORIDE	NR at 70°F	NR at 70°F	NR at 70°F	BC at 70°F	NR at 70°F	NR at 70°F
ETHYLENE GLYCOL	+4% vol 3 days 212°F A to 70°F	+37% vol 3 days 212°F NR at 70°F	+8% vol 3 days 212°F B at 70°F	-1% vol 3 days 212°F A to 185°F	A to 70°F	AB to 70°F
ETHYLENE GLYCOL MONOBUTYL ETHER	+23% vol 3 days 70°F NR at 70°F	+89% vol 3 days 70°F NR at 70°F	+72% vol 3 days 70°F NR at 70°F	A to 70°F AB any conc to 150°F	NR at 70°F	
ETHYLENE GLYCOL MONOETHYL ETHER ACETATE			NR at 70°F	A to 70°F		
ETHYLENE GLYCOL MONOMETHYL ETHER				A to 70°F		
ETHYLENE MONOACETATE			NR at 70°F	A to 70°F		
ETHYLENE OXIDE	NR at 70°F	NR at 70°F	NR at 70°F	C/NR at 70°F	NR at 70°F	
ETHYLENE OXIDE, 12%/FREON 12, 80%	NR at 70°F	NR at 70°F	NR at 70°F	B at 70°F	NR at 70°F	

	SYNTHETIC ISOPRENE IR	NATURAL ISOPRENE NR	CHLOROSULFONATED POLYETHYLENE CSM	FLUOROSILICONE FVMQ	SILICONE VMQ	CHEMRAZ FFKM
ETHYL FORMIC ESTER	NR at 70°F	NR at 70°F	NR at 70°F			
ETHYL HEXANOL	A to 70°F	A to 70°F	A to 70°F	A to 70°F	B at 70°F	A to 70°F
ETHYL HEXOIC ACID	C at 70°F	C at 70°F	B at 70°F			A to 70°F
ETHYL HEXYL ACETATE	NR at 70°F	NR at 70°F	B at 70°F			
ETHYL HEXYL ALCOHOL	A to 70°F	A to 70°F	A to 70°F	A to 70°F	B at 70°F	A to 70°F
ETHYL HEXYL PHTHALATE	NR at 70°F	NR at 70°F				
ETHYL HYDRATE	A to 70°F	A to 70°F	A to 70°F		A to 70°F	
ETHYL HYDROXIDE	A to 70°F	A to 70°F AB to 150°F				
ETHYL IODIDE	NR at 70°F	NR at 70°F	NR at 70°F			
ETHYL ISOBUTYL ETHER	NR at 70°F	NR at 70°F	B at 70°F			
ETHYL ISOBUTYATE	NR at 70°F	NR at 70°F				
ETHYL LACTATE						A to 70°F
ETHYL MERCAPTAN	NR at 70°F	NR at 70°F	B/NR at 70°F		C at 70°F	A to 70°F
ETHYL METHYL CARBINOL	A to 70°F	A to 70°F				
ETHYL METHYL KETONE	NR at 70°F	NR at 70°F	NR at 70°F	NR at 70°F	NR at 70°F	A to 70°F
ETHYL MORPHOLENE, 50% / STANNOUS OCTOATE, 50% MIXTURE						A to 70°F
ETHYL ORTHOSILICATE	C at 70°F	C at 70°F	A to 70°F			
ETHYL OXALATE	A to 70°F	A/NR at 70°F	NR at 70°F	B at 70°F	NR at 70°F	A to 70°F
ETHYL OXIDE	NR at 70°F	NR at 70°F	B at 70°F		C at 70°F	
ETHYL PENTACHLOROBENZENE	NR at 70°F	NR at 70°F	NR at 70°F	B at 70°F	NR at 70°F	A to 70°F
ETHYL PHTHALATE	NR at 70°F	NR at 70°F	NR at 70°F			
ETHYL PROPIONATE	NR at 70°F	NR at 70°F				
ETHYL PROPYL ETHER	NR at 70°F	NR at 70°F	B at 70°F			
ETHYL PROPYL KETONE	NR at 70°F	NR at 70°F	NR at 70°F			
ETHYL PROPYL OXIDE	NR at 70°F	NR at 70°F	B at 70°F			
ETHYL PYRIDENE				NR at 70°F	NR at 70°F	A to 70°F
ETHYL SILICATE	B at 70°F	B at 70°F	AB to 70°F	A to 70°F		A to 70°F
ETHYL SULFATE	NR at 70°F	NR at 70°F	NR at 70°F	C at 70°F	A to 70°F	A to 70°F
ETHYL SULFAHYDRATE	NR at 70°F	NR at 70°F	NR at 70°F			
ETHYLENE			A to 70°F	A to 70°F	B/NR at 70°F	
ETHYLENE ALCOHOL	A to 70°F	A/NR at 70°F	A to 70°F		A to 70°F	
ETHYLENE BROMIDE	NR at 70°F	NR at 70°F	NR at 70°F	C at 70°F	NR at 70°F	A to 70°F
ETHYLENE CHLORIDE	NR at 70°F	NR at 70°F	B/NR at 70°F	C/NR at 70°F	NR at 70°F	A to 70°F
ETHYLENE CHLOROHYDRIN	B at 70°F	B at 70°F	B at 70°F	B at 70°F	C at 70°F	A to 70°F
ETHYLENE DIAMINE	AB to 70°F	AB any conc to 80°F	AB to 70°F	NR at 70°F	A to 70°F	B at 70°F
ETHYLENE DIBROMIDE	NR at 70°F	NR at 70°F	NR at 70°F	C at 70°F	NR at 70°F	A to 70°F
ETHYLENE DICHLORIDE	NR at 70°F	NR at 70°F	B/NR at 70°F	C/NR at 70°F	NR at 70°F	A to 70°F
ETHYLENE GLYCOL	A to 70°F	A to 70°F AB to 150°F	A to 200°F	+1% vol 3 days 212°F +1% vol 120 days 70°F	+3% vol 3 days 212°F A to 70°F	A to 70°F
ETHYLENE GLYCOL MONOBUTYL ETHER	NR at 70°F	NR at 70°F	B/NR at 70°F	+7% vol 3 days 70°F NR at 70°F	+100% vol 3 days 70°F NR at 70°F	A to 70°F
ETHYLENE GLYCOL MONOETHYL ETHER ACETATE	C at 70°F	C at 70°F	NR at 70°F		C at 70°F	
ETHYLENE GLYCOL MONOMETHYL ETHER	NR at 70°F	NR at 70°F	B at 70°F			
ETHYLENE MONOACETATE	C at 70°F	C at 70°F	NR at 70°F		C at 70°F	
ETHYLENE OXIDE	NR at 70°F	NR at 70°F	NR at 70°F	NR at 70°F	NR at 70°F	BC at 70°F C at 70°F white
ETHYLENE OXIDE, 12%/FREON 12, 80%	NR at 70°F	NR at 70°F	NR at 70°F	NR at 70°F	NR at 70°F	A to 70°F

	POLYSULFIDE T	CHLORINATED POLYETHYLENE CM	EPICHLOROHYDRIN CO, ECO	® HYTREL COPOLYESTER TPE	® SANTOPRENE COPOLYMER TPO	® C-FLEX STYRENIC TPE
ETHYL FORMIC ESTER						
ETHYL HEXANOL	B at 70°F				−5% wt 7 days 70°F A to 70°F	
ETHYL HEXOIC ACID						
ETHYL HEXYL ACETATE						
ETHYL HEXYL ALCOHOL	B at 70°F				A to 70°F	
ETHYL HEXYL PHTHALATE						
ETHYL HYDRATE						
ETHYL HYDROXIDE						
ETHYL IODIDE						
ETHYL ISOBUTYL ETHER						
ETHYL ISOBUTYATE						
ETHYL LACTATE						
ETHYL MERCAPTAN	NR at 70°F		NR at 70°F			
ETHYL METHYL CARBINOL						
ETHYL METHYL KETONE	AB to 70°F	AB to 70°F	NR at 70°F	A to 70°F	−4 to −11% vol 3 days 70°F A/NR at 70°F	
ETHYL MORPHOLENE, 50% / STANNOUS OCTOATE, 50% MIXTURE						
ETHYL ORTHOSILICATE						
ETHYL OXALATE	A to 70°F		NR at 70°F			
ETHYL OXIDE						
ETHYL PENTACHLOROBENZENE	NR at 70°F		C at 70°F			
ETHYL PHTHALATE						
ETHYL PROPIONATE						
ETHYL PROPYL ETHER						
ETHYL PROPYL KETONE						
ETHYL PROPYL OXIDE						
ETHYL PYRIDENE						
ETHYL SILICATE	B at 70°F		A to 70°F			
ETHYL SULFATE						
ETHYL SULFAHYDRATE						
ETHYLENE						
ETHYLENE ALCOHOL						
ETHYLENE BROMIDE						
ETHYLENE CHLORIDE	B at 70°F					
ETHYLENE CHLOROHYDRIN	B at 70°F			NR at 70°F	NR at 70°F	AB to 70°F
ETHYLENE DIAMINE	NR at 70°F		A to 70°F			
ETHYLENE DIBROMIDE	NR at 70°F					
ETHYLENE DICHLORIDE	NR at 70°F	NR at 70°F	NR at 70°F	NR at 70°F	NR at 70°F	AB to 70°F
ETHYLENE GLYCOL	+9.5% vol 3 days 212°F C at 70°F	AB to 70°F	A to 70°F	A to 70°F C at 150°F	A to 70°F	AB to 70°F
ETHYLENE GLYCOL MONOBUTYL ETHER	AB to 70°F				*A to 70°F	
ETHYLENE GLYCOL MONOETHYL ETHER ACETATE						
ETHYLENE GLYCOL MONOMETHYL ETHER						
ETHYLENE MONOACETATE						
ETHYLENE OXIDE		NR at 70°F	NR at 70°F	A to 70°F		AB to 70°F
ETHYLENE OXIDE, 12%/FREON 12, 80%	NR at 70°F					

	ETHYLENE ACRYLIC EA	POLYALLOMER LINEAR COPOLYMER	NYLON 11 POLYAMIDE	NYLON 12 POLYAMIDE	ETHYLENE VINYL ACETATE EVA	POLYVINYLCHLORIDE FLEXIBLE PVC
ETHYL FORMIC ESTER						
ETHYL HEXANOL	A to 70°F					
ETHYL HEXOIC ACID						
ETHYL HEXYL ACETATE						
ETHYL HEXYL ALCOHOL	A to 70°F					
ETHYL HEXYL PHTHALATE						
ETHYL HYDRATE						
ETHYL HYDROXIDE						
ETHYL IODIDE						
ETHYL ISOBUTYL ETHER						
ETHYL ISOBUTYATE						
ETHYL LACTATE		A to 122°F				C at 70°F NR at 120°F
ETHYL MERCAPTAN						
ETHYL METHYL CARBINOL						
ETHYL METHYL KETONE	NR at 70°F	A to 70°F NR at 122°F	AB to 105°F NR at 194°F	AB to 70°F	*C at 70°F *NR at 140°F	NR at 70°F
ETHYL MORPHOLENE, 50% / STANNOUS OCTOATE, 50% MIXTURE						
ETHYL ORTHOSILICATE						
ETHYL OXALATE						
ETHYL OXIDE						
ETHYL PENTACHLOROBENZENE						
ETHYL PHTHALATE						
ETHYL PROPIONATE						
ETHYL PROPYL ETHER						
ETHYL PROPYL KETONE						
ETHYL PROPYL OXIDE						
ETHYL PYRIDENE						
ETHYL SILICATE					C at 70°F	C at 70°F
ETHYL SULFATE						
ETHYL SULFAHYDRATE						
ETHYLENE						AB to 70°F
ETHYLENE ALCOHOL						
ETHYLENE BROMIDE						
ETHYLENE CHLORIDE		C at 70°F NR at 122°F	C at 70°F	C at 70°F	*NR at 70°F	NR at 70°F
ETHYLENE CHLOROHYDRIN			NR at 70°F			NR at 70°F
ETHYLENE DIAMINE					C at 70°F	C at 70°F
ETHYLENE DIBROMIDE						
ETHYLENE DICHLORIDE	*NR at 70°F		AC to 70°F		*NR at 70°F	NR at 70°F
ETHYLENE GLYCOL	A to 212°F	A to 122°F	+0.5% vol 180 days 70°F A to 150°F		A to 70°F AB to 140°F	AB to 150°F
ETHYLENE GLYCOL MONOBUTYL ETHER					NR at 70°F	NR at 70°F
ETHYLENE GLYCOL MONOETHYL ETHER ACETATE						
ETHYLENE GLYCOL MONOMETHYL ETHER		A to 122°F				C at 70°F NR at 120°F
ETHYLENE MONOACETATE						
ETHYLENE OXIDE	*NR at 70°F	C to 122°F	AB to 105°F C at 140°F		*AB to 70°F	C/NR at 70°F NR at 122°F
ETHYLENE OXIDE, 12%/FREON 12, 80%						

	POLYETHYLENE LOW DENSITY LDPE	®TEFLON FEP	®KALREZ PERFLUORINATED ELASTOMER	®FLUORAZ FLUORINATED COPOLYMER	®AFLAS FLUORINATED COPOLYMER	®NORPRENE COPOLYMER TPO
ETHYL FORMIC ESTER		A to 70°F				
ETHYL HEXANOL		A to 70°F	A to 212°F	A to 70°F		
ETHYL HEXOIC ACID		A to 70°F	A to 113°F			
ETHYL HEXYL ACETATE		A to 70°F				
ETHYL HEXYL ALCOHOL		A to 70°F		A to 70°F		
ETHYL HEXYL PHTHALATE		A to 70°F				
ETHYL HYDRATE		A to 70°F				
ETHYL HYDROXIDE		A to 70°F				
ETHYL IODIDE		A to 70°F				
ETHYL ISOBUTYL ETHER		A to 70°F				
ETHYL ISOBUTYATE		A to 70°F				
ETHYL LACTATE	A to 122°F	A to 120°F	A to 212°F			
ETHYL MERCAPTAN		A to 70°F	A to 212°F	B at 70°F		
ETHYL METHYL CARBINOL		A to 70°F				
ETHYL METHYL KETONE	AB to 122°F C/NR at 140°F	A to 122°F	+1% vol 7 days 70°F A to 113°F	NR at 70°F	+58% vol 7 days 70°F NR at 70°F	BC at 70°F
ETHYL MORPHOLENE, 50% / STANNOUS OCTOATE, 50% MIXTURE			A to 212°F	C at 70°F		
ETHYL ORTHOSILICATE		A to 70°F				
ETHYL OXALATE		A to 70°F	A to 212°F	A to 70°F		
ETHYL OXIDE		A to 70°F		NR at 70°F		
ETHYL PENTACHLOROBENZENE		A to 70°F	A to 212°F	B at 70°F		
ETHYL PHTHALATE		A to 70°F				
ETHYL PROPIONATE		A to 70°F				
ETHYL PROPYL ETHER		A to 70°F				
ETHYL PROPYL KETONE		A to 70°F				
ETHYL PROPYL OXIDE		A to 70°F				
ETHYL PYRIDENE			A to 212°F	NR at 70°F		
ETHYL SILICATE	A to 70°F	A to 70°F	A to 212°F	A to 70°F		
ETHYL SULFATE		A to 70°F		AB to 70°F		
ETHYL SULFAHYDRATE		A to 70°F				
ETHYLENE	AB to 70°F	A to 70°F	A to 212°F	NR at 70°F		
ETHYLENE ALCOHOL		A to 70°F				
ETHYLENE BROMIDE		A to 300°F	A to 212°F	B at 70°F		
ETHYLENE CHLORIDE	B/NR at 70°F NR at 120°F	A to conc to 300°F	A to 212°F	A/NR at 70°F		
ETHYLENE CHLOROHYDRIN		A to 300°F	A to 70°F *A to 212°F	A to 70°F	+0% vol 7 days 77°F A to 70°F	NR at 70°F
ETHYLENE DIAMINE	A to 70°F	A to 242°F	A to 104°F B at 203°F	AB to 70°F		
ETHYLENE DIBROMIDE		A to 300°F	A to 212°F	B at 70°F		
ETHYLENE DICHLORIDE	B/NR at 70°F NR at 120°F	A to 300°F	A to 212°F	A/NR at 70°F	no swell 7 days 70°F B at 70°F	
ETHYLENE GLYCOL	A to 140°F	A to 300°F	A to 113°F *A to 250°F	A to 70°F	+1% vol 50% conc 14 days 324°F	A to 70°F
ETHYLENE GLYCOL MONOBUTYL ETHER	AB to 70°F	A to 70°F	A to 113°F	C at 70°F		
ETHYLENE GLYCOL MONOETHYL ETHER ACETATE		A to 70°F				
ETHYLENE GLYCOL MONOMETHYL ETHER	A to 120°F	A to 120°F				
ETHYLENE MONOACETATE		A to 70°F				
ETHYLENE OXIDE	AC to 122°F NR at 140°F	A to 300°F	A to 70°F *A to 212°F	NR at 70°F		
ETHYLENE OXIDE, 12%/FREON 12, 80%				NR at 70°F		

	RECOMMENDATIONS	NITRILE NBR	ETHYLENE PROPYLENE EP, EPDM	FLUOROCARBON FKM	CHLOROPRENE CR	HYDROGENATED NITRILE HNBR
ETHYLENE TRICHLORIDE	FKM	NR at 70°F	C/NR at 70°F	A to 70°F	NR at 70°F	NR at 70°F
ETHYLIC ACID	EP, IIR	NR 100% at 70°F B to 30% at 70°F	+16% vol 3 days 158°F AB to 200°F	+112% vol 3 days 70°F NR 90-100% at 70°F	+16% vol 3 days 158°F B to 200°F	B at 70°F
ETHYLIC ALCOHOL		A to 70°F	A to 70°F	B at 70°F	A to 70°F	
ETHYLIC ETHER		B at 70°F	NR at 70°F	NR at 70°F	C at 70°F	
ETHYLIDENE CHLORIDE	FKM	NR at 70°F	C/NR to 140°F	AB to 200°F	NR at 70°F	NR at 70°F
ETHYLIDENE DICHLORIDE		NR at 70°F	NR at 70°F		NR at 70°F	
ETHYLIDENE PERCHLORIDE		NR at 70°F	NR at 70°F		NR at 70°F	
ETHYNE		A to 70°F	A to 70°F	A to 70°F	B at 70°F	
EXSICCATED ALUM		A to 70°F AB any conc to 200°F	A to 70°F		A to 70°F AB any conc to 200°F	
EX-TRI		NR at 70°F		A to 70°F		
EXXON 2300 TURBO OIL			NR at 70°F	A to 392°F		
F60 FLUID (DOW CORNING)	EP, NBR, FKM, CR	A to 70°F	A to 70°F	A to 70°F	A to 70°F	A to 70°F
F61 FLUID (DOW CORNING)	EP, NBR, FKM, CR	A to 70°F	A to 70°F	+1% vol 28 days 300°F A to 300°F	A to 70°F	A to 70°F
FAT LIME		A to 70°F AB any conc to 200°F	A to 70°F		A to 70°F AB any conc to 200°F	
FATTY ACIDS	FKM, NBR, CR	A to 130°F AB to 250°F	C/NR at 70°F	A to 145°F	B to 70°F NR at 140°F	B at 70°F
FC-43 HEPTACOSOFLUOROTRIBUTYLAMINE	EP, NBR, FKM, CR	A to 70°F	A to 70°F	A to 70°F	A to 70°F	A to 70°F
FC75 FLUOROCARBON	NBR, EP, CR	A to 70°F	A to 70°F	B at 70°F	A to 70°F	
FERAN		AB to 70°F			AB to 70°F	
FERMENTATION AMYL ALCOHOL		A to 70°F	A to 70°F	A to 70°F	A to 70°F	
FERRIC ACETATE		NR at 70°F		NR at 70°F		
FERRIC BROMIDE		A to 70°F		A to 70°F		
FERRIC CHLORIDE	NBR, EP, IIR, CR	A to 150°F AB any conc to 200°F	A to conc to 176°F B at 212°F	A to conc to 176°F B at 212°F	A to conc to 140°F B at 176°F	A to 70°F
FERRIC DICHLORIDE		A to 70°F AB any conc to 200°F	A to 70°F	A to 70°F	AB any conc to 140°F	
FERRIC HYDROXIDE		A to 140°F AB 176-200°F	A to 176°F AB to 200°F	A to 176°F AB to 200°F	AB to 200°F	
FERRIC NITRATE	NBR, EP, CR, FKM, IIR	A to 140°F AB 176-200°F	A to 176°F AB to 200°F	A to 212°F	A to 140°F AB any conc to 200°F	A to 70°F
FERRIC PERCHLORIDE		A to 70°F AB any conc to 200°F			B any conc to 80°F	
FERRIC PERSULFATE		A to 70°F AB any conc to 200°F			A to 70°F AB any conc to 200°F	A to 70°F
FERRIC SALTS	NBR, EP, CR, FKM, IIR	AB to 200°F	A to 70°F	A to 70°F	AB to 150°F	
FERRIC SESQUICHLORIDE		A to 70°F AB any conc to 200°F	A to 70°F	A to 70°F	B any conc to 80°F	
FERRIC SESQUISULFATE		A to 70°F AB any conc to 200°F	A to 70°F	A to 70°F	A to 70°F AB any conc to 200°F	
FERRIC SUBSULFATE		A to 70°F AB any conc to 200°F	A to 70°F	A to 70°F	A to 70°F AB any conc to 200°F	
FERRIC SULFATE	NBR, EP, CR, IIR	A to 140°F AB any conc to 200°F	A to 176°F AB any conc to 200°F	A to conc to 176°F B at 212°F	A to conc to 140°F AB to 200°F	A to 70°F
FERRIC SULPHATE	NBR, EP, CR, IIR	A to 140°F AB any conc to 200°F	A to conc to 176°F AB any conc to 200°F	A to conc to 176°F B at 212°F	A to conc to 140°F AB to 200°F	A to 70°F
FERRIC TRICHLORIDE		A to 70°F AB any conc to 200°F	A to 70°F	A to 70°F	B any conc to 80°F	
FERRIC TRISULFATE		A to 70°F AB any conc to 200°F			A to 70°F AB any conc to 200°F	
FERRIFEROUS CHLORIDE		A to 70°F AB any conc to 200°F			B any conc to 80°F	
FERRIFEROUS PERSULFATE		A to 70°F			A to 70°F	
FERROUS ACETATE		NR at 70°F		NR at 70°F		
FERROUS AMMONIUM SULFATE				A 30% to 70°F	A 30% to 70°F	
FERROUS CHLORIDE		A to 140°F AB 176-200°F	A to 176°F B at 212°F	A to 176°F B to 248°F	AB to 200°F	
FERROUS HYDROXIDE		A to 146°F B at 176°F	A to 176°F	A to 212°F	A to 70°F	
FERROUS NITRATE		A to 140°F AB any conc to 200°F	A to 176°F	A to 212°F	A to 70°F AB any conc to 200°F	
FERROUS SALTS	NBR, CR, EP	AB to 70°F	A to 70°F	A to 70°F	AB to 70°F	
FERROUS SULFATE	NBR, CR, EP	A to conc to 160°F AB any conc to 200°F	A to conc to 176°F B at 212°F	A to conc to 176°F B at 212°F	A to conc to 140°F AB any conc to 200°F	

	STYRENE BUTADIENE SBR	POLYACRYLATE ACM	POLYURETHANE AU, EU	ISOBUTYLENE ISOPRENE IIR	POLYBUTADIENE BR	® AEROQUIP AQP
ETHYLENE TRICHLORIDE	NR at 70°F	NR at 70°F	NR at 70°F	C at 70°F	NR at 70°F	
ETHYLIC ACID	+19% vol 3 days 158°F BC at 70°F	+154% vol 3 days 158°F NR at 70°F	+172% vol 3 days 158°F NR at 70°F	+0.5% vol 3 days 70°F B to 125°F	B to 100% at 70°F	NR 30-100% at 70°F
ETHYLIC ALCOHOL			NR at 70°F	A to 70°F		
ETHYLIC ETHER			A to 70°F	C at 70°F		
ETHYLIDENE CHLORIDE				C at 70°F		
ETHYLIDENE DICHLORIDE				C at 70°F		
ETHYLIDENE PERCHLORIDE				C at 70°F		
ETHYNE				A to 70°F		
EXSICCATED ALUM				A to 70°F AB any conc to 185°F		
EX-TRI				NR at 70°F		
EXXON 2300 TURBO OIL						
F60 FLUID (DOW CORNING)	A to 70°F	A to 70°F	A to 70°F	A to 70°F	A to 70°F	
F61 FLUID (DOW CORNING)	A to 70°F	A to 70°F	A to 70°F	A to 70°F	A to 70°F	
FAT LIME				A to 70°F AB any conc to 185°F		
FATTY ACIDS	NR at 70°F	A to 70°F	B/NR at 70°F	C/NR at 70°F	C/NR at 70°F	
FC-43 HEPTACOSOFLUOROTRIBUTYLAMINE	NR at 70°F			A to 70°F		
FC75 FLUOROCARBON	NR at 70°F			A to 70°F		
FERAN				B at 70°F		
FERMENTATION AMYL ALCOHOL			C at 70°F	A to 70°F		
FERRIC ACETATE				A to 70°F		
FERRIC BROMIDE				A to 70°F		
FERRIC CHLORIDE	A to conc to 70°F	A to 70°F	A to conc to 70°F A 20% fumes to 70°F	A to 70°F AB any conc to 185°F	A to 70°F	
FERRIC DICHLORIDE			A to 70°F	A to 70°F AB any conc to 185°F		
FERRIC HYDROXIDE				A to 70°F		
FERRIC NITRATE	A to 70°F	A to 70°F	A to 70°F	A to 70°F AB any conc to 185°F	A to 70°F	
FERRIC PERCHLORIDE				A to 70°F AB any conc to 185°F		
FERRIC PERSULFATE				A to 70°F AB any conc to 185°F		
FERRIC SALTS	AB to 70°F			A to 70°F AB to 150°F		
FERRIC SESQUICHLORIDE			A to 70°F	A to 70°F AB any conc to 150°F		
FERRIC SESQUISULFATE			A to 70°F	A to 70°F AB any conc to 185°F		
FERRIC SUBSULFATE			A to 70°F	A to 70°F AB any conc to 185°F		
FERRIC SULFATE	A to conc to 70°F	A to 70°F	AB conc to 70°F A sol'n to 70°F	A to 70°F AB any conc to 185°F	A to 70°F	NR at 70°F
FERRIC SULPHATE	A to conc to 70°F	A to 70°F	AB conc to 70°F A sol'n to 70°F	A to 70°F AB any conc to 185°F	A to 70°F	NR at 70°F
FERRIC TRICHLORIDE			A to 70°F	A to 70°F AB any conc to 185°F		
FERRIC TRISULFATE				A to 70°F AB any conc to 185°F		
FERRIFEROUS CHLORIDE				A to 70°F		
FERRIFEROUS PERSULFATE				A to 70°F		
FERROUS ACETATE				A to 70°F		
FERROUS AMMONIUM SULFATE						
FERROUS CHLORIDE			AB to 70°F	AB any conc to 185°F		
FERROUS HYDROXIDE				A to 70°F		
FERROUS NITRATE				A to 70°F AB any conc to 185°F		
FERROUS SALTS			AB sol'n to 70°F	AB to 70°F		
FERROUS SULFATE	A sol'n to 70°F		A to conc to 70°F	A to 70°F AB any conc to 185°F		

C153

	SYNTHETIC ISOPRENE IR	NATURAL ISOPRENE NR	CHLOROSULFONATED POLYETHYLENE CSM	FLUOROSILICONE FVMQ	SILICONE VMQ	CHEMRAZ FFKM
ETHYLENE TRICHLORIDE	NR at 70°F	NR at 70°F	NR at 70°F	C at 70°F	C/NR at 70°F	A to 70°F
ETHYLIC ACID	BC 30-100% at 70°F B to 20% at 70°F	B/NR 30-100% at 70°F B to 30% at 70°F	C 100% at 70°F A to 80% to 70°F	+12% vol 3 days 158°F B to 99% at 70°F	+3% vol 3 days 158°F B at 70°F	AB 100% to 70°F A to 30% to 70°F
ETHYLIC ALCOHOL	A to 70°F	A to 70°F	A to 70°F		A to 70°F	
ETHYLIC ETHER	NR at 70°F	NR at 70°F	C at 70°F		NR at 70°F	
ETHYLIDENE CHLORIDE	NR at 70°F	NR at 70°F	B/NR at 70°F	C/NR at 70°F	NR at 70°F	A to 70°F
ETHYLIDENE DICHLORIDE	NR at 70°F	NR at 70°F				
ETHYLIDENE PERCHLORIDE	NR at 70°F	NR at 70°F				
ETHYNE	A to 70°F	A to 70°F	B at 70°F		C at 70°F	
EXSICCATED ALUM	A to 70°F	A to 70°F AB any conc to 150°F	A to 70°F			
EX-TRI	NR at 70°F	NR at 70°F	NR at 70°F			
EXXON 2300 TURBO OIL						
F60 FLUID (DOW CORNING)	A to 70°F	A to 70°F	A to 70°F	A to 70°F	NR at 70°F	A to 70°F
F61 FLUID (DOW CORNING)	A to 70°F	A to 70°F	A to 70°F	A to 70°F	NR at 70°F	A to 70°F
FAT LIME	A to 70°F	A to 70°F AB any conc to 150°F	A to 70°F			
FATTY ACIDS	C/NR at 70°F	C/NR at 70°F	B/NR at 70°F	A to 70°F	C at 70°F	A to 70°F
FC-43 HEPTACOSOFLUOROTRIBUTYLAMINE			A to 70°F	A to 70°F	A to 70°F	A/NR at 70°F
FC75 FLUOROCARBON			A to 70°F	B at 70°F	A to 70°F	A/NR at 70°F
FERAN	B at 70°F	B at 70°F	AB to 70°F			
FERMENTATION AMYL ALCOHOL	A to 70°F	A to 70°F	A to 70°F		NR at 70°F	
FERRIC ACETATE	NR at 70°F	NR at 70°F	A to 70°F		NR at 70°F	A to 70°F
FERRIC BROMIDE	A to 70°F	A to 70°F	A to 70°F			
FERRIC CHLORIDE	A to 70°F	A to 70°F AB any conc to 150°F	A to 200°F	A to conc to 70°F	AB to conc to 70°F	A to 70°F
FERRIC DICHLORIDE	A to 70°F	A to 70°F	A to 70°F		A to 70°F	
FERRIC HYDROXIDE	C at 70°F	C/NR at 70°F	AB to 200°F			A to 70°F
FERRIC NITRATE	A to 70°F	A to 70°F AB any conc to 150°F	A to 70°F AB to 200°F	A to 70°F	BC at 70°F	A to 70°F
FERRIC PERCHLORIDE	A to 70°F	A to 70°F AB any conc to 150°F				
FERRIC PERSULFATE	A to 70°F	A to 70°F AB any conc to 150°F				A to 70°F
FERRIC SALTS	A to 70°F	A to 70°F AB to 150°F	A to 70°F		A to 70°F	
FERRIC SESQUICHLORIDE	A to 70°F	A to 70°F AB any conc to 150°F	A to 70°F		A to 70°F	
FERRIC SESQUISULFATE	A to 70°F	A to 70°F AB any conc to 150°F	A to 80% to 70°F		B at 70°F	
FERRIC SUBSULFATE	A to 70°F	A to 70°F AB any conc to 150°F	A to 70°F		B at 70°F	
FERRIC SULFATE	A to 70°F	A to 70°F AB any conc to 150°F	A to 70°F AB to 200°F	A to conc to 70°F	BC conc at 70°F B to conc at 70°F	A to 70°F
FERRIC SULPHATE	A to 70°F	A to 70°F AB any conc to 150°F	A to 70°F AB to 200°F	A to conc to 70°F	BC conc at 70°F B to conc at 70°F	A to 70°F
FERRIC TRICHLORIDE	A to 70°F	A to 70°F AB any conc to 150°F	A to 70°F		A to 70°F	
FERRIC TRISULFATE	A to 70°F	A to 70°F AB any conc to 150°F				
FERRIFEROUS CHLORIDE	A to 70°F	A to 70°F				
FERRIFEROUS PERSULFATE	A to 70°F	A to 70°F				
FERROUS ACETATE	NR at 70°F	NR at 70°F	A to 70°F			
FERROUS AMMONIUM SULFATE						A to 70°F
FERROUS CHLORIDE	A to 70°F	A to 70°F AB any conc to 150°F	AB to 200°F		BC at 70°F	A to 70°F
FERROUS HYDROXIDE	C at 70°F	C at 70°F	B at 70°F			
FERROUS NITRATE	A to 70°F	A to 70°F AB any conc to 150°F	B at 70°F			
FERROUS SALTS		A to 70°F	AB to 70°F			A to 70°F
FERROUS SULFATE	A to 70°F	A to 70°F AB any conc to 185°F	A to 70°F AB to 200°F	A sol'n to 70°F	C conc at 70°F B sol'n at 70°F	A to 70°F

C154

	POLYSULFIDE T	CHLORINATED POLYETHYLENE CM	EPICHLOROHYDRIN CO, ECO	® HYTREL COPOLYESTER TPE	® SANTOPRENE COPOLYMER TPO	® C-FLEX STYRENIC TPE
ETHYLENE TRICHLORIDE	NR at 70°F					
ETHYLIC ACID	+22% vol 3 days 70°F BC at 70°F	A to 100% to 70°F	NR 100% at 70°F B 30% at 70°F	+16 to +39% vol 100°F AC to 70°F	NR 100% at 70°F C 50% at 70°F	
ETHYLIC ALCOHOL						
ETHYLIC ETHER						
ETHYLIDENE CHLORIDE						
ETHYLIDENE DICHLORIDE						
ETHYLIDENE PERCHLORIDE						
ETHYNE						
EXSICCATED ALUM						
EX-TRI						
EXXON 2300 TURBO OIL				*AB to 70°F		
F60 FLUID (DOW CORNING)	A to 70°F					
F61 FLUID (DOW CORNING)	A to 70°F					
FAT LIME						
FATTY ACIDS	NR at 70°F			AB to 70°F	NR at 70°F	AB to 70°F
FC-43 HEPTACOSOFLUOROTRIBUTYLAMINE	A to 70°F					
FC75 FLUOROCARBON	A to 70°F					
FERAN	AB to 70°F					
FERMENTATION AMYL ALCOHOL						
FERRIC ACETATE						
FERRIC BROMIDE						
FERRIC CHLORIDE	A to 70°F	NR at 70°F	A to 70°F	AB to 140°F		A to 70°F
FERRIC DICHLORIDE						
FERRIC HYDROXIDE						
FERRIC NITRATE	A to 70°F		A to 70°F			
FERRIC PERCHLORIDE						
FERRIC PERSULFATE						
FERRIC SALTS	AB to 70°F					
FERRIC SESQUICHLORIDE						
FERRIC SESQUISULFATE						
FERRIC SUBSULFATE						
FERRIC SULFATE	A to 70°F AB to 150°F	NR at 70°F	A to 70°F	A to 70°F		A to 70°F
FERRIC SULPHATE	A to 70°F AB to 150°F	NR at 70°F	A to 70°F	A to 70°F		A to 70°F
FERRIC TRICHLORIDE						
FERRIC TRISULFATE						
FERRIFEROUS CHLORIDE						
FERRIFEROUS PERSULFATE						
FERROUS ACETATE						
FERROUS AMMONIUM SULFATE						
FERROUS CHLORIDE	AB to 70°F			A to 70°F		A to 70°F
FERROUS HYDROXIDE						
FERROUS NITRATE						
FERROUS SALTS	AB to 70°F			AB sol'n to 70°F		
FERROUS SULFATE	AB to 70°F			A to 70°F		A to 70°F

C 155

	ETHYLENE ACRYLIC EA	POLYALLOMER LINEAR COPOLYMER	NYLON 11 POLYAMIDE	NYLON 12 POLYAMIDE	ETHYLENE VINYL ACETATE EVA	POLYVINYLCHLORIDE FLEXIBLE PVC
ETHYLENE TRICHLORIDE	NR at 70°F					
ETHYLIC ACID	NR 100% at 70°F A to 30% to 70°F	A to 100% to 70°F A to 50% to 122°F	NR 98-100% at 70°F AC 50% to 70°F	NR 40-96% at 70°F AB to 10% to 70°F	B/NR 50-100% at 70°F B 30% at 70°F	C/NR 85-100% at 70°F AB 50% to 70°F
ETHYLIC ALCOHOL						
ETHYLIC ETHER						
ETHYLIDENE CHLORIDE						
ETHYLIDENE DICHLORIDE						
ETHYLIDENE PERCHLORIDE						
ETHYNE						
EXSICCATED ALUM						
EX-TRI						
EXXON 2300 TURBO OIL						
F60 FLUID (DOW CORNING)				AB to 70°F		
F61 FLUID (DOW CORNING)				AB to 70°F		
FAT LIME						
FATTY ACIDS	A to 70°F		A to 70°F AB to 194°F		B at 70°F	AB to 70°F
FC-43 HEPTACOSOFLUOROTRIBUTYLAMINE						
FC75 FLUOROCARBON						
FERAN						
FERMENTATION AMYL ALCOHOL						
FERRIC ACETATE						
FERRIC BROMIDE						
FERRIC CHLORIDE	*AB sol'n to 70°F	A to 70°F	A/NR at 70°F		A to 70°F C at 140°F	A to 150°F
FERRIC DICHLORIDE						
FERRIC HYDROXIDE						
FERRIC NITRATE			A to 70°F			
FERRIC PERCHLORIDE						
FERRIC PERSULFATE						
FERRIC SALTS			AB to 70°F	AB 20% to 70°F		A to 70°F
FERRIC SESQUICHLORIDE						
FERRIC SESQUISULFATE						
FERRIC SUBSULFATE						
FERRIC SULFATE		A to 70°F	A to 70°F		A to 70°F	A to 150°F
FERRIC SULPHATE		A to 70°F	A to 70°F		A to 70°F	A to 150°F
FERRIC TRICHLORIDE						
FERRIC TRISULFATE						
FERRIFEROUS CHLORIDE						
FERRIFEROUS PERSULFATE						
FERROUS ACETATE						
FERROUS AMMONIUM SULFATE						
FERROUS CHLORIDE		A to 70°F	A to 70°F		A to 70°F	A to 150°F
FERROUS HYDROXIDE						
FERROUS NITRATE						
FERROUS SALTS			NR at 70°F			AB to 70°F
FERROUS SULFATE		A to 70°F	A to 70°F		A to 70°F	A to 150°F

	POLYETHYLENE LOW DENSITY LDPE	®TEFLON FEP	® KALREZ PERFLUORINATED ELASTOMER	® FLUORAZ FLUORINATED COPOLYMER	® AFLAS FLUORINATED COPOLYMER	® NORPRENE COPOLYMER TPO
ETHYLENE TRICHLORIDE		A to 70°F	A to 212°F	NR at 70°F		
ETHYLIC ACID	AB 100% to 122°F A 50% to 122°F	A to 100% to boiling	A to 75°F B at 203°F	NR at 70°F B 30% at 70°F	+71% vol 7 days 73°F NR at 70°F	A to 100% to 70°F
ETHYLIC ALCOHOL		A to 70°F				
ETHYLIC ETHER		A to 70°F				
ETHYLIDENE CHLORIDE	B/NR at 70°F NR at 120°F	A to 100% to 300°F	A to 113°F	A/NR at 70°F	A to 70°F	
ETHYLIDENE DICHLORIDE		A to 70°F				
ETHYLIDENE PERCHLORIDE		A to 70°F				
ETHYNE		A to 70°F				
EXSICCATED ALUM		A to 70°F				
EX-TRI		A to 70°F				
EXXON 2300 TURBO OIL		A to 70°F				
F60 FLUID (DOW CORNING)				A to 70°F		
F61 FLUID (DOW CORNING)				A to 70°F		
FAT LIME		A to 70°F				
FATTY ACIDS	B/NR at 70°F	A to 300°F	A to 212°F	A to 200°F		
FC-43 HEPTACOSOFLUOROTRIBUTYLAMINE				C at 70°F		
FC75 FLUOROCARBON				C at 70°F		
FERAN		A to 70°F				
FERMENTATION AMYL ALCOHOL		A to 70°F				
FERRIC ACETATE		A to 70°F	A to 212°F			
FERRIC BROMIDE		A to 70°F				
FERRIC CHLORIDE	AB to 140°F	A to 100% to 300°F	A to 70°F A aqueous to 212°F	A to 160°F		A to 70°F
FERRIC DICHLORIDE		A to 70°F				
FERRIC HYDROXIDE		A to 70°F	A to 212°F			
FERRIC NITRATE	B at 70°F	A to 300°F	A to 70°F A aqueous to 212°F	A to 70°F		A to 70°F
FERRIC PERCHLORIDE		A to 70°F				
FERRIC PERSULFATE		A to 70°F		A to 70°F		
FERRIC SALTS	AB to 140°F	A to 70°F	A to 212°F			A to 70°F
FERRIC SESQUICHLORIDE		A to 70°F				
FERRIC SESQUISULFATE		A to 70°F				
FERRIC SUBSULFATE		A to 70°F				
FERRIC SULFATE	A to 70°F	A to 300°F	A to 70°F A aqueous to 212°F	A to 70°F		
FERRIC SULPHATE	A to 70°F	A to 300°F	A to 70°F A aqueous to 212°F	A to 70°F		
FERRIC TRICHLORIDE		A to 70°F				
FERRIC TRISULFATE		A to 70°F				
FERRIFEROUS CHLORIDE		A to 70°F				
FERRIFEROUS PERSULFATE		A to 70°F				
FERROUS ACETATE		A to 70°F				
FERROUS AMMONIUM SULFATE		A 30% to 70°F	A to 212°F			
FERROUS CHLORIDE	AB to 70°F	A to 300°F	A to 212°F			
FERROUS HYDROXIDE		A to 70°F				
FERROUS NITRATE		A to 70°F				
FERROUS SALTS	AB to 140°F	A to 70°F				
FERROUS SULFATE	A to 70°F	A to 300°F	A to 212°F	A to 70°F		

	RECOMMENDATIONS	NITRILE NBR	ETHYLENE PROPYLENE EP, EPDM	FLUOROCARBON FKM	CHLOROPRENE CR	HYDROGENATED NITRILE HNBR
FERROUS SULFIDE		A to 70°F		A to 70°F		
FERTILIZER SALTS, AQUEOUS		AB to 140°F	AB to 140°F	AB to 212°F	AB to 140°F	
FILTER ALUM		A to 70°F AB any conc to 200°F			A to 70°F AB any conc to 70°F	
FIREDAMP		A to 70°F	NR at 70°F	A to 70°F	B at 70°F	
FIRWOOD OIL		AB to 150°F	NR at 70°F	A to 70°F	NR at 70°F	
FISH OIL	NBR,VMQ,EP	A to 250°F	NR at 70°F	A to 70°F	AB to 70°F	
FLAXSEED OIL	NBR,FKM,FVMQ,ACM	+5% vol 3 days 212°F A to 200°F	+22% vol 3 days 212°F BC at 70°F	no swell 3 days 212°F A to 250°F	+40% vol 3 days 212°F AB to 180°F	
FLORES MARTIS		A to 70°F AB any conc to 200°F			B any conc to 80°F	
FLUOBORIC ACID	EP,CR,NBR,IIR	A to 140°F B to 185°F	A to 140°F B at 176°F	A to 176°F AB to 185°F	A to 140°F AB to 185°F	
FLUORINATED CYCLIC ESTERS	EP,IIR	NR at 70°F	A to 70°F	A to 70°F	NR at 70°F	
FLUORINE GAS		NR at 70°F wet	A to 104°F wet B at 140°F wet	A to 104°F wet B at 140°F wet	NR at 70°F wet	
FLUORINE GAS				A to 300°F dry		
FLUORINE LIQUID		NR at 70°F	C/NR at 70°F	B at 70°F	C/NR at 70°F	
FLUOROBENZENE	FKM,FVMQ	NR at 70°F	NR at 70°F	A to 70°F	NR at 70°F	
FLUOROCARBON OILS	EP,IIR		A to 70°F		A to 70°F	
FLUOROCHLOROETHYLENE	FKM	NR at 70°F		A to 70°F	NR at 70°F	
FLUOROLUBE	EP,NBR,IIR,CR	A to 70°F	A to 70°F	B at 70°F	A to 70°F	A to 70°F
FLUOROMETHANE		NR at 70°F	A to 70°F		NR at 70°F	
FLUOSILICIC ACID (FLUOROSILICIC ACID)	NBR,EP,CR,FKM	A to conc to 140°F B 50% at 212°F	A to conc to 140°F B to 50% 176-212°F	A to conc to 140°F B to 50% at 212°F	A to conc to 158°F B 50-100% 176-212°F	A to 70°F
FORMALDEHYDE (FORMALIN)	EP,IIR,NBR,CR	+2% vol 3 days 70°F BC 50-100% to 140°F	+1% vol 3 days 70°F A to conc to 120°F	no swell 3 days 70°F A to conc to 176°F	+2% vol 3 days 70°F AB to conc to 70°F	B at 70°F
FORMALDEHYDE (FORMALIN)		A to 40% to 140°F B 40% at 212°F	AB conc 140-176°F A to 37% at 212°F	A 37% to 212°F	NR 40-100% at 158°F	
FORMAMIDE (FORMYLAMINE)		A to 70°F	A to 70°F	NR at 70°F	A to 70°F	
FORMIC ACID	EP,CR,IIR	B/NR 50-100% at 70°F NR 100% at 140°F	A to conc to 200°F A to 90% to 212°F	NR 60-100% at 70°F AB to 50% at 104°F	A to conc to 70°F NR conc at 140°F	
FORMIC ACID		B to 50% at 70°F		C 50% at 140°F	B 90% at 140°F C 90% at 176°F	
FORMIC ACID					NR 90% at 212°F A to 40% at 140°F	
FORMIC ACID					B 40% at 176°F NR 40% at 212°F	
FORMIC ALDEHYDE		B 40% to 80°F	A 40% to 70°F	A 40% to 70°F	B 40% at 70°F	
FORMONITRILE		B/NR at 70°F	B at 70°F	A to 70°F	C/NR at 70°F	
FRAUDS REAGENT		NR 10% at 70°F	B 10% at 70°F	A 10% to 70°F	A/NR 10% at 70°F	
FRENCH POLISH				A to 70°F		
FREON 11 (FREON MF)	NBR,T,FKM	AB to 250°F	NR at 70°F	+34% vol 28 days 75°F BC to 140°F	NR at 70°F	B at 70°F
FREON 11 WITH OIL	NBR	A to 200°F	NR at 70°F	NR at 70°F	A to 70°F	
FREON 12	CR,NBR,FKM,T,SBR	A to 250°F dry C at 70°F wet	BC at 70°F dry AB to 130°F wet	+21% vol 28 days 75°F +20% vol 7 days 300°F	A to 250°F dry BC at 70°F wet	A to 70°F
FREON 12				BC to 130°F dry B to 130°F wet		
FREON 12 +ASTM NO. 2 OIL (50/50)	NBR,FKM	A to 250°F	NR at 70°F	A to 70°F B at 200°F	AB to 250°F high & low aniline oil	A to 70°F
FREON 12 +SUNISCO 4G (50/50)	NBR,FKM	A to 70°F	NR at 70°F	+17% vol 7 days 300°F A to 70°F	B at 70°F	
FREON 13	CR,NBR,EP,FKM	A to 70°F	A to 70°F	A to 70°F	A to 250°F	
FREON 13 +HIGH & LOW ANILINE OIL	CR,NBR	AB to 250°F			AB to 250°F	
FREON 13B1	CR,NBR,EP,FKM,T	A to 70°F	A to 70°F	+19% vol 28 days 75°F A to 70°F	A to 70°F	
FREON 14	CR,NBR,EP,FKM,T	A to 250°F	A to 70°F	A to 70°F	A to 250°F	
FREON 14 +HIGH & LOW ANILINE OIL	CR,NBR	A to 250°F			A to 250°F	
FREON 21	CR	NR at 70°F	NR at 70°F	C/NR at 70°F NR at 104°F	B/NR at 70°F	
FREON 22	CR,EP,T,SBR,IIR	NR at 70°F	A to 140°F	A/NR at 70°F	A to 250°F	
FREON 22 + ASTM OIL NO. 2 (50/50)	CR,FKM	NR at 70°F	NR at 70°F	B at 70°F	B at 70°F	

	STYRENE BUTADIENE SBR	POLYACRYLATE ACM	POLYURETHANE AU, EU	ISOBUTYLENE ISOPRENE IIR	POLYBUTADIENE BR	®AEROQUIP AQP
FERROUS SULFIDE				A to 70°F		
FERTILIZER SALTS, AQUEOUS						
FILTER ALUM				A to 70°F AB any conc to 185°F		
FIREDAMP			B at 70°F	NR at 70°F		
FIRWOOD OIL				NR at 70°F		
FISH OIL	NR at 70°F	A to 70°F	AB to 70°F			
FLAXSEED OIL	+56% vol 3 days 212°F NR at 70°F	+4% vol 3 days 212°F A to 70°F	no swell 3 days 212°F AB to 140°F	+11% vol 3 days 212°F AB to 158°F	NR at 70°F	AB to 70°F
FLORES MARTIS				A to 70°F AB any conc to 150°F		
FLUOBORIC ACID	A to 70°F		NR at 70°F	AB conc to 70°F NR 65% at 70°F	A to 70°F	
FLUORINATED CYCLIC ESTERS				AB to 70°F		
FLUORINE GAS			NR at 70°F	NR at 70°F		
FLUORINE GAS						
FLUORINE LIQUID			NR at 70°F	C/NR at 70°F		
FLUOROBENZENE	NR at 70°F			NR at 70°F	NR at 70°F	
FLUOROCARBON OILS				A to 70°F		
FLUOROCHLOROETHYLENE				NR at 70°F		
FLUOROLUBE	NR at 70°F			A to 70°F	NR at 70°F	
FLUOROMETHANE				B at 70°F		
FLUOSILICIC ACID (FLUOROSILICIC ACID)	AB at 70°F		B/NR at 70°F	A to 70°F AB to 65% to 185°F		
FORMALDEHYDE (FORMALIN)	+2% vol 3 days 70°F AC to 70°F	+13% vol 3 days 70°F NR at 70°F	+9% vol 3 days 70°F NR at 70°F	+0.3% vol 3 days 70°F A to conc to 70°F	B at 70°F	AB to 70°F
FORMALDEHYDE (FORMALIN)			BC 40% at 70°F	A to 40% to 150°F		
FORMAMIDE (FORMYLAMINE)				A to 70°F		
FORMIC ACID	A to 70°F		NR 90-100% at 70°F	A to 70°F AB any conc to 150°F	A to 70°F	NR at 70°F
FORMIC ACID						
FORMIC ACID						
FORMIC ACID						
FORMIC ALDEHYDE			NR 40% at 70°F	A to 40% to 70°F AB 40% to 150°F		
FORMONITRILE			C at 70°F	AB any conc to 150°F		
FRAUDS REAGENT				AB 10% to 150°F		
FRENCH POLISH			*C at 70°F *NR at 140°F			
FREON 11 (FREON MF)	NR at 70°F		B/NR to 130°F	NR at 70°F	B/NR at 70°F	
FREON 11 WITH OIL	NR at 70°F	NR at 70°F	NR at 70°F			
FREON 12	A to 70°F		A to 130°F	B/NR at 70°F	NR at 70°F	NR at 70°F
FREON 12						
FREON 12 +ASTM NO. 2 OIL (50/50)	NR at 70°F	NR at 70°F	NR at 70°F	NR at 70°F	NR at 70°F	
FREON 12 +SUNISCO 4G (50/50)	NR at 70°F	NR at 70°F	NR at 70°F	NR at 70°F	NR at 70°F	
FREON 13	A to 70°F		C at 70°F	A to 70°F	A to 70°F	NR at 70°F
FREON 13 +HIGH & LOW ANILINE OIL						
FREON 13B1	A to 70°F		A to 70°F	A to 70°F	A to 70°F	
FREON 14	A to 70°F		A to 70°F	A to 70°F		
FREON 14 +HIGH & LOW ANILINE OIL						
FREON 21	NR at 70°F			NR at 70°F	NR at 70°F	
FREON 22	A to 70°F	B at 70°F	NR at 70°F	C/NR at 70°F	A to 70°F	NR at 70°F
FREON 22 + ASTM OIL NO. 2 (50/50)	NR at 70°F	B at 70°F		NR at 70°F		

	SYNTHETIC ISOPRENE	NATURAL ISOPRENE	CHLOROSULFONATED POLYETHYLENE	FLUOROSILICONE	SILICONE	CHEMRAZ
	IR	NR	CSM	FVMQ	VMQ	FFKM
FERROUS SULFIDE	A to 70°F	A to 70°F	A to 70°F			
FERTILIZER SALTS, AQUEOUS						
FILTER ALUM	A to 70°F	A to 70°F AB any conc to 150°F				
FIREDAMP	NR at 70°F	NR at 70°F	B at 70°F		NR at 70°F	
FIRWOOD OIL	NR at 70°F	NR at 70°F	NR at 70°F			
FISH OIL				A to 70°F	A to 70°F	
FLAXSEED OIL	NR at 70°F	NR at 70°F	AB to 70°F	no swell 3 days 212°F A to 70°F	+3% vol 3 days 212°F NR at 70°F	
FLORES MARTIS	A to 70°F	A to 70°F AB any conc to 150°F				
FLUOBORIC ACID	A to 70°F	A to 70°F AB 65% to 150°F	A 50-100% to 70°F AB conc to 140°F	A to 70°F	A to 70°F	A to 70°F
FLUORINATED CYCLIC ESTERS		NR at 70°F				C at 70°F
FLUORINE GAS	*NR at 70°F	NR at 70°F	NR at 70°F wet or dry		NR at 70°F	B at 70°F
FLUORINE GAS						
FLUORINE LIQUID	NR at 70°F	NR at 70°F	NR at 70°F		NR at 70°F	
FLUOROBENZENE	NR at 70°F	NR at 70°F	NR at 70°F	B at 70°F	NR at 70°F	
FLUOROCARBON OILS						
FLUOROCHLOROETHYLENE	NR at 70°F	NR at 70°F				
FLUOROLUBE	NR at 70°F	NR at 70°F	A to 70°F	B at 70°F	A to 70°F	B at 70°F
FLUOROMETHANE	NR at 70°F	NR at 70°F				
FLUOSILICIC ACID (FLUOROSILICIC ACID)	A to 70°F	A to conc to 70°F AB to 185°F	A to 250°F	A/NR at 70°F	A/NR at 70°F	A to 70°F
FORMALDEHYDE (FORMALIN)	B at 70°F		AB to 70°F A to 40% to 70°F	+2% vol 3 days 70°F NR at 70°F	+2% vol 3 days 70°F B 40-100% at 70°F	A to 70°F
FORMALDEHYDE (FORMALIN)			NR 40% at 158°F	B 40% at 70°F		
FORMAMIDE (FORMYLAMINE)						A to 70°F
FORMIC ACID	AC to 70°F	AC to 70°F	A to 70°F AB to 200°F	C at 70°F	BC at 70°F	B at 70°F
FORMIC ACID						
FORMIC ACID						
FORMIC ACID						
FORMIC ALDEHYDE	B 40% at 70°F	B 40% at 70°F	A 40% to 70°F		A 40% to 70°F	
FORMONITRILE	B at 70°F	B any conc to 80°F	A to 70°F		A to 70°F	
FRAUDS REAGENT	A 10% to 70°F	AB 10% to 150°F	A 10% to 70°F		NR 10% at 70°F	
FRENCH POLISH						
FREON 11 (FREON MF)	NR at 70°F	NR at 70°F	A to 70°F *AB to 140°F	B at 70°F	NR at 70°F	B at 70°F
FREON 11 WITH OIL				NR at 70°F	NR at 70°F	
FREON 12	NR at 70°F	B/NR at 70°F (dry) NR at 70°F (wet)	A to 130°F	C/NR at 70°F	NR at 70°F wet or dry	AB to 70°F
FREON 12						
FREON 12 +ASTM NO. 2 OIL (50/50)	NR at 70°F	NR at 70°F	B at 70°F	B at 70°F	NR at 70°F	AB to 70°F
FREON 12 +SUNISCO 4G (50/50)	NR at 70°F	NR at 70°F	B at 70°F	B at 70°F	NR at 70°F	A to 70°F
FREON 13	A/NR at 70°F	A/NR at 70°F	A to 70°F	NR at 70°F	NR at 70°F	*B at 70°F
FREON 13 +HIGH & LOW ANILINE OIL						
FREON 13B1	A/NR at 70°F	A/NR at 70°F	A to 70°F	B at 70°F	NR at 70°F	*B at 70°F
FREON 14	NR at 70°F	A/NR at 70°F	A/NR at 70°F		NR at 70°F	*B at 70°F
FREON 14 +HIGH & LOW ANILINE OIL						
FREON 21	NR at 70°F	NR at 70°F	NR at 70°F		NR at 70°F	A to 70°F
FREON 22	A/NR at 70°F	A/NR at 70°F	A to 130°F	B/NR at 70°F	NR at 70°F	A to 70°F
FREON 22 + ASTM OIL NO. 2 (50/50)		NR at 70°F		B at 70°F	NR at 70°F	A to 70°F

C160

	POLYSULFIDE T	CHLORINATED POLYETHYLENE CM	EPICHLOROHYDRIN CO, ECO	® HYTREL COPOLYESTER TPE	® SANTOPRENE COPOLYMER TPO	® C-FLEX STYRENIC TPE
FERROUS SULFIDE						
FERTILIZER SALTS, AQUEOUS						
FILTER ALUM						
FIREDAMP						
FIRWOOD OIL						
FISH OIL						
FLAXSEED OIL	+0.5% vol 3 days 212°F AB to 70°F	AB to 70°F		AB to 70°F		NR at 70°F
FLORES MARTIS						
FLUOBORIC ACID	NR at 70°F			NR at 70°F		A to 70°F
FLUORINATED CYCLIC ESTERS						
FLUORINE GAS				NR at 70°F		
FLUORINE GAS						
FLUORINE LIQUID	NR at 70°F			NR at 70°F	NR at 70°F	
FLUOROBENZENE						
FLUOROCARBON OILS						
FLUOROCHLOROETHYLENE						
FLUOROLUBE	A to 70°F					
FLUOROMETHANE						
FLUOSILICIC ACID (FLUOROSILICIC ACID)	AB 50-100% at 70°F			*AB at 70°F		A to 70°F
FORMALDEHYDE (FORMALIN)	+2.8% vol 3 days 70°F B at 70°F	AB to 70°F	B at 70°F	BC conc at 70°F AC 40% at 70°F	A to 70°F	A to 70°F
FORMALDEHYDE (FORMALIN)						
FORMAMIDE (FORMYLAMINE)					*A to 70°F	
FORMIC ACID		NR at 70°F	B at 70°F	BC at 70°F	A to 70°F	AB to 70°F
FORMIC ACID						
FORMIC ACID						
FORMIC ACID						
FORMIC ALDEHYDE						
FORMONITRILE						
FRAUDS REAGENT						
FRENCH POLISH						
FREON 11 (FREON MF)	A to 70°F			A to 70°F	*AB to 70°F	BC at 70°F
FREON 11 WITH OIL						
FREON 12	A to 70°F	C at 70°F	A to 70°F	A to 70°F	NR at 70°F	
FREON 12						
FREON 12 +ASTM NO. 2 OIL (50/50)						
FREON 12 +SUNISCO 4G (50/50)						
FREON 13	A to 70°F		A to 70°F	BC at 70°F		
FREON 13 +HIGH & LOW ANILINE OIL						
FREON 13B1	A to 70°F					
FREON 14	A to 70°F					
FREON 14 +HIGH & LOW ANILINE OIL						
FREON 21	NR at 70°F		B at 70°F			
FREON 22	NR at 70°F	C at 70°F	A to 70°F	C/NR at 70°F	NR at 70°F	
FREON 22 + ASTM OIL NO. 2 (50/50)	B at 70°F					

	ETHYLENE ACRYLIC EA	POLYALLOMER LINEAR COPOLYMER	NYLON 11 POLYAMIDE	NYLON 12 POLYAMIDE	ETHYLENE VINYL ACETATE EVA	POLYVINYLCHLORIDE FLEXIBLE PVC
FERROUS SULFIDE						
FERTILIZER SALTS, AQUEOUS						
FILTER ALUM						
FIREDAMP						
FIRWOOD OIL						
FISH OIL						
FLAXSEED OIL		A to 70°F	A to 70°F *AB to 140°F	AB to 70°F		
FLORES MARTIS						
FLUOBORIC ACID						AB to 70°F
FLUORINATED CYCLIC ESTERS						NR at 70°F
FLUORINE GAS			NR at 70°F			C/NR at 70°F
FLUORINE GAS						
FLUORINE LIQUID		C at 70°F NR at 122°F	NR at 70°F		NR at 70°F	NR at 70°F
FLUOROBENZENE						NR at 70°F
FLUOROCARBON OILS				AB to 70°F		
FLUOROCHLOROETHYLENE						NR at 70°F
FLUOROLUBE						NR at 70°F
FLUOROMETHANE						
FLUOSILICIC ACID (FLUOROSILICIC ACID)	*AB to 70°F					AB to 70°F
FORMALDEHYDE (FORMALIN)		A to conc to 70°F B 40% at 122°F	+2% vol 180 days 70°F A to conc to 70°F	C to 40% at 70°F	A/NR at 70°F AB 40% to 70°F	AB to conc to 70°F C to 40% 122-130°F
FORMALDEHYDE (FORMALIN)		A to 10% to 122°F	C at 105°F NR at 140°F		C 40% at 70°F	
FORMAMIDE (FORMYLAMINE)				C at 70°F		
FORMIC ACID	*AB to 70°F	A to 100% to 70°F B to 100% at 122°F	NR 50-100% at 70°F	NR 85-100% at 70°F B/NR 10-50% at 70°F	A to 70°F *AB to 100% to 140°F	BC 98-100% at 70°F NR 98-100% at 122°F
FORMIC ACID						B 10-50% at 70°F C 10-50% at 122°F
FORMIC ACID						NR 50% at 130°F AB 10% to 130°F
FORMIC ACID						
FORMIC ALDEHYDE						
FORMONITRILE						
FRAUDS REAGENT						
FRENCH POLISH			*AC to 70°F		*AB to 70°F	NR at 70°F
FREON 11 (FREON MF)	BC at 70°F					NR at 70°F
FREON 11 WITH OIL						
FREON 12	AC to 70°F		AB to 70°F			B/NR at 70°F (dry) NR at 70°F (wet)
FREON 12				AB to 70°F		
FREON 12 +ASTM NO. 2 OIL (50/50)						
FREON 12 +SUNISCO 4G (50/50)						
FREON 13			AC to 70°F			NR at 70°F
FREON 13 +HIGH & LOW ANILINE OIL						
FREON 13B1						
FREON 14						
FREON 14 +HIGH & LOW ANILINE OIL						
FREON 21						
FREON 22	NR at 70°F		AB to 70°F			A/NR at 70°F
FREON 22 + ASTM OIL NO. 2 (50/50)	NR at 70°F					

	POLYETHYLENE LOW DENSITY LDPE	®TEFLON FEP	®KALREZ PERFLUORINATED ELASTOMER	®FLUORAZ FLUORINATED COPOLYMER	®AFLAS FLUORINATED COPOLYMER	®NORPRENE COPOLYMER TPO
FERROUS SULFIDE		A to 70°F				
FERTILIZER SALTS, AQUEOUS	A to 140°F aqueous					
FILTER ALUM		A to 70°F				
FIREDAMP		A to 70°F				
FIRWOOD OIL		A to 70°F				
FISH OIL	NR at 70°F (fish solubles)	A to 70°F	A to 212°F			
FLAXSEED OIL	B/NR at 70°F NR at 140°F	A to 70°F	A to 70°F	A to 70°F		A to 70°F
FLORES MARTIS		A to 70°F				
FLUOBORIC ACID	AB to 140°F	A to conc to 200°F	*A to 212°F			A to 70°F
FLUORINATED CYCLIC ESTERS		A to 70°F	*A to 212°F	C at 70°F		
FLUORINE GAS	NR at 70°F	A to 70°F (dry) B to 125°F (wet)	B at 70°F			
FLUORINE GAS		NR at 300°F (dry) A to 300°F (wet)				
FLUORINE LIQUID	C/NR at 70°F NR at 122°F	AB to 70°F B at 122°F	B at 70°F			
FLUOROBENZENE		A to 70°F	A to 212°F			
FLUOROCARBON OILS		A to 70°F	B at 70°F			
FLUOROCHLOROETHYLENE		A to 70°F				
FLUOROLUBE		A to 70°F	B at 70°F			
FLUOROMETHANE		A to 70°F				
FLUOSILICIC ACID (FLUOROSILICIC ACID)	A to conc to 70°F NR conc at 140°F	A to conc to 200°F	A to 212°F	A to 70°F		A to 70°F
FORMALDEHYDE (FORMALIN)	AB to conc to 70°F A to 40% to 70°F	A to conc to 300°F	A to 104°F *A to 212°F	NR 40-100% at 70°F	A to 70°F	A to 70°F
FORMALDEHYDE (FORMALIN)	AB to 40% at 140°F A to 10% to 122°F					
FORMAMIDE (FORMYLAMINE)		A to 70°F	*A to 212°F			A to 70°F
FORMIC ACID	A to 100% to 70°F AB to 100% at 140°F	A to 100% to boiling	A to 70°F B 12% to 212°F	C at 70°F	B at 70°F	A to 70°F
FORMIC ACID						
FORMIC ACID						
FORMIC ACID						
FORMIC ALDEHYDE		A 40% to 70°F				
FORMONITRILE		A to 70°F				
FRAUDS REAGENT		A 10% to 70°F				
FRENCH POLISH	*A to 70°F					
FREON 11 (FREON MF)	BC at 70°F	A to 70°F	BC at 70°F	NR at 70°F		
FREON 11 WITH OIL				NR at 70°F		
FREON 12	AB to 70°F (dry) NR at 140°F (dry)	A to 70°F (wet or dry)	BC at 70°F	NR at 70°F		
FREON 12	C at 70°F (wet)					
FREON 12 +ASTM NO. 2 OIL (50/50)				NR at 70°F		
FREON 12 +SUNISCO 4G (50/50)				NR at 70°F		
FREON 13		A to 70°F	BC at 70°F	NR at 70°F		
FREON 13 +HIGH & LOW ANILINE OIL						
FREON 13B1		A to 70°F	B at 70°F	NR at 70°F		
FREON 14		A to 70°F	B at 70°F	NR at 70°F		
FREON 14 +HIGH & LOW ANILINE OIL						
FREON 21		A to 70°F	A to 130°F *A to 212°F	NR at 70°F		
FREON 22	A to 70°F	A to 300°F	*AB to 212°F	NR at 70°F		
FREON 22 + ASTM OIL NO. 2 (50/50)				C/NR at 70°F		

C 163

	RECOMMENDATIONS	NITRILE NBR	ETHYLENE PROPYLENE EP, EPDM	FLUOROCARBON FKM	CHLOROPRENE CR	HYDROGENATED NITRILE HNBR
FREON 22 +HIGH & LOW ANILINE OIL	CR	NR at 70°F	NR at 70°F	NR at 70°F	A to 200°F AB to 250°F hi aniline	
FREON 31	CR, EP, IIR	NR at 70°F	A to 70°F	NR at 70°F	A to 70°F	
FREON 32	CR, EP, IIR	A to 70°F	A to 70°F	NR at 70°F	A to 70°F	
FREON T-P35		A to 70°F	A to 70°F	A to 70°F	A to 70°F	
FREON 112 WITH OR WITHOUT OIL	CR, NBR, T, FKM	AB to 250°F	NR at 70°F	A to 70°F	B at 70°F	B at 70°F
FREON 113	NBR, CR, AU, T	A to 250°F	NR at 70°F	+18% vol 28 days 75°F AB to 140°F	A to 130°F NR at 140°F	A to 70°F
FREON 113 +HIGH & LOW ANILINE OIL	CR, NBR	A to 250°F			A to 250°F	
FREON 113B1		A to 70°F	A to 70°F	A to 70°F	A to 70°F	
FREON 114	CR, NBR, FKM	A to 250°F	NR at 70°F	A to 104°F	A to 250°F	A to 70°F
FREON 114 +HIGH & LOW ANILINE OIL	CR, NBR	A to 250°F	NR at 70°F	NR at 70°F	A to 250°F	
FREON 114B2	T, CR, FKM, NBR	AB to 250°F	NR at 70°F	B at 70°F	A to 250°F	B at 70°F
FREON 114B2 +HIGH & LOW ANILINE OIL	CR, NBR	A to 250°F	NR at 70°F		A to 250°F	
FREON 115	CR, NBR, EP	A to 250°F	A to 70°F	AB to 70°F	A to 250°F	
FREON 115 +HIGH & LOW ANILINE OIL	CR, NBR	A to 250°F			A to 250°F	
FREON 142B	CR, NBR, EP	A to 70°F	A to 70°F	NR at 70°F	A to 70°F	B at 70°F
FREON 152A	CR, NBR, EP	A to 70°F	A to 70°F	NR at 70°F	A to 70°F	
FREON 218	CR, NBR, EP	A to 250°F	A to 70°F	A to 70°F	A to 250°F	
FREON 218 +HIGH & LOW ANILINE OIL	CR, NBR	A to 250°F			A to 250°F	
FREON C316	CR, NBR, T	A to 250°F	A to 70°F	A to 70°F	A to 250°F	
FREON C316 +HIGH & LOW ANILINE OIL	CR, NBR	A to 250°F hi aniline only			A to 250°F	
FREON C318	CR, NBR, EP	A to 70°F	A to 70°F	A to 70°F	A to 70°F	A to 70°F
FREON 502 (F22 + F316)	CR, EP, SBR, NBR	B to 250°F	A to 70°F	B at 70°F	A to 250°F	
FREON 502 +HIGH & LOW ANILINE OIL	CR				A to 250°F	
FREON T-WD602	FKM	B at 70°F	B at 70°F	A to 70°F	B at 70°F	
FREON BF (F112)	CR, NBR, T, FKM	AB to 250°F	NR at 70°F	A to 70°F	B at 70°F	B at 70°F
FREON MF (F11)	NBR, T, FKM	AB to 250°F	NR at 70°F	+34% vol 28 days 75°F BC to 140°F	NR at 70°F	B at 70°F
FREON PCA	NBR, CR, AU	A to 70°F	NR at 70°F	B at 70°F	A to 70°F	A to 70°F
FREON TA		A to 70°F	A to 70°F	C at 70°F	A to 70°F	
FREON TC		A to 70°F	B at 70°F	A to 70°F	A to 70°F	
FREON TF	NBR, CR, AU, T	A to 250°F	NR at 70°F	+18% vol 28 days 75°F AB to 140°F	A to 130°F NR at 140°F	A to 70°F
FREON TMC		B at 70°F	B at 70°F	A to 70°F	B at 70°F	
FREONS, GENERAL	FKM, NBR	A to 250°F	A to 70°F without oil	A to 70°F wet	A to 70°F	AB to 70°F
FRUIT JUICE	NBR, EP	A to 140°F	A to 250°F	A to 140°F	A to 140°F	
FRUCTOSE		A to 140°F B at 176°F	A to 176°F	A to 212°F	A to 70°F	
FUEL OIL	NBR, FKM, FVMQ, T	A to 250°F	NR at 70°F	A to 200°F	B at 70°F NR at 140°F	A to 70°F
FUEL OIL, ACIDIC	NBR, FKM	A to 70°F AB to 150°F	NR at 70°F	A to 70°F	B to 150°F	A to 70°F
FUEL OIL NO. 6	FKM, ACM	B at 70°F	NR at 70°F	A to 70°F	NR at 70°F	B at 70°F
FUMARIC ACID	NBR, FKM, FVMQ, EP IIR	A to 70°F	B at 70°F	A to 70°F	B at 70°F	A to 70°F
FUMAROLE ACID		A to 70°F	A to 70°F	A to 70°F	A to 70°F	
FUMING SULFURIC ACID (20-25% OLEUM)	FKM	NR at 70°F dissolved	NR at 70°F dissolved	+23% vol 7 days 70°F A to 70°F	NR at 70°F dissolved	B at 70°F
FURALDEHYDE	EP, IIR	NR at 70°F	AB to 160°F NR at 212°F	+86% vol 28 days 158°F NR at 70°F	BC to 104°F C/NR at 140°F	
FURAN	T	NR at 70°F	C/NR at 70°F	A/NR at 70°F	NR at 70°F	NR at 70°F
FURFURAL	EP, IIR	NR at 70°F	AB to 160°F NR at 212°F	+86% vol 28 days 158°F NR at 70°F	BC to 104°F C/NR at 140°F	NR at 70°F
FURFURALDEHYDE	EP, IIR	NR at 70°F	AB to 160°F NR at 212°F	NR at 70°F	BC to 104°F C/NR at 140°F	NR at 70°F

	STYRENE BUTADIENE SBR	POLYACRYLATE ACM	POLYURETHANE AU, EU	ISOBUTYLENE ISOPRENE IIR	POLYBUTADIENE BR	® AEROQUIP AQP
FREON 22 +HIGH & LOW ANILINE OIL	NR at 70°F	NR at 70°F	NR at 70°F			
FREON 31	B at 70°F			A to 70°F	B at 70°F	
FREON 32	A to 70°F			A to 70°F	A to 70°F	
FREON T-P35	A to 70°F		A to 70°F	A to 70°F	A to 70°F	
FREON 112 WITH OR WITHOUT OIL	NR at 70°F		B at 70°F	NR at 70°F	NR at 70°F	
FREON 113	B at 70°F		AB to 70°F	NR at 70°F	B at 70°F	
FREON 113 +HIGH & LOW ANILINE OIL						
FREON 113B1						
FREON 114	A to 70°F		A to 70°F	A/NR at 70°F	A to 70°F	
FREON 114 +HIGH & LOW ANILINE OIL	NR at 70°F	NR at 70°F	NR at 70°F			
FREON 114B2	NR at 70°F			NR at 70°F	C at 70°F	
FREON 114B2 +HIGH & LOW ANILINE OIL						
FREON 115	A to 70°F			A to 70°F	A to 70°F	
FREON 115 +HIGH & LOW ANILINE OIL						
FREON 142B	A to 70°F			A to 70°F	A to 70°F	
FREON 152A	A to 70°F			A to 70°F	A to 70°F	
FREON 218	A to 70°F			A to 70°F	A to 70°F	
FREON 218 +HIGH & LOW ANILINE OIL						
FREON C316	A to 70°F			A to 70°F	A to 70°F	
FREON C316 +HIGH & LOW ANILINE OIL						
FREON C318	A to 70°F			A to 70°F	A to 70°F	
FREON 502 (F22 + F316)	A to 70°F			A to 70°F	A to 70°F	
FREON 502 +HIGH & LOW ANILINE OIL						
FREON T-WD602	B at 70°F		A to 70°F	A to 70°F	B at 70°F	
FREON BF (F112)	NR at 70°F		B at 70°F	NR at 70°F	NR at 70°F	
FREON MF (F11)	NR at 70°F		B/NR to 130°F	NR at 70°F	B/NR at 70°F	
FREON PCA	B at 70°F		A to 70°F	NR at 70°F		
FREON TA	A to 70°F		A to 70°F	A to 70°F	A to 70°F	
FREON TC	B at 70°F		A to 70°F	A to 70°F	B at 70°F	
FREON TF	B at 70°F		AB to 70°F	NR at 70°F	B at 70°F	
FREON TMC	C at 70°F		B at 70°F	B at 70°F	C at 70°F	
FREONS, GENERAL	AB to 70°F	NR at 70°F	AC to 70°F AB fumes to 70°F	A/NR at 70°F	AB to 70°F	NR at 70°F
FRUIT JUICE	B at 70°F		A to 70°F			
FRUCTOSE						
FUEL OIL	NR at 70°F	A to 70°F	B/NR at 70°F	NR at 70°F	NR at 70°F	AB to 70°F
FUEL OIL, ACIDIC	NR at 70°F	A to 70°F	B at 70°F	NR at 70°F	NR at 70°F	
FUEL OIL NO. 6	NR at 70°F	A to 70°F	B at 70°F	NR at 70°F	NR at 70°F	
FUMARIC ACID	AB to 70°F	NR at 70°F		NR at 70°F	AB to 70°F	
FUMAROLE ACID			A to 70°F	A to 70°F		
FUMING SULFURIC ACID (20-25% OLEUM)	NR at 70°F	NR at 70°F	NR at 70°F	NR at 70°F dissolved	NR at 70°F	NR at 70°F
FURALDEHYDE	NR at 70°F	NR at 70°F	C/NR at 70°F	A to 70°F B to 185°F	C/NR at 70°F	AB to 70°F
FURAN	NR at 70°F	NR at 70°F	NR at 70°F	NR at 70°F	NR at 70°F	
FURFURAL	NR at 70°F	NR at 70°F	C/NR at 70°F	A to 70°F B to 185°F	C/NR at 70°F	AB to 70°F
FURFURALDEHYDE	NR at 70°F	NR at 70°F	NR at 70°F	B at 70°F	NR at 70°F	

	SYNTHETIC ISOPRENE IR	NATURAL ISOPRENE NR	CHLOROSULFONATED POLYETHYLENE CSM	FLUOROSILICONE FVMQ	SILICONE VMQ	CHEMRAZ FFKM
FREON 22 +HIGH & LOW ANILINE OIL				NR at 70°F	NR at 70°F	
FREON 31	B at 70°F	B at 70°F	B at 70°F			AB to 70°F
FREON 32	A to 70°F	A to 70°F	A to 70°F			A to 70°F
FREON T-P35	A to 70°F	A to 70°F	A to 70°F		A to 70°F	
FREON 112 WITH OR WITHOUT OIL	NR at 70°F	NR at 70°F	B at 70°F		NR at 70°F	A to 70°F
FREON 113	C/NR at 70°F	C/NR at 70°F	A to 130°F	NR at 70°F	NR at 70°F	C at 70°F
FREON 113 +HIGH & LOW ANILINE OIL	A to 70°F	A to 70°F	A to 70°F		NR at 70°F	C at 70°F
FREON 113B1						
FREON 114	A/NR at 70°F	A/NR at 70°F	A to 70°F *AB to 130°F	B at 70°F	NR at 70°F	C at 70°F
FREON 114 +HIGH & LOW ANILINE OIL				NR at 70°F	NR at 70°F	
FREON 114B2	NR at 70°F	NR at 70°F	A to 70°F		NR at 70°F	C at 70°F
FREON 114B2 +HIGH & LOW ANILINE OIL						
FREON 115	A to 70°F	A to 70°F	A to 70°F		NR at 70°F	C at 70°F
FREON 115 +HIGH & LOW ANILINE OIL						
FREON 142B	A to 70°F	A to 70°F	A to 70°F			C at 70°F
FREON 152A	A to 70°F	A to 70°F	C/NR at 70°F			A to 70°F
FREON 218	A to 70°F	A to 70°F	A to 70°F			
FREON 218 +HIGH & LOW ANILINE OIL						
FREON C316	A to 70°F	A to 70°F	A to 70°F			
FREON C316 +HIGH & LOW ANILINE OIL						
FREON C318	A to 70°F	A to 70°F	A to 70°F			C at 70°F
FREON 502 (F22 + F316)	A to 70°F	A to 70°F				*B at 70°F
FREON 502 +HIGH & LOW ANILINE OIL						
FREON T-WD602	C at 70°F	C at 70°F	B at 70°F		NR at 70°F	
FREON BF (F112)	NR at 70°F	NR at 70°F	B at 70°F		NR at 70°F	B at 70°F
FREON MF (F11)	NR at 70°F	NR at 70°F	A to 70°F *AB to 140°F	B at 70°F	NR at 70°F	B at 70°F
FREON PCA		NR at 70°F	A to 70°F		NR at 70°F	C at 70°F
FREON TA	A to 70°F	A to 70°F	A to 70°F		A to 70°F	
FREON TC	NR at 70°F	NR at 70°F	A to 70°F		NR at 70°F	
FREON TF	C/NR at 70°F	C/NR at 70°F	A to 130°F	NR at 70°F	NR at 70°F	C at 70°F
FREON TMC	B at 70°F	B at 70°F	B at 70°F		C at 70°F	
FREONS, GENERAL	A/NR at 70°F	A/NR at 70°F	AB to 200°F	NR at 70°F	NR at 70°F	AC to 70°F
FRUIT JUICE	C at 70°F	AC to 70°F	AB to 70°F	A to 70°F	A to 70°F	
FRUCTOSE						
FUEL OIL	NR at 70°F	NR at 70°F	BC at 70°F	A to 70°F	NR at 70°F	A to 70°F
FUEL OIL, ACIDIC	NR at 70°F	NR at 70°F	NR at 70°F	A to 70°F	A to 70°F	A to 70°F
FUEL OIL NO. 6	NR at 70°F	NR at 70°F	NR at 70°F	A to 70°F	A to 70°F	A to 70°F
FUMARIC ACID	AB to 70°F	AB any conc to 80°F	B at 70°F	A to 70°F	B at 70°F	A to 70°F
FUMAROLE ACID	A to 70°F	A to 70°F	A to 70°F		A to 70°F	
FUMING SULFURIC ACID (20-25% OLEUM)	NR at 70°F	NR at 70°F	NR at 70°F dissolved		NR at 70°F dissolved	AB to 70°F NR at 70°F white
FURALDEHYDE	NR at 70°F	NR at 70°F	B/NR at 70°F	NR at 70°F	C/NR at 70°F	B at 70°F
FURAN	NR at 70°F	NR at 70°F	NR at 70°F		NR at 70°F (resin)	A to 70°F
FURFURAL	NR at 70°F	NR at 70°F	B/NR at 70°F	NR at 70°F	C/NR at 70°F	B at 70°F
FURFURALDEHYDE	NR at 70°F	NR at 70°F	B/NR at 70°F	NR at 70°F	C/NR at 70°F	B at 70°F

	POLYSULFIDE T	CHLORINATED POLYETHYLENE CM	EPICHLOROHYDRIN CO, ECO	® HYTREL COPOLYESTER TPE	® SANTOPRENE COPOLYMER TPO	® C-FLEX STYRENIC TPE
FREON 22 +HIGH & LOW ANILINE OIL						
FREON 31	C at 70°F					
FREON 32	A to 70°F					
FREON T-P35	A to 70°F					
FREON 112 WITH OR WITHOUT OIL	A to 70°F					
FREON 113	A to 70°F		A to 70°F	A to 130°F	NR at 70°F	
FREON 113 +HIGH & LOW ANILINE OIL						
FREON 113B1						
FREON 114	A to 70°F		A to 70°F	A to 70°F		
FREON 114 +HIGH & LOW ANILINE OIL						
FREON 114B2	AB to 70°F					
FREON 114B2 +HIGH & LOW ANILINE OIL						
FREON 115	A to 70°F					
FREON 115 +HIGH & LOW ANILINE OIL						
FREON 142B	A to 70°F					
FREON 152A	A to 70°F					
FREON 218	A to 70°F					
FREON 218 +HIGH & LOW ANILINE OIL						
FREON C316						
FREON C316 +HIGH & LOW ANILINE OIL						
FREON C318	A to 70°F					
FREON 502 (F22 + F316)						
FREON 502 +HIGH & LOW ANILINE OIL					NR at 70°F	
FREON T-WD602	A to 70°F					
FREON BF (F112)	A to 70°F					
FREON MF (F11)	A to 70°F			A to 70°F	*AB to 70°F	BC at 70°F
FREON PCA	A to 70°F					
FREON TA	A to 70°F					
FREON TC	A to 70°F					
FREON TF	A to 70°F		A to 70°F	A to 130°F	NR at 70°F	
FREON TMC	A to 70°F					
FREONS, GENERAL	A to 70°F	B/NR at 70°F	A to 70°F	A to 70°F	A/NR at 70°F	A/NR at 70°F
FRUIT JUICE				AB to 70°F		
FRUCTOSE						
FUEL OIL	A to 70°F	A to 70°F	A to 70°F	AB to 70°F *AB to 140°F		
FUEL OIL, ACIDIC	A to 70°F					
FUEL OIL NO. 6	A to 70°F					
FUMARIC ACID						
FUMAROLE ACID						
FUMING SULFURIC ACID (20-25% OLEUM)	NR at 70°F	NR at 70°F		no swell 20% conc 70°F NR conc at 70°F		A to 70°F
FURALDEHYDE	B/NR at 70°F	A to 70°F	NR at 70°F	AB to 70°F	A to 70°F	
FURAN	B at 70°F				*A to 70°F	
FURFURAL	B/NR at 70°F	A to 70°F	NR at 70°F	AB to 70°F	A to 70°F	
FURFURALDEHYDE	NR at 70°F					

	ETHYLENE ACRYLIC EA	POLYALLOMER LINEAR COPOLYMER	NYLON 11 POLYAMIDE	NYLON 12 POLYAMIDE	ETHYLENE VINYL ACETATE EVA	POLYVINYLCHLORIDE FLEXIBLE PVC
FREON 22 +HIGH & LOW ANILINE OIL						
FREON 31						
FREON 32						
FREON T-P35						
FREON 112 WITH OR WITHOUT OIL						
FREON 113	BC at 70°F		NR at 70°F	AB to 70°F		NR at 70°F
FREON 113 +HIGH & LOW ANILINE OIL						
FREON 113B1						
FREON 114						NR at 70°F
FREON 114 +HIGH & LOW ANILINE OIL						
FREON 114B2						
FREON 114B2 +HIGH & LOW ANILINE OIL						
FREON 115						
FREON 115 +HIGH & LOW ANILINE OIL						
FREON 142B						
FREON 152A						
FREON 218						
FREON 218 +HIGH & LOW ANILINE OIL						
FREON C316						
FREON C316 +HIGH & LOW ANILINE OIL						
FREON C318						
FREON 502 (F22 + F316)			A to 70°F			
FREON 502 +HIGH & LOW ANILINE OIL						
FREON T-WD602						
FREON BF (F112)						
FREON MF (F11)	BC at 70°F					NR at 70°F
FREON PCA						
FREON TA						
FREON TC						
FREON TF	BC at 70°F		NR at 70°F			NR at 70°F
FREON TMC						
FREONS, GENERAL	C/NR at 70°F	B at 70°F	AB to 70°F		B/NR at 70°F	NR at 70°F
FRUIT JUICE			AB to 105°F	AB to 70°F	AB to 70°F	AB to 70°F
FRUCTOSE						
FUEL OIL	A to 70°F	A to 70°F B at 122°F	AB to 194°F	AB to 140°F	C/NR at 70°F	AB to 122°F C at 150°F
FUEL OIL, ACIDIC	A to 70°F					
FUEL OIL NO. 6	A to 70°F					
FUMARIC ACID						
FUMAROLE ACID						
FUMING SULFURIC ACID (20-25% OLEUM)	NR at 70°F	A to 70°F	NR at 70°F		BC at 70°F	AB to 70°F B at 122°F
FURALDEHYDE	NR at 70°F	NR at 70°F	BC at 70°F NR at 194°F		B/NR at 70°F	A/NR at 70°F NR at 150°F
FURAN						A to 70°F (resin)
FURFURAL	NR at 70°F	NR at 70°F	BC at 70°F NR at 194°F	C at 70°F	B/NR at 70°F	A/NR at 70°F NR at 150°F
FURFURALDEHYDE						

	POLYETHYLENE LOW DENSITY LDPE	®TEFLON FEP	®KALREZ PERFLUORINATED ELASTOMER	®FLUORAZ FLUORINATED COPOLYMER	®AFLAS FLUORINATED COPOLYMER	®NORPRENE COPOLYMER TPO
FREON 22 +HIGH & LOW ANILINE OIL				NR at 70°F		
FREON 31		A to 70°F	AB to 70°F	NR at 70°F		
FREON 32		A to 70°F	AB to 70°F	NR at 70°F		
FREON T-P35		A to 70°F	BC at 70°F	NR at 70°F		
FREON 112 WITH OR WITHOUT OIL		A to 70°F	*AB to 212°F	NR at 70°F		
FREON 113	NR at 70°F	+1.7% wt gain 117°F A to 200°F	BC at 70°F	NR at 70°F	NR at 70°F dissolved	
FREON 113 +HIGH & LOW ANILINE OIL				NR at 70°F		
FREON 113B1			BC at 70°F	NR at 70°F		
FREON 114		A to 70°F	BC at 70°F	NR at 70°F		
FREON 114 +HIGH & LOW ANILINE OIL				NR at 70°F		
FREON 114B2		A to 70°F	BC at 70°F	NR at 70°F		
FREON 114B2 +HIGH & LOW ANILINE OIL						
FREON 115		A to 70°F	BC at 70°F	NR at 70°F		
FREON 115 +HIGH & LOW ANILINE OIL						
FREON 142B		A to 70°F	BC at 70°F	NR at 70°F		
FREON 152A		A to 70°F	BC at 70°F	NR at 70°F		
FREON 218		A to 70°F	BC at 70°F	NR at 70°F		
FREON 218 +HIGH & LOW ANILINE OIL						
FREON C316		A to 70°F	BC at 70°F	NR at 70°F		
FREON C316 +HIGH & LOW ANILINE OIL						
FREON C318		A to 70°F	BC at 70 °F	NR at 70°F		
FREON 502 (F22 + F316)						
FREON 502 +HIGH & LOW ANILINE OIL		A to 70°F	BC at 70°F	NR at 70°F		
FREON T-WD602		A to 70°F	BC at 70°F			
FREON BF (F112)		A to 70°F	BC at 70°F	NR at 70°F		
FREON MF (F11)	BC at 70°F	A to 70°F	BC at 70°F	NR at 70°F		
FREON PCA				NR at 70°F		
FREON TA		A to 70°F	BC at 70°F			
FREON TC		A to 70°F	BC at 70°F			
FREON TF	NR at 70°F	+1.7% wt gain 117°F A to 117°F	BC at 70°F	NR at 70°F	NR at 70°F dissolved	
FREON TMC		A to 70°F	BC at 70°F			
FREONS, GENERAL	AC to 70°F	A to 200°F wet or dry	BC at 70°F	NR at 70°F	NR at 70°F	NR at 70°F
FRUIT JUICE	AB to 140°F	A to 70°F		A to 70°F		
FRUCTOSE	A to 70°F AB to 140°F	A to 70°F				
FUEL OIL	C/NR at 70°F NR at 140°F	A to 300°F	A to 212°F	A to 70°F		
FUEL OIL, ACIDIC				A to 70°F		
FUEL OIL NO. 6				A to 70°F (1,2,3,5A,5B & 6)		
FUMARIC ACID		A to 70°F	A to 212°F	A to 70°F		
FUMAROLE ACID		A to 70°F				
FUMING SULFURIC ACID (20-25% OLEUM)	NR at 70°F	A to 400°F	A to 212°F	A to 70°F	+7.4% vol 180 days 70°F A to 70°F	AB to 70°F
FURALDEHYDE	A/NR at 70°F	A to 70°F	A to 70°F *A to 212°F	A to 70°F		BC at 70°F
FURAN		A to 200°F	A to 212°F	A to 70°F		
FURFURAL	A/NR at 70°F	A to 300°F	A to 70°F *A to 158°F	A to 70°F	B at 70°F	BC at 70°F
FURFURALDEHYDE	A/NR at 70°F	A to 300°F	A to 70°F *A to 158°F	NR at 70°F		BC at 70°F

	RECOMMENDATIONS	NITRILE NBR	ETHYLENE PROPYLENE EP, EPDM	FLUOROCARBON FKM	CHLOROPRENE CR	HYDROGENATED NITRILE HNBR
FURFURAN	T	NR at 70°F	C/NR at 70°F	A/NR at 70°F	NR at 70°F	NR at 70°F
FURFURYL ALCOHOL	EP, IIR	+163% vol 3 days 158°F NR at 70°F	no swell 3 days 158°F B at 70°F	+4% 3 days 158°F A/NR at 75°F	+37% vol 3 days 158°F NR at 70°F	NR at 70°F
FUROL		NR at 70°F	A to 70°F	NR at 70°F	C at 70°F	
FURYLCARBINOL	EP	NR at 70°F	B at 70°F	NR at 70°F	NR at 70°F	NR at 70°F
FUSEL OIL		A to 70°F AB any conc to 180°F	A to 70°F	A to 70°F	A to 70°F AB any conc to 180°F	
FRYQUEL A60		NR at 70°F	B at 70°F	NR at 70°F	NR at 70°F	NR at 70°F
FRYQUEL 90, 100, 150, 220, 300, 500, 550	EP, FKM	NR at 70°F	A to 70°F	A to 212°F	NR at 70°F	NR at 70°F
C.A.!.		NR at 70°F	NR at 70°F	NR at 70°F	NR at 70°F	
GALLIC ACID	FKM, FVMQ, EP	AB to 70°F	AB to 70°F	A to 120°F	B at 70°F	B at 70°F
GALLOTANNIC ACID	NBR, EP, FKM, IIR, CR	A to 100% to 200°F	A to 100% to 140°F AB to 200°F	A to 100% to 140°F AB to 200°F	A to 100% to 70°F AB to 200°F	A to 70°F
GAS OIL	NBR	A to 250°F	NR at 70°F	A to 70°F	NR at 70°F	
GASOLINE, AUTOMOTIVE	NBR, FKM, FVMQ	+6% vol 7 days 70°F A to 250°F	NR at 70°F	+3.1% vol 7 days 70°F AB to 200°F	+55% vol 7 days 70°F NR at 70°F	
GASOLINE, ETHYL & REGULAR	NBR	A to 250°F	NR at 70°F	A to 70°F B at 140°F	B/NR at 70°F	
GASOLINE, REFINED	NBR	A to 250°F	NR at 70°F	A to 70°F	B/NR at 70°F	
GASOLINE, SOUR	NBR	A to 250°F	NR at 70°F	A to 70°F	B/NR at 70°F	
GASOLINE WITH MERCAPTAN	NBR	A to 250°F	NR at 70°F	A to 70°F	NR at 70°F	
GASOLINE 40% AROMATIC	NBR	A to 250°F	NR at 70°F	A to 70°F	B/NR at 70°F	
GASOLINE 65 OCTANE		A to 70°F	NR at 70°F	A to 70°F	AB to 70°F	
GASOLINE 100 OCTANE	NBR	A to 250°F	NR at 70°F	A to 70°F	B/NR at 70°F	
GASOLINE 130 OCTANE	NBR	A to 250°F AB to 70°F aviation	NR at 70°F	A to 70°F	NR at 70°F NR at 70°F aviation	
GAULTHERIA OIL		NR at 70°F	C at 70°F	B at 70°F	NR at 70°F	
GEAR OIL		+1.3% vol 5 days 212°F A to 212°F		+8.7% vol 5 days 212°F A to 350°F		
GELATIN (AQUEOUS SODIUM SULFATE)	NBR, EP, CR, FKM, IIR	A to 200°F	A to 176°F	A to 212°F	A to 140°F AB any conc to 180°F	
GENERATOR GAS		A to 70°F	C at 70°F	A to 70°F	B at 70°F	
GERMAN SALTPETER		A to 70°F	A to 70°F	A to 70°F	A to 70°F	
GIBBSITE		B at 70°F	A to 70°F	C at 70°F	A to 70°F	
GIN		A to 176°F	A to 176°F	A to 212°F		
GIRLING BRAKE FLUID	EP, SBR	C at 70°F	A to 70°F	NR at 70°F	B at 70°F	
GLACIAL ACETIC ACID	EP, IIR	NR at 70°F	AB to 200°F	NR at 70°F	C/NR at 70°F	B at 70°F
GLAUBERS SALT	NBR, EP, FKM, CR, IIR	A to 100% to 200°F B at 212°F	A to 100% to 176°F AB to 200°F	A to 100% to 212°F	A to 100% to 140°F AB to 100% to 200°F	NR at 70°F
GLUCONIC ACID		C at 70°F				
GLUCOSE	NBR, EP, FKM, CR	A to 200°F	A to 176°F B at 212°F	A to 212°F	AB any conc to 140°F	A to 70°F
GLUE	NBR, EP, CR, IIR	A to 160°F B at 176°F	A to 176°F	A to 212°F	A to 158°F	
GLUE SIZING	NBR	A to 160°F	A to 70°F	A to 70°F	A to 70°F	
GLYCERIN	NBR, EP, CR, FKM, IIR	A to 250°F	A to 176°F AB to 200°F	A to 250°F	A to 158°F	
GLYCEROL	NBR, EP, CR, FKM, IIR	A to 250°F	A to 176°F AB to 200°F	A to 250°F	A to 158°F	
GLYCERYL HYDROXIDE		A to 70°F	A to 70°F	A to 70°F	A to 70°F	
GLYCERAL TRIACETATE		A any conc to 80°F	A to 70°F	C at 70°F	A any conc to 80°F	
GLYCERYL TRIOLEATE		AB any conc to 120°F			BC any conc to 80°F	
GLYCOGNENIC ACID		C at 70°F				
GLYCOL, ACETATE		C at 70°F	A to 70°F	A to 70°F	NR at 70°F	
GLYCOL, ALCOHOL		AB any conc to 180°F			AB any conc to 150°F	
GLYCOL, BUTYL ETHER		B at 70°F	A to 70°F	C at 70°F	C at 70°F	
GLYCOL, CHLOROHYDRIN		NR at 70°F	A to 70°F	B at 70°F	B/NR at 70°F	

	STYRENE BUTADIENE SBR	POLYACRYLATE ACM	POLYURETHANE AU, EU	ISOBUTYLENE ISOPRENE IIR	POLYBUTADIENE BR	® AEROQUIP AQP
FURFURAN	NR at 70°F	NR at 70°F	NR at 70°F	NR at 70°F	NR at 70°F	
FURFURYL ALCOHOL	+13% vol 3 days 158°F NR at 70°F	+212% vol 3 days 158°F NR at 70°F	+153% vol 3 days 158°F NR at 70°F	+1.4% vol 3 days 158°F B at 70°F	NR at 70°F	
FUROL			NR at 70°F	A to 70°F		
FURYLCARBINOL	NR at 70°F	NR at 70°F	NR at 70°F	B at 70°F	NR at 70°F	
FUSEL OIL			C at 70°F	A to 70°F AB any conc to 180°F		
FRYQUEL A60	NR at 70°F	NR at 70°F	NR at 70°F	B a 70°F	NR at 70°F	
FRYQUEL 90, 100, 150, 220, 300, 500, 550	NR at 70°F	NR at 70°F	NR at 70°F	A to 70°F	NR at 70°F	AB to 200°F
C.A.J.			NR at 70°F	NR at 70°F		
GALLIC ACID	B at 70°F	NR at 70°F	NR at 70°F	B any conc to 150°F	B at 70°F	
GALLOTANNIC ACID			A to 70°F	B any conc to 185°F		
GAS OIL	NR at 70°F	NR at 70°F	AB to 70°F			
GASOLINE, AUTOMOTIVE	NR at 70°F	NR at 70°F	A to 70°F (ester base) NR at 70°F (ether base)	+240% vol 7 days 70°F NR at 70°F	NR at 70°F	AB to 70°F
GASOLINE, ETHYL & REGULAR	NR at 70°F	NR at 70°F	AB to 70°F			
GASOLINE, REFINED	NR at 70°F	NR at 70°F	BC at 70°F	NR at 70°F		
GASOLINE, SOUR	NR at 70°F	NR at 70°F	B at 70°F	NR at 70°F		
GASOLINE WITH MERCAPTAN	NR at 70°F		B at 70°F			
GASOLINE 40% AROMATIC	NR at 70°F	NR at 70°F	AB to 70°F 40% aromatic AB to 70°F nonaromatic	NR at 70°F		
GASOLINE 65 OCTANE			B at 70°F	NR at 70°F		
GASOLINE 100 OCTANE	NR at 70°F	NR at 70°F	BC at 70°F	NR at 70°F		
GASOLINE 130 OCTANE	NR at 70°F	NR at 70°F	BC at 70°F			
GAULTHERIA OIL				B any conc to 80°F		
GEAR OIL						
GELATIN (AQUEOUS SODIUM SULFATE)	A to 70°F	NR at 70°F	A/NR at 70°F	A to 70°F AB any conc to 185°F	A to 70°F	
GENERATOR GAS			A to 70°F	B at 70°F		
GERMAN SALTPETER			NR at 70°F	A to 70°F		
GIBBSITE				A to 70°F		
GIN						
GIRLING BRAKE FLUID	A to 70°F			B at 70°F		
GLACIAL ACETIC ACID	BC at 70°F	NR at 70°F	BR at 70°F	B to 125°F	B at 70°F NR hot & high pressure	NR at 70°F
GLAUBERS SALT	B to 100% at 70°F	NR any conc at 70°F	A to 100% to 70°F	A to 70°F AB any conc to 150°F	B/NR at 70°F A sol'n to 70°F	
GLUCONIC ACID				C at 70°F		
GLUCOSE	A to 70°F		A/NR at 70°F A sol'n to 70°F	A to 70°F AB any conc to 185°F	A to 70°F	
GLUE	A to 70°F	A to 70°F	A to 70°F	A to 70°F	A to 70°F	TEST
GLUE SIZING	A to 70°F	A to 70°F	A to 70°F			
GLYCERIN	A to 70°F	NR at 70°F	A/NR at 70°F AB fumes to 70°F	A to 70°F AB any conc to 150°F	A to 70°F	AB to 70°F
GLYCEROL	A to 70°F	NR at 70°F	A/NR at 70°F AB fumes to 70°F	A to 70°F AB any conc to 150°F	A to 70°F	AB to 70°F
GLYCERYL HYDROXIDE			A to 70°F	A to 70°F		
GLYCERAL TRIACETATE			NR at 70°F	A any conc to 80°F		
GLYCERYL TRIOLEATE				B any conc to 100°F		
GLYCOGNENIC ACID				C at 70°F		
GLYCOL, ACETATE			NR at 70°F	A to 70°F		
GLYCOL, ALCOHOL				A to 70°F AB any conc to 150°F		
GLYCOL, BUTYL ETHER				A to 70°F		
GLYCOL, CHLOROHYDRIN				B any conc to 80°F		

	SYNTHETIC ISOPRENE IR	NATURAL ISOPRENE NR	CHLOROSULFONATED POLYETHYLENE CSM	FLUOROSILICONE FVMQ	SILICONE VMQ	CHEMRAZ FFKM
FURFURAN	NR at 70°F	NR at 70°F	NR at 70°F		NR at 70°F (resin)	A to 70°F
FURFURYL ALCOHOL	NR at 70°F	NR at 70°F	NR at 70°F	+14% vol 3 days 158°F NR at 70°F	+3% vol 3 days 158°F NR at 70°F	A to 70°F
FUROL	NR at 70°F	NR at 70°F	C at 70°F		C at 70°F	
FURYLCARBINOL	NR at 70°F	NR at 70°F	NR at 70°F	NR at 70°F	NR at 70°F	A to 70°F
FUSEL OIL	A to 70°F	A to 70°F AB any conc to 150°F	A to 70°F		NR at 70°F	
FRYQUEL A60	NR at 70°F	NR at 70°F	NR at 70°F	NR at 70°F	C at 70°F	A to 70°F
FRYQUEL 90, 100, 150, 220, 300, 500, 550	NR at 70°F	NR at 70°F	NR at 70°F	B at 70°F	A to 70°F	A to 70°F
C.A.J.	NR at 70°F	NR at 70°F	NR at 70°F		NR at 70°F	
GALLIC ACID	A to 70°F	A to 70°F AB any conc to 150°F	B at 70°F	A to 70°F		A to 70°F
GALLOTANNIC ACID	A to 100% to 70°F	A to 100% to 70°F AB any conc to 185°F	A to 100% to 140°F AB to 200°F	A 10% to 70°F	B to 100% at 70°F	A to 100% to 70°F
GAS OIL				A to 70°F	NR at 70°F	
GASOLINE, AUTOMOTIVE	NR at 70°F	NR at 70°F	+85% vol 7 days 70°F NR at 70°F	+4% vol 120 days 70°F A to 70°F	+260% vol 7 days 70°F NR at 70°F	A to 70°F
GASOLINE, ETHYL & REGULAR			NR at 70°F	A to 70°F	+130% vol 7 days 70°F NR at 70°F	
GASOLINE, REFINED		NR at 70°F	C/NR at 70°F	A to 70°F	NR at 70°F	
GASOLINE, SOUR		NR at 70°F	NR at 70°F	A to 70°F	NR at 70°F	
GASOLINE WITH MERCAPTAN				A to 70°F	NR at 70°F	
GASOLINE 40% AROMATIC	NR at 70°F	NR at 70°F	NR at 70°F	A to 70°F	NR at 70°F	A to 70°F
GASOLINE 65 OCTANE	NR at 70°F	NR at 70°F	NR at 70°F			
GASOLINE 100 OCTANE	NR at 70°F	NR at 70°F	NR at 70°F	A to 70°F	NR at 70°F	
GASOLINE 130 OCTANE				A to 70°F	NR at 70°F	
GAULTHERIA OIL	NR at 70°F	NR at 70°F				
GEAR OIL					+4.2% vol 5 days 212°F A to 70°F	
GELATIN (AQUEOUS SODIUM SULFATE)	A to 70°F	A to 70°F AB any conc to 150°F	A to 70°F AB to 200°F	A to 70°F	A to 70°F	A to 70°F
GENERATOR GAS	C at 70°F	C at 70°F	B at 70°F		B at 70°F	
GERMAN SALTPETER	A to 70°F	A to 70°F	A to 70°F		B at 70°F	
GIBBSITE	A to 70°F	A to 70°F	B at 70°F			
GIN						
GIRLING BRAKE FLUID			B at 70°F	NR at 70°F		A to 70°F
GLACIAL ACETIC ACID	BC at 70°F NR hot	B/NR at 70°F NR hot	C at 70°F	NR at 70°F	B at 70°F C hot	AB to 70°F
GLAUBERS SALT	AB to 70°F	AB any conc to 150°F	A to 70°F AB to 200°F	A to 100% to 70°F	A to 100% to 70°F	A to 70°F
GLUCONIC ACID	NR at 70°F	NR at 70°F	B at 70°F			A to 70°F
GLUCOSE	AB to 70°F	A to 70°F AB any conc to 120°F	A to 158°F	A to 70°F	A to 70°F	A to 70°F
GLUE	AB to 70°F	AB to 70°F	A to 200°F	A to 70°F	A to 70°F	
GLUE SIZING				A to 70°F	A to 70°F	
GLYCERIN	A to 70°F	A to 70°F AB any conc to 150°F	A to 200°F	A to 70°F	A to 70°F	A to 70°F
GLYCEROL	A to 70°F	A to 70°F AB any conc to 150°F	A to 200°F	A to 70°F	A to 70°F	A to 70°F
GLYCERYL HYDROXIDE	A to 70°F	A to 70°F	A to 70°F		A to 70°F	
GLYCERAL TRIACETATE	B at 70°F	B/NR at 70°F	B at 70°F			A to 70°F
GLYCERYL TRIOLEATE	NR at 70°F	NR at 70°F				
GLYCOGNENIC ACID	NR at 70°F	NR at 70°F	B at 70°F			
GLYCOL, ACETATE	C at 70°F	C at 70°F	NR at 70°F		C at 70°F	
GLYCOL, ALCOHOL	A to 70°F	A to 70°F AB any conc to 150°F				
GLYCOL, BUTYL ETHER	NR at 70°F	NR at 70°F	B at 70°F			
GLYCOL, CHLOROHYDRIN	C at 70°F	C/NR at 70°F	B at 70°F		C at 70°F	

	POLYSULFIDE T	CHLORINATED POLYETHYLENE CM	EPICHLOROHYDRIN CO, ECO	® HYTREL COPOLYESTER TPE	® SANTOPRENE COPOLYMER TPO	® C-FLEX STYRENIC TPE
FURFURAN	B at 70°F				*A to 70°F	
FURFURYL ALCOHOL	disint. 3 days 158°F NR at 70°F			AB to 70°F	A to 70°F	
FUROL						
FURYLCARBINOL	NR at 70°F					
FUSEL OIL						
FRYQUEL A60	NR at 70°F					
FRYQUEL 90, 100, 150, 220, 300, 500, 550	NR at 70°F			AB to 200°F		
C.A.J.						
GALLIC ACID				NR at 70°F		
GALLOTANNIC ACID						
GAS OIL				AB to 70°F		
GASOLINE, AUTOMOTIVE	A to 70°F	A to 70°F	A to 70°F	A to 70°F		NR at 70°F
GASOLINE, ETHYL & REGULAR						
GASOLINE, REFINED		A to 70°F		A to 70°F		
GASOLINE, SOUR						
GASOLINE WITH MERCAPTAN						
GASOLINE 40% AROMATIC				AB to 70°F aromatic & nonaromatic		NR at 70°F
GASOLINE 65 OCTANE	AB to 70°F					
GASOLINE 100 OCTANE	AB to 70°F			A to 70°F	NR at 70°F	
GASOLINE 130 OCTANE	AB to 70°F AB to 70°F (aviation)					
GAULTHERIA OIL						
GEAR OIL						
GELATIN (AQUEOUS SODIUM SULFATE)	B/NR at 70°F		A to 70°F	AB to 70°F		
GENERATOR GAS						
GERMAN SALTPETER						
GIBBSITE						
GIN						
GIRLING BRAKE FLUID	NR at 70°F					
GLACIAL ACETIC ACID	BC at 70°F	A to 70°F	NR at 70°F	AC to 70°F C/NR at 100°F	NR at 70°F	
GLAUBERS SALT	B to 70°F	AB to 70°F	A to 70°F	AB to 70°F		
GLUCONIC ACID						
GLUCOSE	NR at 70°F		A to 70°F	AB to 70°F *AB to 140°F		A to 70°F
GLUE	NR at 70°F		A to 70°F	AB to 70°F		A to 70°F
GLUE SIZING						
GLYCERIN	B at 70°F	A to 70°F	A to 70°F	A to 70°F *AB to 140°F	NR at 70°F	AB to 70°F
GLYCEROL	B at 70°F	A to 70°F	A to 70°F	A to 70°F *AB to 140°F	NR at 70°F	AB to 70°F
GLYCERYL HYDROXIDE						
GLYCERAL TRIACETATE						
GLYCERYL TRIOLEATE						
GLYCOGNENIC ACID						
GLYCOL, ACETATE						
GLYCOL, ALCOHOL						
GLYCOL, BUTYL ETHER						
GLYCOL, CHLOROHYDRIN						

	ETHYLENE ACRYLIC EA	POLYALLOMER LINEAR COPOLYMER	NYLON 11 POLYAMIDE	NYLON 12 POLYAMIDE	ETHYLENE VINYL ACETATE EVA	POLYVINYLCHLORIDE FLEXIBLE PVC
FURFURAN						A to 70°F (resin)
FURFURYL ALCOHOL			AB to 105°F C at 140°F			AB to 70°F
FUROL						
FURYLCARBINOL						
FUSEL OIL						
FRYQUEL A60						
FRYQUEL 90, 100, 150, 220, 300, 500, 550			A to 70°F AB to 200°F			
C.A.J.						
GALLIC ACID			AB to 70°F			A to 150°F
GALLOTANNIC ACID						
GAS OIL			AB to 140°F (yellows)		*C at 70°F *NR at 140°F	AC to 70°F *NR at 140°F
GASOLINE, AUTOMOTIVE	B/NR at 70°F	B at 70°F C at 122°F	AB to 105°F C at 140°F	AB to 70°F	C/NR at 70°F	C at 70°F NR at 130°F
GASOLINE, ETHYL & REGULAR			AB to 70°F			C at 70°F
GASOLINE, REFINED		NR at 70°F	A to 70°F		NR at 70°F	NR at 70°F
GASOLINE, SOUR		NR at 70°F	A to 70°F		NR at 70°F	NR at 70°F
GASOLINE WITH MERCAPTAN						
GASOLINE 40% AROMATIC			AB to 70°F aromatic & nonaromatic			A/NR at 70°F special grades ok
GASOLINE 65 OCTANE						
GASOLINE 100 OCTANE			AB to 140°F yellows at 140°F			C/NR at 70°F NR at 150°F
GASOLINE 130 OCTANE						
GAULTHERIA OIL						NR at 70°F
GEAR OIL						
GELATIN (AQUEOUS SODIUM SULFATE)	A to 70°F	A to 70°F	A to 70°F	AB to 70°F	AB to 70°F	A to 70°F
GENERATOR GAS						
GERMAN SALTPETER						
GIBBSITE						
GIN						
GIRLING BRAKE FLUID	NR at 70°F					
GLACIAL ACETIC ACID	NR at 70°F	A to 70°F	C/NR at 70°F NR at 105°F		B/NR to 140°F	C/NR to 122°F NR at 140°F
GLAUBERS SALT	A sol'n to 70°F	A to 70°F	A to 70°F	AB to 70°F	AB to 70°F	A to 70°F *AB to 140°F
GLUCONIC ACID						
GLUCOSE	A to 70°F	A to 70°F	AB to 194°F		AB to 140°F	A to 70°F AB to 140°F (special)
GLUE		A to 70°F	AC to 70°F		A to 70°F	A to 70°F BC at 70°F (PVA)
GLUE SIZING						
GLYCERIN	A to 70°F	A to 122°F	AB to 105°F NR at 194°F	AB to 70°F	A to 70°F AB to 140°F	A to 122°F AC to 150°F
GLYCEROL	A to 70°F	A to 122°F	AB to 105°F NR at 194°F	AB to 70°F	A to 70°F AB to 140°F	A to 122°F AC to 150°F
GLYCERYL HYDROXIDE						
GLYCERAL TRIACETATE						NR at 70°F
GLYCERYL TRIOLEATE						
GLYCOGNENIC ACID						
GLYCOL, ACETATE						
GLYCOL, ALCOHOL				AB to 70°F		NR at 70°F
GLYCOL, BUTYL ETHER						
GLYCOL, CHLOROHYDRIN						NR at 70°F

	POLYETHYLENE LOW DENSITY LDPE	®TEFLON FEP	®KALREZ PERFLUORINATED ELASTOMER	®FLUORAZ FLUORINATED COPOLYMER	®AFLAS FLUORINATED COPOLYMER	®NORPRENE COPOLYMER TPO
FURFURAN		A to 200°F	A to 212°F	A to 70°F		
FURFURYL ALCOHOL	B/NR at 70°F	A to 70°F	A to 212°F	B at 70°F		
FUROL		A to 70°F				
FURYLCARBINOL		A to 70°F		B at 70°F		
FUSEL OIL		A to 70°F				
FRYQUEL A60			A to 212°F	B at 70°F		
FRYQUEL 90, 100, 150, 220, 300, 500, 550		A to 70°F AB to 400°F	A to 212°F	A to 70°F		
C.A.I.		NR at 70°F				
GALLIC ACID	B/NR at 70°F	A to 200°F	A to 212°F	A to 70°F		
GALLOTANNIC ACID	AB 100% to 70°F A 10% to 70°F	A to 100% to 300°F	A to 70°F	A to 70°F		A to 70°F
GAS OIL	NR at 70°F			A to 70°F		
GASOLINE, AUTOMOTIVE	C/NR at 70°F NR at 122°F	A to 300°F	A to 212°F	C at 70°F	+25% vol 7 days 70°F B at 70°F	BC at 70°F
GASOLINE, ETHYL & REGULAR	NR at 70°F	A to 300°F				NR at 70°F (unleaded)
GASOLINE, REFINED	NR at 70°F	A to 300°F		C at 70°F		
GASOLINE, SOUR	NR at 70°F	A to 70°F				
GASOLINE WITH MERCAPTAN				C at 70°F		
GASOLINE 40% AROMATIC	NR at 70°F aromatic & nonaromatic	A to 70°F		C at 70°F		TEST
GASOLINE 65 OCTANE		A to 70°F				
GASOLINE 100 OCTANE		A to 70°F		C at 70°F		
GASOLINE 130 OCTANE		A to 70°F (aviation gas)		C at 70°F		
GAULTHERIA OIL		A to 70°F				
GEAR OIL					+1.1% vol 5 days 212°F A to 212°F	
GELATIN (AQUEOUS SODIUM SULFATE)	A to 70°F	A to 150°F	A to 212°F	A to 70°F		
GENERATOR GAS		A to 70°F				
GERMAN SALTPETER		A to 70°F				
GIBBSITE		A to 70°F				
GIN	NR at 70°F	A to 70°F				
GIRLING BRAKE FLUID				A to 70°F		
GLACIAL ACETIC ACID	AB to 122°F NR at 140°F	A to boiling	A to 75°F B 203-428°F	NR at 70°F	+71% vol 7 days 70°F NR at 70°F	A to 70°F
GLAUBERS SALT	A to 70°F	A to 300°F	A to 212°F	A to 70°F		
GLUCONIC ACID		A to 70°F	A to 212°F			A to 70°F
GLUCOSE	A to 70°F AB to 140°F	A to 300°F	A to 212°F	A to 70°F		
GLUE	A to 70°F	A to 150°F	A to 212°F	A to 70°F		A to 70°F
GLUE SIZING				A to 70°F		
GLYCERIN	A to 122°F AB to 140°F	A to 300°F	A to 70°F *A to 250°F	A to 70°F	+0% vol 7 days 257°F A to 257°F	A to 70°F
GLYCEROL	A to 122°F AB to 140°F	A to 300°F	A to 70°F *A to 250°F	A to 70°F	A to 257°F	A to 70°F
GLYCERYL HYDROXIDE		A to 70°F				
GLYCERAL TRIACETATE		A to 70°F	A to 212°F			
GLYCERYL TRIOLEATE		A to 70°F				
GLYCOGNENIC ACID		A to 70°F				
GLYCOL, ACETATE		A to 70°F				
GLYCOL, ALCOHOL		A to 70°F				
GLYCOL, BUTYL ETHER		A to 70°F				
GLYCOL, CHLOROHYDRIN		A to 70°F				

	RECOMMENDATIONS	NITRILE NBR	ETHYLENE PROPYLENE EP, EPDM	FLUOROCARBON FKM	CHLOROPRENE CR	HYDROGENATED NITRILE HNBR
GLYCOL, DIBROMIDE		NR at 70°F	C at 70°F	B at 70°F	NR at 70°F	
GLYCOL, DICHLORIDE		NR at 70°F	C at 70°F	B at 70°F	NR at 70°F	
GLYCOL, DIETHYLENE	NBR	A to 250°F	A to 70°F	A to 70°F	A to 70°F	
GLYCOL, ETHYL ETHER	EP, NBR, IIR	AB to 70°F BC 104–150°F	AB to 70°F	B/NR at 70°F	C/NR at 70°F	
GLYCOL, ETHYLENE	EP, NBR, CR, FKM	A to 250°F	A to 200°F	A to 250°F	A to 158°F	A to 70°F
GLYCOL, MONOACETATE		C at 70°F	A to 70°F	A to 70°F	NR at 70°F	
GLYCOL, PROPYLENE	NBR	A to 250°F	A to 70°F	A to 70°F	A to 70°F	
GLYCOLS, GENERAL	NBR, EP, CR, FKM, IIR	A to 70°F AB to 150°F	A to 70°F	A to 70°F	A to 70°F AB to 120°F	
GLYCOLIC ACID	EP	A/NR 100% at 70°F A to 70% to 140°F	A 70–100% to 140°F	A/NR at 70°F A to 70% to 140°F	A/NR at 70°F A to 70% to 140°F	
GRAHAM'S SALT		A to 70°F AB any conc to 200°F			A to 70°F AB any conc to 200°F	
GRAIN ALCOHOL		A to 70°F AB to 150°F	A to 70°F	B at 70°F	A to 70°F AB to 150°F	
GRAIN MASH	EP	A to 70°F	A to 250°F	A to 70°F	A to 70°F	
GRAIN OIL		A to 70°F AB any conc to 150°F	A to 70°F	A to 70°F	A to 100°F	
GRAPE JUICE	EP, NBR, CR	A to 140°F	A to 250°F	A to 140°F	A to 140°F	
GRAPE SUGAR		A to 140°F B at 176°F	A to 176°F	A to 212°F	A to 70°F	
GREASE, PETROLEUM BASE	NBR, FKM	A to 100°F	NR at 70°F	A to 140°F	B/NR at 70°F NR at 140°F	
GREEN COPPERAS (VITRIOL)		A to 70°F AB any conc to 200°F	A to 70°F	A to 70°F	A to 70°F AB any conc to 200°F	
GREEN LIQUOR		AB to 70°F	A to 70°F	A to 70°F	AB to 70°F	B at 70°F
GREEN SULFATE LIQUOR	EP, FKM, NBR, IIR	B to 150°F	A to 200°F	A to 176°F B at 212°F	B to 150°F	
GREY ACETATE		AB any conc to 120°F	A to 70°F	NR at 70°F	B/NR at 70°F	
GULF ENDURANCE OILS	NBR, FKM	A to 70°F	NR at 70°F	A to 70°F	B at 70°F	A to 70°F
GULF FR FLUIDS (EMULSION)	NBR, FKM	A to 70°F	NR at 70°F	A to 70°F	B at 70°F	A to 70°F
GULF FR FLUID G100, G150, G200, G250	EP, NBR, FKM, CR	A to 70°F	A to 70°F	A to 70°F	A to 70°F	
GULF FR FLUID P37, P40, P43, P45, P47	EP	NR at 70°F	B at 70°F	B at 70°F	NR at 70°F	
GULF HARMONY OILS	NBR, FKM	A to 70°F	NR at 70°F	A to 70°F	B at 70°F	A to 70°F
GULF HIGH TEMPERATURE GREASE	NBR, FKM	A to 70°F	NR at 70°F	A to 70°F	B at 70°F	
GULF LEGION OILS	NBR, FKM	A to 70°F	NR at 70°F	A to 70°F	B at 70°F	A to 70°F
GULF PARAMONT OILS	NBR, FKM	A to 70°F	NR at 70°F	A to 70°F	B at 70°F	A to 70°F
GULF SECURITY OILS	NBR, FKM	A to 70°F	NR at 70°F	A to 70°F	B at 70°F	A to 70°F
GULFCROWN GREASE	NBR, FKM	A to 70°F	NR at 70°F	A to 70°F	B at 70°F	A to 70°F
HALITE		A to 70°F AB any conc to 160°F	A to 70°F	A to 70°F	A to 70°F AB any conc to 200°F	
HALOTHANE	FKM	NR at 70°F	NR at 70°F	A to 70°F	NR at 70°F	NR at 70°F
HALOWAX OIL	FKM, T	NR at 70°F	NR at 70°F	A to 70°F	NR at 70°F	NR at 70°F
HANNIFIN LUBE A	NBR, FKM, CR, ACM	A to 70°F	NR at 70°F	A to 70°F	A to 70°F	A to 70°F
HARTSHORN		AB any conc to 200°F	A to 70°F	A to 70°F	AB any conc to 200°F	
HEAVY BENZENE		A to 70°F	NR at 70°F	A to 70°F	B at 70°F	
HEAVY WATER	NBR, EP, SBR	A to 70°F	A to 70°F		B at 70°F	
HEF-2 (HIGH ENERGY FUEL 2)	FKM, FVMQ	B at 70°F	NR at 70°F	+4% vol 3 days 175°F A to 70°F	NR at 70°F	B at 70°F
HEF-3 (HIGH ENERGY FUEL 3)	FKM	BC at 70°F	NR at 70°F	+8% vol 7 days 326°F A to 120°F	NR at 70°F	
HELIUM	EP, IIR, NBR, FKM, CR	A to 200°F	A to 70°F AB to 200°F	A to 70°F	A to 70°F	A to 70°F
HEPAR CALIS		A to 100°F	A to 70°F	A to 70°F	B/NR at 70°F	
HEPTACHLOR, IN PETRO SOLVENTS		C at 70°F		A to 70°F		
HEPTANAL		A to 70°F		NR at 70°F		
N-HEPTANE	NBR, FKM, FVMQ, ACM	A to 250°F	NR at 70°F	A to 140°F AB to 200°F	AB to 70°F NR at 140°F	A to 70°F

	STYRENE BUTADIENE SBR	POLYACRYLATE ACM	POLYURETHANE AU, EU	ISOBUTYLENE ISOPRENE IIR	POLYBUTADIENE BR	®AEROQUIP AQP
GLYCOL, DIBROMIDE				NR at 70°F		
GLYCOL, DICHLORIDE			NR at 70°F	C at 70°F		
GLYCOL, DIETHYLENE	A to 70°F	NR at 70°F	NR at 70°F			
GLYCOL, ETHYL ETHER				A to 70°F		
GLYCOL, ETHYLENE	A to 70°F	NR at 70°F	NR at 70°F			
GLYCOL, MONOACETATE			NR at 70°F	A to 70°F		
GLYCOL, PROPYLENE	A to 70°F	NR at 70°F	NR at 70°F			
GLYCOLS, GENERAL	A to 70°F	NR at 70°F	C/NR at 70°F C 50% at 70°F	A to 70°F AB to 150°F	A to 70°F	
GLYCOLIC ACID	C/NR at 70°F	C/NR at 70°F	NR at 70°F			
GRAHAM'S SALT				A to 70°F AB any conc to 185°F		
GRAIN ALCOHOL			NR at 70°F	A to 70°F AB to 185°F		
GRAIN MASH	A to 70°F	NR at 70°F	NR at 70°F			
GRAIN OIL			C at 70°F	A to 70°F AB any conc to 150°F		
GRAPE JUICE	A to 70°F	NR at 70°F	NR at 70°F			
GRAPE SUGAR			*AB to 70°F			
GREASE, PETROLEUM BASE	NR at 70°F	A to 70°F	A to 70°F *AB to 140°F	NR at 70°F	NR at 70°F	A to 70°F
GREEN COPPERAS (VITRIOL)			A to 70°F	A to 70°F AB any conc to 185°F		
GREEN LIQUOR			A to 70°F	A to 70°F		
GREEN SULFATE LIQUOR	B at 70°F	NR at 70°F	A to 70°F	A to 70°F AB any conc to 150°F	B at 70°F	NR at 70°F
GREY ACETATE				A to 70°F AB any conc to 180°F		
GULF ENDURANCE OILS	NR at 70°F	A to 70°F	A to 70°F	NR at 70°F	NR at 70°F	
GULF FR FLUIDS (EMULSION)	NR at 70°F	A to 70°F	A to 70°F	NR at 70°F	NR at 70°F	
GULF FR FLUID G100, G150, G200, G250	A to 70°F	NR at 70°F	B at 70°F	A to 70°F	A to 70°F	AB to 200°F
GULF FR FLUID P37, P40, P43, P45, P47	NR at 70°F	NR at 70°F	NR at 70°F	B at 70°F	NR at 70°F	AB to 200°F
GULF HARMONY OILS	NR at 70°F	A to 70°F	A to 70°F	NR at 70°F	NR at 70°F	AB to 200°F
GULF HIGH TEMPERATURE GREASE	NR at 70°F	A to 70°F	A to 70°F	NR at 70°F	NR at 70°F	
GULF LEGION OILS	NR at 70°F	A to 70°F	A to 70°F	NR at 70°F	NR at 70°F	
GULF PARAMONT OILS	NR at 70°F	A to 70°F	B at 70°F	NR at 70°F	NR at 70°F	
GULF SECURITY OILS	NR at 70°F	A to 70°F	B at 70°F	NR at 70°F	NR at 70°F	AB to 300°F
GULFCROWN GREASE	NR at 70°F	A to 70°F	A to 70°F	NR at 70°F	NR at 70°F	
HALITE			NR at 70°F	A to 70°F AB any conc to 185°F		
HALOTHANE	NR at 70°F	NR at 70°F	NR at 70°F	NR at 70°F	NR at 70°F	
HALOWAX OIL	NR at 70°F			NR at 70°F	NR at 70°F	
HANNIFIN LUBE A	B at 70°F	A to 70°F	A to 70°F	NR at 70°F	NR at 70°F	
HARTSHORN				A to 70°F AB any conc to 185°F		
HEAVY BENZENE				NR at 70°F		
HEAVY WATER	A to 70°F	NR at 70°F	NR at 70°F	A to 70°F	A to 70°F	
HEF-2 (HIGH ENERGY FUEL 2)	NR at 70°F	NR at 70°F	NR at 70°F	NR at 70°F	NR at 70°F	
HEF-3 (HIGH ENERGY FUEL 3)	NR at 70°F	NR at 70°F	NR at 70°F			
HELIUM	A to 70°F	A to 70°F	A to 70°F	A to 70°F	A to 70°F	
HEPAR CALIS			A to 70°F	A to 70°F AB any conc to 185°F		
HEPTACHLOR, IN PETRO SOLVENTS			NR at 70°F			
HEPTANAL				NR at 70°F		
N-HEPTANE	NR at 70°F	A to 70°F	AB to 70°F	NR at 70°F	NR at 70°F	AB to 70°F

C177

	SYNTHETIC ISOPRENE IR	NATURAL ISOPRENE NR	CHLOROSULFONATED POLYETHYLENE CSM	FLUOROSILICONE FVMQ	SILICONE VMQ	CHEMRAZ FFKM
GLYCOL, DIBROMIDE	NR at 70°F	NR at 70°F	NR at 70°F			
GLYCOL, DICHLORIDE	NR at 70°F	NR at 70°F	NR at 70°F		C at 70°F	
GLYCOL, DIETHYLENE				A to 70°F	A to 70°F	
GLYCOL, ETHYL ETHER	B/NR at 70°F	B/NR at 70°F	B/NR at 70°F	B/NR at 70°F	NR at 70°F	
GLYCOL, ETHYLENE	A to 70°F	A to 70°F AB to 150°F	A to 200°F	A to 212°F	A to 70°F	A to 70°F
GLYCOL, MONOACETATE	C at 70°F	C at 70°F	NR at 70°F		C at 70°F	
GLYCOL, PROPYLENE				A to 70°F	A to 70°F	
GLYCOLS, GENERAL	A to 70°F	A to 70°F AB to 120°F	A to 70°F	A to 70°F	A to 70°F	A to 70°F
GLYCOLIC ACID	NR 10% at 70°F	NR 10% at 70°F		B/NR at 70°F	A/NR at 70°F	A to 70°F
GRAHAM'S SALT	A to 70°F	A to 70°F AB any conc to 150°F				
GRAIN ALCOHOL	A to 70°F	A to 70°F AB any conc to 150°F	A to 70°F		A to 70°F	
GRAIN MASH				A to 70°F	A to 70°F	
GRAIN OIL	A to 70°F	A/NR at 70°F	A to 70°F		NR at 70°F	
GRAPE JUICE				A to 70°F	A to 70°F	
GRAPE SUGAR						
GREASE, PETROLEUM BASE	NR at 70°F	NR at 70°F	C/NR at 70°F	A to 70°F	NR at 70°F	
GREEN COPPERAS (VITRIOL)	A to 70°F	A to 70°F AB any conc to 150°F	A to 70°F		B at 70°F	
GREEN LIQUOR	A to 70°F	A to 70°F	AB to 70°F		A to 70°F	
GREEN SULFATE LIQUOR	AB to 70°F	AB to 120°F	AB to 70°F	B at 70°F	A to 70°F	B at 70°F NR at 70°F white
GREY ACETATE	B at 70°F	B any conc to 80°F	A to 70°F			
GULF ENDURANCE OILS	NR at 70°F	NR at 70°F	NR at 70°F	A to 70°F	NR at 70°F	A to 70°F
GULF FR FLUIDS (EMULSION)	NR at 70°F	NR at 70°F	NR at 70°F	A to 70°F	NR at 70°F	A to 70°F
GULF FR FLUID G100, G150, G200, G250	A to 70°F	A to 70°F	A to 70°F	A to 70°F	A to 70°F	A to 70°F
GULF FR FLUID P37, P40, P43, P45, P47	NR at 70°F	NR at 70°F	NR at 70°F	B at 70°F	A to 70°F	A to 70°F
GULF HARMONY OILS	NR at 70°F	NR at 70°F	NR at 70°F	A to 70°F	NR at 70°F	A to 70°F
GULF HIGH TEMPERATURE GREASE	NR at 70°F	NR at 70°F	NR at 70°F	A to 70°F	NR at 70°F	A to 70°F
GULF LEGION OILS	NR at 70°F	NR at 70°F	NR at 70°F	A to 70°F	NR at 70°F	A to 70°F
GULF PARAMONT OILS	NR at 70°F	NR at 70°F	NR at 70°F	A to 70°F	NR at 70°F	A to 70°F
GULF SECURITY OILS	NR at 70°F	NR at 70°F	NR at 70°F	A to 70°F	NR at 70°F	A to 70°F
GULFCROWN GREASE	NR at 70°F	NR at 70°F	NR at 70°F	A to 70°F	NR at 70°F	A to 70°F
HALITE	A to 70°F	A to 70°F AB any conc to 150°F	A to 70°F		A to 70°F	
HALOTHANE	NR at 70°F	NR at 70°F	NR at 70°F	B at 70°F	NR at 70°F	B at 70°F
HALOWAX OIL	NR at 70°F	NR at 70°F	NR at 70°F	A to 70°F	NR at 70°F	B at 70°F
HANNIFIN LUBE A	NR at 70°F	NR at 70°F	A to 70°F	A to 70°F	B at 70°F	A to 70°F
HARTSHORN	A to 70°F	A to 70°F	B at 70°F		B at 70°F	
HEAVY BENZENE	NR at 70°F	NR at 70°F	NR at 70°F		A to 70°F	
HEAVY WATER	A to 70°F	A to 70°F	A to 70°F	A to 70°F	A to 70°F	A to 70°F
HEF-2 (HIGH ENERGY FUEL 2)	NR at 70°F	NR at 70°F	NR at 70°F	B at 70°F	NR at 70°F	A to 70°F
HEF-3 (HIGH ENERGY FUEL 3)				B at 70°F	NR at 70°F	
HELIUM	A to 70°F	A to 70°F	A to 70°F	A to 70°F	A to 70°F	A to 70°F
HEPAR CALIS	A to 70°F	A to 70°F	A to 70°F		B at 70°F	
HEPTACHLOR, IN PETRO SOLVENTS						A to 70°F
HEPTANAL	NR at 70°F	NR at 70°F	NR at 70°F			
N-HEPTANE	NR at 70°F	NR at 70°F	B to 200°F	A to 70°F	NR at 70°F	

	POLYSULFIDE T	CHLORINATED POLYETHYLENE CM	EPICHLOROHYDRIN CO, ECO	® HYTREL COPOLYESTER TPE	® SANTOPRENE COPOLYMER TPO	® C-FLEX STYRENIC TPE
GLYCOL, DIBROMIDE						
GLYCOL, DICHLORIDE						
GLYCOL, DIETHYLENE						
GLYCOL, ETHYL ETHER						
GLYCOL, ETHYLENE						
GLYCOL, MONOACETATE						
GLYCOL, PROPYLENE						
GLYCOLS, GENERAL	A to 70°F		A to 70°F	BC at 70°F	B at 70°F	
GLYCOLIC ACID						
GRAHAM'S SALT						
GRAIN ALCOHOL						
GRAIN MASH						
GRAIN OIL						
GRAPE JUICE						
GRAPE SUGAR				*AB to 140°F		
GREASE, PETROLEUM BASE	A to 70°F	A to 70°F		A to 70°F	NR at 70°F	
GREEN COPPERAS (VITRIOL)						
GREEN LIQUOR						
GREEN SULFATE LIQUOR	NR at 70°F		A to 70°F	NR at 70°F		
GREY ACETATE						
GULF ENDURANCE OILS	C at 70°F					
GULF FR FLUIDS (EMULSION)	A to 70°F					
GULF FR FLUID G100, G150, G200, G250	C at 70°F			AB to 150°F		
GULF FR FLUID P37, P40, P43, P45, P47	C at 70°F			AB to 150°F		
GULF HARMONY OILS	C at 70°F			AB to 200°F		
GULF HIGH TEMPERATURE GREASE	C at 70°F					
GULF LEGION OILS	C at 70°F					
GULF PARAMONT OILS	C at 70°F					
GULF SECURITY OILS	C at 70°F			AB to 200°F		
GULFCROWN GREASE	C at 70°F					
HALITE						
HALOTHANE	NR at 70°F					
HALOWAX OIL	A to 70°F					
HANNIFIN LUBE A	A to 70°F					
HARTSHORN						
HEAVY BENZENE						
HEAVY WATER	NR at 70°F			AB to 70°F		
HEF-2 (HIGH ENERGY FUEL 2)	B at 70°F					
HEF-3 (HIGH ENERGY FUEL 3)						
HELIUM	A to 70°F					
HEPAR CALIS						
HEPTACHLOR, IN PETRO SOLVENTS				NR at 70°F		
HEPTANAL						
N-HEPTANE	A to 70°F	A to 70°F		AB to 70°F	*A to 70°F	

	ETHYLENE ACRYLIC EA	POLYALLOMER LINEAR COPOLYMER	NYLON 11 POLYAMIDE	NYLON 12 POLYAMIDE	ETHYLENE VINYL ACETATE EVA	POLYVINYLCHLORIDE FLEXIBLE PVC
GLYCOL, DIBROMIDE						
GLYCOL, DICHLORIDE						
GLYCOL, DIETHYLENE						
GLYCOL, ETHYL ETHER						
GLYCOL, ETHYLENE						
GLYCOL, MONOACETATE						
GLYCOL, PROPYLENE						
GLYCOLS, GENERAL	A to 70°F		AB to 105°F NR at 194°F	AB to 70°F		AB 70°F
GLYCOLIC ACID						A 100% to 70°F NR 10% at 70°F
GRAHAM'S SALT						
GRAIN ALCOHOL						
GRAIN MASH						
GRAIN OIL						
GRAPE JUICE						A to 150°F
GRAPE SUGAR			AB to 140°F		AB to 140°F	AB to 130°F
GREASE, PETROLEUM BASE			A to 194°F		C/NR at 70°F *NR at 140°F	AB to 70°F C at 150°F
GREEN COPPERAS (VITRIOL)						
GREEN LIQUOR					C/NR at 70°F	A to 70°F
GREEN SULFATE LIQUOR		A to 70°F	NR at 70°F		A to 70°F	A to 150°F
GREY ACETATE						
GULF ENDURANCE OILS						
GULF FR FLUIDS (EMULSION)						
GULF FR FLUID G100, G150, G200, G250			A to 70°F AB to 160°F			
GULF FR FLUID P37, P40, P43, P45, P47			A to 70°F AB to 200°F			
GULF HARMONY OILS			AB to 200°F			
GULF HIGH TEMPERATURE GREASE	A to 70°F					
GULF LEGION OILS						
GULF PARAMONT OILS						
GULF SECURITY OILS			AB to 200°F			
GULFCROWN GREASE	A to 70°F					
HALITE						
HALOTHANE						
HALOWAX OIL						
HANNIFIN LUBE A	A to 70°F					
HARTSHORN						
HEAVY BENZENE						
HEAVY WATER	A to 70°F		AB to 70°F			
HEF-2 (HIGH ENERGY FUEL 2)						
HEF-3 (HIGH ENERGY FUEL 3)						
HELIUM	A to 70°F		A to 70°F			AB to 70°F
HEPAR CALIS						
HEPTACHLOR, IN PETRO SOLVENTS			AB to 70°F			A to 70°F C at 150°F
HEPTANAL						
N-HEPTANE	A to 70°F	C to 122°F	A to 70°F	AB to 70°F	NR at 70°F	B/NR at 70°F NR at 150°F

	POLYETHYLENE LOW DENSITY LDPE	® TEFLON FEP	® KALREZ PERFLUORINATED ELASTOMER	® FLUORAZ FLUORINATED COPOLYMER	® AFLAS FLUORINATED COPOLYMER	® NORPRENE COPOLYMER TPO
GLYCOL, DIBROMIDE		A to 70°F				
GLYCOL, DICHLORIDE		A to 70°F				
GLYCOL, DIETHYLENE						
GLYCOL, ETHYL ETHER		A to 70°F	A to 70°F	A/NR at 70°F		
GLYCOL, ETHYLENE	A to 140°F	A to 300°F	A to 113°F *A to 250°F	A to 70°F	A 50% to 324°F	A to 70°F
GLYCOL, MONOACETATE		A to 70°F				
GLYCOL, PROPYLENE						
GLYCOLS, GENERAL	AB to 70°F	A to 70°F	A to 212°F	A to 70°F		
GLYCOLIC ACID	A 70-100% to 70°F	A to 100% to 70°F	A to 212°F	NR at 70°F	no swell 67% conc 2 days 158°F	
GRAHAM'S SALT		A to 70°F				
GRAIN ALCOHOL		A to 70°F				
GRAIN MASH						
GRAIN OIL		A to 70°F				
GRAPE JUICE	B at 70°F			A to 70°F		
GRAPE SUGAR	A to 70°F AB to 140°F	A to 70°F				
GREASE, PETROLEUM BASE	AB to 70°F NR at 140°F	A to 70°F		A to 70°F	AB to 70°F	
GREEN COPPERAS (VITRIOL)		A to 70°F				
GREEN LIQUOR	A to 70°F	A to 70°F				
GREEN SULFATE LIQUOR	A to 70°F	A to 70°F	A to 212°F	A to 70°F		
GREY ACETATE		A to 70°F				
GULF ENDURANCE OILS				A to 70°F		
GULF FR FLUIDS (EMULSION)				A to 70°F		
GULF FR FLUID G100, G150, G200, G250		A to 70°F AB to 250°F		A to 70°F		
GULF FR FLUID P37, P40, P43, P45, P47		A to 70°F AB to 400°F		A to 70°F		
GULF HARMONY OILS		AB to 450°F		A to 70°F		
GULF HIGH TEMPERATURE GREASE				A to 70°F		
GULF LEGION OILS				A to 70°F		
GULF PARAMONT OILS				A to 70°F		
GULF SECURITY OILS		AB to 450°F		A to 70°F		
GULFCROWN GREASE				A to 70°F		
HALITE		A to 70°F				
HALOTHANE			A to 212°F	A to 70°F		
HALOWAX OIL		A to 70°F	A to 212°F	A to 70°F		
HANNIFIN LUBE A				A to 70°F		
HARTSHORN		A to 70°F				
HEAVY BENZENE		A to 70°F				
HEAVY WATER			A to 212°F	A to 70°F		
HEF-2 (HIGH ENERGY FUEL 2)				A to 70°F		
HEF-3 (HIGH ENERGY FUEL 3)						
HELIUM	AB to 70°F	A to 70°F	A to 212°F	A to 70°F		
HEPAR CALIS		A to 70°F				
HEPTACHLOR, IN PETRO SOLVENTS		A to 70°F	A to 212°F			
HEPTANAL		A to 70°F	*A to 212°F			
N-HEPTANE	NR at 70°F	A to 300°F	A to 212°F	C/NR at 70°F	BC at 70°F	

C 181

	RECOMMENDATIONS	NITRILE NBR	ETHYLENE PROPYLENE EP, EPDM	FLUOROCARBON FKM	CHLOROPRENE CR	HYDROGENATED NITRILE HNBR
HEPTANE CARBOXYLIC ACID		C at 70°F				
HEPTYL ALDEHYDE		A to 70°F		A to 70°F		
HEPTYL CARBINOL		A to 70°F	A to 70°F	B at 70°F	B at 70°F	
HEPTYL HYDRIDE		A to 70°F	NR at 70°F	A to 70°F	A to 70°F	
HEXACHLORO ACETONE	EP	NR at 70°F	A to 200°F	NR at 70°F	NR at 70°F	
HEXACHLORODIPHENYLMETHANE		NR at 70°F			NR at 70°F	
HEXAFLUOROPROPENE					AB to 70°F	
HEXAHYDROBENZENE	NBR,FKM,FVMQ,T,AU	+11% vol 3 days 70°F AB to 250°F	+148% vol 3 days 70°F NR at 70°F	+4% vol 7 days 75°F A to 200°F	+73% vol 3 days 70°F NR at 70°F	A to 70°F
HEXAHYDROPHENOL		B at 70°F	C at 70°F	A to 70°F	A to 70°F	
HEXAHYDROPYRIDINE		NR at 70°F	NR at 70°F	C at 70°F	NR at 70°F	
n-HEXALDEHYDE	EP,CR,IIR	NR at 70°F	A to 70°F	NR at 70°F	A to 80°F	
HEXALIN		B at 70°F	C at 70°F	A to 70°F	A to 70°F	
HEXAMETHYLENE		B to 80°F	NR at 70°F	A to 70°F	NR at 70°F	
HEXAMINE	EP	NR at 70°F	A to 250°F	NR at 70°F	NR at 70°F	
HEXANAPHTHALENE		A to 80°F	NR at 70°F	A to 70°F	B/NR at 70°F	
HEXANE	NBR,FKM,FVMQ	A to 250°F	NR at 70°F	+1% vol 7 days 70°F A to 200°F	B to 80°F NR at 140°F	
n-HEXANE	NBR,FKM,ACM	A to 70°F	NR at 70°F	+1.3% vol 21 days 75°F A to 75°F	A to 70°F	A to 70°F
HEXADECANOIC ACID	NBR,FKM,FVMQ T,CR,IIR	A to 100% to 160°F	B to 100% at 70°F	A to 100% to 70°F	B at 70°F BC at 158°F	A to 70°F
HEXANOL	NBR,FKM,FVMQ,IIR,T SBR	+31% vol 3 days 158°F A to 104°F	+12% vol 3 days 158°F AB to 200°F	+1% vol 3 days 158°F A to 212°F	+14% vol 3 days 158°F B to 140°F	
HEXAN-3-ONE		NR at 70°F		NR at 70°F		
n-HEXENE-1	FKM,FVMQ,T,ACM	B at 70°F	NR at 70°F	A to 70°F	B at 70°F	B at 70°F
HEXONE	EP,T,IIR	NR at 70°F	AB to 130°F C at 140°F	NR at 70°F	NR at 70°F	NR at 70°F
HEXYL ACETIC ACID		C at 70°F				
HEXYL ALCOHOL	NBR,FKM,FVMQ,T,SBR	+31% vol 3 days 158°F B at 140°F	+12% vol 3 days 158°F AB to 200°F	+1% vol 3 days 158°F B at 248°F	+14% vol 3 days 158°F B to 140°F	
HEXYL HYDRIDE		A to 70°F	NR at 70°F	A to 70°F	B at 70°F	
HEXYL METHYL KETONE		NR at 70°F		NR at 70°F		
HEXYLAMINE		C at 70°F		NR at 70°F		
HEXYLENE		A to 70°F	NR at 70°F	A to 70°F	B at 70°F	
HEXYLENE GLYCOL	NBR	A to 70°F	C at 70°F	A to 70°F	A to 70°F	
HI-LO MS NO. 1	EP	NR at 70°F	A to 70°F	NR at 70°F	NR at 70°F	NR at 70°F
HI-TRI		NR at 70°F		A to 70°F		
HIGH VISCOSITY LUBRICANT H2	NBR,EP,FKM,SBR	A to 70°F	A to 70°F	A to 70°F	B at 70°F	A to 70°F
HIGH VISCOSITY LUBRICANT U4	NBR,EP,FKM,SBR	A to 70°F	A to 70°F	A to 70°F	B at 70°F	A to 70°F
HOME HEATING OIL		A to 70°F	NR at 70°F	AB to 70°F	C at 70°F	
HONEY		A to 140°F	A to 140°F	A to 140°F	A to 140°F	
HOUGHTO-SAFE 271 (WATER & GLYCOL)	NBR,EP,FKM,SBR	A to 70°F	A to 70°F	+14% vol 14 days 158°F B at 70°F	B at 70°F	A to 70°F
HOUGHTO-SAFE 416 & 500	NBR,EP	A to 70°F	A to 70°F			
HOUGHTO-SAFE 620 (WATER & GLYCOL)	NBR,EP,FKM,SBR	A to 250°F	A to 70°F	+8% vol 14 days 158°F B at 70°F	B at 70°F	A to 70°F
HOUGHTO-SAFE 1010 PHOSPHATE ESTER	EP,FKM,IIR	NR at 70°F	A to 250°F	+13% vol 7 days 158°F A to 70°F	NR at 70°F	
HOUGHTO-SAFE 1055 PHOSPHATE ESTER	EP,FKM,IIR	NR at 70°F	A to 250°F	+20% vol 7 days 158°F A to 70°F	NR at 70°F	
HOUGHTO-SAFE 1120 PHOSPHATE ESTER	EP,FKM,IIR	NR at 70°F	A to 250°F	+9% vol 7 days 158°F A to 70°F	NR at 70°F	
HOUGHTO-SAFE 5040 (WATER & OIL)	NBR,FKM,FVMQ	A to 70°F	NR at 70°F	+2% vol 7 days 158°F A to 70°F	B at 70°F	
HYDRARGYLITE		B at 70°F	A to 70°F	C at 70°F	A to 70°F	
HYDRARGYRUM		A to 70°F	A to 70°F	A to 70°F	A to 70°F	

	STYRENE BUTADIENE SBR	POLYACRYLATE ACM	POLYURETHANE AU, EU	ISOBUTYLENE ISOPRENE IIR	POLYBUTADIENE BR	® AEROQUIP AQP
HEPTANE CARBOXYLIC ACID				C at 70°F		
HEPTYL ALDEHYDE				NR at 70°F		
HEPTYL CARBINOL			NR at 70°F	B /NR at 70°F		
HEPTYL HYDRIDE				NR at 70°F		
HEXACHLORO ACETONE	B at 70°F	NR at 70°F	NR at 70°F	A to 70°F		
HEXACHLORODIPHENYLMETHANE				NR at 70°F		
HEXAFLUOROPROPENE				AB to 70°F		
HEXAHYDROBENZENE	+143% vol 3 days 70°F NR at 70°F	+24% vol 3 days 70°F B at 70°F	+21% vol 3 days 70°F AB to 70°F	NR at 70°F	NR at 70°F	
HEXAHYDROPHENOL				NR at 70°F		
HEXAHYDROPYRIDINE				NR at 70°F		
n-HEXALDEHYDE	NR at 70°F		B at 70°F	AB to 150°F	NR at 70°F	
HEXALIN				NR at 70°F		
HEXAMETHYLENE			B at 70°F	NR at 70°F		
HEXAMINE	NR at 70°F	NR at 70°F	NR at 70°F			
HEXANAPHTHALENE			A to 70°F	NR at 70°F		
HEXANE	NR at 70°F	A to 70°F	AB to 70°F	NR at 70°F	NR at 70°F	AB to 70°F
n-HEXANE	NR at 70°F	A to 70°F	B to 122°F	NR at 70°F	NR at 70°F	
HEXADECANOIC ACID				NR at 70°F		
HEXANOL	+17% vol 3 days 158°F A to 200°F	+104% vol 3 days 158°F NR at 70°F	+77% vol 3 days 158°F NR at 70°F	+9.6% vol 3 days 158°F BC at 70°F	A to 70°F	
HEXAN-3-ONE				B at 70°F		
n-HEXENE-1	NR at 70°F	A to 70°F	AB to 70°F	NR at 70°F	NR at 70°F	
HEXONE	A to 70°F	NR at 70°F	NR at 70°F	B any conc to 75°F		
HEXYL ACETIC ACID				C at 70°F		
HEXYL ALCOHOL	+17% vol 3 days 158°F A to 200°F	+104% vol 3 days 158°F NR at 70°F	+77% vol 3 days 158°F NR at 70°F	+9.6% vol 3 days 158°F BC at 70°F	A to 70°F	
HEXYL HYDRIDE			B at 70°F	NR at 70°F		
HEXYL METHYL KETONE				B at 70°F		
HEXYLAMINE				B at 70°F		
HEXYLENE			A to 70°F	NR at 70°F		
HEXYLENE GLYCOL				A to 70°F		
HI-LO MS NO. 1	NR at 70°F	NR at 70°F	NR at 70°F	B at 70°F	NR at 70°F	
HI-TRI				NR at 70°F		
HIGH VISCOSITY LUBRICANT H2	A to 70°F	NR at 70°F	NR at 70°F	A to 70°F	B at 70°F	
HIGH VISCOSITY LUBRICANT U4	A to 70°F	NR at 70°F	NR at 70°F	A to 70°F	B at 70°F	
HOME HEATING OIL	NR at 70°F	AB to 70°F	AB to 70°F			
HONEY						
HOUGHTO-SAFE 271 (WATER & GLYCOL)	A to 70°F	NR at 70°F	NR at 70°F	B at 70°F		AB to 250°F
HOUGHTO-SAFE 416 & 500						AB to 250°F
HOUGHTO-SAFE 620 (WATER & GLYCOL)	A to 70°F	NR at 70°F	NR at 70°F	B at 70°F		AB to 250°F
HOUGHTO-SAFE 1010 PHOSPHATE ESTER	NR at 70°F	NR at 70°F	NR at 70°F	A to 70°F	NR at 70°F	A to 70°F AB to 200°F
HOUGHTO-SAFE 1055 PHOSPHATE ESTER	NR at 70°F	NR at 70°F	NR at 70°F	A to 70°F	NR at 70°F	A to 70°F AB to 200°F
HOUGHTO-SAFE 1120 PHOSPHATE ESTER	NR at 70°F	NR at 70°F	NR at 70°F	A to 70°F	NR at 70°F	A to 70°F AB to 200°F
HOUGHTO-SAFE 5040 (WATER & OIL)	NR at 70°F	NR at 70°F	NR at 70°F	NR at 70°F	NR at 70°F	AB to 200°F
HYDRARGYLITE				A to 70°F		
HYDRARGYRUM			A to 70°F	A to 70°F		

C 183

	SYNTHETIC ISOPRENE IR	NATURAL ISOPRENE NR	CHLOROSULFONATED POLYETHYLENE CSM	FLUOROSILICONE FVMQ	SILICONE VMQ	CHEMRAZ FFKM
HEPTANE CARBOXYLIC ACID	NR at 70°F	NR at 70°F	B at 70°F			
HEPTYL ALDEHYDE	NR at 70°F	NR at 70°F	NR at 70°F			
HEPTYL CARBINOL	B at 70°F	B/NR at 70°F	A to 70°F		B at 70°F	
HEPTYL HYDRIDE	NR at 70°F	NR at 70°F	B at 70°F			
HEXACHLORO ACETONE				NR at 70°F	NR at 70°F	A to 70°F
HEXACHLORODIPHENYLMETHANE	NR at 70°F	NR at 70°F				
HEXAFLUOROPROPENE						
HEXAHYDROBENZENE	NR at 70°F	NR at 70°F	B/NR at 70°F	+15% vol 3 days 70°F AB to 70°F	+102% vol 3 days 70°F NR at 70°F	A to 70°F
HEXAHYDROPHENOL	C at 70°F	C at 70°F	C at 70°F			
HEXAHYDROPYRIDINE	A to 70°F	A to 70°F	NR at 70°F			
n-HEXALDEHYDE	NR at 70°F	NR at 70°F	C at 70°F	NR at 70°F	B at 70°F	A to 70°F
HEXALIN	C at 70°F	C at 70°F	C at 70°F			
HEXAMETHYLENE	NR at 70°F	NR at 70°F	NR at 70°F		NR at 70°F	A to 70°F
HEXAMINE				NR at 70°F	NR at 70°F	A to 70°F
HEXANAPHTHALENE	NR at 70°F	NR at 70°F	B at 70°F		NR at 70°F	
HEXANE	NR at 70°F	NR at 70°F	B to 200°F	A to 70°F	NR at 70°F	
n-HEXANE	NR at 70°F	NR at 70°F	A to 70°F AB to 200°F	A to 70°F	NR at 70°F	A to 70°F
HEXADECANOIC ACID	AB to 70°F	A any conc to 80°F	C to 100% at 70°F	A to 70°F	NR at 70°F	A to 70°F
HEXANOL	A to 70°F	A to 70°F	B at 70°F	+7% vol 3 days 158°F AB to 70°F	+27% vol 3 days 158°F B at 70°F	A to 70°F
HEXAN-3-ONE	NR at 70°F	NR at 70°F	NR at 70°F			
n-HEXENE-1	NR at 70°F	NR at 70°F	B at 70°F	A to 70°F	NR at 70°F	A to 70°F
HEXONE	NR at 70°F	NR at 70°F	NR at 70°F	NR at 70°F	C/NR at 70°F	A to 70°F
HEXYL ACETIC ACID	C at 70°F	C at 70°F	B at 70°F			
HEXYL ALCOHOL	A to 70°F	A to 70°F	B at 70°F	+7% vol 3 days 158°F AB to 70°F	+27% vol 3 days 158°F B at 70°F	A to 70°F
HEXYL HYDRIDE	NR at 70°F	NR at 70°F	B at 70°F			
HEXYL METHYL KETONE	NR at 70°F	NR at 70°F	NR at 70°F			
HEXYLAMINE	C at 70°F	C at 70°F	C at 70°F			
HEXYLENE	NR at 70°F	NR at 70°F	B at 70°F		NR at 70°F	
HEXYLENE GLYCOL	A to 70°F	A to 70°F	A to 70°F			A to 70°F
HI-LO MS NO. 1	NR at 70°F	NR at 70°F	NR at 70°F	C at 70°F	C at 70°F	A to 70°F
HI-TRI	NR at 70°F	NR at 70°F	NR at 70°F			
HIGH VISCOSITY LUBRICANT H2				B at 70°F	A to 70°F	A to 70°F
HIGH VISCOSITY LUBRICANT U4				B at 70°F	A to 70°F	A to 70°F
HOME HEATING OIL					AB to 70°F	
HONEY		A to 70°F			A to 70°F	
HOUGHTO-SAFE 271 (WATER & GLYCOL)		A to 70°F		B at 70°F	B at 70°F	A to 70°F
HOUGHTO-SAFE 416 & 500						
HOUGHTO-SAFE 620 (WATER & GLYCOL)		A to 70°F		B at 70°F	B at 70°F	A to 70°F
HOUGHTO-SAFE 1010 PHOSPHATE ESTER	NR at 70°F	NR at 70°F	NR at 70°F	B at 70°F	C at 70°F	A to 70°F
HOUGHTO-SAFE 1055 PHOSPHATE ESTER	NR at 70°F	NR at 70°F	NR at 70°F	B at 70°F	C at 70°F	A to 70°F
HOUGHTO-SAFE 1120 PHOSPHATE ESTER	NR at 70°F	NR at 70°F	NR at 70°F	B at 70°F	C at 70°F	A to 70°F
HOUGHTO-SAFE 5040 (WATER & OIL)	NR at 70°F	NR at 70°F	NR at 70°F	B at 70°F	C at 70°F	A to 70°F
HYDRARGYLITE	A to 70°F	A to 70°F	B at 70°F			
HYDRARGYRUM	A to 70°F	A to 70°F	A to 70°F		A to 70°F	

	POLYSULFIDE T	CHLORINATED POLYETHYLENE CM	EPICHLOROHYDRIN CO, ECO	® HYTREL COPOLYESTER TPE	® SANTOPRENE COPOLYMER TPO	® C-FLEX STYRENIC TPE
HEPTANE CARBOXYLIC ACID						
HEPTYL ALDEHYDE						
HEPTYL CARBINOL						
HEPTYL HYDRIDE						
HEXACHLORO ACETONE						
HEXACHLORODIPHENYLMETHANE						
HEXAFLUOROPROPENE						
HEXAHYDROBENZENE	A to 70°F			A to 70°F	+20 to +62% vol 70°F C/NR at 70°F	NR at 70°F
HEXAHYDROPHENOL						
HEXAHYDROPYRIDINE						
n-HEXALDEHYDE	B at 70°F					
HEXALIN						
HEXAMETHYLENE						
HEXAMINE						
HEXANAPHTHALENE						
HEXANE	A to 70°F	A to 70°F	A to 70°F	A to 70°F		
n-HEXANE	A to 70°F		A to 70°F	A to 70°F	A to 70°F	
HEXADECANOIC ACID						
HEXANOL	+13.7% vol 3 days 158°F A to 70°F					
HEXAN-3-ONE						
n-HEXENE-1	A to 70°F					
HEXONE						
HEXYL ACETIC ACID						
HEXYL ALCOHOL	+13.7% vol 3 days 158°F A to 70°F					
HEXYL HYDRIDE						
HEXYL METHYL KETONE						
HEXYLAMINE						
HEXYLENE						
HEXYLENE GLYCOL						
HI-LO MS NO. 1	NR at 70°F					
HI-TRI						
HIGH VISCOSITY LUBRICANT H2						
HIGH VISCOSITY LUBRICANT U4						
HOME HEATING OIL						
HONEY						
HOUGHTO-SAFE 271 (WATER & GLYCOL)				AB to 150°F		
HOUGHTO-SAFE 416 & 500				AB to 150°F		
HOUGHTO-SAFE 620 (WATER & GLYCOL)		A to 70°F		+1% vol 7 days 158°F A to 158°F		
HOUGHTO-SAFE 1010 PHOSPHATE ESTER	C at 70°F	A to 70°F		AB to 200°F		
HOUGHTO-SAFE 1055 PHOSPHATE ESTER	C at 70°F	A to 70°F		AB to 200°F		
HOUGHTO-SAFE 1120 PHOSPHATE ESTER	C at 70°F			AB to 200°F		
HOUGHTO-SAFE 5040 (WATER & OIL)				AB to 150°F		
HYDRARGYLITE						
HYDRARGYRUM						

	ETHYLENE ACRYLIC EA	POLYALLOMER LINEAR COPOLYMER	NYLON 11 POLYAMIDE	NYLON 12 POLYAMIDE	ETHYLENE VINYL ACETATE EVA	POLYVINYLCHLORIDE FLEXIBLE PVC
HEPTANE CARBOXYLIC ACID						
HEPTYL ALDEHYDE						
HEPTYL CARBINOL						NR at 70°F
HEPTYL HYDRIDE						
HEXACHLORO ACETONE						
HEXACHLORODIPHENYLMETHANE						NR at 70°F
HEXAFLUOROPROPENE						
HEXAHYDROBENZENE	BC at 70°F	C at 70°F NR at 122°F	AB to 105°F C at 140°F		NR at 70°F	C/NR to 150°F
HEXAHYDROPHENOL						
HEXAHYDROPYRIDINE						
n-HEXALDEHYDE						NR at 70°F
HEXALIN						
HEXAMETHYLENE						
HEXAMINE						
HEXANAPHTHALENE						
HEXANE		B at 70°F C at 122°F	A to 70°F	AB to 70°F		BC at 70°F NR at 122°F
n-HEXANE	A to 70°F					
HEXADECANOIC ACID						
HEXANOL	A to 70°F				A to 70°F	AB to 70°F
HEXAN-3-ONE						
n-HEXENE-1						
HEXONE						NR at 70°F
HEXYL ACETIC ACID						
HEXYL ALCOHOL	A to 70°F				A to 70°F	AB to 70°F
HEXYL HYDRIDE						
HEXYL METHYL KETONE						
HEXYLAMINE						
HEXYLENE						
HEXYLENE GLYCOL						
HI-LO MS NO. 1						
HI-TRI						
HIGH VISCOSITY LUBRICANT H2	A to 70°F		A to 70°F			
HIGH VISCOSITY LUBRICANT U4	A to 70°F					
HOME HEATING OIL						
HONEY						A to 150°F
HOUGHTO-SAFE 271 (WATER & GLYCOL)	A to 70°F	A to 70°F	A to 70°F AB to 160°F		C at 70°F	AB to 70°F
HOUGHTO-SAFE 416 & 500						
HOUGHTO-SAFE 620 (WATER & GLYCOL)	A to 70°F	A to 70°F	A to 70°F AB to 160°F		C at 70°F	AB to 70°F
HOUGHTO-SAFE 1010 PHOSPHATE ESTER		A to 70°F	A to 70°F AB to 200°F		C at 70°F	AC to 70°F
HOUGHTO-SAFE 1055 PHOSPHATE ESTER		A to 70°F	A to 70°F AB to 200°F		C at 70°F	AC to 70°F
HOUGHTO-SAFE 1120 PHOSPHATE ESTER		A to 70°F	A to 70°F AB to 200°F		C at 70°F	AC to 70°F
HOUGHTO-SAFE 5040 (WATER & OIL)		A to 70°F	A to 70°F AB to 180°F		C at 70°F	
HYDRARGYLITE						
HYDRARGYRUM						

	POLYETHYLENE LOW DENSITY LDPE	®TEFLON FEP	®KALREZ PERFLUORINATED ELASTOMER	®FLUORAZ FLUORINATED COPOLYMER	®AFLAS FLUORINATED COPOLYMER	®NORPRENE COPOLYMER TPO
HEPTANE CARBOXYLIC ACID		A to 70°F				
HEPTYL ALDEHYDE		A to 70°F				
HEPTYL CARBINOL		A to 70°F				
HEPTYL HYDRIDE		A to 70°F				
HEXACHLORO ACETONE			A to 212°F	A to 70°F		
HEXACHLORODIPHENYLMETHANE		A to 70°F				
HEXAFLUOROPROPENE	AB to 70°F	AB to 70°F				
HEXAHYDROBENZENE	C at 70°F NR at 122°F	A to 300°F	+1% vol 7 days 70°F A to 70°F	A to 70°F	+13% vol 7 days 70°F AB to 70°F	NR at 70°F
HEXAHYDROPHENOL		A to 70°F				
HEXAHYDROPYRIDINE		A to 70°F				
n-HEXALDEHYDE		A to 70°F	A to 70°F *A to 212°F	NR at 70°F		
HEXALIN		A to 70°F				
HEXAMETHYLENE		A to 70°F	A to 212°F			
HEXAMINE						
HEXANAPHTHALENE		A to 70°F				
HEXANE	NR at 70°F	A to 300°F	+1% vol 7 days 70°F A to 212°F	C at 70°F		
n-HEXANE		A to 70°F	A to 212°F	B at 70°F	+24% vol 7 days 70°F BC at 70°F	
HEXADECANOIC ACID	B to 100% at 70°F NR 100% at 140°F	A to 100% to 300°F	A to 70°F	A to 70°F		
HEXANOL	A to 70°F	A to 200°F	A to 113°F *A to 158°F	A to 70°F		
HEXAN-3-ONE		A to 70°F				
n-HEXENE-1		A to 70°F	A to 70°F	C/NR at 70°F		
HEXONE	C/NR at 70°F NR at 122°F	A to 300°F	A to 212°F	NR at 70°F	NR at 70°F	BC at 70°F
HEXYL ACETIC ACID		A to 70°F				
HEXYL ALCOHOL	A to 70°F	A to 200°F	A to 212°F	A to 70°F		
HEXYL HYDRIDE		A to 70°F				
HEXYL METHYL KETONE		A to 70°F				
HEXYLAMINE		A to 70°F				
HEXYLENE		A to 70°F				
HEXYLENE GLYCOL		A to 70°F	A to 212°F			
HI-LO MS NO. 1				C at 70°F		
HI-TRI		A to 70°F				
HIGH VISCOSITY LUBRICANT H2		A to 70°F		A to 70°F		
HIGH VISCOSITY LUBRICANT U4				A to 70°F		
HOME HEATING OIL						
HONEY	A to 70°F	A to 70°F				
HOUGHTO-SAFE 271 (WATER & GLYCOL)	A to 70°F	A to 70°F AB to 250°F		A to 70°F		
HOUGHTO-SAFE 416 & 500		AB to 250°F		A to 70°F		
HOUGHTO-SAFE 620 (WATER & GLYCOL)	A to 70°F	A to 70°F AB to 250°F		A to 70°F		
HOUGHTO-SAFE 1010 PHOSPHATE ESTER	NR at 70°F	A to 70°F AB to 400°F		A to 70°F		
HOUGHTO-SAFE 1055 PHOSPHATE ESTER	NR at 70°F	A to 70°F AB to 400°F		A to 70°F		
HOUGHTO-SAFE 1120 PHOSPHATE ESTER	A/NR at 70°F	A to 70°F AB to 400°F		A to 70°F .		
HOUGHTO-SAFE 5040 (WATER & OIL)	A to 70°F	A to 70°F AB to 250°F		A to 70°F		
HYDRARGYLITE		A to 70°F				
HYDRARGYRUM		A to 70°F				

C 187

	RECOMMENDATIONS	NITRILE NBR	ETHYLENE PROPYLENE EP, EPDM	FLUOROCARBON FKM	CHLOROPRENE CR	HYDROGENATED NITRILE HNBR
HYDRATED BARYTA		A to 70°F	A to 70°F	A to 70°F	A to 70°F	
HYDRATED LIME		A to 70°F AB any conc to 200°F	A to 70°F	A to 70°F	A to 70°F AB any conc to 200°F	
HYDRATITE LIQUID		A to 70°F		A to 70°F	A to 70°F	
HYDRAULIC FLUIDS (PYDRAULS)	FKM, EP, IIR	C/NR at 70°F	AC to 70°F	A to 70°F AB to 400°F	NR at 70°F	NR at 70°F
HYDRAULIC OIL (PETROLEUM)	NBR, FKM, FVMQ, ACM AU	A to 250°F	NR at 70°F	A to 70°F NR at 140°F	B to 140°F	A to 70°F
HYDRAULIC OIL (PETROLEUM, AIRCRAFT)	NBR, FKM	AC to 70°F	NR at 70°F	A to 70°F	B/NR at 70°F	A to 70°F
HYDRAULIC OIL (PHOSPHATE ESTER)	EP, IIR	NR at 70°F	A to 250°F Skydrol 500 & 7000	A/NR at 70°F	NR at 70°F	NR at 70°F
HYDRAZINE	EP, IIR	B to 140°F	A to 200°F	NR at 70°F	B to 140°F	
HYDRAZINO BENZENE		NR at 70°F	C at 70°F	A to 70°F	NR at 70°F	
HYDRIODIC ACID			A to 37% to 104°F	A to 70°F A to 37% to 104°F	A to 37% to 104°F	
HYDRO-DRIVE MIH10 (PETRO BASED)	NBR, FKM, ACM	A to 70°F	NR at 70°F	+1% vol 14 days 212°F A to 212°F	B at 70°F	
HYDRO-DRIVE MIH50 (PETRO BASED)	NBR, FKM, ACM	A to 70°F	NR at 70°F	no swell 14 days 212°F A to 212°F	B at 70°F	
HYDROBROMIC ACID	EP, FKM, IIR	NR 40-100% at 70°F C/NR 20% to 104°F	A to 100% to 200°F	A to 100% to 140°F B 20% at 176°F	C/NR 100% at 70°F AB 40-50% to 70°F	NR at 70°F
HYDROBROMIC ACID		NR 20% at 140°F			C/NR 20% at 70°F	
HYDROBROMIC ACID, GAS	FKM	C/NR at 70°F	A to 70°F	A to 120°F	B/NR at 70°F	
HYDROCARBONS, ALICYCLIC		B at 70°F	NR at 70°F	A to 70°F	NR at 70°F	
HYDROCARBONS, ALIPHATIC		A to 70°F	NR at 70°F	A to 70°F	B at 70°F	
HYDROCARBONS, AROMATIC		C at 70°F	NR at 70°F	A to 70°F	NR at 70°F	
HYDROCARBONS, CHLORINATED		NR at 70°F	NR at 70°F	AC at 70°F	NR at 70°F	
HYDROCARBONS, LIGHT	NBR	A to 250°F	NR at 70°F	A to 70°F	NR at 70°F	
HYDROCARBONS, NORMAL		A to 70°F	NR at 70°F	A to 70°F	B at 70°F	
HYDROCARBONS, OLEFINIC		A to 70°F		A to 70°F		
HYDROCARBONS, SATURATED	NBR, FKM, ACM	A to 70°F	NR at 70°F	A to 70°F	B at 70°F	
HYDROCHLORIC ACID	FKM, EP, IIR	+2-14% vol 3 days 158°F NR conc at 70°F	+2-6% vol 3 days 158°F BC 50-100% at 70°F	+5% vol 50% conc 7 days 75°F	+2-18% vol 38% conc 3 days 158°F	
HYDROCHLORIC ACID		C/NR 50% at 70°F AB 20-37% to 70°F	NR 100% at 140°F AB to 37% to 130°F	+24% vol 37% conc 3 days 158°F	NR 50-100% at 70°F AB 25-37% to 70°F	
HYDROCHLORIC ACID		C/NR 25-37% at 104°F NR 37% at 140°F	C/NR 37% at 140°F A to 25% to 140°F	+331% vol 37% conc 28 days 158°F	BC 38% at 140°F NR 38% at 158°F	
HYDROCHLORIC ACID		AB to 15% to 150°F NR 0.5% at 175°F	C 25% at 176°F NR 25% at 212°F	+91% vol 37% conc 7 days 180°F	A to 25% to 140°F B/NR 25% at 176°F	
HYDROCHLORIC ACID			A to 3 molar to 158°F	AB 50-100% to 70°F A to 37% to 130°F	NR 15% at 212°F	
HYDROCHLORIC ACID				A/NR 38% at 140°F A to 25% to 140°F		
HYDROCHLORIC ACID				A to 20% to 230°F		
HYDROCHLORIC ETHER		C at 70°F	C at 70°F	A to 70°F	B at 70°F	
HYDROCYANIC ACID	EP, IIR, FKM	B any conc to 140°F	A to 100% to 140°F	A to 100% to 140°F	B to 100% to 140°F	B at 70°F
HYDROFLUORIC ACID	EP, FKM, IIR	+138% vol 50% conc 7 days 70°F	+13% vol 50% conc 7 days 70°F	+22% vol 3 days 70°F blistered 5 days 158°F	+62% vol 50% conc 7 days 70°F	
HYDROFLUORIC ACID		NR 40-100% at 70°F C 20-30% to 130°F	NR 75-100% at 70°F AB to 65% to 70°F	-20% T.S. 75% conc 5 days 158°F	BC 75-100% at 70°F NR 75-100% at 125°F	
HYDROFLUORIC ACID		AB 10% to 70°F	NR 65% at 140°F AB to 60% to 130°F	-40% T.S. 75% conc 5 days 212°F	AC 65% at 70°F B 50-55% to 104°F	
HYDROFLUORIC ACID			B 55% at 130°F C 55% at 140°F	+1% vol 48% conc 28 days 70°F	C 50-55% at 140°F NR 50-55% at 176°F	
HYDROFLUORIC ACID			NR 55% at 176°F AB 50% to 140°F	AB to conc to 120°F A to 60% to 130°F	AB 30% at 140°F A to 25% to 104°F	
HYDROFLUORIC ACID			B/NR 40-50% at 176°F B 30% at 176°F	A to 50% to 176°F B 50% at 212°F	AB 10% to 185°F	
HYDROFLUORIC ACID			NR 30% at 212°F A dilute to 212°F	C 50% at 248°F A to 30% to 212°F		
HYDROFLUOSILICIC ACID	EP, FKM, NBR	AB to 70°F AB to 50% to 150°F	A to 100% to 120°F A to 20% to 140°F	A to 100% to 70°F A to 20% to 140°F	B to 100% to 70°F B to 50% to 150°F	A to 70°F
HYDROGEN GAS	NBR, EP, FKM, CR, IIR	A to 200°F	A to 140°F AB to 200°F	A to 176°F	A to 140°F	
HYDROGEN BROMIDE	FKM	C/NR at 70°F	A to 70°F	A to 120°F	B/NR at 70°F	
HYDROGEN CARBOXYLIC ACID		B at 70°F			B at 70°F	
HYDROGEN CHLORIDE, GAS	EP, IIR	A/NR at 70°F	A to 200°F	A to 70°F wet or dry	AB to 70°F	

	STYRENE BUTADIENE SBR	POLYACRYLATE ACM	POLYURETHANE AU, EU	ISOBUTYLENE ISOPRENE IIR	POLYBUTADIENE BR	® AEROQUIP AQP
HYDRATED BARYTA			A to 70°F	A to 70°F		
HYDRATED LIME			A to 70°F	A to 70°F / AB any conc to 185°F		
HYDRATITE LIQUID						
HYDRAULIC FLUIDS (PYDRAULS)	NR at 70°F	NR at 70°F	NR at 70°F	AB to 70°F	NR at 70°F	AB to 200°F
HYDRAULIC OIL (PETROLEUM)						A to 70°F / AB to 300°F
HYDRAULIC OIL (PETROLEUM, AIRCRAFT)			AB to 70°F			A to 70°F
HYDRAULIC OIL (PHOSPHATE ESTER)	NR at 70°F	NR at 70°F	NR at 70°F	A to 70°F / AB any conc to 120°F	NR at 70°F	AB to 200°F
HYDRAZINE	B at 70°F		NR at 70°F	A any conc to 100°F		
HYDRAZINO BENZENE				C /NR at 70°F		
HYDRIODIC ACID						
HYDRO-DRIVE MIH10 (PETRO BASED)	NR at 70°F	A to 70°F	B at 70°F	NR at 70°F	NR at 70°F	
HYDRO-DRIVE MIH50 (PETRO BASED)	NR at 70°F	A to 70°F	B at 70°F	NR at 70°F	NR at 70°F	
HYDROBROMIC ACID	NR 40-100% at 70°F	NR 40-100% at 70°F	NR 40-100% at 70°F	A 40-100% to 70°F / AB 100% to 150°F	NR 40-100% at 70°F	
HYDROBROMIC ACID						
HYDROBROMIC ACID, GAS	C at 70°F	NR at 70°F	NR at 70°F	A to 70°F / AB any conc to 150°F		
HYDROCARBONS, ALICYCLIC			B at 70°F	NR at 70°F		
HYDROCARBONS, ALIPHATIC			B at 70°F	NR at 70°F		
HYDROCARBONS, AROMATIC			NR at 70°F	NR at 70°F		
HYDROCARBONS, CHLORINATED						
HYDROCARBONS, LIGHT	NR at 70°F	A to 70°F	A to 70°F			
HYDROCARBONS, NORMAL				NR at 70°F		
HYDROCARBONS, OLEFINIC						
HYDROCARBONS, SATURATED	NR at 70°F	A to 70°F	B at 70°F	NR at 70°F	NR at 70°F	
HYDROCHLORIC ACID	+3% vol 3 days 158°F / C/NR 100% at 70°F	+4% vol 3 days 158°F / C/NR 100% at 70°F	+50% vol 3 days 158°F / NR 38-100% at 70°F	A to 70°F / B at 125°F	B/NR 38% at 70°F / NR 38% at 158°F	
HYDROCHLORIC ACID	B/NR 38% at 70°F / NR 38% at 158°F	NR 38% at 70°F / B 3 molar at 70°F	C 25% at 70°F / B 20% at 70°F	BC 50% at 70°F / AB to 38% at 70°F	C 3 molar to 158°F	
HYDROCHLORIC ACID	C 3 molar to 158°F	BC 3 molar at 158°F	NR 20% at 140°F / AB 10% to 70°F	AB 38% to 140°F / C 38% at 158°F		
HYDROCHLORIC ACID			NR 3 molar at 70°F	A 3 molar to 158°F		
HYDROCHLORIC ACID						
HYDROCHLORIC ACID						
HYDROCHLORIC ACID						
HYDROCHLORIC ETHER			C at 70°F	A to 70°F		
HYDROCYANIC ACID	B at 70°F	NR at 70°F	C/NR 50-100% at 70°F	AB to 100% to 150°F	B at 70°F	NR at 70°F
HYDROFLUORIC ACID	NR 75-100% at 70°F / B 10-65% at 70°F	NR 10-100% at 70°F	NR 10-100% at 70°F / *C dilute at 70°F	no swell 7 days 70°F / B 100% to 125°F	NR 65-100% at 70°F / NR 50% at 140°F	
HYDROFLUORIC ACID	NR 65% at 140°F		*NR dilute at 140°F	NR 100% at 140°F / C 75% at 70°F		
HYDROFLUORIC ACID				NR 65% at 140°F / AB 50% to 150°F		
HYDROFLUORIC ACID				AB to 20% at 185°F / A 10% to 70°F		
HYDROFLUORIC ACID						
HYDROFLUORIC ACID						
HYDROFLUORIC ACID						
HYDROFLUOSILICIC ACID	B at 70°F		B /NR at 70°F	A to 70°F / AB to 50% to 185°F	B at 70°F	
HYDROGEN GAS	B to 140°F	B to 140°F	A to 140°F	A to 140°F	A to 140°F	TEST
HYDROGEN BROMIDE	C at 70°F	NR at 70°F	NR at 70°F	A to 70°F / AB any conc to 150°F		
HYDROGEN CARBOXYLIC ACID				A to 70°F / AB any conc to 150°F		
HYDROGEN CHLORIDE, GAS	NR at 70°F			A to 70°F		

	SYNTHETIC ISOPRENE IR	NATURAL ISOPRENE NR	CHLOROSULFONATED POLYETHYLENE CSM	FLUOROSILICONE FVMQ	SILICONE VMQ	CHEMRAZ FFKM
HYDRATED BARYTA	A to 70°F	A to 70°F	A to 70°F		A to 70°F	
HYDRATED LIME	A to 70°F	A to 70°F AB any conc to 150°F	B at 70°F		A to 70°F	
HYDRATITE LIQUID						
HYDRAULIC FLUIDS (PYDRAULS)	NR at 70°F	NR at 70°F	NR at 70°F	A/NR at 70°F	B/NR at 70°F	
HYDRAULIC OIL (PETROLEUM)	NR at 70°F	NR at 70°F	B at 70°F	A to 70°F	+4-12% vol 7 days 70°F BC at 70°F	A to 70°F
HYDRAULIC OIL (PETROLEUM, AIRCRAFT)		NR at 70°F				A to 70°F
HYDRAULIC OIL (PHOSPHATE ESTER)	NR at 70°F	NR at 70°F	NR at 70°F	A/NR at 70°F	B/NR at 70°F	
HYDRAZINE	NR at 70°F	NR at 70°F	B at 70°F	NR at 70°F	B/NR at 70°F	B to 70°F
HYDRAZINO BENZENE	A to 70°F	A any conc to 80°F	C at 70°F			
HYDRIODIC ACID						A to 70°F
HYDRO-DRIVE MIH10 (PETRO BASED)	NR at 70°F	NR at 70°F	NR at 70°F	A to 70°F	B at 70°F	
HYDRO-DRIVE MIH50 (PETRO BASED)	NR at 70°F	NR at 70°F	NR at 70°F	A to 70°F	B at 70°F	
HYDROBROMIC ACID	A 40-100% to 70°F	A 20-100% to 70°F AB 100% to 100°F	A to 100% to 70°F A to 20% to 200°F	C 40-100% at 70°F	NR 20-100% at 70°F	A 40-100% to 70°F
HYDROBROMIC ACID						
HYDROBROMIC ACID, GAS	B at 70°F	B at 70°F		NR at 70°F	NR at 70°F	
HYDROCARBONS, ALICYCLIC	NR at 70°F	NR at 70°F	NR at 70°F		NR at 70°F	
HYDROCARBONS, ALIPHATIC	NR at 70°F	NR at 70°F	B at 70°F		NR at 70°F	
HYDROCARBONS, AROMATIC	NR at 70°F	NR at 70°F	NR at 70°F		NR at 70°F	
HYDROCARBONS, CHLORINATED	NR at 70°F	NR at 70°F	NR at 70°F			
HYDROCARBONS, LIGHT				A to 70°F	NR at 70°F	
HYDROCARBONS, NORMAL	NR at 70°F	NR at 70°F	B at 70°F			
HYDROCARBONS, OLEFINIC						
HYDROCARBONS, SATURATED	NR at 70°F	NR at 70°F	C at 70°F	A to 70°F	NR at 70°F	A to 70°F
HYDROCHLORIC ACID	B 100% at 70°F A to 50% to 70°F	A/NR 100% at 70°F A to 50% to 70°F	A/NR 100% at 70°F AB to 50% to 70°F	+8% vol 7 days 70°F +5% vol 3 days 158°F	+15% vol 7 days 70°F +12% vol 3 days 158°F	A to 100% to 70°F
HYDROCHLORIC ACID	NR 37% at 158°F C 3 molar at 158°F	AB to 38% at 150°F NR 38% at 158°F	A to 38% to 122°F BC 38% at 158°F	+1% vol 10% conc 7 days 70°F	+1% vol 38% conc 7 days 70°F	
HYDROCHLORIC ACID		C 3 molar at 158°F	A to 25% to 200°F AB 3 molar at 158°F	C 100% at 70°F B 38% at 70°F	-29% vol 38% conc 3 days 104°F	
HYDROCHLORIC ACID				A 10% to 70°F BC 3 molar at 158°F	+4% vol 10% conc 7 days 70°F	
HYDROCHLORIC ACID					NR 100% at 70°F BC 50% at 70°F	
HYDROCHLORIC ACID					B/NR 38% at 70°F AB to 20% to 125°F	
HYDROCHLORIC ACID					NR 3 molar at 158°F	
HYDROCHLORIC ETHER	B at 70°F	B at 70°F	B at 70°F		C at 70°F	
HYDROCYANIC ACID	AB to 70°F	AB to 100% to 80°F AB to 20% to 150°F	A to 70°F	B at 70°F	AC to 70°F	A to 70°F
HYDROFLUORIC ACID	NR 75-100% at 70°F BC 50-65% at 70°F	NR 75-100% at 70°F BC 20-65% at 70°F	+4% vol 50% conc 7 days 158°F	NR 65-100% at 70°F	NR 65-100% at 70°F B/NR 20-50% at 70°F	A 100% to 70°F A to 65% to 140°F
HYDROFLUORIC ACID	NR 50-65% at 140°F B 10% at 70°F	NR 50-65% at 140°F B 10-20% at 70°F	A to 100% to 70°F NR 65% at 140°F			
HYDROFLUORIC ACID			C 50% at 140°F			
HYDROFLUORIC ACID						
HYDROFLUORIC ACID						
HYDROFLUORIC ACID						
HYDROFLUORIC ACID						
HYDROFLUOSILICIC ACID	A to 70°F	A to 70°F AB 10% to 100°F	A to 250°F	A/NR at 70°F	A/NR at 70°F	A to 70°F
HYDROGEN GAS	AB to 140°F	B to 150°F	A to 140°F	C to 140°F	C to 140°F	A to 140°F
HYDROGEN BROMIDE	B at 70°F	B at 70°F		NR at 70°F	NR at 70°F	A 100% to 70°F
HYDROGEN CARBOXYLIC ACID	NR at 70°F	NR at 70°F				
HYDROGEN CHLORIDE, GAS			A to 70°F	C/NR at 70°F	NR at 70°F	A to 70°F B at 70°F white

	POLYSULFIDE T	CHLORINATED POLYETHYLENE CM	EPICHLOROHYDRIN CO, ECO	® HYTREL COPOLYESTER TPE	® SANTOPRENE COPOLYMER TPO	® C-FLEX STYRENIC TPE
HYDRATED BARYTA						
HYDRATED LIME						
HYDRATITE LIQUID						
HYDRAULIC FLUIDS (PYDRAULS)	NR at 70°F	AB to 70°F	NR at 70°F	A/NR to 212°F	AB to 212°F	AB to 212°F B at 250°F Pydraul 312
HYDRAULIC OIL (PETROLEUM)	A to 70°F	A to 70°F	A to 70°F	+5% vol 3 days 212°F A to 212°F	NR at 70°F	
HYDRAULIC OIL (PETROLEUM, AIRCRAFT)				A to 70°F	NR at 70°F	
HYDRAULIC OIL (PHOSPHATE ESTER)	NR at 70°F	AB to 70°F	NR at 70°F	A/NR to 212°F	A to 70°F Skydrol AB to 212°F Pydraul	AB to 212°F
HYDRAZINE	NR at 70°F			NR at 70°F		
HYDRAZINO BENZENE						
HYDRIODIC ACID						AB to 70°F
HYDRO-DRIVE MIH10 (PETRO BASED)	A to 70°F					
HYDRO-DRIVE MIH50 (PETRO BASED)	A to 70°F					
HYDROBROMIC ACID		NR at 70°F				
HYDROBROMIC ACID						
HYDROBROMIC ACID, GAS						
HYDROCARBONS, ALICYCLIC			AB to 125°F hydrocarbons general			
HYDROCARBONS, ALIPHATIC						
HYDROCARBONS, AROMATIC						
HYDROCARBONS, CHLORINATED						
HYDROCARBONS, LIGHT						
HYDROCARBONS, NORMAL						
HYDROCARBONS, OLEFINIC						
HYDROCARBONS, SATURATED	A to 70°F		C at 70°F			
HYDROCHLORIC ACID	NR 38-100% at 70°F NR 3 molar at 70°F	NR 38-100% at 70°F NR 3 molar at 70°F	NR 38% at 70°F NR 3 molar at 70°F	NR 38-100% at 70°F BC 20% at 70°F	A to 100% to 70°F	AB to 100% to 70°F
HYDROCHLORIC ACID				NR 20% at 140°F B 10% at 70°F		
HYDROCHLORIC ACID				NR 3 molar at 158°F		
HYDROCHLORIC ACID						
HYDROCHLORIC ACID						
HYDROCHLORIC ACID						
HYDROCHLORIC ACID						
HYDROCHLORIC ETHER						
HYDROCYANIC ACID	NR at 70°F	NR at 70°F		NR at 70°F	A to 70°F	A to 70°F
HYDROFLUORIC ACID	NR 50-100% at 70°F	NR 50-100% at 70°F		NR 45-100% at 70°F	NR 100% at 70°F	A 75-100% at 70°F
HYDROFLUORIC ACID						
HYDROFLUORIC ACID						
HYDROFLUORIC ACID						
HYDROFLUORIC ACID						
HYDROFLUORIC ACID						
HYDROFLUORIC ACID						
HYDROFLUOSILICIC ACID	NR at 70°F	NR at 70°F		AB to 70°F		
HYDROGEN GAS	C at 70°F NR at 140°F	C at 70°F		A to 70°F AB to 140°F		
HYDROGEN BROMIDE						
HYDROGEN CARBOXYLIC ACID						
HYDROGEN CHLORIDE, GAS						

	ETHYLENE ACRYLIC EA	POLYALLOMER LINEAR COPOLYMER	NYLON 11 POLYAMIDE	NYLON 12 POLYAMIDE	ETHYLENE VINYL ACETATE EVA	POLYVINYLCHLORIDE FLEXIBLE PVC
HYDRATED BARYTA						
HYDRATED LIME						
HYDRATITE LIQUID						
HYDRAULIC FLUIDS (PYDRAULS)	NR at 70°F		A to 70°F AB to 200°F			B/NR at 70°F
HYDRAULIC OIL (PETROLEUM)	A to 70°F	A to 70°F	A to 70°F AB to 200°F	AB to 70°F	C at 70°F	A to 70°F
HYDRAULIC OIL (PETROLEUM, AIRCRAFT)			A to 70°F			C at 70°F
HYDRAULIC OIL (PHOSPHATE ESTER)	NR at 70°F		A to 70°F AB to 200°F		NR at 70°F	B/NR at 70°F
HYDRAZINE			NR at 70°F		C at 70°F	B/NR at 70°F
HYDRAZINO BENZENE						NR at 70°F
HYDRIODIC ACID			NR at 70°F			AB to 70°F
HYDRO-DRIVE MIH10 (PETRO BASED)						
HYDRO-DRIVE MIH50 (PETRO BASED)						
HYDROBROMIC ACID			NR at 70°F		B at 70°F	AB to 100% to 150°F
HYDROBROMIC ACID						
HYDROBROMIC ACID, GAS						
HYDROCARBONS, ALICYCLIC						
HYDROCARBONS, ALIPHATIC						
HYDROCARBONS, AROMATIC						
HYDROCARBONS, CHLORINATED						
HYDROCARBONS, LIGHT						
HYDROCARBONS, NORMAL						
HYDROCARBONS, OLEFINIC						
HYDROCARBONS, SATURATED	A to 70°F					A to 70°F
HYDROCHLORIC ACID	NR 38% at 70°F A 20% to 70°F	A to 50% to 70°F B 35% at 122°F	+0.8% vol 6M conc 28 days 70°F	NR 10-100% at 70°F C 1% at 70°F	C/NR at 70°F A to 50% to 70°F	+1% vol 10% conc 14 days 70°F
HYDROCHLORIC ACID		A to 20% to 122°F	+1.7% vol 6M conc 180 days 70°F		C 38% at 140°F AB dilute to 140°F	AB 100% to 70°F C 100% at 140°F
HYDROCHLORIC ACID			C/NR 50-100% at 70°F NR 37% at 70°F			A to 70% to 70°F C 37-70% at 150°F
HYDROCHLORIC ACID			BC 20-30% at 70°F AB 10% to 70°F			AB to 25% at 131°F A 10% to 150°F
HYDROCHLORIC ACID			C 10% at 105°F NR 1% at 140°F			
HYDROCHLORIC ACID						
HYDROCHLORIC ACID						
HYDROCHLORIC ETHER						
HYDROCYANIC ACID		A to 70°F	C/NR at 70°F		AB to 70°F	AB 50-100% to 70°F
HYDROFLUORIC ACID		A to 50% to 70°F AB to 48% to 122°F	NR 38-100% at 70°F C dilute at 70°F		NR 38-100% at 70°F C dilute at 70°F	NR 60-100% at 70°F AB to 50% to 70°F
HYDROFLUORIC ACID						C 20-50% at 122°F
HYDROFLUORIC ACID						
HYDROFLUORIC ACID						
HYDROFLUORIC ACID						
HYDROFLUORIC ACID						
HYDROFLUORIC ACID						
HYDROFLUOSILICIC ACID			NR at 70°F		C/NR at 70°F AB 31% to 140°F	A/NR 20-100% to 150°
HYDROGEN GAS	*A to 70°F	A to 70°F	A to 70°F AB to 194°F		A to 70°F AB to 140°F	AB to 130°F
HYDROGEN BROMIDE						
HYDROGEN CARBOXYLIC ACID						
HYDROGEN CHLORIDE, GAS				NR 3-100% at 70°F C 2% at 70°F		

	POLYETHYLENE LOW DENSITY LDPE	®TEFLON FEP	®KALREZ PERFLUORINATED ELASTOMER	®FLUORAZ FLUORINATED COPOLYMER	®AFLAS FLUORINATED COPOLYMER	®NORPRENE COPOLYMER TPO
HYDRATED BARYTA		A to 70°F				
HYDRATED LIME		A to 70°F				
HYDRATITE LIQUID	TEST	A to 70°F				
HYDRAULIC FLUIDS (PYDRAULS)	NR at 70°F	A to 70°F AB to 400°F	A to 212°F	A to 70°F	+14-17% vol 3 days 212°F A to 70°F	
HYDRAULIC OIL (PETROLEUM)	A/NR at 70°F	A to 70°F AB to 450°F	A to 212°F	A to 70°F		
HYDRAULIC OIL (PETROLEUM, AIRCRAFT)	A to 70°F	A to 70°F				
HYDRAULIC OIL (PHOSPHATE ESTER)	NR at 70°F	A to 70°F AB to 400°F	A to 212°F	A to 70°F	+14-17% vol 3 days 212°F A to 70°F	BC at 212°F Skydrol 500-B4
HYDRAZINE	AB to 70°F	A to boiling	A to 75°F *A to 212°F	A to 70°F	AB to 70°F	
HYDRAZINO BENZENE		A to 70°F				
HYDRIODIC ACID			*A to 212°F			AB to 70°F
HYDRO-DRIVE MIH10 (PETRO BASED)						
HYDRO-DRIVE MIH50 (PETRO BASED)						
HYDROBROMIC ACID	AB to 100% to 70°F AB to 50% to 140°F	A to 300°F	A 40-100% to 212°F	A 40-100% to 70°F		TEST
HYDROBROMIC ACID						
HYDROBROMIC ACID, GAS	A to 70°F	A to 70°F				
HYDROCARBONS, ALICYCLIC		A to 70°F				
HYDROCARBONS, ALIPHATIC		A to 70°F				
HYDROCARBONS, AROMATIC		A to 70°F				
HYDROCARBONS, CHLORINATED		A to 70°F				
HYDROCARBONS, LIGHT						
HYDROCARBONS, NORMAL		A to 70°F	A to 212°F			
HYDROCARBONS, OLEFINIC		A to 70°F	A to 212°F			
HYDROCARBONS, SATURATED		A to 70°F		C at 70°F		
HYDROCHLORIC ACID	A to 100% to 70°F A to 35% to 122°F	A to 100% to 248°F	A to fuming to 200°F A to 20% to 230°F	A 37-100% to 70°F	+0.2% vol 37% conc 7 days 73°F	AB to 100% to 70°F
HYDROCHLORIC ACID	AB 37% at 140°F		A 100% fumes to 70°F		+4.5% vol 37% conc 180 days 73°F	
HYDROCHLORIC ACID					+2.3% vol 37% conc 3 days 104°F	
HYDROCHLORIC ACID					+7% vol 37% conc 3 days 158°F	
HYDROCHLORIC ACID					+7.4% vol 20% conc 3 days 158°F	
HYDROCHLORIC ACID					A to conc to 194°F BC 20% at 212°F	
HYDROCHLORIC ACID						
HYDROCHLORIC ETHER		A to 70°F				
HYDROCYANIC ACID	A to 100% to 70°F	A to 100% to 300°F	A to 212°F	A to 70°F		AB to 70°F
HYDROFLUORIC ACID	C/NR 75-100% at 70°F AB to 60% to 70°F	A to 100% to 122°F A to 50% to 300°F	A to 100% to 140°F A 100% to 212°F dry	A to 100% to 70°F A to 65% to 150°F	+1.5% vol 50% conc 7 days 70°F	TEST
HYDROFLUORIC ACID	AC to 60% at 140°F A to 50% to 122°F				+4.1% vol 50% conc 180 days 70°F	
HYDROFLUORIC ACID					A to 50% to 70°F	
HYDROFLUORIC ACID						
HYDROFLUORIC ACID						
HYDROFLUORIC ACID						
HYDROFLUORIC ACID						
HYDROFLUOSILICIC ACID	A to conc to 70°F AB to 32% at 140°F	A to 100% to 200°F	A to 212°F	A to 70°F		
HYDROGEN GAS	A to 70°F AB at 140°F	A to 300°F	A to 212°F dry			
HYDROGEN BROMIDE	A to 70°F	A to 70°F				
HYDROGEN CARBOXYLIC ACID		A to 70°F				
HYDROGEN CHLORIDE, GAS	A dry to 70°F AB dry to 140°F	A to 300°F wet or dry	A to 212°F dry	A to 150°F		

C 193

	RECOMMENDATIONS	NITRILE NBR	ETHYLENE PROPYLENE EP, EPDM	FLUOROCARBON FKM	CHLOROPRENE CR	HYDROGENATED NITRILE HNBR
HYDROGEN CYANIDE	EP, IIR, FKM	B any conc to 140°F	A to 100% to 140°F	A to 100% to 140°F	B to 100% to 140°F	B at 70°F
HYDROGEN DIOXIDE		NR 90% at 70°F C 30% at 70°F	C 90% at 70°F B 3-30% at 70°F	A to 90% to 70°F	NR 30-90% at 70°F C/NR 10% at 70°F	
HYDROGEN DIOXIDE		BC 10% to 80°F B 3% at 70°F			B 3% at 70°F	
HYDROGEN FLUORIDE	FKM	NR 100% at 70°F AB 20% to 80°F	AB to 70°F	A to 120°F	NR 20-100% at 70°F	
HYDROGEN OXIDE	NBR, EP, SBR, IIR	A to 180°F	A to 275°F	A to 70°F B at 275°F	AC to 212°F	A to 70°F
HYDROGEN PEROXIDE	FKM, FVMQ, EP, CR, IIR	NR 30-90% at 70°F BC 10% to 80°F	BC 90-100% at 70°F AC 50-80% to 70°F	A to 100% to 104°F AB 100% at 160°F	NR 20-90% at 70°F C/NR 10% at 70°F	B at 70°F
HYDROGEN PEROXIDE		NR 10% at 104°F B 3% at 70°F	B 3-30% at 70°F NR 30% at 104°F	NR 100% at 270°F A 50% to 200°F	C 5% at 104°F B 3% at 70°F	
HYDROGEN PEROXIDE			C 10% at 104°F B 5% to 140°F	C 10-30% at 104°F A 5% to 176°F		
HYDROGEN PEROXIDE			C 5% at 176°F A 0.5% to 70°F			
HYDROGEN SULFIDE, DRY COLD	EP, NBR, SBR	A to 140°F	A to 140°F	NR at 70°F	A to 140°F	A to 70°F
HYDROGEN SULFIDE, DRY HOT	EP, IIR	B 176-212°F	B 176-212°F	NR at 140°F	B at 150°F	NR at 140°F
HYDROGEN SULFIDE, WET COLD	EP, CR, IIR	B/NR at 70°F B/NR sol'n at 130°F	A to 140°F A sol'n to 130°F	A/NR at 70°F NR sol'n at 130°F	A to 70°F	NR at 70°F
HYDROGEN SULFIDE, WET HOT	EP, IIR	NR at 140°F	B at 212°F	B/NR to 270°F	B at 140°F NR sol'n at 140°F	NR at 140°F
HYDROLUBE (WATER/ETHYLENE GLYCOL)	NBR, FKM, EP, SBR	A to 70°F	A to 70°F	A to 70°F	B at 70°F	
HYDROQUINOL		C/NR at 70°F		C at 70°F	NR at 70°F	NR at 70°F
HYDROQUINONE	FKM, FVMQ, EP, IIR	C/NR at 70°F	A/NR to 140°F	AB to 140°F	NR at 70°F	NR at 70°F
HYDROXYPROPANE TRICARBOXYLIC ACID		B any conc to 110°F			B any conc to 150°F	
HYDROXYACETIC ACID	EP	A/NR at 70°F A 70% to 140°F	A 70-100% to 200°F	A/NR at 70°F A 70% to 140°F	B/NR at 70°F A 70% to 140°F	
HYDROXYBENZENE		NR at 70°F	C at 70°F	A to 70°F	C/NR at 70°F	
HYDROXYBUTANE		A to 70°F	A to 70°F	A to 70°F	A to 70°F	
HYDROXYBUTANEDIOIC ACID		B at 70°F	NR at 70°F	A to 70°F	C at 70°F	
HYDROXYETHER		C at 70°F	A to 70°F	B at 70°F	C at 70°F	
HYDROXYETHYL ACETATE		C at 70°F	A to 70°F	A to 70°F	NR at 70°F	
HYDROXYETHYL AMINE		B any conc to 80°F	B at 70°F	C at 70°F	B any conc to 80°F	
HYDROXYFORMIC ACID		A to 70°F	A to 70°F	A to 70°F	A to 70°F	
HYDROXYOCTANE		A to 70°F	NR at 70°F	B at 70°F	A to 70°F	
HYDROXYSUCCINIC ACID	NBR, FKM, FVMQ	A to 70°F	NR at 70°F	A to 70°F	A to 140°F	A to 70°F
HYDYNE	EP	B at 70°F	A to 70°F	NR at 70°F	B at 70°F	
HYJET	EP	NR at 70°F	A to 70°F	NR at 70°F	NR at 70°F	NR at 70°F
HYJET S	EP	NR at 70°F	A to 70°F	NR at 70°F	NR at 70°F	
HYJET W	EP	NR at 70°F	A to 70°F	NR at 70°F	NR at 70°F	NR at 70°F
HYJET III	EP	NR at 70°F	A to 70°F	NR at 70°F	NR at 70°F	
HYKIL NO. 6 33%, H$_2$O 67%		C at 70°F			NR at 70°F	
HYPNONE		NR at 70°F	A to 70°F	NR at 70°F	NR at 70°F	
HYPO	NBR, FKM, CR, EP	A to 70°F AB any conc to 200°F	A to 104°F	A to 140°F	A to 140°F AB any conc to 200°F	A to 70°F
HYPOCHLOROUS ACID	FKM, EP, IIR	NR at 70°F C/NR 5-10% at 70°F	B 100% to 160°F A 10% to 104°F	A to 100% to 70°F B 10% at 176°F	NR 5-100% at 70°F	
HYPOID LUBES (GEAR LUBES)	ACM	+9% vol 3 days 300°F AB to 70°F	+124% vol 3 days 300°F NR at 70°F	+4% vol 3 days 300°F +23% vol 14 days 300°F	softened 3 days 300°F C/NR at 70°F	
ICE SPAR (ICE STONE)		B at 70°F	A to 70°F	A to 70°F	A to 70°F	
IDOBUTANE		NR at 70°F			NR at 70°F	
IDOETHANE		NR at 70°F	C at 70°F	B at 70°F	NR at 70°F	
IDOFORM	EP, IIR		A to 120°F	AC to 70°F	NR at 70°F	
IDOPENTANE		NR at 70°F	NR at 70°F		NR at 70°F	
INDUSTRON FF44	NBR, FKM, ACM	A to 70°F	NR at 70°F	A to 70°F	B at 70°F	A to 70°F
INDUSTRON FF48	NBR, FKM, ACM	A to 70°F	NR at 70°F	A to 70°F	B at 70°F	A to 70°F

	STYRENE BUTADIENE SBR	POLYACRYLATE ACM	POLYURETHANE AU, EU	ISOBUTYLENE ISOPRENE IIR	POLYBUTADIENE BR	® AEROQUIP AQP
HYDROGEN CYANIDE	B at 70°F	NR at 70°F	C/NR 50-100% at 70°F	AB any conc to 150°F	B at 70°F	NR at 70°F
HYDROGEN DIOXIDE			C 90% at 70°F	NR 90% at 70°F C 10-30% at 70°F		
HYDROGEN DIOXIDE				B 3% at 70°F		
HYDROGEN FLUORIDE	NR at 70°F	NR at 70°F	NR at 70°F	NR 20% at 70°F		
HYDROGEN OXIDE			A to 70°F	A to 70°F		
HYDROGEN PEROXIDE	NR 90% at 70°F B/NR to 90% at 70°F	NR to 90% at 70°F	B/NR 90-100% at 70°F AB to 30% to 70°F	C/NR 90% at 70°F B to 50% at 70°F	NR 90% at 70°F B to 90% at 70°F	NR 3-90% at 70°F
HYDROGEN PEROXIDE						
HYDROGEN PEROXIDE						
HYDROGEN PEROXIDE						
HYDROGEN SULFIDE, DRY COLD	A to 70°F	NR at 70°F	A to 70°F	A to 70°F	A to 70°F	NR at 70°F
HYDROGEN SULFIDE, DRY HOT	NR at 140°F	NR at 140°F		AB any conc to 150°F	NR at 140°F	NR at 70°F
HYDROGEN SULFIDE, WET COLD	NR at 70°F	NR at 70°F	A/NR at 70°F	A to 70°F	NR at 70°F	NR at 70°F
HYDROGEN SULFIDE, WET HOT	NR at 140°F	NR at 140°F		AB any conc to 150°F	NR at 140°F	NR at 70°F
HYDROLUBE (WATER/ETHYLENE GLYCOL)	A to 70°F	NR at 70°F	NR at 70°F	B at 70°F		*AB to 200°F
HYDROQUINOL				NR at 70°F		
HYDROQUINONE	NR at 70°F	NR at 70°F		NR at 70°F	NR at 70°F	
HYDROXYPROPANE TRICARBOXYLIC ACID				A to 70°F AB any conc to 185°F		
HYDROXYACETIC ACID	C/NR at 70°F	C/NR at 70°F	NR at 70°F			
HYDROXYBENZENE			C at 70°F	B any conc to 100°F		
HYDROXYBUTANE			NR at 70°F	A to 70°F		
HYDROXYBUTANEDIOIC ACID				NR at 70°F		
HYDROXYETHER				A to 70°F		
HYDROXYETHYL ACETATE			NR at 70°F	A to 70°F		
HYDROXYETHYL AMINE			C at 70°F	A to 70°F AB any conc to 140°F		
HYDROXYFORMIC ACID			A to 70°F	A to 70°F		
HYDROXYOCTANE				B at 70°F		
HYDROXYSUCCINIC ACID				NR at 70°F		
HYDYNE	B at 70°F	NR at 70°F		B at 70°F	B at 70°F	
HYJET	NR at 70°F	NR at 70°F	NR at 70°F	B at 70°F	NR at 70°F	
HYJET S	NR at 70°F	NR at 70°F	NR at 70°F	B at 70°F	NR at 70°F	
HYJET W	NR at 70°F	NR at 70°F	NR at 70°F	B at 70°F	NR at 70°F	
HYJET III	NR at 70°F	NR at 70°F	NR at 70°F	B at 70°F	NR at 70°F	
HYKIL NO. 6 33%, H_2O 67%				NR at 70°F		
HYPNONE			NR at 70°F	A any conc to 80°F		
HYPO			A to 70°F	A to 70°F AB any conc to 185°F		
HYPOCHLOROUS ACID	NR at 70°F	NR at 70°F	NR at 70°F	B at 70°F	NR at 70°F	
HYPOID LUBES (GEAR LUBES)	+154% vol 3 days 300°F NR at 70°F	+30% vol 3 days 300°F A to 300°F	dissolved 3 days 300°F B/NR at 70°F	dissolved 3 days 300°F NR at 300°F		
ICE SPAR (ICE STONE)				A to 70°F		
IDOBUTANE						
IDOETHANE				C at 70°F		
IDOFORM				A to 70°F		
IDOPENTANE				NR at 70°F		
INDUSTRON FF44	NR at 70°F	A to 70°F	B at 70°F	NR at 70°F	NR at 70°F	
INDUSTRON FF48	NR at 70°F	A to 70°F	B at 70°F	NR at 70°F	NR at 70°F	

C195

	SYNTHETIC ISOPRENE IR	NATURAL ISOPRENE NR	CHLOROSULFONATED POLYETHYLENE CSM	FLUOROSILICONE FVMQ	SILICONE VMQ	CHEMRAZ FFKM
HYDROGEN CYANIDE	AB to 70°F	AB to 100% to 80°F AB 20% to 150°F	A to 70°F	B at 70°F	AC to 70°F	A to 70°F
HYDROGEN DIOXIDE	NR 90% at 70°F C 30% at 70°F	NR 90% at 70°F C 30% at 70°F	C 30-90% at 70°F B to 10% at 70°F		A to 90% to 70°F	
HYDROGEN DIOXIDE	B to 10% at 70°F	B to 10% to 80°F				
HYDROGEN FLUORIDE	NR at 70°F	NR 20-100% at 70°F		NR at 70°F	NR at 70°F	A to 100% to 70°F
HYDROGEN OXIDE	A to 70°F	A to 70°F	A to 70°F	A to 70°F	A to 70°F	A to 70°F B at 70°F white
HYDROGEN PEROXIDE	NR 90% at 70°F BC 30% at 70°F	NR 80-90% at 70°F C 30% at 70°F	AB to 100% to 70°F A to 80% to 70°F	+5% vol 7 days 150°F B 90% at 70°F	+1% vol 3% conc 7 days 70°F	AB 90-100% at 70°F A to 70°F white
HYDROGEN PEROXIDE	B to 10% to 70°F	NR 30-50% at 200°F B to 10% at 80°F	AB to 50% at 200°F	A to 30% to 70°F	BC 100% at 70°F B 90% at 160°F	
HYDROGEN PEROXIDE					AB to 50% to 125°F A to 30% to 70°F	
HYDROGEN PEROXIDE						
HYDROGEN SULFIDE, DRY COLD	A to 70°F	A to 70°F	A to 70°F	C to 140°F	C/NR at 70°F	A to 70°F
HYDROGEN SULFIDE, DRY HOT	NR at 140°F	NR at 70°F	C at 140°F	C to 140°F	C/NR at 70°F	A to 140°F
HYDROGEN SULFIDE, WET COLD	C/NR to 140°F	C/NR at 70°F	A to 70°F	C to 140°F	C/NR at 70°F	A to 70°F
HYDROGEN SULFIDE, WET HOT	NR at 140°F	NR at 140°F	C at 140°F	C to 140°F	C/NR at 70°F	A to 140°F
HYDROLUBE (WATER/ETHYLENE GLYCOL)				B at 70°F	B at 70°F	A to 70°F
HYDROQUINOL	B at 70°F	B any conc to 80°F	A to 70°F			B at 70°F
HYDROQUINONE	B at 70°F	B any conc to 80°F	A/NR at 70°F	B at 70°F		B at 70°F
HYDROXYPROPANE TRICARBOXYLIC ACID	A to 70°F	A to 70°F AB any conc to 150°F				
HYDROXYACETIC ACID	NR 10% at 70°F	NR 10% at 70°F		B/NR 100% at 70°F	A/NR 100% at 70°F	A to 70°F
HYDROXYBENZENE	NR at 70°F	NR at 70°F	C at 70°F		NR at 80°F	
HYDROXYBUTANE	A to 70°F	A to 70°F	A to 70°F		B at 70°F	
HYDROXYBUTANEDIOIC ACID	A to 70°F	A to 70°F	B at 70°F		B at 70°F	
HYDROXYETHER	NR at 70°F	NR at 70°F	B at 70°F			
HYDROXYETHYL ACETATE	C at 70°F	C at 70°F	NR at 70°F		C at 70°F	
HYDROXYETHYL AMINE	B at 70°F	B any conc to 80°F	B at 70°F		B at 70°F	
HYDROXYFORMIC ACID	A to 70°F	A to 70°F	A to 70°F		A to 70°F	
HYDROXYOCTANE	B at 70°F	B at 70°F	A to 70°F			
HYDROXYSUCCINIC ACID	A to 70°F	A any conc to 80°F	B at 70°F	A to 70°F	B at 70°F	A to 70°F
HYDYNE	B at 70°F	B at 70°F		NR at 70°F	NR at 70°F	A to 70°F
HYJET	NR at 70°F	NR at 70°F	NR at 70°F			A to 70°F
HYJET S	NR at 70°F	NR at 70°F	NR at 70°F			A to 70°F
HYJET W	NR at 70°F	NR at 70°F	NR at 70°F			A to 70°F
HYJET III	NR at 70°F	NR at 70°F	NR at 70°F			A to 70°F
HYKIL NO. 6 33%, H₂O 67%	NR at 70°F	NR at 70°F				
HYPNONE	C at 70°F	C/NR at 70°F	NR at 70°F			
HYPO	A to 70°F	A to 70°F AB any conc to 150°F	A to 70°F	A to 70°F	AB to 70°F	A to 70°F
HYPOCHLOROUS ACID	B at 70°F	B at 70°F AB 5% to 120°F	NR at 70°F	A to 70°F	NR at 70°F	A to 70°F
HYPOID LUBES (GEAR LUBES)				dissolved 3 days 300°F NR at 70°F	+16% vol 3 days 300°F C/NR at 70°F	
ICE SPAR (ICE STONE)						
IDOBUTANE	NR at 70°F	NR at 70°F				
IDOETHANE	NR at 70°F	NR at 70°F				
IDOFORM		B at 70°F				*B at 70°F
IDOPENTANE	NR at 70°F	NR at 70°F	NR at 70°F			
INDUSTRON FF44	NR at 70°F	NR at 70°F	NR at 70°F	A to 70°F	NR at 70°F	A to 70°F
INDUSTRON FF48	NR at 70°F	NR at 70°F	NR at 70°F	A to 70°F	NR at 70°F	A to 70°F

	POLYSULFIDE T	CHLORINATED POLYETHYLENE CM	EPICHLOROHYDRIN CO, ECO	® HYTREL COPOLYESTER TPE	® SANTOPRENE COPOLYMER TPO	® C-FLEX STYRENIC TPE
HYDROGEN CYANIDE	NR at 70°F	NR at 70°F		NR at 70°F	A to 70°F	A to 70°F
HYDROGEN DIOXIDE						
HYDROGEN DIOXIDE						
HYDROGEN FLUORIDE						
HYDROGEN OXIDE						
HYDROGEN PEROXIDE	NR 90% at 70°F C 30% at 70°F	NR 30–90% at 70°F	AB to 50% to 125°F	NR 3–90% at 70°F AB dilute at 70°F		A to 90% to 70°F
HYDROGEN PEROXIDE	AB dilute at 70°F					
HYDROGEN PEROXIDE						
HYDROGEN PEROXIDE						
HYDROGEN SULFIDE, DRY COLD	B at 70°F	NR at 70°F		A to 70°F		A to 70°F
HYDROGEN SULFIDE, DRY HOT	A to 140°F	NR at 140°F				
HYDROGEN SULFIDE, WET COLD	A to 140°F	NR at 70°F	B at 70°F	A to 70°F		A to 70°F
HYDROGEN SULFIDE, WET HOT	A to 140°F	NR at 140°F	B at 140°F			
HYDROLUBE (WATER/ETHYLENE GLYCOL)				no swell 7 days 70°F *AB to 150°F	+4.2% vol 3 days 240°F A to 250°F	
HYDROQUINOL						
HYDROQUINONE	C at 70°F					
HYDROXYPROPANE TRICARBOXYLIC ACID						
HYDROXYACETIC ACID						
HYDROXYBENZENE						
HYDROXYBUTANE						
HYDROXYBUTANEDIOIC ACID						
HYDROXYETHER						
HYDROXYETHYL ACETATE						
HYDROXYETHYL AMINE						
HYDROXYFORMIC ACID						
HYDROXYOCTANE						
HYDROXYSUCCINIC ACID						
HYDYNE	NR at 70°F					
HYJET	NR at 70°F					
HYJET S	NR at 70°F					
HYJET W	NR at 70°F		B at 70°F			
HYJET III	NR at 70°F					
HYKIL NO. 6 33%, H$_2$O 67%	AB to 70°F					
HYPNONE						
HYPO						
HYPOCHLOROUS ACID	NR at 70°F		B at 70°F			A to 70°F
HYPOID LUBES (GEAR LUBES)	softened 3 days 300°F					
ICE SPAR (ICE STONE)						
IDOBUTANE						
IDOETHANE						
IDOFORM						
IDOPENTANE						
INDUSTRON FF44						
INDUSTRON FF48						

	ETHYLENE ACRYLIC EA	POLYALLOMER LINEAR COPOLYMER	NYLON 11 POLYAMIDE	NYLON 12 POLYAMIDE	ETHYLENE VINYL ACETATE EVA	POLYVINYLCHLORIDE FLEXIBLE PVC
HYDROGEN CYANIDE		A to 70°F	C/NR at 70°F		AB to 70°F	AB 50-100% to 70°F
HYDROGEN DIOXIDE						
HYDROGEN DIOXIDE						
HYDROGEN FLUORIDE				NR 40% at 70°F		
HYDROGEN OXIDE						
HYDROGEN PEROXIDE		A to 90% to 70°F B 30-90% at 122°F	NR 25-90% at 70°F AB to 20% to 70°F	NR 30% at 70°F C to 10% at 70°F	AC 90% to 70°F AB 30% to 140°F	A/NR 50-100% at 70°F AB 25-30% at 122°F
HYDROGEN PEROXIDE		A to 3% to 122°F	C 20% at 105°F			C 30% at 140°F AB 10% to 70°F
HYDROGEN PEROXIDE						B 10% at 150°F A 3% to 122°F
HYDROGEN PEROXIDE						no swell 3% conc 14 days 70°F
HYDROGEN SULFIDE, DRY COLD	A to 70°F	A to 70°F	C/NR at 70°F	AB to 5% to 70°F	A to 70°F	A to 70°F
HYDROGEN SULFIDE, DRY HOT	A to 450°F					
HYDROGEN SULFIDE, WET COLD		A to 70°F	C/NR at 70°F AB to 5% to 70°F		A to 70°F	A to 70°F
HYDROGEN SULFIDE, WET HOT						
HYDROLUBE (WATER/ETHYLENE GLYCOL)			A to 70°F *AB to 160°F		A to 70°F	AB to 70°F
HYDROQUINOL						
HYDROQUINONE				AB 5% to 70°F		AB to 70°F
HYDROXYPROPANE TRICARBOXYLIC ACID						
HYDROXYACETIC ACID						A 100% to 70°F NR 10% at 70°F
HYDROXYBENZENE						NR at 70°F
HYDROXYBUTANE						
HYDROXYBUTANEDIOIC ACID						
HYDROXYETHER						
HYDROXYETHYL ACETATE						
HYDROXYETHYL AMINE						
HYDROXYFORMIC ACID						
HYDROXYOCTANE						
HYDROXYSUCCINIC ACID						
HYDYNE						
HYJET						
HYJET S						
HYJET W						
HYJET III						
HYKIL NO. 6 33%, H_2O 67%						
HYPNONE						NR at 70°F
HYPO						AB to 70°F
HYPOCHLOROUS ACID						AB to 70°F *NR at 140°F
HYPOID LUBES (GEAR LUBES)						
ICE SPAR (ICE STONE)						
IDOBUTANE						
IDOETHANE						
IDOFORM						C at 70°F
IDOPENTANE						
INDUSTRON FF44						
INDUSTRON FF48						

C 198

	POLYETHYLENE LOW DENSITY LDPE	® TEFLON FEP	® KALREZ PERFLUORINATED ELASTOMER	® FLUORAZ FLUORINATED COPOLYMER	® AFLAS FLUORINATED COPOLYMER	® NORPRENE COPOLYMER TPO
HYDROGEN CYANIDE	A to 100% to 70°F	A to 100% to 300°F	A to 212°F	A to 70°F		AB to 70°F
HYDROGEN DIOXIDE		A to 90% to 70°F	A 90% to 70°F *A 90% to 270°F			
HYDROGEN DIOXIDE						
HYDROGEN FLUORIDE	AB to 70°F	AB to 70°F	B at 70°F *A to 212°F dry			
HYDROGEN OXIDE	A to 140°F NR at 212°F	A to boiling	A to 194°F	A to 70°F	A to 212°F	A to 212°F
HYDROGEN PEROXIDE	B 100% at 70°F AB 30-90% to 70°F	A to 100% to boiling A to 90% to 300°F	A to 212°F	NR 90-100% at 70°F	-1%vol 30% conc 7 days 212°F	A 3-90% at 70°F
HYDROGEN PEROXIDE	B 30-90% at 122°F NR 30-90% at 140°F				A to 30% to 212°F	
HYDROGEN PEROXIDE	A to 30% to 70°F AB to 30% 122°F to 140°F					
HYDROGEN PEROXIDE	A to 3% to 140°F					
HYDROGEN SULFIDE, DRY COLD	A to 104°F	A to 70°F		A to 70°F		A to 70°F
HYDROGEN SULFIDE, DRY HOT	AC to 140°F	A to 300°F	*A to 270°F	A to 140°F		
HYDROGEN SULFIDE, WET COLD	A to 70°F	A to 70°F	A to 70°F	A to 70°F		A to 70°F
HYDROGEN SULFIDE, WET HOT	AB to 140°F	A to300°F	A to 140°F	B at 140°F	+3.5% vol 35% conc 4 days 400°F	
HYDROLUBE (WATER/ETHYLENE GLYCOL)	A to 70°F	A to 70°F *AB to 250°F		A to 70°F	+ 3.3% vol 14 days 324°F A to 70°F	
HYDROQUINOL		A to 70°F				
HYDROQUINONE	A to 70°F AB to 140°F	A to 70°F	A to 212°F	A to 70°F		
HYDROXYPROPANE TRICARBOXYLIC ACID		A to 70°F				
HYDROXYACETIC ACID	A 70-100% to 70°F	A to 100% to 70°F	A to 212°F	NR at 70°F	no swell 67% conc 2 days 158°F	
HYDROXYBENZENE		A to 70°F				
HYDROXYBUTANE		A to 70°F				
HYDROXYBUTANEDIOIC ACID		A to 70°F				
HYDROXYETHER		A to 70°F				
HYDROXYETHYL ACETATE		A to 70°F				
HYDROXYETHYL AMINE		A to 70°F				
HYDROXYFORMIC ACID		A to 70°F				
HYDROXYOCTANE		A to 70°F				
HYDROXYSUCCINIC ACID	B /NR at 70°F	A to 300°F	A to 70°F	A to 70°F		TEST
HYDYNE			A to 212°F	NR at 70°F		
HYJET				B at 70°F		
HYJET S				B at 70°F		
HYJET W				B at 70°F		
HYJET III				B at 70°F		
HYKIL NO. 6 33%, H_2O 67%		A to 70°F				
HYPNONE		A to 70°F				
HYPO	A to 104°F B at 140°F	A to 300°F	A to 70°F	AC to 70°F		
HYPOCHLOROUS ACID	A to 70°F AB to 140°F	A to 300°F	A to 70°F *A to 212°F	A to 70°F		A to 70°F
HYPOID LUBES (GEAR LUBES)						
ICE SPAR (ICE STONE)		A to 70°F				
IDOBUTANE		A to 70°F				
IDOETHANE		A to 70°F				
IDOFORM		A to 70°F	A to 212°F	NR at 70°F		
IDOPENTANE		A to 70°F				
INDUSTRON FF44				A to 70°F		
INDUSTRON FF48				A to 70°F		

	RECOMMENDATIONS	NITRILE NBR	ETHYLENE PROPYLENE EP, EPDM	FLUOROCARBON FKM	CHLOROPRENE CR	HYDROGENATED NITRILE HNBR
INDUSTRON FF53	NBR, FKM, ACM	A to 70°F	NR at 70°F	A to 70°F	B at 70°F	A to 70°F
INDUSTRON FF80	NBR, FKM, ACM	A to 70°F	NR at 70°F	A to 70°F	B at 70°F	A to 70°F
INK	NBR	A to 140°F	A to 70°F	A to 70°F	A to 70°F	
IODINE	FKM, EP	B to 140°F A 6.5% to 70°F	AB to 160°F A 6.5% to 70°F	A to conc to 140°F	NR 6-100% at 70°F	A to 70°F
IODINE PENTAFLUORIDE		NR at 70°F	NR at 70°F	NR at 70°F	NR at 70°F	NR at 70°F
I.P.A.	EP, IIR, FKM, NBR	A to 70°F B any conc to 120°F	A to 160°F B at 176°F	A to 160°F B at 176°F	AB any conc to 120°F C at 140°F	B at 70°F
IRON ACETATE		NR at 70°F		NR at 70°F		
IRON CHLORIDE	NBR, EP, IIR, CR	A to 70°F AB any conc to 200°F	A to conc to 176°F B at 212°F	A to conc to 176°F B at 212°F	AB to conc to 176°F	
IRON DICHLORIDE		A to 70°F AB any conc to 200°F	A to 70°F	A to 70°F	AB any conc to 140°F	
IRON HYDROXIDE		A to 140°F AB 176-200°F	A to 176°F AB to 200°F	A to 176°F AB to 200°F	AB to 200°F	
IRON MONOSULFIDE		A to 70°F		A to 70°F		
IRON NITRATE	NBR, EP, CR, FKM, IIR	A to 140°F AB any conc to 200°F	A to 176°F AB to 212°F	A to 212°F	A to 140°F AB any conc to 200°F	
IRON PERCHLORIDE		A to 70°F AB any conc to 200°F			B any conc to 80°F	
IRON PERSULFATE		A to 70°F AB any conc to 200°F			A to 70°F AB any conc to 200°F	
IRON PROTOCHLORIDE		A to 70°F AB any conc to 200°F	A to 70°F	A to 70°F	B any conc to 80°F	
IRON SALTS	NBR, EP, CR, FKM, IIR	A to 70°F AB to 200°F	A to 70°F	A to 70°F	AB to 150°F	
IRON SESQUICHLORIDE		A to 70°F AB any conc to 200°F	A to 70°F	A to 70°F	B any conc to 80°F	
IRON SESQUISULFATE		A to 70°F AB any conc to 200°F	A to 70°F	A to 70°F	A to 70°F AB any conc to 200°F	
IRON SULFATE	NBR, EP, CR, IIR	A to 70°F AB any conc to 200°F	A to conc to 70°F AB to conc to 200°F	A to conc to 140°F AB to 200°F	A to 70°F AB any conc to 200°F	
IRON SULFIDE		A to 70°F		A to 70°F		
IRON TERSULFATE		A to 70°F AB any conc to 200°F	A to 70°F	A to 70°F	A to 70°F AB any conc to 200°F	
IRON TRICHLORIDE		A to 70°F AB any conc to 200°F	A to 70°F	A to 70°F	B any conc to 80°F	
IRON VITRIOL		A to 70°F AB any conc to 200°F	A to 70°F	A to 70°F	A to 70°F AB any conc to 200°F	
IRUS 902		A to 70°F			A to 70°F	
ISOAMYL ACETATE		NR at 70°F	B at 70°F	NR at 70°F	NR at 70°F	
ISOAMYL ACETIC ESTER		NR at 70°F	B at 70°F	NR at 70°F	NR at 70°F	
ISOAMYL ALCOHOL		A to 70°F	A to 70°F	A to 158°F	A to 70°F	
ISOAMYL BROMIDE		NR at 70°F	NR at 70°F	B at 70°F	NR at 70°F	
ISOAMYL BUTYRATE		NR at 70°F		NR at 70°F		
ISOAMYL CHLORIDE		NR at 70°F	NR at 70°F	A to 70°F	NR at 70°F	
ISOAMYL ETHER		C at 70°F	NR at 70°F		NR at 70°F	
ISOAMYL PHTHALATE		NR at 70°F		C at 70°F		
ISOBUTANE	NBR, SBR	A to 250°F	NR at 70°F	A to 70°F	B/NR at 70°F	
ISOBUTANOL	EP, FKM, CR, IIR, SBR	B any conc to 80°F	A to 160°F	+1% vol 21 days 75°F A to 75°F	A any conc to 80°F	B at 70°F
ISOBUTYL ACETATE		NR at 70°F	C at 70°F	NR at 70°F	NR at 70°F	
ISOBUTYL ALCOHOL	EP, FKM, CR, IIR, SBR	B any conc to 80°F	A to 160°F	+1% vol 21 days 75°F A to 75°F	A any conc to 80°F	B at 70°F
ISOBUTYL ALDEHYDE		NR at 70°F	B at 70°F	NR at 70°F	C at 70°F	
ISOBUTYL AMINE		NR at 70°F		NR at 70°F		
ISOBUTYL BROMIDE		NR at 70°F		B at 70°F		
ISOBUTYL n-BUTYRATE	EP, FKM, IIR	NR at 70°F	A to 70°F	A to 70°F	NR at 70°F	
ISOBUTYL CARBINOL		A to 70°F	A to 70°F	A to 70°F	A to 70°F	
ISOBUTYL CHLORIDE		NR at 70°F		B at 70°F		NR at 70°F
ISOBUTYL ETHER		B at 70°F				B at 70°F
ISOBUTYLENE	NBR	A to 250°F	NR at 70°F	A to 70°F	A/NR at 70°F	

	STYRENE BUTADIENE SBR	POLYACRYLATE ACM	POLYURETHANE AU, EU	ISOBUTYLENE ISOPRENE IIR	POLYBUTADIENE BR	® AEROQUIP AQP
INDUSTRON FF53	NR at 70°F	A to 70°F	B at 70°F	NR at 70°F	NR at 70°F	AB to 300°F
INDUSTRON FF80	NR at 70°F	A to 70°F	B at 70°F	NR at 70°F	NR at 70°F	
INK			A to 70°F	A to 70°F		
IODINE	B at 70°F		NR at 70°F B sol'n at 70°F discolor	B at 70°F		
IODINE PENTAFLUORIDE	NR at 70°F	NR at 70°F	NR at 70°F	NR at 70°F	NR at 70°F	
I.P.A.				B any conc to 120°F		
IRON ACETATE				A to 70°F		
IRON CHLORIDE	A to conc to 70°F	A to 70°F	A to conc to 70°F A 20% fumes to 70°F	A to 70°F AB any conc to 185°F	A to 70°F	
IRON DICHLORIDE			A to 70°F	A to 70°F AB any conc to 185°F		
IRON HYDROXIDE				A to 70°F		
IRON MONOSULFIDE				A to 70°F		
IRON NITRATE	A to 70°F	A to 70°F	A to 70°F	A to 70°F AB any conc to 185°F	A to 70°F	
IRON PERCHLORIDE				A to 70°F AB any conc to 185°F		
IRON PERSULFATE				A to 70°F AB any conc to 185°F		
IRON PROTOCHLORIDE			A to 70°F	A to 70°F AB to 150°F		
IRON SALTS	AB to 70°F			A to 70°F AB to 150°F		
IRON SESQUICHLORIDE			A to 70°F	A to 70°F AB any conc to 185°F		
IRON SESQUISULFATE			A to 70°F	A to 70°F AB any conc to 185°F		
IRON SULFATE	A to conc to 70°F	A to 70°F	AB conc to 70°F AB sol'n to 70°F	A to 70°F AB any conc to 185°F	A to 70°F	NR at 70°F
IRON SULFIDE				A to 70°F		
IRON TERSULFATE			A to 70°F	A to 70°F AB any conc to 185°F		
IRON TRICHLORIDE			A to 70°F	A to 70°F AB any conc to 185°F		
IRON VITRIOL			A to 70°F	A to 70°F AB any conc to 185°F		
IRUS 902			AB to 70°F	NR at 70°F		
ISOAMYL ACETATE			NR at 70°F	B at 70°F		
ISOAMYL ACETIC ESTER			NR at 70°F	B at 70°F		
ISOAMYL ALCOHOL			C at 70°F	A to 70°F		
ISOAMYL BROMIDE				NR at 70°F		
ISOAMYL BUTYRATE				C at 70°F		
ISOAMYL CHLORIDE				C at 70°F		
ISOAMYL ETHER				NR at 70°F		
ISOAMYL PHTHALATE				A to 70°F		
ISOBUTANE	NR at 70°F	A to 70°F	A to 70°F	A/NR at 70°F		
ISOBUTANOL	A to 160°F	NR at 70°F	NR at 70°F	A to 70°F AB any conc to 100°F	B at 70°F	
ISOBUTYL ACETATE				B at 70°F		
ISOBUTYL ALCOHOL	A to 160°F	NR at 70°F	NR at 70°F	A to 70°F AB any conc to 100°F	B at 70°F	
ISOBUTYL ALDEHYDE			C at 70°F	B at 70°F		
ISOBUTYL AMINE				A to 70°F		
ISOBUTYL BROMIDE				NR at 70°F		
ISOBUTYL n-BUTYRATE	NR at 70°F	NR at 70°F		A to 70°F	NR at 70°F	
ISOBUTYL CARBINOL			C at 70°F	A to 70°F		
ISOBUTYL CHLORIDE				NR at 70°F		
ISOBUTYL ETHER				NR at 70°F		
ISOBUTYLENE	NR at 70°F	NR at 70°F	NR at 70°F	A/NR at 70°F		

C201

	SYNTHETIC ISOPRENE IR	NATURAL ISOPRENE NR	CHLOROSULFONATED POLYETHYLENE CSM	FLUOROSILICONE FVMQ	SILICONE VMQ	CHEMRAZ FFKM
INDUSTRON FF53	NR at 70°F	NR at 70°F	NR at 70°F	A to 70°F	NR at 70°F	A to 70°F
INDUSTRON FF80	NR at 70°F	NR at 70°F	NR at 70°F	A to 70°F	NR at 70°F	A to 70°F
INK		B /NR at 70°F				
IODINE	NR at 70°F	NR at 70°F	AB to 70°F	A to 70°F	C at 70°F	A to 70°F
IODINE PENTAFLUORIDE	NR at 70°F	NR at 70°F	NR at 70°F	NR at 70°F	NR at 70°F	AB to 70°F B at 70°F white
I.P.A.	AB to 70°F	A to 70°F AB any conc to 120°F	A to 200°F	B at 70°F	A to 70°F	A to 70°F
IRON ACETATE	NR at 70°F	NR at 70°F	A to 70°F		NR at 70°F	
IRON CHLORIDE	A to 70°F	A to 70°F AB any conc to 150°F	A to 200°F	A to conc to 70°F	AB to conc to 70°F	
IRON DICHLORIDE	A to 70°F	A to 70°F AB any conc to 150°F	A to 70°F		A to 70°F	
IRON HYDROXIDE	C at 70°F	C /NR at 70°F	AB to 200°F			
IRON MONOSULFIDE	A to 70°F	A to 70°F	A to 70°F			
IRON NITRATE	A to 70°F	A to 70°F AB any conc to 150°F	A to 70°F AB to 200°F	A to 70°F	BC at 70°F	
IRON PERCHLORIDE	A to 70°F	A to 70°F AB any conc to 150°F				
IRON PERSULFATE	A to 70°F	A to 70°F AB any conc to 150°F				
IRON PROTOCHLORIDE	A to 70°F	A to 70°F AB any conc to 150°F	A to 70°F		A to 70°F	
IRON SALTS	A to 70°F	A to 70°F AB any conc to 150°F	A to 70°F		AB to 70°F	
IRON SESQUICHLORIDE	A to 70°F	A to 70°F AB any conc to 150°F	A to 70°F		A to 70°F	
IRON SESQUISULFATE	A to 70°F	A to 70°F AB any conc to 150°F	A to 70°F		B at 70°F	
IRON SULFATE	A to 70°F	A to 70°F AB any conc to 150°F	A to 70°F	A to 70°F	BC conc at 70°F B to conc at 70°F	
IRON SULFIDE	A to 70°F	A to 70°F	A to 70°F			
IRON TERSULFATE	A to 70°F	A to 70°F AB any conc to 150°F	A to 70°F		B at 70°F	
IRON TRICHLORIDE	A to 70°F	A to 70°F AB any conc to 150°F	A to 70°F		A to 70°F	
IRON VITRIOL	A to 70°F	A to 70°F AB any conc to 150°F	A to 70°F		B at 70°F	
IRUS 902		NR at 70°F				
ISOAMYL ACETATE	NR at 70°F	NR at 70°F	C at 70°F		NR at 70°F	
ISOAMYL ACETIC ESTER	C at 70°F	C at 70°F	C at 70°F		NR at 70°F	A to 70°F
ISOAMYL ALCOHOL	A to 70°F	A to 70°F	A to 70°F		NR at 70°F	
ISOAMYL BROMIDE	NR at 70°F	NR at 70°F	NR at 70°F			
ISOAMYL BUTYRATE	NR at 70°F	NR at 70°F	NR at 70°F			A to 70°F
ISOAMYL CHLORIDE	NR at 70°F	NR at 70°F	NR at 70°F			
ISOAMYL ETHER	NR at 70°F	NR at 70°F	C at 70°F			
ISOAMYL PHTHALATE	NR at 70°F	NR at 70°F	NR at 70°F			
ISOBUTANE	NR at 70°F	NR at 70°F	NR at 70°F	A to 70°F	NR at 70°F	A to 70°F
ISOBUTANOL	A to 70°F	A any conc to 80°F	A to 70°F	B at 70°F	A to 70°F	A to 70°F
ISOBUTYL ACETATE	NR at 70°F	NR at 70°F	C at 70°F			A to 70°F
ISOBUTYL ALCOHOL	A to 70°F	A any conc to 80°F	A to 70°F	B at 70°F	A to 70°F	A to 70°F
ISOBUTYL ALDEHYDE	C at 70°F	C at 70°F	NR at 70°F		C at 70°F	
ISOBUTYL AMINE	C at 70°F	C at 70°F	C at 70°F			
ISOBUTYL BROMIDE	NR at 70°F	NR at 70°F	NR at 70°F			
ISOBUTYL n-BUTYRATE	NR at 70°F	NR at 70°F	NR at 70°F	A to 70°F		
ISOBUTYL CARBINOL	A to 70°F	A to 70°F	A to 70°F		NR at 70°F	
ISOBUTYL CHLORIDE	NR at 70°F	NR at 70°F	NR at 70°F			A to 70°F
ISOBUTYL ETHER	NR at 70°F	NR at 70°F	B at 70°F			A to 70°F
ISOBUTYLENE	NR at 70°F	NR at 70°F	NR at 70°F	A to 70°F	NR at 70°F	A to 70°F

	POLYSULFIDE T	CHLORINATED POLYETHYLENE CM	EPICHLOROHYDRIN CO, ECO	® HYTREL COPOLYESTER TPE	® SANTOPRENE COPOLYMER TPO	® C-FLEX STYRENIC TPE
INDUSTRON FF53				AB to 250°F		
INDUSTRON FF80						
INK						
IODINE				*AB IMS to 70°F		NR at 70°F
IODINE PENTAFLUORIDE	NR at 70°F		NR at 70°F			NR at 70°F
I.P.A.						
IRON ACETATE						
IRON CHLORIDE	A to 70°F	NR at 70°F	A to 70°F	AB to 140°F		A to 70°F
IRON DICHLORIDE						
IRON HYDROXIDE						
IRON MONOSULFIDE						
IRON NITRATE	A to 70°F		A to 70°F			
IRON PERCHLORIDE						
IRON PERSULFATE						
IRON PROTOCHLORIDE						
IRON SALTS	AB to 70°F					
IRON SESQUICHLORIDE						
IRON SESQUISULFATE						
IRON SULFATE	A to 70°F AB to 150°F	NR at 70°F	A to 70°F	A to 70°F		A to 70°F
IRON SULFIDE						
IRON TERSULFATE						
IRON TRICHLORIDE						
IRON VITRIOL						
IRUS 902				AB to 70°F		
ISOAMYL ACETATE						
ISOAMYL ACETIC ESTER						
ISOAMYL ALCOHOL	AB to 70°F					
ISOAMYL BROMIDE						
ISOAMYL BUTYRATE						
ISOAMYL CHLORIDE						
ISOAMYL ETHER						
ISOAMYL PHTHALATE						
ISOBUTANE						
ISOBUTANOL	B at 70°F					
ISOBUTYL ACETATE	AB to 70°F					
ISOBUTYL ALCOHOL	B at 70°F					
ISOBUTYL ALDEHYDE	AB to 70°F					
ISOBUTYL AMINE						
ISOBUTYL BROMIDE						
ISOBUTYL n-BUTYRATE	NR at 70°F					
ISOBUTYL CARBINOL						
ISOBUTYL CHLORIDE						
ISOBUTYL ETHER						
ISOBUTYLENE	AB to 70°F					

	ETHYLENE ACRYLIC EA	POLYALLOMER LINEAR COPOLYMER	NYLON 11 POLYAMIDE	NYLON 12 POLYAMIDE	ETHYLENE VINYL ACETATE EVA	POLYVINYLCHLORIDE FLEXIBLE PVC
INDUSTRON FF53			AB to 200°F			
INDUSTRON FF80						
INK				AB to 70°F	A to 70°F	AC to 70°F
IODINE	A to 70°F	A to 70°F in alcohol	A to 70°F in alcohol AC to 70°F methylated		C at 70°F B at 70°F in alcohol	A/NR at 70°F A to 70°F in alcohol
IODINE PENTAFLUORIDE					NR at 70°F	NR at 70°F
I.P.A.						
IRON ACETATE						
IRON CHLORIDE	*AB sol'n to 70°F	A to 70°F	A/NR at 70°F		A to 70°F C at 140°F	A to 150°F
IRON DICHLORIDE						
IRON HYDROXIDE						
IRON MONOSULFIDE						
IRON NITRATE			A to 70°F			
IRON PERCHLORIDE						
IRON PERSULFATE						
IRON PROTOCHLORIDE						
IRON SALTS			AB to 70°F	AB 20% to 70°F		A to 70°F
IRON SESQUICHLORIDE						
IRON SESQUISULFATE						
IRON SULFATE		A to 70°F	A to 70°F		A to 70°F	A to 150°F
IRON SULFIDE						
IRON TERSULFATE						
IRON TRICHLORIDE						
IRON VITRIOL						
IRUS 902			AB to 70°F			AB to 70°F
ISOAMYL ACETATE						
ISOAMYL ACETIC ESTER						
ISOAMYL ALCOHOL						
ISOAMYL BROMIDE						
ISOAMYL BUTYRATE						
ISOAMYL CHLORIDE						
ISOAMYL ETHER						
ISOAMYL PHTHALATE						
ISOBUTANE						AB to 70°F
ISOBUTANOL	A to 70°F	A to 122°F				A to 70°F B at 122°F
ISOBUTYL ACETATE						
ISOBUTYL ALCOHOL	A to 70°F	A to 122°F				A to 70°F B at 122°F
ISOBUTYL ALDEHYDE						
ISOBUTYL AMINE						
ISOBUTYL BROMIDE						
ISOBUTYL n-BUTYRATE						
ISOBUTYL CARBINOL						
ISOBUTYL CHLORIDE						
ISOBUTYL ETHER						
ISOBUTYLENE						AB to 70°F

	POLYETHYLENE LOW DENSITY LDPE	® TEFLON FEP	® KALREZ PERFLUORINATED ELASTOMER	® FLUORAZ FLUORINATED COPOLYMER	® AFLAS FLUORINATED COPOLYMER	® NORPRENE COPOLYMER TPO
INDUSTRON FF53		AB to 450°F		A to 70°F		
INDUSTRON FF80				A to 70°F		
INK	A/NR at 70°F	A to 70°F		NR at 70°F		
IODINE	NR conc at 70°F A 6.5% conc at 70°F	A to 300°F	A to 212°F	A to 70°F (wet)		A to 70°F
IODINE PENTAFLUORIDE	C/NR at 70°F	A to 70°F	B at 70°F	NR at 70°F		
I.P.A.	A to 122°F AB at 140°F	A to 300°F	A to 212°F	A to 70°F		
IRON ACETATE		A to 70°F				
IRON CHLORIDE	AB to 140°F	A to boiling	A to 70°F	A to 160°F		A to 70°F
IRON DICHLORIDE		A to 70°F				
IRON HYDROXIDE		A to 70°F				
IRON MONOSULFIDE		A to 70°F				
IRON NITRATE	B at 70°F	A to 70°F	A to 70°F			A to 70°F
IRON PERCHLORIDE		A to 70°F				
IRON PERSULFATE		A to 70°F				
IRON PROTOCHLORIDE		A to 70°F				
IRON SALTS	AB to 140°F	A to 70°F				A to 70°F
IRON SESQUICHLORIDE		A to 70°F				
IRON SESQUISULFATE		A to 70°F				
IRON SULFATE	A to 70°F	A to 150°F	A to 70°F	A to 70°F		
IRON SULFIDE		A to 70°F				
IRON TERSULFATE		A to 70°F				
IRON TRICHLORIDE		A to 70°F				
IRON VITRIOL		A to 70°F				
IRUS 902	C at 70°F					
ISOAMYL ACETATE		A to 70°F	A to 212°F		NR at 70°F	
ISOAMYL ACETIC ESTER		A to 70°F				
ISOAMYL ALCOHOL		A to 70°F			no swell 7 days 70°F A to 70°F	
ISOAMYL BROMIDE		A to 70°F				
ISOAMYL BUTYRATE		A to 70°F	A to 212°F			
ISOAMYL CHLORIDE		A to 70°F				
ISOAMYL ETHER		A to 70°F				
ISOAMYL PHTHALATE		A to 70°F				
ISOBUTANE	AB to 70°F	A to 70°F	A to 212°F	NR at 70°F		
ISOBUTANOL	A to 122°F	A to 200°F	A to 122°F	A to 70°F		
ISOBUTYL ACETATE		A to 70°F	A to 241°F			
ISOBUTYL ALCOHOL	A to 122°F	A to 200°F	A to 212°F	A to 70°F		
ISOBUTYL ALDEHYDE		A to 70°F		NR at 70°F		
ISOBUTYL AMINE		A to 70°F				
ISOBUTYL BROMIDE		A to 70°F				
ISOBUTYL n-BUTYRATE		A to 70°F		C at 70°F		
ISOBUTYL CARBINOL		A to 70°F				
ISOBUTYL CHLORIDE		A to 70°F	A to 212°F	NR at 70°F		
ISOBUTYL ETHER		A to 70°F		NR at 70°F		
ISOBUTYLENE	AB to 70°F	A to 70°F	A to 212°F	NR at 70°F		

	RECOMMENDATIONS	NITRILE NBR	ETHYLENE PROPYLENE EP, EPDM	FLUOROCARBON FKM	CHLOROPRENE CR	HYDROGENATED NITRILE HNBR
ISOBUTYL METHYL KETONE	EP	NR at 70°F	A to 200°F	NR at 70°F	NR at 70°F	
ISOBUTYRALDEHYDE	EP	NR at 70°F	A to 160°F	NR at 70°F	NR at 70°F	B at 70°F
ISOBUTYRIC ACID		NR at 70°F	A to 70°F		B at 70°F	B at 70°F
ISOCYANATES		AB to 70°F		AB to 70°F		
ISODODECANE	NBR,FKM,FVMQ	A to 70°F	NR at 70°F	A to 70°F	B at 70°F	A to 70°F
ISOOCTANE	NBR,FKM,ACM,FVMQ	+2% vol 3 days 70°F A to 250°F	+103% vol 3 days 70°F NR at 70°F	no swell 3 days 70°F A to 70°F	+14% vol 3 days 70°F B to 80°F	A to 70°F
ISOPAR G		AB to 70°F	NR at 70°F	A to 70°F	AB to 70°F	
ISOPENTANE	NBR	A to 250°F	NR at 70°F	A to 70°F	NR at 70°F	
ISOPHORONE	EP,IIR	NR at 70°F	A to 70°F	NR at 70°F	NR at 70°F	NR at 70°F
ISOPROPANOL	EP,IIR,FKM,NBR	A to 70°F B to 120°F	A to 160°F B at 176°F	A to 170°F B at 212°F	AB any conc to 120°F C at 140°F	B at 70°F
ISOPROPANOL AMINE		B at 70°F		*NR at 70°F		
ISOPROPYL ACETATE	EP,IIR,T	NR at 70°F	AB to 160°F	NR at 70°F	NR at 70°F	NR at 70°F
ISOPROPYL ALCOHOL	EP,IIR,FKM,NBR	A to 70°F B any conc to 130°F	A to 160°F B at 176°F	A to 170°F B at 212°F	AB any conc to 120°F C at 140°F	B at 70°F
ISOPROPYL AMINE		NR at 70°F		NR at 70°F		
ISOPROPYL BENZENE	FKM,FVMQ	NR at 70°F	A/NR at 70°F	A to 200°F	B/NR at 70°F	NR at 70°F
ISOPROPYL CARBINOL		B any conc to 80°F	A to 70°F	B at 70°F	B any conc to 80°F	
ISOPROPYL CHLORIDE	FKM,FVMQ	NR at 70°F	NR at 70°F	A to 160°F	NR at 70°F	NR at 70°F
ISOPROPYL DIENACETONE		NR at 70°F			NR at 70°F	
ISOPROPYL ETHER	NBR,T,FVMQ	AB any conc to 140°F	NR at 70°F	NR at 70°F	NR at 70°F	B at 70°F
ISOPROPYL METHYL BENZENE		NR at 70°F	NR at 70°F	A to 70°F	NR at 70°F	
ISOPROPYL TOLUENE		NR at 70°F	NR at 70°F	A to 70°F	NR at 70°F	
ISOTANE		A to 70°F		A to 70°F	NR at 70°F	
JAPANESE GELATIN		AB any conc to 180°F			AB any conc to 180°F	
JET FUEL (JP1 TO JP6, A & A1)	NBR,FKM,FVMQ	A to 70°F AB to 250°F	NR at 70°F	A to 400°F	NR at 70°F	A to 70°F
JET FUEL TYPE B (GASOLINE & KEROSENE)		AB to 100°F			NR at 70°F	
JEWS PITCH		A to 70°F	NR at 70°F	A to 70°F	B at 70°F	
JP3 (MIL-J-5624)	NBR,FKM,FVMQ	A to 70°F B at 140°F	NR at 70°F	A to 140°F	NR at 70°F	A to 70°F
JP4 (MIL-J-5624)	NBR,FKM,FVMQ	A to 200°F ground or airborn	NR at 70°F	+12% vol 3 days 400°F A to 400°F	NR at 70°F	A to 70°F
JP5 (MIL-J-5624)	NBR,FKM,FVMQ	A to 200°F ground or airborn	NR at 70°F	+4% vol 3 days 400°F NR at 500°F	NR at 70°F	A to 70°F
JP6 (MIL-J-25656)	NBR,FKM,FVMQ	A to 70°F AB to 120°F	NR at 70°F	+9% vol 7 days 400°F NR at 550°F	NR at 70°F	A to 70°F
JPX (MIL-F-25604)	NBR	A to 70°F AB to 120°F	NR at 70°F	NR at 70°F	B at 70°F	A to 70°F
KALILAUGE		C at 70°F	B at 70°F	C at 70°F	B at 70°F	
KALINITE		A to 70°F	A to 70°F	A to 70°F	A to 70°F	
KANDOL		A to 70°F	NR at 70°F	A to 70°F	B at 70°F	
KARO SYRUP						
KEL-F LIQUIDS	EP,NBR,SBR,IIR	A to 70°F	A to 70°F	B at 70°F		
KEROSENE	NBR,FKM,FVMQ,ACM AU	A to 250°F	NR at 70°F	+1.3% vol 7 days 70°F no swell 7 days 158°F	BC at 70°F NR at 140°F	A to 70°F
KEROSENE				+20% vol 28 days 300°F A to 158°F		
KEROSENE + NAPHTHA		A to 70°F	NR at 70°F	A to 70°F	NR at 70°F	
KETCHUP	NBR	A to 200°F	A to 70°F	A to 70°F	A to 70°F	
KETO HEXAMETHYLENE	EP	NR at 70°F	BC at 70°F	NR at 70°F	NR at 70°F	NR at 70°F
KETONES	EP,SBR,IIR,NR	NR at 70°F	A to 200°F	NR at 70°F	NR at 70°F	
KETONES: ALIPHATIC, SATURATED		NR at 70°F	A to 70°F	NR at 70°F	C at 70°F	
KETONES: ALIPHATIC, UNSATURATED		NR at 70°F	A to 70°F	NR at 70°F	NR at 70°F	

	STYRENE BUTADIENE SBR	POLYACRYLATE ACM	POLYURETHANE AU, EU	ISOBUTYLENE ISOPRENE IIR	POLYBUTADIENE BR	®AEROQUIP AQP
ISOBUTYL METHYL KETONE	B at 70°F	NR at 70°F	NR at 70°F	A to 70°F		
ISOBUTYRALDEHYDE	B at 70°F	NR at 70°F	NR at 70°F			
ISOBUTYRIC ACID				A to 70°F		
ISOCYANATES			AB to 70°F			
ISODODECANE	NR at 70°F	NR at 70°F	B at 70°F	NR at 70°F	NR at 70°F	
ISOOCTANE	+52% vol 3 days 70°F NR at 70°F	–2% vol 3 days 70°F A to 70°F	no swell 3 days 70°F B to 158°F	+111% vol 3 days 70°F NR at 70°F	NR at 70°F	
ISOPAR G		AB to 70°F	AB to 70°F	NR at 70°F		
ISOPENTANE	NR at 70°F	A to 70°F	B at 70°F	NR at 70°F		
ISOPHORONE	NR at 70°F	NR at 70°F	NR at 70°F	A to 70°F	NR at 70°F	
ISOPROPANOL	A to 160°F	NR at 70°F	NR at 70°F	A to 70°F AB any conc to 185°F	B at 70°F	
ISOPROPANOL AMINE				A to 70°F		
ISOPROPYL ACETATE	NR at 70°F	NR at 70°F	NR at 70°F	AB any conc to 80°F	NR at 70°F	
ISOPROPYL ALCOHOL	A to 160°F	NR at 70°F	NR at 70°F	A to 70°F AB any conc to 185°F	B at 70°F	
ISOPROPYL AMINE				B at 70°F		
ISOPROPYL BENZENE	NR at 70°F	NR at 70°F	NR at 70°F	NR at 70°F	NR at 70°F	
ISOPROPYL CARBINOL			NR at 70°F	A any conc to 100°F		
ISOPROPYL CHLORIDE	NR at 70°F	NR at 70°F	NR at 70°F	NR at 70°F	NR at 70°F	
ISOPROPYL DIENACETONE				NR at 70°F		
ISOPROPYL ETHER	NR at 70°F	C at 70°F	B at 70°F	NR at 70°F	NR at 70°F	
ISOPROPYL METHYL BENZENE				NR at 70°F		
ISOPROPYL TOLUENE				NR at 70°F		
ISOTANE						
JAPANESE GELATIN				A to 70°F AB any conc to 185°F		
JET FUEL (JP1 TO JP6, A & A1)	NR at 70°F	AB to 70°F	BC at 70°F	NR at 70°F	NR at 70°F	
JET FUEL TYPE B (GASOLINE & KEROSENE)	*NR at 70°F			NR at 70°F	*NR at 70°F	
JEWS PITCH			B at 70°F	NR at 70°F	NR at 70°F	
JP3 (MIL-J-5624)	NR at 70°F	B at 70°F	BC at 70°F	NR at 70°F	NR at 70°F	
JP4 (MIL-J-5624)	NR at 70°F	AB to 70°F ground NR at 70°F airborn	BC at 70°F	NR at 70°F	NR at 70°F	
JP5 (MIL-J-5624)	NR at 70°F	AB to 70°F ground NR at 70°F airborn	B at 70°F	NR at 70°F	NR at 70°F	
JP6 (MIL-J-25656)	NR at 70°F	B at 70°F	BC at 70°F	NR at 70°F	NR at 70°F	
JPX (MIL-F-25604)	NR at 70°F			NR at 70°F	NR at 70°F	
KALILAUGE			B at 70°F	A to 70°F		
KALINITE				A to 70°F		
KANDOL			B at 70°F	NR at 70°F		
KARO SYRUP						
KEL-F LIQUIDS	A to 70°F			A to 70°F		
KEROSENE	NR at 70°F	A to 70°F	AB to 110°F C at 140°F	NR at 70°F	NR at 70°F	AB to 70°F
KEROSENE						
KEROSENE + NAPHTHA			A to 70°F			
KETCHUP	A to 70°F	NR at 70°F	AB to 70°F			
KETO HEXAMETHYLENE				C/NR at 70°F		
KETONES	AB to 200°F	NR at 70°F	NR at 70°F	AB to 70°F	NR at 70°F	AB to 70°F
KETONES: ALIPHATIC, SATURATED			NR at 70°F	A to 70°F		
KETONES: ALIPHATIC, UNSATURATED			NR at 70°F	B at 70°F		

C 207

	SYNTHETIC ISOPRENE IR	NATURAL ISOPRENE NR	CHLOROSULFONATED POLYETHYLENE CSM	FLUOROSILICONE FVMQ	SILICONE VMQ	CHEMRAZ FFKM
ISOBUTYL METHYL KETONE				NR at 70°F	NR at 70°F	A to 70°F
ISOBUTYRALDEHYDE				NR at 70°F	NR at 70°F	B/NR at 70°F NR at 70°F white
ISOBUTYRIC ACID	A to 70°F	A to 70°F				A to 70°F
ISOCYANATES						
ISODODECANE	NR at 70°F	NR at 70°F	B at 70°F	A to 70°F	NR at 70°F	A to 70°F
ISOOCTANE	NR at 70°F	NR at 70°F	AB to 70°F	+3% vol 3 days 70°F +15% vol 7 days 70°F	+93% vol 3 days 70°F +150% vol 7 days 70°F	A to 70°F
ISOPAR G			C at 70°F		NR at 70°F	
ISOPENTANE	NR at 70°F	NR at 70°F	NR at 70°F	A to 70°F	NR at 70°F	A to 70°F
ISOPHORONE	NR at 70°F	NR at 70°F	NR at 70°F	NR at 70°F	NR at 70°F	A to 70°F
ISOPROPANOL	A to 70°F	A to 70°F AB any conc to 120°F	A to 200°F	B at 70°F	A to 70°F	A to 70°F
ISOPROPANOL AMINE	B at 70°F	B at 70°F	C at 70°F			
ISOPROPYL ACETATE	NR at 70°F	NR at 70°F	NR at 70°F	NR at 70°F	NR at 70°F	A to 70°F
ISOPROPYL ALCOHOL	AB to 70°F	A to 70°F AB any conc to 120°F	A to 200°F	B at 70°F	A to 70°F	A to 70°F
ISOPROPYL AMINE	B at 70°F	B at 70°F	C at 70°F			A to 70°F
ISOPROPYL BENZENE	NR at 70°F	NR at 70°F	NR at 70°F	AB to 70°F	NR at 70°F	A to 70°F
ISOPROPYL CARBINOL	B at 70°F	B any conc to 80°F	A to 70°F		A to 70°F	
ISOPROPYL CHLORIDE	NR at 70°F	NR at 70°F	NR at 70°F	B at 70°F	NR at 70°F	A to 70°F
ISOPROPYL DIENACETONE	NR at 70°F	NR at 70°F				
ISOPROPYL ETHER	NR at 70°F	NR at 70°F	BC at 70°F	C at 70°F	NR at 70°F	A to 70°F
ISOPROPYL METHYL BENZENE	NR at 70°F	NR at 70°F	NR at 70°F		NR at 70°F	
ISOPROPYL TOLUENE	NR at 70°F	NR at 70°F	NR at 70°F		NR at 70°F	
ISOTANE		NR at 70°F				
JAPANESE GELATIN	A to 70°F	A to 70°F AB any conc to 150°F				
JET FUEL (JP1 TO JP6, A & A1)	NR at 70°F	NR at 70°F	NR at 70°F	AB to 70°F	NR at 70°F	A to 70°F
JET FUEL TYPE B (GASOLINE & KEROSENE)	NR at 70°F	NR at 70°F	*C /NR at 70°F	*A to 70°F	*NR at 70°F	
JEWS PITCH	NR at 70°F	NR at 70°F	B at 70°F		NR at 70°F	
JP3 (MIL-J-5624)	NR at 70°F	NR at 70°F	NR at 70°F	AB to 70°F	NR at 70°F	A to 70°F
JP4 (MIL-J-5624)	NR at 70°F	NR at 70°F	NR at 70°F	+10% vol 7 days 70°F B at 70°F	+150% vol 7 days 70°F NR at 70°F	A to 70°F
JP5 (MIL-J-5624)	NR at 70°F	NR at 70°F	NR at 70°F	B at 70°F	NR at 70°F	A to 70°F
JP6 (MIL-J-25656)	NR at 70°F	NR at 70°F	NR at 70°F	B at 70°F	NR at 70°F	A to 70°F
JPX (MIL-F-25604)	NR at 70°F	NR at 70°F	NR at 70°F	NR at 70°F	NR at 70°F	A to 70°F
KALILAUGE	B at 70°F	B at 70°F	A to 70°F		A to 70°F	
KALINITE	A to 70°F	A to 70°F	A to 70°F		A to 70°F	
KANDOL	NR at 70°F	NR at 70°F	NR at 70°F		C at 70°F	
KARO SYRUP						
KEL-F LIQUIDS			A to 70°F	B at 70°F	A to 70°F	C at 70°F
KEROSENE	NR at 70°F	NR at 70°F	C /NR at 70°F	A to 70°F	NR at 70°F	A to 70°F
KEROSENE						
KEROSENE + NAPHTHA						
KETCHUP		AB to 70°F		A to 70°F	A to 70°F	
KETO HEXAMETHYLENE	NR at 70°F	NR at 70°F	NR at 70°F	NR at 70°F	NR at 70°F	B at 70°F
KETONES	A/NR at 70°F	A/NR at 70°F	B/NR at 70°F	NR at 70°F	A/NR at 70°F	
KETONES: ALIPHATIC, SATURATED	B at 70°F	B at 70°F	B at 70°F		A to 70°F	
KETONES: ALIPHATIC, UNSATURATED	NR at 70°F	NR at 70°F	NR at 70°F		C at 70°F	

C 208

	POLYSULFIDE T	CHLORINATED POLYETHYLENE CM	EPICHLOROHYDRIN CO, ECO	® HYTREL COPOLYESTER TPE	® SANTOPRENE COPOLYMER TPO	® C-FLEX STYRENIC TPE
ISOBUTYL METHYL KETONE						
ISOBUTYRALDEHYDE						
ISOBUTYRIC ACID						
ISOCYANATES				AB to 70°F		
ISODODECANE	A to 70°F					
ISOOCTANE	-0.2% vol 3 days 70°F A to 70°F	A to 70°F	A to 70°F	+1 to 8% vol 7 days 70°F A to 158°F	NR at 70°F	
ISOPAR G			C at 70°F			
ISOPENTANE						
ISOPHORONE	B at 70°F					
ISOPROPANOL	A to 70°F		A to 70°F	A to 70°F		
ISOPROPANOL AMINE						
ISOPROPYL ACETATE	B at 70°F			C at 70°F		
ISOPROPYL ALCOHOL	A to 70°F		A to 70°F	A to 70°F		
ISOPROPYL AMINE						
ISOPROPYL BENZENE	B at 70°F					
ISOPROPYL CARBINOL						
ISOPROPYL CHLORIDE	NR at 70°F					
ISOPROPYL DIENACETONE						
ISOPROPYL ETHER	A to 200°F					
ISOPROPYL METHYL BENZENE						
ISOPROPYL TOLUENE						
ISOTANE						
JAPANESE GELATIN						
JET FUEL (JP1 TO JP6, A & A1)	B at 70°F				NR at 70°F	
JET FUEL TYPE B (GASOLINE & KEROSENE)						
JEWS PITCH						
JP3 (MIL-J-5624)	B at 70°F		A to 70°F			
JP4 (MIL-J-5624)	B at 70°F		A to 70°F	+4% vol 7 days 212°F A to 100°F	NR at 70°F	
JP5 (MIL-J-5624)	B at 70°F		A to 70°F			
JP6 (MIL-J-25656)	B at 70°F		A to 70°F			
JPX (MIL-F-25604)						
KALILAUGE						
KALINITE						
KANDOL						
KARO SYRUP						
KEL-F LIQUIDS	A to 70°F					
KEROSENE	B at 70°F	AB to 70°F	A to 70°F	A to 70°F *AB to 140°F	NR at 70°F	NR at 70°F
KEROSENE						
KEROSENE + NAPHTHA						
KETCHUP						
KETO HEXAMETHYLENE						
KETONES	AB to 70°F	AB to 70°F	NR at 70°F	C/NR at 70°F	NR at 70°F	NR at 70°F A to 70°F water soluble
KETONES: ALIPHATIC, SATURATED						
KETONES: ALIPHATIC, UNSATURATED						

	ETHYLENE ACRYLIC EA	POLYALLOMER LINEAR COPOLYMER	NYLON 11 POLYAMIDE	NYLON 12 POLYAMIDE	ETHYLENE VINYL ACETATE EVA	POLYVINYLCHLORIDE FLEXIBLE PVC
ISOBUTYL METHYL KETONE						
ISOBUTYRALDEHYDE						
ISOBUTYRIC ACID						
ISOCYANATES			AB to 70°F		*NR at 70°F	NR at 70°F
ISODODECANE	A to 70°F					
ISOOCTANE	A to 70°F		A to 70°F	AB to 70°F	NR at 70°F	AB to 70°F C at 150°F
ISOPAR G						A to 70°F
ISOPENTANE						
ISOPHORONE						
ISOPROPANOL	A to 70°F	A to 122°F	A to 70°F NR at 140°F	C /NR at 70°F	B at 70°F	A to 70°F B at 122°F
ISOPROPANOL AMINE						
ISOPROPYL ACETATE	NR at 70°F	B at 70°F C at 122°F	AB to 70°F			NR at 70°F
ISOPROPYL ALCOHOL	A to 70°F	A to 122°F	A to 70°F NR at 140°F	C /NR at 70°F	B at 70°F	A to 70°F B at 122°F
ISOPROPYL AMINE						
ISOPROPYL BENZENE		C at 70°F NR at 122°F				NR at 70°F
ISOPROPYL CARBINOL						
ISOPROPYL CHLORIDE						
ISOPROPYL DIENACETONE						
ISOPROPYL ETHER	*NR at 70°F					A to 70°F
ISOPROPYL METHYL BENZENE						
ISOPROPYL TOLUENE						NR at 70°F
ISOTANE						
JAPANESE GELATIN						
JET FUEL (JP1 TO JP6, A & A1)			AC to 70°F			AC to 70°F
JET FUEL TYPE B (GASOLINE & KEROSENE)						
JEWS PITCH						
JP3 (MIL-J-5624)			C at 70°F			AC to 70°F NR at 150°F
JP4 (MIL-J-5624)			AC to 70°F		NR at 70°F	AC to 70°F NR at 130°F
JP5 (MIL-J-5624)						AC to 70°F NR at 150°F
JP6 (MIL-J-25656)						
JPX (MIL-F-25604)						NR at 70°F
KALILAUGE						
KALINITE						
KANDOL						
KARO SYRUP		A to 70°F	A to 70°F		A to 70°F	A to 70°F
KEL-F LIQUIDS						
KEROSENE	BC at 70°F	B at 70°F C at 122°F	A to 70°F swells at 140°F		NR at 70°F	AC to 70°F A to 122°F special
KEROSENE						
KEROSENE + NAPHTHA						
KETCHUP						
KETO HEXAMETHYLENE						NR at 70°F
KETONES	NR at 70°F	A to 70°F	A to 70°F		C /NR at 70°F	NR at 70°F
KETONES: ALIPHATIC, SATURATED						
KETONES: ALIPHATIC, UNSATURATED						

C210

	POLYETHYLENE LOW DENSITY LDPE	®TEFLON FEP	®KALREZ PERFLUORINATED ELASTOMER	® FLUORAZ FLUORINATED COPOLYMER	® AFLAS FLUORINATED COPOLYMER	®NORPRENE COPOLYMER TPO
ISOBUTYL METHYL KETONE			A to 212°F	NR at 70°F		
ISOBUTYRALDEHYDE				NR at 70°F		
ISOBUTYRIC ACID		A to 70°F	A to 212°F	C at 70°F		
ISOCYANATES	AB to 70°F					
ISODODECANE		A to 70°F		A to 70°F		
ISOOCTANE	B at 70°F	+1% wt gain 3 days 212°F A to 300°F	A to 212°F	BC at 70°F	+19% vol 7 days 70°F BC at 70°F	
ISOPAR G						
ISOPENTANE		A to 70°F	A to 212°F			
ISOPHORONE		A to 70°F	A to 212°F	B at 70°F	+15% vol 7 days 70°F B at 70°F	*NR at 70°F
ISOPROPANOL	A to 122°F AB to 140°F	A to 300°F	A to 212°F	A to 70°F		
ISOPROPANOL AMINE		A to 70°F	*A to 212°F			
ISOPROPYL ACETATE	B at 70°F C at 122°F	A to 125°F	A to 212°F	NR at 70°F		
ISOPROPYL ALCOHOL	A to 122°F AB at 140°F	A to 300°F	A to 212°F	A to 70°F		
ISOPROPYL AMINE		A to 70°F		C at 70°F		
ISOPROPYL BENZENE	C at 70°F NR at 122°F	A to 122°F	A to 70°F	C at 70°F		
ISOPROPYL CARBINOL		A to 70°F				
ISOPROPYL CHLORIDE		A to 70°F	A to 212°F	NR at 70°F		
ISOPROPYL DIENACETONE		A to 70°F				
ISOPROPYL ETHER	B/NR at 70°F NR at 140°F	A to 300°F	A to 212°F	NR at 70°F	+40% vol 7 days 70°F NR at 70°F	
ISOPROPYL METHYL BENZENE		A to 70°F				
ISOPROPYL TOLUENE		A to 70°F				
ISOTANE						
JAPANESE GELATIN		A to 70°F				
JET FUEL (JP1 TO JP6, A & A1)	NR at 70°F	A to 300°F	A to 212°F	BC at 70°F	+30% vol 3 days 212°F	
JET FUEL TYPE B (GASOLINE & KEROSENE)						
JEWS PITCH		A to 70°F				
JP3 (MIL-J-5624)	NR at 70°F	A to 300°F	A to 212°F	BC at 70°F	AB to 70°F	
JP4 (MIL-J-5624)	A/NR at 70°F NR at 140°F	A to 300°F	A to 212°F	BC at 70°F	+15% vol 3 days 70°F AB to 70°F	
JP5 (MIL-J-5624)	NR at 70°F	A to 300°F	A to 75°F B at 392°F	BC at 70°F	AB to 70°F	
JP6 (MIL-J-25656)			A to 212°F	BC at 70°F	AB to 70°F	
JPX (MIL-F-25604)				BC at 70°F		
KALILAUGE		A to 70°F				
KALINITE		A to 70°F				
KANDOL		A to 70°F				
KARO SYRUP	A to 70°F					
KEL-F LIQUIDS				C at 70°F		
KEROSENE	C at 70°F NR at 122°F	A to 300°F	A to 212°F *A to 400°F	AC to 70°F	+2-5% vol 7 days 70°F A to 70°F	NR at 70°F
KEROSENE						
KEROSENE + NAPHTHA						
KETCHUP						
KETO HEXAMETHYLENE	C/NR at 70°F	A to 312°F	A to 70°F	C at 70°F	B at 70°F	
KETONES	AC to 70°F B/NR at 122°F	A to 300°F	A to 212°F	NR at 70°F	NR at 70°F	B/NR at 70°F
KETONES: ALIPHATIC, SATURATED		A to 70°F				
KETONES: ALIPHATIC, UNSATURATED		A to 70°F				

	RECOMMENDATIONS	NITRILE NBR	ETHYLENE PROPYLENE EP, EPDM	FLUOROCARBON FKM	CHLOROPRENE CR	HYDROGENATED NITRILE HNBR
KETONES, AROMATIC		NR at 70°F	A to 70°F	NR at 70°F	NR at 70°F	
KETOPROPANE		NR at 70°F			NR at 70°F	
KEYSTONE NO. 87HX GREASE	NBR, FKM, ACM, AU	A to 70°F	NR at 70°F	A to 70°F	NR at 70°F	A to 70°F
KRYPTON					AB to 70°F	
KRYSTALLIN		NR at 70°F	B at 70°F	A to 70°F	C at 70°F	
KURROL'S SALT		A to 70°F AB any conc to 200°F			A to 70°F AB any conc to 200°F	
KYANOL		NR at 70°F	B at 70°F	A to 70°F	C at 70°F	
LABARRAQUE'S SOLUTION		NR 20% at 70°F			NR 20% at 70°F	
LACQUER SOLVENTS	T, IIR	NR at 70°F	NR at 70°F	NR at 70°F	NR at 70°F	NR at 70°F
LACQUER SOLVENTS, SYNTHETIC		NR at 70°F	NR at 70°F	NR at 70°F	NR at 70°F	NR at 70°F
LACQUERS	T, IIR	NR at 70°F	NR at 70°F	+91% vol 7 days 75°F NR at 70°F	NR at 70°F	NR at 70°F
LACQUERS, SYNTHETIC		NR at 70°F	NR at 70°F	NR at 70°F	NR at 70°F	
LACQUERS, WITH KETONE SOLVENTS	EP, SBR	NR at 70°F	A to 160°F	NR at 70°F	NR at 70°F	NR at 70°F
LACTAMS (AMINO ACIDS)	EP	NR at 70°F	B at 70°F	NR at 70°F	B at 70°F	NR at 70°F
LACTIC ACID	NBR, EP, FKM, CR	A to 100% to 70°F A/NR 100% at 200°F	A to 100% to 140°F AB 100% 176-200°F	A to 100% to 140°F A to 80% to 176°F	A to 100% to 70°F AC 100% at 140°F	
LACTIC ACID		B 25-80% at 104°F C 25-80% at 140°F	A to 80% to 176°F B 80% at 212°F	B 80% at 212°F A to 25% to 212°F	C/NR 100% at 180°F NR 10% at 212°F	
LACTIC ACID		NR 5% at 150°F	A to 25% to 212°F			
LACTOL		AB to 100°F		A to 70°F	B/NR at 70°F	
LACTONES (CYCLIC ESTERS)	EP	NR at 70°F	B at 70°F	NR at 70°F	NR at 70°F	
LARD, ANIMAL FAT	NBR, FKM, ACM, AU, EP VMQ	A to 200°F	AB to 140°F	A to 140°F	B at 70°F C at 140°F	A to 70°F
LARVACIDE		NR at 70°F		NR at 70°F	NR at 70°F	
LATEX	NBR	A to 200°F	AB to 70°F	A to 140°F	A to 70°F C at 140°F	
LAUGHING GAS (NITROUS OXIDE)	NBR	A to 70°F	A to 140°F B at 176°F	A to 176°F B 212-248°F	AB to 70°F	
LAURIC ACID		NR at 70°F	C at 70°F	A to 70°F	A to 70°F	
LAURYL ALCOHOL		A to 70°F		B at 70°F		
LAVENDER OIL	FKM	B at 70°F	NR at 70°F	A to 70°F	NR at 70°F	B at 70°F
LAYOR CARANGA		B any conc to 180°F			B any conc to 180°F	
LEAD ACETATE	EP, IIR, NBR	AB any conc to 140°F B at 176°F	A to 200°F	A to 140°F B at 176°F	AB any conc to 80°F	B at 70°F
LEAD ARSENATE		AB to 70°F			AB to 70°F	
LEAD CHLORIDE		A to 140°F B at 176°F	A to 176°F	A to 212°F	B at 70°F	
LEAD NITRATE	NBR, EP, IIR, CR, SBR	A any conc to 120°F AB 176-200°F	A to 176°F	A to 212°F	A to 70°F AB any conc to 120°F	A to 70°F
LEAD SALTS		AB to 70°F	A to 176°F	A to 212°F	AB to 70°F	
LEAD STYPHNATE		B at 70°F			A to 70°F	
LEAD SULFAMATE	EP, CR, FKM, IIR	B any conc to 140°F	A to 140°F	A to 140°F	A to 140°F AB any conc to 180°F	
LEAD SULFATE		A any conc to 120°F B at 176°F	A to 176°F	A to 212°F	AB any conc to 180°F	
LEAD TETRAETHYL		B at 70°F	NR at 70°F	A to 70°F	NR at 70°F	
LEAD TETRAMETHYL						
LEAD TRINITRORESORCINOL		B any conc to 120°F			A to 70°F AB any conc to 180°F	
LEHIGH X1169	NBR, FKM, ACM, AU	A to 70°F	NR at 70°F	+11.3% vol 28 days 300°F A to 70°F	B at 70°F	A to 70°F
LEHIGH X1170	NBR, FKM, ACM, AU	A to 70°F	NR at 70°F	+4.8% vol 28 days 300°F A to 70°F	B at 70°F	A to 70°F
LEMON OIL		A to 70°F	C at 70°F	A to 70°F	AC to 70°F	
LEUCOGEN		A to 70°F AB any conc to 160°F	A to 70°F	A to 70°F	A to 70°F AB any conc to 200°F	
LICHENIC ACID		NR at 70°F				
LIGHT ANILINE		NR at 70°F	B at 70°F	A to 70°F	C at 70°F	

	STYRENE BUTADIENE SBR	POLYACRYLATE ACM	POLYURETHANE AU, EU	ISOBUTYLENE ISOPRENE IIR	POLYBUTADIENE BR	® AEROQUIP AQP
KETONES, AROMATIC			NR at 70°F	A to 70°F		
KETOPROPANE				A to 70°F AB any conc to 150°F		
KEYSTONE NO. 87HX GREASE	NR at 70°F	A to 70°F	A to 70°F	NR at 70°F	NR at 70°F	
KRYPTON				AB to 70°F		
KRYSTALLIN			C at 70°F	A to 70°F		
KURROL'S SALT				A to 70°F AB any conc to 185°F		
KYANOL			C at 70°F	A to 70°F		
LABARRAQUE'S SOLUTION			NR 20% at 70°F	C/NR 20% at 70°F		
LACQUER SOLVENTS	NR at 70°F	NR at 70°F	NR at 70°F	NR at 70°F	NR at 70°F	TEST
LACQUER SOLVENTS, SYNTHETIC			NR at 70°F	NR at 70°F		TEST
LACQUERS	NR at 70°F	NR at 70°F	NR at 70°F	NR at 70°F	NR at 70°F	TEST
LACQUERS, SYNTHETIC	NR at 70°F	NR at 70°F	NR at 70°F	NR at 70°F		TEST
LACQUERS, WITH KETONE SOLVENTS	A to 160°F	NR at 70°F	NR at 70°F			
LACTAMS (AMINO ACIDS)	NR at 70°F			B at 70°F	NR at 70°F	
LACTIC ACID	A to 70°F NR at 140°F	NR at 70°F	B at 70°F NR at 140°F	A to 70°F NR at 150°F	A to 70°F NR at 140°F	NR at 70°F
LACTIC ACID			AB 10% to 70°F liquid or fumes	AB 50% to 150°F		
LACTIC ACID			NR 5% at 150°F			
LACTOL				NR at 70°F		
LACTONES (CYCLIC ESTERS)	NR at 70°F	NR at 70°F	NR at 70°F	B at 70°F	NR at 70°F	
LARD, ANIMAL FAT	NR at 70°F	A to 70°F	AC to 70°F	BC at 70°F	NR at 70°F	
LARVACIDE			NR at 70°F			
LATEX	A to 70°F	NR at 70°F	NR at 70°F			
LAUGHING GAS (NITROUS OXIDE)			AB to 70°F	A to 70°F		
LAURIC ACID						
LAURYL ALCOHOL				A to 70°F		
LAVENDER OIL	NR at 70°F	B at 70°F	NR at 70°F	NR at 70°F	NR at 70°F	
LAYOR CARANGA				A to 70°F AB any conc to 185°F		
LEAD ACETATE	NR at 70°F	NR at 70°F	B/NR at 70°F	A to 70°F AB any conc to 120°F	NR at 70°F	
LEAD ARSENATE			AB to 70°F			
LEAD CHLORIDE						
LEAD NITRATE	A to 70°F			A to 70°F AB any conc to 180°F		
LEAD SALTS	A to 70°F		AB to 70°F	A to 70°F AB any conc to 180°F		
LEAD STYPHNATE				A to 70°F		
LEAD SULFAMATE	B at 70°F	NR at 70°F		A to 70°F AB any conc to 185°F	B at 70°F	
LEAD SULFATE			AB to 70°F	A to 70°F AB any conc to 185°F		
LEAD TETRAETHYL				NR at 70°F		
LEAD TETRAMETHYL			AB to 70°F			
LEAD TRINITRORESORCINOL				A to 70°F AB any conc to 185°F		
LEHIGH X1169	NR at 70°F	A to 70°F	A to 70°F	NR at 70°F	NR at 70°F	
LEHIGH X1170	NR at 70°F	A to 70°F	A to 70°F	NR at 70°F	NR at 70°F	
LEMON OIL						
LEUCOGEN				A to 70°F AB any conc to 185°F		
LICHENIC ACID						
LIGHT ANILINE			C at 70°F	A to 70°F		

C213

	SYNTHETIC ISOPRENE IR	NATURAL ISOPRENE NR	CHLOROSULFONATED POLYETHYLENE CSM	FLUOROSILICONE FVMQ	SILICONE VMQ	CHEMRAZ FFKM
KETONES, AROMATIC	C at 70°F	C at 70°F	NR at 70°F			
KETOPROPANE	NR at 70°F	NR at 70°F				
KEYSTONE NO. 87HX GREASE	NR at 70°F	NR at 70°F	NR at 70°F	A to 70°F	NR at 70°F	A to 70°F
KRYPTON						
KRYSTALLIN	NR at 70°F	NR at 70°F	C at 70°F		B at 70°F	
KURROL'S SALT	A to 70°F	A to 70°F AB any conc to 150°F				
KYANOL	NR at 70°F	NR at 70°F	C at 70°F		B at 70°F	
LABARRAQUE'S SOLUTION	NR 20% at 70°F	NR 20% at 70°F				
LACQUER SOLVENTS	NR at 70°F	NR at 70°F	NR at 70°F	NR at 70°F	NR at 70°F	A to 70°F
LACQUER SOLVENTS, SYNTHETIC	NR at 70°F	NR at 70°F	NR at 70°F		NR at 70°F	
LACQUERS	NR at 70°F	NR at 70°F	NR at 70°F	NR at 70°F	NR at 70°F	A to 70°F
LACQUERS, SYNTHETIC	NR at 70°F	NR at 70°F	NR at 70°F		NR at 70°F	
LACQUERS, WITH KETONE SOLVENTS				NR at 70°F	NR at 70°F	
LACTAMS (AMINO ACIDS)	NR at 70°F	NR at 70°F	B at 70°F	NR at 70°F		A to 70°F
LACTIC ACID	A to 70°F NR at 140°F	A to 70°F NR at 140°F	A to 70°F BC at 200°F	A to 70°F B at 140°F	A to 70°F B at 140°F	A to 140°F
LACTIC ACID		AB 50% to 120°F				
LACTIC ACID						
LACTOL	NR at 70°F	NR at 70°F	B at 70°F			
LACTONES (CYCLIC ESTERS)	NR at 70°F	NR at 70°F	NR at 70°F	NR at 70°F	B at 70°F	
LARD, ANIMAL FAT	NR at 70°F	NR at 70°F	B at 70°F	A to 70°F	+1% vol 7 days 70°F AB to 70°F	A to 70°F
LARVACIDE						
LATEX		AB to 70°F	C at 70°F	A to 70°F	A to 70°F	
LAUGHING GAS (NITROUS OXIDE)	A to 70°F	A to 70°F	A to 70°F			A to 70°F
LAURIC ACID		NR at 70°F	C at 70°F			A to 70°F
LAURYL ALCOHOL	A to 70°F	A to 70°F	A to 70°F			
LAVENDER OIL	NR at 70°F	NR at 70°F	NR at 70°F	B at 70°F	NR at 70°F	A to 70°F
LAYOR CARANGA	A to 70°F	A to 70°F AB any conc to 150°F				
LEAD ACETATE	AB to 70°F	A any conc to 80°F AB to 185°F	A/NR at 70°F	NR at 70°F	NR at 70°F	A to 70°F
LEAD ARSENATE			AB to 70°F			A to 70°F
LEAD CHLORIDE	B at 70°F	B at 70°F	A to 70°F			A to 70°F
LEAD NITRATE	A to 70°F	A to 70°F AB any conc to 120°F	A/NR at 70°F	A to 70°F	B at 70°F	A to 70°F
LEAD SALTS	AB to 70°F	AB to 70°F	AB to 70°F	A to 70°F	B at 70°F	A to 70°F
LEAD STYPHNATE	B at 70°F	B any conc to 120°F				
LEAD SULFAMATE	B at 70°F	B at 70°F	A to 70°F	A to 70°F	B at 70°F	A to 70°F
LEAD SULFATE	A to 70°F	A to 70°F AB any conc to 120°F	A to 70°F			
LEAD TETRAETHYL	NR at 70°F	NR at 70°F	NR at 70°F			
LEAD TETRAMETHYL						
LEAD TRINITRORESORCINOL	B at 70°F	B any conc to 120°F				
LEHIGH X1169	NR at 70°F	NR at 70°F	B at 70°F	A to 70°F	NR at 70°F	A to 70°F
LEHIGH X1170	NR at 70°F	NR at 70°F	B at 70°F	A to 70°F	NR at 70°F	A to 70°F
LEMON OIL		C/NR at 70°F	A to 70°F			
LEUCOGEN	A to 70°F	A to 70°F AB any conc to 150°F	A to 70°F			
LICHENIC ACID	C at 70°F	BC any conc to 80°F				
LIGHT ANILINE	NR at 70°F	NR at 70°F	C at 70°F		B at 70°F	

	POLYSULFIDE T	CHLORINATED POLYETHYLENE CM	EPICHLOROHYDRIN CO, ECO	® HYTREL COPOLYESTER TPE	® SANTOPRENE COPOLYMER TPO	® C-FLEX STYRENIC TPE
KETONES, AROMATIC						
KETOPROPANE						
KEYSTONE NO. 87HX GREASE	A to 70°F					
KRYPTON						
KRYSTALLIN						
KURROL'S SALT						
KYANOL						
LABARRAQUE'S SOLUTION						
LACQUER SOLVENTS	A to 70°F	C at 70°F	NR at 70°F	B/NR at 70°F		NR at 70°F
LACQUER SOLVENTS, SYNTHETIC		C at 70°F		BC at 70°F		NR at 70°F
LACQUERS	A to 70°F	C at 70°F	NR at 70°F	NR at 70°F		
LACQUERS, SYNTHETIC		C at 70°F		NR at 70°F		
LACQUERS, WITH KETONE SOLVENTS						
LACTAMS (AMINO ACIDS)	NR at 70°F					
LACTIC ACID	NR at 70°F	NR at 70°F		NR at 70°F		A to 70°F
LACTIC ACID						
LACTIC ACID						
LACTOL						
LACTONES (CYCLIC ESTERS)	B at 70°F					
LARD, ANIMAL FAT	NR at 70°F		A to 70°F	AB to 70°F	+1% vol 7 days 212°F A to 212°F	
LARVACIDE						
LATEX						
LAUGHING GAS (NITROUS OXIDE)						
LAURIC ACID						
LAURYL ALCOHOL						
LAVENDER OIL	B at 70°F					
LAYOR CARANGA						
LEAD ACETATE	NR at 70°F		B at 70°F			
LEAD ARSENATE				AB to 70°F		
LEAD CHLORIDE						
LEAD NITRATE	NR at 70°F					
LEAD SALTS	NR at 70°F			AB to 70°F		
LEAD STYPHNATE						
LEAD SULFAMATE	NR at 70°F					
LEAD SULFATE				AB to 70°F		
LEAD TETRAETHYL						
LEAD TETRAMETHYL				AB to 70°F		
LEAD TRINITRORESORCINOL						
LEHIGH X1169	A to 70°F					
LEHIGH X1170	A to 70°F					
LEMON OIL						
LEUCOGEN						
LICHENIC ACID						
LIGHT ANILINE						

	ETHYLENE ACRYLIC EA	POLYALLOMER LINEAR COPOLYMER	NYLON 11 POLYAMIDE	NYLON 12 POLYAMIDE	ETHYLENE VINYL ACETATE EVA	POLYVINYLCHLORIDE FLEXIBLE PVC
KETONES, AROMATIC						
KETOPROPANE						NR at 70°F
KEYSTONE NO. 87HX GREASE						
KRYPTON						AB to 70°F
KRYSTALLIN						
KURROL'S SALT						
KYANOL						
LABARRAQUE'S SOLUTION						
LACQUER SOLVENTS	*NR at 70°F	A to 70°F	A to 70°F		B/NR at 70°F	NR at 70°F
LACQUER SOLVENTS, SYNTHETIC	*NR at 70°F	A to 70°F	A to 70°F		B/NR at 70°F	NR at 70°F
LACQUERS		A to 70°F	A to 70°F		B/NR at 70°F	NR at 70°F
LACQUERS, SYNTHETIC		A to 70°F	A to 70°F		B/NR at 70°F	NR at 70°F
LACQUERS, WITH KETONE SOLVENTS						
LACTAMS (AMINO ACIDS)						
LACTIC ACID		A 85-100% to 70°F B 85% at 122°F	A to 70°F AB at 140°F	C 50-90% at 70°F AB 5% to 70°F	B/NR 100% at 70°F AB 90% to 140°F	AB to 100% to 70°F C to 85% at 122°F
LACTIC ACID			C at 194°F			
LACTIC ACID						
LACTOL						
LACTONES (CYCLIC ESTERS)						
LARD, ANIMAL FAT	A to 70°F	A to 70°F	A to 70°F		B at 70°F	AB to 70°F C at 150°F
LARVACIDE						
LATEX						A to 70°F
LAUGHING GAS (NITROUS OXIDE)			B/NR at 70°F			AB to 70°F
LAURIC ACID						
LAURYL ALCOHOL						AB to 70°F
LAVENDER OIL						
LAYOR CARANGA						
LEAD ACETATE		A to 70°F	A to 70°F		A to 70°F AB to 140°F	AB to 70°F *AB to 140°F
LEAD ARSENATE			AB to 70°F			A to 70°F
LEAD CHLORIDE						A to 70°F
LEAD NITRATE						A to 70°F
LEAD SALTS			AB to 70°F	AB to 70°F		A to 70°F
LEAD STYPHNATE						
LEAD SULFAMATE						A to 70°F
LEAD SULFATE			AB to 70°F			A to 70°F
LEAD TETRAETHYL			AB to 70°F		*AB to 70°F *NR at 70°F	*AB to 70°F
LEAD TETRAMETHYL			AB to 70°F			
LEAD TRINITRORESORCINOL						
LEHIGH X1169						
LEHIGH X1170						
LEMON OIL						A to 70°F B at 150°F
LEUCOGEN						
LICHENIC ACID						
LIGHT ANILINE						

C216

	POLYETHYLENE LOW DENSITY LDPE	®TEFLON FEP	®KALREZ PERFLUORINATED ELASTOMER	®FLUORAZ FLUORINATED COPOLYMER	®AFLAS FLUORINATED COPOLYMER	®NORPRENE COPOLYMER TPO
KETONES, AROMATIC		A to 70°F				
KETOPROPANE		A to 70°F				
KEYSTONE NO. 87HX GREASE				A to 70°F		
KRYPTON	AB to 70°F	A to 70°F				
KRYSTALLIN		A to 70°F				
KURROL'S SALT		A to 70°F				
KYANOL		A to 70°F				
LABARRAQUE'S SOLUTION		A 20% to 70°F				
LACQUER SOLVENTS	AB to 70°F	A to 70°F	A to 212°F	NR at 70°F	+53% vol 7 days 70°F C/NR at 70°F	NR at 70°F
LACQUER SOLVENTS, SYNTHETIC	AB to 70°F	A to 70°F				NR at 70°F
LACQUERS	AB to 70°F	A to 70°F	A to 212°F	NR at 70°F		
LACQUERS, SYNTHETIC	AB to 70°F	A to 70°F				
LACQUERS, WITH KETONE SOLVENTS		A to 70°F	A to 212°F	NR at 70°F		
LACTAMS (AMINO ACIDS)				C at 70°F		
LACTIC ACID	AB to 100% to 140°F A to 85% to 122°F	A to 100% to 300°F	A to 212°F	A to 100% to 140°F		A to 70°F
LACTIC ACID						
LACTIC ACID						
LACTOL		A to 70°F				
LACTONES (CYCLIC ESTERS)						
LARD, ANIMAL FAT	A to 70°F	A to 300°F	A to 212°F	A to 70°F		
LARVACIDE			A to 212°F			
LATEX	A/NR at 70°F	A to 70°F		A to 70°F		
LAUGHING GAS (NITROUS OXIDE)	BC at 70°F	A to 300°F	A to 212°F			
LAURIC ACID			A to 212°F			
LAURYL ALCOHOL		A to 70°F				
LAVENDER OIL		A to 70°F	A to 212°F	A to 70°F		
LAYOR CARANGA		A to 70°F				
LEAD ACETATE	A to 70°F AB at 140°F	A to 300°F	A to 212°F	NR at 70°F		A to 70°F
LEAD ARSENATE	AB to 70°F	A to 70°F	A to 212°F			
LEAD CHLORIDE		A to 70°F	A to 212°F			A to 70°F
LEAD NITRATE		A to 125°F	A to 212°F	BC at 70°F		A to 70°F
LEAD SALTS	AB to 70°F	A to 70°F	A to 212°F	C at 70°F		A to 70°F
LEAD STYPHNATE		A to 70°F				
LEAD SULFAMATE		A to 125°F	A to 212°F	A to 70°F		A to 70°F
LEAD SULFATE	AB to 70°F	A to 70°F				
LEAD TETRAETHYL		A to 70°F				
LEAD TETRAMETHYL						
LEAD TRINITRORESORCINOL		A to 70°F				
LEHIGH X1169				A to 70°F		
LEHIGH X1170				A to 70°F		
LEMON OIL		NR at 70°F				
LEUCOGEN		A to 70°F				
LICHENIC ACID		A to 70°F				
LIGHT ANILINE		A to 70°F				

	RECOMMENDATIONS	NITRILE NBR	ETHYLENE PROPYLENE EP, EPDM	FLUOROCARBON FKM	CHLOROPRENE CR	HYDROGENATED NITRILE HNBR
LIGHT GREASE		A to 70°F	NR at 70°F	A to 70°F	NR at 70°F	A to 70°F
LIGROIN (PETROLEUM ETHER; BENZINE)	NBR,FKM,FVMQ,ACM	A to 140°F	NR at 70°F	A to 140°F AB to 260°F	B/NR at 70°F	A to 70°F
LIME (CALCIUM OXIDE)	EP,IIR,NBR	A to 140°F AB any conc to 200°F	A to 140°F	A to 100% to 140°F	A to 140°F AB any conc to 200°F	
LIME ACETATE		B any conc to 120°F			NR at 70°F	
LIME, AGRICULTURAL		A to 70°F	A to 70°F	A to 70°F	A to 70°F	
LIME BISULFITE		A to 70°F	C at 70°F	A to 70°F	A to 70°F	
LIME BLEACH	NBR,EP,IIR,FKM	A to 70°F	A to 70°F	A to 70°F	B at 70°F	A to 70°F
LIME, CAUSTIC		A to 70°F	A to 70°F	A to 70°F	A to 70°F	
LIME HYDRATE		A to 70°F AB any conc to 200°F	A to 70°F	A to 70°F	A to 70°F AB any conc to 200°F	
LIME NITRATE		A to 70°F	A to 70°F	A to 70°F	A to 70°F	
LIME SALTPETER		A to 70°F AB any conc to 200°F	A to 70°F	A to 70°F	A to 70°F AB any conc to 200°F	
LIME, SODA		B at 70°F	A to 70°F	B at 70°F	B at 70°F	
LIME SLURRIES		A to 70°F	AC to 70°F	A/NR at 70°F	A to 70°F	
LIME SULFUR-DRY	EP,FKM,IIR,CR	A to 140°F AB to 200°F	A to 176°F	A to 212°F	A to 70°F AB to 150°F	A to 70°F
LIME SULFUR-WET	EP	A to 140°F AB any conc to 200°F	C at 70°F	A to 70°F	A to 70°F AB any conc to 150°F	
LIME + WATER (CALCIUM HYDROXIDE)	NBR,EP,CR,FKM,SBR	A to 140°F AB any conc to 200°F	A to 176°F B at 212°F	A to 212°F	A to 158°F AB any conc to 200°F	A to 70°F
LIME WATER (MILK OF LIME)	NBR	A to 160°F	A to 70°F	A to 70°F	AB to 70°F	
LIMESTONE		A to 70°F	A to 70°F	A to 70°F	A to 70°F	
LIMONENE		C/NR at 70°F	NR at 70°F	A to 70°F	NR at 70°F	
LINDOIL		NR at 70°F			NR at 70°F	
LINDOL (HYDR. FLUID, PHOSPHATE ESTER)	EP,IIR,FKM	NR at 70°F	A to 250°F	+66% vol 28 days 250°F B at 70°F	NR at 70°F	A to 70°F
LINOLEIC ACID (LINOLENIC ACID)	VMQ	B to 80°F	NR at 70°F	AB to 70°F	B/NR at 70°F	B at 70°F
LINSEED CAKE						
LINSEED OIL	NBR,FKM,FVMQ,ACM	+5% vol 3 days 212°F A to 200°F	+22% vol 3 days 212°F BC at 70°F	no swell 3 days 212°F A to 250°F	+40% vol 3 days 212°F A to 70°F	A to 70°F
LIQUID OXYGEN	VMQ,FKM,IIR	NR	NR	NR	NR	NR
LIQUID ROSIN		A to 70°F AB to 150°F		A to 70°F	A to 70°F	
LIQUID SOAP		A to 70°F	A to 70°F	A to 70°F	A to 70°F	
LIQUIFIED PETROLEUM GAS	NBR,FKM,FVMQ,T	A to 250°F gas or liquid	NR at 70°F gas or liquid	A to 176°F gas or liquid	AC to 140°F gas or liquid	A to 70°F
LIQUIMOLY	NBR,FKM,ACM	A to 70°F	NR at 70°F	A to 70°F	B at 70°F	A to 70°F
LITHIUM BROMIDE BRINE	NBR	A to 160°F	A to 70°F	A to 70°F A to 60% to 212°F	A/NR at 70°F	
LITHIUM CHLORIDE	NBR	A to 160°F	A to 70°F	A to 70°F	A to 70°F	
LITHIUM HYDROXIDE	EP	B/NR at 70°F	A to 160°F	AC to 70°F A 5% to 70°F	A/NR at 70°F	
LUBOIL (LUBE OILS)		A to 140°F	NR at 70°F	A to 140°F	B to 120°F NR at 140°F	
LUBRICATING OILS (CRUDE & REFINED)	NBR,FKM,FVMQ	A to 250°F	NR at 70°F	A to 158°F B 176-212°F	B to 158°F	B at 70°F
LUBRICATING OILS, DIESTER	FKM,NBR	B at 70°F	NR at 70°F	A to 70°F	C at 70°F	
LUBRICATING OILS, PETROLEUM	NBR,FKM,ACM	A to 70°F	NR at 70°F	A to 158°F	B to 158°F	NR at 70°F
LUBRICATING OILS, SAE 10, 20, 30, 40 & 50	NBR,FKM,ACM	A to 250°F	NR at 70°F	A to 70°F	B/NR at 70°F	NR at 70°F
LYE	EP,IIR,NBR,CR	NR 80-100% at 70°F A to 50% to 176°F	A to 100% to 70°F A to 50% to 176°F	B 100% at 70°F AC to 50% to 140°F	A to 100% to 70°F A to 50% to 140°F	B at 70°F
LYSOL		AB to 150°F			AB to 150°F	
MACASSAR GUM		AB any conc to 180°F			AB any conc to 180°F	
MACHINE OIL		A to 104°F B at 140°F	NR at 70°F	A to 140°F	NR at 70°F	
MADDRELL'S SALT		AB any conc to 200°F	A to 70°F	A to 70°F	AB any conc to 200°F	
MAGNESIUM ACETATE	EP	NR any conc at 70°F	A sol'n to 120°F	NR any conc at 70°F	NR sol'n at 70°F	
MAGNESIUM AMMONIUM SULFATE			A to 70°F	A to 70°F	A to 70°F	

	STYRENE BUTADIENE SBR	POLYACRYLATE ACM	POLYURETHANE AU, EU	ISOBUTYLENE ISOPRENE IIR	POLYBUTADIENE BR	® AEROQUIP AQP
LIGHT GREASE	NR at 70°F	A to 70°F	A to 70°F	NR at 70°F	NR at 70°F	
LIGROIN (PETROLEUM ETHER; BENZINE)	NR at 70°F	A to 70°F	BC at 70°F	NR at 70°F	NR at 70°F	AB to 70°F
LIME (CALCIUM OXIDE)			B to 100% at 70°F	A to 70°F AB any conc to 185°F		
LIME ACETATE				A to 70°F AB any conc to 180°F		
LIME, AGRICULTURAL			A to 70°F	A to 70°F		
LIME BISULFITE			A to 70°F	B at 70°F		
LIME BLEACH	A to 70°F	NR at 70°F		A to 70°F AB any conc to 150°F	AB to 70°F	
LIME, CAUSTIC			A to 70°F	A to 70°F		
LIME HYDRATE			A to 70°F	A to 70°F AB any conc to 185°F		
LIME NITRATE			A to 70°F	A to 70°F		
LIME SALTPETER			A to 70°F	A to 70°F AB any conc to 185°F		
LIME, SODA			C at 70°F	A to 70°F		
LIME SLURRIES			B at 70°F			
LIME SULFUR-DRY	B/NR at 70°F	NR at 70°F	A to 70°F	A to 70°F AB to 185°F	B/NR at 70°F	
LIME SULFUR-WET				A to 70°F AB any conc to 185°F		
LIME + WATER (CALCIUM HYDROXIDE)	A to 70°F	NR at 70°F	A/NR at 70°F NR 10% boiling	A to 70°F AB any conc to 185°F	A to 70°F	AB to 70°F
LIME WATER (MILK OF LIME)	NR at 70°F	NR at 70°F				
LIMESTONE				A to 70°F		
LIMONENE				NR at 70°F		
LINDOIL				NR at 70°F		
LINDOL (HYDR. FLUID, PHOSPHATE ESTER)	NR at 70°F	NR at 70°F	NR at 70°F	A to 70°F	NR at 70°F	
LINOLEIC ACID (LINOLENIC ACID)	NR at 70°F			NR at 70°F	NR at 70°F	
LINSEED CAKE			AB to 70°F			
LINSEED OIL	+56% vol 3 days 212°F NR at 70°F	+4% vol 3 days 212°F A to 70°F	no swell 3 days 212°F AB to 140°F	+11% vol 3 days 212°F AB to 150°F	NR at 70°F	AB to 70°F
LIQUID OXYGEN	NR	NR	NR	NR	NR	
LIQUID ROSIN				NR at 70°F		
LIQUID SOAP				A to 70°F		
LIQUIFIED PETROLEUM GAS	NR at 70°F	A/NR at 70°F	AC to 70°F	NR at 70°F	NR at 70°F	
LIQUIMOLY	NR at 70°F	A to 70°F	B at 70°F	NR at 70°F	NR at 70°F	
LITHIUM BROMIDE BRINE	A to 70°F	NR at 70°F	NR at 70°F			
LITHIUM CHLORIDE	A to 70°F	NR at 70°F	NR at 70°F			
LITHIUM HYDROXIDE	NR at 70°F	NR at 70°F	NR at 70°F			
LUBOIL (LUBE OILS)			B at 70°F	NR at 70°F		
LUBRICATING OILS (CRUDE & REFINED)	NR at 70°F	A to 70°F	AB to 70°F	NR at 70°F		A to 70°F
LUBRICATING OILS, DIESTER	NR at 70°F	B at 70°F	NR at 70°F	NR at 70°F	NR at 70°F	A to 70°F
LUBRICATING OILS, PETROLEUM	NR at 70°F	A to 70°F	B at 70°F	NR at 70°F	NR at 70°F	A to 70°F
LUBRICATING OILS, SAE 10, 20, 30, 40 & 50	NR at 70°F	A to 70°F	A to 70°F AB to 158°F	NR at 70°F	NR at 70°F	A to 70°F
LYE	A conc to 70°F B sol'n at 70°F	A/NR conc at 70°F NR sol'n at 70°F	B to 100% at 70°F	A to 100% to 70°F A to 80% to 175°F	A to 100% to 70°F	C 50% at 70°F
LYSOL				A to 70°F AB to 150°F		
MACASSAR GUM				A to 70°F AB any conc to 185°F		
MACHINE OIL			AB to 70°F			
MADDRELL'S SALT			A to 70°F	A to 70°F AB any conc to 185°F		
MAGNESIUM ACETATE	NR sol'n at 70°F	NR sol'n at 70°F	NR sol'n at 70°F	A to 70°F		
MAGNESIUM AMMONIUM SULFATE						

	SYNTHETIC ISOPRENE IR	NATURAL ISOPRENE NR	CHLOROSULFONATED POLYETHYLENE CSM	FLUOROSILICONE FVMQ	SILICONE VMQ	CHEMRAZ FFKM
LIGHT GREASE	NR at 70°F	NR at 70°F	NR at 70°F	A to 70°F	NR at 70°F	A to 70°F
LIGROIN (PETROLEUM ETHER; BENZINE)	NR at 70°F	NR at 70°F	C at 70°F	A to 70°F	NR at 70°F	A to 70°F
LIME (CALCIUM OXIDE)	A to 70°F	A to 70°F AB any conc to 150°F	A to 100% to 70°F AB 30% to 125°F		C conc at 70°F A dilute to 70°F	
LIME ACETATE	C at 70°F	BC any conc to 80°F				
LIME, AGRICULTURAL	A to 70°F	A to 70°F	B at 70°F		A to 70°F	
LIME BISULFITE	C at 70°F	C at 70°F	A to 70°F		C at 70°F	
LIME BLEACH	A to 70°F	A to 70°F AB any conc to 120°F	B at 70°F	A to 70°F	B at 70°F	A to 70°F B at 70°F white
LIME, CAUSTIC	A to 70°F	A to 70°F	B at 70°F		A to 70°F	
LIME HYDRATE	A to 70°F	A to 70°F AB any conc to 150°F	B at 70°F		A to 70°F	
LIME NITRATE	A to 70°F	A to 70°F	A to 70°F		B at 70°F	
LIME SALTPETER	A to 70°F	A to 70°F AB any conc to 150°F	A to 70°F		B at 70°F	
LIME, SODA	A to 70°F	A to 70°F	B at 70°F		C at 70°F	
LIME SLURRIES		A to 70°F	B at 70°F			
LIME SULFUR-DRY	B /NR at 70°F	B /NR to 150°F	A to 70°F	A to 70°F	AB to 70°F	A to 70°F
LIME SULFUR-WET	B /NR at 70°F	B /NR to 150°F	AB to 70°F			A to 70°F
LIME + WATER (CALCIUM HYDROXIDE)	A to 70°F	A to 70°F AB any conc to 150°F	AB to 200°F	A to 70°F	A to 70°F	BC at 70°F B at 70°F white
LIME WATER (MILK OF LIME)			C at 70°F	A to 70°F	B at 70°F	
LIMESTONE	A to 70°F	A to 70°F	A to 70°F			
LIMONENE	NR at 70°F	NR at 70°F	NR at 70°F			
LINDOIL	NR at 70°F	NR at 70°F				
LINDOL (HYDR. FLUID, PHOSPHATE ESTER)	NR at 70°F	NR at 70°F	C /NR at 70°F	C at 70°F	C at 70°F	A to 70°F
LINOLEIC ACID (LINOLENIC ACID)	NR at 70°F	NR at 70°F	NR at 70°F		B at 70°F	A to 70°F
LINSEED CAKE						
LINSEED OIL	NR at 70°F	NR at 70°F	AB to 70°F	no swell 3 days 212°F A to 70°F	+1% vol 7 days 70°F +3% vol 7 days 70°F	A to 70°F
LIQUID OXYGEN	NR	NR	NR	NR	NR	BC B white
LIQUID ROSIN	NR at 70°F	NR at 70°F	B at 70°F			
LIQUID SOAP	A to 70°F	A to 70°F	A to 70°F			
LIQUIFIED PETROLEUM GAS	NR at 70°F gas or liquid	NR at 70°F gas or liquid	A /NR at 70°F	AB to 70°F	NR at 70°F	A to 70°F
LIQUIMOLY	NR at 70°F	NR at 70°F	NR at 70°F	A to 70°F	NR at 70°F	A to 70°F
LITHIUM BROMIDE BRINE		NR at 70°F		A to 70°F	A to 70°F	A to 70°F
LITHIUM CHLORIDE		AB to 70°F	AB to 70°F	A to 70°F	A to 70°F	A aqueous sol'n to 70°F
LITHIUM HYDROXIDE		AB to 70°F		NR at 70°F	NR 5-100% at 70°F	A to 70°F
LUBOIL (LUBE OILS)	NR at 70°F	NR at 70°F	BC at 70°F		C at 70°F	
LUBRICATING OILS (CRUDE & REFINED)	NR at 70°F	NR at 70°F	BC to 158°F	A to 70°F	B /NR at 70°F	A to 70°F
LUBRICATING OILS, DIESTER	NR at 70°F	NR at 70°F	NR at 70°F			A to 70°F
LUBRICATING OILS, PETROLEUM	NR at 70°F	NR at 70°F	BC to 158°F	A to 70°F	B /NR at 70°F	A to 70°F
LUBRICATING OILS, SAE 10, 20, 30, 40 & 50	NR at 70°F	NR at 70°F	C /NR at 70°F	A to 70°F	A /NR at 70°F	A to 70°F
LYE	A to conc to 70°F	AB any conc to 150°F A sol'n to 70°F	A to conc to 70°F A to 73% to 280°F	B to conc at 70°F	A to conc to 70°F A 20% to 212°F	AB to 100% to 70°F C at 70°F white
LYSOL	A to 70°F	A to 70°F AB to 150°F				
MACASSAR GUM	A to 70°F	A to 70°F AB any conc to 150°F				
MACHINE OIL						
MADDRELL'S SALT	A to 70°F	A to 70°F AB any conc to 150°F	A to 70°F		C at 70°F	
MAGNESIUM ACETATE	NR at 70°F	NR at 70°F	A to 70°F	NR sol'n at 70°F	NR sol'n at 70°F	
MAGNESIUM AMMONIUM SULFATE						

C220

	POLYSULFIDE T	CHLORINATED POLYETHYLENE CM	EPICHLOROHYDRIN CO, ECO	® HYTREL COPOLYESTER TPE	® SANTOPRENE COPOLYMER TPO	® C-FLEX STYRENIC TPE
LIGHT GREASE	A to 70°F					
LIGROIN (PETROLEUM ETHER; BENZINE)	A to 70°F					
LIME (CALCIUM OXIDE)				AB to 70°F		AB dilute to 70°F
LIME ACETATE						
LIME, AGRICULTURAL						
LIME BISULFITE						
LIME BLEACH	NR at 70°F					
LIME, CAUSTIC						
LIME HYDRATE						
LIME NITRATE						
LIME SALTPETER						
LIME, SODA						
LIME SLURRIES						
LIME SULFUR-DRY	NR at 70°F		B at 70°F			
LIME SULFUR-WET						
LIME + WATER (CALCIUM HYDROXIDE)	NR at 70°F	AB to 70°F	A to 70°F	AB to 70°F		
LIME WATER (MILK OF LIME)						
LIMESTONE						
LIMONENE						
LINDOIL						
LINDOL (HYDR. FLUID, PHOSPHATE ESTER)	NR at 70°F					
LINOLEIC ACID (LINOLENIC ACID)	NR at 70°F					
LINSEED CAKE				AB to 70°F		
LINSEED OIL	-0.5% vol 3 days 212°F AB to 70°F	AB to 70°F		AB to 70°F		NR at 70°F
LIQUID OXYGEN	NR					
LIQUID ROSIN						
LIQUID SOAP						
LIQUIFIED PETROLEUM GAS	A to 70°F		A to 70°F	AB to 70°F		
LIQUIMOLY						
LITHIUM BROMIDE BRINE						
LITHIUM CHLORIDE	AB to 70°F					
LITHIUM HYDROXIDE						A to conc to 70°F
LUBOIL (LUBE OILS)						
LUBRICATING OILS (CRUDE & REFINED)	AB to 70°F	AB to 70°F		A to 70°F		
LUBRICATING OILS, DIESTER	C at 70°F			C/NR at 70°F		
LUBRICATING OILS, PETROLEUM	C at 70°F	AB to 70°F	A to 70°F	A to 70°F		
LUBRICATING OILS, SAE 10, 20, 30, 40 & 50	C at 70°F	AB to 70°F	A to 70°F	A to 70°F	NR at 70°F	
LYE	NR conc at 70°F AB 20-50% at 70°F	C at 70°F	B at 70°F	BC 46% at 70°F AC to 20% to 70°F	A to 50% to 70°F	A to conc to 70°F
LYSOL						
MACASSAR GUM						
MACHINE OIL						
MADDRELL'S SALT						
MAGNESIUM ACETATE						
MAGNESIUM AMMONIUM SULFATE						

	ETHYLENE ACRYLIC EA	POLYALLOMER LINEAR COPOLYMER	NYLON 11 POLYAMIDE	NYLON 12 POLYAMIDE	ETHYLENE VINYL ACETATE EVA	POLYVINYLCHLORIDE FLEXIBLE PVC
LIGHT GREASE	A to 70°F					
LIGROIN (PETROLEUM ETHER; BENZINE)	NR at 70°F				NR at 70°F	AC to 70°F
LIME (CALCIUM OXIDE)			AB to 70°F		AB to 140°F	A to conc to 70°F AB at 130°F
LIME ACETATE						
LIME, AGRICULTURAL						
LIME BISULFITE						
LIME BLEACH						
LIME, CAUSTIC						
LIME HYDRATE						
LIME NITRATE						
LIME SALTPETER						
LIME, SODA						
LIME SLURRIES						
LIME SULFUR-DRY		A to 70°F	A to 70°F		A to 70°F	
LIME SULFUR-WET						
LIME + WATER (CALCIUM HYDROXIDE)		A to 122°F	A to 70°F		A to 70°F AB to 140°F	A to conc to 150°F
LIME WATER (MILK OF LIME)						
LIMESTONE						
LIMONENE						NR at 70°F
LINDOIL						
LINDOL (HYDR. FLUID, PHOSPHATE ESTER)	NR at 70°F		A to 70°F			
LINOLEIC ACID (LINOLENIC ACID)						
LINSEED CAKE			AB to 194°F			AB to 70°F
LINSEED OIL		A to 70°F	A to 70°F *AB to 140°F	AB to 70°F		
LIQUID OXYGEN						NR
LIQUID ROSIN						
LIQUID SOAP						
LIQUIFIED PETROLEUM GAS	A to 70°F	B /NR at 70°F	AB to 140°F	AB to 70°F	NR at 70°F	A to 70°F B at 122°F
LIQUIMOLY						
LITHIUM BROMIDE BRINE						
LITHIUM CHLORIDE					NR at 70°F	NR at 70°F
LITHIUM HYDROXIDE						
LUBOIL (LUBE OILS)						
LUBRICATING OILS (CRUDE & REFINED)	A to 200°F		AB to 140°F	AB to 70°F	NR at 70°F	B at 70°F
LUBRICATING OILS, DIESTER			A to 70°F			
LUBRICATING OILS, PETROLEUM	A to 200°F		A to 70°F AB to 200°F	AB to 70°F	NR at 70°F	AB to 150°F
LUBRICATING OILS, SAE 10, 20, 30, 40 & 50	A to 70°F		A to 70°F AB to 140°F	AB to 70°F	NR at 70°F	AB to 70°F
LYE	NR 73% at 70°F A 20-46% to 70°F	A to conc to 122°F	AB to 100% to 70°F NR 50% at 140°F	AB 40% to 70°F	A to 70°F AB to 140°F	A to conc to 70°F B 50% to conc at 122°
LYSOL						
MACASSAR GUM						
MACHINE OIL		A to 70°F	A to 70°F		NR at 70°F	AB to 70°F
MADDRELL'S SALT						
MAGNESIUM ACETATE						
MAGNESIUM AMMONIUM SULFATE						

	POLYETHYLENE LOW DENSITY LDPE	®TEFLON FEP	®KALREZ PERFLUORINATED ELASTOMER	®FLUORAZ FLUORINATED COPOLYMER	®AFLAS FLUORINATED COPOLYMER	®NORPRENE COPOLYMER TPO
LIGHT GREASE		A to 70°F		A to 70°F	+5% vol 7 days 70°F A to 70°F	
LIGROIN (PETROLEUM ETHER; BENZINE)	NR at 70°F	A to 70°F	A to 212°F	BC at 70°F		
LIME (CALCIUM OXIDE)	A to 70°F AB to 140°F	A to 200°F				A to 70°F dilute
LIME ACETATE		A to 70°F				
LIME, AGRICULTURAL		A to 70°F				
LIME BISULFITE		A to 70°F				
LIME BLEACH		A to 70°F	A to 212°F	A to 70°F		
LIME, CAUSTIC		A to 70°F				
LIME HYDRATE		A to 70°F				
LIME NITRATE		A to 70°F				
LIME SALTPETER		A to 70°F				
LIME, SODA		A to 70°F				
LIME SLURRIES						
LIME SULFUR-DRY	A to 70°F	A to 300°F	A to 212°F	A to 70°F		
LIME SULFUR-WET		A to 70°F		A to 70°F		
LIME + WATER (CALCIUM HYDROXIDE)	A to 100% to 122°F A conc to 140°F	A to conc to 125°F	A to 70°F	A to 70°F		
LIME WATER (MILK OF LIME)						
LIMESTONE		A to 70°F				
LIMONENE		A to 70°F				
LINDOIL		A to 70°F				
LINDOL (HYDR. FLUID, PHOSPHATE ESTER)		A to 70°F	A to 212°F	A to 70°F	+14% vol 3 days 212°F A to 70°F	
LINOLEIC ACID (LINOLENIC ACID)		A to 300°F	A to 212°F	A to 70°F		
LINSEED CAKE	NR at 70°F				A to 70°F	
LINSEED OIL	B/NR at 70°F NR at 140°F	A to 300°F	A to 212°F	A to 70°F	A to 70°F	A to 70°F
LIQUID OXYGEN		A		NR		
LIQUID ROSIN		A to 70°F				
LIQUID SOAP		A to 70°F				
LIQUIFIED PETROLEUM GAS	AC to 70°F	A to 200°F gas or liquid	A to 212°F	AC to 70°F		
LIQUIMOLY						
LITHIUM BROMIDE BRINE		A to 70°F	A to 212°F	A to 70°F	+0.3% vol 58% conc 11 days 320°F	
LITHIUM CHLORIDE	A to 70°F	A to 125°F	A to 212°F	A to 70°F aqueous		
LITHIUM HYDROXIDE		A to 125°F	A to 212°F			AB to 70°F
LUBOIL (LUBE OILS)	NR at 70°F	A to 300°F	A to 212°F	AB to 70°F	AB to 70°F	
LUBRICATING OILS (CRUDE & REFINED)	A to 104°F B at 140°F	A to 300°F	A to 70°F	A to 70°F	AB to 70°F	
LUBRICATING OILS, DIESTER				B at 70°F		
LUBRICATING OILS, PETROLEUM	AC to 70°F	A to 300°F	A to 212°F	A to 70°F	AB to 70°F	
LUBRICATING OILS, SAE 10, 20, 30, 40 & 50	AB to 70°F	A to 300°F		A to 70°F		
LYE	A to conc to 122°F AB to conc to 140°F	A to 100% to 300°F	A to 100% to 212°F A to 50% to 303°F	A to conc to 250°F dry A to 50% to 150°F wet	+2% vol 50% conc 3 days 212°F	A 50% to conc to 70°F BC 46% at 70°F
LYSOL		A to 70°F				
MACASSAR GUM		A to 70°F				
MACHINE OIL	C at 70°F	A to 70°F				
MADDRELL'S SALT		A to 70°F				
MAGNESIUM ACETATE		A to 70°F				
MAGNESIUM AMMONIUM SULFATE		A to 70°F				

C223

	RECOMMENDATIONS	NITRILE NBR	ETHYLENE PROPYLENE EP, EPDM	FLUOROCARBON FKM	CHLOROPRENE CR	HYDROGENATED NITRILE HNBR
MAGNESIUM BISULFITE		B any conc to 120°F			B any conc to 120°F	
MAGNESIUM CARBONATE		A to 140°F B at 176°F	AB to 176°F	A to 212°F	A to 140°F	
MAGNESIUM CHLORIDE	NBR,EP,CR,FKM	A to 100% to 176°F B at 212°F	A to 100% to 176°F B at 212°F	A to 100% to 176°F B at 212°F	A to 158°F AB any conc to 200°F	A to 70°F
MAGNESIUM CITRATE		A to 140°F B at 176°F	A to 176°F	A to 212°F		
MAGNESIUM HYDRATE		AB any conc to 200°F	A to 70°F	B at 70°F	AB any conc to 200°F	
MAGNESIUM HYDROXIDE	EP,FKM,IIR,NBR,CR	A to 140°F AB any conc to 200°F	A to 176°F AB to 200°F	A to conc to 212°F	A to conc to 158°F AB any conc to 200°F	B at 70°F
MAGNESIUM NITRATE	IIR,CR,EP	A to 140°F AB any conc to 200°F	A to 176°F AB to 200°F	A to 212°F	A to 140°F AB any conc to 200°F	
MAGNESIUM OXIDE		A to 140°F	A to 70°F	A to 70°F	A to 140°F	
MAGNESIUM SALTS	NBR	A to 70°F	A to 70°F	A to 70°F	A to 70°F	A to 70°F
MAGNESIUM SULFATE (EPSOM SALTS)	NBR,EP,IIR,CR,FKM	A to 176°F AB any conc to 200°F	A to 176°F B at 212°F	A to boiling	A to 176°F AB any conc to 200°F	
MAGNESIUM SULFIDE		NR at 70°F	C at 70°F	NR at 70°F	B at 70°F	
MAGNESIUM SULFITE	NBR,EP,FKM,CR,IIR	A to 70°F	A to 70°F	A to 70°F	A to 70°F	
MAIZE OIL	FKM,NBR,ACM,VMQ	A to 250°F	B/NR to 104°F	A to 140°F	AC to 70°F	A to 70°F
MALATHION	FKM	B 100% at 70°F A 50% at 70°F	NR at 70°F	A to 70°F		
MALEIC ACID	NBR,FKM,T,NR	B/NR to 176°F	A to 120°F AB 140-212°F	A to 140°F B at 176°F	NR 25-100% at 70°F	NR at 70°F
MALEIC ANHYDRIDE	FKM	NR at 70°F	NR at 70°F	+69% vol 7 days 180°F A to 140°F	NR at 70°F	NR at 70°F
MALEINIC ACID		NR 25% at 70°F			NR 25% at 70°F	
MALIC ACID	NBR,FKM,FVMQ	A to 70°F	NR at 70°F	A to 70°F	AB to 140°F	A to 70°F
MALONYL NITRILE		A to 70°F	A to 70°F		A to 70°F	
MALT BEVERAGE	EP	A to 70°F	A to 250°F	A to 70°F	A to 70°F	
MALT SALT		B at 70°F			B at 70°F	
MANGANESE CHLORIDE	NBR	A 50-100% to 160°F	AC to 70°F	A to 70°F	AB to 70°F	
MANGANESE NITRATE	NBR,CR	A to 70°F		C at 70°F	A to 70°F	
MANGANESE SALTS		A to 70°F		A to 70°F	A to 70°F	
MANGANESE SULFATE		A to 140°F AB any conc to 200°F	A to 176°F	A to 212°F	A to 70°F AB any conc to 200°F	
MANGANESE SULFIDE		A to 70°F		A to 70°F		
MANGANESE SULFITE		A to 70°F		A to 70°F		
MAPLE SUGAR LIQUORS		A to 70°F	A to 70°F	A to 70°F	A to 70°F	
MARSH GAS	NBR,FKM,ACM	A to 250°F	NR at 70°F	A to 176°F	A to 140°F	A to 70°F
MASH	NBR,CR	A to 140°F			A to 140°F	
MASTER KILL EMULSION		A to 70°F		C at 70°F	A to 70°F	
MAYONNAISE	NBR	A to 250°F	NR at 70°F	A to 140°F	A to 70°F	
MCS 312	FKM	NR at 70°F	NR at 70°F	A to 70°F	NR at 70°F	
MCS 352	EP	NR at 70°F	A to 70°F	NR at 70°F	NR at 70°F	
MCS 463	EP	NR at 70°F	A to 70°F	NR at 70°F	NR at 70°F	
M.E.A.	EP,IIR,NBR,CR	B any conc to 80°F	AB to 200°F	NR at 70°F	B any conc to 80°F	
M.E.K.	EP,IIR	NR any conc at 70°F	A to 140°F AB to 200°F	+240% vol 7 days 70°F NR at 70°F	NR any conc at 70°F	
MELAMINE RESINS	EP	BC to 140°F NR 0.5% boiling	A to 120°F	A to 140°F	B/NR at 70°F	
MERCAPTANS	EP	NR at 70°F	A to 160°F	A/NR at 70°F	NR at 70°F	
MERCURIC CHLORIDE	NBR,EP,CR,FKM,IIR	A to 140°F A sol'n to 160°F	A any conc to 140°F	A any conc to 140°F A 0.5% to boiling	A any conc to 140°F	A to 70°F
MERCURIC CYANIDE		A to 140°F AB any conc to 150°F	A to 70°F	A to 70°F	B at 70°F	
MERCURIC SULFATE		A to 70°F	A to 70°F	A to 70°F	B at 70°F	
MERCUROUS NITRATE		AB any conc to 80°F	A to 70°F	A to 70°F	AB any conc to 80°F	
MERCUROUS SALTS		A to 70°F	A to 70°F	A to 70°F	A to 70°F	

	STYRENE BUTADIENE SBR	POLYACRYLATE ACM	POLYURETHANE AU, EU	ISOBUTYLENE ISOPRENE IIR	POLYBUTADIENE BR	® AEROQUIP AQP
MAGNESIUM BISULFITE				AB any conc to 120°F		
MAGNESIUM CARBONATE			AB to 70°F	A to 70°F		
MAGNESIUM CHLORIDE	A to conc to 70°F		A to conc to 70°F	A to 70°F / AB any conc to 185°F	A to 70°F	AB to 70°F
MAGNESIUM CITRATE						
MAGNESIUM HYDRATE			A to 70°F	A to 70°F / AB any conc to 185°F		
MAGNESIUM HYDROXIDE	B to conc at 70°F	NR any conc at 70°F	A/NR at 70°F	A to 70°F / AB any conc to 185°F	B at 70°F	AB to 70°F
MAGNESIUM NITRATE			AB to 70°F	A to 70°F / AB any conc to 185°F		
MAGNESIUM OXIDE						
MAGNESIUM SALTS	A to 70°F	A to 70°F	A to 70°F	A to 70°F	A to 70°F	
MAGNESIUM SULFATE (EPSOM SALTS)	B at 70°F	NR at 70°F	B/NR at 70°F	A to 70°F / AB any conc to 185°F	B at 70°F	AB to 70°F
MAGNESIUM SULFIDE			A to 70°F	B at 70°F		
MAGNESIUM SULFITE	B at 70°F	NR at 70°F		A to 70°F	B at 70°F	
MAIZE OIL	NR at 70°F	A to 70°F	A to 70°F	AB to 100°F	NR at 70°F	
MALATHION	NR at 70°F		NR at 70°F	NR at 70°F	NR at 70°F	
MALEIC ACID	NR at 70°F	NR at 70°F	B/NR at 70°F / AB fumes to 70°F	C/NR 25-100% at 70°F	NR at 70°F	
MALEIC ANHYDRIDE	NR at 70°F	NR at 70°F		C/NR at 70°F	NR at 70°F	
MALEINIC ACID				NR 25% at 70°F		
MALIC ACID	B at 70°F	NR at 70°F		NR any conc at 70°F	B at 70°F	
MALONYL NITRILE				A to 70°F		
MALT BEVERAGE	A to 70°F	NR at 70°F	B/NR at 70°F			
MALT SALT				A to 70°F		
MANGANESE CHLORIDE	A to 70°F	NR at 70°F	AB to 70°F	B at 70°F		
MANGANESE NITRATE						
MANGANESE SALTS						
MANGANESE SULFATE			A to 70°F	AB any conc to 185°F		
MANGANESE SULFIDE				A to 70°F		
MANGANESE SULFITE				A to 70°F		
MAPLE SUGAR LIQUORS				A to 70°F		
MARSH GAS	NR at 70°F	A to 70°F	BC at 70°F	NR at 70°F	NR at 70°F	
MASH			A to 70°F			
MASTER KILL EMULSION				A to 70°F		
MAYONNAISE	NR at 70°F		NR at 70°F			
MCS 312	NR at 70°F	NR at 70°F		NR at 70°F	NR at 70°F	
MCS 352	NR at 70°F	NR at 70°F	NR at 70°F	B at 70°F	NR at 70°F	
MCS 463	NR at 70°F	NR at 70°F	NR at 70°F	B at 70°F	NR at 70°F	
M.E.A.	B at 70°F	NR at 70°F	NR at 70°F	B any conc to 140°F	B at 70°F	
M.E.K.	NR at 70°F	NR at 70°F	NR at 70°F / AB fumes to 70°F	A any conc to 100°F	NR at 70°F	A to 70°F
MELAMINE RESINS		NR at 70°F	NR at 70°F			
MERCAPTANS	NR at 70°F	NR at 70°F	NR at 70°F			
MERCURIC CHLORIDE	A to 70°F		A to 70°F	A to 70°F / AB any conc to 150°F	A to 70°F	NR at 70°F
MERCURIC CYANIDE				A to 70°F / AB any conc to 150°F		
MERCURIC SULFATE				B at 70°F		
MERCUROUS NITRATE				A to 70°F / AB any conc to 150°F		
MERCUROUS SALTS				A to 70°F / AB any conc to 150°F		

	SYNTHETIC ISOPRENE IR	NATURAL ISOPRENE NR	CHLOROSULFONATED POLYETHYLENE CSM	FLUOROSILICONE FVMQ	SILICONE VMQ	CHEMRAZ FFKM
MAGNESIUM BISULFITE	B at 70°F	AB any conc to 120°F				
MAGNESIUM CARBONATE	A to 70°F	A to 70°F	A to 70°F			
MAGNESIUM CHLORIDE	A to 70°F	A to 70°F AB any conc to 185°F	A to 100% to 158°F AB to 200°F	A to conc to 70°F	A to conc to 70°F	A to 70°F
MAGNESIUM CITRATE						
MAGNESIUM HYDRATE	A to 70°F	A to 70°F AB any conc to 150°F	A to 70°F		B at 70°F	
MAGNESIUM HYDROXIDE	A to 70°F	A to 70°F AB any conc to 185°F	A to 100% to 158°F AB to 200°F		AB to 70°F	AB to 70°F C at 70°F white
MAGNESIUM NITRATE	A to 70°F	AB any conc to 185°F	A to 70°F AB to 200°F		AB to 70°F	
MAGNESIUM OXIDE		AB to 70°F				
MAGNESIUM SALTS	A to 70°F	A to 70°F	A to 70°F	A to 70°F	A to 70°F	A to 70°F
MAGNESIUM SULFATE (EPSOM SALTS)	AB to 70°F	AB any conc to 185°F	A to 70°F AB to 200°F	A to 70°F	A to 70°F	
MAGNESIUM SULFIDE		B at 70°F	B at 70°F		C at 70°F	A to 70°F
MAGNESIUM SULFITE	B at 70°F	B at 70°F	A to 70°F	A to 70°F	A to 70°F	A to 70°F
MAIZE OIL	NR at 70°F	NR at 70°F	B at 70°F	A to 70°F	A to 70°F	A to 70°F
MALATHION	NR at 70°F	NR at 70°F		B at 70°F	NR at 70°F	A to 70°F
MALEIC ACID	NR at 70°F	NR at 70°F AB 25% to 80°F	C/NR at 70°F		NR at 70°F	A to 70°F
MALEIC ANHYDRIDE	NR at 70°F	NR at 70°F	NR at 70°F			A to 70°F
MALEINIC ACID	C 25% at 70°F	BC 25% to 80°F				
MALIC ACID	A to 70°F	A any conc to 80°F	B at 70°F	A to 70°F	B at 70°F	A to 70°F
MALONYL NITRILE	A to 70°F	A to 70°F				
MALT BEVERAGE				A to 70°F	A to 70°F	
MALT SALT	B at 70°F	B at 70°F				
MANGANESE CHLORIDE		B at 70°F	B at 70°F	A to 70°F	AC to 70°F	A aqueous sol'n to 70°F
MANGANESE NITRATE		AB to 70°F				
MANGANESE SALTS		AB to 70°F	AB to 70°F		AB to 70°F	AB to 70°F
MANGANESE SULFATE	B at 70°F	B any conc to 150°F	A to 70°F		C at 70°F	A to 70°F
MANGANESE SULFIDE	C at 70°F	C at 70°F	A to 70°F			
MANGANESE SULFITE	C at 70°F	C at 70°F	A to 70°F			
MAPLE SUGAR LIQUORS	A to 70°F	A to 70°F	A to 70°F		A to 70°F	
MARSH GAS	B/NR at 70°F	BC at 70°F AB to 70°F dry	A to 70°F wet or dry	AC to 70°F	A/NR at 70°F	A to 70°F
MASH						
MASTER KILL EMULSION			AB to 70°F			
MAYONNAISE						
MCS 312	NR at 70°F	NR at 70°F		A to 70°F	A to 70°F	A to 70°F
MCS 352	NR at 70°F	NR at 70°F	NR at 70°F	C at 70°F	C at 70°F	A to 70°F
MCS 463	NR at 70°F	NR at 70°F	NR at 70°F	C at 70°F	C at 70°F	A to 70°F
M.E.A.	B at 70°F	B any conc to 80°F	B/NR at 70°F	NR at 70°F	B at 70°F	A to 70°F
M.E.K.	NR at 70°F	NR at 70°F	NR at 70°F	NR at 70°F	NR at 70°F	A to 70°F
MELAMINE RESINS		NR at 70°F			C at 70°F	
MERCAPTANS						A to 70°F
MERCURIC CHLORIDE	A to 70°F	A any conc to 70°F AB any conc to 150°F	A to conc to 70°F	A sol'n to 70°F	A to conc to 70°F	A to 70°F
MERCURIC CYANIDE	A to 70°F	A to 70°F AB any conc to 150°F	A to 70°F			A to 70°F
MERCURIC SULFATE		B at 70°F	B at 70°F		AB to 70°F	A to 70°F
MERCUROUS NITRATE	A to 70°F	A to 70°F AB any conc to 150°F	A to 70°F			A to 70°F
MERCUROUS SALTS	A to 70°F	A to 70°F	A to 70°F			A to 70°F

	POLYSULFIDE T	CHLORINATED POLYETHYLENE CM	EPICHLOROHYDRIN CO, ECO	® HYTREL COPOLYESTER TPE	® SANTOPRENE COPOLYMER TPO	® C-FLEX STYRENIC TPE
MAGNESIUM BISULFITE						
MAGNESIUM CARBONATE	AB to 70°F					
MAGNESIUM CHLORIDE	C at 70°F	AB to 70°F	A to 70°F	AB to 70°F *AB to 140°F		A to 70°F
MAGNESIUM CITRATE						
MAGNESIUM HYDRATE						
MAGNESIUM HYDROXIDE	C at 70°F	AB to 70°F	A to 70°F	BC at 70°F		
MAGNESIUM NITRATE	AB to 70°F					
MAGNESIUM OXIDE						
MAGNESIUM SALTS	A to 70°F					
MAGNESIUM SULFATE (EPSOM SALTS)	B at 70°F	AB to 70°F	A to 70°F	AB to 70°F		A to 70°F
MAGNESIUM SULFIDE						
MAGNESIUM SULFITE	B at 70°F		A to 70°F			
MAIZE OIL	NR at 70°F		A to 70°F	no swell at 212°F A to 212°F	+0.8% vol 7 days 212°F A to 212°F	
MALATHION				NR at 70°F		
MALEIC ACID	B at 70°F					
MALEIC ANHYDRIDE						
MALEINIC ACID						
MALIC ACID						A to 70°F
MALONYL NITRILE						
MALT BEVERAGE						
MALT SALT						
MANGANESE CHLORIDE						
MANGANESE NITRATE						
MANGANESE SALTS						AB to 70°F
MANGANESE SULFATE						
MANGANESE SULFIDE						
MANGANESE SULFITE						
MAPLE SUGAR LIQUORS						
MARSH GAS	A to 70°F	AB to 70°F	A to 70°F	AB to 70°F		NR at 70°F
MASH						
MASTER KILL EMULSION						
MAYONNAISE						
MCS 312						
MCS 352	NR at 70°F					
MCS 463	NR at 70°F					
M.E.A.	NR at 70°F					A to 70°F
M.E.K.	AB to 70°F	AB to 70°F	NR at 70°F	A to 70°F	-4 to -11% vol 7 days 70°F A/NR at 70°F	
MELAMINE RESINS						
MERCAPTANS						
MERCURIC CHLORIDE		NR at 70°F	A to 70°F			
MERCURIC CYANIDE						
MERCURIC SULFATE						
MERCUROUS NITRATE						
MERCUROUS SALTS						*AB to 70°F

	ETHYLENE ACRYLIC EA	POLYALLOMER LINEAR COPOLYMER	NYLON 11 POLYAMIDE	NYLON 12 POLYAMIDE	ETHYLENE VINYL ACETATE EVA	POLYVINYLCHLORIDE FLEXIBLE PVC
MAGNESIUM BISULFITE						
MAGNESIUM CARBONATE						*AB to 140°F
MAGNESIUM CHLORIDE	A to 70°F	A to 70°F	A to 70°F AB 50% to 194°F	AB 15% to 70°F	A to 70°F AB at 140°F	A to 150°F
MAGNESIUM CITRATE						
MAGNESIUM HYDRATE						
MAGNESIUM HYDROXIDE	A to 70°F	A to 70°F	A to 70°F	AB 10% to 70°F	A to 70°F AB at 140°F	A to 150°F
MAGNESIUM NITRATE						*AB to 140°F
MAGNESIUM OXIDE						
MAGNESIUM SALTS	A to 70°F		AB to 70°F	AB 10% to 70°F		AB to 70°F
MAGNESIUM SULFATE (EPSOM SALTS)		A to 70°F	A to 70°F		A to 70°F	A to 150°F
MAGNESIUM SULFIDE					B at 70°F	BC at 70°F
MAGNESIUM SULFITE						
MAIZE OIL	A to 70°F		AB to 70°F		B at 70°F	AB to 70°F B at 150°F
MALATHION			AB to 70°F			
MALEIC ACID	A to 70°F		AC to 70°F			
MALEIC ANHYDRIDE						AB to 70°F A/NR at 150°F
MALEINIC ACID						
MALIC ACID		B at 70°F	A to 70°F		C at 70°F	A to 70°F
MALONYL NITRILE						
MALT BEVERAGE						
MALT SALT						
MANGANESE CHLORIDE					B at 70°F	BC at 70°F
MANGANESE NITRATE						
MANGANESE SALTS		A to 70°F	A to 70°F		A to 70°F	A to 70°F
MANGANESE SULFATE					B at 70°F	BC at 70°F
MANGANESE SULFIDE						
MANGANESE SULFITE						
MAPLE SUGAR LIQUORS						
MARSH GAS			A to 70°F AB to 140°F			AB to 70°F
MASH					B at 70°F	AB to 70°F
MASTER KILL EMULSION						
MAYONNAISE		A to 70°F	A to 70°F	A to 70°F	C at 70°F	A to 150°F food grades
MCS 312						
MCS 352						
MCS 463						
M.E.A.						
M.E.K.	NR at 70°F	A to 70°F NR at 122°F	AB to 105°F NR at 140°F	AB to 70°F	*C at 70°F *NR at 140°F	NR at 70°F
MELAMINE RESINS						NR at 70°F
MERCAPTANS						
MERCURIC CHLORIDE		A to 70°F	A/NR at 70°F		A to 70°F	A/NR at 70°F
MERCURIC CYANIDE						NR at 70°F
MERCURIC SULFATE						B at 70°F
MERCUROUS NITRATE						
MERCUROUS SALTS				*AB to 70°F		

	POLYETHYLENE LOW DENSITY LDPE	®TEFLON FEP	®KALREZ PERFLUORINATED ELASTOMER	®FLUORAZ FLUORINATED COPOLYMER	®AFLAS FLUORINATED COPOLYMER	®NORPRENE COPOLYMER TPO
MAGNESIUM BISULFITE		A to 70°F				
MAGNESIUM CARBONATE	B at 70°F	A to 300°F				
MAGNESIUM CHLORIDE	AB to 140°F	A to 300°F	A to 212°F	A to 70°F	+0% vol 7 days 212°F 18% conc	AB to 70°F A sol'n to 70°F
MAGNESIUM CITRATE		A to 70°F				
MAGNESIUM HYDRATE		A to 70°F				
MAGNESIUM HYDROXIDE	A to 70°F AB to 140°F	A to 300°F	A to 212°F	A to 70°F		
MAGNESIUM NITRATE	AB to 70°F	A to 300°F		A to 70°F		
MAGNESIUM OXIDE		A to 70°F				
MAGNESIUM SALTS	A to 70°F AB to 140°F	A to 70°F	A to 212°F	A to 70°F		
MAGNESIUM SULFATE (EPSOM SALTS)	A to 70°F	A to 300°F	A to 212°F	A to 50% to 140°F		AB to 70°F
MAGNESIUM SULFIDE	A to 70°F					
MAGNESIUM SULFITE			A to 212°F	A to 70°F aqueous		
MAIZE OIL	A to 70°F	A to 70°F	A to 212°F	A to 70°F		
MALATHION			A to 212°F	A to 70°F		
MALEIC ACID	A to 70°F NR at 140°F	A to 300°F	A to 212°F	A to 70°F		A to 70°F
MALEIC ANHYDRIDE	NR at 70°F	A to 70°F	A to 336°F	A to 70°F		
MALEINIC ACID		A 25% to 70°F				
MALIC ACID	B/NR at 70°F	A to 300°F	A to 212°F	A to 70°F		TEST
MALONYL NITRILE		A to 70°F				
MALT BEVERAGE		A to 70°F				
MALT SALT						
MANGANESE CHLORIDE	A to 70°F	A to 300°F	A to 212°F	A to 70°F aqueous		
MANGANESE NITRATE						
MANGANESE SALTS	A to 70°F	A to 70°F				A to 70°F
MANGANESE SULFATE	A to 70°F	A to 300°F	A to 212°F aqueous	A to 70°F aqueous		
MANGANESE SULFIDE		A to 70°F				
MANGANESE SULFITE		A to 70°F				
MAPLE SUGAR LIQUORS		A to 70°F				
MARSH GAS	AC to 70°F	A to 300°F	A to 212°F	B/NR at 70°F		A to 70°F
MASH	A to 70°F					
MASTER KILL EMULSION		A to 70°F				
MAYONNAISE	A to 70°F	A to 70°F		NR at 70°F		
MCS 312				A to 70°F		
MCS 352				C at 70°F		
MCS 463		A to 70°F	A to 150°F	A to 70°F		
M.E.A.	A to 70°F B at 122°F	A to boiling	+1% vol 7 days 70°F A to 150°F	A to 70°F	+58% vol 7 days 70°F NR at 70°F	TEST
M.E.K.	AB to 122°F C/NR at 140°F	A to 122°F	+1% vol 7 days 70°F A to 212°F	NR at 70°F	+56% vol 7 days 70°F NR at 70°F	BC at 70°F
MELAMINE RESINS				NR at 70°F		
MERCAPTANS			A to 212°F	NR at 70°F		
MERCURIC CHLORIDE	A to 70°F	A to 300°F	A to 212°F	A to 70°F		
MERCURIC CYANIDE	A to 70°F	A to 300°F	A to 212°F			
MERCURIC SULFATE		A to 70°F	A to 212°F			
MERCUROUS NITRATE	A to 70°F	A to 300°F	A to 212°F			
MERCUROUS SALTS	AB to 140°F	A to 70°F	A to 212°F	A to 70°F		A to 70°F

C 229

	RECOMMENDATIONS	NITRILE NBR	ETHYLENE PROPYLENE EP, EPDM	FLUOROCARBON FKM	CHLOROPRENE CR	HYDROGENATED NITRILE HNBR
MERCURY	NBR,EP,CR,FKM,IIR	A to 140°F	A to 140°F	A to 140°F	A to 140°F	A to 70°F
MERCURY BICHLORIDE		A to 70°F AB any conc to 150°F	A to 70°F	A to 70°F	B/NR at 70°F	
MERCURY CHLORIDE		A to 70°F AB any conc to 150°F	A to 70°F	A to conc to 70°F	B/NR at 70°F	
MERCURY SALTS		A to 70°F	AB to 70°F	A to 70°F	AB to 70°F	
MERCURY VAPOR	NBR,EP,FKM,CR	A to 70°F	A to 70°F	A to 70°F	A/NR at 70°F	A to 70°F
MESITYL OXIDE	EP,IIR,T	NR at 70°F	AB to 120°F	NR at 70°F	NR at 70°F	NR at 70°F
METER-CRESOL		NR at 70°F	NR at 70°F	AB to 70°F	AB to 70°F	
METHADIENE		C at 70°F	NR at 70°F	A to 70°F	NR at 70°F	
METHALLYL ALCOHOL		A to 70°F		B at 70°F		
METHANAL	EP,IIR,NBR,CR	BC 50-100% to 140°F AB 40% to 80°F	A to 100% to 120°F	A to 40% to 212°F	NR 40-100% at 158°F AB 40% to 70°F	B at 70°F
METHANAMIDE		A to 70°F	A to 70°F	NR at 70°F	A to 70°F	
METHANE	NBR,FKM,CR,ACM	A to 250°F	NR at 70°F	A to 176°F	AB to 140°F	A to 70°F
METHANOL	NBR,EP,CR,IIR,SBR	+18% vol 3 days 158°F A to 70°F	+1% vol 3 days 158°F A to 160°F	+18% vol 21 days 75°F +49% vol 3 days 158°F	+8% vol 3 days 158°F A to 100% to 140°F	A to 70°F
METHANOL		AB any conc to 150°F	AB at 176°F	BC at 70°F C/NR at 140°F	B at 176°F NR at 212°F	
METHYL ACETATE	EP,T,IIR	NR at 70°F	AB to 160°F	NR at 70°F	B/NR at 70°F NR at 140°F	NR at 70°F
METHYL ACETOACETATE	EP	NR at 70°F	B at 70°F	NR at 70°F	NR at 70°F	NR at 70°F
METHYL ACETONE	EP,IIR	NR at 70°F	A to 70°F	NR at 70°F	NR at 70°F	
METHYL ACETYLENE					AB to 70°F	NR at 70°F
METHYL ACRYLATE	EP,T,IIR	NR at 70°F	AB to 160°F	+214% vol 7 days 75°F NR at 70°F	B/NR at 70°F NR at 140°F	
METHYL ACRYIC ACID		NR at 70°F	B at 70°F	BC at 70°F	B at 70°F	
METHYL ALCOHOL	NBR,EP,CR,IIR,SBR	+18% vol 3 days 158°F A to 70°F	+1% vol 3 days 158°F A to 160°F	+18% vol 21 days 75°F +49% vol 3 days 158°F	+8% vol 3 days 158°F A to 100% to 140°F	A to 70°F
METHYL ALCOHOL		AB any conc to 150°F	AB at 176°F	BC at 70°F C/NR at 140°F	B at 176°F NR at 212°F	
METHYL ALCOHOL-WOOD	NBR,EP,CR,IIR	A to 70°F AB any conc to 150°F	A to 160°F AB to 176°F	BC at 70°F C/NR at 140°F	A to 100% at 140°F NR at 212°F	A to 70°F
METHYL AMINE		B at 140°F	A to 70°F	AC to 70°F	A/NR at 70°F	
METHYL AMYL AMINE		NR at 70°F	NR at 70°F	NR at 70°F	NR at 70°F	
METHYL AMYL ALCOHOL		AB to 70°F	A to 70°F	A/NR at 70°F	A to 70°F	
METHYL AMYL CARBINOL		A to 70°F		B at 70°F		
METHYL AMYL KETONE		NR at 70°F		NR at 70°F		
METHYL ANILINE	FKM	A/NR at 70°F	A/NR at 70°F	B at 70°F	A/NR at 70°F	
METHYL BENZENE	FKM	NR 30-100% at 70°F	NR 30-100% at 70°F	A to 100°F BC to 200°F	NR 30-100% at 70°F	NR at 70°F
METHYL BENZOATE	FKM	NR at 70°F	NR at 70°F	A to 70°F	NR at 70°F	NR at 70°F
METHYL BICHLORIDE		NR at 70°F	C at 70°F	B at 70°F	NR at 70°F	
METHYL BROMIDE	FKM,FVMQ	B at 70°F NR at 140°F	A/NR at 70°F	A to 160°F	NR at 70°F	B at 70°F
METHYL BUTANOL	EP,IIR,NBR	A to 70°F AB any conc to 180°F	A to 200°F	A to 212°F	A to 158°F AB any conc to 180°F	B at 70°F
METHYL BUTANONE		NR at 70°F		NR at 70°F		
METHYL BUTYL KETONE	EP,IIR	NR at 70°F	A to 140°F	NR at 70°F	NR at 70°F	NR at 70°F
METHYL BUTYRATE		NR at 70°F	NR at 70°F		NR at 70°F	
METHYL CARBITOL		C at 70°F				
METHYL CARBONATE	FKM	NR at 70°F	NR at 70°F	A to 70°F	NR at 70°F	NR at 70°F
METHYL CELLOSOLVE	EP,IIR	C at 70°F NR at 140°F	B to 200°F	NR at 70°F	B at 70°F NR at 140°F	C at 70°F
METHYL CELLULOSE	NBR	B at 70°F	B at 70°F	NR at 70°F	B at 70°F	B at 70°F
METHYL CHLORIDE	FKM,FVMQ	NR at 70°F	NR to 140°F	A to 140°F AB to 200°F	NR at 70°F	NR at 70°F
METHYL CHLOROFORM	FKM,FVMQ	+279% vol 3 days 70°F NR at 70°F	+113% vol 3 days 70°F NR at 70°F	+11% vol 7 days 75°F AB to 200°F	+191% vol 3 days 70°F NR at 70°F	NR at 70°F
METHYL CHLOROFORMATE	FKM	NR at 70°F	NR at 70°F	A to 70°F	NR at 70°F	NR at 70°F

	STYRENE BUTADIENE SBR	POLYACRYLATE ACM	POLYURETHANE AU, EU	ISOBUTYLENE ISOPRENE IIR	POLYBUTADIENE BR	® AEROQUIP AQP
MERCURY	A to 70°F		A to 70°F *AB to 140°F	A to 70°F	A to 70°F	A to 70°F
MERCURY BICHLORIDE				A to 70°F AB any conc to 150°F		
MERCURY CHLORIDE				A to 70°F AB any conc to 150°F		
MERCURY SALTS				A to 70°F AB any conc to 150°F		
MERCURY VAPOR	A to 70°F			A to 70°F	A to 70°F	
MESITYL OXIDE	NR at 70°F	NR at 70°F	NR at 70°F	B/NR at 70°F	NR at 70°F	
METER-CRESOL	NR at 70°F	NR at 70°F	NR at 70°F			
METHADIENE				NR at 70°F		
METHALLYL ALCOHOL				A to 70°F		
METHANAL	AC to 70°F	NR at 70°F	NR 40% at 70°F	A to conc to 70°F AB 40% to 150°F	B at 70°F	AB to 70°F
METHANAMIDE				A to 70°F		
METHANE	B/NR at 70°F	AB to 70°F	BC at 70°F	NR at 70°F	B/NR at 70°F	
METHANOL	+2% vol 3 days 158°F A to 70°F	+140% vol 3 days 158°F NR at 70°F	+28% vol 3 days 158°F NR 50-100% at 70°F	-1.7% vol 3 days 158°F A to 70°F	A to 70°F	A to 70°F
METHANOL			AB to 70°F fumes	AB any conc to 185°F		
METHYL ACETATE	NR at 70°F	NR at 70°F	NR at 70°F	B at 70°F	NR at 70°F	
METHYL ACETOACETATE		NR at 70°F	NR at 70°F	B at 70°F		
METHYL ACETONE				B at 70°F		
METHYL ACETYLENE				AB to 70°F		
METHYL ACRYLATE	NR at 70°F	NR at 70°F	NR at 70°F	B at 70°F	NR at 70°F	
METHYL ACRYIC ACID	NR at 70°F	NR at 70°F	NR at 70°F	B at 70°F	NR at 70°F	
METHYL ALCOHOL	+2% vol 3 days 158°F A to 70°F	+140% vol 3 days 158°F NR at 70°F	+28% vol 3 days 158°F NR 50-100% at 70°F	-1.7% vol 3 days 158°F A to 70°F	A to 70°F	A to 70°F
METHYL ALCOHOL			AB to 70°F fumes	AB any conc to 185°F		
METHYL ALCOHOL-WOOD	A to 70°F	NR at 70°F	NR 50-100% at 70°F	A to 70°F	A to 70°F	A to 70°F
METHYL AMINE				A/NR at 70°F		
METHYL AMYL AMINE				C at 70°F		
METHYL AMYL ALCOHOL				A to 70°F		
METHYL AMYL CARBINOL				A to 70°F		
METHYL AMYL KETONE				B at 70°F		
METHYL ANILINE	NR at 70°F	NR at 70°F	NR at 70°F	A/NR at 70°F	NR at 70°F	
METHYL BENZENE	NR 30-100% at 70°F	NR 30-100% at 70°F	NR 100% at 70°F	NR at 70°F	NR at 70°F	TEST
METHYL BENZOATE	NR at 70°F	NR at 70°F	NR at 70°F	NR at 70°F	NR at 70°F	
METHYL BICHLORIDE			NR at 70°F	NR at 70°F		
METHYL BROMIDE	NR at 70°F	C at 70°F	NR at 70°F	B/NR at 70°F	NR at 70°F	
METHYL BUTANOL	B at 70°F	NR at 70°F	C/NR at 70°F	A to 70°F AB any conc to 180°F	B at 70°F	AB to 70°F
METHYL BUTANONE				B at 70°F		
METHYL BUTYL KETONE	NR at 70°F	NR at 70°F	NR at 70°F	A to 70°F AB any conc to 150°F	NR at 70°F	
METHYL BUTYRATE				NR at 70°F		
METHYL CARBITOL				A to 70°F		
METHYL CARBONATE	NR at 70°F	NR at 70°F	NR at 70°F	NR at 70°F	NR at 70°F	
METHYL CELLOSOLVE	NR at 70°F	NR at 70°F	NR at 70°F	B at 70°F	NR at 70°F	
METHYL CELLULOSE	B at 70°F	NR at 70°F	B at 70°F	B at 70°F	B at 70°F	
METHYL CHLORIDE	NR at 70°F	NR at 70°F	NR at 70°F	NR at 70°F	NR at 70°F	TEST
METHYL CHLOROFORM	+217% vol 3 days 70°F NR at 70°F	+296% vol 3 days 70°F NR at 70°F	+263% vol 3 days 70°F NR at 70°F	NR at 70°F	NR at 70°F	NR at 70°F
METHYL CHLOROFORMATE	NR at 70°F	NR at 70°F	NR at 70°F	NR at 70°F	NR at 70°F	

	SYNTHETIC ISOPRENE IR	NATURAL ISOPRENE NR	CHLOROSULFONATED POLYETHYLENE CSM	FLUOROSILICONE FVMQ	SILICONE VMQ	CHEMRAZ FFKM
MERCURY	A to 70°F	A to 70°F	A to 70°F	AC to 70°F	A to 70°F	A to 70°F
MERCURY BICHLORIDE	B at 70°F	B any conc to 150°F	A to 70°F		A to 70°F	
MERCURY CHLORIDE	B at 70°F	B any conc to 150°F	A to 70°F		A to 70°F	
MERCURY SALTS			A to 70°F			A to 70°F
MERCURY VAPOR	A/NR at 70°F	A/NR at 70°F	A to 70°F			A to 70°F
MESITYL OXIDE	NR at 70°F	NR at 70°F	NR at 70°F	NR at 70°F	NR at 70°F	A to 70°F
METER-CRESOL					NR at 70°F	A to 70°F
METHADIENE	NR at 70°F	NR at 70°F	NR at 70°F			
METHALLYL ALCOHOL	A to 70°F	A to 70°F	A to 70°F			
METHANAL	B 40-100% at 70°F	B 100% at 70°F A to 40% to 80°F	AB 50-100% to 70°F A 40% to 70°F	NR 100% at 70°F B 40% at 70°F	B 40-100% at 70°F A to 40% to 70°F	A to 70°F
METHANAMIDE	A to 70°F	A to 70°F				
METHANE	B/NR at 70°F	BC at 70°F AB to 70°F dry	A to 70°F wet or dry	AC to 70°F	A/NR at 70°F	A to 70°F
METHANOL	A to 70°F	A to 100% to 70°F AB any conc to 100°F	A to 100% to 70°F A 100% to 158°F	no swell 120 days 70°F +8% vol 3 days 158°F	+6% vol 120 days 70°F no swell 3 days 158°F	A to 70°F
METHANOL				A to 158°F	A to 100% to 70°F A to 158°F	
METHYL ACETATE	NR at 70°F	NR at 70°F	B/NR at 70°F	NR at 70°F	NR at 70°F	A to 70°F
METHYL ACETOACETATE	NR at 70°F	NR at 70°F	NR at 70°F	NR at 70°F	B at 70°F	A to 70°F
METHYL ACETONE	C at 70°F	AC to 70°F	NR at 70°F			
METHYL ACETYLENE						
METHYL ACRYLATE	NR at 70°F	NR at 70°F	NR at 70°F	NR at 70°F	NR at 70°F	A to 70°F
METHYL ACRYIC ACID	NR at 70°F	NR at 70°F	NR at 70°F	NR at 70°F	NR at 70°F	A to 70°F
METHYL ALCOHOL	A to 70°F	A to 100% to 70°F AB any conc to 100°F	A to 100% to 70°F A 100% to 158°F	no swell 120 days 70°F +8% vol 3 days 158°F	+6% vol 120 days 70°F no swell 3 days 158°F	A to 70°F
METHYL ALCOHOL				A to 158°F	A to 100% to 70°F A to 158°F	
METHYL ALCOHOL-WOOD	A to 70°F	A to 100% to 70°F AB any conc to 100°F	A to 100% to 70°F A 100% to 158°F	A to 158°F	A to 100% to 70°F A to 158°F	A to 70°F
METHYL AMINE	A/NR at 70°F	A/NR at 70°F	A/NR at 70°F			A to 70°F
METHYL AMYL AMINE	NR at 70°F	NR at 70°F	B at 70°F			
METHYL AMYL ALCOHOL	AB to 70°F	AB to 70°F	A to 70°F			A to 70°F
METHYL AMYL CARBINOL	A to 70°F	A to 70°F	A to 70°F			
METHYL AMYL KETONE	NR at 70°F	NR at 70°F	NR at 70°F			A to 70°F
METHYL ANILINE	A/NR at 70°F	A/NR at 70°F	NR at 70°F			
METHYL BENZENE	NR at 70°F	NR at 70°F	NR at 70°F	B at 70°F	NR 30-100% at 70°F	A to 70°F
METHYL BENZOATE	NR at 70°F	NR at 70°F	NR at 70°F	A to 70°F	NR at 70°F	A to 70°F
METHYL BICHLORIDE	NR at 70°F	NR at 70°F	NR at 70°F			
METHYL BROMIDE	C/NR at 70°F	C/NR at 70°F	NR at 70°F	A to 70°F		A to 70°F
METHYL BUTANOL	AB to 70°F	A to 70°F AB any conc to 150°F	A to 200°F	A to 70°F	NR at 70°F	A to 70°F
METHYL BUTANONE	NR at 70°F	NR at 70°F	NR at 70°F			
METHYL BUTYL KETONE	NR at 70°F	NR at 70°F	NR at 70°F	NR at 70°F	NR at 70°F	A to 70°F
METHYL BUTYRATE	NR at 70°F	NR at 70°F				
METHYL CARBITOL	NR at 70°F	NR at 70°F	A to 70°F			
METHYL CARBONATE	NR at 70°F	NR at 70°F	NR at 70°F	B at 70°F	NR at 70°F	A to 70°F
METHYL CELLOSOLVE	NR at 70°F	NR at 70°F	B/NR at 70°F	NR at 70°F	NR at 70°F	A to 70°F
METHYL CELLULOSE	B at 70°F	B at 70°F	B at 70°F	NR at 70°F	B at 70°F	A to 70°F
METHYL CHLORIDE	NR at 70°F	NR at 70°F	NR at 70°F	B at 70°F	NR at 70°F	A to 70°F
METHYL CHLOROFORM	NR at 70°F	NR at 70°F	NR at 70°F	+36% vol 3 days 70°F B/NR at 70°F	+99% vol 3 days 70°F NR at 70°F	A to 70°F
METHYL CHLOROFORMATE	NR at 70°F	NR at 70°F	NR at 70°F	B at 70°F	NR at 70°F	A to 70°F

	POLYSULFIDE T	CHLORINATED POLYETHYLENE CM	EPICHLOROHYDRIN CO, ECO	® HYTREL COPOLYESTER TPE	® SANTOPRENE COPOLYMER TPO	® C-FLEX STYRENIC TPE
MERCURY	A to 70°F	AB to 70°F	A to 70°F	A to 70°F *AB to 140°F		
MERCURY BICHLORIDE						
MERCURY CHLORIDE						AB to 70°F
MERCURY SALTS						
MERCURY VAPOR	B at 70°F					
MESITYL OXIDE						
METER-CRESOL						
METHADIENE						
METHALLYL ALCOHOL						
METHANAL	B at 70°F	AB to 70°F	B at 70°F	BC 100% at 70°F AC 40% at 70°F	A to 70°F	A to 70°F
METHANAMIDE						
METHANE	AB to 70°F wet or dry	AB to 70°F	A to 70°F	AB to 70°F	NR at 70°F	NR at 70°F
METHANOL	+14.5% vol 3 days 158°F B at 70°F	AB to 70°F	B at 70°F	A to 70°F	+1% vol 7 days 70°F A to 70°F	A to 70°F
METHANOL						
METHYL ACETATE	B at 70°F		NR at 70°F	C at 70°F		
METHYL ACETOACETATE	B at 70°F					
METHYL ACETONE	AB to 70°F					
METHYL ACETYLENE						
METHYL ACRYLATE	B at 70°F				NR at 70°F	
METHYL ACRYIC ACID						
METHYL ALCOHOL	+14.5% vol 3 days 158°F B at 70°F	AB to 70°F	B at 70°F	A to 70°F	+1% vol 7 days 70°F A to 70°F	A to 70°F
METHYL ALCOHOL						
METHYL ALCOHOL-WOOD	B at 70°F	AB to 70°F	B at 70°F	A to 70°F	A to 70°F	A to 70°F
METHYL AMINE	A to 70°F					
METHYL AMYL AMINE						
METHYL AMYL ALCOHOL						
METHYL AMYL CARBINOL						
METHYL AMYL KETONE						
METHYL ANILINE	NR at 70°F					
METHYL BENZENE	B/NR at 70°F	C at 70°F	NR at 70°F	BC at 70°F NR at 140°F	NR at 70°F	NR at 70°F
METHYL BENZOATE	B at 70°F					
METHYL BICHLORIDE						
METHYL BROMIDE				NR at 70°F	NR at 70°F	
METHYL BUTANOL	B at 70°F	A to 70°F	A to 70°F	A to 70°F	A to 70°F	NR at 70°F
METHYL BUTANONE						
METHYL BUTYL KETONE	A to 70°F					
METHYL BUTYRATE						
METHYL CARBITOL						
METHYL CARBONATE	NR at 70°F					
METHYL CELLOSOLVE					*A to 70°F	
METHYL CELLULOSE	NR at 70°F					
METHYL CHLORIDE	NR at 70°F	C at 70°F		NR at 70°F	NR at 70°F	B/NR at 70°F
METHYL CHLOROFORM	NR at 70°F	C at 70°F		NR at 70°F	NR at 70°F	B/NR at 70°F
METHYL CHLOROFORMATE	NR at 70°F					

	ETHYLENE ACRYLIC EA	POLYALLOMER LINEAR COPOLYMER	NYLON 11 POLYAMIDE	NYLON 12 POLYAMIDE	ETHYLENE VINYL ACETATE EVA	POLYVINYLCHLORIDE FLEXIBLE PVC
MERCURY	A to 70°F	A to 70°F	A to 70°F AB 140–194°F	AB to 70°F	A to 70°F *AB to 140°F	AB to 70°F *AB to 140°F
MERCURY BICHLORIDE						
MERCURY CHLORIDE	*AB sol'n to 70°F					
MERCURY SALTS				AB to 70°F		A to 70°F
MERCURY VAPOR						NR at 70°F
MESITYL OXIDE						
METER-CRESOL						
METHADIENE						
METHALLYL ALCOHOL						
METHANAL		A to conc to 70°F B 40% at 122°F	A to conc to 70°F C conc at 105°F	C to 40% at 70°F	A/NR 100% at 70°F AB 40% to 70°F	AB to conc to 70°F C to 40% 122–130°F
METHANAMIDE						
METHANE	A to 70°F	B at 70°F	A to 70°F AB 140°F		C at 70°F	AB to 70°F
METHANOL	A to 70°F	A to 70°F	+11% vol 28 days 70°F +14% vol 180 days 70°F	C 50–100% at 70°F AB 10% to 70°F	B at 70°F NR at 140°F	AB to 100% to 70°F C/NR 122–150°F
METHANOL			AB to 70°F NR at 140°F		*C 6% at 140°F	
METHYL ACETATE	NR at 70°F		A to 70°F AB to 140°F		*NR at 70°F	NR at 70°F
METHYL ACETOACETATE	NR at 70°F					
METHYL ACETONE						A to 70°F
METHYL ACETYLENE						AB to 70°F
METHYL ACRYLATE	NR at 70°F		C at 70°F			
METHYL ACRYIC ACID	NR at 70°F					
METHYL ALCOHOL	A to 70°F	A to 70°F	+11% vol 28 days 70°F +14% vol 180 days 70°F	C 50–100% at 70°F AB 10% to 70°F	B at 70°F NR at 140°F	AB to 100% to 70°F C/NR 122–150°F
METHYL ALCOHOL			AB to 70°F NR at 140°F		*C 6% at 140°F	
METHYL ALCOHOL-WOOD	A to 70°F	A to 70°F	AB to 70°F NR at 140°F	C 50–100% at 70°F AB 10% to 70°F	B at 70°F NR at 140°F	AB to 100% to 70°F C/NR 122–150°F
METHYL AMINE						NR at 70°F
METHYL AMYL AMINE						
METHYL AMYL ALCOHOL						
METHYL AMYL CARBINOL						
METHYL AMYL KETONE						
METHYL ANILINE						
METHYL BENZENE	NR at 70°F	B at 70°F NR at 122°F	A to 70°F C at 194°F	AB to 70°F	NR at 70°F	NR at 70°F
METHYL BENZOATE						
METHYL BICHLORIDE						
METHYL BROMIDE			AB to 70°F NR at 105°F			C/NR at 70°F
METHYL BUTANOL	B/NR at 70°F	B at 70°F	A to 70°F *BC at 140°F	AB to 70°F	A to 70°F *NR at 140°F	AB to 70°F C at 70°F food grades
METHYL BUTANONE						
METHYL BUTYL KETONE	NR at 70°F					NR at 70°F
METHYL BUTYRATE						
METHYL CARBITOL						
METHYL CARBONATE						
METHYL CELLOSOLVE	NR at 70°F					A to 70°F
METHYL CELLULOSE	A to 70°F					
METHYL CHLORIDE	NR at 70°F	NR at 70°F	C at 70°F NR at 140°F		NR at 70°F	NR at 70°F
METHYL CHLOROFORM	NR at 70°F	NR at 70°F	C at 70°F NR at 140°F		NR at 70°F	NR at 70°F
METHYL CHLOROFORMATE						

	POLYETHYLENE LOW DENSITY LDPE	®TEFLON FEP	® KALREZ PERFLUORINATED ELASTOMER	® FLUORAZ FLUORINATED COPOLYMER	® AFLAS FLUORINATED COPOLYMER	®NORPRENE COPOLYMER TPO
MERCURY	A to 70°F AB to 140°F	A to 300°F	A to 212°F	A to 70°F		
MERCURY BICHLORIDE		A to 70°F				
MERCURY CHLORIDE		A to 70°F	A to 212°F			
MERCURY SALTS	AB to 140°F	A to 70°F	A to 212°F	A to 70°F		A to 70°F
MERCURY VAPOR		A to 70°F		A to 70°F		
MESITYL OXIDE		A to 70°F	A to 212°F	NR at 70°F	NR at 70°F	
METER-CRESOL			A to 212°F			
METHADIENE		A to 70°F				
METHALLYL ALCOHOL		A to 70°F				
METHANAL	AB to 100% to 70°F A to 40% to 70°F	A to conc to 300°F	A to 104°F	NR 40-100% at 70°F		A to 70°F
METHANAMIDE		A to 70°F				
METHANE	AC to 70°F	A to 300°F	A to 212°F	B/NR at 70°F		A to 70°F
METHANOL	AB to 100% to 122°F B/NR at 140°F	A to 300°F	A 50-100% to 212°F	A to 70°F	+0.2% vol 7 days 70°F A to 70°F	A to 70°F
METHANOL						
METHYL ACETATE	B/NR at 70°F	A to 70°F	A to 212°F	NR at 70°F		
METHYL ACETOACETATE		A to 70°F	A to 212°F	NR at 70°F		
METHYL ACETONE	NR at 70°F	A to 70°F				
METHYL ACETYLENE	AB to 70°F	A to 70°F				
METHYL ACRYLATE		A to 70°F	A to 212°F	NR at 70°F		
METHYL ACRYIC ACID		A to boiling	A to 150°F	B at 70°F		
METHYL ALCOHOL	AB to 100% to 122°F B/NR at 140°F	A to 300°F	A 50-100% to 212°F	A to 70°F	+0.2% vol 7 days 70°F A to 70°F	A to 70°F
METHYL ALCOHOL						
METHYL ALCOHOL-WOOD	AB to 100% to 122°F B/NR at 140°F	A to 300°F	A 50-100% to 212°F	A to 70°F	A to 70°F	A to 70°F
METHYL AMINE	AB to 70°F	A to 70°F	*A to 212°F			
METHYL AMYL AMINE		A to 70°F				
METHYL AMYL ALCOHOL		A to 70°F				
METHYL AMYL CARBINOL		A to 70°F				
METHYL AMYL KETONE		A to 70°F	A to 212°F			
METHYL ANILINE		A to 70°F	A to 70°F			
METHYL BENZENE	B/NR at 70°F NR at 122°F	A to 300°F	A to 75°F	NR at 70°F	NR at 70°F	NR at 70°F
METHYL BENZOATE			A to 212°F	A to 70°F		
METHYL BICHLORIDE		A to 70°F				
METHYL BROMIDE	NR at 70°F	A to 300°F	A to 70°F	AC at 70°F		
METHYL BUTANOL	AB to 70°F B/NR at 140°F	A to 70°F	A to 212°F	A to 70°F	C at 70°F	BC at 70°F
METHYL BUTANONE		A to 70°F				
METHYL BUTYL KETONE		A to 70°F	A to 212°F	NR at 70°F		
METHYL BUTYRATE		A to 70°F				
METHYL CARBITOL		A to 70°F				
METHYL CARBONATE			A to 212°F	A to 70°F		
METHYL CELLOSOLVE		A to 300°F	A to 212°F	A to 70°F	+1.4% vol 7 days 70°F A to 70°F	
METHYL CELLULOSE			A to 212°F	A to 70°F		
METHYL CHLORIDE	NR at 70°F	A to 200°F wet or dry	A to 212°F	NR at 70°F wet or dry		NR at 70°F
METHYL CHLOROFORM	NR at 70°F	A to boiling	A to 70°F *AB to 212°F	NR at 70°F wet or dry	+125% vol 7 days 70°F NR at 70°F	NR at 70°F
METHYL CHLOROFORMATE			A to 212°F	A to 70°F		

	RECOMMENDATIONS	NITRILE NBR	ETHYLENE PROPYLENE EP, EPDM	FLUOROCARBON FKM	CHLOROPRENE CR	HYDROGENATED NITRILE HNBR
METHYL CYANIDE	CR,NR	C at 70°F	A to 70°F	C/NR at 70°F	A to 70°F	
METHYL CYCLOHEXANE		NR at 70°F		B at 70°F		
METHYL CYCLOPENTANE	FKM,FVMQ,T	NR at 70°F	NR at 70°F	A to 250°F	NR at 70°F	NR at 70°F
METHYL D-BROMIDE	FKM	NR at 70°F		A to 70°F	NR at 70°F	
METHYL DICHLORIDE		NR at 70°F	NR at 70°F	A to 140°F	NR at 70°F	
METHYL DIPHENYL DIISOCYANATE					NR at 70°F	
METHYL ETHER	EP,NBR,FKM,IIR	A to 70°F	A to 70°F	A to 70°F	C at 70°F	A to 70°F
METHYL ETHYL KETONE	EP,IIR	NR any conc at 70°F	A to 140°F AB to 200°F	+240% vol 7 days 70°F NR at 70°F	NR any conc at 70°F	
METHYL ETHYL KETONE PEROXIDE	VMQ	NR at 70°F	NR at 70°F	NR at 70°F	NR at 70°F	
METHYL FORMATE	CR,EP,IIR	NR at 70°F	AB to 120°F	NR at 70°F	B at 70°F	NR at 70°F
METHYL HEXANE		A to 70°F	NR at 70°F	A to 70°F	AB to 70°F	
METHYL HEXANOL		A to 70°F		B at 70°F		
METHYL HEXANONE		NR at 70°F		NR at 70°F		
METHYL HEXYL CARBINOL		AB to 100°F			NR at 70°F	
METHYL HEXYL KETONE		NR at 70°F		NR at 70°F		
METHYL HYDRATE		A to 70°F	B at 70°F	C at 70°F	A to 70°F	
METHYL HYDRIDE		A to 70°F	NR at 70°F	A to 70°F	B at 70°F	
METHYL HYDROXIDE		A to 70°F	B at 70°F	C at 70°F	A to 70°F	
METHYL IODIDE		NR at 70°F	A to 70°F		NR at 70°F	
METHYL ISOBUTENYL KETONE		NR at 70°F			NR at 70°F	
METHYL ISOBUTYL CARBINOL		AB to 70°F	A to 70°F	A/NR at 70°F	A to 70°F	
METHYL ISOBUTYL KETONE	EP,T,IIR	NR at 70°F	AB to 130°F C at 140°F	+203% vol 7 days 70°F NR at 70°F	NR at 70°F	NR at 70°F
METHYL ISOPROPYL KETONE	EP,T,IIR,VMQ	NR at 70°F	BC to 140°F	+203% vol 7 days 70°F NR at 70°F	NR at 70°F	
METHYL MERCAPTAN	EP,IIR		A to 70°F		variable	
METHYL METHACRYLATE	T	NR at 70°F	NR at 70°F	NR at 70°F	NR at 70°F	NR at 70°F
METHYL METHANE		A to 70°F	NR at 70°F	A to 70°F	B at 70°F	
METHYL NORMAL AMYL KETONE		NR at 70°F		NR at 70°F		
METHYL OLEATE	FKM,EP,IIR	NR at 70°F	B at 70°F	A to 70°F	NR at 70°F	NR at 70°F
METHYL PHENOL	FKM,FVMQ	C/NR at 70°F	NR any conc at 70°F	A to 104°F AB to 350°F	NR at 70°F	
METHYL POLYSILOXANES	EP,NBR,FKM,IIR,CR	A to 140°F AB to 180°F	A to 140°F	A to 400°F	A to 140°F	A to 70°F
METHYL PROPANOL		A any conc to 80°F	AB to 70°F	A to 70°F	A any conc to 80°F	
METHYL PROPYL BENZENE		NR at 70°F			NR at 70°F	
METHYL PROPYL CARBINOL		A to 70°F		B at 70°F		
METHYL PROPYL ETHER		NR at 70°F				
METHYL PROPYL KETONE		NR at 70°F	B at 70°F	NR at 70°F	NR at 70°F	
METHYL PROPYL SALICYLATE	EP,IIR	NR at 70°F	B at 70°F	B at 70°F	NR at 70°F	
METHYL SULFATE					C/NR at 70°F	
METHYLACETAL		NR at 70°F	A to 70°F	NR at 70°F	C at 70°F	
METHYLACRYLIC ACID			B at 70°F	B at 70°F	B at 70°F	
METHYLALLYL ACETATE		NR at 70°F		NR at 70°F		
METHYLALLYL CHLORIDE		NR at 70°F		C at 70°F		
METHYLATED SPIRITS		A to 70°F	A to 70°F	B at 70°F	AB to 70°F	
METHYLENE BROMIDE		NR at 70°F	NR at 70°F	AC to 70°F	NR at 70°F	
METHYLENE CHLORIDE	FKM,FVMQ	NR at 70°F	BC to 130°F NR at 140°F	+29% vol 28 days 100°F B at 70°F	NR at 70°F	

	STYRENE BUTADIENE SBR	POLYACRYLATE ACM	POLYURETHANE AU, EU	ISOBUTYLENE ISOPRENE IIR	POLYBUTADIENE BR	® AEROQUIP AQP
METHYL CYANIDE				A to 70°F		
METHYL CYCLOHEXANE				NR at 70°F		
METHYL CYCLOPENTANE	NR at 70°F	NR at 70°F	NR at 70°F	NR at 70°F	NR at 70°F	
METHYL D-BROMIDE	NR at 70°F	NR at 70°F	NR at 70°F		NR at 70°F	
METHYL DICHLORIDE						
METHYL DIPHENYL DIISOCYANATE				AB to 70°F		
METHYL ETHER	A to 70°F	NR at 70°F		AC to 70°F	A to 70°F	
METHYL ETHYL KETONE	NR at 70°F	NR at 70°F	NR at 70°F	A any conc to 100°F	NR at 70°F	A to 70°F
METHYL ETHYL KETONE PEROXIDE	NR at 70°F	NR at 70°F	NR at 70°F	NR at 70°F	NR at 70°F	
METHYL FORMATE	NR at 70°F	NR at 70°F	NR at 70°F	B at 70°F	NR at 70°F	
METHYL HEXANE				NR at 70°F		
METHYL HEXANOL				A to 70°F		
METHYL HEXANONE				B at 70°F		
METHYL HEXYL CARBINOL				NR at 70°F		
METHYL HEXYL KETONE				B at 70°F		
METHYL HYDRATE			NR at 70°F	A to 70°F		
METHYL HYDRIDE			B at 70°F	NR at 70°F		
METHYL HYDROXIDE			NR at 70°F	A to 70°F		
METHYL IODIDE				A to 70°F		
METHYL ISOBUTENYL KETONE				NR at 70°F		
METHYL ISOBUTYL CARBINOL				A to 70°F		
METHYL ISOBUTYL KETONE	NR at 70°F	NR at 70°F	NR at 70°F	C 100% at 70°F AB to 75% to 70°F	NR at 70°F	
METHYL ISOPROPYL KETONE	NR at 70°F	NR at 70°F	NR at 70°F	B at 70°F	NR at 70°F	NR at 70°F
METHYL MERCAPTAN				B at 70°F		
METHYL METHACRYLATE	NR at 70°F	NR at 70°F	NR at 70°F	B /NR at 70°F	NR at 70°F	
METHYL METHANE			B at 70°F	NR at 70°F		
METHYL NORMAL AMYL KETONE				B at 70°F		
METHYL OLEATE	NR at 70°F			B at 70°F	NR at 70°F	
METHYL PHENOL			NR at 70°F	NR at 70°F		
METHYL POLYSILOXANES	A to 70°F	A to 70°F	A to 70°F	A to 70°F	A to 70°F	
METHYL PROPANOL			NR at 70°F	A any conc to 100°F		
METHYL PROPYL BENZENE				NR at 70°F		
METHYL PROPYL CARBINOL				A to 70°F		
METHYL PROPYL ETHER				NR at 70°F		
METHYL PROPYL KETONE				B at 70°F		
METHYL PROPYL SALICYLATE				B any conc to 80°F		
METHYL SULFATE			AB to 70°F			
METHYLACETAL			NR at 70°F	A to 70°F		
METHYLACRYLIC ACID				B at 70°F		
METHYLALLYL ACETATE				A to 70°F		
METHYLALLYL CHLORIDE				C at 70°F		
METHYLATED SPIRITS			C /NR at 70°F *NR at 140°F	A to 70°F		
METHYLENE BROMIDE				NR at 70°F		
METHYLENE CHLORIDE	NR at 70°F	NR at 70°F	NR at 70°F	NR at 70°F	NR at 70°F	

C 237

	SYNTHETIC ISOPRENE IR	NATURAL ISOPRENE NR	CHLOROSULFONATED POLYETHYLENE CSM	FLUOROSILICONE FVMQ	SILICONE VMQ	CHEMRAZ FFKM
METHYL CYANIDE	B at 70°F	B at 70°F	B at 70°F		TEST	A to 70°F
METHYL CYCLOHEXANE	NR at 70°F	NR at 70°F	NR at 70°F			
METHYL CYCLOPENTANE	NR at 70°F	NR at 70°F	NR at 70°F	B at 70°F	B at 70°F	A to 70°F
METHYL D-BROMIDE	NR at 70°F	NR at 70°F	NR at 70°F	B at 70°F	NR at 70°F	
METHYL DICHLORIDE		NR at 70°F				
METHYL DIPHENYL DIISOCYANATE					NR at 70°F	
METHYL ETHER	A/NR at 70°F	A/NR at 70°F	BC at 70°F	A to 70°F	A to 70°F	A to 70°F
METHYL ETHYL KETONE	NR at 70°F	NR at 70°F	NR at 70°F	NR at 70°F	NR at 70°F	A to 70°F
METHYL ETHYL KETONE PEROXIDE	NR at 70°F	NR at 70°F	NR at 70°F	NR at 70°F	B at 70°F	A to 70°F
METHYL FORMATE	NR at 70°F	NR at 70°F	B at 70°F	NR at 70°F	B at 70°F	A to 70°F
METHYL HEXANE	NR at 70°F	NR at 70°F	B/NR at 70°F			
METHYL HEXANOL	A to 70°F	A to 70°F	A to 70°F			
METHYL HEXANONE	NR at 70°F	NR at 70°F	NR at 70°F			
METHYL HEXYL CARBINOL	NR at 70°F	NR at 70°F				
METHYL HEXYL KETONE	NR at 70°F	NR at 70°F	NR at 70°F			
METHYL HYDRATE	A to 70°F	A to 70°F	A to 70°F		A to 70°F	
METHYL HYDRIDE	NR at 70°F	NR at 70°F	B at 70°F		NR at 70°F	
METHYL HYDROXIDE	A to 70°F	A to 70°F	A to 70°F		A to 70°F	
METHYL IODIDE	A to 70°F	A to 70°F				A to 70°F
METHYL ISOBUTENYL KETONE		NR at 70°F				A to 70°F
METHYL ISOBUTYL CARBINOL	AB to 70°F	AB to 70°F	A to 70°F			A to 70°F
METHYL ISOBUTYL KETONE	NR at 70°F	NR at 70°F	NR at 70°F	NR at 70°F	C/NR at 70°F	A to 70°F
METHYL ISOPROPYL KETONE	NR at 70°F	NR at 70°F	NR at 70°F	NR at 70°F	B/NR at 70°F	
METHYL MERCAPTAN						
METHYL METHACRYLATE	NR at 70°F	NR at 70°F	NR at 70°F	NR at 70°F	C/NR at 70°F	A to 70°F
METHYL METHANE	NR at 70°F	NR at 70°F	B at 70°F		NR at 70°F	
METHYL NORMAL AMYL KETONE	NR at 70°F	NR at 70°F	NR at 70°F			A to 70°F
METHYL OLEATE	NR at 70°F	NR at 70°F	NR at 70°F	B at 70°F		A to 70°F
METHYL PHENOL	NR at 70°F	NR at 70°F	B/NR at 70°F	AB to 70°F	B/NR at 70°F	A to 70°F (m)
METHYL POLYSILOXANES	AC to 70°F	AC to 70°F	A to 70°F	A to 70°F	C/NR at 70°F	A to 70°F
METHYL PROPANOL	A to 70°F	A any conc to 80°F	A to 70°F		B at 70°F	
METHYL PROPYL BENZENE	NR at 70°F	NR at 70°F				
METHYL PROPYL CARBINOL	A to 70°F	A to 70°F	A to 70°F			
METHYL PROPYL ETHER	NR at 70°F	NR at 70°F	B at 70°F			
METHYL PROPYL KETONE	NR at 70°F	NR at 70°F	NR at 70°F			
METHYL PROPYL SALICYLATE	NR at 70°F	NR at 70°F	NR at 70°F			
METHYL SULFATE						
METHYLACETAL	B at 70°F	B at 70°F	B at 70°F		A to 70°F	
METHYLACRYLIC ACID	NR at 70°F	NR at 70°F				A to 70°F
METHYLALLYL ACETATE	NR at 70°F	NR at 70°F	B at 70°F			
METHYLALLYL CHLORIDE	NR at 70°F	NR at 70°F	NR at 70°F			
METHYLATED SPIRITS	A to 70°F	A to 70°F	A to 70°F		A to 70°F	
METHYLENE BROMIDE	NR at 70°F	NR at 70°F	NR at 70°F			A to 70°F
METHYLENE CHLORIDE	NR at 70°F	NR at 70°F	NR at 70°F	+14% vol 120 days 70°F B at 70°F	-4% vol 120 days 70°F NR at 70°F	A to 70°F

	POLYSULFIDE T	CHLORINATED POLYETHYLENE CM	EPICHLOROHYDRIN CO, ECO	® HYTREL COPOLYESTER TPE	® SANTOPRENE COPOLYMER TPO	® C-FLEX STYRENIC TPE
METHYL CYANIDE					*A to 70°F	AC to 70°F
METHYL CYCLOHEXANE						
METHYL CYCLOPENTANE	B at 70°F					
METHYL D-BROMIDE	NR at 70°F					
METHYL DICHLORIDE						
METHYL DIPHENYL DIISOCYANATE						
METHYL ETHER	A to 70°F					
METHYL ETHYL KETONE	AB to 70°F	AB to 70°F	NR at 70°F	A to 70°F	−4% to −11% vol 3 days 70°F A/NR at 70°F	
METHYL ETHYL KETONE PEROXIDE	NR at 70°F					
METHYL FORMATE	B/NR at 70°F		NR at 70°F			
METHYL HEXANE						
METHYL HEXANOL						
METHYL HEXANONE						
METHYL HEXYL CARBINOL						
METHYL HEXYL KETONE						
METHYL HYDRATE						
METHYL HYDRIDE						
METHYL HYDROXIDE						
METHYL IODIDE					*A to 70°F	
METHYL ISOBUTENYL KETONE						
METHYL ISOBUTYL CARBINOL						
METHYL ISOBUTYL KETONE	B at 70°F	NR at 70°F	NR at 70°F	+6−42% vol 7 days 70°F C/NR at 70°F	NR at 70°F	
METHYL ISOPROPYL KETONE	B at 70°F	NR at 70°F		NR at 70°F		
METHYL MERCAPTAN	B at 70°F					
METHYL METHACRYLATE	AB to 160°F		NR at 70°F		C at 70°F	
METHYL METHANE						
METHYL NORMAL AMYL KETONE						
METHYL OLEATE						
METHYL PHENOL						
METHYL POLYSILOXANES	A to 70°F softened		A to 70°F	AB to 70°F		A to 70°F
METHYL PROPANOL						
METHYL PROPYL BENZENE						
METHYL PROPYL CARBINOL						
METHYL PROPYL ETHER						
METHYL PROPYL KETONE						
METHYL PROPYL SALICYLATE						
METHYL SULFATE				AB to 70°F		
METHYLACETAL						
METHYLACRYLIC ACID						
METHYLALLYL ACETATE						
METHYLALLYL CHLORIDE						
METHYLATED SPIRITS	*AB to 70°F					
METHYLENE BROMIDE						
METHYLENE CHLORIDE	NR at 70°F			NR at 70°F	NR at 70°F	

C 239

	ETHYLENE ACRYLIC EA	POLYALLOMER LINEAR COPOLYMER	NYLON 11 POLYAMIDE	NYLON 12 POLYAMIDE	ETHYLENE VINYL ACETATE EVA	POLYVINYLCHLORIDE FLEXIBLE PVC
METHYL CYANIDE		C at 70°F NR at 122°F				NR at 70°F
METHYL CYCLOHEXANE						
METHYL CYCLOPENTANE	NR at 70°F				NR at 70°F	NR at 70°F
METHYL D-BROMIDE						
METHYL DICHLORIDE						
METHYL DIPHENYL DIISOCYANATE						NR at 70°F
METHYL ETHER						
METHYL ETHYL KETONE	NR at 70°F	A to 70°F NR at 122°F	AB to 105°F NR at 194°F	AB to 70°F	*C at 70°F *NR at 140°F	NR at 70°F
METHYL ETHYL KETONE PEROXIDE	NR at 70°F					
METHYL FORMATE						
METHYL HEXANE						
METHYL HEXANOL						
METHYL HEXANONE						
METHYL HEXYL CARBINOL						NR at 70°F
METHYL HEXYL KETONE						
METHYL HYDRATE						
METHYL HYDRIDE						
METHYL HYDROXIDE						
METHYL IODIDE						
METHYL ISOBUTENYL KETONE						
METHYL ISOBUTYL CARBINOL						
METHYL ISOBUTYL KETONE	NR at 70°F	B at 70°F C at 122°F	AB to 105°F NR at 194°F		*NR at 70°F	NR at 70°F
METHYL ISOPROPYL KETONE	NR at 70°F		A to 70°F			NR at 70°F
METHYL MERCAPTAN						
METHYL METHACRYLATE	NR at 70°F		C at 70°F			NR at 70°F
METHYL METHANE						
METHYL NORMAL AMYL KETONE						
METHYL OLEATE						
METHYL PHENOL						
METHYL POLYSILOXANES	A to 70°F		A to 70°F	AB to 70°F	B at 70°F	AB to 70°F
METHYL PROPANOL						
METHYL PROPYL BENZENE						NR at 70°F
METHYL PROPYL CARBINOL						
METHYL PROPYL ETHER						
METHYL PROPYL KETONE		B at 70°F C at 122°F				NR at 70°F
METHYL PROPYL SALICYLATE						NR at 70°F
METHYL SULFATE			AB to 70°F C 105–140°F			*C at 70°F *NR at 130°F
METHYLACETAL						
METHYLACRYLIC ACID						
METHYLALLYL ACETATE						
METHYLALLYL CHLORIDE						
METHYLATED SPIRITS			*AC to 70°F *NR at 140°F		*C at 70°F *NR at 140°F	AB to 70°F
METHYLENE BROMIDE						
METHYLENE CHLORIDE	NR at 70°F	C at 70°F NR at 122°F	AC to 70°F NR at 140°F	NR at 70°F	NR at 70°F	NR at 70°F

	POLYETHYLENE LOW DENSITY LDPE	®TEFLON FEP	®KALREZ PERFLUORINATED ELASTOMER	®FLUORAZ FLUORINATED COPOLYMER	®AFLAS FLUORINATED COPOLYMER	®NORPRENE COPOLYMER TPO
METHYL CYANIDE		A to 200°F	A to 212°F	A to 70°F		
METHYL CYCLOHEXANE		A to 70°F				
METHYL CYCLOPENTANE	A to 70°F	A to 70°F	A to 212°F	NR at 70°F		
METHYL D-BROMIDE						
METHYL DICHLORIDE			A to 212°F	NR at 70°F		
METHYL DIPHENYL DIISOCYANATE						
METHYL ETHER	AB to 70°F	A to 70°F	A to 212°F	NR at 70°F		
METHYL ETHYL KETONE	AB to 122°F C/NR at 140°F	A to 300°F	+1% vol 7 days 70°F A to 212°F	NR at 70°F	+58% vol 7 days 70°F NR at 70°F	BC at 70°F
METHYL ETHYL KETONE PEROXIDE			A to 212°F	NR at 70°F		
METHYL FORMATE		A to 70°F	A to 212°F	NR at 70°F		
METHYL HEXANE		A to 70°F				
METHYL HEXANOL		A to 70°F				
METHYL HEXANONE		A to 70°F				
METHYL HEXYL CARBINOL		A to 70°F				
METHYL HEXYL KETONE		A to 70°F	A to 212°F			
METHYL HYDRATE		A to 70°F				
METHYL HYDRIDE		A to 70°F				
METHYL HYDROXIDE		A to 70°F				
METHYL IODIDE		A to 70°F	A to 212°F			
METHYL ISOBUTENYL KETONE						
METHYL ISOBUTYL CARBINOL		A to 70°F	A to 212°F			
METHYL ISOBUTYL KETONE	C/NR at 70°F NR at 122°F	A to 300°F	A to 212°F	NR at 70°F	NR at 70°F	BC at 70°F
METHYL ISOPROPYL KETONE		A to 70°F	A to 212°F	NR at 70°F		
METHYL MERCAPTAN	AB to 70°F	A to 70°F				
METHYL METHACRYLATE		A to boiling	A to 320°F	NR at 70°F		
METHYL METHANE		A to 70°F				
METHYL NORMAL AMYL KETONE		A to 70°F	A to 212°F			
METHYL OLEATE		A to 70°F	A to 212°F	A to 70°F		
METHYL PHENOL	NR at 70°F	A to 376°F	A to 70°F			
METHYL POLYSILOXANES	A to 70°F	A to 300°F	A to 70°F	A to 70°F		TEST
METHYL PROPANOL		A to 70°F				
METHYL PROPYL BENZENE		A to 70°F				
METHYL PROPYL CARBINOL		A to 70°F				
METHYL PROPYL ETHER		A to 70°F				
METHYL PROPYL KETONE	B at 70°F C at 122°F	A to 122°F				
METHYL PROPYL SALICYLATE		A to 70°F	A to 70°F			
METHYL SULFATE						
METHYLACETAL		A to 70°F				
METHYLACRYLIC ACID		A to 70°F	A to 212°F			
METHYLALLYL ACETATE		A to 70°F				
METHYLALLYL CHLORIDE		A to 70°F				
METHYLATED SPIRITS	C at 70°F NR at 140°F	A to 70°F				
METHYLENE BROMIDE		A to 70°F	A to 212°F			
METHYLENE CHLORIDE	BC at 70°F NR at 122°F	A to 300°F	A to 300°F	B/NR at 70°F	B at 70°F	

	RECOMMENDATIONS	NITRILE NBR	ETHYLENE PROPYLENE EP, EPDM	FLUOROCARBON FKM	CHLOROPRENE CR	HYDROGENATED NITRILE HNBR
METHYLENE CHLOROBROMIDE		NR at 70°F		NR at 70°F	NR at 70°F	
METHYLENE DICHLORIDE	FKM, FVMQ	+243% vol 3 days 70°F NR at 70°F	NR at 70°F	+21% vol 3 days 75°F B to 160°F	+168% vol 3 days 70°F NR at 70°F	
M.I.B.K.	EP, T, IIR	NR at 70°F	AB to 130°F C at 140°F	+203% vol 7 days 75°F NR at 70°F	NR at 70°F	NR at 70°F
MILK	NBR, EP, FKM, CR	A to 140°F	A to 250°F	A to 212°F	A to 140°F	
MILK ACID		BC 50% to 80°F	B 50% at 70°F	A 50% to 70°F	B 50% to 80°F	
MILK OF MAGNESIA	EP, FKM, IIR, NBR, CR	A to 140°F AB any conc to 200°F	A to 176°F AB to 200°F	A to conc to 212°F	A to conc to 158°F AB any conc to 200°F	B at 70°F
MINE GUARD FR	NBR, CR	A to 70°F	variable		A to 70°F	
MINE WATER	FKM, EP, IIR	A to 140°F	A to 70°F	A to 180°F	C to 140°F	
MINERAL NAPHTHA		A to 70°F	NR at 70°F	A to 70°F	C at 70°F	
MINERAL OIL	NBR, FKM, ACM, AU IIR	A to 250°F	NR at 70°F	A to 70°F	A to 140°F	A to 70°F
MINERAL PITCH		A to 70°F AB to 150°F	NR at 70°F	A to 70°F	B to 80°F	
MINERAL SPIRITS	NBR	A to 250°F	NR at 70°F	A to 70°F	C/NR at 70°F	
MINERAL THINNER		NR at 70°F			NR at 70°F	
MINERS OIL		AB to 70°F			AB to 70°F	
MIRIBILITE		A to 70°F AB any conc to 200°F	A to 70°F	A to 70°F	A to 70°F AB any conc to 200°F	
MITTEL H.N.A.		NR at 70°F	A to 70°F	NR at 70°F	NR at 70°F	
MLO-7277	FKM	C/NR at 70°F	NR at 70°F	A to 70°F	NR at 70°F	C at 70°F
MLO-7557	FKM	C/NR at 70°F	NR at 70°F	A to 70°F	NR at 70°F	C at 70°F
MLO-8200	FKM, CR, AU	B at 70°F	NR at 70°F	no swell 28 days 350°F +2% vol 28 days 400°F	A to 70°F	B at 70°F
MLO-8515	FKM, CR, AU	B at 70°F	NR at 70°F	+3.6% vol 21 days 350°F A to 70°F	AB to 70°F	B at 70°F
M.N.V.		A to 70°F	A to 70°F	A to 70°F	A to 70°F	
MOBIL DELVAC 1100, 1110, 1120, 1130	NBR, FKM, ACM	A to 70°F	NR at 70°F	A to 70°F	B at 70°F	A to 70°F
MOBIL 24DTE	NBR, FKM, ACM	A to 70°F	NR at 70°F	A to 70°F	B at 70°F	A to 70°F
MOBIL HF	NBR, FKM, ACM	A to 70°F	NR at 70°F	A to 70°F	B at 70°F	A to 70°F
MOBIL NIVAC 20, 30	NBR, EP, FKM, CR	A to 70°F	A to 70°F	A to 70°F	A to 70°F	A to 70°F
MOBIL THERM 600	NBR, FKM, ACM	A to 70°F	NR at 70°F	A to 70°F	B at 70°F	A to 70°F
MOBIL VELOCITE C	NBR, FKM, ACM	A to 70°F	NR at 70°F	A to 70°F	B at 70°F	A to 70°F
MOBIL XRM 206A				A to 350°F		
MOBILGAS WA200 ATF		A to 70°F	NR at 70°F	A to 70°F	B at 70°F	A to 70°F
MOBILOIL SAE 20	NBR, FKM, ACM	A to 70°F	NR at 70°F	A to 70°F	B at 70°F	A to 70°F
MOBILUX	NBR, FKM, ACM	A to 70°F	NR at 70°F	A to 70°F	B at 70°F	A to 70°F
MOLASSES	NBR	A any conc to 200°F	A to 70°F	A to 140°F	A to 140°F AB any conc to 200°F	
MOLYBDIC ACID	NBR, EP, FKM	A to 70°F	A to 70°F	A to 70°F		
MOLTEN SULFUR		B	A	A	A	
MOLYSITE		A to 70°F AB any conc to 200°F	A to 70°F	A to 70°F	B any conc to 80°F	
MONOAMMONIUM PHOSPHATE	NBR, EP, IIR, CR, SBR	A any conc to 200°F	A to 176°F AB to 200°F	A to 176°F	A to 140°F AB to 200°F	
MONOBROMOBENZENE	FKM, T, FVMQ	NR at 70°F	NR at 70°F	A to 70°F	NR at 70°F	NR at 70°F
MONOBROMO TRIFLUOROMETHANE	CR, NBR, EP, FKM	A to 70°F	A to 70°F	A to 70°F	A to 70°F	
MONOBUTYL ETHER	T	AC to 150°F	C/NR at 70°F	NR at 70°F	B/NR at 70°F	
MONOCHLOROETHANE		C at 70°F	C at 70°F	A to 70°F	B/NR at 70°F	
MONOCHLOROACETIC ACID	EP, IIR	NR 10-100% at 70°F	AC to 200°F	B 50-100% at 70°F C 100% at 104°F	A to 70°F C at 104°F	
MONOCHLOROACETONE	EP, IIR	NR at 70°F	A to 70°F	B/NR at 70°F	C at 70°F	
MONOCHLOROBENZENE	FKM, T, FVMQ	NR at 70°F	NR at 70°F	A to 200°F	NR at 70°F	NR at 70°F
MONOCHLORODIFLUOROMETHANE (F 22)	CR, EP, T, SBR, IIR	NR at 70°F	A to 140°F	A/NR at 70°F	A to 250°F	

	STYRENE BUTADIENE SBR	POLYACRYLATE ACM	POLYURETHANE AU, EU	ISOBUTYLENE ISOPRENE IIR	POLYBUTADIENE BR	® AEROQUIP AQP
METHYLENE CHLOROBROMIDE				NR at 70°F		
METHYLENE DICHLORIDE	+164% vol 3 days 70°F NR at 70°F	+254% vol 3 days 70°F NR at 70°F	NR at 70°F	+74% vol 3 days 70°F NR at 70°F	NR at 70°F	
M.I.B.K.	NR at 70°F	NR at 70°F	NR at 70°F	C 100% at 70°F AB to 75% to 70°F	NR at 70°F	
MILK	A to 70°F	NR at 70°F	A/NR at 70°F	A to 70°F AB to 100°F	A to 70°F	
MILK ACID			B 50% at 70°F	AB 50% to 150°F		
MILK OF MAGNESIA	B to conc at 70°F	NR any conc at 70°F	A/NR at 70°F	A to 70°F AB any conc to 185°F	B at 70°F	AB to 70°F
MINE GUARD FR						
MINE WATER	NR at 70°F	NR at 70°F	C/NR at 70°F	A to 70°F		
MINERAL NAPHTHA			C at 70°F	NR at 70°F		
MINERAL OIL	NR at 70°F	A to 70°F	A to 70°F	C/NR at 70°F	NR at 70°F	A to 70°F
MINERAL PITCH			B at 70°F	NR at 70°F		
MINERAL SPIRITS	NR at 70°F	A to 70°F	B at 70°F	NR at 70°F		
MINERAL THINNER				NR at 70°F		
MINERS OIL						
MIRIBILITE			A to 70°F	A to 70°F AB any conc to 185°F		
MITTEL H.N.A.			NR at 70°F	A to 70°F		
MLO-7277	NR at 70°F	C at 70°F	C/NR at 70°F	NR at 70°F	NR at 70°F	
MLO-7557	NR at 70°F	C at 70°F	C/NR at 70°F	NR at 70°F	NR at 70°F	
MLO-8200	NR at 70°F		A to 70°F	NR at 70°F	NR at 70°F	
MLO-8515	NR at 70°F	C at 70°F	A/NR at 70°F	NR at 70°F	NR at 70°F	
M.N.V.			A to 70°F	A to 70°F		
MOBIL DELVAC 1100, 1110, 1120, 1130	NR at 70°F	A to 70°F	B at 70°F	NR at 70°F	NR at 70°F	
MOBIL 24DTE	NR at 70°F	A to 70°F	B at 70°F	NR at 70°F	NR at 70°F	
MOBIL HF	NR at 70°F	A to 70°F	B at 70°F	NR at 70°F	NR at 70°F	
MOBIL NIVAC 20, 30	A to 70°F			A to 70°F	A to 70°F	
MOBIL THERM 600	NR at 70°F	A to 70°F	B at 70°F	NR at 70°F	NR at 70°F	
MOBIL VELOCITE C	NR at 70°F	A to 70°F	B at 70°F	NR at 70°F	NR at 70°F	
MOBIL XRM 206A						
MOBILGAS WA200 ATF	NR at 70°F	A to 70°F	A to 70°F	NR at 70°F	NR at 70°F	
MOBILOIL SAE 20	NR at 70°F	A to 70°F	A to 70°F	NR at 70°F	NR at 70°F	
MOBILUX	NR at 70°F	A to 70°F	B at 70°F	NR at 70°F	NR at 70°F	
MOLASSES	A to 70°F	NR at 70°F	AB to 70°F	A to 70°F AB any conc to 185°F		
MOLYBDIC ACID						
MOLTEN SULFUR			B	B		
MOLYSITE			A to 70°F	A to 70°F AB any conc to 180°F		
MONOAMMONIUM PHOSPHATE	A to 70°F		B at 70°F	A any conc to 185°F	A to 70°F	NR at 70°F
MONOBROMOBENZENE	NR at 70°F	NR at 70°F	NR at 70°F	NR at 70°F	NR at 70°F	
MONOBROMO TRIFLUOROMETHANE				NR at 70°F		
MONOBUTYL ETHER	NR at 70°F	NR at 70°F	B at 70°F	C/NR at 70°F	NR at 70°F	
MONOCHLOROETHANE			C at 70°F	A/NR at 70°F		
MONOCHLOROACETIC ACID	NR at 70°F	NR at 70°F	NR at 70°F	B at 70°F AB 10% to 150°F	NR at 70°F	NR at 70°F
MONOCHLOROACETONE	NR at 70°F	NR at 70°F	NR at 70°F	BC at 70°F	NR at 70°F	
MONOCHLOROBENZENE	NR at 70°F	NR at 70°F	NR at 70°F	NR at 70°F	NR at 70°F	
MONOCHLORODIFLUOROMETHANE (F 22)	A to 70°F	B at 70°F	NR at 70°F	C/NR at 70°F	A to 70°F	NR at 70°F

	SYNTHETIC ISOPRENE IR	NATURAL ISOPRENE NR	CHLOROSULFONATED POLYETHYLENE CSM	FLUOROSILICONE FVMQ	SILICONE VMQ	CHEMRAZ FFKM
METHYLENE CHLOROBROMIDE	NR at 70°F	NR at 70°F				
METHYLENE DICHLORIDE	NR at 70°F	NR at 70°F	NR at 70°F	B at 70°F	+72% vol 3 days 70°F NR at 70°F	A to 70°F
M.I.B.K.	NR at 70°F	NR at 70°F	NR at 70°F	NR at 70°F	C/NR at 70°F	A to 70°F
MILK	A to 70°F	A to 70°F AB to 100°F	A to 70°F	A to 70°F	A to 70°F	
MILK ACID	C 50% at 70°F	BC 50% to 120°F	A 50% to 70°F		A 50% at 70°F	
MILK OF MAGNESIA	A to 70°F	A to 70°F AB any conc to 150°F	A to 100% to 158°F AB to 200°F		AB to 70°F	A to 70°F C at 70°F white
MINE GUARD FR						
MINE WATER	A to 70°F	AB to 70°F	A to 70°F		AB to 70°F	
MINERAL NAPHTHA	NR at 70°F	NR at 70°F	C at 70°F		C at 70°F	
MINERAL OIL	NR at 70°F	NR at 70°F	A to 70°F	A to 70°F	+6% vol 7 days 70°F B at 70°F	A to 70°F
MINERAL PITCH	NR at 70°F	NR at 70°F	B at 70°F		NR at 70°F	
MINERAL SPIRITS	NR at 70°F	NR at 70°F	B at 70°F	A to 70°F	+110% vol 7 days 70°F NR at 70°F	
MINERAL THINNER	NR at 70°F	NR at 70°F				
MINERS OIL						
MIRIBILITE	A to 70°F	A to 70°F AB any conc to 150°F	A to 70°F		A to 70°F	
MITTEL H.N.A.	C at 70°F	C at 70°F	NR at 70°F			
MLO-7277	NR at 70°F	NR at 70°F	NR at 70°F	C at 70°F	NR at 70°F	A to 70°F
MLO-7557	NR at 70°F	NR at 70°F	NR at 70°F	C at 70°F	NR at 70°F	A to 70°F
MLO-8200	NR at 70°F	NR at 70°F	NR at 70°F	A to 70°F	NR at 70°F	A to 70°F
MLO-8515	NR at 70°F	NR at 70°F	C at 70°F	A to 70°F	NR at 70°F	A to 70°F
M.N.V.	A to 70°F	A to 70°F	A to 70°F		A to 70°F	
MOBIL DELVAC 1100, 1110, 1120, 1130	NR at 70°F	NR at 70°F	NR at 70°F	A to 70°F	NR at 70°F	A to 70°F
MOBIL 24DTE	NR at 70°F	NR at 70°F	NR at 70°F	A to 70°F	NR at 70°F	A to 70°F
MOBIL HF	NR at 70°F	NR at 70°F	NR at 70°F	A to 70°F	NR at 70°F	A to 70°F
MOBIL NIVAC 20, 30	A to 70°F	A to 70°F	A to 70°F	A to 70°F	A to 70°F	A to 70°F
MOBIL THERM 600	NR at 70°F	NR at 70°F	NR at 70°F	A to 70°F	NR at 70°F	A to 70°F
MOBIL VELOCITE C	NR at 70°F	NR at 70°F	NR at 70°F	A to 70°F	NR at 70°F	A to 70°F
MOBIL XRM 206A						
MOBILGAS WA200 ATF	NR at 70°F	NR at 70°F	NR at 70°F	A to 70°F	NR at 70°F	A to 70°F
MOBILOIL SAE 20	NR at 70°F	NR at 70°F	B at 70°F	A to 70°F	NR at 70°F	A to 70°F
MOBILUX	NR at 70°F	NR at 70°F	NR at 70°F	A to 70°F	NR at 70°F	A to 70°F
MOLASSES	A to 70°F	A to 70°F AB any conc to 150°F	A to 70°F	A to 70°F	A to 70°F	
MOLYBDIC ACID						
MOLTEN SULFUR	B	B	A		B	
MOLYSITE	A to 70°F	A to 70°F AB any conc to 150°F	A to 70°F		A to 70°F	
MONOAMMONIUM PHOSPHATE	A to 70°F	A any conc to 150°F	A to 70°F AB to 200°F	NR at 70°F	A to 70°F	A to 70°F
MONOBROMOBENZENE	NR at 70°F	NR at 70°F	NR at 70°F	AB to 70°F	NR at 70°F	A to 70°F
MONOBROMO TRIFLUOROMETHANE	A/NR at 70°F	A/NR at 70°F	A to 70°F	B at 70°F	NR at 70°F	*B at 70°F
MONOBUTYL ETHER	NR at 70°F	NR at 70°F	C/NR at 70°F	C at 70°F	NR at 70°F	
MONOCHLOROETHANE	BC at 70°F	B/NR at 70°F	B/NR at 70°F		C at 70°F	
MONOCHLOROACETIC ACID	B/NR at 70°F	B/NR 10-100% at 70°F	AB to 200°F	NR at 70°F	B/NR at 70°F	A to 70°F
MONOCHLOROACETONE	B/NR at 70°F	B/NR at 70°F	B/NR at 70°F	NR at 70°F	NR at 70°F	
MONOCHLOROBENZENE	NR at 70°F	NR at 70°F	NR at 70°F	B at 70°F	NR at 70°F	A to 70°F
MONOCHLORODIFLUOROMETHANE (F 22)	A/NR at 70°F	A/NR at 70°F	A to 130°F	B/NR at 70°F	NR at 70°F	A to 70°F

	POLYSULFIDE T	CHLORINATED POLYETHYLENE CM	EPICHLOROHYDRIN CO, ECO	® HYTREL COPOLYESTER TPE	® SANTOPRENE COPOLYMER TPO	® C-FLEX STYRENIC TPE
METHYLENE CHLOROBROMIDE						
METHYLENE DICHLORIDE	+250% vol 3 days 70°F NR at 70°F					
M.I.B.K.	B at 70°F	NR at 70°F	NR at 70°F	+6-42% vol 7 days 70°F C/NR at 70°F	NR at 70°F	
MILK	B at 70°F			AB to 140°F		
MILK ACID						
MILK OF MAGNESIA	C at 70°F	AB to 70°F	A to 70°F	BC at 70°F		
MINE GUARD FR						
MINE WATER						
MINERAL NAPHTHA						
MINERAL OIL	B at 70°F	AB to 70°F	A to 70°F	A to 70°F	NR at 70°F	AB to 70°F
MINERAL PITCH						
MINERAL SPIRITS	AB to 70°F				*AB to 70°F	
MINERAL THINNER						
MINERS OIL	AB to 70°F					
MIRIBILITE						
MITTEL H.N.A.						
MLO-7277	C at 70°F					
MLO-7557	C at 70°F					
MLO-8200						
MLO-8515	BC at 70°F		C at 70°F			
M.N.V.						
MOBIL DELVAC 1100, 1110, 1120, 1130						
MOBIL 24DTE						
MOBIL HF						
MOBIL NIVAC 20, 30						
MOBIL THERM 600	C at 70°F					
MOBIL VELOCITE C						
MOBIL XRM 206A				*AB to 70°F		
MOBILGAS WA200 ATF	A to 70°F					
MOBILOIL SAE 20	A to 70°F					
MOBILUX	C at 70°F					
MOLASSES				AB to 70°F		
MOLYBDIC ACID						
MOLTEN SULFUR						
MOLYSITE						
MONOAMMONIUM PHOSPHATE	A to 70°F	AB to 70°F	A to 70°F	AB to 70°F		AB to 70°F
MONOBROMOBENZENE	AC to 70°F		NR at 70°F		NR at 70°F	
MONOBROMO TRIFLUOROMETHANE						
MONOBUTYL ETHER	A to 70°F					
MONOCHLOROETHANE						
MONOCHLOROACETIC ACID	NR at 70°F	NR at 70°F		NR at 70°F	NR at 70°F	AB to 70°F
MONOCHLOROACETONE	NR at 70°F					
MONOCHLOROBENZENE	NR at 70°F		NR at 70°F	C at 70°F	NR at 70°F	
MONOCHLORODIFLUOROMETHANE (F 22)	NR at 70°F	C at 70°F	A to 70°F	C/NR at 70°F	NR at 70°F	

	ETHYLENE ACRYLIC EA	POLYALLOMER LINEAR COPOLYMER	NYLON 11 POLYAMIDE	NYLON 12 POLYAMIDE	ETHYLENE VINYL ACETATE EVA	POLYVINYLCHLORIDE FLEXIBLE PVC
METHYLENE CHLOROBROMIDE						NR at 70°F
METHYLENE DICHLORIDE						NR at 70°F
M.I.B.K.	NR at 70°F	B at 70°F C at 122°F	AB to 105°F NR at 194°F		*NR at 70°F	NR at 70°F
MILK	A to 70°F	A to 70°F	A to 194°F	AB to 70°F	A to 70°F *AB to 140°F	A to 150°F food grades
MILK ACID						
MILK OF MAGNESIA	A to 70°F	A to 70°F	A to 70°F	AB 10% to 70°F	A to 70°F AB at 140°F	A to 150°F food grades
MINE GUARD FR			A to 70°F			
MINE WATER					C at 70°F acidic	B/NR at 70°F acidic
MINERAL NAPHTHA						
MINERAL OIL	A to 70°F	A to 122°F	A to 70°F	AB to 70°F	NR at 70°F	A to 70°F B/NR 140-150°F
MINERAL PITCH						
MINERAL SPIRITS						NR at 70°F
MINERAL THINNER						NR at 70°F
MINERS OIL						
MIRIBILITE						
MITTEL H.N.A.						
MLO-7277						
MLO-7557						
MLO-8200						
MLO-8515						
M.N.V.						
MOBIL DELVAC 1100, 1110, 1120, 1130						
MOBIL 24DTE						
MOBIL HF						
MOBIL NIVAC 20, 30						
MOBIL THERM 600						
MOBIL VELOCITE C						
MOBIL XRM 206A						
MOBILGAS WA200 ATF	A to 70°F					
MOBILOIL SAE 20	A to 70°F					
MOBILUX						
MOLASSES		A to 70°F	A to 70°F		A to 70°F	A to 150°F food grades
MOLYBDIC ACID						
MOLTEN SULFUR						
MOLYSITE						
MONOAMMONIUM PHOSPHATE		A to 70°F	A to 70°F		AB to 70°F	AB to 70°F *AB to 140°F
MONOBROMOBENZENE		NR at 70°F				NR at 70°F
MONOBROMO TRIFLUOROMETHANE						NR at 70°F
MONOBUTYL ETHER						
MONOCHLOROETHANE						
MONOCHLOROACETIC ACID		A to 70°F B at 122°F	NR at 70°F		C/NR at 70°F	NR 10-100% at 70°F
MONOCHLOROACETONE						
MONOCHLOROBENZENE			B at 70°F			NR at 70°F
MONOCHLORODIFLUOROMETHANE (F 22)						NR at 70°F

	POLYETHYLENE LOW DENSITY LDPE	®TEFLON FEP	®KALREZ PERFLUORINATED ELASTOMER	®FLUORAZ FLUORINATED COPOLYMER	®AFLAS FLUORINATED COPOLYMER	®NORPRENE COPOLYMER TPO
METHYLENE CHLOROBROMIDE		A to 70°F				
METHYLENE DICHLORIDE		A to 70°F		NR at 70°F		
M.I.B.K.	C/NR at 70°F NR at 122°F	A to 300°F	A to 212°F	NR at 70°F	NR at 70°F	BC at 70°F
MILK	A to 70°F AB to 140°F	A to 212°F	A to 70°F	A to 70°F		
MILK ACID		A 50% to 70°F				
MILK OF MAGNESIA	A to 70°F AB to 140°F	A to 150°F	A to 70°F	A to 70°F		
MINE GUARD FR		A to 70°F				
MINE WATER	A to 70°F	A to 70°F		A to 70°F		
MINERAL NAPHTHA		A to 70°F				
MINERAL OIL	AC to 70°F NR at 122°F	A to 356°F	A to 212°F	A to 70°F	A to 70°F	A to 70°F
MINERAL PITCH		A to 70°F				
MINERAL SPIRITS		A to 70°F		A to 70°F		
MINERAL THINNER		A to 70°F				
MINERS OIL		A to 70°F				
MIRIBILITE		A to 70°F				
MITTEL H.N.A.		A to 70°F				
MLO-7277				A to 70°F		
MLO-7557				A to 70°F		
MLO-8200				A to 70°F		
MLO-8515				A to 70°F		
M.N.V.		A to 70°F				
MOBIL DELVAC 1100, 1110, 1120, 1130				A to 70°F		
MOBIL 24DTE				A to 70°F		
MOBIL HF						
MOBIL NIVAC 20, 30				A to 70°F		
MOBIL THERM 600				A to 70°F		
MOBIL VELOCITE C				A to 70°F		
MOBIL XRM 206A		A to 70°F				
MOBILGAS WA200 ATF				A to 70°F		
MOBILOIL SAE 20				A to 70°F		
MOBILUX				A to 70°F		
MOLASSES	A to 70°F AB to 140°F	A to 300°F		A to 70°F		
MOLYBDIC ACID			A to 212°F			
MOLTEN SULFUR		A				
MOLYSITE		A to 70°F				
MONOAMMONIUM PHOSPHATE	AB to 70°F	A to 70°F	A to 70°F	NR at 70°F		
MONOBROMOBENZENE	NR at 70°F	A to 122°F	A to 212°F	C at 70°F	B at 70°F	
MONOBROMO TRIFLUOROMETHANE		A to 70°F				
MONOBUTYL ETHER		A to 70°F				
MONOCHLOROETHANE		A to 70°F				
MONOCHLOROACETIC ACID	A/NR to 122°F	A to 200°F	A 100% to 212°F A 10% to 104°F	NR at 70°F		A to 70°F
MONOCHLOROACETONE		A to 70°F	A to 70°F	NR at 70°F	NR at 70°F	
MONOCHLOROBENZENE	NR at 70°F	A to 300°F	A to 212°F	C/NR at 70°F		NR at 70°F
MONOCHLORODIFLUOROMETHANE (F 22)	A to 70°F	A to 300°F	AB to 70°F	NR at 70°F		

	RECOMMENDATIONS	NITRILE NBR	ETHYLENE PROPYLENE EP, EPDM	FLUOROCARBON FKM	CHLOROPRENE CR	HYDROGENATED NITRILE HNBR
MONOCHLOROPHENOL	FKM, FVMQ	NR at 70°F	NR at 70°F	AB to 70°F	NR at 70°F	
MONOCHLOROTRIFLUOROMETHANE (F 13)	CR, NBR, EP, FKM	A to 70°F	A to 70°F	A to 70°F	A to 250°F	
MONOETHANOLAMINE	EP, IIR, NBR, CR	B any conc to 80°F	AB to 200°F	NR at 70°F	B any conc to 80°F	
MONOETHYLAMINE	EP	C/NR at 70°F	A to 250°F	NR at 70°F	NR at 70°F	
MONOISOPROPANOL AMINE		B at 70°F		*NR at 70°F		
MONOMETHYL AMINE		B to 140°F	A to 70°F	C at 70°F	A/NR at 70°F	
MONOMETHYL ANILINE	FKM	A/NR at 70°F	A/NR at 70°F	B at 70°F	NR at 70°F	NR at 70°F
MONOMETHYL ETHER	NBR, CR	A to 100°F	A to 70°F	A to 70°F	AB to 70°F	
MONOMETHYL HYDRAZINE	EP	B at 70°F	A to 70°F		B at 70°F	B at 70°F
MONONITROCHLOROBENZENE	FKM	NR at 70°F	NR at 70°F	A to 160°F	NR at 70°F	
MONONITROTOLUENE 40% + DINITRO-TOLUENE 60%		NR at 70°F	NR at 70°F	C at 70°F	NR at 70°F	NR at 70°F
MONOSODIUM ACID METHANEARSENATE					A to 70°F AB 25% to 180°F	
MONOVINYL ACETATE	EP	NR at 70°F	B to 160°F	+7% vol 7 days 75°F A to 70°F	NR at 70°F	
MONOVINYL ARESNATE	EP, NBR, IIR, FKM	A to 70°F	A to 70°F	A to 70°F	B at 70°F	
MONSEL'S SALT		A to 70°F AB any conc to 200°F	A to 70°F	A to 70°F	A to 70°F AB any conc to 200°F	
MOPAR BRAKE FLUID	EP, SBR	C at 70°F	A to 70°F	NR at 70°F	B at 70°F	
MOREA PREMIC						
MORPHOLINE		NR at 70°F	NR at 70°F		NR at 70°F	
MORRHUA OIL		AB to 120°F	A to 70°F	A to 70°F	B at 70°F	
MOSAIC GOLD		A to 70°F				
MOTOR OIL		A to 70°F	NR at 70°F	A to 70°F	AB to 70°F	
MOTOR SPIRITS		AB to 100°F	NR at 70°F	A to 70°F	B/NR at 70°F	
MOULD OIL		AB to 70°F			AB to 70°F	
MURIATE OF AMONIA		A to 70°F	A to 70°F	A to 70°F	A to 70°F	
MURIATIC ACID	FKM, EP, IIR	NR conc at 70°F AB 20-37% to 70°F	BC 50-100% at 70°F AB to 37% to 130°F	AB 50-100% to 70°F A to 37% to 130°F	NR 50-100% at 70°F AB 25-37% to 70°F	
MUSTARD	NBR	AB to 160°F	A to 70°F	A to 140°F	A to 70°F C at 140°F	
MUSTARD GAS			A to 70°F	A to 70°F	A to 70°F	
MUTHMANN'S LIQUID		NR at 70°F			NR at 70°F	
NAPALM	NBR	B at 70°F	NR at 70°F	A to 70°F		
NAPHTHA	FKM, NBR, FVMQ	A to 250°F	NR at 70°F	+7.4% vol 28 days 158°F A to 158°F	NR at 70°F	B at 70°F
NAPHTHA-COAL TAR		A/NR at 70°F	NR at 70°F	A to 70°F	NR at 70°F	
NAPHTHA, SOUR	NBR	A to 250°F	NR at 70°F	A to 70°F	NR at 70°F	
NAPHTHALENE	FKM, FVMQ, T	NR at 70°F	NR at 70°F	A to 176°F	NR at 70°F	NR at 70°F
NAPHTHALENIC	FKM, FVMQ, NBR	B at 70°F	NR at 70°F	A to 160°F	NR at 70°F	
NAPHTHYLBENBENE		NR at 70°F			NR at 70°F	
NATURAL GAS	NBR, FKM, CR, SBR	A to 250°F	NR at 70°F	A to 176°F	A to 140°F	A to 70°F
NAVEE		BC at 70°F				
NEATSFOOT OIL	NBR, FKM, ACM, AU	A to 250°F	B at 70°F	A to 70°F	NR at 70°F	A to 70°F
NEOHEXANE		A to 70°F		A to 70°F		
NEON	NBR, FKM, FVMQ, EP CR, IIR	A to 70°F	A to 70°F	A to 70°F	A to 70°F	A to 70°F
NEOSOL		A to 70°F		C at 70°F	A to 70°F	
NEU-TRI		NR at 70°F		A to 70°F		
NEVILLE ACID	FKM	NR at 70°F	B at 70°F	A to 70°F	NR at 70°F	NR at 70°F
NEVOLI		C at 70°F		B at 70°F	C at 70°F	

	STYRENE BUTADIENE SBR	POLYACRYLATE ACM	POLYURETHANE AU, EU	ISOBUTYLENE ISOPRENE IIR	POLYBUTADIENE BR	® AEROQUIP AQP
MONOCHLOROPHENOL	NR at 70°F	NR at 70°F	NR at 70°F	NR at 70°F	NR at 70°F	
MONOCHLOROTRIFLUOROMETHANE (F 13)	A to 70°F		C at 70°F	A to 70°F	A to 70°F	NR at 70°F
MONOETHANOLAMINE	B at 70°F	NR at 70°F	NR at 70°F	B any conc to 140°F	B at 70°F	
MONOETHYLAMINE	B/NR at 70°F	NR at 70°F	C/NR at 70°F	B any conc to 140°F	B at 70°F	
MONOISOPROPANOL AMINE				A to 70°F		
MONOMETHYL AMINE				A/NR at 70°F		
MONOMETHYL ANILINE	NR at 70°F	NR at 70°F	NR at 70°F	A/NR at 70°F	NR at 70°F	
MONOMETHYL ETHER	B at 70°F			A to 80°F	B at 70°F	
MONOMETHYL HYDRAZINE	B at 70°F			A to 70°F		
MONONITROCHLOROBENZENE	NR at 70°F	NR at 70°F	NR at 70°F			
MONONITROTOLUENE 40% + DINITRO-TOLUENE 60%	NR at 70°F	NR at 70°F	NR at 70°F	NR at 70°F	NR at 70°F	
MONOSODIUM ACID METHANEARSENATE				A to 70°F AB 25% to 200°F		
MONOVINYL ACETATE	NR at 70°F	NR at 70°F	NR at 70°F	A to 70°F		
MONOVINYL ARESNATE	B at 70°F			A to 70°F	B at 70°F	
MONSEL'S SALT			A to 70°F	A to 70°F AB any conc to 185°F		
MOPAR BRAKE FLUID	A to 70°F			B at 70°F		
MOREA PREMIC				A to 70°F AB to 150°F		
MORPHOLINE						
MORRHUA OIL			A to 70°F	A to 70°F AB to 150°F		
MOSAIC GOLD				A to 70°F		
MOTOR OIL			AB to 70°F			
MOTOR SPIRITS			B at 70°F	NR at 70°F		
MOULD OIL						
MURIATE OF AMONIA			A to 70°F	A to 70°F		
MURIATIC ACID	C/NR 100% at 70°F B/NR 38% at 70°F	C/NR 100% at 70°F NR 38% at 70°F	NR 38-100% at 70°F B 20% at 70°F	A 100% to 70°F BC 50% at 70°F	B/NR 38% at 70°F NR 38% at 158°F	
MUSTARD	A to 70°F		AB to 70°F	B at 70°F		
MUSTARD GAS				A to 70°F		
MUTHMANN'S LIQUID				NR at 70°F		
NAPALM			B at 70°F			
NAPHTHA	NR at 70°F	AB to 70°F	AC to 70°F NR at 140°F	NR at 70°F	NR at 70°F	AB to 70°F
NAPHTHA-COAL TAR			AB to 70°F	NR at 70°F		
NAPHTHA, SOUR	NR at 70°F	A to 70°F	A to 70°F			
NAPHTHALENE	NR at 70°F		B at 70°F	NR at 70°F	NR at 70°F	AB to 70°F
NAPHTHALENIC	NR at 70°F			NR at 70°F	NR at 70°F	
NAPHTHYLBENBENE				NR at 70°F		
NATURAL GAS	BC at 70°F	AB to 70°F	BC at 70°F	NR at 70°F	B/NR at 70°F	
NAVEE						
NEATSFOOT OIL	NR at 70°F	A to 70°F	A to 70°F	B at 70°F	NR at 70°F	
NEOHEXANE				NR at 70°F		
NEON	A to 70°F	A to 70°F	A to 70°F	A to 70°F	A to 70°F	
NEOSOL				A to 70°F		
NEU-TRI				NR at 70°F		
NEVILLE ACID	NR at 70°F	NR at 70°F		B at 70°F	NR at 70°F	
NEVOLI				A to 70°F		

	SYNTHETIC ISOPRENE IR	NATURAL ISOPRENE NR	CHLOROSULFONATED POLYETHYLENE CSM	FLUOROSILICONE FVMQ	SILICONE VMQ	CHEMRAZ FFKM
MONOCHLOROPHENOL	NR at 70°F	NR at 70°F	NR at 70°F	B at 70°F	NR at 70°F	
MONOCHLOROTRIFLUOROMETHANE (F 13)	A/NR at 70°F	A/NR at 70°F	A to 70°F	NR at 70°F	NR at 70°F	*B at 70°F
MONOETHANOLAMINE	B at 70°F	B any conc to 80°F	B/NR at 70°F	NR at 70°F	B at 70°F	A to 70°F
MONOETHYLAMINE	BC at 70°F	BC at 70°F	BC at 70°F	NR at 70°F	BC at 70°F	A to 70°F
MONOISOPROPANOL AMINE	B at 70°F	B at 70°F	C at 70°F			A to 70°F
MONOMETHYL AMINE	A/NR at 70°F	A/NR at 70°F	A/NR at 70°F			A to 70°F
MONOMETHYL ANILINE	A/NR at 70°F	A/NR at 70°F	NR at 70°F			
MONOMETHYL ETHER	BC at 70°F	B/NR at 70°F				
MONOMETHYL HYDRAZINE			B at 70°F		NR at 70°F	B at 70°F
MONONITROCHLOROBENZENE				A to 70°F	NR at 70°F	
MONONITROTOLUENE 40% + DINITRO-TOLUENE 60%	NR at 70°F	NR at 70°F	NR at 70°F	C at 70°F	NR at 70°F	A to 70°F
MONOSODIUM ACID METHANEARSENATE						
MONOVINYL ACETATE	NR at 70°F	NR at 70°F	C at 70°F		NR at 70°F	
MONOVINYL ARESNATE	B at 70°F	B at 70°F	B at 70°F	NR at 70°F	B at 70°F	
MONSEL'S SALT	A to 70°F	A to 70°F AB any conc to 150°F	A to 70°F		B at 70°F	
MOPAR BRAKE FLUID			B at 70°F	NR at 70°F	C at 70°F	A to 70°F
MOREA PREMIC	A to 70°F	A to 70°F AB to 150°F				
MORPHOLINE						A to 70°F
MORRHUA OIL	NR at 70°F	NR at 70°F	B at 70°F		B at 70°F	
MOSAIC GOLD	A to 70°F	A to 70°F	A to 70°F			
MOTOR OIL			AB to 70°F			
MOTOR SPIRITS	NR at 70°F	NR at 70°F	B at 70°F		NR at 70°F	
MOULD OIL			AB to 70°F			
MURIATE OF AMONIA	A to 70°F	A to 70°F	A to 70°F		B at 70°F	
MURIATIC ACID	B 100% at 70°F A to 50% to 70°F	A/NR 100% at 70°F A to 50% to 70°F	A/NR 100% at 70°F AB to 50% to 70°F	C 100% at 70°F B 38% at 70°F	NR 100% at 70°F BC 50% at 70°F	A to 100% to 70°F
MUSTARD		C/NR at 70°F	B at 70°F	A to 70°F	A to 70°F	
MUSTARD GAS	A to 70°F	A to 70°F	A to 70°F		A to 70°F	
MUTHMANN'S LIQUID	NR at 70°F	NR at 70°F				
NAPALM						
NAPHTHA	NR at 70°F	NR at 70°F	C/NR at 70°F	AB to 70°F	NR at 70°F	A to 70°F
NAPHTHA-COAL TAR	NR at 70°F	NR at 70°F	NR at 70°F		NR at 70°F	
NAPHTHA, SOUR				A to 70°F	NR at 70°F	
NAPHTHALENE	NR at 70°F	NR at 70°F	NR at 70°F	A to 70°F	NR at 70°F	A to 70°F
NAPHTHALENIC	NR at 70°F	NR at 70°F	NR at 70°F	A to 70°F	NR at 70°F	A to 70°F
NAPHTHYLBENBENE	NR at 70°F	NR at 70°F				
NATURAL GAS	BC at 70°F	BC at 70°F AB to 70°F (dry)	A to 70°F wet or dry	AC to 70°F	A/NR at 70°F	A to 70°F
NAVEE						
NEATSFOOT OIL	NR at 70°F	NR at 70°F	NR at 70°F	A to 70°F	B at 70°F	A to 70°F
NEOHEXANE	NR at 70°F	NR at 70°F	NR at 70°F			
NEON	A to 70°F	A to 70°F	A to 70°F	A to 70°F	A to 70°F	A to 70°F
NEOSOL	A to 70°F	A to 70°F				
NEU-TRI	NR at 70°F	NR at 70°F	NR at 70°F			
NEVILLE ACID	NR at 70°F	NR at 70°F	NR at 70°F	B at 70°F	NR at 70°F	A to 70°F
NEVOLI	NR at 70°F	NR at 70°F				

	POLYSULFIDE T	CHLORINATED POLYETHYLENE CM	EPICHLOROHYDRIN CO, ECO	® HYTREL COPOLYESTER TPE	® SANTOPRENE COPOLYMER TPO	® C-FLEX STYRENIC TPE
MONOCHLOROPHENOL	NR at 70°F					
MONOCHLOROTRIFLUOROMETHANE (F 13)	A to 70°F		A to 70°F	BC at 70°F		
MONOETHANOLAMINE	NR at 70°F					A to 70°F
MONOETHYLAMINE	B at 70°F		B at 70°F			A to 70°F
MONOISOPROPANOL AMINE						
MONOMETHYL AMINE	A to 70°F					
MONOMETHYL ANILINE	NR at 70°F					
MONOMETHYL ETHER	B at 70°F					
MONOMETHYL HYDRAZINE	NR at 70°F					
MONONITROCHLOROBENZENE						
MONONITROTOLUENE 40% + DINITRO-TOLUENE 60%	NR at 70°F					
MONOSODIUM ACID METHANEARSENATE						
MONOVINYL ACETATE						
MONOVINYL ARESNATE	C at 70°F					
MONSEL'S SALT						
MOPAR BRAKE FLUID	NR at 70°F					
MOREA PREMIC						
MORPHOLINE						
MORRHUA OIL						
MOSAIC GOLD						
MOTOR OIL	AB to 70°F			AB to 70°F		
MOTOR SPIRITS	AB to 70°F					
MOULD OIL						
MURIATE OF AMONIA						
MURIATIC ACID	NR 38-100% at 70°F	NR 38-100% at 70°F	NR 38-100% at 70°F	NR 38-100% at 70°F BC 20% at 70°F	A to 100% to 70°F	AB to conc to 70°F
MUSTARD				AB to 70°F		
MUSTARD GAS						
MUTHMANN'S LIQUID						
NAPALM						
NAPHTHA	B at 70°F	A to 70°F	A to 70°F	A to 70°F	NR at 70°F	NR at 70°F
NAPHTHA-COAL TAR						
NAPHTHA, SOUR						
NAPHTHALENE	B at 70°F	C at 70°F		A to 70°F hard BC at 70°F soft	B/NR at 70°F	
NAPHTHALENIC	B at 70°F					
NAPHTHYLBENBENE						
NATURAL GAS	AB to 70°F wet or dry	AB to 70°F	A to 70°F	AB to 70°F	NR at 70°F	NR at 70°F
NAVEE						
NEATSFOOT OIL	NR at 70°F					
NEOHEXANE						
NEON	A to 70°F				A to 70°F	
NEOSOL						
NEU-TRI						
NEVILLE ACID	A to 70°F					
NEVOLI						

	ETHYLENE ACRYLIC EA	POLYALLOMER LINEAR COPOLYMER	NYLON 11 POLYAMIDE	NYLON 12 POLYAMIDE	ETHYLENE VINYL ACETATE EVA	POLYVINYLCHLORIDE FLEXIBLE PVC
MONOCHLOROPHENOL						
MONOCHLOROTRIFLUOROMETHANE (F 13)			AC to 70°F			NR at 70°F
MONOETHANOLAMINE						
MONOETHYLAMINE						NR at 70°F
MONOISOPROPANOL AMINE						
MONOMETHYL AMINE						NR at 70°F
MONOMETHYL ANILINE						
MONOMETHYL ETHER						NR at 70°F
MONOMETHYL HYDRAZINE						
MONONITROCHLOROBENZENE						
MONONITROTOLUENE 40% + DINITRO-TOLUENE 60%						
MONOSODIUM ACID METHANEARSENATE						
MONOVINYL ACETATE						
MONOVINYL ARESNATE						
MONSEL'S SALT						
MOPAR BRAKE FLUID	NR at 70°F					
MOREA PREMIC						
MORPHOLINE			NR at 70°F			
MORRHUA OIL						
MOSAIC GOLD						
MOTOR OIL			AB to 70°F			AB to 70°F
MOTOR SPIRITS						
MOULD OIL						
MURIATE OF AMONIA						
MURIATIC ACID	NR 38% at 70°F A 20% to 70°F	A to 50% to 70°F A to 20% to 122°F	C/NR 50-100% at 70°F NR 37% at 70°F	NR 10-100% at 70°F C 1% at 70°F	C/NR 100% at 70°F A to 50% at 70°F	AB 100% at 70°F A to 70% to 70°F
MUSTARD			AB to 70°F		B at 70°F	B at 70°F
MUSTARD GAS						
MUTHMANN'S LIQUID						NR at 70°F
NAPALM						
NAPHTHA	*NR at 70°F		A to 105°F BC at 140°F		C at 70°F NR at 140°F	C at 70°F NR at 140°F
NAPHTHA-COAL TAR			B/NR at 70°F		NR at 70°F	C/NR at 70°F
NAPHTHA, SOUR			AB to 70°F crude			C/NR at 70°F crude
NAPHTHALENE			AB to 140°F C at 194°F	AB to 70°F	*NR at 70°F	C/NR at 70°F NR at 150°F
NAPHTHALENIC						
NAPHTHYLBENBENE						
NATURAL GAS	A to 70°F	B at 70°F	A to 70°F AB to 140°F		C at 70°F	AB to 70°F
NAVEE						
NEATSFOOT OIL						
NEOHEXANE						
NEON	A to 70°F		A to 70°F			AB to 70°F
NEOSOL						
NEU-TRI						
NEVILLE ACID						
NEVOLI						

	POLYETHYLENE LOW DENSITY LDPE	® TEFLON FEP	® KALREZ PERFLUORINATED ELASTOMER	® FLUORAZ FLUORINATED COPOLYMER	® AFLAS FLUORINATED COPOLYMER	® NORPRENE COPOLYMER TPO
MONOCHLOROPHENOL		A to 70°F				
MONOCHLOROTRIFLUOROMETHANE (F 13)		A to 70°F	BC at 70°F	NR at 70°F		
MONOETHANOLAMINE	A to 70°F B at 122°F	A to 300°F	+1% vol 7 days 70°F *A to 212°F	A to 70°F	+58% vol 7 days 70°F NR at 70°F	TEST
MONOETHYLAMINE	AB to 70°F	A to 70°F	A to 113°F *A to 212°F	A to 70°F		
MONOISOPROPANOL AMINE		A to 70°F				
MONOMETHYL AMINE	AB to 70°F	A to 70°F	A to 212°F			
MONOMETHYL ANILINE		A to 70°F	A to 212°F	B at 70°F		
MONOMETHYL ETHER		A to 70°F	A to 212°F			
MONOMETHYL HYDRAZINE			*A to 212°F	A to 70°F		
MONONITROCHLOROBENZENE						
MONONITROTOLUENE 40% + DINITRO-TOLUENE 60%			A to 212°F	C at 70°F		
MONOSODIUM ACID METHANEARSENATE		A to 70°F				
MONOVINYL ACETATE		A to 70°F	A to 113°F			
MONOVINYL ARESNATE		A to 70°F	A to 70°F			
MONSEL'S SALT		A to 70°F				
MOPAR BRAKE FLUID				A to 70°F		
MOREA PREMIC		A to 70°F				
MORPHOLINE		A to 300°F	A to 212°F			
MORRHUA OIL		A to 70°F				
MOSAIC GOLD	AB to 140°F	A to 70°F				
MOTOR OIL		A to 300°F	A to 212°F			
MOTOR SPIRITS		A to 70°F				
MOULD OIL		A to 70°F				
MURIATE OF AMONIA		A to 70°F				
MURIATIC ACID	A to 100% to 70°F A to 35% to 122°F	A to 100% to 248°F	A to fuming to 200°F	A 37-100% to 70°F		AB to 100% to 70°F
MUSTARD	*AB to 70°F	A to 300°F		A to 70°F		
MUSTARD GAS		A to 70°F	A to 212°F			
MUTHMANN'S LIQUID		A to 70°F				
NAPALM						
NAPHTHA	NR at 70°F	A to 300°F	A to 70°F *A to 212°F	A to 70°F	+4% vol 7 days 70°F A to 70°F	NR at 70°F
NAPHTHA-COAL TAR	NR at 70°F	A to 70°F				
NAPHTHA, SOUR	NR at 70°F			A to 70°F crude		
NAPHTHALENE	NR at 70°F	A to 424°F	A to 212°F	C/NR at 70°F	B at 70°F	
NAPHTHALENIC		A to 70°F	A to 212°F	A to 550°F		
NAPHTHYLBENBENE		A to 70°F				
NATURAL GAS	C at 70°F	A to 70°F wet or dry	A to 212°F sweet or sour	A to 70°F		A to 70°F
NAVEE		A to 70°F				
NEATSFOOT OIL		A to 70°F	A to 212°F	A to 70°F		
NEOHEXANE		A to 70°F				
NEON	A to 70°F	A to 70°F	A to 212°F	A to 70°F		
NEOSOL		A to 70°F				
NEU-TRI		A to 70°F				
NEVILLE ACID		A to 70°F	A to 212°F	A to 70°F		
NEVOLI		A to 70°F				

C 253

	RECOMMENDATIONS	NITRILE NBR	ETHYLENE PROPYLENE EP, EPDM	FLUOROCARBON FKM	CHLOROPRENE CR	HYDROGENATED NITRILE HNBR
NICKEL		A to 70°F	A to 70°F	A to 70°F	A to 70°F	
NICKEL ACETATE	EP,IIR,NBR	AB to 70°F	A to 70°F	NR at 70°F	B at 70°F	B at 70°F
NICKEL AMMONIUM SULFATE				A to 70°F	A to 70°F	
NICKEL CARBONYL					NR at 70°F	
NICKEL CHLORIDE	NBR,EP,IIR,FKM,CR	A to 176°F B at 212°F	A to 176°F B at 212°F	A to 212°F	AB any conc to 200°F	A to 70°F
NICKEL NITRATE		A to 140°F AB any conc to 200°F	A to 212°F	A to 248°F	A to 140°F AB any conc to 200°F	
NICKEL SALTS	NBR,EP,FKM,SBR	A to 70°F AB any conc to 200°F	A to 70°F	A to 70°F	AB any conc to 200°F	A to 70°F
NICKEL SULFATE	NBR,EP,CR,IIR	A to 176°F B at 212°F	A to 176°F B at 212°F	A to boiling	A to 140°F B 176-200°F	A to 70°F
NICKELOUS SULFATE	NBR,EP,FKM,CR	A to 70°F	A to 70°F	A to 70°F	A to 70°F	
NICOTINE				A to 70°F	AC to 70°F	
NICOTINE BENTONITE		A to 70°F AB any conc to 200°F		C at 70°F	A to 70°F AB any conc to 200°F	
NICOTINE SULFATE		A to 70°F AB any conc to 200°F		C at 70°F	A to 70°F AB any conc to 200°F	
NICOTINIC ACID (NIACIN)		A to 70°F	A to 70°F		A to 70°F	
NITER-POTASSIUM NITRATE	NBR	A to 140°F AB any conc to 200°F	A to 100% to 176°F	A to 100% to 212°F	A to 100% to 140°F AB to 100% to 200°F	A to 70°F
NITER-SODIUM NITRATE	EP,IIR,NBR,CR,NR	AB any conc to 150°F	A to 100% to 200°F	A to 212°F	B to 100% to 200°F	
NITER CAKE	NBR,EP,FKM,CR	A to 70°F AB any conc to 160°F	A to 70°F	A to 70°F	A to 70°F AB any conc to 200°F	A to 70°F
NITRANA 2 & NITRANA 3		B at 70°F		C at 70°F	B at 70°F	
NITRATINE		C at 70°F	A to 70°F	A to 70°F	B at 70°F	
NITRATING ACIDS	EP	NR at 70°F	A to 120°F	NR at 70°F	NR at 70°F	
NITRIC ACID	FKM,IIR	dissolved 100% conc 3 days 158°F	+2% vol 100% conc 3 days 158°F	+6% vol 100% conc 3 days 158°F	dissolved 100% conc 3 days 158°F	NR 95-100% at 70°F
NITRIC ACID		dissolved 60% conc 7 days 70°F	dissolved 60% conc 7 days 70°F	+13% vol 90% conc 7 days 77°F	dissolved 60% conc 7 days 70°F	
NITRIC ACID		NR 0-100% at 70°F	NR 40-100% at 70°F C 35% at 70°F	+27% vol 69% conc 7 days 75°F	NR 15-100% at 70°F B 10% at 70°F	
NITRIC ACID			B 25-30% to 104°F NR 25-30% at 140°F	+13% vol 60% conc 7 days 70°F	C 10% at 104°F NR 10% at 140°F	
NITRIC ACID			A to 25% to 70°F A to 10% to 104°F	+43% vol 60% conc 28 days 75°F	NR gas at 70°F	
NITRIC ACID			B 10% at 140°F NR 10% at 176°F	+94% vol 10% conc 28 days 158°F		
NITRIC ACID			NR gas at 70°F	A 90-100% to 158°F NR 100% boiling		
NITRIC ACID				AC 60-70% to 70°F BC 70% at 100°F		
NITRIC ACID				C 70% boiling A to 50% to 140°F		
NITRIC ACID				B 20% boiling NR 10% at 158°F		
NITRIC ACID-RED FUMING	FKM	NR at 70°F	NR at 70°F	+23% vol 7 days 75°F +42% vol 7 days 125°F	NR at 70°F	NR at 70°F
NITRIC ACID-RED FUMING				+60% vol 7 days 158°F BC at 70°F		
NITRIC ACID-RED FUMING				C/NR at 130°F NR at 158°F		
NITRIC ACID-RED FUMING, INHIBITED	FKM	NR at 70°F	NR at 70°F	+70% vol 3 days 77°F BC at 70°F	NR at 70°F	NR at 70°F
NITRIC ACID-WHITE FUMING	FKM	NR at 70°F	NR at 70°F	B to 120°F	NR at 70°F	*NR at 70°F
NITRIC ACID, CRUDE		NR at 70°F		AC to 70°F	NR at 70°F	
NITRIC OXIDES		NR at 70°F	C at 70°F	NR at 70°F	C at 70°F variable	
NITROBENZENE	FKM,EP,IIR	NR at 70°F	A to 70°F B at 140°F	+40% vol 21 days 75°F +77% vol 28 days 158°F	NR at 70°F	NR at 70°F
NITROBENZINE	NBR,FKM,FVMQ	A to 140°F	NR at 70°F	A to 140°F AB to 260°F	B/NR at 70°F	A to 70°F
NITROCALCITE		A to 70°F AB any conc to 200°F	A to 70°F	A to 70°F	A to 70°F AB any conc to 200°F	
NITROCARBOL		NR at 70°F	A to 70°F	C at 70°F	C at 70°F	
NITROETHANE	CR,EP,IIR,T	NR at 70°F	B at 70°F	NR at 70°F	BC at 70°F	
NITROGEN, GAS	NBR,IIR,CR,EP,FKM	A to 200°F	A to 70°F	A to 70°F	A to 70°F AB to 200°F	
NITROGEN FERTILIZER SOLUTION		B to 150°F			B to 150°F	
NITROGEN OXIDE		NR at 70°F	B/NR at 70°F	C/NR at 70°F	NR at 70°F	

C254

	STYRENE BUTADIENE SBR	POLYACRYLATE ACM	POLYURETHANE AU, EU	ISOBUTYLENE ISOPRENE IIR	POLYBUTADIENE BR	®AEROQUIP AQP
NICKEL						
NICKEL ACETATE	NR at 70°F	NR at 70°F	NR at 70°F	A to 70°F	NR at 70°F	
NICKEL AMMONIUM SULFATE						
NICKEL CARBONYL						
NICKEL CHLORIDE	A to 70°F	C at 70°F	B/NR at 70°F	A to 70°F AB any conc to 150°F	A to 70°F	
NICKEL NITRATE				A to 70°F AB any conc to 150°F		
NICKEL SALTS	A to 70°F	C at 70°F	C/NR at 70°F	A to 70°F AB any conc to 185°F	A to 70°F	
NICKEL SULFATE	B at 70°F	NR at 70°F	AC to 70°F	A to 70°F AB any conc to 185°F	B at 70°F	NR at 70°F
NICKELOUS SULFATE			A to 70°F	A to 70°F		
NICOTINE			AB to 70°F			
NICOTINE BENTONITE				A to 70°F AB any conc to 185°F		
NICOTINE SULFATE			A to 70°F	A to 70°F AB any conc to 185°F		
NICOTINIC ACID (NIACIN)						
NITER-POTASSIUM NITRATE	A to 70°F	A to 70°F	A to 70°F	A to 70°F AB any conc to 185°F	A to 70°F	
NITER-SODIUM NITRATE	B at 70°F		C/NR at 70°F	A to 70°F AB any conc to 185°F	A to 70°F	A to 70°F
NITER CAKE	A to 70°F	NR at 70°F	A to 70°F	A to 70°F AB any conc to 185°F	A to 70°F	
NITRANA 2 & NITRANA 3				B at 70°F		
NITRATINE				A to 70°F		
NITRATING ACIDS	NR at 70°F	NR at 70°F	NR at 70°F			
NITRIC ACID	dissolved 100% conc 3 days 70°F	+52% vol 100% conc 3 days 70°F	dissolved 100% conc 3 days 158°F	softened 100% conc 3 days 70°F	NR 10-100% at 70°F	NR 10-70% at 70°F
NITRIC ACID	dissolved 100% conc 3 days 158°F	NR 20-100% at 70°F	-43% T.S. 10% conc 3 days 70°F	dissolved 60% conc 7 days 70°F		
NITRIC ACID	NR 2-100% at 70°F		NR 2-100% at 70°F AB 2% to 70°F	NR 40-100% at 70°F C 35% at 70°F		
NITRIC ACID			AB 10% fumes to 70°F	B 25% to 100°F AB 20% to 70°F		
NITRIC ACID				A to 10% to 150°F		
NITRIC ACID						
NITRIC ACID						
NITRIC ACID						
NITRIC ACID						
NITRIC ACID						
NITRIC ACID-RED FUMING	NR at 70°F	NR at 70°F	NR at 70°F	B/NR at 70°F	NR at 70°F	*NR at 70°F
NITRIC ACID-RED FUMING						
NITRIC ACID-RED FUMING						
NITRIC ACID-RED FUMING, INHIBITED	NR at 70°F	NR at 70°F	NR at 70°F	B/NR at 70°F	NR at 70°F	*NR at 70°F
NITRIC ACID-WHITE FUMING	NR at 70°F	NR at 70°F	NR at 70°F	NR at 70°F	NR at 70°F	*NR at 70°F
NITRIC ACID, CRUDE				C/NR at 70°F		NR at 70°F
NITRIC OXIDES			NR at 70°F	NR at 70°F		
NITROBENZENE	NR at 70°F	NR at 70°F	NR at 70°F	B any conc to 80°F	NR at 70°F	TEST
NITROBENZINE				C at 70°F		
NITROCALCITE			NR at 70°F	A to 70°F AB any conc to 185°F		
NITROCARBOL				A to 70°F		
NITROETHANE	B at 70°F	NR at 70°F	NR at 70°F	B any conc to 80°F	B at 70°F	
NITROGEN, GAS	A to 70°F	A to 70°F	A to 70°F *AB to 140°F	A to 70°F AB any conc to 185°F	A to 70°F	
NITROGEN FERTILIZER SOLUTION				B to 150°F		
NITROGEN OXIDE	*NR at 70°F	*NR at 70°F	AB to 50% to 70°F	B/NR at 70°F	NR at 70°F	

C 255

	SYNTHETIC ISOPRENE IR	NATURAL ISOPRENE NR	CHLOROSULFONATED POLYETHYLENE CSM	FLUOROSILICONE FVMQ	SILICONE VMQ	CHEMRAZ FFKM
NICKEL						
NICKEL ACETATE	A to 70°F	A any conc to 80°F	NR at 70°F	NR at 70°F	NR at 70°F	A to 70°F
NICKEL AMMONIUM SULFATE						A to 70°F
NICKEL CARBONYL						
NICKEL CHLORIDE	A to 70°F	A to 70°F AB to 185°F	A to 70°F AB to 185°F	A to 70°F	A to 70°F	A to 70°F
NICKEL NITRATE	A to 70°F	A to 70°F AB any conc to 150°F	A to 70°F			A to 70°F
NICKEL SALTS	A to 70°F	A to 70°F AB any conc to 150°F	A to 70°F	A to 70°F	A to 70°F	A to 70°F
NICKEL SULFATE	AB to 70°F	AB to 185°F	A to 70°F AB to 200°F	A to 70°F	A to 70°F	A to 70°F
NICKELOUS SULFATE	A to 70°F	A to 70°F	A to 70°F		A to 70°F	
NICOTINE						
NICOTINE BENTONITE	A to 70°F	A to 70°F AB any conc to 150°F				
NICOTINE SULFATE	A to 70°F	A to 70°F AB any conc to 150°F				
NICOTINIC ACID (NIACIN)						
NITER-POTASSIUM NITRATE	A to 70°F	A to 70°F AB any conc to 150°F	A at 70°F AB to 185°F	A to 70°F	A to 70°F	A to 70°F
NITER-SODIUM NITRATE	AB to 70°F	B any conc at 70°F	A to 70°F AB to 200°F		C/NR conc at 70°F NR sol'n at 70°F	A to 70°F
NITER CAKE	A to 70°F	A to 70°F AB any conc to 150°F	A to 70°F	A to 70°F	A to 70°F	A to 70°F
NITRANA 2 & NITRANA 3	B at 70°F	B at 70°F				
NITRATINE	B at 70°F	B at 70°F	A to 70°F		C at 70°F	
NITRATING ACIDS		C at 70°F		NR at 70°F	NR at 70°F	
NITRIC ACID	NR 2-100% at 70°F	NR 2-100% at 70°F	dissolved 60% conc 7 days 70°F	+4% vol 100% conc 7 days 70°F	-10% to +10% vol 100% conc 7 days 70°F	B 50-100% at 70°F A vapors to 70°F
NITRIC ACID			NR 60-100% at 70°F B/NR 50% at 70°F	+2% vol 100% conc 3 days 158°F	deteriorated 100% conc 7 days 70°F	B to conc to 70°F whit
NITRIC ACID			A/NR 30-35% at 70°F NR 30% at 158°F	+5% vol 70% conc 7 days 75°F	+25% vol 100% conc 3 days 158°F	
NITRIC ACID			AB to 20% to 140°F AB to 10% to 200°F	-40% T.S. 70% conc 7 days 75°F	+13% vol 50% conc 7 days 70°F	
NITRIC ACID				+1% vol 10% conc 7 days 70°F	+1-8% vol 10% conc 7 days 70°F	
NITRIC ACID				NR 10-100% at 70°F B 10% at 70°F	+1% vol 7% conc 7 days 70°F	
NITRIC ACID				B dilute at 70°F	NR 50-100% at 70°F C/NR 10-100% at 70°F	
NITRIC ACID					BC 5-10% at 70°F B dilute at 70°F	
NITRIC ACID						
NITRIC ACID						
NITRIC ACID-RED FUMING	NR at 70°F	NR at 70°F	NR at 70°F	NR at 70°F	NR at 70°F	AB to 70°F C at 70°F white
NITRIC ACID-RED FUMING						
NITRIC ACID-RED FUMING						
NITRIC ACID-RED FUMING, INHIBITED	NR at 70°F	NR at 70°F	NR at 70°F	NR at 70°F	NR at 70°F	
NITRIC ACID-WHITE FUMING	NR at 70°F	NR at 70°F	NR at 70°F	NR at 70°F	NR at 70°F	AB to 70°F
NITRIC ACID, CRUDE	NR at 70°F	NR at 70°F	C at 70°F		NR at 70°F	
NITRIC OXIDES		C at 70°F	NR at 70°F		NR at 70°F	
NITROBENZENE	NR at 70°F	NR at 70°F	NR at 70°F	NR at 70°F	C/NR at 70°F	A to 70°F
NITROBENZINE	NR at 70°F	NR at 70°F	C at 70°F	A to 70°F	NR at 70°F	A to 70°F
NITROCALCITE	A to 70°F	A to 70°F AB any conc to 150°F	A to 70°F		B at 70°F	
NITROCARBOL	A to 70°F	A to 70°F	C at 70°F		C at 70°F	
NITROETHANE	B at 70°F	B any conc to 80°F	BC at 70°F	NR at 70°F	NR at 70°F	A to 70°F
NITROGEN, GAS	A to 70°F	A to 70°F AB to 150°F	A to 70°F	A to 70°F	A to 70°F	A to 70°F
NITROGEN FERTILIZER SOLUTION	A to 70°F	A to 70°F AB to 150°F				
NITROGEN OXIDE	*NR at 70°F	*NR at 70°F	*NR at 70°F	*NR at 70°F	NR at 70°F	A to 70°F

C 256

	POLYSULFIDE T	CHLORINATED POLYETHYLENE CM	EPICHLOROHYDRIN CO, ECO	® HYTREL COPOLYESTER TPE	® SANTOPRENE COPOLYMER TPO	® C-FLEX STYRENIC TPE
NICKEL						
NICKEL ACETATE	NR at 70°F					
NICKEL AMMONIUM SULFATE						
NICKEL CARBONYL						
NICKEL CHLORIDE	A to 70°F AB to 150°F	NR at 70°F		A/NR at 70°F		
NICKEL NITRATE	AB to 70°F					
NICKEL SALTS	AB to 70°F					A to 70°F
NICKEL SULFATE	BC to 150°F	NR at 70°F		A/NR at 70°F		
NICKELOUS SULFATE						
NICOTINE				AB to 70°F		
NICOTINE BENTONITE						
NICOTINE SULFATE						
NICOTINIC ACID (NIACIN)						
NITER-POTASSIUM NITRATE	A to 70°F		A to 70°F			
NITER-SODIUM NITRATE		AB to 70°F	A to 70°F	AB to 70°F		A to 70°F
NITER CAKE	C at 70°F					
NITRANA 2 & NITRANA 3						
NITRATINE						
NITRATING ACIDS						
NITRIC ACID	dissolved 100% conc 3 days 70°F	NR 10–100% at 70°F	NR 2–100% at 70°F	NR 20–100% at 70°F B/NR 10% at 70°F	NR 100% at 70°F A 10% to 70°F	A to 100% to 70°F
NITRIC ACID	NR 2–100% at 70°F					
NITRIC ACID						
NITRIC ACID						
NITRIC ACID						
NITRIC ACID						
NITRIC ACID						
NITRIC ACID						
NITRIC ACID						
NITRIC ACID						
NITRIC ACID-RED FUMING	NR at 70°F	NR at 70°F	NR at 70°F	NR at 70°F	NR at 70°F	
NITRIC ACID-RED FUMING						
NITRIC ACID-RED FUMING						
NITRIC ACID-RED FUMING, INHIBITED	NR at 70°F	NR at 70°F	NR at 70°F	NR at 70°F	NR at 70°F	
NITRIC ACID-WHITE FUMING	NR at 70°F	NR at 70°F	NR at 70°F	NR at 70°F		
NITRIC ACID, CRUDE				NR at 70°F		
NITRIC OXIDES						
NITROBENZENE	NR at 70°F	C at 70°F	NR at 70°F	NR at 70°F		NR at 70°F
NITROBENZINE						
NITROCALCITE						
NITROCARBOL						
NITROETHANE					no swell 7 days 70°F A to 70°F	
NITROGEN, GAS	A to 70°F		A to 70°F	*AB to 140°F		
NITROGEN FERTILIZER SOLUTION						
NITROGEN OXIDE	*NR at 70°F			AB to 50% to 70°F		AB to 70°F

C 257

	ETHYLENE ACRYLIC EA	POLYALLOMER LINEAR COPOLYMER	NYLON 11 POLYAMIDE	NYLON 12 POLYAMIDE	ETHYLENE VINYL ACETATE EVA	POLYVINYLCHLORIDE FLEXIBLE PVC
NICKEL						
NICKEL ACETATE						
NICKEL AMMONIUM SULFATE						
NICKEL CARBONYL						
NICKEL CHLORIDE		A to 70°F	A/NR at 70°F		A to 70°F	AB to 70°F *AB to 140°F
NICKEL NITRATE						
NICKEL SALTS		A to 70°F	A to 70°F	AB to 70°F aqueous	A to 70°F AB to 140°F	A to 70°F *AB to 130°F
NICKEL SULFATE		A to 70°F	A/NR at 70°F		A to 70°F	A to 70°F *AB to 140°F
NICKELOUS SULFATE						
NICOTINE			AB to 70°F			A to 70°F
NICOTINE BENTONITE						
NICOTINE SULFATE			B at 70°F		B at 70°F	AB to 70°F
NICOTINIC ACID (NIACIN)						
NITER-POTASSIUM NITRATE	A to 70°F		A to 70°F NR at 140°F		A/NR at 70°F	AC to 70°F
NITER-SODIUM NITRATE		A to 70°F	AB to 70°F		A/NR at 70°F	AC to 70°F
NITER CAKE					NR at 70°F	C at 70°F
NITRANA 2 & NITRANA 3						
NITRATINE						
NITRATING ACIDS						NR at 70°F
NITRIC ACID	NR 100% at 70°F *NR 50% at 70°F	A to 100% to 70°F	NR 69-100% at 70°F C/NR 20-50% at 70°F	NR conc at 70°F C 10% at 70°F	NR 100% at 70°F C 30-50% to 140°F	no swell 10% conc 14 days 70°F
NITRIC ACID	*AB 10% to 70°F		NR 50% at 122°F A/NR to 10% to 122°F		B dilute to 70°F	C/NR 69-100% at 70°F NR 100% at 122°F
NITRIC ACID						AB 20-50% to 70°F C 50% 122-140°F
NITRIC ACID						A to 10% to 70°F B 10% 122-150°F
NITRIC ACID						
NITRIC ACID						
NITRIC ACID						
NITRIC ACID						
NITRIC ACID						
NITRIC ACID						
NITRIC ACID-RED FUMING	NR at 70°F	A to 70°F	NR at 70°F			NR at 70°F
NITRIC ACID-RED FUMING						
NITRIC ACID-RED FUMING						
NITRIC ACID-RED FUMING, INHIBITED	NR at 70°F		NR at 70°F			NR at 70°F
NITRIC ACID-WHITE FUMING			NR at 70°F			NR at 70°F
NITRIC ACID, CRUDE			NR at 70°F			NR at 70°F
NITRIC OXIDES					C/NR at 70°F	A to 70°F
NITROBENZENE	*NR at 70°F	B at 70°F	B/NR at 70°F	BC at 70°F	NR at 70°F	NR at 70°F
NITROBENZINE						
NITROCALCITE						
NITROCARBOL						
NITROETHANE						NR at 70°F
NITROGEN, GAS	A to 70°F		*AB to 70°F		*AB to 70°F	AB to 70°F
NITROGEN FERTILIZER SOLUTION						
NITROGEN OXIDE		A to 70°F	AB to 70°F		A to 70°F	A to 70°F

	POLYETHYLENE LOW DENSITY LDPE	® TEFLON FEP	® KALREZ PERFLUORINATED ELASTOMER	® FLUORAZ FLUORINATED COPOLYMER	® AFLAS FLUORINATED COPOLYMER	® NORPRENE COPOLYMER TPO
NICKEL	A to 70°F	A to 70°F				
NICKEL ACETATE		A to 70°F	A to 70°F A aqueous to 212°F	NR at 70°F		
NICKEL AMMONIUM SULFATE		A to 70°F	A to 212°F			
NICKEL CARBONYL	AB to 70°F	A to 70°F				
NICKEL CHLORIDE	AB to 70°F	A to 300°F	A to 70°F A aqueous to 212°F	A to 100% to 70°F A to 30% to 150°F		
NICKEL NITRATE	A to 70°F	A to 300°F	A to 212°F			
NICKEL SALTS	A to 70°F AB to 140°F	A to 70°F	A to 212°F	A to 70°F		A to 70°F
NICKEL SULFATE	A to 70°F	A to 300°F	A to 70°F A aqueous to 212°F	A to 70°F		
NICKELOUS SULFATE		A to 70°F				
NICOTINE	B/NR at 70°F	A to 70°F	A to 212°F			
NICOTINE BENTONITE		A to 70°F				
NICOTINE SULFATE	A to 70°F	A to 70°F		A to 70°F		
NICOTINIC ACID (NIACIN)	NR at 70°F	A to 70°F				
NITER-POTASSIUM NITRATE	A to 70°F AB to 140°F	A to 100% to 300°F	A to 212°F	A to 100% to 70°F		*A to 70°F
NITER-SODIUM NITRATE	A to 70°F AB to 140°F	A to 70°F	A to 212°F	A/NR at 70°F		AB to 70°F
NITER CAKE	A to 70°F	A to 70°F	A to 212°F	A to 70°F		
NITRANA 2 & NITRANA 3		A to 70°F				
NITRATINE		A to 70°F				
NITRATING ACIDS						
NITRIC ACID	NR 95-100% at 70°F C/NR 50-70% at 70°F	no swell 10% conc 365 days 158°F	A to 100% to 70°F A to 69% to 158°F	B 100% at 70°F C 100% at 150°F	+15-19% vol 100% conc 180 days 70°F	TEST 100% at 70°F NR 70% at 70°F
NITRIC ACID	NR 50-70% at 140°F NR 66% at 104°F	A to 100% to 248°F	B to 69% at 212°F	A to 85% to 70°F	+21% vol 98% conc 30 days 70°F	AB 50% at 70°F BC 30% at 70°F
NITRIC ACID	B 50% to 122°F AB to 30% to 140°F				+5% vol 60% conc 180 days 70°F	A 10% to 70°F
NITRIC ACID	NR 30-50% at 140°F				+10% vol 60% conc 3 days 158°F	
NITRIC ACID					no swell 20% conc 7 days 70°F	
NITRIC ACID					+25% vol 20% conc 3 days 158°F	
NITRIC ACID					B 98-100% at 70°F A to 60% to 70°F	
NITRIC ACID					AB 60% at 158°F C 10% at 158°F	
NITRIC ACID					NR 100% at 168°F	
NITRIC ACID						
NITRIC ACID-RED FUMING	NR at 70°F	A to boiling	A to 70°F *A to 212°F	C 70-150°F	+ 15% vol 180 days 77°F B at 70°F	*NR at 70°F
NITRIC ACID-RED FUMING						
NITRIC ACID-RED FUMING						
NITRIC ACID-RED FUMING, INHIBITED	NR at 70°F		A to 70°F *A to 212°F			*NR at 70°F
NITRIC ACID-WHITE FUMING	NR at 70°F	A to boiling	B at 70°F			*NR at 70°F
NITRIC ACID, CRUDE		A to 70°F				
NITRIC OXIDES	A to 70°F	A to 70°F				
NITROBENZENE	NR at 70°F	A to 410°F	+1% vol 7 days 70°F A to 410°F	A to 70°F	+5.6% vol 7 days 70°F A to 70°F	NR at 70°F
NITROBENZINE	NR at 70°F	A to 70°F	A to 212°F	BC at 70°F		
NITROCALCITE		A to 70°F				
NITROCARBOL		A to 70°F				
NITROETHANE	A to 70°F	A to 70°F	A to 70°F	NR at 70°F	+16% vol 7 days 70°F B at 70°F	A to 70°F
NITROGEN, GAS	AB to 70°F	A to 300°F		A to 70°F		
NITROGEN FERTILIZER SOLUTION		A to 70°F				
NITROGEN OXIDE	C/NR 50% at 70°F	A to 70°F dioxide & trioxide	A to 350°F			A to 70°F

	RECOMMENDATIONS	NITRILE NBR	ETHYLENE PROPYLENE EP, EPDM	FLUOROCARBON FKM	CHLOROPRENE CR	HYDROGENATED NITRILE HNBR
NITROGEN TETROXIDE	IIR, EP	NR at 70°F	B/NR at 70°F	+100% vol 1 day 77°F NR at 70°F	NR at 70°F	C/NR at 70°F
NITROHYDROCHLORIC ACID	FKM, EP, FVMQ	NR at 70°F	B to 104°F C at 140°F	A to 70°F B to 185°F	NR at 70°F	
NITROMETHANE	EP, T, SBR, IIR	NR at 70°F	AB to 120°F	NR at 70°F	C/NR at 70°F	C/NR at 70°F
NITROMURIATRIC ACID	FKM, EP, FVMQ	NR at 70°F	B to 104°F C at 140°F	A to 70°F B to 185°F	NR at 70°F	
1-NITROPROPANE	EP, T, IIR	NR at 70°F	B at 70°F	NR at 70°F	NR at 70°F	
NITROUS ACIDS		NR 10% at 70°F	AB 10-100% to 70°F	A 10-100% to 70°F		
NITROUS MONOXIDE	NBR	A to 70°F	A to 140°F B at 176°F	A to 176°F B 212-248°F	AB to 70°F	
NITROUS OXIDE	NBR	A to 70°F	A to 140°F B at 176°F	A to 176°F B 212-248°F	AB to 70°F	
NITROUS OXIDES			A to 70°F	A to 70°F		
NITROXANTHIC ACID		C/NR at 70°F	B at 70°F	A to 70°F	C/NR at 70°F	
NONANOIC ACID		A to 70°F				
NONENES		A to 70°F		A to 70°F		
NORGE NITER		A to 70°F AB any conc to 200°F	A to 70°F	A to 70°F	A to 70°F AB any conc to 200°F	
NORGE SALTPETER		A to 70°F	A to 70°F	A to 70°F	A to 70°F	
NORWAY SALTPETER		A to 70°F AB any conc to 200°F	A to 70°F	A to 70°F	A to 70°F AB any conc to 200°F	
NORWEGIAN SALTPETER		A to 70°F AB any conc to 200°F	A to 70°F	A to 70°F	A to 70°F AB any conc to 200°F	
NORYL-GENERAL ELECTRIC	NBR, EP	A to 70°F	A to 70°F			
N.R.R.O.	NBR, FKM	B at 70°F	NR at 70°F	A to 70°F	C at 70°F	
NUT OIL	NBR	A to 250°F	NR at 70°F	A to 70°F	NR at 70°F	
O-A-548A		A to 70°F	A to 70°F	B at 70°F	B at 70°F	
O-T-634B		C at 70°F	NR at 70°F	A to 70°F	NR at 70°F	
OAKITE ALKALINE MATERIALS		AB any conc to 150°F	A to 70°F		A to 70°F AB any conc to 200°F	
OAKITE-O.C. 31 & 32					BC at 70°F	
OAKITE-O.C. 33 & 34					NR at 70°F	
OAKITE-O.C. 36, 84H & 84M					C to 85°F	
OAKITE-O.C. 85					NR at 70°F	
OAKITE-O.C. 88					C at 70°F	
OAKITE-O.C. 131					BC to 85°F	
OAKITE CRYSCOAT FH RINSE					NR at 70°F	
OAKITE O DRYCID & OMF 184					BC to 85°F	
OAKITE O STRIPPER S.A.					NR at 70°F	
OAKITE SOLVENT MATERIALS		NR at 70°F			NR at 70°F	
OCTACHLOROTOLUENE	FKM	NR at 70°F	NR at 70°F	A to 70°F	NR at 70°F	
OCTADECANE	NBR, FKM, FVMQ, AU	A to 70°F	NR at 70°F	A to 70°F	B at 70°F	NR at 70°F
OCTADECANOIC ACID	NBR, EP, IIR	A to 70°F B to 250°F	B to 140°F	A to 140°F C at 176°F	B to 140°F BC at 158°F	B at 70°F
OCTADECATRIENOIC ACID		BC to 80°F			NR at 70°F	
OCTAFLUOROCYCLOBUTANE		NR at 70°F			NR at 70°F	
OCTANE	NBR, FKM, T	A to 70°F	NR at 70°F	A to 70°F	B/NR at 70°F	
n-OCTANE	FKM	B at 70°F	NR at 70°F	A to 70°F	NR at 70°F	
OCTANOL	NBR, EP, FKM, CR, AU	AB to 100°F	A to 160°F	A to 70°F	AB to 70°F	B at 70°F
2-OCTANONE		NR at 70°F		NR at 70°F		
n-OCTENE-2		A to 70°F	NR at 70°F	A to 70°F	C at 70°F	
OCTOIC ACID		C at 70°F				
OCTYL ACETATE		NR at 70°F		NR at 70°F		

	STYRENE BUTADIENE SBR	POLYACRYLATE ACM	POLYURETHANE AU, EU	ISOBUTYLENE ISOPRENE IIR	POLYBUTADIENE BR	®AEROQUIP AQP
NITROGEN TETROXIDE	NR at 70°F	NR at 70°F	NR at 70°F	BC at 70°F	NR at 70°F	
NITROHYDROCHLORIC ACID	NR at 70°F	NR at 70°F	NR at 70°F	C/NR at 70°F	NR at 70°F	
NITROMETHANE	B/NR at 70°F	NR at 70°F	NR at 70°F	B any conc to 80°F	B at 70°F	
NITROMURIATRIC ACID	NR at 70°F	NR at 70°F	NR at 70°F	C/NR at 70°F	NR at 70°F	
1-NITROPROPANE	NR at 70°F	NR at 70°F	NR at 70°F AB fumes to 70°F	B any conc to 80°F	NR at 70°F	
NITROUS ACIDS						
NITROUS MONOXIDE			AB to 70°F	A to 70°F		
NITROUS OXIDE			AB to 70°F	A to 70°F		
NITROUS OXIDES						
NITROXANTHIC ACID			B at 70°F	C at 70°F		
NONANOIC ACID				A to 70°F		
NONENES				NR at 70°F		
NORGE NITER			NR at 70°F	A to 70°F AB any conc to 185°F		
NORGE SALTPETER			NR at 70°F	A to 70°F AB any conc to 185°F		
NORWAY SALTPETER			NR at 70°F	A to 70°F AB any conc to 185°F		
NORWEGIAN SALTPETER			NR at 70°F	A to 70°F AB any conc to 185°F		
NORYL-GENERAL ELECTRIC						
N.R.R.O.	NR at 70°F	B at 70°F	B at 70°F	NR at 70°F		
NUT OIL	NR at 70°F	A to 70°F	A to 70°F			
O-A-548A	A to 70°F	C at 70°F	C at 70°F	A to 70°F	A to 70°F	
O-T-634B	NR at 70°F	NR at 70°F	NR at 70°F	NR at 70°F	NR at 70°F	
OAKITE ALKALINE MATERIALS				A to 70°F A to 185°F		
OAKITE-O.C. 31 & 32						
OAKITE-O.C. 33 & 34						
OAKITE-O.C. 36, 84H & 84M						
OAKITE-O.C. 85						
OAKITE-O.C. 88						
OAKITE-O.C. 131						
OAKITE CRYSCOAT FH RINSE						
OAKITE O DRYCID & OMF 184						
OAKITE O STRIPPER S.A.						
OAKITE SOLVENT MATERIALS				NR at 70°F		
OCTACHLOROTOLUENE	NR at 70°F	NR at 70°F	NR at 70°F	NR at 70°F	NR at 70°F	
OCTADECANE	NR at 70°F	B at 70°F	A to 70°F	NR at 70°F	NR at 70°F	
OCTADECANOIC ACID			A to 70°F	B/NR at 70°F		
OCTADECATRIENOIC ACID				AB to 150°F		
OCTAFLUOROCYCLOBUTANE				NR at 70°F		
OCTANE	C/NR at 70°F	NR at 70°F	NR at 70°F	NR at 70°F		
n-OCTANE	NR at 70°F	NR at 70°F	NR at 70°F	NR at 70°F	NR at 70°F	
OCTANOL	AB to 160°F	B/NR at 70°F	NR at 70°F	B at 70°F	B at 70°F	
2-OCTANONE				B at 70°F		
n-OCTENE-2						
OCTOIC ACID				C at 70°F		
OCTYL ACETATE				A to 70°F		

	SYNTHETIC ISOPRENE IR	NATURAL ISOPRENE NR	CHLOROSULFONATED POLYETHYLENE CSM	FLUOROSILICONE FVMQ	SILICONE VMQ	CHEMRAZ FFKM
NITROGEN TETROXIDE	NR at 70°F	NR at 70°F	NR at 70°F	NR at 70°F	NR at 70°F	A/NR at 70°F B at 70°F white
NITROHYDROCHLORIC ACID	NR at 70°F	NR at 70°F	B/NR at 70°F	C at 70°F	C/NR at 70°F	AB to 70°F NR at 70°F white
NITROMETHANE	B at 70°F	B any conc to 80°F	BC at 70°F	NR at 70°F	NR at 70°F	A to 70°F
NITROMURIATRIC ACID	NR at 70°F	NR at 70°F	B/NR at 70°F	C at 70°F	C/NR at 70°F	AB to 70°F NR at 70°F white
1-NITROPROPANE	NR at 70°F	NR at 70°F	NR at 70°F	NR at 70°F	NR at 70°F	A to 70°F
NITROUS ACIDS		C at 70°F				A to 70°F
NITROUS MONOXIDE	A to 70°F	A to 70°F	A to 70°F			A to 70°F
NITROUS OXIDE	A to 70°F	A to 70°F	A to 70°F			A to 70°F
NITROUS OXIDES						A to 70°F
NITROXANTHIC ACID	C at 70°F	C/NR at 70°F	A to 70°F		NR at 70°F	
NONANOIC ACID	NR at 70°F	NR at 70°F	NR at 70°F			
NONENES	NR at 70°F	NR at 70°F	NR at 70°F			
NORGE NITER	A to 70°F	A to 70°F AB any conc to 150°F	A to 70°F		B at 70°F	
NORGE SALTPETER	A to 70°F	A to 70°F	A to 70°F		B at 70°F	
NORWAY SALTPETER	A to 70°F	A to 70°F AB any conc to 150°F	A to 70°F		B at 70°F	
NORWEGIAN SALTPETER	A to 70°F	A to 70°F AB any conc to 150°F	A to 70°F		B at 70°F	
NORYL-GENERAL ELECTRIC						
N.R.R.O.	NR at 70°F	C at 70°F			B at 70°F	
NUT OIL				A to 70°F	NR at 70°F	
O-A-548A	B at 70°F	B at 70°F	B at 70°F	B at 70°F	B at 70°F	
O-T-634B	NR at 70°F	NR at 70°F	NR at 70°F	B at 70°F	NR at 70°F	
OAKITE ALKALINE MATERIALS	A to 70°F	A to 70°F AB to 150°F				
OAKITE-O.C. 31 & 32	C at 70°F	C at 70°F				
OAKITE-O.C. 33 & 34	NR at 70°F	NR at 70°F				
OAKITE-O.C. 36, 84H & 84M	C at 70°F	C at 70°F				
OAKITE-O.C. 85	NR at 70°F	NR at 70°F				
OAKITE-O.C. 88	NR at 70°F	NR at 70°F				
OAKITE-O.C. 131	C at 70°F	C at 70°F				
OAKITE CRYSCOAT FH RINSE	NR at 70°F	NR at 70°F				
OAKITE O DRYCID & OMF 184	C at 70°F	C at 70°F				
OAKITE O STRIPPER S.A.	NR at 70°F	NR at 70°F				
OAKITE SOLVENT MATERIALS	NR at 70°F	NR at 70°F				
OCTACHLOROTOLUENE	NR at 70°F	NR at 70°F	NR at 70°F	B at 70°F	NR at 70°F	B at 70°F
OCTADECANE	NR at 70°F	NR at 70°F	B at 70°F	A to 70°F	NR at 70°F	A to 70°F
OCTADECANOIC ACID	B/NR at 70°F	B/NR 70°F	BC to 158°F	BC at 70°F	B at 70°F	A to 70°F
OCTADECATRIENOIC ACID	NR at 70°F	NR at 70°F				
OCTAFLUOROCYCLOBUTANE	NR at 70°F	NR at 70°F				
OCTANE	NR at 70°F	NR at 70°F	NR at 70°F		NR at 70°F	
n-OCTANE	NR at 70°F	NR at 70°F	NR at 70°F	B at 70°F	NR at 70°F	A to 70°F
OCTANOL	B at 70°F	B at 70°F	AB at 70°F	B at 70°F	B at 70°F	A to 70°F
2-OCTANONE	NR at 70°F	NR at 70°F	NR at 70°F			
n-OCTENE-2						
OCTOIC ACID	C at 70°F	C at 70°F	B at 70°F			
OCTYL ACETATE	NR at 70°F	NR at 70°F	A to 70°F			A to 70°F

	POLYSULFIDE T	CHLORINATED POLYETHYLENE CM	EPICHLOROHYDRIN CO, ECO	® HYTREL COPOLYESTER TPE	® SANTOPRENE COPOLYMER TPO	® C-FLEX STYRENIC TPE
NITROGEN TETROXIDE	NR at 70°F			*AB to 50% to 70°F		*AB to 70°F
NITROHYDROCHLORIC ACID	NR at 70°F					AB to 70°F
NITROMETHANE				+13-33% vol 7 days 70°F B/NR at 70°F		
NITROMURIATRIC ACID	NR at 70°F					AB to 70°F
1-NITROPROPANE					*B/NR at 70°F	
NITROUS ACIDS						A to 70°F
NITROUS MONOXIDE						
NITROUS OXIDE						
NITROUS OXIDES						
NITROXANTHIC ACID						
NONANOIC ACID						
NONENES						
NORGE NITER						
NORGE SALTPETER						
NORWAY SALTPETER						
NORWEGIAN SALTPETER						
NORYL-GENERAL ELECTRIC						
N.R.R.O.						
NUT OIL						
O-A-548A	C at 70°F		B at 70°F			
O-T-634B	C at 70°F		C at 70°F			
OAKITE ALKALINE MATERIALS						
OAKITE-O.C. 31 & 32						
OAKITE-O.C. 33 & 34						
OAKITE-O.C. 36, 84H & 84M						
OAKITE-O.C. 85						
OAKITE-O.C. 88						
OAKITE-O.C. 131						
OAKITE CRYSCOAT FH RINSE						
OAKITE O DRYCID & OMF 184						
OAKITE O STRIPPER S.A.						
OAKITE SOLVENT MATERIALS						
OCTACHLOROTOLUENE	NR at 70°F					
OCTADECANE	A to 70°F					
OCTADECANOIC ACID						
OCTADECATRIENOIC ACID						
OCTAFLUOROCYCLOBUTANE						
OCTANE	AB to 70°F					
n-OCTANE	B at 70°F				*AB to 70°F	
OCTANOL	B at 70°F					
2-OCTANONE						
n-OCTENE-2						
OCTOIC ACID						
OCTYL ACETATE	AB to 70°F					

	ETHYLENE ACRYLIC EA	POLYALLOMER LINEAR COPOLYMER	NYLON 11 POLYAMIDE	NYLON 12 POLYAMIDE	ETHYLENE VINYL ACETATE EVA	POLYVINYLCHLORIDE FLEXIBLE PVC
NITROGEN TETROXIDE					NR at 70°F	NR at 70°F
NITROHYDROCHLORIC ACID	NR at 70°F					NR at 70°F
NITROMETHANE				AB to 70°F	NR at 70°F	NR at 70°F
NITROMURIATRIC ACID	NR at 70°F					NR at 70°F
1-NITROPROPANE					NR at 70°F	NR at 70°F
NITROUS ACIDS		C at 70°F	B/NR at 70°F		B at 70°F	A to 70°F
NITROUS MONOXIDE			B/NR at 70°F			AB to 70°F
NITROUS OXIDE			B/NR at 70°F			AB to 70°F
NITROUS OXIDES						NR at 70°F
NITROXANTHIC ACID						NR at 70°F
NONANOIC ACID						
NONENES						
NORGE NITER						
NORGE SALTPETER						
NORWAY SALTPETER						
NORWEGIAN SALTPETER						
NORYL-GENERAL ELECTRIC						
N.R.R.O.						
NUT OIL						
O-A-548A						
O-T-634B						
OAKITE ALKALINE MATERIALS						
OAKITE-O.C. 31 & 32						
OAKITE-O.C. 33 & 34						NR at 70°F
OAKITE-O.C. 36, 84H & 84M						
OAKITE-O.C. 85						
OAKITE-O.C. 88						
OAKITE-O.C. 131						
OAKITE CRYSCOAT FH RINSE						
OAKITE O DRYCID & OMF 184						
OAKITE O STRIPPER S.A.						
OAKITE SOLVENT MATERIALS						NR at 70°F
OCTACHLOROTOLUENE						
OCTADECANE						
OCTADECANOIC ACID						
OCTADECATRIENOIC ACID						
OCTAFLUOROCYCLOBUTANE						NR at 70°F
OCTANE				AB to 70°F		
n-OCTANE	A to 70°F	A to 122°F		AB to 70°F	C/NR at 70°F	NR at 70°F
OCTANOL	A to 70°F					NR at 70°F
2-OCTANONE						
n-OCTENE-2						
OCTOIC ACID						
OCTYL ACETATE						

	POLYETHYLENE LOW DENSITY LDPE	®TEFLON FEP	®KALREZ PERFLUORINATED ELASTOMER	®FLUORAZ FLUORINATED COPOLYMER	®AFLAS FLUORINATED COPOLYMER	®NORPRENE COPOLYMER TPO
NITROGEN TETROXIDE	A to 70°F	A to boiling	A to 212°F	C at 70°F		
NITROHYDROCHLORIC ACID	AB to 70°F BC at 104°F	A to 248°F	A to 212°F	C at 70°F		BC at 70°F
NITROMETHANE	A to 70°F	A to 300°F	A to 212°F	C/NR at 70°F		
NITROMURIATRIC ACID	AB to 70°F BC at 104°F	A to 248°F	A to 212°F	C at 70°F		BC at 70°F
1-NITROPROPANE	A to 70°F	A to 70°F	A to 212°F	BC at 70°F	B at 70°F	
NITROUS ACIDS	TEST	A to 300°F	A to 212°F			A to 70°F
NITROUS MONOXIDE	BC at 70°F	A to 300°F	*A to 212°F			
NITROUS OXIDE	BC at 70°F	A to 300°F	*A to 212°F			
NITROUS OXIDES		A to 300°F	*A to 212°F			
NITROXANTHIC ACID		A to 70°F	*A to 212°F			
NONANOIC ACID		A to 70°F				
NONENES		A to 70°F				
NORGE NITER		A to 70°F				
NORGE SALTPETER		A to 70°F				
NORWAY SALTPETER		A to 70°F				
NORWEGIAN SALTPETER		A to 70°F				
NORYL-GENERAL ELECTRIC						
N.R.R.O.						
NUT OIL						
O-A-548A						
O-T-634B						
OAKITE ALKALINE MATERIALS		A to 70°F				
OAKITE-O.C. 31 & 32		A to 70°F				
OAKITE-O.C. 33 & 34		A to 70°F				
OAKITE-O.C. 36, 84H & 84M		A to 70°F				
OAKITE-O.C. 85		A to 70°F				
OAKITE-O.C. 88		A to 70°F				
OAKITE-O.C. 131		A to 70°F				
OAKITE CRYSCOAT FH RINSE		A to 70°F				
OAKITE O DRYCID & OMF 184		A to 70°F				
OAKITE O STRIPPER S.A.		A to 70°F				
OAKITE SOLVENT MATERIALS		A to 70°F				
OCTACHLOROTOLUENE		A to 70°F	A to 212°F	B at 70°F		
OCTADECANE		A to 70°F	A to 212°F	A to 70°F		
OCTADECANOIC ACID	AC to 70°F A crystals to 122°F	A to 300°F A crystals to 125°F	A to 70°F	AC to 70°F		A to 70°F
OCTADECATRIENOIC ACID		A to 70°F				
OCTAFLUOROCYCLOBUTANE		A to 70°F				
OCTANE		A to 70°F				
n-OCTANE	A to 122°F	A to 300°F	A to 212°F	B at 70°F		
OCTANOL		A to 70°F	A to 212°F	A to 70°F		
2-OCTANONE		A to 70°F	A to 212°F			
n-OCTENE-2		A to 70°F				
OCTOIC ACID		A to 70°F				
OCTYL ACETATE		A to 70°F	A to 212°F			

	RECOMMENDATIONS	NITRILE NBR	ETHYLENE PROPYLENE EP, EPDM	FLUOROCARBON FKM	CHLOROPRENE CR	HYDROGENATED NITRILE HNBR
OCTYL ALCOHOL	EP,NBR,FKM,CR,AU	AB to 100°F	AB to 160°F	AB to 70°F	AB to 70°F	B at 70°F
n-OCTYL ALCOHOL	EP,NBR,FKM,CR	A to 70°F AB to 100°F	A to 70°F	AB to 70°F	AB to 70°F	
OCTYL ALDEHYDE		NR at 70°F		NR at 70°F		
OCTYL AMINE		C at 70°F		NR at 70°F		
OCTYL CARBINOL		A to 70°F		B at 70°F		
OCTYLENE GLYCOL		A to 70°F		A to 70°F		
OCTYLIC ACID		C at 70°F				
OCTYLIC ALCOHOL		A to 70°F	A to 70°F	B at 70°F	B at 70°F	
OIL-SAE		A to 70°F AB to 150°F	NR at 70°F	A to 70°F	BC at 70°F	
OIL OF ACETONE		NR at 70°F	A to 70°F	NR at 70°F	C at 70°F	
OIL OF BIRCHWOOD				A to 70°F	NR at 70°F	
OIL OF BITTER ALMONDS. ARTICICIAL		NR at 70°F	C at 70°F	A to 70°F	NR at 70°F	
OIL OF HARTSHORN		A to 70°F		A to 70°F		
OIL OF MIRBANE	FKM,EP,IIR	NR at 70°F	A to 70°F B at 140°F	A to 70°F	NR at 70°F	
OIL OF PALMA CHRISTI		A to 70°F	B at 70°F	A to 70°F	A to 70°F	
OIL OF TURPENTINE	NBR,FKM,FVMQ	A to 200°F	NR at 70°F	A to 158°F	NR at 70°F	A to 70°F
OIL OF VITRIOL	FKM	NR at 70°F	C at 70°F	A to 158°F	NR at 70°F	NR at 70°F
OKONITE		A to 70°F		A to 70°F		
OLEFIANT GAS		A to 70°F	C at 70°F	A to 70°F		
OLEFIN CRUDE	NBR	A to 250°F	NR at 70°F	A to 70°F	NR at 70°F	
OLEIC ACID	NBR,FKM,CR,FVMQ	A to 275°F	BC at 70°F NR at 140°F	A to 212°F B at 248°F	BC at 70°F NR at 140°F	A to 70°F
OLEIN		AB to 120°F			BC to 80°F	
OLEUM (RED FUMING SULFURIC ACID)	FKM,IIR	dissolved 7 days 70°F NR at 70°F	dissolved 7 days 70°F NR at 70°F	+23% vol 7 days 70°F B at 140°F	dissolved 7 days 70°F NR at 70°F	B at 70°F
OLEUM LINI		A to 70°F	B at 70°F	A to 70°F	B at 70°F	
OLEUM SPIRITS	NBR,FKM,CR	B at 70°F	NR at 70°F	A to 160°F	C/NR at 70°F	
OLIVE OIL	NBR,FKM,CR,AU	A to 250°F	B at 70°F	+4% vol 7 days 75°F A to 176°F	B to 80°F NR at 140°F	A to 70°F
ORANGE OIL		A to 140°F		A to 140°F	C/NR at 70°F NR at 140°F	
ORONITE 8200, 8515	FKM,CR,NBR,AU	B at 70°F	NR at 70°F	A to 100°F NR at 400°F	A to 70°F	B at 70°F
ORTHOBORIC ACID		A to 70°F AB any conc to 200°F	A to 70°F	A to 70°F	A to 70°F AB any conc to 200°F	
ORTHOCHLOROETHYLBENZENE	FKM	NR at 70°F	NR at 70°F	A to 70°F	NR at 70°F	NR at 70°F
ORTHODICHLOROBENZENE	FKM,FVMQ	NR at 70°F	NR at 70°F	A to 200°F	NR at 70°F	NR at 70°F
ORTHODICHLOROBENZOL	FKM,FVMQ	NR at 70°F	NR at 70°F	A to 70°F	NR at 70°F	NR at 70°F
ORTHO-HYDROXYBENZOIC ACID		NR at 70°F			NR at 70°F	
ORTHOPHOSPHORIC ACID	FKM	NR at 70°F		AB to 70°F	NR at 70°F	
ORTHOXYLENE	FKM,FVMQ,T	NR at 70°F	A/NR at 70°F	A to 140°F	NR at 70°F	
OS45 TYPE III, SILICATE ESTER BASED	CR,FKM,FVMQ	+3% vol 3 days 300°F B at 70°F	+36% vol 3 days 300°F NR at 70°F	+1% vol 3 days 300°F +11% vol 7 days 400°F	A to 250°F	B at 70°F
OS45 TYPE IV	CR,FKM,FVMQ	B at 70°F	NR at 70°F	+7% vol 21 days 350°F +9% vol 3 days 500°F	+34% vol 3 days 300°F A to 250°F	B at 70°F
OS45 + OS45-1	FKM	NR at 70°F	NR at 70°F	A to 250°F	A to 70°F	
OS70	FKM,CR	B at 70°F	NR at 70°F	+9% vol 28 days 158°F A to 70°F	A to 70°F	B at 70°F
OXALIC ACID		AB to 100% to 140°F NR 10% boiling	A to 100% to 250°F	A to 100% to 140°F A to 50% to 176°F	B to 100% to 140°F NR 20-50% at 176°F	B at 70°F
OXYETHYLENE SUCCINIC ACID		B at 70°F	NR at 70°F	A to 70°F	C at 70°F	
OXYGEN, COLD		B to 100°F	A to 70°F	A to 70°F	A to 200°F	NR at 70°F
OXYGEN, HOT		NR at 250°F	NR at 250°F	B 250-400°F	NR at 250°F	NR
OXYGEN, LIQUID		NR	NR	NR	NR	NR

	STYRENE BUTADIENE SBR	POLYACRYLATE ACM	POLYURETHANE AU, EU	ISOBUTYLENE ISOPRENE IIR	POLYBUTADIENE BR	®AEROQUIP AQP
OCTYL ALCOHOL	AB to 160°F	B/NR at 70°F	NR at 70°F	B at 70°F	B at 70°F	
n-OCTYL ALCOHOL			NR at 70°F	B at 70°F		
OCTYL ALDEHYDE				C at 70°F		
OCTYL AMINE				A to 70°F		
OCTYL CARBINOL				A to 70°F		
OCTYLENE GLYCOL				A to 70°F		
OCTYLIC ACID				C at 70°F		
OCTYLIC ALCOHOL			NR at 70°F	B at 70°F		
OIL-SAE			A to 70°F	C/NR at 70°F		AB to 70°F
OIL OF ACETONE			NR at 70°F	A to 70°F		
OIL OF BIRCHWOOD						
OIL OF BITTER ALMONDS. ARTICICIAL			NR at 70°F	B any conc to 75°F		
OIL OF HARTSHORN				B at 70°F		
OIL OF MIRBANE	NR at 70°F	NR at 70°F	NR at 70°F	B any conc to 80°F	NR at 70°F	TEST
OIL OF PALMA CHRISTI			A to 70°F	A to 70°F		
OIL OF TURPENTINE	NR at 70°F	B at 70°F	NR at 70°F	NR at 70°F	NR at 70°F	AB to 70°F
OIL OF VITRIOL				NR at 70°F		
OKONITE				A to 70°F		
OLEFIANT GAS						
OLEFIN CRUDE	NR at 70°F	A to 70°F	A to 70°F			
OLEIC ACID	NR at 70°F	AB to 70°F	B at 70°F AB to 70°F fumes	NR at 70°F	NR at 70°F	NR at 70°F
OLEIN				AB to 100°F		
OLEUM (RED FUMING SULFURIC ACID)	dissolved 7 days 70°F NR at 70°F	dissolved 7 days 70°F NR at 70°F	NR at 70°F	dissolved 7 days 70°F NR at 70°F	NR at 70°F	NR at 70°F
OLEUM LINI			B at 70°F	A to 70°F		
OLEUM SPIRITS	NR at 70°F		BC at 70°F	NR at 70°F	NR at 70°F	AB to 70°F
OLIVE OIL	NR at 70°F	A to 70°F	A to 70°F	B to 100°F	NR at 70°F	
ORANGE OIL						
ORONITE 8200, 8515	NR at 70°F		A to 70°F	NR at 70°F	NR at 70°F	
ORTHOBORIC ACID	A to 70°F	NR at 70°F	A to 70°F	A to 70°F AB any conc to 185°F	A to 70°F	
ORTHOCHLOROETHYLBENZENE	NR at 70°F	NR at 70°F	NR at 70°F	NR at 70°F	NR at 70°F	
ORTHODICHLOROBENZENE	NR at 70°F	NR at 70°F	NR at 70°F	NR at 70°F	NR at 70°F	
ORTHODICHLOROBENZOL	NR at 70°F	NR at 70°F	NR at 70°F	NR at 70°F	NR at 70°F	
ORTHO-HYDROXYBENZOIC ACID				B any conc to 80°F		
ORTHOPHOSPHORIC ACID			C/NR at 70°F	AB to 120°F	NR at 70°F	NR at 70°F
ORTHOXYLENE	NR at 70°F	NR at 70°F	C/NR at 70°F	NR at 70°F	NR at 70°F	TEST
OS45 TYPE III, SILICATE ESTER BASED	dissolved 3 days 300°F NR at 70°F	+8% vol 3 days 300°F NR at 70°F	-6% vol 3 days 300°F A/NR at 70°F	dissolved 3 days 300°F A/NR at 70°F	NR at 70°F	
OS45 TYPE IV	NR at 70°F	NR at 70°F	NR at 70°F	NR 100% at 70°F AB 25% to 100°F	NR at 70°F	
OS45 + OS45-1	NR at 70°F	NR at 70°F	NR at 70°F			
OS70	NR at 70°F		NR at 70°F	NR at 70°F	NR at 70°F	
OXALIC ACID	B at 70°F		A/NR 100% at 70°F A 5% to 70°F	A to 70°F AB 10% to 185°F	B at 70°F	NR at 70°F
OXYETHYLENE SUCCINIC ACID				NR at 70°F		
OXYGEN, COLD	B/NR at 70°F	AB to 70°F	A to 70°F	A to 70°F AB to 100°F	B at 70°F	NR at 70°F
OXYGEN, HOT	NR at 250°F	NR at 250°F	NR at 250°F	NR at 250°F	NR at 250°F	
OXYGEN, LIQUID	NR	NR	NR			

	SYNTHETIC ISOPRENE IR	NATURAL ISOPRENE NR	CHLOROSULFONATED POLYETHYLENE CSM	FLUOROSILICONE FVMQ	SILICONE VMQ	CHEMRAZ FFKM
OCTYL ALCOHOL	B at 70°F	B at 70°F	AB to 70°F	B at 70°F	B at 70°F	A to 70°F
n-OCTYL ALCOHOL	B at 70°F	B at 70°F	A to 70°F		B at 70°F	
OCTYL ALDEHYDE	NR at 70°F	NR at 70°F	NR at 70°F			
OCTYL AMINE	C at 70°F	C at 70°F	C at 70°F			
OCTYL CARBINOL	A to 70°F	A to 70°F	A to 70°F			
OCTYLENE GLYCOL	A to 70°F	A to 70°F	A to 70°F			
OCTYLIC ACID	C at 70°F	C at 70°F	B at 70°F			
OCTYLIC ALCOHOL	B at 70°F	B at 70°F	A to 70°F		B at 70°F	
OIL-SAE		NR at 70°F	BC at 70°F		B at 70°F	
OIL OF ACETONE	B at 70°F	B at 70°F	B at 70°F		A to 70°F	
OIL OF BIRCHWOOD		NR at 70°F				
OIL OF BITTER ALMONDS. ARTICICIAL	NR at 70°F	NR at 70°F	NR at 70°F		C at 70°F	
OIL OF HARTSHORN	NR at 70°F	NR at 70°F	NR at 70°F			
OIL OF MIRBANE	NR at 70°F	NR at 70°F	NR at 70°F	NR at 70°F	C/NR at 70°F	A to 70°F
OIL OF PALMA CHRISTI	B at 70°F	B at 70°F	A to 70°F		A to 70°F	
OIL OF TURPENTINE	NR at 70°F	NR at 70°F	NR at 70°F	B at 70°F	NR at 70°F	A to 70°F
OIL OF VITRIOL	NR at 70°F	NR at 70°F	NR at 70°F	NR at 70°F	NR at 70°F	A to 70°F
OKONITE	A to 70°F	A to 70°F	A to 70°F			
OLEFIANT GAS			A to 70°F			
OLEFIN CRUDE				A to 70°F	NR at 70°F	A to 70°F
OLEIC ACID	NR at 70°F	NR at 70°F	B at 70°F	A to 400°F	NR at 70°F	A to 70°F
OLEIN	NR at 70°F	NR at 70°F				
OLEUM (RED FUMING SULFURIC ACID)	NR at 70°F	NR at 70°F	dissolved 7 days 70°F C/NR 20% at 70°F	NR at 70°F	dissolved 7 days 70°F NR 25-100% at 70°F	AB to 70°F NR at 70°F white
OLEUM LINI	NR at 70°F	NR at 70°F	B at 70°F		NR at 70°F	
OLEUM SPIRITS	NR at 70°F	NR at 70°F	B/NR at 70°F	B at 70°F	NR at 70°F	A to 70°F
OLIVE OIL	NR at 70°F	NR at 70°F	B at 70°F	A to 70°F	A/NR at 70°F	A to 70°F
ORANGE OIL					NR at 70°F	
ORONITE 8200, 8515	NR at 70°F	NR at 70°F	NR at 70°F	+6% vol 3 days 300°F A to 70°F	+150% vol 3 days 300°F NR at 70°F	A to 70°F
ORTHOBORIC ACID	A to 70°F	A to 70°F AB any conc to 150°F	A to 70°F	A to 70°F	A to 70°F	
ORTHOCHLOROETHYLBENZENE	NR at 70°F	NR at 70°F	NR at 70°F	B at 70°F	NR at 70°F	A to 70°F
ORTHODICHLOROBENZENE	NR at 70°F	NR at 70°F	NR at 70°F	B at 70°F	NR at 70°F	A to 70°F
ORTHODICHLOROBENZOL	NR at 70°F	NR at 70°F	NR at 70°F	B at 70°F	NR at 70°F	A to 70°F
ORTHO-HYDROXYBENZOIC ACID	C at 70°F	BC any conc to 80°F				
ORTHOPHOSPHORIC ACID		NR at 70°F	AB to 200°F		TEST	A to 70°F
ORTHOXYLENE	NR at 70°F	NR at 70°F	NR at 70°F	AC to 70°F	NR at 70°F	
OS45 TYPE III, SILICATE ESTER BASED	NR at 70°F	NR at 70°F	B at 70°F	+5% vol 3 days 300°F AB to 70°F	+35% vol 3 days 300°F NR at 70°F	A to 70°F
OS45 TYPE IV	NR at 70°F	NR at 70°F	B at 70°F	B at 70°F	NR at 70°F	A to 70°F
OS45 OS45-1				B at 70°F	NR at 70°F	
OS70	NR at 70°F	NR at 70°F	B at 70°F	B at 70°F	NR at 70°F	A to 70°F
OXALIC ACID	B at 70°F	BC at 70°F AB 10% to 150°F	A to 70°F AB to 185°F	A to 70°F	B at 70°F	A to 70°F
OXYETHYLENE SUCCINIC ACID	A to 70°F	A to 70°F	B at 70°F		B at 70°F	
OXYGEN, COLD	B at 70°F	B at 70°F	AB to 70°F	A to 70°F	A to 70°F	AC to 70°F A to 70°F white
OXYGEN, HOT	NR at 70°F	NR at 70°F	NR at 70°F	NR at 70°F	B at 70°F	A to 400°F
OXYGEN, LIQUID				NR at 70°F	NR at 70°F	B/NR at 70°F B at 70°F white

	POLYSULFIDE T	CHLORINATED POLYETHYLENE CM	EPICHLOROHYDRIN CO, ECO	® HYTREL COPOLYESTER TPE	® SANTOPRENE COPOLYMER TPO	® C-FLEX STYRENIC TPE
OCTYL ALCOHOL	B at 70°F					
n-OCTYL ALCOHOL						
OCTYL ALDEHYDE						
OCTYL AMINE						
OCTYL CARBINOL						
OCTYLENE GLYCOL						
OCTYLIC ACID						
OCTYLIC ALCOHOL						
OIL-SAE	AB to 70°F			AB to 70°F	A/NR at 70°F	TEST
OIL OF ACETONE						
OIL OF BIRCHWOOD						
OIL OF BITTER ALMONDS. ARTICICIAL						
OIL OF HARTSHORN						
OIL OF MIRBANE	NR at 70°F	C at 70°F	NR at 70°F	NR at 70°F	C at 70°F	NR at 70°F
OIL OF PALMA CHRISTI						
OIL OF TURPENTINE	B at 70°F	AB at 70°F	A to 70°F	AB to 70°F	NR at 70°F	NR at 70°F
OIL OF VITRIOL						
OKONITE						
OLEFIANT GAS						
OLEFIN CRUDE						
OLEIC ACID	A to 70°F	NR at 70°F	A to 70°F	A to 70°F		A to 70°F
OLEIN						
OLEUM (RED FUMING SULFURIC ACID)	dissolved 7 days 70°F NR at 70°F	NR at 70°F	NR at 70°F	NR 20–100% at 70°F		
OLEUM LINI						
OLEUM SPIRITS		AB to 70°F		AB to 70°F		
OLIVE OIL	NR at 70°F		B at 70°F			
ORANGE OIL						
ORONITE 8200, 8515				+4% vol 14 days 302°F grade 63D		
ORTHOBORIC ACID	NR at 70°F	NR at 70°F	A to 70°F	A to 70°F	A to 70°F	AB to 70°F
ORTHOCHLOROETHYLBENZENE	B/NR at 70°F					
ORTHODICHLOROBENZENE	A/NR at 70°F			NR at 70°F	NR at 70°F	
ORTHODICHLOROBENZOL	A/NR at 70°F			NR at 70°F	NR at 70°F	
ORTHO-HYDROXYBENZOIC ACID						
ORTHOPHOSPHORIC ACID	NR at 70°F	NR at 70°F		NR at 70°F		A to 70°F
ORTHOXYLENE	B at 70°F	C at 70°F	NR at 70°F	BC at 70°F	NR at 70°F	NR at 70°F
OS45 TYPE III, SILICATE ESTER BASED	softened 3 days 300°F			AC to 70°F		
OS45 TYPE IV						
OS45 + OS45-1						
OS70						
OXALIC ACID	NR at 70°F	NR at 70°F	C at 70°F	NR at 70°F	A to 70°F	A to 70°F
OXYETHYLENE SUCCINIC ACID						
OXYGEN, COLD	NR at 70°F	NR at 70°F	B at 70°F	AB to 70°F	A to 70°F	A to 70°F
OXYGEN, HOT	NR at 250°F	NR at 250°F	NR at 250°F			
OXYGEN, LIQUID						

	ETHYLENE ACRYLIC EA	POLYALLOMER LINEAR COPOLYMER	NYLON 11 POLYAMIDE	NYLON 12 POLYAMIDE	ETHYLENE VINYL ACETATE EVA	POLYVINYLCHLORIDE FLEXIBLE PVC
OCTYL ALCOHOL	A to 70°F					NR at 70°F
n-OCTYL ALCOHOL						
OCTYL ALDEHYDE						
OCTYL AMINE						
OCTYL CARBINOL						
OCTYLENE GLYCOL						
OCTYLIC ACID						
OCTYLIC ALCOHOL						
OIL-SAE			A to 70°F AB to 194°F		C/NR at 70°F	AB to 70°F
OIL OF ACETONE						
OIL OF BIRCHWOOD						
OIL OF BITTER ALMONDS. ARTICICIAL						
OIL OF HARTSHORN						
OIL OF MIRBANE	*NR at 70°F	B at 70°F	B/NR at 70°F	C at 70°F	NR at 70°F	NR at 70°F
OIL OF PALMA CHRISTI						
OIL OF TURPENTINE	B/NR at 70°F	AB to 70°F C at 122°F	AB to 105°F BC at 140°F	AB to 70°F	NR at 70°F	AB to 70°F NR at 130°F
OIL OF VITRIOL						
OKONITE						
OLEFIANT GAS						
OLEFIN CRUDE						
OLEIC ACID		A to 70°F	A to 70°F C at 194°F	AB to 70°F	B/NR at 70°F	no swell 14 days 70°F C/NR at 150°F
OLEIN						
OLEUM (RED FUMING SULFURIC ACID)	NR at 70°F		NR at 70°F	NR at 70°F	NR at 70°F	NR at 70°F
OLEUM LINI						
OLEUM SPIRITS	NR at 70°F		A to 70°F		NR at 70°F	B/NR at 70°F
OLIVE OIL		A to 70°F	A to 70°F	AB to 70°F	C/NR at 70°F	BC at 70°F A to 150°F olives
ORANGE OIL		B at 70°F C at 122°F				C at 70°F NR at 122°F
ORONITE 8200, 8515						
ORTHOBORIC ACID	A to 70°F	A to 70°F	AC to 70°F		AB to 140°F	A to 150°F
ORTHOCHLOROETHYLBENZENE						
ORTHODICHLOROBENZENE		C at 70°F NR at 122°F	A to 70°F			NR at 70°F
ORTHODICHLOROBENZOL		C at 70°F	A to 70°F			NR at 70°F
ORTHO-HYDROXYBENZOIC ACID						NR at 70°F
ORTHOPHOSPHORIC ACID			NR at 70°F	NR at 70°F	C/NR at 70°F	AB to 70°F
ORTHOXYLENE	NR at 70°F	C at 70°F NR at 122°F	AB to 105°F C at 194°F	AB to 70°F	NR at 70°F	NR at 70°F
OS45 TYPE III, SILICATE ESTER BASED			AB to 70°F		NR at 70°F	NR at 70°F
OS45 TYPE IV						
OS45 + OS45-1						
OS70						
OXALIC ACID		A to 70°F	AB to 105°F NR at 194°F	AB 10% to 70°F	AB to 140°F	AB to 150°F
OXYETHYLENE SUCCINIC ACID						
OXYGEN, COLD		A to 70°F	AB to 105°F NR at 194°F		A to 70°F AB to 140°F	A to 70°F AB to 131°F
OXYGEN, HOT			NR at 250°F			
OXYGEN, LIQUID						

	POLYETHYLENE LOW DENSITY LDPE	® TEFLON FEP	® KALREZ PERFLUORINATED ELASTOMER	® FLUORAZ FLUORINATED COPOLYMER	® AFLAS FLUORINATED COPOLYMER	® NORPRENE COPOLYMER TPO
OCTYL ALCOHOL		A to 70°F	A to 212°F	A to 70°F		
n-OCTYL ALCOHOL		A to 70°F				
OCTYL ALDEHYDE		A to 70°F				
OCTYL AMINE		A to 70°F				
OCTYL CARBINOL		A to 70°F				
OCTYLENE GLYCOL		A to 70°F				
OCTYLIC ACID		A to 70°F				
OCTYLIC ALCOHOL		A to 70°F				
OIL-SAE	C at 70°F	A to 70°F	A to 70°F		AB to 212°F	A to 70°F
OIL OF ACETONE		A to 70°F				
OIL OF BIRCHWOOD	.					
OIL OF BITTER ALMONDS. ARTICICIAL		A to 70°F				
OIL OF HARTSHORN		A to 70°F				
OIL OF MIRBANE	NR at 70°F	A to 410°F	+1% vol 7 days 70°F A to 410°F	A to 70°F	+5.6% vol 7 days 70°F A to 70°F	NR at 70°F
OIL OF PALMA CHRISTI		A to 70°F				
OIL OF TURPENTINE	NR at 70°F	A to 300°F	A to 70°F	A to 70°F		BC at 70°F
OIL OF VITRIOL	NR at 70°F	A to 400°F	A to 133°F	AB to 70°F	A to 70°F	TEST
OKONITE		A to 70°F				
OLEFIANT GAS		A to 70°F				
OLEFIN CRUDE			A to 212°F	A to 70°F		
OLEIC ACID	B /NR at 70°F	A to 300°F	A to 212°F	A to 70°F	A to 70°F	BC at 70°F
OLEIN		A to 70°F				
OLEUM (RED FUMING SULFURIC ACID)	NR at 70°F	A to 400°F	A to 212°F	B at 70°F	+4.4% vol 7 days 70°F A to 70°F	NR at 70°F
OLEUM LINI		A to 70°F		NR at 70°F		
OLEUM SPIRITS	A to 70°F	A to 70°F	A to 70°F	B /NR at 70°F		
OLIVE OIL	A to 70°F	A to boiling	A to 212°F	A to 70°F		
ORANGE OIL	C at 70°F NR at 122°F	A to 122°F				
ORONITE 8200, 8515				A to 70°F		
ORTHOBORIC ACID	A to 70°F	A to boiling	A to 70°F	A to 70°F		
ORTHOCHLOROETHYLBENZENE		A to 70°F		NR at 70°F		
ORTHODICHLOROBENZENE	C at 70°F NR at 140°F	A to 122°F	A to 338°F	NR at 70°F		
ORTHODICHLOROBENZOL	C at 70°F	A to 122°F	A to 338°F	NR at 70°F		
ORTHO-HYDROXYBENZOIC ACID		A to 70°F				
ORTHOPHOSPHORIC ACID	B at 70°F		A to 105% to 400°F	A to 115% to 175°F		A to 70°F
ORTHOXYLENE	B/NR at 70°F NR at 122°F	A to boiling	A to 450°F	NR at 70°F	C /NR at 70°F	NR at 70°F
OS45 TYPE III, SILICATE ESTER BASED	C /NR at 70°F	A to 70°F	A to 70°F	A to 70°F	AB to 70°F	
OS45 TYPE IV				A/NR at 70°F		
OS45 + OS45-1						
OS70				A to 70°F		
OXALIC ACID	A to 70°F AB any conc to 150°F	A to 200°F	A to 212°F	A to 70°F		A to 70°F
OXYETHYLENE SUCCINIC ACID		A to 70°F				
OXYGEN, COLD	A to 70°F AB to 140°F	A to 70°F	A to 70°F	A/NR at 70°F		A to 70°F
OXYGEN, HOT		A 250-400°F	A 250-400°F	C /NR 200-400°F		
OXYGEN, LIQUID				NR		

	RECOMMENDATIONS	NITRILE NBR	ETHYLENE PROPYLENE EP, EPDM	FLUOROCARBON FKM	CHLOROPRENE CR	HYDROGENATED NITRILE HNBR
OXYMETHYLENE		C/NR 40% at 70°F			BC 40% to 80°F	
OXYMURIATE OF TIN		A to 70°F	B at 70°F	A to 70°F	A to 70°F	
OZONE	EP, FKM, IIR, VMQ FVMQ	NR 2% to sat'd at 70°F	A to sat'd to 70°F NR sat'd at 140°F	A to sat'd to 70°F NR sat'd at 140°F	BC to sat'd at 70°F NR sat'd at 140°F	NR at 70°F
OZONE-HIGH CONCENTRATION, DYNAMIC	FKM	NR at 70°F	A to 70°F	A to 400°F	B at 70°F	
P-D-680		A to 70°F	NR at 70°F	A to 70°F	C at 70°F	
P-S-661B		A to 70°F	NR at 70°F	A to 70°F	C at 70°F	
PAINT		BC at 70°F AB to 70°F oil based		C at 70°F		
PAINT THINNER-DUCO	FKM	NR at 70°F	NR at 70°F	AB to 200°F	NR at 70°F	NR at 70°F
PAINTERS NAPHTHA		NR at 70°F			NR at 70°F	
PALM OIL	NBR	A to 250°F	NR at 70°F	A to 140°F	C/NR at 70°F NR at 140°F	
PALMITIC ACID (HEXADECANOIC ACID)	NBR, FKM, FVMQ, T CR, IIR	A to 100% to 160°F	B to 100% at 70°F	A to 100% to 70°F	B at 70°F BC at 158°F	A to 70°F
PAPERMAKERS ALUM		A to 70°F AB any conc to 200°F		A to 70°F	A to 70°F AB any conc to 200°F	
PAR-AL-KETONE	T	NR at 70°F	NR at 70°F	NR at 70°F	NR at 70°F	
PARA DICHLOROBENZENE	FKM, FVMQ	NR at 70°F	NR at 70°F	A to 70°F	NR at 70°F	NR at 70°F
PARA DICHLOROBENZINE		NR at 70°F	NR at 70°F	A to 70°F	NR at 70°F	
PARA DICHLOROBENZOL	FKM, FVMQ	NR at 70°F	NR at 70°F	A to 70°F	NR at 70°F	
PARA SAN		B 10% at 70°F		C 10% at 70°F	A 10% to 70°F	
PARADIHYDROXYBENZENE		NR at 70°F			NR at 70°F	
PARAFIN, MOLTEN	NBR	A to 250°F	NR at 250°F	A to 250°F	B at 250°F	
PARAFINS	NBR	A to 140°F	NR at 70°F	A to 150°F	AB to 70°F	
PARAFORM		B/NR at 70°F		C at 70°F	B/NR at 70°F	
PARAFORMALDEHYDE		B at 70°F	A to 70°F	C at 70°F	B at 70°F	
PARALDEHYDE		B/NR at 70°F	A to 70°F	NR at 70°F	B/NR at 70°F	
PARALKETONE		NR at 70°F	NR at 70°F	NR at 70°F	NR at 70°F	NR at 70°F
PARAPLEX G62	NBR	A to 70°F			NR at 70°F	
PARAPOID 10-C (HYPOID GEAR OIL)	NBR	+9% vol 3 days 300°F A to 70°F	NR at 70°F	dissolved 3 days 300°F	softened 3 days 300°F	
PARAXYLENE		C at 70°F	NR at 70°F	A to 70°F	NR at 70°F	
PARIS GREEN & LIME		A 37% to 70°F		A 37% to 70°F	A 37% to 70°F	
PARKER O-LUBE	NBR, FKM, CR, ACM	A to 70°F	NR at 70°F	A to 70°F	A to 70°F	A to 70°F
PARKER SUPER O-LUBE	EP, NBR, FKM, IIR, CR FVMQ	A to 140°F AB to 180°F	A to 140°F	A to 140°F	A to 140°F	
PATENT ALUM		A to 70°F AB any conc to 200°F	A to 70°F	A to 70°F	A to 70°F AB any conc to 200°F	
PEANUT OIL	NBR, FKM, ACM	A to 250°F	C at 70°F	A to 70°F	B/NR at 70°F	
PEAR ALUM		A to 70°F AB any conc to 200°F			A to 70°F AB any conc to 200°F	
PEAR OIL		NR at 70°F	B at 70°F	NR at 70°F	NR at 70°F	
PEARL ASH	NBR	A to 70°F AB any conc to 180°F	A to 176°F AB to 200°F	A to 212°F	A to 140°F AB any conc to 200°F	
PECTIN LIQUOR	NBR	A to 200°F	NR at 70°F	A to 70°F	C at 70°F	
PELARGONIC ACID		A to 70°F				
PELLA OIL		A to 70°F		A to 70°F	NR at 70°F	
PENDA OIL (DIBASIC ESTER)		+27% vol 3 days 300°F			+87% vol 3 days 300°F	
PENICILLIN LIQUID	FVMQ, VMQ				AB to 70°F	
PENTACHLORODIPHENYL		NR at 70°F			NR at 70°F	
PENTACHLORODIPHENYL KETONE		NR at 70°F			NR at 70°F	
PENTACHLORODIPHENYL OXIDE		NR at 70°F			NR at 70°F	
PENTACHLOROETHANE		NR at 70°F		A to 70°F	NR at 70°F	

	STYRENE BUTADIENE SBR	POLYACRYLATE ACM	POLYURETHANE AU, EU	ISOBUTYLENE ISOPRENE IIR	POLYBUTADIENE BR	® AEROQUIP AQP
OXYMETHYLENE				B 40% to 80°F		
OXYMURIATE OF TIN			B at 70°F	B at 70°F		
OZONE	C/NR high conc at 70°F	AB high conc to 70°F	A high conc to 70°F	B high conc at 70°F	NR high conc at 70°F	
OZONE-HIGH CONCENTRATION, DYNAMIC	C/NR at 70°F	AB to 70°F	A to 70°F		NR at 70°F	
P-D-680	NR at 70°F		C at 70°F	NR at 70°F	NR at 70°F	
P-S-661B	NR at 70°F		C at 70°F	NR at 70°F	NR at 70°F	
PAINT			B at 70°F AB to 70°F oil base	C/NR at 70°F		TEST
PAINT THINNER-DUCO	NR at 70°F	NR at 70°F	NR at 70°F C at 70°F oil base	NR at 70°F	NR at 70°F	
PAINTERS NAPHTHA				NR at 70°F		
PALM OIL	NR at 70°F	A to 70°F	A to 70°F	B/NR at 70°F		
PALMITIC ACID (HEXADECANOIC ACID)	B at 70°F	NR at 70°F	A to 70°F	B at 70°F	B at 70°F	AB to 70°F
PAPERMAKERS ALUM				A to 70°F AB any conc to 185°F		
PAR-AL-KETONE	NR at 70°F	NR at 70°F	NR at 70°F	NR at 70°F	NR at 70°F	
PARA DICHLOROBENZENE	NR at 70°F	NR at 70°F	NR at 70°F	NR at 70°F	NR at 70°F	
PARA DICHLOROBENZINE				NR at 70°F		
PARA DICHLOROBENZOL				NR at 70°F		
PARA SAN				A to 70°F		
PARADIHYDROXYBENZENE				NR at 70°F		
PARAFIN, MOLTEN	A to 250°F	A to 250°F	AB to 250°F	NR at 250°F		
PARAFINS			A to 70°F	NR at 70°F		
PARAFORM				B/NR at 70°F		
PARAFORMALDEHYDE				B/NR at 70°F		
PARALDEHYDE				A to 70°F		
PARALKETONE			NR at 70°F			
PARAPLEX G62						
PARAPOID 10-C (HYPOID GEAR OIL)	+154% vol 3 days 300°F NR at 70°F	+30% vol 3 days 300°F		dissolved 3 days 300°F NR at 70°F		
PARAXYLENE			C at 70°F	NR at 70°F		
PARIS GREEN & LIME				A 37% to 70°F		
PARKER O-LUBE	B at 70°F	A to 70°F	A to 70°F	NR at 70°F	NR at 70°F	
PARKER SUPER O-LUBE	A to 70°F	A to 70°F	A to 70°F	A to 70°F	A to 70°F	
PATENT ALUM				A to 70°F AB any conc to 185°F		
PEANUT OIL	NR at 70°F	A to 70°F	B at 70°F	C at 70°F	NR at 70°F	
PEAR ALUM				A to 70°F AB any conc to 185°F		
PEAR OIL			NR at 70°F	B/NR at 70°F		
PEARL ASH				A to 70°F AB any conc to 180°F		
PECTIN LIQUOR	NR at 70°F	A to 70°F	A to 70°F			
PELARGONIC ACID				A to 70°F		
PELLA OIL						
PENDA OIL (DIBASIC ESTER)	dissolved 3 days 300°F NR at 70°F	+36% vol 3 days 300°F		softened 3 days 300°F		
PENICILLIN LIQUID						
PENTACHLORODIPHENYL				NR at 70°F		
PENTACHLORODIPHENYL KETONE				NR at 70°F		
PENTACHLORODIPHENYL OXIDE				NR at 70°F		
PENTACHLOROETHANE				NR at 70°F		

	SYNTHETIC ISOPRENE IR	NATURAL ISOPRENE NR	CHLOROSULFONATED POLYETHYLENE CSM	FLUOROSILICONE FVMQ	SILICONE VMQ	CHEMRAZ FFKM
OXYMETHYLENE	C 40% at 70°F	BC 40% to 80°F				
OXYMURIATE OF TIN	A to 70°F	A to 70°F	A to 70°F		B at 70°F	
OZONE	NR high conc at 70°F	NR high conc at 70°F	A to 70°F	A high conc to 400°F	A high conc to 400°F	AB to 70°F A to 70°F white
OZONE-HIGH CONCENTRATION, DYNAMIC	NR high conc at 70°F	NR high conc at 70°F		NR high conc at 70°F	NR high conc at 70°F	
P-D-680	NR at 70°F	NR at 70°F	C at 70°F	A to 70°F	NR at 70°F	
P-S-661B	NR at 70°F	NR at 70°F	C at 70°F	A to 70°F	NR at 70°F	
PAINT		NR at 70°F	AB to 70°F			
PAINT THINNER-DUCO	NR at 70°F	NR at 70°F	NR at 70°F	B at 70°F	NR at 70°F	A to 70°F
PAINTERS NAPHTHA	NR at 70°F	NR at 70°F				
PALM OIL	NR at 70°F	NR at 70°F	AB to 70°F	A to 70°F	NR at 70°F	
PALMITIC ACID (HEXADECANOIC ACID)	AB to 70°F	A to 100% to 80°F	C to 100% at 70°F	A to 70°F	NR at 70°F	A to 70°F
PAPERMAKERS ALUM	A to 70°F	A to 70°F AB any conc to 150°F	A to 70°F			
PAR-AL-KETONE	NR at 70°F	NR at 70°F	NR at 70°F	NR at 70°F	NR at 70°F	
PARA DICHLOROBENZENE	NR at 70°F	NR at 70°F	NR at 70°F	B at 70°F	NR at 70°F	A to 70°F
PARA DICHLOROBENZINE	NR at 70°F	NR at 70°F	NR at 70°F			
PARA DICHLOROBENZOL	NR at 70°F	NR at 70°F	NR at 70°F			
PARA SAN	A 10% to 70°F	A 10% to 70°F	A 10% to 70°F			
PARADIHYDROXYBENZENE	C at 70°F	BC any conc to 80°F				
PARAFIN, MOLTEN	NR at 250°F	NR at 250°F	NR at 250°F	A to 250°F	B at 250°F	
PARAFINS	NR at 70°F	B/NR at 70°F	NR at 70°F		NR at 70°F	
PARAFORM	NR at 70°F	NR at 70°F				
PARAFORMALDEHYDE	NR at 70°F	NR at 70°F	AB to 70°F			*B at 70°F
PARALDEHYDE	C at 70°F	C at 70°F	NR at 70°F			*B at 70°F
PARALKETONE						A to 70°F
PARAPLEX G62		NR at 70°F				
PARAPOID 10-C (HYPOID GEAR OIL)					+16% vol 3 days 300°F	
PARAXYLENE	NR at 70°F	NR at 70°F	NR at 70°F		NR at 70°F	
PARIS GREEN & LIME			AB 37% at 70°F			
PARKER O-LUBE	NR at 70°F	NR at 70°F	A to 70°F	A to 70°F	B at 70°F	A to 70°F
PARKER SUPER O-LUBE	A to 70°F	A to 70°F	A to 70°F	A to 70°F	BC to 70°F	
PATENT ALUM	A to 70°F	A to 70°F AB any conc to 150°F	A to 70°F		A to 70°F	
PEANUT OIL	NR at 70°F	NR at 70°F	B at 70°F	A to 70°F	A to 70°F	A to 70°F
PEAR ALUM	A to 70°F	A to 70°F AB any conc to 150°F				
PEAR OIL	C at 70°F	C/NR at 70°F	C at 70°F		NR at 70°F	
PEARL ASH	A to 70°F	A to 70°F AB any conc to 185°F	A to 70°F AB to 200°F	A to 70°F	AC to 70°F	A aqueous sol'n to 70°F
PECTIN LIQUOR				A to 70°F	A to 70°F	
PELARGONIC ACID	NR at 70°F	NR at 70°F	B/NR at 70°F			
PELLA OIL						
PENDA OIL (DIBASIC ESTER)					+16% vol 3 days 300°F	
PENICILLIN LIQUID				AB to 200°F	AB to 200°F	
PENTACHLORODIPHENYL	NR at 70°F	NR at 70°F				
PENTACHLORODIPHENYL KETONE		NR at 70°F				
PENTACHLORODIPHENYL OXIDE	NR at 70°F	NR at 70°F				
PENTACHLOROETHANE	NR at 70°F	NR at 70°F	NR at 70°F			A to 70°F

	POLYSULFIDE T	CHLORINATED POLYETHYLENE CM	EPICHLOROHYDRIN CO, ECO	® HYTREL COPOLYESTER TPE	® SANTOPRENE COPOLYMER TPO	® C-FLEX STYRENIC TPE
OXYMETHYLENE						
OXYMURIATE OF TIN						
OZONE	A high conc to 70°F		A high conc to 70°F	C high conc to 70°F		
OZONE-HIGH CONCENTRATION, DYNAMIC	B at 70°F			C at 70°F oil based	NR at 70°F	
P-D-680	A to 70°F		A to 70°F			
P-S-661B	A to 70°F		A to 70°F	AB to 70°F AB to 70°F oil based		
PAINT						
PAINT THINNER-DUCO						
PAINTERS NAPHTHA						
PALM OIL						
PALMITIC ACID (HEXADECANOIC ACID)	NR at 70°F	AB to 70°F	B at 70°F	A to 70°F	A to 70°F	
PAPERMAKERS ALUM						
PAR-AL-KETONE	B at 70°F					
PARA DICHLOROBENZENE	NR at 70°F					
PARA DICHLOROBENZINE						
PARA DICHLOROBENZOL						
PARA SAN	AB 10% to 70°F					
PARADIHYDROXYBENZENE						
PARAFIN, MOLTEN						
PARAFINS						
PARAFORM						
PARAFORMALDEHYDE						
PARALDEHYDE						
PARALKETONE						
PARAPLEX G62						
PARAPOID 10-C (HYPOID GEAR OIL)	softened 3 days 300°F					
PARAXYLENE						
PARIS GREEN & LIME	AB 37% to 70°F					
PARKER O-LUBE	A to 70°F					
PARKER SUPER O-LUBE	A to 70°F		A to 70°F	A to 70°F		
PATENT ALUM						
PEANUT OIL	NR at 70°F		A to 70°F			
PEAR ALUM						
PEAR OIL						
PEARL ASH						
PECTIN LIQUOR						
PELARGONIC ACID						
PELLA OIL						
PENDA OIL (DIBASIC ESTER)	−8% vol 3 days 300°F					
PENICILLIN LIQUID						
PENTACHLORODIPHENYL						
PENTACHLORODIPHENYL KETONE						
PENTACHLORODIPHENYL OXIDE						
PENTACHLOROETHANE					*B /NR at 70°F	

	ETHYLENE ACRYLIC EA	POLYALLOMER LINEAR COPOLYMER	NYLON 11 POLYAMIDE	NYLON 12 POLYAMIDE	ETHYLENE VINYL ACETATE EVA	POLYVINYLCHLORIDE FLEXIBLE PVC
OXYMETHYLENE						
OXYMURIATE OF TIN						
OZONE	A high conc to 70°F	A high conc to 70°F B high conc at 122°F	C high conc at 70°F NR high conc at 105°F	NR high conc at 70°F AB 1 PPM to 70°F	A high conc to 70°F	A high conc to 70°F B high conc at 122°F
OZONE-HIGH CONCENTRATION, DYNAMIC				NR at 70°F		
P-D-680						
P-S-661B						
PAINT			AB to 70°F AB to 70°F, oil base		B at 70°F	B at 70°F C/NR at 70°F, oil base
PAINT THINNER-DUCO			AB to 70°F, oil base			NR at 70°F NR at 70°F, oil base
PAINTERS NAPHTHA						NR at 70°F
PALM OIL					NR at 70°F	B at 70°F
PALMITIC ACID (HEXADECANOIC ACID)		A to 70°F	A/NR at 70°F		C/NR at 70°F	AC to 70°F
PAPERMAKERS ALUM						
PAR-AL-KETONE						
PARA DICHLOROBENZENE						
PARA DICHLOROBENZINE						
PARA DICHLOROBENZOL						
PARA SAN						
PARADIHYDROXYBENZENE						
PARAFIN, MOLTEN					NR at 250°F	NR at 250°F
PARAFINS					NR at 70°F	A to 70°F
PARAFORM						
PARAFORMALDEHYDE						
PARALDEHYDE						
PARALKETONE						
PARAPLEX G62						
PARAPOID 10-C (HYPOID GEAR OIL)						
PARAXYLENE						
PARIS GREEN & LIME						
PARKER O-LUBE						
PARKER SUPER O-LUBE	A to 158°F		A to 70°F			A to 70°F
PATENT ALUM						
PEANUT OIL				AB to 70°F		
PEAR ALUM						
PEAR OIL						NR at 70°F
PEARL ASH						
PECTIN LIQUOR						
PELARGONIC ACID						
PELLA OIL						
PENDA OIL (DIBASIC ESTER)						
PENICILLIN LIQUID						
PENTACHLORODIPHENYL						NR at 70°F
PENTACHLORODIPHENYL KETONE						NR at 70°F
PENTACHLORODIPHENYL OXIDE						NR at 70°F
PENTACHLOROETHANE						

	POLYETHYLENE LOW DENSITY LDPE	® TEFLON FEP	® KALREZ PERFLUORINATED ELASTOMER	® FLUORAZ FLUORINATED COPOLYMER	® AFLAS FLUORINATED COPOLYMER	® NORPRENE COPOLYMER TPO
OXYMETHYLENE		A 40% to 70°F				
OXYMURIATE OF TIN		A to 70°F				
OZONE	BC to sat'd at 70°F NR sat'd at 104°F	A high conc to 125°F	A high conc to 70°F *A to 212°F	A high conc to 70°F		
OZONE-HIGH CONCENTRATION, DYNAMIC						
P-D-680						
P-S-661B						
PAINT	A to 70°F AC to 70°F, oil base	A to 70°F	A to 212°F			
PAINT THINNER-DUCO	AB to 70°F, oil base	A to 70°F	A to 70°F	C at 70°F		
PAINTERS NAPHTHA		A to 70°F				
PALM OIL	A to 70°F	A to 70°F		A to 70°F		
PALMITIC ACID (HEXADECANOIC ACID)	B to 100% at 70°F NR at 140°F	A to 100% to 300°F	A to 212°F	A to 70°F		
PAPERMAKERS ALUM		A to 70°F				
PAR-AL-KETONE						
PARA DICHLOROBENZENE		A to 70°F	A to 212°F	C/NR at 70°F	BC at 70°F	
PARA DICHLOROBENZINE		A to 70°F				
PARA DICHLOROBENZOL		A to 70°F				
PARA SAN		A 10% to 70°F				
PARADIHYDROXYBENZENE		A to 70°F				
PARAFIN, MOLTEN						
PARAFINS	C at 70°F	A to 300°F	A to 212°F	A liquid to 70°F		
PARAFORM		A to 70°F				
PARAFORMALDEHYDE		A to 70°F	*A to 212°F			
PARALDEHYDE		A to 70°F	*A to 212°F			
PARALKETONE				NR at 70°F		
PARAPLEX G62						
PARAPOID 10-C (HYPOID GEAR OIL)						
PARAXYLENE		A to 70°F				
PARIS GREEN & LIME		A 37% to 70°F				
PARKER O-LUBE				A to 70°F		
PARKER SUPER O-LUBE		A to 70°F	A to 70°F	A to 70°F		TEST
PATENT ALUM		A to 70°F				
PEANUT OIL		A to 70°F	A to 212°F	A to 70°F		
PEAR ALUM		A to 70°F				
PEAR OIL		A to 70°F				
PEARL ASH	AB to 104°F	A to 200°F		A to conc to 70°F		AB to 70°F
PECTIN LIQUOR			A to 212°F	A to 70°F		
PELARGONIC ACID		A to 70°F	A to 212°F			
PELLA OIL						
PENDA OIL (DIBASIC ESTER)						
PENICILLIN LIQUID			A to 212°F	NR at 70°F		
PENTACHLORODIPHENYL		A to 70°F				
PENTACHLORODIPHENYL KETONE						
PENTACHLORODIPHENYL OXIDE		A to 70°F				
PENTACHLOROETHANE		A to 70°F	A to 212°F			

	RECOMMENDATIONS	NITRILE NBR	ETHYLENE PROPYLENE EP, EPDM	FLUOROCARBON FKM	CHLOROPRENE CR	HYDROGENATED NITRILE HNBR
PENTACHLOROETHYLBENZENE		NR at 70°F			NR at 70°F	
PENTACHLOROPHENOL	FKM	C/NR at 70°F AB 10% to 70°F	B/NR at 70°F	A to 250°F	B/NR at 70°F AB 10% to 70°F	
PENTACHLOROPHENYLBENZOATE		NR at 70°F			NR at 70°F	
PENTADICHLORODIPHENYL KETONE		NR at 70°F			NR at 70°F	
PENTAHYDROXY HEXOIC ACID		C at 70°F				
PENTAMETHYLENE AMINE		NR at 70°F	NR at 70°F	C at 70°F	NR at 70°F	
PENTANE	NBR, FKM	A to 250°F	NR at 70°F	A to 140°F	A/NR at 70°F	
N-PENTANE	NBR, FKM, CR, ACM	A to 70°F	NR at 70°F	A to 70°F	A to 70°F	
n-PENTANE, 2 METHYL, 2-4 METHYL & 3 METHYL	NBR, FKM, ACM	A to 70°F	NR at 70°F	A to 70°F	B at 70°F	
2, 4 PENTANE DIONE	EP, IIR	NR at 70°F	A to 70°F	NR at 70°F	A/NR at 70°F	
PENTANOIC ACID		NR at 70°F	A to 70°F		NR at 70°F	
PENTANOL		A to 70°F	A to 70°F	AB to 70°F	A to 70°F	
PENTANONE		NR at 70°F	B at 70°F	NR at 70°F	NR at 70°F	
PENTASOL	EP	AB to 70°F	A to 160°F	AB to 70°F	A to 70°F	
PENTASODIUM TRIPHOSPHATE		NR at 70°F			NR at 70°F	
PENTENE		B at 70°F	NR at 70°F	A to 70°F	A to 70°F	
PENTENE-2, 4-METHYL				A to 70°F	B at 70°F	
PENTYL AMINE		C at 70°F	NR at 70°F	NR at 70°F	NR at 70°F	
PENTYL OXYPENTANE		C at 70°F	NR at 70°F		NR at 70°F	
PEPPERMINT OIL	FKM	NR at 70°F		A to 70°F	NR at 70°F	
PERCHLORIC ACID	FKM, FVMQ, CR, EP	NR 10-100% at 70°F	B to 100% at 70°F B to 70% to 140°F	A to 100% to 70°F AB to 70% to 140°F	AB to 100% to 70°F A 10% to 140°F	NR at 70°F
PERCHLORIC ACID			C to 70% at 176°F	BC to 70% at 200°F	B 10% at 176°F NR 10% at 212°F	
PERCHLOROETHYLENE	FKM, FVMQ	B/NR at 70°F NR at 140°F	NR at 70°F	+11% vol 28 days 212°F A to 212°F	NR at 70°F	B/NR at 70°F
PERCHLOROMETHANE		NR at 70°F			NR at 70°F	
PERMACHLOR (DEGREASING FLUID)		NR at 70°F	NR at 70°F	AC to 70°F		
PEROXIDE	FKM	NR 90% at 70°F C/NR 30% at 70°F	C 90% at 70°F B to 30% at 70°F	A to 90% to 70°F	NR 30-90% at 70°F C/NR 10% at 70°F	
PEROXIDE		BC 10% to 80°F B 3% at 70°F			B 3% at 70°F	
PEROXYDOL		B any conc to 200°F	A to 70°F	A to 70°F	B any conc to 200°F	
PEROXYHYDRATE		A to 70°F AB any conc to 200°F			A to 70°F AB any conc to 200°F	
PERPHOSPHATE	NBR, EP, FKM	A to 70°F	A to 70°F	A to 70°F		
PERU SALTPETER		C at 70°F	A to 70°F	A to 70°F	B at 70°F	
PETROL	NBR, FKM, FVMQ	A to 250°F	NR at 70°F	AB to 200°F	NR at 70°F	
PETROLATUM (PETROLEUM JELLY)	NBR, FKM, FVMQ, ACM	A to 140°F	NR at 70°F	A to 140°F	AB to 140°F	
PETROLENE		A to 70°F	NR at 70°F	A to 70°F	B at 70°F	
PETROLEUM	NBR, FKM	A to 250°F C/NR at 260°F	NR at 70°F	A to 250°F B at 260°F	B at 70°F B/NR at 104°F	
PETROLEUM-CRUDE	NBR, FKM, ACM, AU	A to 250°F B at 70°F sour	NR at 70°F	+3% vol 28 days 300°F A to 70°F	B at 70°F B at 70°F sour	
PETROLEUM ETHER	T	A to 70°F	NR at 70°F	+3% vol 7 days 75°F A to 70°F	B/NR at 70°F	
PETROLEUM ETHER, LIQUIFIED	NBR, FKM	A to 200°F	NR at 70°F	A to 70°F AB to 260°F	NR at 70°F	A to 70°F
PETROLEUM NAPHTHA		AC to 70°F	NR at 70°F	A to 70°F	B/NR at 70°F	
PETROLEUM OILS	NBR, FKM, FVMQ	A to 250°F	NR at 70°F	A to 70°F	B at 70°F B/NR to 180°F	
PETROLEUM PITCH		A to 70°F	NR at 70°F	A to 70°F	B at 70°F	
PETROLEUM SPIRIT		A/NR at 70°F	NR at 70°F	A to 70°F	B/NR at 70°F	
PETROLEUM THINNER		NR at 70°F			NR at 70°F	
PHENE		NR at 70°F	NR at 70°F	A to 70°F	NR at 70°F	

	STYRENE BUTADIENE SBR	POLYACRYLATE ACM	POLYURETHANE AU, EU	ISOBUTYLENE ISOPRENE IIR	POLYBUTADIENE BR	® AEROQUIP AQP
PENTACHLOROETHYLBENZENE				NR at 70°F		
PENTACHLOROPHENOL	NR at 70°F		NR at 70°F	A to 70°F		
PENTACHLOROPHENYLBENZOATE				NR at 70°F		
PENTADICHLORODIPHENYL KETONE				NR at 70°F		
PENTAHYDROXY HEXOIC ACID				C at 70°F		
PENTAMETHYLENE AMINE	NR at 70°F	A to 70°F	NR at 70°F	NR at 70°F	NR at 70°F	
PENTANE	C/NR at 70°F	A to 70°F	B/NR at 70°F	C/NR at 70°F		
N-PENTANE	C at 70°F	A to 70°F	NR at 70°F	NR at 70°F	NR at 70°F	
n-PENTANE, 2 METHYL, 2-4 METHYL & 3 METHYL	NR at 70°F	A to 70°F	NR at 70°F	NR at 70°F	NR at 70°F	
2, 4 PENTANE DIONE			B at 70°F	B at 70°F		
PENTANOIC ACID				A to 70°F		
PENTANOL			C at 70°F	A to 70°F		
PENTANONE				B at 70°F		
PENTASOL	B at 70°F	NR at 70°F	NR at 70°F	A to 70°F		
PENTASODIUM TRIPHOSPHATE						
PENTENE				NR at 70°F		
PENTENE-2, 4-METHYL						
PENTYL AMINE				B at 70°F		
PENTYL OXYPENTANE				NR at 70°F		
PEPPERMINT OIL						
PERCHLORIC ACID	NR 100% at 70°F NR 2N at 70°F	NR 2N at 70°F	NR 100% at 70°F NR 2N at 70°F	B to 100% at 70°F AB 10% to 150°F	NR 2N at 70°F	
PERCHLORIC ACID						
PERCHLOROETHYLENE	NR at 70°F	NR at 70°F	+60% vol 7 days 70°F NR at 70°F	NR at 70°F	NR at 70°F	TEST
PERCHLOROMETHANE				NR at 70°F		
PERMACHLOR (DEGREASING FLUID)						
PEROXIDE			C 90% at 70°F	NR 90% at 70°F C/NR 10-30% at 70°F		
PEROXIDE				B 3% at 70°F		
PEROXYDOL				A to 70°F AB any conc to 185°F		
PEROXYHYDRATE				A to 70°F AB any conc to 185°F		
PERPHOSPHATE						
PERU SALTPETER				A to 70°F		
PETROL	NR at 70°F	NR at 70°F	B at 70°F BC at 70°F high octane	NR at 70°F	NR at 70°F	AB to 70°F
PETROLATUM (PETROLEUM JELLY)	NR at 70°F	A to 70°F	A/NR at 70°F	NR at 70°F	NR at 70°F	
PETROLENE			B at 70°F	NR at 70°F		
PETROLEUM	C at 70°F NR at 250°F	AB to 250°F C/NR at 260°F	AB to 70°F NR at 250°F	NR at 70°F	NR at 250°F	
PETROLEUM-CRUDE	NR at 70°F	A to 70°F	A to 70°F AB to 70°F sour	NR at 70°F sweet or sour	NR at 70°F	
PETROLEUM ETHER	NR at 70°F	A/NR at 70°F	AB to 70°F	NR at 70°F		AB to 70°F
PETROLEUM ETHER, LIQUIFIED	NR at 70°F	B at 70°F	A to 70°F			
PETROLEUM NAPHTHA				NR at 70°F		
PETROLEUM OILS	C/NR at 70°F NR at 250°F	AB to 70°F	AB to 70°F	NR at 70°F	NR at 250°F	
PETROLEUM PITCH			B at 70°F	NR at 70°F		
PETROLEUM SPIRIT				NR at 70°F		
PETROLEUM THINNER				NR at 70°F		
PHENE			NR at 70°F	NR at 70°F		

	SYNTHETIC ISOPRENE IR	NATURAL ISOPRENE NR	CHLOROSULFONATED POLYETHYLENE CSM	FLUOROSILICONE FVMQ	SILICONE VMQ	CHEMRAZ FFKM
PENTACHLOROETHYLBENZENE	NR at 70°F	NR at 70°F				
PENTACHLOROPHENOL	NR at 70°F	NR at 70°F		B at 70°F		A to 70°F
PENTACHLOROPHENYLBENZOATE	NR at 70°F	NR at 70°F				
PENTADICHLORODIPHENYL KETONE	NR at 70°F	NR at 70°F				
PENTAHYDROXY HEXOIC ACID	NR at 70°F	NR at 70°F	B at 70°F			
PENTAMETHYLENE AMINE	NR at 70°F	NR at 70°F	NR at 70°F			
PENTANE	NR at 70°F	NR at 70°F	BC at 70°F	A to 70°F	NR at 70°F	
N-PENTANE	NR at 70°F	NR at 70°F	B at 70°F	C at 70°F	NR at 70°F	A to 70°F
n-PENTANE, 2 METHYL, 2-4 METHYL & 3 METHYL	NR at 70°F	NR at 70°F	B at 70°F	C at 70°F	NR at 70°F	A to 70°F
2, 4 PENTANE DIONE	C at 70°F	C at 70°F	NR at 70°F			
PENTANOIC ACID	A to 70°F	A to 70°F				
PENTANOL	A to 70°F	A to 70°F	A to 70°F		NR at 70°F	
PENTANONE	NR at 70°F	NR at 70°F	NR at 70°F			
PENTASOL	A to 70°F	A to 70°F	A to 70°F	A to 70°F	NR at 70°F	
PENTASODIUM TRIPHOSPHATE	NR at 70°F	NR at 70°F				
PENTENE	NR at 70°F	NR at 70°F	NR at 70°F			
PENTENE-2, 4-METHYL						
PENTYL AMINE	C at 70°F	C at 70°F	C at 70°F			
PENTYL OXYPENTANE	NR at 70°F	NR at 70°F	C at 70°F			
PEPPERMINT OIL		NR at 70°F				
PERCHLORIC ACID	NR at 70°F NR 2N at 70°F	NR 70-100% at 70°F AB 10% to 150°F	AB 100% to 70°F B 2N at 70°F	A to 100% to 70°F	NR 100% at 70°F NR 2N at 70°F	AB to 70°F A to 70°F white
PERCHLORIC ACID						
PERCHLOROETHYLENE	NR at 70°F	NR at 70°F	NR at 70°F	+3% vol 120 days 70°F B at 70°F	+3% vol 120 days 70°F B/NR at 70°F	A to 70°F
PERCHLOROMETHANE	NR at 70°F	NR at 70°F				
PERMACHLOR (DEGREASING FLUID)						
PEROXIDE	NR 90% at 70°F C 30% at 70°F	NR 90% at 70°F C/NR 30% at 70°F	C 30-90% at 70°F B 3-10% at 70°F		A to 90% to 70°F	A to 70°F
PEROXIDE	B 3-10% at 70°F	B 3-10% to 80°F				
PEROXYDOL	B at 70°F	B any conc to 150°F	A to 70°F		B at 70°F	
PEROXYHYDRATE	A to 70°F	A to 70°F AB any conc to 150°F				
PERPHOSPHATE						
PERU SALTPETER	B at 70°F	B at 70°F	A to 70°F		C at 70°F	
PETROL	NR at 70°F	NR at 70°F	NR at 70°F	A to 70°F	NR at 70°F	A to 70°F
PETROLATUM (PETROLEUM JELLY)	NR at 70°F	C/NR at 70°F	B at 70°F	A to 70°F	NR at 70°F	A to 70°F
PETROLENE	NR at 70°F	NR at 70°F	B at 70°F		NR at 70°F	
PETROLEUM	NR at 70°F	NR at 70°F	B to 200°F NR at 250°F	B to 250°F NR at 260°F	NR at 70°F	
PETROLEUM-CRUDE	NR at 70°F sweet or sour	NR at 70°F sweet or sour	B to 200°F	A to 70°F	NR at 70°F	A to 250°F
PETROLEUM ETHER	NR at 70°F	NR at 70°F	NR at 70°F	A to 70°F	C/NR at 70°F	
PETROLEUM ETHER, LIQUIFIED	NR at 70°F	NR at 70°F	C at 70°F	A to 70°F	NR at 70°F	A to 70°F
PETROLEUM NAPHTHA	NR at 70°F	NR at 70°F	NR at 70°F		A to 70°F	
PETROLEUM OILS	NR at 70°F	NR at 70°F	B to 200°F NR at 250°F	A to 70°F B to 250°F	NR at 70°F	A to 250°F
PETROLEUM PITCH	NR at 70°F	NR at 70°F	B at 70°F		NR at 70°F	
PETROLEUM SPIRIT	NR at 70°F	NR at 70°F	NR at 70°F		A to 70°F	
PETROLEUM THINNER	NR at 70°F	NR at 70°F				
PHENE	NR at 70°F	NR at 70°F	NR at 70°F		NR at 70°F	

	POLYSULFIDE T	CHLORINATED POLYETHYLENE CM	EPICHLOROHYDRIN CO, ECO	® HYTREL COPOLYESTER TPE	® SANTOPRENE COPOLYMER TPO	® C-FLEX STYRENIC TPE
PENTACHLOROETHYLBENZENE						
PENTACHLOROPHENOL	AB to 100% to 70°F					
PENTACHLOROPHENYLBENZOATE						
PENTADICHLORODIPHENYL KETONE						
PENTAHYDROXY HEXOIC ACID						
PENTAMETHYLENE AMINE						
PENTANE				AB to 70°F		
N-PENTANE	A to 70°F				*AB to 70°F	
n-PENTANE, 2 METHYL, 2-4 METHYL & 3 METHYL	A to 70°F					
2, 4 PENTANE DIONE						
PENTANOIC ACID						
PENTANOL						
PENTANONE						
PENTASOL						
PENTASODIUM TRIPHOSPHATE						
PENTENE						
PENTENE-2, 4-METHYL						
PENTYL AMINE						
PENTYL OXYPENTANE						
PEPPERMINT OIL						
PERCHLORIC ACID	A/NR 100% at 70°F NR 2N at 70°F		C at 70°F	NR at 70°F	NR at 70°F	A to 70°F
PERCHLORIC ACID						
PERCHLOROETHYLENE	AB to 70°F	C at 70°F	BC at 70°F	+10-81% vol 7 days 70°F B/NR at 70°F	NR at 70°F	AB to 70°F
PERCHLOROMETHANE						
PERMACHLOR (DEGREASING FLUID)						
PEROXIDE						
PEROXIDE						
PEROXYDOL						
PEROXYHYDRATE						
PERPHOSPHATE						
PERU SALTPETER						
PETROL	A to 70°F	A to 70°F	A to 70°F	A to 70°F		NR at 70°F
PETROLATUM (PETROLEUM JELLY)	B at 70°F					
PETROLENE						
PETROLEUM	NR at 250°F	AB to 70°F	A to 250°F B at 260°F	A to 70°F AB to 300°F	AC at 212°F	
PETROLEUM-CRUDE	NR at 70°F		A to 70°F	BC at 158°F sweet AB to 70°F sour		
PETROLEUM ETHER	A to 160°F					
PETROLEUM ETHER, LIQUIFIED						
PETROLEUM NAPHTHA						
PETROLEUM OILS	NR at 250°F	AB to 70°F	A to 250°F	A to 70°F	AC at 212°F BC at 257°F	
PETROLEUM PITCH						
PETROLEUM SPIRIT						
PETROLEUM THINNER						
PHENE						

C281

	ETHYLENE ACRYLIC EA	POLYALLOMER LINEAR COPOLYMER	NYLON 11 POLYAMIDE	NYLON 12 POLYAMIDE	ETHYLENE VINYL ACETATE EVA	POLYVINYLCHLORIDE FLEXIBLE PVC
PENTACHLOROETHYLBENZENE						NR at 70°F
PENTACHLOROPHENOL						B at 70°F in oil C at 150°F in oil
PENTACHLOROPHENYLBENZOATE						NR at 70°F
PENTADICHLORODIPHENYL KETONE						
PENTAHYDROXY HEXOIC ACID						
PENTAMETHYLENE AMINE						
PENTANE			AB to 70°F			AC to 70°F NR at 150°F
N-PENTANE						
n-PENTANE, 2 METHYL, 2-4 METHYL & 3 METHYL						
2, 4 PENTANE DIONE						
PENTANOIC ACID						
PENTANOL						
PENTANONE						
PENTASOL						
PENTASODIUM TRIPHOSPHATE						
PENTENE						
PENTENE-2, 4-METHYL						
PENTYL AMINE						
PENTYL OXYPENTANE						
PEPPERMINT OIL						
PERCHLORIC ACID		AB to 70°F NR at 122°F	NR at 70°F		B at 70°F	NR at 70°F
PERCHLORIC ACID						
PERCHLOROETHYLENE	NR at 70°F	NR at 70°F	C at 70°F NR at 105°F	AB to 70°F	*NR at 70°F	C/NR at 70°F
PERCHLOROMETHANE						
PERMACHLOR (DEGREASING FLUID)						
PEROXIDE			AB 20% to 70°F			
PEROXIDE						
PEROXYDOL						
PEROXYHYDRATE						
PERPHOSPHATE						
PERU SALTPETER						
PETROL	B/NR at 70°F	B at 70°F C at 122°F	AB to 105°F all types C at 140°F all types	AB to 70°F	C/NR at 70°F	C at 70°F hi & lo octane NR at 130°F all types
PETROLATUM (PETROLEUM JELLY)	A to 70°F			AB to 70°F		B at 70°F
PETROLENE						
PETROLEUM		B at 70°F	AB to 70°F	AB to 70°F	NR at 70°F	AB to 70°F
PETROLEUM-CRUDE	A to 70°F	B at 70°F sweet or sour	AB to 105°F sweet & sour C at 140°F sweet		NR at 70°F sweet or sour	AB to 70°F sour
PETROLEUM ETHER			A to 70°F	AB to 70°F	*NR at 70°F	C/NR at 70°F
PETROLEUM ETHER, LIQUIFIED						
PETROLEUM NAPHTHA						
PETROLEUM OILS		B at 70°F	AB to 70°F	AB to 70°F	NR at 70°F	AB to 70°F
PETROLEUM PITCH						
PETROLEUM SPIRIT						NR at 70°F
PETROLEUM THINNER						NR at 70°F
PHENE						

	POLYETHYLENE LOW DENSITY LDPE	® TEFLON FEP	® KALREZ PERFLUORINATED ELASTOMER	® FLUORAZ FLUORINATED COPOLYMER	® AFLAS FLUORINATED COPOLYMER	® NORPRENE COPOLYMER TPO
PENTACHLOROETHYLBENZENE		A to 70°F				
PENTACHLOROPHENOL		A to 100% to 70°F	A to 212°F	A to 70°F		
PENTACHLOROPHENYLBENZOATE		A to 70°F				
PENTADICHLORODIPHENYL KETONE		A to 70°F				
PENTAHYDROXY HEXOIC ACID		A to 70°F				
PENTAMETHYLENE AMINE		A to 70°F				
PENTANE	NR at 70°F	A to 70°F	A to 212°F	A to 70°F		
N-PENTANE				B at 70°F		
n-PENTANE, 2 METHYL, 2-4 METHYL & 3 METHYL		A to 70°F		B at 70°F		
2, 4 PENTANE DIONE		A to 70°F				
PENTANOIC ACID		A to 70°F				
PENTANOL		A to 70°F				
PENTANONE	B at 70°F C at 122°F	A to 122°F				
PENTASOL		A to 70°F		A to 70°F		
PENTASODIUM TRIPHOSPHATE		A to 70°F				
PENTENE		A to 70°F				
PENTENE-2, 4-METHYL		A to 70°F				
PENTYL AMINE		A to 70°F				
PENTYL OXYPENTANE		A to 70°F				
PEPPERMINT OIL						
PERCHLORIC ACID	AB 70-100% to 70°F NR 100% at 122°F	A to 200°F AC to 300°F	A to 70°F *A to 212°F	A to 70°F		NR at 70°F
PERCHLORIC ACID	NR 70% at 140°F AB 10% at 140°F					
PERCHLOROETHYLENE	NR at 70°F	+2% vol 7 days 250°F A to 250°F	+2% vol 7 days 70°F A to 212°F	NR at 70°F		NR at 70°F
PERCHLOROMETHANE		A to 70°F				
PERMACHLOR (DEGREASING FLUID)		A to 70°F				
PEROXIDE		A to 90% to 70°F		NR 90-100% at 70°F		
PEROXIDE		A to 70°F				
PEROXYDOL		A to 70°F				
PEROXYHYDRATE		A to 70°F				
PERPHOSPHATE		A to 70°F				
PERU SALTPETER		A to 70°F				
PETROL	C/NR at 70°F NR at 122°F	A to boiling	A to 212°F		+25% vol 7 days 70°F B at 70°F	BC at 70°F
PETROLATUM (PETROLEUM JELLY)		A to 70°F	A to 212°F	A to 70°F		
PETROLENE		A to 70°F				
PETROLEUM	AB to 105°F C at 140°F	A to 250°F	A to 250°F	A to 70°F	A to 212°F	A to 70°F
PETROLEUM-CRUDE	C at 70°F sour	A to 200°F	A to 70°F *A to 212°F	A to 250°F	+13% vol 3 days 212°F AB to 212°F	
PETROLEUM ETHER	NR at 70°F	A to 300°F		NR at 70°F		
PETROLEUM ETHER, LIQUIFIED	NR at 70°F	A to 70°F		BC at 70°F		
PETROLEUM NAPHTHA		A to 70°F				
PETROLEUM OILS	AB to 105°F	A to 200°F	A to 250°F	A to 250°F	+2-8% vol 3 days 212°F +10% vol 3 days 400°F	A to 70°F
PETROLEUM PITCH		A to 70°F				
PETROLEUM SPIRIT		A to 70°F				
PETROLEUM THINNER		A to 70°F				
PHENE		A to 70°F				

	RECOMMENDATIONS	NITRILE NBR	ETHYLENE PROPYLENE EP, EPDM	FLUOROCARBON FKM	CHLOROPRENE CR	HYDROGENATED NITRILE HNBR
PHENETOLE		NR at 70°F	NR at 70°F	C/NR at 70°F	NR at 70°F	NR at 70°F
PHENIC ACID		NR at 70°F	C at 70°F	A to 70°F	C at 70°F	
PHENMETHYLOL		NR at 70°F	C at 70°F	A to 70°F	C at 70°F	
PHENOL	FKM, FVMQ, IIR	NR 10-100% at 70°F	NR 5-100% at 70°F	no swell 3 days 75°F +7% vol 28 days 158°F	NR 5-100% at 70°F	NR 90-100% at 70°F
PHENOL				+10% vol 28 days 212°F A to 100% to 140°F		
PHENOL POLYSILOXANE		A to 70°F AB to 180°F			NR at 70°F	
PHENOL SULFONIC ACID		NR at 70°F		NR at 70°F A 65% to 158°F		
PHENOL TRINITRATE		NR at 70°F			NR at 70°F	
PHENOLATES		NR at 70°F		B at 70°F	NR at 70°F	
PHENOLS		NR at 70°F		B at 70°F	B at 70°F	
PHENOXIDE		NR at 70°F		B at 70°F	NR at 70°F	
PHENOXIN		C at 70°F	NR at 70°F	A to 70°F	NR at 70°F	
PHENOXYBENZENE		NR at 70°F	NR at 70°F	A to 70°F	NR at 70°F	
PHENYL ACETATE		NR at 70°F	B at 70°F	C/NR at 70°F	NR at 70°F	
PHENYL ACETIC ACID	EP	NR 75% at 70°F	A 75% to 200°F	C/NR 75% at 70°F	NR 75% at 70°F	
PHENYL ALDEHYDE		NR at 70°F	B at 70°F	C/NR at 70°F	NR at 70°F	
PHENYL AMINE	EP, IIR	NR at 70°F	AB to 200°F	A to 70°F C at 158°F	NR at 70°F	
PHENYL BENZENE	FKM, FVMQ, T	NR at 70°F	NR at 70°F	A to 300°F	NR at 70°F	NR at 70°F
PHENYL BISULFIDE	FKM	C to 104°F NR at 140°F	NR at 70°F	A to 70°F		
PHENYL BROMIDE		NR at 70°F	NR at 70°F	B at 70°F	NR at 70°F	
PHENYL CARBINOL		NR at 70°F	C at 70°F	A to 70°F	C/NR at 70°F	
PHENYL CELLOSOLVE (2 PHENOXYETHALOL)	EP	NR at 70°F	A to 200°F	A to 70°F	A to 70°F	
PHENYL CHLORIDE	FKM, FVMQ, T	NR at 70°F	NR at 70°F	A to 200°F	NR at 70°F	NR at 70°F
PHENYL ETHANE	FKM, FVMQ	NR at 70°F	NR at 70°F	A to 70°F	NR at 70°F	NR at 70°F
PHENYL ETHER		NR at 70°F	NR at 70°F	A to 70°F	NR at 70°F	
PHENYL ETHYL ETHER		NR at 70°F	NR at 70°F	C/NR at 70°F	NR at 70°F	NR at 70°F
PHENYL FORMIC ACID	T	NR at 70°F	B at 70°F	A to 70°F	A to 70°F AB any conc to 150°F	
PHENYL HYDRAZINE	FKM, NR	NR at 70°F	C/NR at 70°F	A to 70°F	NR at 70°F	NR at 70°F
PHENYL HYDRIDE		NR at 70°F	NR at 70°F	A to 70°F	NR at 70°F	
PHENYL HYDROXIDE		NR at 70°F	C at 70°F	A to 70°F	C at 70°F	
PHENYL METHANE		C/NR at 70°F	NR at 70°F	A to 70°F	NR at 70°F	
PHENYL METHYL KETONE		NR at 70°F	A to 70°F	NR at 70°F	NR at 70°F	
PHENYLIC ACID		NR at 70°F	C at 70°F	A to 70°F	C/NR at 70°F	
PHORONE	EP, IIR	NR at 70°F	AB to 70°F	NR at 70°F	NR at 70°F	NR at 70°F
PHOSGENE GAS		NR at 70°F gas or liquid	A to 70°F gas or liquid	NR at 70°F gas or liquid	C at 70°F gas or liquid	
PHOSPHATE ESTERS	EP, IIR	NR at 70°F	A to 250°F	A/NR at 70°F	NR at 70°F	NR at 70°F
PHOSPHOLEUM		NR at 70°F			NR 100-106% at 70°F	
PHOSPHORIC ACID	EP, FKM, IIR, SBR	NR 50-100% at 70°F AB to 50% to 104°F	A to 100% to 130°F AB 100% 140-250°F	no swell 85% conc 7 days 77°F	B/NR 100% at 70°F NR 100% at 140°F	
PHOSPHORIC ACID		C/NR 30-50% at 140°F AB 30% to 104°F	A to 85% to 176°F B to 30% at 212°F	+0.5% vol 60% conc 7 days 212°F	A to 85% to 104°F B 85% to 150°F	
PHOSPHORIC ACID		A to 10% to 104°F B 10% at 140°F		+4.2% vol 60% conc 28 days 212°F	NR 85% at 176°F C 80% at 200°F	
PHOSPHORIC ACID		C 10% at 176°F NR 10% boiling		A to 100% to 140°F B 100% at 185°F	A to 50% to 140°F B 20-50% at 176°F	
PHOSPHORIC ACID				A to 85% to 176°F B 85% at 212°F	C 20% at 212°F	
PHOSPHORIC ACID				A to 75% to 212°F		
PHOSPHORIC ACID, CRUDE	FKM	C to 140°F	B to 140°F	A to 140°F	BC at 70°F NR at 140°F	

	STYRENE BUTADIENE SBR	POLYACRYLATE ACM	POLYURETHANE AU, EU	ISOBUTYLENE ISOPRENE IIR	POLYBUTADIENE BR	® AEROQUIP AQP
PHENETOLE				NR at 70°F		
PHENIC ACID			C at 70°F	B at 70°F		
PHENMETHYLOL				A to 70°F		
PHENOL	NR 70-100% at 70°F	NR 70-100% at 70°F	NR 70-100% at 70°F	B any conc to 100°F	NR 70-100% at 70°F	AB 70-90% to 70°F
PHENOL						
PHENOL POLYSILOXANE				NR at 70°F		
PHENOL SULFONIC ACID				C at 70°F		
PHENOL TRINITRATE				NR at 70°F		
PHENOLATES			NR at 70°F			
PHENOLS						
PHENOXIDE						
PHENOXIN			C at 70°F	NR at 70°F		
PHENOXYBENZENE				NR at 70°F		
PHENYL ACETATE			NR at 70°F	B at 70°F		
PHENYL ACETIC ACID	NR 75% at 70°F	NR 75% at 70°F	NR 75% at 70°F			
PHENYL ALDEHYDE			NR at 70°F	B at 70°F		
PHENYL AMINE			C at 70°F	A to 70°F		
PHENYL BENZENE	NR at 70°F	NR at 70°F	NR at 70°F	NR at 70°F	NR at 70°F	
PHENYL BISULFIDE						
PHENYL BROMIDE			NR at 70°F	NR at 70°F		
PHENYL CARBINOL				A to 70°F AB any conc to 185°F		
PHENYL CELLOSOLVE (2 PHENOXYETHALOL)	NR at 70°F	NR at 70°F	NR at 70°F	A to 70°F		
PHENYL CHLORIDE			NR at 70°F	NR at 70°F		
PHENYL ETHANE			NR at 70°F	NR at 70°F		
PHENYL ETHER				NR at 70°F		
PHENYL ETHYL ETHER	NR at 70°F	NR at 70°F	NR at 70°F	NR at 70°F	NR at 70°F	
PHENYL FORMIC ACID				A to 70°F		
PHENYL HYDRAZINE	B at 70°F	NR at 70°F		NR at 70°F	B at 70°F	
PHENYL HYDRIDE			NR at 70°F	NR at 70°F		
PHENYL HYDROXIDE			C at 70°F	B at 70°F		
PHENYL METHANE			NR at 70°F	NR at 70°F		
PHENYL METHYL KETONE			NR at 70°F	A any conc to 80°F		
PHENYLIC ACID			C at 70°F	B any conc to 100°F		–
PHORONE	NR at 70°F	NR at 70°F	NR at 70°F	AB to 70°F		
PHOSGENE GAS						
PHOSPHATE ESTERS	NR at 70°F	NR at 70°F	NR at 70°F	A to 70°F AB any conc to 120°F	NR at 70°F	AB to 200°F
PHOSPHOLEUM			NR at 70°F	A 100% to 70°F AB 106% to 185°F	C/NR at 70°F	NR at 70°F
PHOSPHORIC ACID	AB to 100% to 70°F B 100% at 158°F	BC to 100% at 70°F C 100% at 158°F	C/NR 85-100% at 70°F AC 20-70% at 70°F	AB to 100% at 158°F AB to 50% at 185°F	C 100% at 70°F NR 45% at 70°F	NR 85-100% at 70°F
PHOSPHORIC ACID			A to 10% to 140°F		C 20% at 70°F B 3 molar at 70°F	
PHOSPHORIC ACID						
PHOSPHORIC ACID						
PHOSPHORIC ACID						
PHOSPHORIC ACID						
PHOSPHORIC ACID, CRUDE				C at 70°F		

C 285

	SYNTHETIC ISOPRENE IR	NATURAL ISOPRENE NR	CHLOROSULFONATED POLYETHYLENE CSM	FLUOROSILICONE FVMQ	SILICONE VMQ	CHEMRAZ FFKM
PHENETOLE	NR at 70°F	NR at 70°F	NR at 70°F	NR at 70°F	NR at 70°F	A to 70°F
PHENIC ACID	NR at 70°F	NR at 70°F	C at 70°F		NR at 70°F	
PHENMETHYLOL	C at 70°F	C at 70°F	B at 70°F			
PHENOL	NR 70-100% at 70°F	NR 70-100% at 70°F C/NR 10% at 70°F	C/NR any conc at 70°F	AB 70-100% at 70°F	NR 10-100% at 70°F	A 70-100% to 70°F
PHENOL						
PHENOL POLYSILOXANE	NR at 70°F	NR at 70°F				
PHENOL SULFONIC ACID	NR at 70°F	NR at 70°F	NR at 70°F			A to 70°F
PHENOL TRINITRATE	NR at 70°F	NR at 70°F				
PHENOLATES	NR at 70°F	NR at 70°F	NR at 70°F			
PHENOLS	NR at 70°F	NR at 70°F	B at 70°F			
PHENOXIDE	NR at 70°F	NR at 70°F				
PHENOXIN	NR at 70°F	NR at 70°F	NR at 70°F		NR at 70°F	
PHENOXYBENZENE	NR at 70°F	NR at 70°F	C at 70°F		C at 70°F	
PHENYL ACETATE	C at 70°F	C/NR at 70°F	C at 70°F		NR at 70°F	A to 70°F
PHENYL ACETIC ACID		NR 75% at 100°F		NR 75% at 70°F	NR 75% at 70°F	A to 70°F
PHENYL ALDEHYDE	NR at 70°F	NR at 70°F	NR at 70°F		C at 70°F	
PHENYL AMINE	NR at 70°F	NR at 70°F	C/NR at 70°F NR at 100°F	BC at 70°F	AB to 70°F	A to 70°F
PHENYL BENZENE	NR at 70°F	NR at 70°F	NR at 70°F	B at 70°F	NR at 70°F	A to 70°F
PHENYL BISULFIDE						
PHENYL BROMIDE	NR at 70°F	NR at 70°F	NR at 70°F		NR at 70°F	
PHENYL CARBINOL	C at 70°F	C/NR at 70°F	B at 70°F			
PHENYL CELLOSOLVE (2 PHENOXYETHALOL)				A to 70°F	A to 70°F	
PHENYL CHLORIDE	NR at 70°F	NR at 70°F	NR at 70°F	B at 70°F	NR at 70°F	A to 70°F
PHENYL ETHANE	NR at 70°F	NR at 70°F	NR at 70°F	A to 70°F	NR at 70°F	A to 70°F
PHENYL ETHER	NR at 70°F	NR at 70°F	C at 70°F		C at 70°F	
PHENYL ETHYL ETHER	NR at 70°F	NR at 70°F	NR at 70°F	NR at 70°F	NR at 70°F	A to 70°F
PHENYL FORMIC ACID	B at 70°F	AB any conc to 150°F	B at 70°F		B at 70°F	
PHENYL HYDRAZINE	A to 70°F	A any conc to 80°F	C/NR at 70°F			A to 70°F
PHENYL HYDRIDE	NR at 70°F	NR at 70°F	NR at 70°F		NR at 70°F	
PHENYL HYDROXIDE	NR at 70°F	NR at 70°F	C at 70°F		NR at 70°F	
PHENYL METHANE	NR at 70°F	NR at 70°F	NR at 70°F		NR at 70°F	
PHENYL METHYL KETONE	C at 70°F	C/NR at 70°F	NR at 70°F			
PHENYLIC ACID	NR at 70°F	NR at 70°F	C at 70°F		NR at 70°F	
PHORONE	NR at 70°F	NR at 70°F	NR at 70°F	NR at 70°F	NR at 70°F	A to 70°F
PHOSGENE GAS						A to 70°F
PHOSPHATE ESTERS	NR at 70°F	NR at 70°F	NR at 70°F	A/NR at 70°F	B/NR at 70°F	
PHOSPHOLEUM	NR at 70°F	NR 100-106% at 70°F				
PHOSPHORIC ACID	C 50-100% at 70°F B 20% at 70°F	C/NR 100% at 70°F AB to 85% at 150°F	AB 100% to 200°F A to 85% to 200°F	BC 100% at 70°F C 100% at 158°F	C/NR 50-100% at 70°F NR 100% at 158°F	A to conc to 70°F
PHOSPHORIC ACID	A 10% to 70°F			B 20-50% at 70°F B 3 molar to 158°F	BC 20% at 70°F C 10% at 70°F	
PHOSPHORIC ACID						
PHOSPHORIC ACID						
PHOSPHORIC ACID						
PHOSPHORIC ACID						
PHOSPHORIC ACID, CRUDE	C at 70°F	C/NR at 70°F	A to 70°F		C at 70°F	

	POLYSULFIDE T	CHLORINATED POLYETHYLENE CM	EPICHLOROHYDRIN CO, ECO	® HYTREL COPOLYESTER TPE	® SANTOPRENE COPOLYMER TPO	® C-FLEX STYRENIC TPE
PHENETOLE						
PHENIC ACID						
PHENMETHYLOL						
PHENOL	NR 70-100% at 70°F	AB to 70°F	C 70% at 70°F	NR at 70°F	NR at 70°F	NR at 70°F
PHENOL						
PHENOL POLYSILOXANE						
PHENOL SULFONIC ACID						
PHENOL TRINITRATE						
PHENOLATES				NR at 70°F		
PHENOLS						
PHENOXIDE						
PHENOXIN						
PHENOXYBENZENE						
PHENYL ACETATE						
PHENYL ACETIC ACID						
PHENYL ALDEHYDE						
PHENYL AMINE						
PHENYL BENZENE	B at 70°F					
PHENYL BISULFIDE						
PHENYL BROMIDE						
PHENYL CARBINOL						
PHENYL CELLOSOLVE (2 PHENOXYETHALOL)						
PHENYL CHLORIDE						
PHENYL ETHANE						
PHENYL ETHER						
PHENYL ETHYL ETHER	B at 70°F					
PHENYL FORMIC ACID						
PHENYL HYDRAZINE	NR at 70°F					
PHENYL HYDRIDE						
PHENYL HYDROXIDE						
PHENYL METHANE						
PHENYL METHYL KETONE						
PHENYLIC ACID						
PHORONE	C/NR at 70°F					
PHOSGENE GAS						
PHOSPHATE ESTERS	NR at 70°F	AB to 70°F	NR at 70°F	A/NR at 70°F	A to 70°F	AB to 212°F
PHOSPHOLEUM	NR at 70°F	NR at 70°F		NR at 70°F		
PHOSPHORIC ACID	NR 20-100% at 70°F NR 3 molar at 70°F	NR 100% at 70°F		NR 60-100% at 70°F	A 40% to 70°F	A 100% to 70°F
PHOSPHORIC ACID						
PHOSPHORIC ACID						
PHOSPHORIC ACID						
PHOSPHORIC ACID						
PHOSPHORIC ACID						
PHOSPHORIC ACID, CRUDE						

	ETHYLENE ACRYLIC EA	POLYALLOMER LINEAR COPOLYMER	NYLON 11 POLYAMIDE	NYLON 12 POLYAMIDE	ETHYLENE VINYL ACETATE EVA	POLYVINYLCHLORIDE FLEXIBLE PVC
PHENETOLE						
PHENIC ACID						
PHENMETHYLOL						
PHENOL	*NR at 70°F C 70% at 70°F	A to 70°F NR crystals at 122°F	NR at 70°F	NR any conc at 70°F	C/NR at 70°F	C at 70°F NR at 150°F
PHENOL						
PHENOL POLYSILOXANE						
PHENOL SULFONIC ACID						
PHENOL TRINITRATE						NR at 70°F
PHENOLATES			NR at 70°F			B/NR at 70°F
PHENOLS			NR at 70°F			B/NR at 70°F
PHENOXIDE						
PHENOXIN						
PHENOXYBENZENE						
PHENYL ACETATE						
PHENYL ACETIC ACID						
PHENYL ALDEHYDE						
PHENYL AMINE						
PHENYL BENZENE						NR at 70°F
PHENYL BISULFIDE						
PHENYL BROMIDE						
PHENYL CARBINOL						
PHENYL CELLOSOLVE (2 PHENOXYETHALOL)						
PHENYL CHLORIDE						NR at 70°F
PHENYL ETHANE						NR at 70°F
PHENYL ETHER						
PHENYL ETHYL ETHER						
PHENYL FORMIC ACID						
PHENYL HYDRAZINE						NR at 70°F
PHENYL HYDRIDE						
PHENYL HYDROXIDE						
PHENYL METHANE						
PHENYL METHYL KETONE						NR at 70°F
PHENYLIC ACID						NR at 70°F
PHORONE						
PHOSGENE GAS						
PHOSPHATE ESTERS	NR at 70°F		A to 70°F AB to 200°F		NR at 70°F	B/NR at 70°F
PHOSPHOLEUM			NR at 70°F	NR at 70°F	NR at 70°F	
PHOSPHORIC ACID	*AB 20% to 70°F	A to 85% to 70°F B 85% to 122°F	NR 100% at 70°F C 50-85% at 70°F	NR 50% at 70°F C 10% at 70°F	C/NR 100% at 70°F BC 50-85% to 140°F	AB 100% to 70°F A to 85% to 70°F
PHOSPHORIC ACID		A to 5% to 122°F	AB 50% to 70°F C 50% at 105°F		B 25-50% at 70°F A to 25% to 70°F	AB 50-85% to 150°F A to 5% to 122°F
PHOSPHORIC ACID			NR 50% at 140°F A to 25% to 70°F			
PHOSPHORIC ACID						
PHOSPHORIC ACID						
PHOSPHORIC ACID						
PHOSPHORIC ACID						
PHOSPHORIC ACID, CRUDE						B/NR at 70°F

	POLYETHYLENE LOW DENSITY LDPE	®TEFLON FEP	®KALREZ PERFLUORINATED ELASTOMER	®FLUORAZ FLUORINATED COPOLYMER	®AFLAS FLUORINATED COPOLYMER	®NORPRENE COPOLYMER TPO
PHENETOLE		A to 70°F	A to 212°F	NR at 70°F		
PHENIC ACID		A to 70°F				
PHENMETHYLOL		A to 70°F				
PHENOL	AB to 70°F	A to 300°F A crystals to 122°F	A to 212°F *A to 300°F	AB to 100% to 400°F	+9% vol 7 days 158°F A to 70°F	B/NR at 70°F
PHENOL	B crystals at 70°F NR crystals at 122°F		A 90% to 203°F			
PHENOL POLYSILOXANE		A to 70°F				
PHENOL SULFONIC ACID		A to 70°F	A to 212°F			
PHENOL TRINITRATE		A to 70°F				
PHENOLATES		A to 70°F				
PHENOLS		A to 70°F				
PHENOXIDE		A to 70°F				
PHENOXIN		A to 70°F				
PHENOXYBENZENE		A to 70°F				
PHENYL ACETATE		A to 70°F	A to 212°F			
PHENYL ACETIC ACID			A to 212°F	NR at 70°F		
PHENYL ALDEHYDE		A to 70°F				
PHENYL AMINE	A to 70°F AB to 122°F	A to 365°F	A to 300°F	A to 70°F	A to 70°F	
PHENYL BENZENE	C at 70°F	A to 300°F	A to 500°F	C/NR at 70°F		
PHENYL BISULFIDE						
PHENYL BROMIDE		A to 70°F				
PHENYL CARBINOL		A to 70°F				
PHENYL CELLOSOLVE (2 PHENOXYETHALOL)						
PHENYL CHLORIDE	NR at 70°F	A to 300°F	A to 70°F	C/NR at 70°F		NR at 70°F
PHENYL ETHANE	A to 70°F NR at 122°F	A to 125°F	A to 70°F	B/NR at 70°F	BC at 70°F	
PHENYL ETHER		A to 70°F				
PHENYL ETHYL ETHER		A to 70°F	A to 212°F	NR at 70°F		
PHENYL FORMIC ACID		A to 70°F				
PHENYL HYDRAZINE		A to 70°F	*A to 212°F	A to 70°F		
PHENYL HYDRIDE		A to 70°F				
PHENYL HYDROXIDE		A to 70°F				
PHENYL METHANE		A to 70°F				
PHENYL METHYL KETONE		A to 70°F				
PHENYLIC ACID		A to 70°F				
PHORONE		A to 70°F	A to 212°F	NR at 70°F		
PHOSGENE GAS	TEST	AB to 70°F	A to 212°F			
PHOSPHATE ESTERS	NR at 70°F	A to 70°F AB to 400°F	A to 212°F	A to 70°F	A to 70°F B at 212°F	BC at 212°F Skydrol 500-B4
PHOSPHOLEUM		A to 70°F				
PHOSPHORIC ACID	AC 95-100% at 70°F NR 95-100% at 140°F	no swell 7 days 212°F A to 100% to boiling	A to 100% to 400°F	A to 100% to 175°F	+1% vol 7 days 300°F A to 300°F	A 85-100% to 70°F
PHOSPHORIC ACID	AB 90% to 70°F C 90% at 140°F					
PHOSPHORIC ACID	A to 85% to 122°F B/NR 85% at 140°F					
PHOSPHORIC ACID	A to 50% to 140°F					
PHOSPHORIC ACID						
PHOSPHORIC ACID						
PHOSPHORIC ACID, CRUDE	C at 70°F	A to 70°F				

	RECOMMENDATIONS	NITRILE NBR	ETHYLENE PROPYLENE EP, EPDM	FLUOROCARBON FKM	CHLOROPRENE CR	HYDROGENATED NITRILE HNBR
PHOSPHORIC ACID, ORTHO		NR at 70°F		AB to 70°F	NR at 70°F	
PHOSPHORIC ANHYDRIDE		NR at 70°F wet or dry C molten		NR at 70°F wet or dry NR molten	NR at 70°F wet or dry NR molten	
PHOSPHOROUS OXYCHLORIDE	IIR				NR at 70°F	
PHOSPHOROUS PENTOXIDE		NR 36-100% at 70°F AB to 36% to 70°F	A to 72.5% to 130°F A to 61% to 176°F	A to 72.5% to 140°F A to 61% to 176°F	NR 100% at 70°F A to 61% to 104°F	
PHOSPHOROUS TRICHLORIDE ACID	EP,FKM,IIR	NR at 70°F	A to 70°F	A to 70°F	NR at 70°F	NR at 70°F
PHOTOGEN		A to 70°F	NR at 70°F	A to 70°F	C at 70°F	
PHOTOGRAPHIC DEVELOPER	NBR,FKM,EP,CR,IIR	A to 70°F AB any conc to 200°F	AB to 104°F	A to 140°F	A to 140°F	
PHOTOGRAPHIC EMULSION		BC at 70°F	A to 104°F	A to 104°F	A to 104°F	
PHOTOGRAPHIC FIXER		AC to 104°F	A to 104°F	AB to 104°F	AB to 104°F	
PHOTOGRAPHIC SOLUTIONS	NBR,FKM,EP,CR,IIR	AB to 70°F	A to 104°F	A to 70°F	A to 70°F	
PHTHALIC ACID		AC to 70°F	A to 70°F	A to 70°F	C at 70°F	
PHTHALIC ANHYDRIDE	EP	BC at 70°F	A to 200°F	A to 70°F BC at 400°F	A to 70°F	
PICKLE ALUM		A to 70°F AB any conc to 200°F	A to 70°F	A to 70°F	A to 70°F AB any conc to 200°F	
PICKLING SOLUT'N, EG 20% NITRIC, 4% HF	FKM,EP,IIR	NR at 70°F	C at 70°F	B at 70°F NR at 225°F	NR at 70°F	
PICOLINE, ALPHA	EP	A to 70°F	A to 250°F	NR at 70°F		
PICRIC ACID	NBR,EP,FKM,CR,IIR	NR conc at 70°F AB to 10% to 160°F	BC conc at 70°F AB to 10% to 200°F	A to conc to 70°F A to 10% to 140°F	BC conc at 70°F A to 10% to 70°F	
PICRIC ACID	FKM (molten)	NR 10% at 176°F BC molten	BC molten	B 10% at 176°F C 10% at 212°F	C 10% at 104°F NR 10% at 140°F	
PICRONITRIC ACID		C/NR at 70°F	B at 70°F	A to 70°F	C/NR at 70°F	
PIMELIC KETONE	EP	NR at 70°F	BC at 70°F	NR at 70°F	NR at 70°F	NR at 70°F
PINOCOL (PINOCOLIN)		A to 70°F	C at 70°F	A to 70°F	A to 70°F	
PINE OIL	NBR,FKM,FVMQ,T	AB to 250°F	NR at 70°F	A to 70°F	NR at 70°F	
PINENE	FKM,FVMQ,NBR	B at 70°F	NR at 70°F	A to 158°F	C/NR at 70°F	B at 70°F
PINK LIQUOR						
PIPE LINE CLEANER, DIVAN IND.		NR at 70°F		A to 70°F	A to 70°F	
PIPERIDINE		NR at 70°F	NR at 70°F	NR at 70°F	NR at 70°F	
PITCH		A to 70°F	A/NR at 70°F	A to 70°F	B/NR at 70°F	
PLASTICIZER		NR at 70°F		A to 70°F	C/NR at 70°F	
PLATING SOLUTION, ANTIMONY		A to 140°F		A to 140°F	A to 140°F	
PLATING SOLUTION, ARSENIC		A to 140°F		A to 140°F	A to 140°F	
PLATING SOLUTION, BRASS		A to 140°F regular A to 110°F high speed	A to 70°F	A to 140°F	A to 110°F regular & high speed	
PLATING SOLUTION, BRONZE		A to 140°F A to 160°F Cu/Sn	A to 160°F Cu/Sn A to 70°F Cu/Cd	A to 160°F Cu/Sn A to 70°F Cu/Cd	AB to 160°F Cu/Sn A to 70°F Cu/Cd	
PLATING SOLUTION, CADMIUM		A to 140°F AB at 150°F	A to 70°F	A to 140°F	A to 140°F AB to 150°F	
PLATING SOLUTION, CHROME	EP,FKM,IIR	NR at 70°F all types	A to 70°F	A to 140°F	NR at 70°F all types	NR at 70°F
PLATING SOLUTION, COBALT		AB to 150°F			AB to 150°F	A to 70°F
PLATING SOLUTION, COPPER	EP	A to 140°F	A to 200°F	A to 140°F all types	AB to 140°F NR at 70°F electroless	A to 70°F
PLATING SOLUTION, GOLD		A to 150°F cyanide A to 75°F acid & neut	A to 70°F	A to 150°F cyanide A to 75°F acid & neut	A to 150°F cyanide A to 75°F acid & neut	A to 70°F
PLATING SOLUTION, INDIUM		A to 140°F		A to 140°F	A to 70°F	A to 70°F
PLATING SOLUTION, IRON		A to 140°F AB to 180°F		A to 140°F all types	A to 140°F	A to 70°F
PLATING SOLUTION, LEAD		AB to 150°F	A to 70°F	A to 140°F	AB to 150°F	A to 70°F
PLATING SOLUTION, NICKEL		A to 140°F NR at 70°F electroless	A to 70°F	A to 140°F all types	A to 140°F NR at 70°F electroless	A to 70°F
PLATING SOLUTION, OTHERS	EP,NBR,FKM,IIR	A to 70°F	A to 70°F	A to 70°F	A to 70°F	
PLATING SOLUTION, PLATINUM				A to 70°F	A to 70°F AB to 200°F	
PLATING SOLUTION, RHODIUM		A to 70°F	A to 120°F	A to 70°F	AB to 70°F	
PLATING SOLUTION, SILVER		A to 140°F	A to 120°F	A to 140°F	A to 140°F	A to 70°F

	STYRENE BUTADIENE SBR	POLYACRYLATE ACM	POLYURETHANE AU, EU	ISOBUTYLENE ISOPRENE IIR	POLYBUTADIENE BR	® AEROQUIP AQP
PHOSPHORIC ACID, ORTHO			C/NR at 70°F	AB to 120°F	NR at 70°F	NR at 70°F
PHOSPHORIC ANHYDRIDE				NR at 70°F		
PHOSPHOROUS OXYCHLORIDE						
PHOSPHOROUS PENTOXIDE						
PHOSPHOROUS TRICHLORIDE ACID	NR at 70°F	NR at 70°F		A to 70°F	NR at 70°F	
PHOTOGEN			C at 70°F	NR at 70°F		
PHOTOGRAPHIC DEVELOPER	B at 70°F		A/NR at 70°F	B any conc to 185°F	B at 70°F	
PHOTOGRAPHIC EMULSION						
PHOTOGRAPHIC FIXER				B at 70°F		
PHOTOGRAPHIC SOLUTIONS	B at 70°F		A/NR at 70°F	B at 70°F	B at 70°F	
PHTHALIC ACID						
PHTHALIC ANHYDRIDE			A to 70°F			
PICKLE ALUM			A to 70°F	A to 70°F AB any conc to 185°F		
PICKLING SOLUT'N, EG 20% NITRIC, 4% HF	NR at 70°F	NR at 70°F	NR at 70°F	C at 70°F	NR at 70°F	
PICOLINE, ALPHA						
PICRIC ACID	B sol'n at 70°F B molten		B to conc at 70°F B molten	C conc at 70°F AB sol'n to 70°F	B sol'n at 70°F B molten	NR sol'n at 70°F NR molten
PICRIC ACID						
PICRONITRIC ACID			B at 70°F	C at 70°F		
PIMELIC KETONE	NR at 70°F	NR at 70°F	NR at 70°F	B/NR at 70°F	NR at 70°F	
PINOCOL (PINOCOLIN)				A to 70°F		
PINE OIL	NR at 70°F	A to 70°F	A to 70°F	NR at 70°F	NR at 70°F	
PINENE	NR at 70°F	NR at 70°F	B at 70°F	NR at 70°F	NR at 70°F	
PINK LIQUOR			C/NR at 70°F			
PIPE LINE CLEANER, DIVAN IND.			NR at 70°F			
PIPERIDINE	NR at 70°F	NR at 70°F	NR at 70°F	NR at 70°F	NR at 70°F	
PITCH			AB to 70°F	NR at 70°F		
PLASTICIZER						
PLATING SOLUTION, ANTIMONY				A to 70°F AB to 185°F		
PLATING SOLUTION, ARSENIC				*AB to 70°F		
PLATING SOLUTION, BRASS				A to 70°F AB to 185°F		
PLATING SOLUTION, BRONZE				*AB to 70°F		
PLATING SOLUTION, CADMIUM				A to 70°F AB to 150°F		
PLATING SOLUTION, CHROME	NR at 70°F			A/NR at 70°F	NR at 70°F	
PLATING SOLUTION, COBALT				AB to 150°F		
PLATING SOLUTION, COPPER	B at 70°F	NR at 70°F	NR at 70°F	A to 70°F AB to 180°F		
PLATING SOLUTION, GOLD				A to 70°F AB to 180°F		
PLATING SOLUTION, INDIUM						
PLATING SOLUTION, IRON						
PLATING SOLUTION, LEAD				A to 70°F AB to 150°F		
PLATING SOLUTION, NICKEL						
PLATING SOLUTION, OTHERS	A to 70°F			A to 70°F		
PLATING SOLUTION, PLATINUM						
PLATING SOLUTION, RHODIUM						
PLATING SOLUTION, SILVER				A to 70°F AB to 150°F		

C 291

	SYNTHETIC ISOPRENE IR	NATURAL ISOPRENE NR	CHLOROSULFONATED POLYETHYLENE CSM	FLUOROSILICONE FVMQ	SILICONE VMQ	CHEMRAZ FFKM
PHOSPHORIC ACID, ORTHO		NR at 70°F	AB to 200°F		TEST	
PHOSPHORIC ANHYDRIDE		A to 70°F wet or dry NR molten				
PHOSPHOROUS OXYCHLORIDE	NR at 70°F	NR at 70°F	NR at 70°F			
PHOSPHOROUS PENTOXIDE						
PHOSPHOROUS TRICHLORIDE ACID	NR at 70°F	NR at 70°F	NR at 70°F	A to 70°F		
PHOTOGEN	NR at 70°F	NR at 70°F	C at 70°F		C at 70°F	
PHOTOGRAPHIC DEVELOPER	A to 70°F	A to 70°F AB any conc to 150°F	A to 70°F	A to 70°F	AB to 70°F	
PHOTOGRAPHIC EMULSION						
PHOTOGRAPHIC FIXER			C at 70°F		A to 70°F	
PHOTOGRAPHIC SOLUTIONS	A to 70°F	A to 70°F	AC to 70°F	A to 70°F	A to 70°F	
PHTHALIC ACID					AB to 70°F	
PHTHALIC ANHYDRIDE		C at 70°F				
PICKLE ALUM	A to 70°F	A to 70°F AB any conc to 150°F	A to 70°F		A to 70°F	
PICKLING SOLUT'N, EG 20% NITRIC, 4% HF	C/NR at 70°F	NR at 70°F	A to 150°F	NR at 70°F	NR at 70°F	
PICOLINE, ALPHA						
PICRIC ACID	C conc at 70°F AB sol'n to 70°F	NR conc at 70°F AB sol'n to 70°F	A to conc to 70°F AB molten	B sol'n at 70°F B molten	NR any conc at 70°F NR molten	A to 100% to molten
PICRIC ACID	BC molten	BC molten				
PICRONITRIC ACID	C at 70°F	C at 70°F	A to 70°F		NR at 70°F	
PIMELIC KETONE	NR at 70°F	NR at 70°F	NR at 70°F	NR at 70°F	NR at 70°F	B at 70°F
PINOCOL (PINOCOLIN)	A to 70°F	A to 70°F	A to 70°F			
PINE OIL	NR at 70°F	NR at 70°F	NR at 70°F	A to 70°F	NR at 70°F	*A to 70°F
PINENE	NR at 70°F	NR at 70°F	NR at 70°F	AB at 70°F	NR at 70°F	A to 70°F
PINK LIQUOR						
PIPE LINE CLEANER, DIVAN IND.						
PIPERIDINE	NR at 70°F	NR at 70°F	NR at 70°F	NR at 70°F	NR at 70°F	A to 70°F
PITCH	NR at 70°F	NR at 70°F	B at 70°F		NR at 70°F	
PLASTICIZER						
PLATING SOLUTION, ANTIMONY	B at 70°F	B to 130°F			NR at 130°F	
PLATING SOLUTION, ARSENIC	B at 70°F	B to 100°F			NR at 110°F	
PLATING SOLUTION, BRASS	B at 70°F	B to 100°F	A to 70°F		NR at 70°F regular or hi speed	
PLATING SOLUTION, BRONZE					*NR at 70°F	
PLATING SOLUTION, CADMIUM	A to 70°F	A to 70°F AB to 150°F	A to 70°F		*NR at 70°F	
PLATING SOLUTION, CHROME	NR at 70°F	NR at 70°F	C/NR at 70°F		NR at 70°F	A to 70°F
PLATING SOLUTION, COBALT		AB to 150°F			*NR at 70°F	A to 70°F
PLATING SOLUTION, COPPER	B at 70°F	B to 100°F	A to 70°F	C at 70°F	C/NR at 70°F	A to 70°F
PLATING SOLUTION, GOLD	B at 70°F	B to 100°F				A to 70°F
PLATING SOLUTION, INDIUM						A to 70°F
PLATING SOLUTION, IRON	B at 70°F	B to 100°F				A to 70°F
PLATING SOLUTION, LEAD	A to 70°F	A to 70°F AB to 150°F				A to 70°F
PLATING SOLUTION, NICKEL	B at 70°F	B to 100°F				A to 70°F
PLATING SOLUTION, OTHERS	A to 70°F	A to 70°F	A to 70°F		NR at 70°F	A to 70°F
PLATING SOLUTION, PLATINUM						
PLATING SOLUTION, RHODIUM						
PLATING SOLUTION, SILVER	B at 70°F	B to 100°F				A to 70°F

	POLYSULFIDE T	CHLORINATED POLYETHYLENE CM	EPICHLOROHYDRIN CO, ECO	® HYTREL COPOLYESTER TPE	® SANTOPRENE COPOLYMER TPO	® C-FLEX STYRENIC TPE
PHOSPHORIC ACID, ORTHO	NR at 70°F	NR at 70°F		NR at 70°F		A to 70°F
PHOSPHORIC ANHYDRIDE						
PHOSPHOROUS OXYCHLORIDE						
PHOSPHOROUS PENTOXIDE						
PHOSPHOROUS TRICHLORIDE ACID						
PHOTOGEN						
PHOTOGRAPHIC DEVELOPER	A to 70°F			NR at 70°F		
PHOTOGRAPHIC EMULSION						
PHOTOGRAPHIC FIXER				B at 70°F		
PHOTOGRAPHIC SOLUTIONS	A to 70°F			B/NR at 70°F		
PHTHALIC ACID						NR at 70°F
PHTHALIC ANHYDRIDE						
PICKLE ALUM						
PICKLING SOLUT'N, EG 20% NITRIC, 4% HF	NR at 70°F		NR at 70°F	*NR at 70°F		
PICOLINE, ALPHA						
PICRIC ACID	NR sol'n at 70°F NR molten	NR sol'n at 70°F NR molten		NR sol'n at 70°F NR molten	NR sol'n at 70°F	
PICRIC ACID						
PICRONITRIC ACID						
PIMELIC KETONE					NR at 70°F	NR at 70°F
PINOCOL (PINOCOLIN)						
PINE OIL	B at 70°F					
PINENE	B at 70°F					
PINK LIQUOR						
PIPE LINE CLEANER, DIVAN IND.						
PIPERIDINE	NR at 70°F				*A to 70°F	
PITCH						
PLASTICIZER						
PLATING SOLUTION, ANTIMONY						
PLATING SOLUTION, ARSENIC						
PLATING SOLUTION, BRASS						
PLATING SOLUTION, BRONZE						
PLATING SOLUTION, CADMIUM						
PLATING SOLUTION, CHROME	NR at 70°F					
PLATING SOLUTION, COBALT						
PLATING SOLUTION, COPPER						
PLATING SOLUTION, GOLD						
PLATING SOLUTION, INDIUM						
PLATING SOLUTION, IRON						
PLATING SOLUTION, LEAD						
PLATING SOLUTION, NICKEL						
PLATING SOLUTION, OTHERS					A to 70°F	
PLATING SOLUTION, PLATINUM						
PLATING SOLUTION, RHODIUM						
PLATING SOLUTION, SILVER						

C 293

	ETHYLENE ACRYLIC EA	POLYALLOMER LINEAR COPOLYMER	NYLON 11 POLYAMIDE	NYLON 12 POLYAMIDE	ETHYLENE VINYL ACETATE EVA	POLYVINYLCHLORIDE FLEXIBLE PVC
PHOSPHORIC ACID, ORTHO			NR at 70°F	NR at 70°F	C/NR at 70°F	A to 70°F
PHOSPHORIC ANHYDRIDE						NR at 70°F wet or dry
PHOSPHOROUS OXYCHLORIDE						
PHOSPHOROUS PENTOXIDE			C at 70°F		*AB to 140°F	AB to 70°F
PHOSPHOROUS TRICHLORIDE ACID						
PHOTOGEN						
PHOTOGRAPHIC DEVELOPER					A to 70°F	AB to 70°F *AB to 140°F
PHOTOGRAPHIC EMULSION						*AB to 140°F
PHOTOGRAPHIC FIXER						A to 70°F *AB to 140°F
PHOTOGRAPHIC SOLUTIONS					A to 70°F	A to 70°F
PHTHALIC ACID						A to 70°F
PHTHALIC ANHYDRIDE						B at 70°F
PICKLE ALUM						
PICKLING SOLUT'N, EG 20% NITRIC, 4% HF	*NR at 70°F					
PICOLINE, ALPHA						
PICRIC ACID		A sol'n to 70°F	C sol'n at 70°F NR sol'n at 105°F		C/NR sol'n at 70°F C 1% to 140°F	C/NR sol'n at 70°F AB 1% to 130°F
PICRIC ACID			NR molten			
PICRONITRIC ACID						NR at 70°F
PIMELIC KETONE		B at 70°F	AB to 70°F NR at 140°F	AB to 70°F	NR at 70°F	NR at 70°F
PINOCOL (PINOCOLIN)						
PINE OIL		A to 70°F B at 122°F				C at 70°F NR at 122°F
PINENE						
PINK LIQUOR					C/NR at 70°F	A to 70°F
PIPE LINE CLEANER, DIVAN IND.						
PIPERIDINE						
PITCH						B at 70°F C at 150°F
PLASTICIZER				AB to 70°F phthalete & phosphate		
PLATING SOLUTION, ANTIMONY						
PLATING SOLUTION, ARSENIC						
PLATING SOLUTION, BRASS						
PLATING SOLUTION, BRONZE						
PLATING SOLUTION, CADMIUM						
PLATING SOLUTION, CHROME				NR at 70°F		AB to 70°F
PLATING SOLUTION, COBALT						
PLATING SOLUTION, COPPER						NR at 70°F electroless
PLATING SOLUTION, GOLD						
PLATING SOLUTION, INDIUM						
PLATING SOLUTION, IRON						
PLATING SOLUTION, LEAD						
PLATING SOLUTION, NICKEL						
PLATING SOLUTION, OTHERS						A to 70°F
PLATING SOLUTION, PLATINUM						
PLATING SOLUTION, RHODIUM						
PLATING SOLUTION, SILVER						

	POLYETHYLENE LOW DENSITY LDPE	® TEFLON FEP	® KALREZ PERFLUORINATED ELASTOMER	® FLUORAZ FLUORINATED COPOLYMER	® AFLAS FLUORINATED COPOLYMER	® NORPRENE COPOLYMER TPO
PHOSPHORIC ACID, ORTHO	B at 70°F		A 105% to 400°F	A to 115% to 175°F		AB to 70°F
PHOSPHORIC ANHYDRIDE	NR molten	A molten				
PHOSPHOROUS OXYCHLORIDE		A to 70°F	A to 212°F			
PHOSPHOROUS PENTOXIDE	AB to 140°F					
PHOSPHOROUS TRICHLORIDE ACID	A to 70°F AB to 140°F	A to 300°F	A to 212°F	A to 70°F dry		
PHOTOGEN		A to 70°F				
PHOTOGRAPHIC DEVELOPER	A to 70°F B at 140°F	A to 300°F	A to 70°F	C at 70°F		
PHOTOGRAPHIC EMULSION	A to 104°F					
PHOTOGRAPHIC FIXER	A to 104°F	A to 300°F				
PHOTOGRAPHIC SOLUTIONS	A to 70°F	A to 70°F	A to 70°F	C at 70°F		
PHTHALIC ACID		A to 300°F	A to 212°F			A to 70°F
PHTHALIC ANHYDRIDE		A to 300°F	A to 212°F *A to 400°F	C at 70°F dry		
PICKLE ALUM		A to 70°F				
PICKLING SOLUT'N, EG 20% NITRIC, 4% HF	NR at 70°F	A to 70°F	A to 212°F			BC at 70°F
PICOLINE, ALPHA						
PICRIC ACID	B/NR sol'n at 70°F	A to 300°F	A to 100% to 212°F	A sol'n to 70°F A molten		
PICRIC ACID						
PICRONITRIC ACID		A to 70°F				
PIMELIC KETONE	C/NR to 140°F	A to 312°F	A to 70°F	C at 70°F	B at 70°F	
PINOCOL (PINOCOLIN)		A to 70°F				
PINE OIL	B at 70°F NR at 122°F	A to 122°F	A to 212°F	A to 70°F	A to 70°F	
PINENE		A to boiling	A to 212°F	A to 70°F		
PINK LIQUOR	A to 70°F					
PIPE LINE CLEANER, DIVAN IND.						`
PIPERIDINE		A to boiling	A to 212°F			
PITCH		A to 70°F				
PLASTICIZER		A to 70°F				
PLATING SOLUTION, ANTIMONY		A to 70°F				
PLATING SOLUTION, ARSENIC		A to 70°F				
PLATING SOLUTION, BRASS	A to 70°F	A to 300°F				
PLATING SOLUTION, BRONZE		A to 70°F				
PLATING SOLUTION, CADMIUM	A to 70°F	A to 300°F				
PLATING SOLUTION, CHROME	A to 70°F	A to 300°F	A to 212°F	A to 70°F		
PLATING SOLUTION, COBALT				A to 70°F		
PLATING SOLUTION, COPPER	A to 70°F	A to 300°F		A to 70°F		
PLATING SOLUTION, GOLD	A to 70°F	A to 300°F		A to 70°F		
PLATING SOLUTION, INDIUM	A to 70°F	A to 70°F		A to 70°F		
PLATING SOLUTION, IRON		A to 70°F		A to 70°F		
PLATING SOLUTION, LEAD	A to 70°F	A to 300°F		A to 70°F		
PLATING SOLUTION, NICKEL	A to 70°F	A to 70°F		A to 70°F		
PLATING SOLUTION, OTHERS		A to 70°F	A to 212°F	A to 70°F		A to 70°F
PLATING SOLUTION, PLATINUM		A to 70°F				
PLATING SOLUTION, RHODIUM	A to 70°F	A to 70°F				
PLATING SOLUTION, SILVER	A to 70°F	A to 70°F		A to 70°F		

	RECOMMENDATIONS	NITRILE NBR	ETHYLENE PROPYLENE EP, EPDM	FLUOROCARBON FKM	CHLOROPRENE CR	HYDROGENATED NITRILE HNBR
PLATING SOLUTION, TIN		AB to 140°F	A to 104°F B at 140°F	A to 140°F B at 176°F	A to 140°F C at 150°F	A to 70°F
PLATING SOLUTION, ZINC		A to 140°F B at 70°F fluoborate	A to 70°F	A to 140°F all types	A to 140°F B to 150°F sulfate	A to 70°F
PNEUMATIC SERVICE	NBR,EP,FKM,CR	A to 70°F	A to 70°F	A to 70°F	A to 70°F	A to 70°F
POLYESTER MONOMERS						
POLYETHYLENE GLYCOL		A to 70°F	A to 176°F	A to 212°F	A to 104°F B 140-176°F	B at 70°F
POLYFORMALDEHYDE		NR at 70°F			NR at 70°F	
POLYGLYCOLS	NBR	A to 250°F	A to 70°F	A to 70°F	A to 70°F	
POLYOL ESTERS (HYDRAULIC FLUID)					AB to 70°F	
POLYOXYMETHYLENE		B/NR at 70°F		C at 70°F	B/NR at 70°F	
POLYPROPYLENE GLYCOL		A to 70°F		A to 70°F		
POLYSTYRENE EMULSION						
POLYVINYL ACETATE EMULSION	EP,IIR	A to 70°F	A to 200°F	A/NR at 70°F	B at 70°F	
POLYVINYL ALCOHOL		A to 104°F	A to 104°F	A to 140°F		
POTASH (POTASSIUM CARBONATE)	NBR	A to 200°F A any conc to 180°F	A to 176°F AB to 200°F	A to 212°F	A to 140°F AB any conc to 200°F	
POTASH ALUM		A to 70°F AB any conc to 180°F	A to 70°F	A to 70°F	A to 140°F AB any conc to 200°F	A to 70°F
POTASH CAUSTIC	EP,IIR,CR,NBR	AB any conc to 150°F NR 36% at 176°F	A to 100% to 200°F B 25% at 212°F	AB to 70°F NR at 140°F	AB to 70°F NR at 140°F	B at 70°F
POTASSA		B any conc to 150°F	B at 70°F	C at 70°F	B any conc to 200°F	
POTASSIUM ACETATE	EP,NBR,IIR	B any conc to 120°F	A to 70°F	A/NR at 70°F	B at 70°F	
POTASSIUM ALUM	NBR,EP,CR,IIR	A to 140°F AB at 176°F	A to 176°F B at 212°F	A to 212°F	A to 140°F AB any conc to 200°F	A to 70°F
POTASSIUM ALUMINUM SULFATE	NBR,EP,CR,IIR	A to 140°F AB to 176°F	A to 176°F B at 212°F	A to 212°F	A to 140°F AB any conc to 200°F	A to 70°F
POTASSIUM BICARBONATE	NBR	A to 200°F	A to 176°F	A to 212°F	A to 140°F	
POTASSIUM BICHROMATE	NBR,EP,IIR,FKM,CR	A to 140°F B at 176°F	A to 140°F AC to 176°F	A to 212°F	A to 104°F B/NR at 140°F	A to 70°F
POTASSIUM BISULFATE		A to 140°F B at 176°F	A to 176°F	A to 212°F	A to 70°F	
POTASSIUM BISULFITE		A to 70°F	A to 70°F	A to 70°F	A to 70°F	
POTASSIUM BORATES		A to 176°F	A to 176°F	A to 212°F	A to 70°F	
POTASSIUM BROMATE		AB to 140°F NR at 176°F	A to 70°F AB at 212°F	A to 70°F AB to 212°F	AB to 140°F C 176-212°F	
POTASSIUM BROMIDE	NBR	A to 200°F	A to 176°F AB to 200°F	A to 212°F	A to 140°F AB to 200°F	
POTASSIUM CARBONATE	NBR	A to 200°F A any conc to 180°F	A to 176°F AB to 200°F	A to 212°F	A to 140°F AB any conc to 200°F	
POTASSIUM CHLORATE		A to 70°F AC to 130°F	A to 130°F AB 140-200°F	A to 140°F AB to 200°F	A to 140°F AB to 200°F	
POTASSIUM CHLORIDE	NBR,EP,CR,FKM,IIR	A to 176°F B at 212°F	A to 176°F AB to 212°F	A to 212°F	A any conc to 150°F AB to 200°F	A to 70°F
POTASSIUM CHROMATE		A to 140°F B at 176°F	A to 176°F B at 212°F	A to 212°F	A to 140°F	
POTASSIUM CHROMIC SULFATE		A to 70°F	AB at 70°F	AB to 70°F	A to 70°F	
POTASSIUM CUPRO CYANIDE	NBR,EP,CR,FKM	A to 140°F	A to 176°F B at 212°F	A to 212°F	A to 70°F	A to 70°F
POTASSIUM CYANIDE	NBR,EP,CR,FKM	A to 140°F AB to 200°F	A to 140°F AB to 200°F	A to 70°F A/NR at 140°F	AB to 150°F B/NR at 200°F	A to 70°F
POTASSIUM DICHROMATE	NBR,EP,IIR,FKM,CR	A to 140°F B at 176°F	A to 140°F AC at 176°F	A to 212°F	A to 104°F B/NR at 140°F	A to 70°F
POTASSIUM FERRICYANIDE		A/NR at 70°F	A to 70°F	A to 70°F	A to 70°F	
POTASSIUM FERROCYANIDE		AC to 70°F	A to 140°F	A to 140°F	A to 70°F	
POTASSIUM FLUORIDE		A to 140°F B at 176°F	A to 176°F	A to 212°F	A to 70°F	
POTASSIUM HYDRATE		B any conc to 150°F	B at 70°F	C at 70°F	B any conc to 200°F	
POTASSIUM HYDROXIDE	EP,IIR,CR,NBR	AB any conc to 150°F NR 30% at 175°F	A to 100% to 200°F B 25% at 212°F	AB to 100% to 70°F NR conc at 140°F	AB to 100% to 70°F B/NR 100% at 140°F	B at 70°F
POTASSIUM HYDROXIDE		A to 5% to 150°F		AB to 70% to 140°F C 30-50% at 175°F	AB to 50% to 212°F A to 25% to 140°F	
POTASSIUM HYDROXIDE				NR 30% at 212°F A 5% to 150°F	A to 5% to 158°F	
POTASSIUM HYPOCHLORITE		B to conc at 70°F	A to 70°F	A/NR conc at 70°F A dilute to 70°F	B to 140°F C at 176°F	
POTASSIUM IODIDE		A to 70°F B/NR at 140°F	A to 176°F	A to 212°F	A to 70°F A/NR at 140°F	

	STYRENE BUTADIENE SBR	POLYACRYLATE ACM	POLYURETHANE AU, EU	ISOBUTYLENE ISOPRENE IIR	POLYBUTADIENE BR	® AEROQUIP AQP
PLATING SOLUTION, TIN				A to 70°F AB to 150°F		
PLATING SOLUTION, ZINC				A to 70°F AB to 150°F		
PNEUMATIC SERVICE	NR at 70°F	NR at 70°F	A to 70°F	A to 70°F	NR at 70°F	
POLYESTER MONOMERS						
POLYETHYLENE GLYCOL				A to 70°F		
POLYFORMALDEHYDE				NR at 70°F		
POLYGLYCOLS						
POLYOL ESTERS (HYDRAULIC FLUID)			NR at 70°F			
POLYOXYMETHYLENE				B/NR at 70°F		
POLYPROPYLENE GLYCOL				A to 70°F		
POLYSTYRENE EMULSION						
POLYVINYL ACETATE EMULSION				A to 70°F emulsion A to 70°F liquid		
POLYVINYL ALCOHOL						
POTASH (POTASSIUM CARBONATE)	A to 70°F	NR at 70°F	B/NR at 70°F	A to 70°F AB any conc to 180°F		
POTASH ALUM			C/NR at 70°F	A to 70°F AB any conc to 185°F		
POTASH CAUSTIC	B 50-100% at 70°F	NR 50-100% at 70°F	B/NR 50-100% at 70°F AB 10% to 70°F	A 50-100% to 70°F AB any conc to 185°F	B 50-100% at 70°F	NR at 70°F
POTASSA			B at 70°F	A to 70°F AB any conc to 185°F		
POTASSIUM ACETATE	NR at 70°F	NR at 70°F	NR at 70°F	A to 70°F AB any conc to 150°F	NR at 70°F	
POTASSIUM ALUM	AB to 70°F			A to 70°F AB any conc to 185°F		
POTASSIUM ALUMINUM SULFATE	AB to 70°F			A to 70°F AB any conc to 185°F		
POTASSIUM BICARBONATE	A to 70°F	NR at 70°F	B/NR at 70°F	AB to 70°F		
POTASSIUM BICHROMATE	AB to 70°F	A to 70°F	AB to 70°F	A to 70°F AB any conc to 150°F	AB to 70°F	
POTASSIUM BISULFATE	AB to 70°F			A to 70°F		
POTASSIUM BISULFITE	AB to 70°F		A to 70°F	A to 70°F		
POTASSIUM BORATES	AB to 70°F			A to 70°F		
POTASSIUM BROMATE	AB to 70°F		AB to 70°F	AB to 70°F		
POTASSIUM BROMIDE	A to 70°F	NR at 70°F	A/NR at 70°F	A to 70°F		
POTASSIUM CARBONATE	A to 70°F	NR at 70°F	B/NR at 70°F	A to 70°F AB any conc to 180°F		
POTASSIUM CHLORATE	AB to 70°F		A to 70°F	AB to 70°F		
POTASSIUM CHLORIDE	A to 70°F	A to 70°F	A to 100% to 70°F	A to 70°F AB any conc to 185°F	A to 70°F	AB to 70°F
POTASSIUM CHROMATE	AB to 70°F		AB to 70°F	B at 70°F		
POTASSIUM CHROMIC SULFATE	AB to 70°F			A to 70°F		
POTASSIUM CUPRO CYANIDE	A to 70°F	A to 70°F	A to 70°F	A to 70°F AB any conc to 150°F	A to 70°F	
POTASSIUM CYANIDE	A to 70°F	A to 70°F	A to 70°F	A to 70°F	A to 70°F	AB to 70°F
POTASSIUM DICHROMATE	AB to 70°F	A to 70°F	AB to 70°F	A to 70°F AB any conc to 150°F	AB to 70°F	
POTASSIUM FERRICYANIDE	AB to 70°F			AB to 70°F		
POTASSIUM FERROCYANIDE	AB to 70°F		AB to 70°F	AB to 70°F		
POTASSIUM FLUORIDE	AB to 70°F		AB to 70°F	AB to 70°F		
POTASSIUM HYDRATE	AB to 70°F		B at 70°F	A to 70°F AB any conc to 185°F		
POTASSIUM HYDROXIDE	B 50-100% at 70°F	NR 50-100% at 70°F	B/NR 50-100% at 70°F AB 10% to 70°F	A 50-100% to 70°F AB any conc to 180°F	B 50-100% at 70°F	NR at 70°F
POTASSIUM HYDROXIDE						
POTASSIUM HYDROXIDE						
POTASSIUM HYPOCHLORITE	AB to 70°F		AB 25-100% to 70°F A 5% to 60°F	C at 70°F		
POTASSIUM IODIDE	AB to 70°F			AB to 70°F		

	SYNTHETIC ISOPRENE IR	NATURAL ISOPRENE NR	CHLOROSULFONATED POLYETHYLENE CSM	FLUOROSILICONE FVMQ	SILICONE VMQ	CHEMRAZ FFKM
PLATING SOLUTION, TIN						A to 70°F
PLATING SOLUTION, ZINC	B at 70°F	B to 100°F				A to 70°F
PNEUMATIC SERVICE	NR at 70°F	NR at 70°F	A to 70°F	NR at 70°F	NR at 70°F	A to 70°F
POLYESTER MONOMERS						A to 70°F
POLYETHYLENE GLYCOL	A to 70°F	A to 70°F	A to 70°F			A to 70°F
POLYFORMALDEHYDE	NR at 70°F	NR at 70°F				
POLYGLYCOLS				A to 70°F	A to 70°F	A to 70°F
POLYOL ESTERS (HYDRAULIC FLUID)						
POLYOXYMETHYLENE	NR at 70°F	NR at 70°F				
POLYPROPYLENE GLYCOL	A to 70°F	A to 70°F	A to 70°F			
POLYSTYRENE EMULSION						A to 70°F
POLYVINYL ACETATE EMULSION			B at 70°F		NR at 70°F	A to 70°F
POLYVINYL ALCOHOL						
POTASH (POTASSIUM CARBONATE)	A to 70°F	A to 70°F AB to 185°F	A to 70°F AB to 200°F	A to 70°F	AC to 70°F	A aqueous sol'n to 70°F
POTASH ALUM	A to 70°F	A to 70°F AB any conc to 150°F	A to 70°F		A to 70°F	A sol'n to 70°F
POTASH CAUSTIC	B at 70°F	B any conc to 150°F	A to 100% to 70°F AB to 200°F	C 50-100% at 70°F	C 50-100% at 70°F AB 1% to 70°F	AB to 70°F C at 70°F white
POTASSA	B at 70°F	B any conc to 150°F	A to 70°F		A to 70°F	
POTASSIUM ACETATE	A to 70°F	A to 70°F AB any conc to 150°F	B/NR at 70°F	NR at 70°F	NR at 70°F	A to 70°F
POTASSIUM ALUM	A to 70°F	A to 70°F AB any conc to 150°F	A to 70°F		A to 70°F	A to 70°F
POTASSIUM ALUMINUM SULFATE	A to 70°F	A to 70°F AB any conc to 150°F	A to 70°F		A to 70°F	A to 70°F
POTASSIUM BICARBONATE	A to 70°F	A to 70°F	A to 70°F	A to 70°F	A to 70°F AB to 125°F	A to 70°F
POTASSIUM BICHROMATE	A/NR at 70°F	A/NR at 70°F	A to 200°F AB 40% at 125°F	A to 70°F	A to 100% to 70°F	A to 70°F
POTASSIUM BISULFATE	A to 70°F	A to 70°F	A to 70°F			A sol'n to 70°F
POTASSIUM BISULFITE	A to 70°F	A to 70°F	A to 70°F			A to 70°F
POTASSIUM BORATES	A to 70°F	A to 70°F	A to 70°F			
POTASSIUM BROMATE		AB to 70°F	AB to 70°F			
POTASSIUM BROMIDE	A to 70°F	A to 70°F AB to 185°F	A to 70°F AB to 200°F	A to 70°F	A to 70°F	A to 70°F
POTASSIUM CARBONATE	A to 70°F	A to 70°F AB to 185°F	A to 70°F AB to 200°F	A to 70°F	AC to 70°F	A aqueous sol'n to 70°F
POTASSIUM CHLORATE	A to 70°F	B/NR at 70°F	A to 70°F AB to 200°F		B to 125°F	A to 70°F
POTASSIUM CHLORIDE	A to 70°F	A to 70°F AB to 185°F	A to 70°F AB to 200°F	A to 70°F	A to 70°F	A to 70°F
POTASSIUM CHROMATE		AB to 70°F	C at 70°F			A to 70°F
POTASSIUM CHROMIC SULFATE	A to 70°F	A to 70°F	AB to 70°F			A to 70°F
POTASSIUM CUPRO CYANIDE	A to 70°F	A to 70°F AB to 185°F	A to 70°F AB to 200°F	A to 70°F	A to 70°F	A to 70°F
POTASSIUM CYANIDE	A to 70°F	A to 70°F AB any conc to 150°F	A to 70°F	A to 70°F	A to 70°F	A to 70°F
POTASSIUM DICHROMATE	A/NR at 70°F	A/NR at 70°F	A to 200°F AB 40% at 125°F	A to 70°F	A to 100% to 70°F	A to 70°F
POTASSIUM FERRICYANIDE		AB to 70°F	A to 70°F			A to 70°F
POTASSIUM FERROCYANIDE		A to 70°F	AB to 70°F			A to 70°F
POTASSIUM FLUORIDE		AB to 70°F	A to 70°F			A to 70°F
POTASSIUM HYDRATE	B at 70°F	B any conc to 150°F	A to 70°F		A to 70°F	
POTASSIUM HYDROXIDE	B 50-100% at 70°F	B any conc to 150°F	A to 100% to 70°F AB at 200°F	C 50-100% at 70°F	C 50-100% at 70°F AB 1% to 70°F	AB to 70°F C at 70°F white
POTASSIUM HYDROXIDE						
POTASSIUM HYDROXIDE						
POTASSIUM HYPOCHLORITE	C at 70°F	C at 70°F	B at 70°F			A to 70°F
POTASSIUM IODIDE		AB to 70°F	A to 70°F			A to 70°F

C 298

	POLYSULFIDE T	CHLORINATED POLYETHYLENE CM	EPICHLOROHYDRIN CO, ECO	® HYTREL COPOLYESTER TPE	® SANTOPRENE COPOLYMER TPO	® C-FLEX STYRENIC TPE
PLATING SOLUTION, TIN						
PLATING SOLUTION, ZINC						
PNEUMATIC SERVICE	NR at 70°F					
POLYESTER MONOMERS				*AB to 70°F		
POLYETHYLENE GLYCOL						
POLYFORMALDEHYDE						
POLYGLYCOLS						AB to 70°F
POLYOL ESTERS (HYDRAULIC FLUID)				NR at 70°F		
POLYOXYMETHYLENE						
POLYPROPYLENE GLYCOL						
POLYSTYRENE EMULSION				*AB to 70°F		
POLYVINYL ACETATE EMULSION						
POLYVINYL ALCOHOL						
POTASH (POTASSIUM CARBONATE)	AB to 70°F			NR at 70°F		A to 70°F
POTASH ALUM						
POTASH CAUSTIC	B 50-100% at 70°F	NR at 70°F	A 50-100% to 70°F	C/NR 50-100% at 70°F A sol'n to 70°F		A 50-100% to 70°F
POTASSA						
POTASSIUM ACETATE	NR at 70°F					
POTASSIUM ALUM						
POTASSIUM ALUMINUM SULFATE						
POTASSIUM BICARBONATE	AB to 70°F					
POTASSIUM BICHROMATE	A to 70°F	NR at 70°F		*AB to 70°F		
POTASSIUM BISULFATE						
POTASSIUM BISULFITE						
POTASSIUM BORATES	AB to 70°F					
POTASSIUM BROMATE						
POTASSIUM BROMIDE	AB to 70°F					
POTASSIUM CARBONATE	AB to 70°F			NR at 70°F		A to 70°F
POTASSIUM CHLORATE						A to 70°F
POTASSIUM CHLORIDE	A to 70°F	AB to 70°F	A to 70°F	AB to 70°F		
POTASSIUM CHROMATE						
POTASSIUM CHROMIC SULFATE						
POTASSIUM CUPRO CYANIDE	A to 70°F					
POTASSIUM CYANIDE	A to 70°F	AB to 70°F	A to 70°F	AB to 70°F		
POTASSIUM DICHROMATE	A to 70°F	NR at 70°F		*AB to 70°F		
POTASSIUM FERRICYANIDE						
POTASSIUM FERROCYANIDE						
POTASSIUM FLUORIDE						
POTASSIUM HYDRATE						
POTASSIUM HYDROXIDE	B 50-100% at 70°F	NR at 70°F	A 50-100% to 70°F	C/NR 50-100% at 70°F A solution to 70°F	AB to conc to 70°F	A to 100% to 70°F
POTASSIUM HYDROXIDE						
POTASSIUM HYDROXIDE						
POTASSIUM HYPOCHLORITE						
POTASSIUM IODIDE						A to 70°F

	ETHYLENE ACRYLIC EA	POLYALLOMER LINEAR COPOLYMER	NYLON 11 POLYAMIDE	NYLON 12 POLYAMIDE	ETHYLENE VINYL ACETATE EVA	POLYVINYLCHLORIDE FLEXIBLE PVC
PLATING SOLUTION, TIN						
PLATING SOLUTION, ZINC						
PNEUMATIC SERVICE						
POLYESTER MONOMERS			*AB to 70°F emulsions	AB to 70°F		NR at 70°F emulsions
POLYETHYLENE GLYCOL			AB to 70°F			AB to 70°F
POLYFORMALDEHYDE						
POLYGLYCOLS						
POLYOL ESTERS (HYDRAULIC FLUID)			AB to 70°F			NR at 70°F
POLYOXYMETHYLENE						
POLYPROPYLENE GLYCOL						
POLYSTYRENE EMULSION			*AB to 70°F	AB to 70°F		NR at 70°F
POLYVINYL ACETATE EMULSION	A to 70°F					
POLYVINYL ALCOHOL						
POTASH (POTASSIUM CARBONATE)		A to 70°F	AC to 70°F	AB to 70°F aqueous	AB to 70°F	AB to 70°F *AB to 140°F
POTASH ALUM					B at 70°F	A to 70°F
POTASH CAUSTIC		A to conc to 122°F	AB to 100% to 70°F NR 50% at 140°F	AB to 50% to 70°F	AB to 70°F	AB to 70°F B at 122°F
POTASSA						
POTASSIUM ACETATE			AB to 70°F			AB to 70°F
POTASSIUM ALUM			AB to 70°F			AB to 100% to 70°F
POTASSIUM ALUMINUM SULFATE			AB to 70°F			AB to 100% to 70°F
POTASSIUM BICARBONATE			AB to 70°F			A to 70°F *AB to 140°F
POTASSIUM BICHROMATE			AB to 70°F		B 100% at 70°F AB 40% to 140°F	BC 100% to 130°F A 40% to 70°F
POTASSIUM BISULFATE			AB to 70°F			AB to 70°F
POTASSIUM BISULFITE			AB to 70°F			AB to 70°F *AB to 140°F
POTASSIUM BORATES			AB to 70°F			AB to 70°F
POTASSIUM BROMATE			AB to 70°F			AB to 70°F
POTASSIUM BROMIDE			AB to 70°F	AB to 10% to 70°F		AB to 70°F
POTASSIUM CARBONATE		A to 70°F	AC to 70°F	AB to 70° aqueous	AB to 70°F	AB to 70°F *AB to 140°F
POTASSIUM CHLORATE		A to 70°F	A to 70°F	C 7% at 70°F AB to 5% to 70°F	A to 70°F	AB to 70°F
POTASSIUM CHLORIDE	A to 70°F	A to 70°F	A to 70°F		A to 70°F	A to 70°F AB to 140°F
POTASSIUM CHROMATE			AB to 70°F			*AB to 140°F
POTASSIUM CHROMIC SULFATE			AB to 70°F			AB to 70°F
POTASSIUM CUPRO CYANIDE			AB to 70°F			AB to 140°F
POTASSIUM CYANIDE	A to 70°F		A to 70°F		B at 70°F	A to 70°F
POTASSIUM DICHROMATE			AB to 70°F		B 100% at 70°F AB 40% to 140°F	BC 100% to 130°F A 40% to 70°F
POTASSIUM FERRICYANIDE			AB to 70°F			AB to 70°F
POTASSIUM FERROCYANIDE			AB to 70°F			*AB to 140°F
POTASSIUM FLUORIDE			AB to 70°F			*AB to 140°F
POTASSIUM HYDRATE			AB to 70°F			AB to 70°F
POTASSIUM HYDROXIDE		A to conc to 122°F	AB to 100% to 70°F C 50% at 105°F	AB to 50% to 70°F	AB to 100% to 70°F AB 20% to 140°F	AB to 100% to 70°F B 100% at 122°F
POTASSIUM HYDROXIDE			NR 50% at 140°F			NR 100% at 130°F C 50% at 130°F
POTASSIUM HYDROXIDE						AB to 30% to 130°F AB 10% to 140°F
POTASSIUM HYPOCHLORITE			AB to 70°F			AB to 70°F
POTASSIUM IODIDE		A to 70°F	A to 70°F	AB 10% to 70°F	B at 70°F	AB to 70°F

C 300

	POLYETHYLENE LOW DENSITY LDPE	®TEFLON FEP	® KALREZ PERFLUORINATED ELASTOMER	® FLUORAZ FLUORINATED COPOLYMER	® AFLAS FLUORINATED COPOLYMER	® NORPRENE COPOLYMER TPO
PLATING SOLUTION, TIN	A to 70°F	A to 70°F		A to 70°F		
PLATING SOLUTION, ZINC	A to 70°F	A to 70°F		A to 70°F		
PNEUMATIC SERVICE				A to 70°F		
POLYESTER MONOMERS			A to 464°F NR at 600°F			
POLYETHYLENE GLYCOL	AB to 70°F	A to 70°F	A to 212°F	A to 70°F		
POLYFORMALDEHYDE		A to 70°F				
POLYGLYCOLS			A to 212°F	A to 70°F		AB to 70°F
POLYOL ESTERS (HYDRAULIC FLUID)						
POLYOXYMETHYLENE		A to 70°F				
POLYPROPYLENE GLYCOL		A to 70°F		A to 70°F		
POLYSTYRENE EMULSION						
POLYVINYL ACETATE EMULSION		A to 70°F	A to 212°F	A/NR at 70°F		
POLYVINYL ALCOHOL		A to 70°F				
POTASH (POTASSIUM CARBONATE)	AB to 104°F	A to 200°F	A to 212°F	A to conc to 70°F		AB to 70°F
POTASH ALUM	A to 70°F	A to 70°F		A to 70°F		
POTASH CAUSTIC	A to conc to 122°F AB to conc at 140°F	A to conc to 300°F	A to sat'd to 176°F A 42% to 250°F	A to conc to 70°F		A to 100% to 70°F
POTASSA		A to 70°F				
POTASSIUM ACETATE	AB to 70°F	A to 390°F	A to 212°F		A sat'd to 70°F	
POTASSIUM ALUM	B 100% to 175°F A 10% to 70°F	A to 300°F	A to 212°F	A to 70°F	A 30% to 70°F	
POTASSIUM ALUMINUM SULFATE	B 100% to 175°F A 10% to 70°F	A to 300°F	A to 212°F	A to 70°F		
POTASSIUM BICARBONATE	AB to 70°F	A to 300°F	A to 212°F	A to 70°F		
POTASSIUM BICHROMATE	A to 70°F AB to 140°F	A to 200°F	A to 212°F	A to conc to 70°F		
POTASSIUM BISULFATE	AB to 70°F	A to 70°F	A to 212°F			
POTASSIUM BISULFITE	AB to 70°F	A to 70°F	A to 212°F			
POTASSIUM BORATES	A to 70°F	A to 70°F				
POTASSIUM BROMATE	AB to 104°F C at 140°F					
POTASSIUM BROMIDE	B at 70°F	A to 200°F	A to 212°F	A to 70°F		
POTASSIUM CARBONATE	AB to 104°F	A to 200°F	A to 212°F	A to conc to 70°F		AB to 70°F
POTASSIUM CHLORATE	AB to 100% to 70°F	A to 100% to 200°F	A to 212°F	A to 70°F		AB to 70°F
POTASSIUM CHLORIDE	AB to 140°F	A to 70°F	A to 212°F	A to 70°F aqueous	A sat'd to 70°F	
POTASSIUM CHROMATE	B at 70°F	A to 70°F	A to 212°F			
POTASSIUM CHROMIC SULFATE	AB to 70°F	A to 70°F		A to 70°F		
POTASSIUM CUPRO CYANIDE	AB to 70°F	A to 70°F	A to 212°F	A to conc to 70°F		
POTASSIUM CYANIDE	AB to 140°F	A to 200°F	A to 212°F	A to conc to 70°F		
POTASSIUM DICHROMATE	A to 70°F AB to 140°F	A to 200°F	A to 212°F	A to conc to 70°F		
POTASSIUM FERRICYANIDE	A to 70°F	A to 300°F	A to 212°F			
POTASSIUM FERROCYANIDE	A to 70°F	A to 300°F				
POTASSIUM FLUORIDE	A to 70°F	A to 70°F	A to 212°F			
POTASSIUM HYDRATE	AB to 70°F	A to 70°F				
POTASSIUM HYDROXIDE	A to conc to 122°F AB to conc at 140°F	A to conc to 300°F	A to sat'd to 212°F A 42% to 250°F	A to conc to 70°F	A 30% to 70°F	A to 100% to 70°F
POTASSIUM HYDROXIDE						
POTASSIUM HYDROXIDE						
POTASSIUM HYPOCHLORITE	AB to 70°F	A to 300°F	A to 212°F			
POTASSIUM IODIDE	AB to 140°F	A to 300°F	A to 212°F			AB to 70°F

C 301

	RECOMMENDATIONS	NITRILE NBR	ETHYLENE PROPYLENE EP, EPDM	FLUOROCARBON FKM	CHLOROPRENE CR	HYDROGENATED NITRILE HNBR
POTASSIUM MURIATE		A to 70°F AB any conc to 200°F	A to 70°F AB to 185°F	A to 70°F AB to 185°F	A to 70°F AB any conc to 200°F	
POTASSIUM NITRATE	NBR	A to 140°F AB any conc to 200°F	A to 100% to 176°F	A to 100% to 212°F	A to 100% to 140°F AB to 100% to 200°F	A to 70°F
POTASSIUM NITRITE	NBR,EP,CR,FKM,IIR	A to 70°F AB any conc to 180°F	A to 70°F	A to 70°F	A to 70°F AB any conc to 200°F	
POTASSIUM OXIDE		AB any conc to 150°F		AB to 70°F	A to 70°F AB any conc to 200°F	
POTASSIUM PERBORATE		AB to 70°F		AB to 70°F	A to 70°F	
POTASSIUM PERCHLORATE		AB to 70°F		A to 70°F	C at 70°F	
POTASSIUM PERFLUOROACETATE	EP	AB to 70°F	A to 200°F	NR at 70°F	B at 70°F	
POTASSIUM PERMANGANATE	EP	AC any conc to 150°F	A to 100% to 200°F	A to 100% to 140°F BC 30% at 158°F	A to 100% to 140°F	
POTASSIUM PERSULFATE		A to 70°F	A to 70°F	AB to 70°F	C at 70°F	
POTASSIUM PHOSPHATE	NBR	A to 200°F	A to 70°F	A to 70°F	A to 70°F	
POTASSIUM SALTS	NBR,EP,FKM,IIR	A to 70°F	A to 70°F	A to 70°F	A to 70°F	A to 70°F
POTASSIUM SILICATE	NBR	A to 160°F	A to 70°F	A to 70°F	A to 70°F	
POTASSIUM SULFATE	NBR,EP,CR,IIR	A to 160°F AB to 200°F	A to 140°F AB 176-200°F	A to 212°F B at 248°F	A to 100% to 140°F AB to 100% to 200°F	A to 70°F
POTASSIUM SULFIDE		A to 70°F	A to 70°F	A to 70°F	A to 70°F	
POTASSIUM SULFITE	NBR,EP,FKM,CR,IIR	A to 70°F AB any conc to 150°F	A to 70°F	A to 70°F	A to 70°F AB any conc to 150°F	
POTASSIUM SULPHATE	NBR,EP,FKM,CR	A to 70°F	A to 100% to 70°F	A to 70°F	A to 100% to 70°F	
POTASSIUM THIOSULPHATE		A to 70°F		A to 70°F	A to 70°F	
POTASSIUM TRIPHOSPHATE	NBR	A to 200°F	A to 70°F	A to 70°F	A to 70°F	
POTATO OIL		A to 70°F AB any conc to 180°F	A to 70°F	A to 70°F	A to 70°F AB any conc to 180°F	
POTATO SPIRIT		A to 70°F AB any conc to 180°F	A to 70°F	A to 70°F	A to 70°F AB any conc to 180°F	
PRESTONE ANTIFREEZE	NBR,EP,FKM,CR,IIR	A to 250°F	A to 70°F	A to 70°F	AC to 70°F	A to 70°F
PRL-HIGH TEMP. HYDRAULIC OIL	FKM,ACM	B at 70°F	NR at 70°F	A to 70°F	B at 70°F	B at 70°F
PRODUCER GAS		A to 250°F	NR at 70°F	A to 70°F	B at 70°F	A to 70°F
PROPANE	NBR,FKM,FVMQ,T	A to 250°F gas or liquid	NR at 70°F gas or liquid	A to 176°F gas or liquid	AC to 140°F gas or liquid	A to 70°F
PROPANE PROPIONITRILE		A to 70°F	NR at 70°F	A to 70°F	B at 70°F	
PROPANEDIAMINE		B at 70°F				
PROPANEDIOL	NBR,CR,EP,FKM	A to 250°F	A to 70°F	A to 140°F	AC to 70°F C at 140°F	
PROPANETRIOL	NBR,EP,CR,FKM,IIR	A to 250°F	A to 170°F AB to 200°F	A to 250°F	A to 158°F	
PROPANOL	NBR,EP,CR,FKM,IIR	+2% vol 21 days 75°F AB to 200°F	A to 176°F AB to 200°F	A to 212°F	A to 140°F AB 176-200°F	A to 70°F
2-PROPANONE	EP,IIR,NR	NR at 70°F	A any conc to 200°F	NR 100% at 70°F C 10% to 104°F	C/NR at 70°F NR at 140°F	NR at 70°F
PROPELLER OIL		AB to 70°F			AB to 70°F	
PROPENOL		B at 70°F		A to 70°F		
PROPENE	FKM,FVMQ,T	NR at 70°F	NR at 70°F	A to 70°F	NR at 70°F	NR at 70°F
PROPENE NITRILE		NR at 70°F			NR at 70°F	
PROPENE OXIDE			B at 70°F		NR at 70°F	
PROPENEOL		A to 70°F	A to 70°F	B at 70°F	A to 70°F	
PROPENYL HYDRATE		A to 70°F	A to 70°F	A to 70°F	A to 70°F	
PROPENYL ANISOLE		NR at 70°F		B at 70°F		
PROPION ALDEHYDE		NR at 70°F	A to 200°F	NR at 70°F	NR at 70°F	
PROPIONIC ACID		A/NR at 70°F	A to 200°F	NR at 70°F	B/NR at 70°F	
PROPIONITRILE		NR at 70°F	A to 70°F	NR at 70°F	B at 70°F	A to 70°F
PROPYL ACETATE		NR at 70°F	B at 70°F	NR at 70°F	NR at 70°F	NR at 70°F
n-PROPYL ACETATE		NR at 70°F	A to 70°F	NR at 70°F	NR at 70°F	
PROPYL ACETONE		NR at 70°F			NR at 70°F	

	STYRENE BUTADIENE SBR	POLYACRYLATE ACM	POLYURETHANE AU, EU	ISOBUTYLENE ISOPRENE IIR	POLYBUTADIENE BR	® AEROQUIP AQP
POTASSIUM MURIATE	AB to 70°F		A to 70°F	A to 70°F		
POTASSIUM NITRATE	A to 70°F	A to 70°F	A to 100% to 70°F	A to 70°F AB any conc to 185°F	A to 70°F	
POTASSIUM NITRITE	AB to 70°F		A to 70°F	A to 70°F		
POTASSIUM OXIDE	AB to 70°F			A to 70°F AB any conc to 185°F		
POTASSIUM PERBORATE	AB to 70°F			AB to 70°F		
POTASSIUM PERCHLORATE	AB to 70°F			AB to 70°F		
POTASSIUM PERFLUOROACETATE	AB to 70°F			AB to 70°F		
POTASSIUM PERMANGANATE	AB to 70°F		B 100% at 70°F NR 5% at 70°F	A to 70°F AB any conc to 130°F		
POTASSIUM PERSULFATE	AB to 70°F		AB to 70°F	AB to 70°F		
POTASSIUM PHOSPHATE	A to 70°F	NR at 70°F	B /NR at 70°F	AB to 70°F		
POTASSIUM SALTS	A to 70°F	A to 70°F	A to 70°F	A to 70°F	A to 70°F	
POTASSIUM SILICATE	A to 70°F	A to 70°F	A to 70°F	A to 70°F		
POTASSIUM SULFATE	B at 70°F	NR at 70°F	A to 100% to 70°F	A to 70°F AB any conc to 185°F	B at 70°F	AB to 70°F
POTASSIUM SULFIDE	AB to 70°F		A to 70°F	A to 70°F		
POTASSIUM SULFITE	B at 70°F	NR at 70°F	A to 70°F	A to 70°F AB any conc to 180°F	A to 70°F	
POTASSIUM SULPHATE	B at 70°F	NR at 70°F	A to 70°F	A to 70°F	A to 70°F	
POTASSIUM THIOSULPHATE	AB to 70°F		AB to 70°F	AB to 70°F		
POTASSIUM TRIPHOSPHATE	A to 70°F	NR at 70°F	NR at 70°F	AB to 70°F		
POTATO OIL			C at 70°F	A to 70°F AB any conc to 180°F		
POTATO SPIRIT			C at 70°F	A to 70°F AB any conc to 180°F		
PRESTONE ANTIFREEZE	A to 70°F	NR at 70°F	NR at 70°F	A to 70°F	A to 70°F	AB to 70°F
PRL-HIGH TEMP. HYDRAULIC OIL	NR at 70°F	A to 70°F	B at 70°F	NR at 70°F	NR at 70°F	
PRODUCER GAS	NR at 70°F	B at 70°F	A to 70°F	NR at 70°F	NR at 70°F	
PROPANE	NR at 70°F	A /NR at 70°F	AC to 70°F	NR at 70°F	NR at 70°F	
PROPANE PROPIONITRILE	NR at 70°F	A to 70°F	NR at 70°F	NR at 70°F	NR at 70°F	
PROPANEDIAMINE				A to 70°F		
PROPANEDIOL	A to 70°F	NR at 70°F	AB to 70°F	AC to 70°F		
PROPANETRIOL	A to 70°F	NR at 70°F	A /NR at 70°F	A to 70°F AB any conc to 150°F	A to 70°F	AB to 70°F
PROPANOL	A to 250°F	NR at 70°F	NR at 70°F	A to 70°F AB any conc to 120°F	A to 70°F	
2-PROPANONE	BC at 70°F	NR at 70°F	NR at 70°F	A to 70°F AB any conc to 150°F	NR at 70°F	AB to 70°F
PROPELLER OIL						
PROPENOL				A to 70°F		
PROPENE	NR at 70°F	NR at 70°F	NR at 70°F	NR at 70°F	NR at 70°F	
PROPENE NITRILE				NR at 70°F		
PROPENE OXIDE				B at 70°F		
PROPENEOL				A to 70°F		
PROPENYL HYDRATE			A to 70°F	A to 70°F		
PROPENYL ANISOLE				NR at 70°F		
PROPION ALDEHYDE	NR at 70°F	NR at 70°F	NR at 70°F	A to 70°F		
PROPIONIC ACID	NR at 70°F	NR at 70°F	NR at 70°F	A to 70°F		
PROPIONITRILE				A to 70°F		
PROPYL ACETATE	NR at 70°F	NR at 70°F	NR at 70°F	AB to 80°F	NR at 70°F	
n-PROPYL ACETATE	NR at 70°F	NR at 70°F		AB to 70°F	NR at 70°F	
PROPYL ACETONE				A to 70°F AB any conc to 150°F		

	SYNTHETIC ISOPRENE IR	NATURAL ISOPRENE NR	CHLOROSULFONATED POLYETHYLENE CSM	FLUOROSILICONE FVMQ	SILICONE VMQ	CHEMRAZ FFKM
POTASSIUM MURIATE	A to 70°F	A to 70°F AB any conc to 150°F	A to 70°F		A to 70°F	
POTASSIUM NITRATE	A to 70°F	A to 70°F AB to 185°F	A to 70°F AB to 185°F	A to 70°F	A to 70°F	A to 70°F
POTASSIUM NITRITE	A to 70°F	A to 70°F	A to 70°F		A to 70°F	A to 70°F
POTASSIUM OXIDE	A to 70°F	A to 70°F AB any conc to 150°F	AB to 70°F			A to 70°F
POTASSIUM PERBORATE		AB to 70°F	AB to 70°F			
POTASSIUM PERCHLORATE		AB to 70°F	AB to 70°F			A to 70°F
POTASSIUM PERFLUOROACETATE		AB to 70°F	AB to 70°F	NR at 70°F		A to 70°F
POTASSIUM PERMANGANATE	A to 70°F	A to 70°F	AB to 140°F			A to 70°F
POTASSIUM PERSULFATE		AB to 70°F	AB to 70°F			A to 70°F
POTASSIUM PHOSPHATE		AB to 70°F	A to 70°F	A to 70°F	A to 70°F	A to 70°F
POTASSIUM SALTS	A to 70°F	A to 70°F	A to 70°F	A to 70°F	A to 70°F	A to 70°F
POTASSIUM SILICATE	A to 70°F	A to 70°F	A to 70°F	A to 70°F	A to 70°F	A to 70°F
POTASSIUM SULFATE	A to 70°F	A to 70°F AB to 185°F	A to 70°F AB to 200°F	A to 70°F	A to 70°F	A to 70°F
POTASSIUM SULFIDE	A to 70°F	AB to 70°F	AB to 70°F		A to 70°F	A to 70°F
POTASSIUM SULFITE	AB to 70°F	AB any conc to 150°F	AB to 70°F	A to 70°F	A to 70°F	A to 100% to 70°F
POTASSIUM SULPHATE	B at 70°F	B at 70°F	B at 70°F	A to 70°F	A to 70°F	A to 70°F
POTASSIUM THIOSULPHATE	A to 70°F	A to 70°F	A to 70°F			A to 70°F
POTASSIUM TRIPHOSPHATE		AB to 70°F	AB to 70°F	A to 70°F	A to 70°F	A to 70°F
POTATO OIL	A to 70°F	A to 70°F AB any conc to 150°F	A to 70°F		NR at 70°F	
POTATO SPIRIT	A to 70°F	A to 70°F AB any conc to 150°F	A to 70°F		NR at 70°F	
PRESTONE ANTIFREEZE	A to 70°F	A to 70°F	A to 70°F	A to 70°F	A to 70°F	A to 70°F
PRL-HIGH TEMP. HYDRAULIC OIL	NR at 70°F	NR at 70°F	NR at 70°F	A to 70°F	+7% vol 7 days 70°F +9% vol 3 days 212°F	A to 70°F
PRODUCER GAS	NR at 70°F	NR at 70°F	B at 70°F	B at 70°F	B at 70°F	A to 70°F
PROPANE	NR at 70°F gas or liquid	NR at 70°F gas or liquid	A/NR at 70°F	AB to 70°F	NR at 70°F	A to 70°F
PROPANE PROPIONITRILE	NR at 70°F	NR at 70°F	B at 70°F	C at 70°F	NR at 70°F	
PROPANEDIAMINE	B at 70°F	B at 70°F	C at 70°F			
PROPANEDIOL	A to 70°F	A to 70°F	A to 70°F	A to 70°F	A to 70°F	A to 70°F
PROPANETRIOL	A to 70°F	A to 70°F AB any conc to 150°F	A to 200°F	A to 70°F	A to 70°F	
PROPANOL	A to 70°F	A to 70°F AB any conc to 150°F	A to 70°F AB to 200°F	A to 70°F	A to 70°F	A to 70°F
2-PROPANONE	NR at 70°F	B/NR at 70°F	NR at 70°F	NR at 70°F	BC at 70°F	A to 70°F
PROPELLER OIL			AB to 70°F			
PROPENOL	B at 70°F	B at 70°F	B at 70°F			A to 70°F
PROPENE	NR at 70°F	NR at 70°F	NR at 70°F	B at 70°F	NR at 70°F	A to 70°F
PROPENE NITRILE		AB any conc to 80°F				
PROPENE OXIDE			NR at 70°F		NR at 70°F	
PROPENEOL	A to 70°F	A to 70°F	A to 70°F			
PROPENYL HYDRATE	A to 70°F	A to 70°F	A to 70°F		A to 70°F	
PROPENYL ANISOLE	NR at 70°F	NR at 70°F	NR at 70°F			
PROPION ALDEHYDE				NR at 70°F	NR at 70°F	A to 70°F
PROPIONIC ACID	A to 70°F	A to 70°F	AB to 70°F		A to 70°F	A to 70°F
PROPIONITRILE	A to 70°F	A to 70°F				A to 70°F
PROPYL ACETATE	NR at 70°F	NR at 70°F	NR at 70°F	NR at 70°F	NR at 70°F	A to 70°F
n-PROPYL ACETATE	NR at 70°F	NR at 70°F		NR at 70°F		
PROPYL ACETONE	NR at 70°F	NR at 70°F				A to 70°F

C 304

	POLYSULFIDE T	CHLORINATED POLYETHYLENE CM	EPICHLOROHYDRIN CO, ECO	® HYTREL COPOLYESTER TPE	® SANTOPRENE COPOLYMER TPO	® C-FLEX STYRENIC TPE
POTASSIUM MURIATE						
POTASSIUM NITRATE	A to 70°F		A to 70°F	AB to 70°F		
POTASSIUM NITRITE						
POTASSIUM OXIDE						
POTASSIUM PERBORATE						
POTASSIUM PERCHLORATE						
POTASSIUM PERFLUOROACETATE						
POTASSIUM PERMANGANATE				NR 5% at 70°F		
POTASSIUM PERSULFATE						
POTASSIUM PHOSPHATE						
POTASSIUM SALTS	A to 70°F					
POTASSIUM SILICATE						
POTASSIUM SULFATE	B at 70°F		A to 70°F	AB to 70°F		
POTASSIUM SULFIDE						
POTASSIUM SULFITE	B at 70°F					
POTASSIUM SULPHATE	B at 70°F	AB to 70°F	A to 70°F			
POTASSIUM THIOSULPHATE						
POTASSIUM TRIPHOSPHATE						
POTATO OIL						
POTATO SPIRIT						
PRESTONE ANTIFREEZE	C at 70°F	AB to 70°F		AB to 70°F	A to 200°F A to 50% to 250°F	
PRL-HIGH TEMP. HYDRAULIC OIL	B at 70°F					
PRODUCER GAS	NR at 70°F					
PROPANE	A to 70°F gas or liquid		A to 70°F	AB to 70°F		NR at 70°F
PROPANE PROPIONITRILE	A to 70°F					
PROPANEDIAMINE						
PROPANEDIOL						
PROPANETRIOL	B at 70°F	A to 70°F	A to 70°F	A to 70°F	NR at 70°F	TEST
PROPANOL	A to 70°F		A to 70°F		-1% to -8% vol 7 days 70°F A to 70°F	
2-PROPANONE	AC to 70°F	AB to 70°F	NR at 70°F	BC at 70°F	A to 70°F	AB to 70°F
PROPELLER OIL	AB to 70°F					
PROPENOL						
PROPENE	B at 70°F					
PROPENE NITRILE						
PROPENE OXIDE						
PROPENEOL						
PROPENYL HYDRATE						
PROPENYL ANISOLE						
PROPION ALDEHYDE						
PROPIONIC ACID					*A to 70°F	
PROPIONITRILE					*NR at 70°F	
PROPYL ACETATE	B at 70°F		NR at 70°F			
n-PROPYL ACETATE	B at 70°F		NR at 70°F			
PROPYL ACETONE						

C 305

	ETHYLENE ACRYLIC EA	POLYALLOMER LINEAR COPOLYMER	NYLON 11 POLYAMIDE	NYLON 12 POLYAMIDE	ETHYLENE VINYL ACETATE EVA	POLYVINYLCHLORIDE FLEXIBLE PVC
POTASSIUM MURIATE			AB to 70°F			AB to 70°F
POTASSIUM NITRATE	A to 70°F		AB to 70°F NR at 140°F	AB to 70°F	A to 70°F AB to 140°F	A to 70°F *AB to 140°F
POTASSIUM NITRITE			AB to 70°F			AB to 70°F
POTASSIUM OXIDE			AB to 70°F			AB to 70°F
POTASSIUM PERBORATE			AB to 70°F			AB to 70°F
POTASSIUM PERCHLORATE			AB to 70°F			AB to 70°F
POTASSIUM PERFLUOROACETATE			AB to 70°F			AB to 70°F
POTASSIUM PERMANGANATE			NR 5-100% at 70°F	C/NR 1-5% at 70°F	NR 20% at 70°F	AB 5-100% at 70°F
POTASSIUM PERSULFATE			AB to 70°F			*AB to 140°F
POTASSIUM PHOSPHATE			AB to 70°F			*AB to 140°F
POTASSIUM SALTS			AB to 70°F			AB to 70°F
POTASSIUM SILICATE			AB to 70°F			AB to 70°F
POTASSIUM SULFATE		A to 70°F	A to 70°F AB to 140°F	AB 10% to 70°F	AB to 140°F	AB to 70°F *AB to 140°F
POTASSIUM SULFIDE			AB to 70°F			AB to 140°F
POTASSIUM SULFITE			AB to 70°F			AB to 70°F
POTASSIUM SULPHATE		A to 70°F	A to 70°F AB 140-194°F	AB 10% to 70°F	AB to 70°F	AB to 70°F
POTASSIUM THIOSULPHATE			AB to 70°F			AB to 140°F
POTASSIUM TRIPHOSPHATE			AB to 70°F			AB to 70°F
POTATO OIL						
POTATO SPIRIT						
PRESTONE ANTIFREEZE	A to 70°F		AB to 70°F			
PRL-HIGH TEMP. HYDRAULIC OIL						
PRODUCER GAS					NR at 70°F	A to 70°F
PROPANE	A to 70°F	B/NR at 70°F	AB to 140°F	AB to 70°F	NR at 70°F	A to 70°F B at 122°F
PROPANE PROPIONITRILE						
PROPANEDIAMINE						
PROPANEDIOL		A to 122°F			B at 70°F	BC at 70°F NR at 122°F
PROPANETRIOL	A to 70°F	A to 122°F	AB to 105°F NR at 194°F	AB to 70°F	A to 70°F AB to 140°F	A to 122°F AC to 150°F
PROPANOL	A to 70°F			C/NR at 70°F		A to 70°F C at 150°F
2-PROPANONE	NR at 70°F	A to 122°F	A to 70°F C/NR at 140°F	NR 40-100% at 70°F C 10% at 70°F	C/NR at 70°F	NR any conc at 70°F
PROPELLER OIL						
PROPENOL						
PROPENE						AB to 70°F
PROPENE NITRILE						
PROPENE OXIDE						
PROPENEOL						
PROPENYL HYDRATE						
PROPENYL ANISOLE						
PROPION ALDEHYDE						
PROPIONIC ACID						NR at 70°F
PROPIONITRILE						
PROPYL ACETATE						
n-PROPYL ACETATE						
PROPYL ACETONE	*NR at 70°F					NR at 70°F

C 306

	POLYETHYLENE LOW DENSITY LDPE	®TEFLON FEP	®KALREZ PERFLUORINATED ELASTOMER	®FLUORAZ FLUORINATED COPOLYMER	®AFLAS FLUORINATED COPOLYMER	®NORPRENE COPOLYMER TPO
POTASSIUM MURIATE	AB to 70°F	A to 70°F				
POTASSIUM NITRATE	A to 70°F AB to 140°F	A to 100% to 300°F	A to 212°F	A to 100% to 70°F	A sat'd to 70°F	
POTASSIUM NITRITE	AB to 70°F	A to 70°F	A to 212°F			
POTASSIUM OXIDE	AB to 70°F	A to 70°F				
POTASSIUM PERBORATE	A to 70°F	A to 70°F				
POTASSIUM PERCHLORATE	A to 70°F	A to 70°F	A to 212°F			
POTASSIUM PERFLUOROACETATE	AB to 70°F		A to 212°F	NR at 70°F		
POTASSIUM PERMANGANATE	AB to 140°F	A to 300°F	A to 212°F	A to 70°F		
POTASSIUM PERSULFATE	A to 70°F	A to 70°F	A to 212°F			
POTASSIUM PHOSPHATE	AB to 70°F	A to 70°F	A to 212°F	A to 70°F		
POTASSIUM SALTS	A to 70°F AB to 140°F	A to 70°F	A to 212°F	A to 70°F		A to 70°F
POTASSIUM SILICATE	AB to 70°F	A to 70°F	A to 212°F			
POTASSIUM SULFATE	AB to 140°F	A to 300°F	A to 212°F	A to 70°F		
POTASSIUM SULFIDE	A to 70°F	A to 70°F	A to 212°F			
POTASSIUM SULFITE	AB to 70°F	A to 70°F	A to 212°F			
POTASSIUM SULPHATE	A to 70°F	A to 100% to 300°F	A to 212°F	A to 70°F aqueous		
POTASSIUM THIOSULPHATE	AB to 70°F	A to 70°F	A to 212°F			
POTASSIUM TRIPHOSPHATE	AB to 70°F			A to 70°F		
POTATO OIL		A to 70°F				
POTATO SPIRIT		A to 70°F				
PRESTONE ANTIFREEZE		A to 70°F	A to 212°F	B at 70°F		
PRL-HIGH TEMP. HYDRAULIC OIL				A to 70°F		
PRODUCER GAS	C/NR at 70°F	A to 70°F	A to 212°F	A to 70°F		
PROPANE	AC to 70°F	A to 300°F gas or liquid	A to 212°F	AC to 70°F		
PROPANE PROPIONITRILE						
PROPANEDIAMINE		A to 70°F				
PROPANEDIOL	A to 122°F	A to 122°F	A to 104°F	A to 70°F		
PROPANETRIOL	A to 122°F AB to 140°F	A to 300°F	A to 70°F *A to 250°F	A to 70°F		A to 70°F
PROPANOL	A/NR at 70°F	A to 300°F	A to 212°F	A to 70°F	A to 70°F	BC at 70°F
2-PROPANONE	C/NR at 70°F NR at 122°F	A to boiling	A to 113°F	NR at 70°F	NR at 70°F	BC at 70°F
PROPELLER OIL						
PROPENOL		A to 70°F				
PROPENE	AB to 70°F	A to 70°F	A to 70°F	A to 70°F		
PROPENE NITRILE		A to 70°F				
PROPENE OXIDE		A to 70°F				
PROPENEOL		A to 70°F				
PROPENYL HYDRATE		A to 70°F				
PROPENYL ANISOLE		A to 70°F				
PROPION ALDEHYDE			A to 212°F	NR at 70°F		
PROPIONIC ACID		A to 70°F	A to 212°F			
PROPIONITRILE		A to 70°F	A to 212°F	A to 70°F		
PROPYL ACETATE		A to 70°F	A to 212°F	NR at 70°F		
n-PROPYL ACETATE		A to 70°F	A to 212°F	NR at 70°F		
PROPYL ACETONE		A to 70°F	A to 212°F	NR at 70°F		

	RECOMMENDATIONS	NITRILE NBR	ETHYLENE PROPYLENE EP, EPDM	FLUOROCARBON FKM	CHLOROPRENE CR	HYDROGENATED NITRILE HNBR
n-PROPYL ACETONE	EP, IIR	NR at 70°F	A to 70°F	NR at 70°F	NR at 70°F	
PROPYL ALCOHOL	NBR, EP, CR, FKM, IIR	+2% vol 21 days 75°F AB to 200°F	A to 176°F AB to 200°F	A to 212°F	A to 140°F AB 176–200°F	A to 70°F
PROPYL ALDEHYDE		NR at 70°F		NR at 70°F		
PROPYL AMINE	EP	NR at 70°F	A to 250°F	NR at 70°F	NR at 70°F	
PROPYL CHLORIDE		NR at 70°F		B at 70°F		
PROPYL CYANIDE		NR at 70°F	A to 70°F		NR at 70°F	
PROPYL ETHYLENE		C at 70°F	NR at 70°F		A to 70°F	
PROPYL FORMIC ACID		NR at 70°F			NR at 70°F	
PROPYL NITRATE	EP, IIR	+141% vol 7 days 75°F NR at 70°F	B to 160°F	NR at 70°F	NR at 70°F	
PROPYLENE	FKM, FVMQ, T	NR at 70°F	NR at 70°F	A to 70°F	NR at 70°F	NR at 70°F
PROPYLENE ALDEHYDE		NR at 70°F		NR at 70°F		
PROPYLENE CHLORIDE		NR at 70°F	NR at 70°F	B at 70°F	NR at 70°F	
PROPYLENE DIAMINE		B at 70°F				
PROPYLENE DICHLORIDE	FKM	NR at 70°F	NR at 70°F	A to 300°F	NR at 70°F	
PROPYLENE GLYCOL	NBR, CR, EP, FKM	A to 250°F	A to 70°F	A to 140°F	AC to 70°F C at 140°F	
PROPYLENE OXIDE	EP, IIR	NR at 70°F	B to 120°F	NR at 70°F	NR at 70°F	NR at 70°F
PROTOCHLORIDE					AB to 80°F	
PRUSSIC ACID	EP, IIR, FKM	B any conc to 140°F	A to 100% to 140°F	A to 100% to 140°F	B to 100% to 140°F	B at 70°F
PULP STOCK	NBR	A to 200°F	A to 70°F	A to 70°F	A to 70°F	
PURINA WOODY PLANT SPRAY		NR 10% at 70°F		C 10% at 70°F	NR 10% at 70°F	
PURPLE SALT		AB any conc to 150°F			NR at 70°F	
PYRACETIC ACID		C at 70°F	B at 70°F	A to 70°F	C at 70°F	
PYDRAUL 10E	EP, IIR	NR at 70°F	AB to 70°F	A to 70°F BC at 158°F	NR at 70°F	NR at 70°F
PYDRAUL 29ELT, 30E, 50E, 65E & 90E	EP, FKM, IIR	NR at 70°F	AB to 70°F	A to 158°F	NR at 70°F	NR at 70°F
PYDRAUL 115E	EP. FKM, IIR	NR at 70°F	AB to 70°F	A to 70°F	NR at 70°F	NR at 70°F
PYDRAUL 150	EP, IIR, FKM	C/NR at 70°F	AC to 70°F	A to 212°F	C/NR at 70°F	
PYDRAUL 230C, 312C & 540C	FKM	NR at 70°F	NR at 70°F	A to 212°F	NR at 70°F	NR at 70°F
PYDRAUL 600 & 625	EP, FKM, IIR(625)	C at 70°F (600) NR at 70°F (625)	C at 70°F (600) A to 70°F (625)	AB to 70°F (625)	C at 70°F	
PYDRAUL F-9	EP, FKM, IIR	NR at 70°F	AC to 70°F	+3% vol 14 days 180°F A to 180°F	C/NR at 70°F	
PYDRAULS, GENERAL	FKM, EP, IIR	C/NR at 70°F	AB to 70°F	A to 70°F AB to 400°F	NR at 70°F	NR at 70°F
PYRANOL	NBR, FKM, ACM, FVMQ	A to 70°F	NR at 70°F	+5% vol 7 days 300°F A to 212°F	NR at 70°F	
PYRANOL TRANSFORMER OIL	NBR, FKM, ACM	A to 70°F	NR at 70°F	A to 70°F	B at 70°F	A to 70°F
PYRANOL 1467		A to 70°F	NR at 70°F		BC at 70°F	
PYRANOL 1476		A to 70°F	NR at 70°F	A to 70°F	B at 70°F	
PYRENE		C at 70°F	NR at 70°F	A to 70°F	B/NR at 70°F	
PYRETHRUM						
PYREX		C at 70°F	NR at 70°F	A to 70°F	NR at 70°F	
PYRIDINE	EP, IIR	NR at 70°F	B to 160°F	+118% vol 3 days 75°F NR at 70°F	NR at 70°F	NR at 70°F
PYRIDINE OIL	EP, IIR	NR at 70°F	B at 70°F	NR at 70°F	NR at 70°F	
PYRITE		A to 70°F		A to 70°F		
PYROACETIC ETHER		NR at 70°F			NR at 70°F	
PYROACETIC SPIRIT		NR at 70°F	A to 70°F	NR at 70°F	C at 70°F	
PYROGALLIC ACID (PYROGALLOL)		A/NR at 70°F	B at 70°F	A to 140°F	A to 70°F	
PYROGALLOLCARBOXYLIC ACID		NR at 70°F	B at 70°F	A to 70°F	C at 70°F	

	STYRENE BUTADIENE SBR	POLYACRYLATE ACM	POLYURETHANE AU, EU	ISOBUTYLENE ISOPRENE IIR	POLYBUTADIENE BR	® AEROQUIP AQP
n-PROPYL ACETONE	NR at 70°F	NR at 70°F	NR at 70°F	A to 70°F	NR at 70°F	
PROPYL ALCOHOL	A to 250°F	NR at 70°F	NR at 70°F	A to 70°F AB any conc to 120°F	A to 70°F	
PROPYL ALDEHYDE				B at 70°F		
PROPYL AMINE	NR at 70°F	NR at 70°F	NR at 70°F			
PROPYL CHLORIDE				C at 70°F		
PROPYL CYANIDE				A to 70°F		
PROPYL ETHYLENE				NR at 70°F		
PROPYL FORMIC ACID				NR at 70°F		
PROPYL NITRATE	NR at 70°F	NR at 70°F	NR at 70°F	B to 120°F	NR at 70°F	
PROPYLENE	NR at 70°F	NR at 70°F	NR at 70°F	NR at 70°F	NR at 70°F	
PROPYLENE ALDEHYDE				A to 70°F		
PROPYLENE CHLORIDE				NR at 70°F		
PROPYLENE DIAMINE				A to 70°F		
PROPYLENE DICHLORIDE	NR at 70°F	NR at 70°F	NR at 70°F	NR at 70°F		
PROPYLENE GLYCOL	A to 70°F	NR at 70°F	AB to 70°F	AC to 70°F		
PROPYLENE OXIDE	NR at 70°F	NR at 70°F	NR at 70°F	B at 70°F	NR at 70°F	
PROTOCHLORIDE				A to 70°F AB any conc to 185°F		
PRUSSIC ACID	B at 70°F	NR at 70°F	C/NR 50-100% at 70°F	AB any conc to 150°F	B at 70°F	NR at 70°F
PULP STOCK	B at 70°F	NR at 70°F	NR at 70°F			
PURINA WOODY PLANT SPRAY				NR at 70°F		
PURPLE SALT				AB any conc to 130°F		
PYRACETIC ACID				A to 70°F		
PYDRAUL 10E	NR at 70°F	NR at 70°F	NR at 70°F	AB to 70°F	NR at 70°F	AB to 200°F
PYDRAUL 29ELT, 30E, 50E, 65E & 90E	NR at 70°F	NR at 70°F	NR at 70°F	AB to 70°F	NR at 70°F	AB to 200°F
PYDRAUL 115E	NR at 70°F	NR at 70°F	NR at 70°F	AB to 70°F	NR at 70°F	AB to 200°F
PYDRAUL 150			NR at 70°F	AC to 70°F		
PYDRAUL 230C, 312C & 540C	NR at 70°F	NR at 70°F	NR at 70°F	NR at 70°F	NR at 70°F	AB to 200°F
PYDRAUL 600 & 625			NR at 70°F 600 & 625	AC at 70°F 600 A to 70°F 625		
PYDRAUL F-9			NR at 70°F	C/NR at 70°F		
PYDRAULS, GENERAL	NR at 70°F	NR at 70°F	NR at 70°F	AB to 70°F	NR at 70°F	AB to 200°F
PYRANOL	NR at 70°F	A to 70°F	B at 70°F	NR at 70°F	NR at 70°F	
PYRANOL TRANSFORMER OIL	NR at 70°F	A to 70°F	B at 70°F	NR at 70°F	NR at 70°F	
PYRANOL 1467						
PYRANOL 1476						
PYRENE			C at 70°F	NR at 70°F		
PYRETHRUM			AB to 70°F			
PYREX			C at 70°F	NR at 70°F		
PYRIDINE	B/NR at 70°F	NR at 70°F	NR at 70°F	AB any conc to 100°F	NR at 70°F	
PYRIDINE OIL	NR at 70°F	NR at 70°F		B at 70°F	NR at 70°F	
PYRITE				A to 70°F		
PYROACETIC ETHER				A to 70°F AB any conc to 150°F		
PYROACETIC SPIRIT			NR at 70°F	A to 70°F		
PYROGALLIC ACID (PYROGALLOL)			B/NR at 70°F			
PYROGALLOLCARBOXYLIC ACID			NR at 70°F	B at 70°F		

	SYNTHETIC ISOPRENE IR	NATURAL ISOPRENE NR	CHLOROSULFONATED POLYETHYLENE CSM	FLUOROSILICONE FVMQ	SILICONE VMQ	CHEMRAZ FFKM
n-PROPYL ACETONE	NR at 70°F	NR at 70°F	NR at 70°F	NR at 70°F	NR at 70°F	A to 70°F
PROPYL ALCOHOL	A to 70°F	A to 70°F AB any conc to 120°F	A to 70°F AB to 200°F	A to 70°F	A to 70°F	A to 70°F
PROPYL ALDEHYDE	C at 70°F	C at 70°F	NR at 70°F			
PROPYL AMINE				NR at 70°F	C at 70°F	A to 70°F
PROPYL CHLORIDE	NR at 70°F	NR at 70°F	NR at 70°F			
PROPYL CYANIDE	NR at 70°F	NR at 70°F				
PROPYL ETHYLENE	NR at 70°F	NR at 70°F				
PROPYL FORMIC ACID	NR at 70°F	NR at 70°F				
PROPYL NITRATE	NR at 70°F	NR at 70°F	C/NR at 70°F	NR at 70°F	C/NR at 70°F	A to 70°F
PROPYLENE	NR at 70°F	NR at 70°F	NR at 70°F	B at 70°F	NR at 70°F	A to 70°F
PROPYLENE ALDEHYDE	NR at 70°F	NR at 70°F	NR at 70°F			
PROPYLENE CHLORIDE	NR at 70°F	NR at 70°F	NR at 70°F			A to 70°F
PROPYLENE DIAMINE	B at 70°F	B at 70°F	C at 70°F			
PROPYLENE DICHLORIDE	NR at 70°F	NR at 70°F	NR at 70°F	B at 70°F	NR at 70°F	A to 70°F
PROPYLENE GLYCOL	A to 70°F	A to 70°F	A to 70°F	A to 70°F	A to 70°F	A to 70°F
PROPYLENE OXIDE	NR at 70°F	NR at 70°F	NR at 70°F	NR at 70°F	NR at 70°F	A to 70°F
PROTOCHLORIDE	A to 70°F	A to 70°F AB any conc to 150°F				
PRUSSIC ACID	AB to 70°F	AB any conc to 80°F AB to 20% to 150°F	A to 70°F	B at 70°F	AC to 70°F	A to 70°F
PULP STOCK				A to 70°F	A to 70°F	
PURINA WOODY PLANT SPRAY	NR at 70°F	NR at 70°F			NR at 70°F	
PURPLE SALT	NR at 70°F	NR at 70°F				
PYRACETIC ACID	C at 70°F	C at 70°F	B at 70°F		A to 70°F	
PYDRAUL 10E	NR at 70°F	NR at 70°F	NR at 70°F	NR at 70°F	NR at 70°F	B at 70°F
PYDRAUL 29ELT, 30E, 50E, 65E & 90E	NR at 70°F	NR at 70°F	NR at 70°F	A to 70°F	A to 70°F	B at 70°F
PYDRAUL 115E	NR at 70°F	NR at 70°F	NR at 70°F	C at 70°F	NR at 70°F	B at 70°F
PYDRAUL 150		NR at 70°F				
PYDRAUL 230C, 312C & 540C	NR at 70°F	NR at 70°F	NR at 70°F	NR at 70°F	NR at 70°F	B at 70°F
PYDRAUL 600 & 625		NR at 70°F 600 & 625				
PYDRAUL F-9	NR at 70°F	NR at 70°F				
PYDRAULS, GENERAL	NR at 70°F	NR at 70°F	NR at 70°F	A/NR at 70°F	B/NR at 70°F	B at 70°F
PYRANOL	NR at 70°F	NR at 70°F	NR at 70°F	A to 70°F	B/NR at 70°F	
PYRANOL TRANSFORMER OIL	NR at 70°F	NR at 70°F	B at 70°F	A to 70°F	NR at 70°F	A to 70°F
PYRANOL 1467			AB to 70°F		+4% vol 7 days 70°F A to 70°F	
PYRANOL 1476			AB to 70°F			
PYRENE	NR at 70°F	NR at 70°F	NR at 70°F		NR at 70°F	
PYRETHRUM						
PYREX	NR at 70°F	NR at 70°F	NR at 70°F		NR at 70°F	
PYRIDINE	NR at 70°F	NR at 70°F	NR at 70°F	NR at 70°F	NR at 70°F	A to 70°F
PYRIDINE OIL	NR at 70°F	NR at 70°F	NR at 70°F	NR at 70°F	NR at 70°F	A to 70°F
PYRITE	A to 70°F	A to 70°F	A to 70°F			
PYROACETIC ETHER	NR at 70°F	NR at 70°F				
PYROACETIC SPIRIT	B at 70°F	B at 70°F	B at 70°F		A to 70°F	
PYROGALLIC ACID (PYROGALLOL)						A to 70°F
PYROGALLOLCARBOXYLIC ACID	A to 70°F	A to 70°F	B at 70°F			

	POLYSULFIDE T	CHLORINATED POLYETHYLENE CM	EPICHLOROHYDRIN CO, ECO	® HYTREL COPOLYESTER TPE	® SANTOPRENE COPOLYMER TPO	® C-FLEX STYRENIC TPE
n-PROPYL ACETONE	B at 70°F		NR at 70°F			
PROPYL ALCOHOL	A to 70°F		A to 70°F		−1% to −8% vol 7 days 70°F A to 70°F	
PROPYL ALDEHYDE						
PROPYL AMINE						
PROPYL CHLORIDE						
PROPYL CYANIDE						
PROPYL ETHYLENE						
PROPYL FORMIC ACID						
PROPYL NITRATE						
PROPYLENE	B at 70°F					
PROPYLENE ALDEHYDE						
PROPYLENE CHLORIDE						
PROPYLENE DIAMINE						
PROPYLENE DICHLORIDE						
PROPYLENE GLYCOL						
PROPYLENE OXIDE					*A to 70°F	
PROTOCHLORIDE						
PRUSSIC ACID	NR at 70°F	NR at 70°F		NR at 70°F	A to 70°F	A to 70°F
PULP STOCK						
PURINA WOODY PLANT SPRAY						
PURPLE SALT						
PYRACETIC ACID						
PYDRAUL 10E			NR at 70°F	AB to 200°F		
PYDRAUL 29ELT, 30E, 50E, 65E & 90E			NR at 70°F	A to 70°F AB to 200°F		
PYDRAUL 115E			NR at 70°F	AB to 200°F		
PYDRAUL 150		AB to 70°F		C/NR at 70°F		
PYDRAUL 230C, 312C & 540C	NR at 70°F		NR at 70°F	AC to 70°F B/NR at 200–212°F	A to 212°F C at 302°F	
PYDRAUL 600 & 625		AB to 70°F		C/NR at 70°F 600 & 625		
PYDRAUL F-9		AB to 70°F		C/NR at 70°F		
PYDRAULS, GENERAL	NR at 70°F	AB to 70°F	NR at 70°F	A/NR at 70°F	AB to 212°F	AB to 212°F
PYRANOL	NR at 70°F		NR at 70°F			
PYRANOL TRANSFORMER OIL	NR at 70°F		NR at 70°F			
PYRANOL 1467	AB to 70°F					
PYRANOL 1476	AB to 70°F					
PYRENE						
PYRETHRUM				AB to 70°F		
PYREX						
PYRIDINE			NR at 70°F	C at 70°F	A to 70°F	A to 70°F
PYRIDINE OIL	NR at 70°F		NR at 70°F			
PYRITE						
PYROACETIC ETHER						
PYROACETIC SPIRIT						
PYROGALLIC ACID (PYROGALLOL)						
PYROGALLOLCARBOXYLIC ACID						

	ETHYLENE ACRYLIC EA	POLYALLOMER LINEAR COPOLYMER	NYLON 11 POLYAMIDE	NYLON 12 POLYAMIDE	ETHYLENE VINYL ACETATE EVA	POLYVINYLCHLORIDE FLEXIBLE PVC
n-PROPYL ACETONE	NR at 70°F					*NR at 70°F
PROPYL ALCOHOL	A to 70°F			C/NR at 70°F		A to 70°F C at 150°F
PROPYL ALDEHYDE						
PROPYL AMINE						
PROPYL CHLORIDE						
PROPYL CYANIDE						
PROPYL ETHYLENE						
PROPYL FORMIC ACID						
PROPYL NITRATE						
PROPYLENE						AB to 70°F
PROPYLENE ALDEHYDE						
PROPYLENE CHLORIDE						
PROPYLENE DIAMINE						
PROPYLENE DICHLORIDE						
PROPYLENE GLYCOL		A to 122°F			B at 70°F	BC at 70°F NR at 122°F
PROPYLENE OXIDE		A to 70°F B at 122°F				C at 70°F NR at 122°F
PROTOCHLORIDE						
PRUSSIC ACID		A to 70°F	C/NR at 70°F		AB to 70°F	AB 50-100% to 70°F
PULP STOCK						
PURINA WOODY PLANT SPRAY						
PURPLE SALT						
PYRACETIC ACID						NR at 70°F
PYDRAUL 10E	NR at 70°F		AB to 200°F			
PYDRAUL 29ELT, 30E, 50E, 65E & 90E	NR at 70°F		A to 70°F AB to 200°F			
PYDRAUL 115E	NR at 70°F		AB to 200°F			
PYDRAUL 150			A to 70°F AB to 150°F			NR at 70°F
PYDRAUL 230C, 312C & 540C	NR at 70°F		A to 70°F AB to 200°F			NR at 70°F
PYDRAUL 600 & 625			AB to 150°F 600 AB to 150°F 625			NR at 70°F 600 NR at 70°F 625
PYDRAUL F-9			A to 70°F AB to 150°F			NR at 70°F
PYDRAULS, GENERAL	NR at 70°F		A to 70°F AB to 200°F			B/NR at 70°F
PYRANOL						
PYRANOL TRANSFORMER OIL	A to 70°F			AB to 70°F		
PYRANOL 1467						
PYRANOL 1476						
PYRENE						
PYRETHRUM			AB to 70°F			A to 70°F
PYREX						
PYRIDINE	*NR at 70°F	A to 70°F	C at 70°F NR at 105°F	AB to 70°F	B/NR at 70°F	NR at 70°F
PYRIDINE OIL						
PYRITE						
PYROACETIC ETHER						
PYROACETIC SPIRIT						
PYROGALLIC ACID (PYROGALLOL)					C/NR at 70°F	C at 70°F
PYROGALLOLCARBOXYLIC ACID						

	POLYETHYLENE LOW DENSITY LDPE	®TEFLON FEP	®KALREZ PERFLUORINATED ELASTOMER	®FLUORAZ FLUORINATED COPOLYMER	®AFLAS FLUORINATED COPOLYMER	®NORPRENE COPOLYMER TPO
n-PROPYL ACETONE		*A to 70°F	A to 212°F	NR at 70°F		
PROPYL ALCOHOL	A/NR at 70°F	A to 300°F	A to 212°F	A to 70°F	A to 70°F	BC at 70°F
PROPYL ALDEHYDE		A to 70°F				
PROPYL AMINE			A to 212°F			
PROPYL CHLORIDE		A to 70°F				
PROPYL CYANIDE		A to 70°F				
PROPYL ETHYLENE		A to 70°F				
PROPYL FORMIC ACID		A to 70°F				
PROPYL NITRATE		A to 70°F	A to 212°F	C at 70°F		
PROPYLENE	AB to 70°F	A to 70°F	A to 212°F	A to 70°F		
PROPYLENE ALDEHYDE		A to 70°F				
PROPYLENE CHLORIDE		A to 70°F	A to 212°F			
PROPYLENE DIAMINE		A to 70°F				
PROPYLENE DICHLORIDE	C at 70°F	A to 70°F	A to 212°F			
PROPYLENE GLYCOL	A to 122°F	A to 125°F	A to 212°F	A to 70°F		
PROPYLENE OXIDE	A to 70°F B at 122°F	A to 300°F	A to 212°F	NR at 70°F		
PROTOCHLORIDE		A to 70°F				
PRUSSIC ACID	A to 100% to 70°F	A to 100% to 70°F	A to 70°F	A to 70°F		AB to 70°F
PULP STOCK				A to 70°F		
PURINA WOODY PLANT SPRAY		A to 70°F				
PURPLE SALT		A to 70°F				
PYRACETIC ACID		A to 70°F				
PYDRAUL 10E		A to 70°F AB to 400°F	A to 212°F	A to 70°F		
PYDRAUL 29ELT, 30E, 50E, 65E & 90E		A to 70°F AB to 400°F	A to 212°F	A to 70°F		
PYDRAUL 115E		AB to 400°F	A to 212°F	A to 70°F		
PYDRAUL 150	NR at 70°F	A to 70°F				
PYDRAUL 230C, 312C & 540C		A to 200°F AB to 400°F	A to 212°F	A to 70°F	A to 70°F	
PYDRAUL 600 & 625	NR at 70°F 600 & 625	A to 70°F 600 & 625				
PYDRAUL F-9	NR at 70°F	A to 70°F	A to 70°F			
PYDRAULS, GENERAL	NR at 70°F	A to 70°F AB to 400°F	A to 212°F	A to 70°F	A to 70°F	
PYRANOL		A to 70°F	A to 212°F			
PYRANOL TRANSFORMER OIL			A to 212°F	A to 70°F		
PYRANOL 1467		A to 70°F				
PYRANOL 1476		A to 70°F				
PYRENE		A to 70°F				
PYRETHRUM	AB to 70°F					
PYREX		A to 70°F				
PYRIDINE	C at 70°F	A to 300°F	A to 212°F	B/NR at 70°F	B at 70°F	NR at 70°F
PYRIDINE OIL		A to 70°F		B at 70°F		
PYRITE		A to 70°F				
PYROACETIC ETHER		A to 70°F				
PYROACETIC SPIRIT		A to 70°F				
PYROGALLIC ACID (PYROGALLOL)	B at 70°F	A to 200°F	A to 212°F	NR at 70°F		
PYROGALLOLCARBOXYLIC ACID		A to 70°F				

	RECOMMENDATIONS	NITRILE NBR	ETHYLENE PROPYLENE EP, EPDM	FLUOROCARBON FKM	CHLOROPRENE CR	HYDROGENATED NITRILE HNBR
PYROGARD C & D (WATER & PETROLEUM)	NBR, FKM	A to 70°F	NR at 70°F	A to 70°F	B at 70°F	A to 70°F
PYROGARD 42, 43, 53 & 55	EP, FKM, IIR	NR at 70°F	A to 70°F	A to 70°F	NR at 70°F	NR at 70°F
PYROLIGNEOUS ACID	EP, IIR, VMQ	NR at 70°F	B at 70°F	NR at 70°F	NR at 70°F	NR at 70°F
PYROLIGNEOUS SPIRIT		A to 70°F	B at 70°F	C at 70°F	A to 70°F	
PYROLUBE	FKM, EP, IIR	NR at 70°F	B at 70°F	A to 70°F	NR at 70°F	NR at 70°F
PYROMUCIC ALDEHYDE		NR at 70°F			NR at 70°F	
PYROXYLIC SPIRIT		A to 70°F	B at 70°F	C at 70°F	A to 70°F	
PYRROLE	SBR, VMQ	NR at 70°F	NR at 70°F	NR at 70°F	NR at 70°F	
QUATERNARY AMMONIUM SALTS		A to 70°F		A to 70°F	A to 70°F	
QUENCHING OIL	NBR	A to 250°F	NR at 70°F	A to 70°F	B/NR at 70°F	
QUICKLIME		A to 70°F	A to 70°F		A to 70°F	
QUICKSILVER		A to 70°F	A to 70°F	A to 70°F	A to 70°F	
QUININE BISULPHATE		A to 70°F dry	A to 70°F dry	A to 70°F dry	A to 70°F dry	
QUININE SULPHATE		A to 70°F dry	A to 70°F dry	A to 70°F dry	A to 70°F dry	
QUINOL		C/NR at 70°F		C at 70°F	NR at 70°F	
RADIATION	EP, VMQ	BC at 70°F	BC at 70°F	NR at 70°F	BC at 70°F	C at 70°F
RANGE OIL		AB to 70°F			AB to 70°F	
RAPE OIL		B at 70°F	A to 70°F	A to 70°F	B at 70°F	
RAPESEED OIL	EP, FKM, IIR, NBR, AU	AB to 250°F	A to 70°F	A to 70°F	B at 70°F	B at 70°F
RAW LINSEED OIL		A to 70°F AB to 180°F	B at 70°F	A to 70°F	AB to 180°F	
RED LINE 100 OIL	NBR, FKM, ACM, AU	A to 70°F	NR at 70°F	A to 70°F	B at 70°F	A to 70°F
RED OIL	NBR, FKM, FVMQ, CR	A to 275°F	BC at 70°F NR at 140°F	A to 212°F B at 248°F	BC at 70°F NR at 140°F	A to 70°F
RED POTASSIUM CHROMATE		AB any conc to 180°F			NR at 70°F	
REFINED LINSEED OIL		AB to 180°F			AB to 180°F	
REFINED OIL		AB to 70°F			AB to 70°F	
RESINS		BC at 70°F	NR at 70°F	A to 70°F	NR at 70°F	
RESORCINOL			AB to 70°F	A to 70°F	NR at 70°F	
RETINOL		A to 70°F AB to 150°F		A to 70°F	A to 70°F	
RHIGOLENE		A to 70°F	NR at 70°F	A to 70°F	B at 70°F	
RHODAN SALTS		C at 70°F	A to 140°F	A to 176°F		
RICHFIELD A (WEED KILLER)		BC 100% at 70°F	NR 100% at 70°F	C 100% at 70°F	NR 100% at 70°F	
RICHFIELD B (WEED KILLER)		B 33% at 70°F	NR 33% at 70°F	C 33% at 70°F	B 33% at 70°F	
RICINUS OIL (RICININE)		A to 70°F AB to 150°F	B at 70°F	A to 70°F	A to 70°F AB to 120°F	
RJ-1 (MIL-F-25558)	NBR, FKM, ACM, FVMQ AU	A to 70°F	NR at 70°F	A to 70°F	B at 70°F	A to 70°F
ROCK SALT		AB any conc to 160°F	A to 70°F	A to 70°F	A to 70°F AB any conc to 200°F	
ROMAN VITRIOL		A to 70°F	A to 70°F	A to 70°F	A to 70°F	
ROSE OIL				A to 70°F	C at 70°F	
ROSIN OIL		A to 70°F		A to 70°F	A to 70°F	
ROSIN, PAPER MILL	NBR	A to 200°F	A to 70°F	A to 70°F	A to 70°F	
ROSINS	NBR	A to 200°F NR molten	A/NR at 70°F	A to molten	A to 70°F	
ROTENONE & WATER		A to 70°F	A to 70°F	A to 70°F	A to 70°F	
RP-1 (MIL-R-25576)	NBR, FKM, ACM FVMQ, AU	A to 70°F	NR at 70°F	A to 70°F	BC at 70°F	A to 70°F
RUM		A to 140°F	A to 70°F	A to 140°F	A to 70°F	
RUST INHIBITORS		A to 140°F		A to 140°F	C to 140°F	

	STYRENE BUTADIENE SBR	POLYACRYLATE ACM	POLYURETHANE AU, EU	ISOBUTYLENE ISOPRENE IIR	POLYBUTADIENE BR	® AEROQUIP AQP
PYROGARD C & D (WATER & PETROLEUM)	NR at 70°F	NR at 70°F	NR at 70°F	NR at 70°F	NR at 70°F	AB to 200°F
PYROGARD 42, 43, 53 & 55	NR at 70°F	NR at 70°F	NR at 70°F	A to 70°F	NR at 70°F	AB to 200°F
PYROLIGNEOUS ACID	NR at 70°F	NR at 70°F	NR at 70°F	B any conc to 100°F	NR at 70°F	
PYROLIGNEOUS SPIRIT			NR at 70°F	A to 70°F		
PYROLUBE	NR at 70°F	NR at 70°F	NR at 70°F	B at 70°F	NR at 70°F	
PYROMUCIC ALDEHYDE				A to 70°F AB to 185°F		
PYROXYLIC SPIRIT			NR at 70°F	A to 70°F		
PYRROLE	BC at 70°F	NR at 70°F		NR at 70°F	BC at 70°F	
QUATERNARY AMMONIUM SALTS				A to 70°F		
QUENCHING OIL	NR at 70°F	A to 70°F	A to 70°F			
QUICKLIME				A to 70°F		
QUICKSILVER			A to 70°F	A to 70°F		
QUININE BISULPHATE			A to 70°F dry			
QUININE SULPHATE			A to 70°F dry			
QUINOL				NR at 70°F		
RADIATION	C at 70°F	BC at 70°F	AC to 70°F	NR at 70°F	BC at 70°F	
RANGE OIL						
RAPE OIL			B at 70°F	A to 70°F		
RAPESEED OIL	NR at 70°F	B at 70°F	B at 70°F	A to 70°F AB to 150°F	NR at 70°F	
RAW LINSEED OIL			B at 70°F	B at 70°F		
RED LINE 100 OIL	NR at 70°F	A to 70°F	A to 70°F	NR at 70°F	NR at 70°F	
RED OIL	NR at 70°F	AB to 70°F	B at 70°F AB to 70°F fumes	NR at 70°F	NR at 70°F	NR at 70°F
RED POTASSIUM CHROMATE				A to 70°F AB any conc to 150°F		
REFINED LINSEED OIL	NR at 70°F			AB to 150°F	NR at 70°F	
REFINED OIL						
RESINS			A to 70°F			
RESORCINOL	AB to 70°F		NR at 70°F			
RETINOL				NR at 70°F		
RHIGOLENE			NR at 70°F	NR at 70°F		
RHODAN SALTS						
RICHFIELD A (WEED KILLER)				NR 100% at 70°F		
RICHFIELD B (WEED KILLER)				B 33% at 70°F		
RICINUS OIL (RICININE)			A to 70°F	A to 70°F AB to 150°F		
RJ-1 (MIL-F-25558)	NR at 70°F	A to 70°F	AB to 70°F	NR at 70°F	NR at 70°F	
ROCK SALT			A to 70°F	A to 70°F AB any conc to 150°F		
ROMAN VITRIOL			A to 70°F	A to 70°F		
ROSE OIL			A to 70°F			
ROSIN OIL				NR at 70°F		
ROSIN, PAPER MILL	A to 70°F	NR at 70°F	NR at 70°F			
ROSINS	A to 70°F	NR at 70°F	A/NR at 70°F			
ROTENONE & WATER						
RP-1 (MIL-R-25576)	NR at 70°F	A to 70°F	AC to 70°F	NR at 70°F	NR at 70°F	
RUM			NR at 70°F	A to 70°F		
RUST INHIBITORS			A to 70°F			

	SYNTHETIC ISOPRENE IR	NATURAL ISOPRENE NR	CHLOROSULFONATED POLYETHYLENE CSM	FLUOROSILICONE FVMQ	SILICONE VMQ	CHEMRAZ FFKM
PYROGARD C & D (WATER & PETROLEUM)	NR at 70°F	NR at 70°F	NR at 70°F	B at 70°F	B at 70°F	A to 70°F
PYROGARD 42, 43, 53 & 55	NR at 70°F	NR at 70°F	NR at 70°F	NR at 70°F	NR at 70°F	A to 70°F
PYROLIGNEOUS ACID	NR at 70°F	NR at 70°F	B/NR at 70°F	NR at 70°F	A to 70°F	A to 70°F
PYROLIGNEOUS SPIRIT	A to 70°F	A to 70°F	A to 70°F		A to 70°F	
PYROLUBE	NR at 70°F	NR at 70°F	NR at 70°F	B at 70°F	B at 70°F	A to 70°F
PYROMUCIC ALDEHYDE	NR at 70°F	NR at 70°F				
PYROXYLIC SPIRIT	A to 70°F	A to 70°F	A to 70°F		A to 70°F	
PYRROLE	BC at 70°F	BC at 70°F	NR at 70°F	NR at 70°F	B at 70°F	
QUATERNARY AMMONIUM SALTS	A to 70°F	A to 70°F	A to 70°F			
QUENCHING OIL			AB to 70°F	A to 70°F	NR at 70°F	
QUICKLIME	A to 70°F	A to 70°F	A to 70°F			
QUICKSILVER	A to 70°F	A to 70°F	A to 70°F		A to 70°F	
QUININE BISULPHATE						A to 70°F
QUININE SULPHATE						A to 70°F
QUINOL	B at 70°F	B any conc to 80°F	A to 70°F			
RADIATION	BC at 70°F	BC at 70°F	BC at 70°F	NR at 70°F	AB to 77°F 55 megarads AB to 390°F 8 megarads	B at 70°F
RANGE OIL			AB to 70°F			
RAPE OIL	NR at 70°F	NR at 70°F	B at 70°F		NR at 70°F	
RAPESEED OIL	NR at 70°F	NR at 70°F	B/NR at 70°F	A to 70°F	NR at 70°F	A to 70°F
RAW LINSEED OIL	NR at 70°F	NR at 70°F	B at 70°F		NR at 70°F	
RED LINE 100 OIL	NR at 70°F	NR at 70°F	B at 70°F	A to 70°F	NR at 70°F	A to 70°F
RED OIL	NR at 70°F	NR at 70°F	B at 70°F	A to 400°F	NR at 70°F	A to 70°F
RED POTASSIUM CHROMATE	NR at 70°F	NR at 70°F				
REFINED LINSEED OIL	NR at 70°F	NR at 70°F				
REFINED OIL			AB to 70°F			
RESINS						A to 70°F
RESORCINOL						
RETINOL	NR at 70°F	NR at 70°F	B at 70°F			
RHIGOLENE	NR at 70°F	NR at 70°F	B at 70°F			
RHODAN SALTS						
RICHFIELD A (WEED KILLER)	NR 100% at 70°F	NR 100% at 70°F	NR 100% at 70°F			
RICHFIELD B (WEED KILLER)	NR 33% at 70°F	NR 33% at 70°F				
RICINUS OIL (RICININE)	B at 70°F	B to 80°F	A to 70°F		A to 70°F	
RJ-1 (MIL-F-25558)	NR at 70°F	NR at 70°F	B at 70°F	A to 70°F	NR at 70°F	A to 70°F
ROCK SALT	A to 70°F	A to 70°F AB any conc to 150°F	A to 70°F		A to 70°F	
ROMAN VITRIOL	C at 70°F	C at 70°F	A to 70°F		A to 70°F	
ROSE OIL		C at 70°F				
ROSIN OIL			B at 70°F			
ROSIN, PAPER MILL				A to 70°F	A to 70°F	
ROSINS			B at 70°F	A to 70°F	A to 70°F	A to 70°F
ROTENONE & WATER	A to 70°F	A to 70°F	AB to 70°F			
RP-1 (MIL-R-25576)	NR at 70°F	NR at 70°F	BC at 70°F	A to 70°F	NR at 70°F	A to 70°F
RUM	A to 70°F	A to 70°F	A to 70°F		A to 70°F	
RUST INHIBITORS		A/NR at 70°F				

C 316

	POLYSULFIDE T	CHLORINATED POLYETHYLENE CM	EPICHLOROHYDRIN CO, ECO	® HYTREL COPOLYESTER TPE	® SANTOPRENE COPOLYMER TPO	® C-FLEX STYRENIC TPE
PYROGARD C & D (WATER & PETROLEUM)				AB to 150°F		
PYROGARD 42, 43, 53 & 55	NR at 70°F			AB to 200°F		
PYROLIGNEOUS ACID	B at 70°F					
PYROLIGNEOUS SPIRIT						
PYROLUBE						
PYROMUCIC ALDEHYDE						
PYROXYLIC SPIRIT						
PYRROLE	NR at 70°F					
QUATERNARY AMMONIUM SALTS						
QUENCHING OIL	AB to 70°F					
QUICKLIME						
QUICKSILVER						
QUININE BISULPHATE						
QUININE SULPHATE						
QUINOL						
RADIATION	NR at 70°F					
RANGE OIL	AB to 70°F					
RAPE OIL						
RAPESEED OIL	NR at 70°F		A to 70°F			
RAW LINSEED OIL						
RED LINE 100 OIL	A to 70°F					
RED OIL	A to 70°F	NR at 70°F	A to 70°F	A to 70°F		A to 70°F
RED POTASSIUM CHROMATE						
REFINED LINSEED OIL						NR at 70°F
REFINED OIL						
RESINS						
RESORCINOL				NR at 70°F	NR at 70°F	
RETINOL						
RHIGOLENE						
RHODAN SALTS						
RICHFIELD A (WEED KILLER)	AB 100% to 70°F					
RICHFIELD B (WEED KILLER)	AB 33% to 70°F					
RICINUS OIL (RICININE)						
RJ-1 (MIL-F-25558)	A to 70°F		A to 70°F			
ROCK SALT						
ROMAN VITRIOL						
ROSE OIL						
ROSIN OIL						
ROSIN, PAPER MILL						
ROSINS						
ROTENONE & WATER						
RP-1 (MIL-R-25576)	A to 70°F		A to 70°F			
RUM						
RUST INHIBITORS						

C 317

	ETHYLENE ACRYLIC EA	POLYALLOMER LINEAR COPOLYMER	NYLON 11 POLYAMIDE	NYLON 12 POLYAMIDE	ETHYLENE VINYL ACETATE EVA	POLYVINYLCHLORIDE FLEXIBLE PVC
PYROGARD C & D (WATER & PETROLEUM)			A to 70°F AB to 180°F			
PYROGARD 42, 43, 53 & 55						
PYROLIGNEOUS ACID					C/NR at 70°F	C at 70°F
PYROLIGNEOUS SPIRIT						
PYROLUBE	NR at 70°F		A to 70°F AB to 180°F			
PYROMUCIC ALDEHYDE						NR at 70°F
PYROXYLIC SPIRIT						
PYRROLE						
QUATERNARY AMMONIUM SALTS						
QUENCHING OIL					NR at 70°F	B at 70°F
QUICKLIME						
QUICKSILVER						
QUININE BISULPHATE						
QUININE SULPHATE						
QUINOL						
RADIATION	A to 70°F				NR at 70°F	B at 70°F
RANGE OIL						
RAPE OIL						
RAPESEED OIL						
RAW LINSEED OIL						
RED LINE 100 OIL						
RED OIL		A to 70°F	A to 70°F C at 194°F	AB to 70°F	B/NR at 70°F	C/NR at 70°F
RED POTASSIUM CHROMATE						
REFINED LINSEED OIL						
REFINED OIL						
RESINS						
RESORCINOL		A to sat'd to 122°F	NR at 70°F	NR at 70°F		C at 70°F NR 5% at 122°F
RETINOL						
RHIGOLENE					NR at 70°F	NR at 70°F
RHODAN SALTS						
RICHFIELD A (WEED KILLER)						A 100% to 70°F C 100% at 150°F
RICHFIELD B (WEED KILLER)						
RICINUS OIL (RICININE)						
RJ-1 (MIL-F-25558)						
ROCK SALT						
ROMAN VITRIOL						
ROSE OIL						
ROSIN OIL		A to 70°F	A to 70°F		A to 70°F	
ROSIN, PAPER MILL						
ROSINS					NR at 70°F	C/NR at 70°F
ROTENONE & WATER						
RP-1 (MIL-R-25576)						
RUM				AB to 70°F		
RUST INHIBITORS						

	POLYETHYLENE LOW DENSITY LDPE	®TEFLON FEP	®KALREZ PERFLUORINATED ELASTOMER	®FLUORAZ FLUORINATED COPOLYMER	®AFLAS FLUORINATED COPOLYMER	®NORPRENE COPOLYMER TPO
PYROGARD C & D (WATER & PETROLEUM)		A to 70°F AB to 250°F				
PYROGARD 42, 43, 53 & 55		A to 70°F AB to 400°F		B at 70°F		
PYROLIGNEOUS ACID	B at 70°F	A to boiling	A to 212°F	NR at 70°F		
PYROLIGNEOUS SPIRIT		A to 70°F				
PYROLUBE				A to 70°F		
PYROMUCIC ALDEHYDE		A to 70°F				
PYROXYLIC SPIRIT		A to 70°F				
PYRROLE		A to 70°F	A to 212°F			
QUATERNARY AMMONIUM SALTS		A to 70°F				
QUENCHING OIL	A to 70°F	A to 70°F		A to 70°F		
QUICKLIME		A to 70°F				
QUICKSILVER		A to 70°F				
QUININE BISULPHATE			A to 212°F			
QUININE SULPHATE			A to 212°F			
QUINOL		A to 70°F				
RADIATION		A to 70°F	B at 70°F	A to 70°F	A to 70°F	
RANGE OIL		A to 70°F				
RAPE OIL		A to 70°F				
RAPESEED OIL	A to 70°F	A to 70°F	A to 212°F	A to 70°F		
RAW LINSEED OIL		A to 70°F		A to 70°F		
RED LINE 100 OIL				A to 70°F		
RED OIL	B/NR at 70°F	A to 300°F	A to 212°F	A to 70°F	+12% vol 3 days 212°F	BC at 70°F
RED POTASSIUM CHROMATE		A to 70°F				
REFINED LINSEED OIL						
REFINED OIL		A to 70°F				
RESINS						
RESORCINOL	A to sat'd to 122°F	A to sat'd to 125°F	A to 212°F			
RETINOL		A to 70°F				
RHIGOLENE		A to 70°F				
RHODAN SALTS		A to 70°F				
RICHFIELD A (WEED KILLER)		A 100% to 70°F				
RICHFIELD B (WEED KILLER)		A 33% to 70°F				
RICINUS OIL (RICININE)		A to 70°F				
RJ-1 (MIL-F-25558)			A to 212°F	A to 70°F		
ROCK SALT		A to 70°F				
ROMAN VITRIOL		A to 70°F				
ROSE OIL						
ROSIN OIL		A to 70°F		A to 70°F		
ROSIN, PAPER MILL						
ROSINS	A to 70°F	A to 70°F	A to 212°F	A to 100°F		
ROTENONE & WATER		A to 70°F				
RP-1 (MIL-R-25576)			A to 212°F	A to 70°F		
RUM		A to 70°F				
RUST INHIBITORS					A to 70°F	

C 319

	RECOMMENDATIONS	NITRILE NBR	ETHYLENE PROPYLENE EP, EPDM	FLUOROCARBON FKM	CHLOROPRENE CR	HYDROGENATED NITRIL HNBR
SACCHAROSE	NBR,EP,FKM,SBR	A to 100% to 140°F	A to 250°F	A to 140°F	A to 70°F AB any conc to 120°F	
SACCHARUM AMYLACEUM		B at 70°F	A to 70°F	A to 70°F	B at 70°F	
SACCHARUM SOLUTIONS		A to 70°F AB any conc to 120°F	A to 70°F	A to 70°F	A to 70°F AB any conc to 120°F	
SAE NO. 10 OIL	NBR,FKM,ACM	A to 250°F	NR at 70°F	A to 70°F	B/NR at 70°F	
SAE NO. 10-30 OIL	NBR,FKM,ACM	A to 70°F	NR at 70°F	A to 70°F	BC at 70°F	
SAL AMMONIAC	EP,NBR,FKM,CR,IIR	A to 160°F AB any conc to 200°F	A to 176°F B at 212°F	A to 100% to 212°F	A to 100% to 140°F B 176–200°F	A to 70°F
SAL CHALYBDIS		A to 70°F AB any conc to 200°F	A to 70°F	A to 70°F	A to 70°F AB any conc to 200°F	
SAL SODA	NBR,EP,FKM,CR	A to 100% to 160°F AB any conc to 200°F	A to 100% to 176°F B at 212°F	A to 100% to 212°F	A to 100% to 176°F AB any conc to 200°F	A to 70°F
SAL TARTAN		A to 70°F	A to 70°F	A to 70°F	A to 70°F	
SAL VOLATILE		NR at 70°F	A to 70°F	A to 70°F	B at 70°F	
SALICYLALDEHYDE		A to 70°F	A to 70°F	A to 70°F		
SALICYLIC ACID	EP,FKM,NBR,IIR	AB to 70°F	A to 200°F	A to 70°F	C/NR at 70°F	B at 70°F
SALIMIAK (SALMIAC)		A to 70°F	A to 70°F	A to 70°F	A to 70°F	
SALITER (SALITRE)		C at 70°F	A to 70°F	A to 70°F	B at 70°F	
SALT	NBR,EP,FKM,CR,IIR	−0.8% vol 3 days 212°F A to 140°F	A to 100% to 176°F AB to 200°F	A to 100% to 212°F	+0.4% vol 3 days 212°F A to 140°F	
SALT SOLUTION	NBR	−1% vol 3 days 212°F A to 200°F	+1% vol 3 days 70°F A to 200°F	no swell 3 days 212°F A to 200°F	no swell 3 days 212°F A to 200°F	
SALT CAKE		A to 70°F AB any conc to 200°F	A to 70°F	A to 70°F	A to 70°F AB any conc to 200°F	
SALT OF LEMERY		A to 70°F AB any conc to 200°F	A to 70°F	A to 70°F	A to 70°F AB any conc to 200°F	
SALT OF TARTER		A to 70°F AB any conc to 200°F	A to 70°F	A to 70°F	A to 70°F AB any conc to 200°F	
SALT OF VITRIOL		A to 70°F	A to 70°F	A to 70°F	A to 70°F	
SALT WATER	EP,NBR,FVMQ,SBR	A to 140°F AB to 200°F	A to 176°F AB to 200°F	A to 176°F AB to 200°F	A to 140°F AB to 200°F	A to 70°F
SALTPETER-AMMONIUM NITRATE	NBR	A to 200°F	A to 176°F AB to 200°F	A to 176°F	A to 200°F	A to 70°F
SALTPETER-CALCIUM NITRATE	NBR,EP,CR,FKM	A to 176°F AB to 200°F	A to 176°F AB to 200°F	A to 212°F	A to 70°F AB to 200°F	A to 70°F
SALTPETER-POTASSIUM NITRATE		A to 70°F AB any conc to 200°F	A to 176°F	A to 100% to 212°F	A to 70°F AB any conc to 200°F	A to 70°F
SALTPETER-SODIUM NITRATE	EP,IIR,NBR,CR,NR	AB any conc to 150°F B at 176°F	A to 176°F	A to 212°F	B to 200°F	A to 70°F
SAND ACID (FLUOROSILICIC ACID)	EP,FKM,NBR,CR	AB 100% to 70°F AB 50% to 80°F	A to 100% to 120°F A to 20% to 140°F	A to 100% to 120°F A to 20% to 140°F	B 100% at 70°F AB 50% to 80°F	A to 70°F
SANITIZER 160 (MONSANTO)			A to 70°F	B at 70°F		
SANITIZERS (CHLORINATED TSP)		A to 70°F		A to 70°F	A to 70°F	
SANITIZERS (KLENZADE)		A to 70°F		A to 70°F	A to 70°F	
SANTOSAFE 300	FKM	NR at 70°F	C at 70°F	A to 212°F	NR at 70°F	NR at 70°F
SAUERKRAUT						
SEA SALT		A to 70°F AB any conc to 180°F	A to 70°F	A to 70°F	A to 70°F AB any conc to 200°F	A to 70°F
SEA WATER	NBR,EP,SBR	A to 140°F B to 200°F	A to 180°F AB to 200°F	A to 212°F	AB to 140°F BC to 200°F	A to 70°F
SEED OIL		A to 70°F AB to 150°F	A to 70°F	A to 70°F	A to 70°F AB to 150°F	
SEPARAN NP-10						
SESAME SEED OIL		A to 70°F		A to 70°F	C at 70°F	
SEWAGE	NBR,EP,FKM,CR,SBR IIR	A to 200°F	A to 140°F	A to 176°F	AB to 180°F	A to 70°F
SEWERAGE	EP	A to 70°F	A to 200°F	A to 70°F	AC to 70°F	
SF96 (GE SILICONE OIL)	EP,FKM,CR,SBR	B at 70°F	A to 70°F	A to 70°F	A to 70°F	
SF1147 (GE SILICONE FLUID)	FKM	B at 70°F	C at 70°F	A to 70°F		
SF1153 & SF1154 (GE SILICONE FLUID)	EP,FKM,CR,SBR	B at 70°F	A to 70°F	A to 70°F	A to 70°F	
SHALE NAPHTHA		A to 70°F	NR at 70°F	A to 70°F	B at 70°F	
SHELL ALVANIA GREASE NO. 2	NBR,FKM,ACM,AU	A to 70°F	NR at 70°F	A to 70°F	B at 70°F	A to 70°F
SHELL CARNEA 19 & 29	NBR,FKM,ACM	A to 70°F	NR at 70°F	A to 70°F	NR at 70°F	A to 70°F

	STYRENE BUTADIENE SBR	POLYACRYLATE ACM	POLYURETHANE AU, EU	ISOBUTYLENE ISOPRENE IIR	POLYBUTADIENE BR	® AEROQUIP AQP
SACCHAROSE	A to 70°F	NR at 70°F	A/NR at 70°F	A to 70°F AB any conc to 120°F	A to 70°F	
SACCHARUM AMYLACEUM			A to 70°F	A to 70°F		
SACCHARUM SOLUTIONS			NR at 70°F	A to 70°F AB any conc to 120°F		
SAE NO. 10 OIL	NR at 70°F	A to 70°F	A to 70°F AB to 158°F	NR at 70°F	NR at 70°F	A to 70°F
SAE NO. 10-30 OIL	NR at 70°F	A to 70°F	A to 70°F			
SAL AMMONIAC	A to 70°F	A to 70°F	AB to 70°F	A to 70°F AB any conc to 185°F	A to 70°F	AB to 70°F
SAL CHALYBDIS				A to 70°F AB any conc to 185°F		
SAL SODA				A to 70°F AB any conc to 185°F		
SAL TARTAN				A to 70°F		
SAL VOLATILE				A to 70°F		
SALICYLALDEHYDE						
SALICYLIC ACID	B at 70°F			A any conc to 80°F	B at 70°F	
SALIMIAK (SALMIAC)			A to 70°F	A to 70°F		
SALITER (SALITRE)				A to 70°F		
SALT	+0.1% vol 3 days 212°F A to 70°F	+2% vol 3 days 212°F A to 70°F	A to 100% to 70°F	-0.1% vol 3 days 212°F AB any conc to 185°F	A to 70°F	AB to 70°F
SALT SOLUTION	no swell 3 days 212°F A to 70°F	+2% vol 3 days 212°F A/NR at 70°F	+1% vol 3 days 212°F A/NR at 70°F	no swell 3 days 212°F A any conc to 125°F	A to 70°F	
SALT CAKE			A to 70°F	A to 70°F AB any conc to 185°F		
SALT OF LEMERY			A to 70°F	A to 70°F AB any conc to 185°F		
SALT OF TARTER				A to 70°F AB any conc to 185°F		
SALT OF VITRIOL				A to 70°F		
SALT WATER	A to 70°F	A/NR at 70°F	A/NR at 70°F	A to 70°F AB to 185°F	A to 70°F	
SALTPETER-AMMONIUM NITRATE	A to 70°F	AB to 70°F	B/NR at 70°F	A to 185°F	A to 70°F	AB to 70°F
SALTPETER-CALCIUM NITRATE	A to 70°F	A to 70°F	A/NR at 70°F	A to 70°F AB to 185°F	A to 70°F	
SALTPETER-POTASSIUM NITRATE	A to 70°F	A to 70°F	A to 100% to 70°F	A to 70°F AB any conc to 185°F	A to 70°F	
SALTPETER-SODIUM NITRATE	B to 100% at 70°F		B at 70°F	A to 70°F AB any conc to 185°F	A to 70°F	AB to 70°F
SAND ACID (FLUOROSILICIC ACID)			B at 70°F	A/NR at 70°F		
SANITIZER 160 (MONSANTO)						
SANITIZERS (CHLORINATED TSP)						
SANITIZERS (KLENZADE)						
SANTOSAFE 300	NR at 70°F	NR at 70°F		C at 70°F	NR at 70°F	AB to 250°F W/G15, 20 & 30
SAUERKRAUT			-40% T.S. at 70°F A to 70°F	A to 70°F AB any conc to 185°F		
SEA SALT						
SEA WATER	A to 70°F	A/NR at 70°F	A/NR at 70°F	A to 70°F AB to 120°F	A to 70°F	
SEED OIL			A to 70°F	B to 80°F		
SEPARAN NP-10						
SESAME SEED OIL						
SEWAGE	AB to 70°F	NR at 70°F	NR at 70°F	AB to 150°F	AB to 70°F	A to 70°F
SEWERAGE			NR at 70°F	B at 70°F		
SF96 (GE SILICONE OIL)	A to 70°F	A to 70°F	A to 70°F	A to 70°F	A to 70°F	
SF1147 (GE SILICONE FLUID)				C at 70°F		
SF1153 & SF1154 (GE SILICONE FLUID)	A to 70°F	A to 70°F	A to 70°F	A to 70°F	A to 70°F	
SHALE NAPHTHA				NR at 70°F		
SHELL ALVANIA GREASE NO. 2	NR at 70°F	A to 70°F	A to 70°F	NR at 70°F	NR at 70°F	
SHELL CARNEA 19 & 29	NR at 70°F	A to 70°F	B at 70°F	NR at 70°F	NR at 70°F	

	SYNTHETIC ISOPRENE	NATURAL ISOPRENE	CHLOROSULFONATED POLYETHYLENE	FLUOROSILICONE	SILICONE	CHEMRAZ
	IR	NR	CSM	FVMQ	VMQ	FFKM
SACCHAROSE	A to 70°F	A to 70°F AB any conc to 120°F	AB to 70°F	A to 70°F	A to 70°F	A to 70°F
SACCHARUM AMYLACEUM	B at 70°F	B at 70°F	A to 70°F		A to 70°F	
SACCHARUM SOLUTIONS	A to 70°F	A to 70°F AB any conc to 120°F	A to 70°F		A to 70°F	
SAE NO. 10 OIL	NR to 70°F	NR to 70°F	C/NR at 70°F	A to 70°F	A/NR to 70°F	
SAE NO. 10-30 OIL						
SAL AMMONIAC	A to 70°F	A to 70°F AB any conc to 150°F	A to 70°F	AC to 70°F	BC at 70°F	A to 70°F
SAL CHALYBDIS	A to 70°F	A to 70°F AB any conc to 150°F	A to 70°F			
SAL SODA	A to 100% to 70°F	A to 100% to 70°F AB any conc to 185°F	A to 100% to 70°F AB to 200°F	A to 100% to 70°F	A to 100% to 70°F	A to 70°F
SAL TARTAN	A to 70°F	A to 70°F	A to 70°F			
SAL VOLATILE	A to 70°F	A to 70°F	B at 70°F		B at 70°F	
SALICYLALDEHYDE						
SALICYLIC ACID	A to 70°F	A any conc to 80°F	A to 70°F	A to 70°F		A to 70°F
SALIMIAK (SALMIAC)	A to 70°F	A to 70°F	A to 70°F		B at 70°F	
SALITER (SALITRE)	B at 70°F	B at 70°F	A to 70°F		C at 70°F	
SALT	A to 70°F	A to 70°F AB any conc to 150°F	A to 70°F AB to 200°F	A to 70°F	no swell 3 days 212°F A to 70°F	A to 70°F
SALT SOLUTION	A to 70°F	A to 70°F	A to 70°F	+1% vol 3 days 212°F A to 70°F	no swell 3 days 212°F A to 70°F	
SALT CAKE	A to 70°F	A to 70°F AB any conc to 150°F	A to 70°F		A to 70°F	
SALT OF LEMERY	A to 70°F	A to 70°F AB any conc to 150°F	A to 70°F		A to 70°F	
SALT OF TARTER	A to 70°F	A to 70°F AB any conc to 150°F	A to 70°F			
SALT OF VITRIOL	A to 70°F	A to 70°F	A to 70°F		A to 70°F	
SALT WATER	A to 70°F	A to 70°F AB to 150°F	A to 70°F AB to 200°F	A to 70°F	A to 70°F	A to 70°F B at 70°F white
SALTPETER-AMMONIUM NITRATE	A to 70°F	A to 150°F	A to 70°F AB to 200°F	BC at 70°F	BC at 70°F	A to 70°F
SALTPETER-CALCIUM NITRATE	A to 70°F	A to 70°F AB to 150°F	A to 70°F	A to 70°F	AB to 70°F	A to 70°F
SALTPETER-POTASSIUM NITRATE	A to 70°F	A to 70°F AB any conc to 150°F	A to 70°F AB to 185°F	A to 70°F	A to 70°F	A to 70°F
SALTPETER-SODIUM NITRATE	B at 70°F	B any conc to 150°F	A to 70°F AB to 200°F	A to 70°F	C/NR at 70°F	A to 70°F
SAND ACID (FLUOROSILICIC ACID)	A to 70°F	A to 80°F AB 10% to 100°F	A to 250°F	A/NR at 70°F	A/NR at 70°F	A to 70°F
SANITIZER 160 (MONSANTO)	NR at 70°F	NR at 70°F				
SANITIZERS (CHLORINATED TSP)						
SANITIZERS (KLENZADE)						
SANTOSAFE 300	NR at 70°F	NR at 70°F		A to 70°F	A to 70°F	A to 70°F
SAUERKRAUT						
SEA SALT	A to 70°F	A to 70°F AB to 150°F	A to 70°F		A to 70°F	A to 70°F
SEA WATER	A to 70°F	A to 70°F NR at 200°F	A to 70°F AB to 200°F	A to 70°F	A to 70°F B at 180°F	A to 70°F B at 70°F white
SEED OIL	NR at 70°F	NR at 70°F	B at 70°F		A to 70°F	
SEPARAN NP-10	A to 70°F	A to 70°F AB to 150°F				
SESAME SEED OIL		NR at 70°F				
SEWAGE	AB to 70°F	AB to 70°F	A to 70°F	A to 70°F	AB to 70°F	A to 70°F B at 70°F white
SEWERAGE	C at 70°F	C/NR at 70°F	A to 70°F		A to 70°F	
SF96 (GE SILICONE OIL)	A to 70°F	A to 70°F	A to 70°F	A to 70°F	NR at 70°F	A to 70°F
SF1147 (GE SILICONE FLUID)					NR at 70°F	A to 70°F
SF1153 & SF1154 (GE SILICONE FLUID)	A to 70°F	A to 70°F	A to 70°F	A to 70°F	NR at 70°F	A to 70°F
SHALE NAPHTHA	NR at 70°F	NR at 70°F	NR at 70°F		A to 70°F	
SHELL ALVANIA GREASE NO. 2	NR at 70°F	NR at 70°F	NR at 70°F	A to 70°F	B at 70°F	A to 70°F
SHELL CARNEA 19 & 29	NR at 70°F	NR at 70°F	NR at 70°F	A to 70°F		A to 70°F

	POLYSULFIDE T	CHLORINATED POLYETHYLENE CM	EPICHLOROHYDRIN CO, ECO	® HYTREL COPOLYESTER TPE	® SANTOPRENE COPOLYMER TPO	® C-FLEX STYRENIC TPE
SACCHAROSE	NR at 70°F			*AB to 140°F		
SACCHARUM AMYLACEUM						
SACCHARUM SOLUTIONS						
SAE NO. 10 OIL	C at 70°F	AB to 70°F	A to 70°F	A to 70°F	NR at 70°F	
SAE NO. 10-30 OIL						
SAL AMMONIAC	A to 70°F	AB to 70°F	A to 70°F	A to 70°F *AB to 140°F		AB to 70°F
SAL CHALYBDIS						
SAL SODA						
SAL TARTAN						
SAL VOLATILE						
SALICYLALDEHYDE						
SALICYLIC ACID						
SALIMIAK (SALMIAC)						
SALITER (SALITRE)						
SALT	−0.8% vol 3 days 212°F C at 70°F	AB to 70°F	A to 70°F	A to conc to 70°F	no swell 15% conc 7 days 70°F	A to 70°F
SALT SOLUTION	+1.2% vol 3 days 212°F C/NR at 70°F			*A to 70°F	A to 70°F	
SALT CAKE						
SALT OF LEMERY						
SALT OF TARTER						
SALT OF VITRIOL						
SALT WATER	C/NR at 70°F			A to 70°F	no swell 15% conc 7 days 70°F	
SALTPETER-AMMONIUM NITRATE	AB to 70°F	AB to 70°F	A to 70°F	AB to 70°F		AB to 70°F
SALTPETER-CALCIUM NITRATE	A to 70°F		A to 70°F			
SALTPETER-POTASSIUM NITRATE	A to 70°F		A to 70°F	AB to 70°F		
SALTPETER-SODIUM NITRATE	AB to 70°F	AB to 70°F	A to 70°F	AB to 70°F		A to 70°F
SAND ACID (FLUOROSILICIC ACID)						
SANITIZER 160 (MONSANTO)						
SANITIZERS (CHLORINATED TSP)						
SANITIZERS (KLENZADE)						
SANTOSAFE 300				*AB to 150°F W/G 15, 20 & 30		
SAUERKRAUT						
SEA SALT						
SEA WATER	NR at 70°F			A to 70°F	A to 70°F	
SEED OIL						
SEPARAN NP-10						
SESAME SEED OIL						
SEWAGE	NR at 70°F			AB to 70°F		
SEWERAGE						
SF96 (GE SILICONE OIL)	A to 70°F					
SF1147 (GE SILICONE FLUID)						
SF1153 & SF1154 (GE SILICONE FLUID)	A to 70°F					
SHALE NAPHTHA						
SHELL ALVANIA GREASE NO. 2	A to 70°F					
SHELL CARNEA 19 & 29						

	ETHYLENE ACRYLIC EA	POLYALLOMER LINEAR COPOLYMER	NYLON 11 POLYAMIDE	NYLON 12 POLYAMIDE	ETHYLENE VINYL ACETATE EVA	POLYVINYLCHLORIDE FLEXIBLE PVC
SACCHAROSE			*AB to 140°F	AB to 70°F	A to 70°F *AB to 140°F	AB to 70°F *AB to 140°F
SACCHARUM AMYLACEUM						
SACCHARUM SOLUTIONS						
SAE NO. 10 OIL	A to 70°F		A to 70°F AB to 140°F		NR at 70°F	AB to 70°F
SAE NO. 10-30 OIL			A to 70°F			AB to 70°F
SAL AMMONIAC	A to 70°F	A to 70°F	A/NR at 70°F		AB to 140°F	A to 70°F AB to 140°F
SAL CHALYBDIS						
SAL SODA						
SAL TARTAN						
SAL VOLATILE						
SALICYLALDEHYDE		A to 70°F B at 122°F				C at 70°F NR at 122°F
SALICYLIC ACID		A to 122°F powder or sat'd sol'n		AB to 70°F		B to 70°F C at 122°F
SALIMIAK (SALMIAC)						
SALITER (SALITRE)						
SALT	A to 70°F	A to 70°F	AB to 100% to 194°F	AB to 70°F aqueous		AB to 70°F
SALT SOLUTION		A to 122°F metallic solutions	AB to 70°F	AB to 70°F	A to 70°F	A to 122°F
SALT CAKE					A to 70°F	A to 70°F
SALT OF LEMERY						
SALT OF TARTER						
SALT OF VITRIOL						
SALT WATER		A to 70°F	AB to 70°F		A to 70°F	AB to 70°F *AB to 140°F
SALTPETER-AMMONIUM NITRATE		A to 70°F	AB to 70°F		AB to 70°F	AB to 70°F *AB to 140°F
SALTPETER-CALCIUM NITRATE			A to 70°F			
SALTPETER-POTASSIUM NITRATE	A to 70°F		AB to 70°F NR at 140°F	AB 10% to 70°F	A to 70°F AB to 140°F	A to 70°F *AB to 130°F
SALTPETER-SODIUM NITRATE		A to 70°F	A to 70°F *AB to 140°F	AB 10% to 70°F	A to 70°F AB to 140°F	A to 70°F *AB to 140°F
SAND ACID (FLUOROSILICIC ACID)						
SANITIZER 160 (MONSANTO)						
SANITIZERS (CHLORINATED TSP)						
SANITIZERS (KLENZADE)						
SANTOSAFE 300			AB to 160°F W/G 15,20, & 30			
SAUERKRAUT		A to 70°F	A to 70°F		A to 70°F	A to 70°F
SEA SALT						
SEA WATER	A to 70°F		A to 194°F		A to 70°F AB to 140°F	AB to 70°F *AB to 130°F
SEED OIL						
SEPARAN NP-10						
SESAME SEED OIL						
SEWAGE	A to 70°F		A to 70°F			
SEWERAGE						
SF96 (GE SILICONE OIL)						
SF1147 (GE SILICONE FLUID)						
SF1153 & SF1154 (GE SILICONE FLUID)						
SHALE NAPHTHA						
SHELL ALVANIA GREASE NO. 2						
SHELL CARNEA 19 & 29						

	POLYETHYLENE LOW DENSITY LDPE	®TEFLON FEP	®KALREZ PERFLUORINATED ELASTOMER	®FLUORAZ FLUORINATED COPOLYMER	®AFLAS FLUORINATED COPOLYMER	®NORPRENE COPOLYMER TPO
SACCHAROSE	A to 140°F	A to 200°F	A to 70°F	A to 70°F		
SACCHARUM AMYLACEUM		A to 70°F				
SACCHARUM SOLUTIONS		A to 70°F				
SAE NO. 10 OIL	AB to 70°F	A to 70°F		A to 70°F		
SAE NO. 10-30 OIL	AB to 70°F			A to 70°F		
SAL AMMONIAC	A to 70°F AB to 140°F	A to boiling	A to 212°F	A/NR at 70°F A 2N to 70°F		*A to 70°F
SAL CHALYBDIS		A to 70°F				
SAL SODA	AB to 100% to 140°F	A to 100% to 300°F	A to 70°F	A to 70°F		A to 70°F
SAL TARTAN		A to 70°F				
SAL VOLATILE		A to 70°F				
SALICYLALDEHYDE	A to 70°F B at 122°F	A to 300°F				
SALICYLIC ACID	A to 122°F	A to 300°F	A to 212°F			
SALIMIAK (SALMIAC)		A to 70°F				
SALITER (SALITRE)		A to 70°F				
SALT	AB to 140°F	A to 100% to 300°F	A to 70°F	A to 70°F		A to 70°F
SALT SOLUTION	A to 122°F	A to 122°F	*A to 212°F	A to 70°F		
SALT CAKE	A to 70°F	A to 70°F		A to 70°F		
SALT OF LEMERY		A to 70°F				
SALT OF TARTER		A to 70°F				
SALT OF VITRIOL		A to 70°F				
SALT WATER	A to 122°F AB to 140°F	A to 250°F	*A to 212°F	A to 70°F		
SALTPETER-AMMONIUM NITRATE	A to 70°F	A to 200°F	A to 70°F	A/NR at 70°F		
SALTPETER-CALCIUM NITRATE	A to 70°F	A to 200°F	A to 70°F	A to 70°F		
SALTPETER-POTASSIUM NITRATE	AB to 140°F	A to 100% to 300°F	A to 70°F	A to 100% to 70°F		
SALTPETER-SODIUM NITRATE	A to 70°F AB to 140°F	A to 300°F	A to 70°F	NR at 70°F		AB to 70°F
SAND ACID (FLUOROSILICIC ACID)	A to conc to 70°F AB to 32% at 140°F	A to 100% to 200°F	AB to 70°F	A to 70°F		
SANITIZER 160 (MONSANTO)		A to 70°F				
SANITIZERS (CHLORINATED TSP)						
SANITIZERS (KLENZADE)						
SANTOSAFE 300		A to 70°F AB to 250°F		A to 70°F		
SAUERKRAUT	A to 70°F					
SEA SALT		A to 70°F		A to 70°F		
SEA WATER	A to 70°F AB to 140°F	A to 250°F	A to 70°F	A to 70°F		
SEED OIL		A to 70°F				
SEPARAN NP-10		A to 70°F				
SESAME SEED OIL				A to 70°F		
SEWAGE		A to 70°F	A to 212°F	A to 70°F		
SEWERAGE		A to 70°F				
SF96 (GE SILICONE OIL)					A to 70°F	
SF1147 (GE SILICONE FLUID)					A to 70°F	
SF1153 & SF1154 (GE SILICONE FLUID)					A to 70°F	
SHALE NAPHTHA		A to 70°F				
SHELL ALVANIA GREASE NO. 2				A to 70°F		
SHELL CARNEA 19 & 29				A to 70°F		

C 325

	RECOMMENDATIONS	NITRILE NBR	ETHYLENE PROPYLENE EP, EPDM	FLUOROCARBON FKM	CHLOROPRENE CR	HYDROGENATED NITRILE HNBR
SHELL DD		NR at 70°F		C at 70°F	NR at 70°F	
SHELL DIALA	NBR,FKM,ACM	A to 70°F	NR at 70°F	A to 70°F	B at 70°F	A to 70°F
SHELL IRUS 905	NBR,FKM,ACM,AU	A to 70°F 902 & 905	NR at 70°F	+4.7% vol 28 days 150°F A to 70°F	A to 70°F 902 & 905	A to 70°F
SHELL LO HYDRAX 27 & 29	NBR,FKM,ACM	A to 70°F	NR at 70°F	A to 70°F	B at 70°F	
SHELL MACOME 72	NBR,FKM,ACM	A to 70°F	NR at 70°F	A to 70°F	B at 70°F	
SHELL TELLUS 27 (PETRO BASE)	NBR,FKM,ACM,AU	A to 70°F	NR at 70°F	+1.5% vol 7 days 158°F A to 70°F	B at 70°F	A to 70°F
SHELL TELLUS 33	NBR,FKM,ACM,AU	A to 70°F	NR at 70°F	+1% vol 28 days 158°F A to 70°F	B at 70°F	A to 70°F
SHELL TURBINE OIL 307			*NR at 70°F	BC at 392°F	*AB to 70°F	
SHELL UMF (5% AROMATIC)	NBR,FKM,ACM,AU	A to 70°F	NR at 70°F	A to 70°F	B at 70°F	A to 70°F
SHELL 3XF MINE FLUID	NBR,FKM	A to 70°F	NR at 70°F	A to 300°F	B at 70°F	A to 70°F
SHELLAC	NBR,CR,EP	A to 70°F bleached or orange	A to 200°F	A to 70°F bleached or orange	B/NR at 70°F bleached or orange	
SHERWOOD OIL		A to 70°F	NR at 70°F	A to 70°F	B at 70°F	
SHORTENING		A to 140°F	NR at 70°F	A to 140°F	C at 140°F	
SIEMANS GAS		A to 70°F	C at 70°F	A to 70°F	B at 70°F	
SILANE					AB to 70°F	
SILENT SPIRIT		A to 70°F	A to 70°F	B at 70°F	A to 70°F	
SILICATE ESTERS	FKM,CR,FVMQ,AU	+3% vol 3 days 300°F AB any conc to 180°F	+36% vol 3 days 300°F NR at 70°F	+1% vol 3 days 300°F A to 400°F	+34% vol 3 days 300°F A to 250°F	B at 70°F
SILICATE OF SODA	NBR,EP,IIR,CR	A to 100% to 140°F A soln to 200°F	A to 100% to 176°F AB to 100% to 200°F	A to 100% to 212°F	A to 100% to 140°F AB to 100% to 200°F	A to 70°F
SILICIC ACID		A to 176°F	A to 176°F	A to 212°F	A to 70°F	
SILICOFLUORIC ACID		B 100% at 70°F AB 50% to 80°F	B at 70°F	A to 70°F	B 100% at 70°F AB 50% to 80°F	
SILICONE GREASES	EP,NBR,FKM,IIR,CR FVMQ	A to 140°F AB to 180°F	A to 140°F	A to 140°F	A to 140°F	A to 70°F
SILICONE OILS	EP,NBR,FKM,IIR,CR FVMQ	-4% vol 3 days 300°F A to 140°F	-5% vol 3 days 300°F A to 140°F	no swell 3 days 300°F A to 400°F	-2% vol 3 days 300°F A to 140°F	A to 70°F
SILVER ACETATE				A to 70°F	A to 70°F	
SILVER CYANIDE	NBR,EP,IIR	A to 70°F	A to 140°F	A to 140°F	A to 70°F	
SILVER NITRATE	EP,NBR,FKM,IIR,CR SBR	AB any conc to 180°F	A to 176°F	A to 176°F AB to 200°F	A to 140°F AB any conc to 200°F	B at 70°F
SILVER SALTS	NBR	A to 200°F	A to 70°F	A to 70°F	A to 70°F	
SILVER SULFATE		A to 140°F B at 176°F	A to 176°F	A to 212°F	A to 70°F	
SINCLAIR OPALINE CX-EP LUBE	NBR,FKM,ACM,AU	A to 70°F	NR at 70°F	A to 70°F	B at 70°F	A to 70°F
SPELLY SOLVENT B, C & E	NBR,FKM	A to 70°F	NR at 70°F	A to 70°F	NR at 70°F	
SKYDROL HYDRAULIC FLUID	EP,IIR	NR at 70°F	AB to 70°F	+114% vol 7 days 158°F B/NR at 70°F	NR at 70°F	NR at 70°F
SKYDROL 500	EP,IIR	NR at 70°F	A to 250°F	+262% vol 28 days 158°F NR at 70°F	NR at 70°F	NR at 70°F
SKYDROL 7000	EP,IIR	NR at 70°F	A to 250°F	+29.5% vol 7 days 158°F B at 70°F	NR at 70°F	NR at 70°F
SLAKED LIME	NBR,EP,CR,FKM,SBR	A to 140°F AB any conc to 200°F	A to 176°F B at 212°F	A to 212°F	A to 158°F AB any conc to 200°F	A to 70°F
SLUDGE ACID	EP	A to 70°F	A to 200°F	A to 70°F	AC to 70°F	
SOAP LIQUOR	NBR	A to 200°F	A to 70°F	A to 70°F	A to 70°F	
SOAP OIL		AB to 70°F			AB to 70°F	
SOAP SOLUTIONS	EP,NBR,FKM,SBR,CR	+4% vol 3 days 158°F A to 200°F	+2% vol 3 days 212°F A to 212°F	no swell 3 days 158°F A to 212°F	+6% vol 3 days 158°F AB to 180°F	A to 70°F
SOCONY MOBIL TYPE A	NBR,FKM,ACM	A to 70°F	NR at 70°F	A to 70°F	B at 70°F	A to 70°F
SOCONY PD959B VACUUM	NBR,FKM,ACM,AU	A to 70°F	NR at 70°F	A to 70°F	B at 70°F	A to 70°F
SOCONY VACUUM AMV AC781 GREASE	NBR,FKM,ACM	A to 70°F	NR at 70°F	A to 70°F	B at 70°F	A to 70°F
SODA		A to 70°F AB any conc to 200°F	A to 70°F	A to 70°F	A to 70°F AB any conc to 200°F	
SODA ALUM		A to 70°F AB any conc to 200°F	A to 70°F	A to 70°F	A to 70°F AB any conc to 200°F	
SODA ASH	NBR,EP,FKM,CR,IIR	A to 100% to 160°F AB to 100% to 200°F	A to 100% to 176°F B at 212°F	A to 100% to 212°F	A to 176°F AB any conc to 200°F	A to 70°F
SODA, BAKING	NBR,EP,FKM,CR,IIR	A to 70°F AB to 200°F	A to 100% to 176°F B at 212°F	A to 100% to 212°F	A to 70°F AB to 200°F	A to 70°F

	STYRENE BUTADIENE SBR	POLYACRYLATE ACM	POLYURETHANE AU, EU	ISOBUTYLENE ISOPRENE IIR	POLYBUTADIENE BR	® AEROQUIP AQP
SHELL DD				NR at 70°F		
SHELL DIALA	NR at 70°F	A to 70°F	B at 70°F	NR at 70°F	NR at 70°F	
SHELL IRUS 905	NR at 70°F	A to 70°F	A to 70°F	NR at 70°F	NR at 70°F	
SHELL LO HYDRAX 27 & 29	NR at 70°F	A to 70°F	B at 70°F	NR at 70°F	NR at 70°F	
SHELL MACOME 72	NR at 70°F	A to 70°F	B at 70°F	NR at 70°F	NR at 70°F	
SHELL TELLUS 27 (PETRO BASE)	NR at 70°F	A to 70°F	A to 70°F	NR at 70°F	NR at 70°F	
SHELL TELLUS 33	NR at 70°F	A to 70°F	A to 70°F	NR at 70°F	NR at 70°F	
SHELL TURBINE OIL 307						
SHELL UMF (5% AROMATIC)	NR at 70°F	A to 70°F	A to 70°F	NR at 70°F	NR at 70°F	
SHELL 3XF MINE FLUID	NR at 70°F	NR at 70°F	NR at 70°F	NR at 70°F	NR at 70°F	
SHELLAC	A to 200°F		NR at 70°F bleached or orange			
SHERWOOD OIL			B at 70°F	NR at 70°F		
SHORTENING						
SIEMANS GAS			A to 70°F	B at 70°F		
SILANE				AB to 70°F		
SILENT SPIRIT			NR at 70°F	A to 70°F		
SILICATE ESTERS	dissolved 3 days 300°F NR at 70°F	+8% vol 3 days 300°F NR at 70°F	-6% vol 3 days 300°F A to 70°F	dissolved 3 days 300°F B/NR to 150°F	NR at 70°F	
SILICATE OF SODA				A to 70°F		
SILICIC ACID			AB to 70°F			
SILICOFLUORIC ACID			B at 70°F	A/NR at 70°F		
SILICONE GREASES	A to 70°F	A to 70°F	A to 70°F	A to 70°F	A to 70°F	
SILICONE OILS	-3% vol 3 days 300°F A to 70°F	-3% vol 3 days 300°F A to 70°F	-9% vol 3 days 300°F A to 70°F	-5.3% vol 3 days 300°F A to 70°F	A to 70°F	
SILVER ACETATE						
SILVER CYANIDE	NR at 70°F	NR at 70°F	NR at 70°F			
SILVER NITRATE	A to 70°F	A to 70°F	A to 70°F	A to 70°F AB any conc to 150°F	A to 70°F	
SILVER SALTS	A to 70°F	NR at 70°F	NR at 70°F			
SILVER SULFATE						
SINCLAIR OPALINE CX-EP LUBE	NR at 70°F	A to 70°F	A to 70°F	NR at 70°F	NR at 70°F	
SPELLY SOLVENT B, C & E	NR at 70°F			NR at 70°F	NR at 70°F	
SKYDROL HYDRAULIC FLUID	NR at 70°F	NR at 70°F	NR at 70°F	A to 70°F AB to 130°F	NR at 70°F	AB to 200°F
SKYDROL 500	NR at 70°F	NR at 70°F	NR at 70°F	AB to 70°F	NR at 70°F	AB to 200°F
SKYDROL 7000	NR at 70°F	NR at 70°F	NR at 70°F	A to 70°F	NR at 70°F	
SLAKED LIME			A to 70°F	A to 70°F AB any conc to 185°F		
SLUDGE ACID			NR at 70°F	B at 70°F		
SOAP LIQUOR	A to 70°F	NR at 70°F	C/NR at 70°F			
SOAP OIL						
SOAP SOLUTIONS	+5% vol 3 days 158°F B at 70°F	+33% vol 3 days 158°F NR at 70°F	-1% vol 3 days 158°F A to 70°F	+0.6% vol 3 days 158°F AB to 150°F	AB to 70°F	AB to 70°F
SOCONY MOBIL TYPE A	NR at 70°F	A to 70°F	B at 70°F	NR at 70°F	NR at 70°F	
SOCONY PD959B VACUUM	NR at 70°F	A to 70°F	A to 70°F	NR at 70°F	NR at 70°F	
SOCONY VACUUM AMV AC781 GREASE	NR at 70°F	A to 70°F	B at 70°F	NR at 70°F	NR at 70°F	
SODA				A to 70°F AB any conc to 185°F		
SODA ALUM			A to 70°F	A to 70°F AB any conc to 185°F		
SODA ASH	A to 100% to 70°F	NR at 70°F	AB to 100% to 70°F	A to 70°F AB any conc to 185°F	A to 70°F	AB to 70°F
SODA, BAKING	A to 100% to 70°F	NR sol'n at 70°F	A to 70°F	A to 70°F AB to 185°F	A to 70°F	

	SYNTHETIC ISOPRENE IR	NATURAL ISOPRENE NR	CHLOROSULFONATED POLYETHYLENE CSM	FLUOROSILICONE FVMQ	SILICONE VMQ	CHEMRAZ FFKM
SHELL DD	NR at 70°F	NR at 70°F				
SHELL DIALA	NR at 70°F	NR at 70°F	NR at 70°F	A to 70°F	NR at 70°F	A to 70°F
SHELL IRUS 905	NR at 70°F	NR at 70°F	NR at 70°F	A to 70°F	NR at 70°F	A to 70°F
SHELL LO HYDRAX 27 & 29	NR at 70°F	NR at 70°F	NR at 70°F	A to 70°F	NR at 70°F	A to 70°F
SHELL MACOME 72	NR at 70°F	NR at 70°F	NR at 70°F	A to 70°F	NR at 70°F	A to 70°F
SHELL TELLUS 27 (PETRO BASE)	NR at 70°F	NR at 70°F	NR at 70°F	A to 70°F	NR at 70°F	A to 70°F
SHELL TELLUS 33	NR at 70°F	NR at 70°F	NR at 70°F	A to 70°F	NR at 70°F	A to 70°F
SHELL TURBINE OIL 307			*AB to 70°F			
SHELL UMF (5% AROMATIC)	NR at 70°F	NR at 70°F	NR at 70°F	A to 70°F	NR at 70°F	A to 70°F
SHELL 3XF MINE FLUID	NR at 70°F	NR at 70°F	B at 70°F	A to 70°F		A to 70°F
SHELLAC		NR at 70°F orange				
SHERWOOD OIL	NR at 70°F	NR at 70°F	NR at 70°F		C at 70°F	
SHORTENING		NR at 70°F	C at 140°F			
SIEMANS GAS	C at 70°F	C at 70°F	B at 70°F		B at 70°F	
SILANE						A to 70°F
SILENT SPIRIT	A to 70°F	A to 70°F	A to 70°F		A to 70°F	
SILICATE ESTERS	NR at 70°F	NR at 70°F	B at 70°F	+5% vol 3 days 300°F AB to 70°F	+150% vol 3 days 300°F NR at 70°F	A to 70°F
SILICATE OF SODA	A to 70°F	A to 70°F AB any conc to 185°F	A to 70°F AB to 200°F		A to 70°F	A to 70°F
SILICIC ACID						
SILICOFLUORIC ACID	A to 70°F	A 50-100% to 80°F	A to 70°F		NR at 70°F	
SILICONE GREASES	A to 70°F	A to 70°F	A to 70°F	A to 70°F	BC at 70°F	A to 70°F
SILICONE OILS	AC to 70°F	AC to 70°F	A to 70°F	no swell 3 days 300°F A to 70°F	+10% vol 7 days 300°F C/NR at 70°F	A to 70°F
SILVER ACETATE						
SILVER CYANIDE	A to 70°F	A to 70°F		A to 70°F	NR at 70°F	A to 70°F
SILVER NITRATE	A to 70°F	A to 70°F AB to 185°F	A to 70°F AB to 200°F	A to 70°F	A to 70°F	A to 70°F
SILVER SALTS		AB to 70°F	AB to 70°F	A to 70°F	A to 70°F	
SILVER SULFATE						A to 70°F
SINCLAIR OPALINE CX-EP LUBE	NR at 70°F	NR at 70°F	B at 70°F	A to 70°F	NR at 70°F	A to 70°F
SPELLY SOLVENT B, C & E	NR at 70°F	NR at 70°F	NR at 70°F	A to 70°F		A to 70°F
SKYDROL HYDRAULIC FLUID	NR at 70°F	NR at 70°F	NR at 70°F	C at 70°F	+8% vol 3 days 212°F NR at 70°F	A to 70°F
SKYDROL 500	NR at 70°F	NR at 70°F	NR at 70°F	+25% vol 7 days 120°F +30% vol 3 days 212°F	+10-20% vol 3 days 212°F C at 70°F	A to 70°F
SKYDROL 7000	NR at 70°F	NR at 70°F	NR at 70°F	C at 70°F	B/NR at 70°F	A to 70°F
SLAKED LIME	A to 70°F	A to 70°F AB any conc to 150°F	A to 200°F	A to 70°F	A to 70°F	A to 70°F B at 70°F white
SLUDGE ACID	C at 70°F	C/NR at 70°F	A to 70°F		A to 70°F	
SOAP LIQUOR				A to 70°F	A to 70°F	
SOAP OIL			AB to 70°F			
SOAP SOLUTIONS	AB to 70°F	B to 150°F	A to 200°F	+1% vol 3 days 158°F A to 70°F	+4% vol 3 days 158°F A to 70°F	A to 70°F B at 70°F white
SOCONY MOBIL TYPE A	NR at 70°F	NR at 70°F	NR at 70°F	B at 70°F	NR at 70°F	A to 70°F
SOCONY PD959B VACUUM	NR at 70°F	NR at 70°F	B at 70°F	A to 70°F	NR at 70°F	A to 70°F
SOCONY VACUUM AMV AC781 GREASE	NR at 70°F	NR at 70°F	NR at 70°F	B at 70°F	NR at 70°F	A to 70°F
SODA	A to 70°F	A to 70°F AB any conc to 150°F	A to 70°F		A to 70°F	
SODA ALUM	A to 70°F	A to 70°F AB any conc to 150°F	A to 70°F		A to 70°F	
SODA ASH	A to 100% to 70°F	A to 70°F AB any conc to 150°F	A to 100% to 70°F AB to 200°F	A to 100% to 70°F	A to 100% to 70°F	A to 70°F
SODA, BAKING	A to 70°F	A to 70°F AB to 150°F	A to 70°F AB to 200°F	A to 100% to 70°F	A to 70°F	A to 70°F

	POLYSULFIDE T	CHLORINATED POLYETHYLENE CM	EPICHLOROHYDRIN CO, ECO	® HYTREL COPOLYESTER TPE	® SANTOPRENE COPOLYMER TPO	® C-FLEX STYRENIC TPE
SHELL DD						
SHELL DIALA						
SHELL IRUS 905	A to 70°F					
SHELL LO HYDRAX 27 & 29						
SHELL MACOME 72						
SHELL TELLUS 27 (PETRO BASE)	B at 70°F					
SHELL TELLUS 33	B at 70°F					
SHELL TURBINE OIL 307				*AB to 70°F		
SHELL UMF (5% AROMATIC)	B at 70°F					
SHELL 3XF MINE FLUID						
SHELLAC						
SHERWOOD OIL						
SHORTENING						
SIEMANS GAS						
SILANE						
SILENT SPIRIT						
SILICATE ESTERS	softened 3 days 300°F			AC to 70°F		
SILICATE OF SODA						
SILICIC ACID						
SILICOFLUORIC ACID						
SILICONE GREASES	A to 70°F		A to 70°F	A to 70°F		A to 70°F
SILICONE OILS	softened 3 days 300°F A to 70°F		A to 70°F	AB to 70°F		A to 70°F
SILVER ACETATE						
SILVER CYANIDE						
SILVER NITRATE	B at 70°F		NR at 70°F			A to 70°F
SILVER SALTS	AB to 70°F					
SILVER SULFATE						
SINCLAIR OPALINE CX-EP LUBE	A to 70°F					
SPELLY SOLVENT B, C & E						
SKYDROL HYDRAULIC FLUID	NR at 70°F	AB to 70°F	NR at 70°F	AB to 200°F	A to 70°F	
SKYDROL 500	NR at 70°F	AB to 70°F	NR at 70°F	AB to 200°F BC at 250°F	-1% to -13% vol 7 days 212°F A to 70°F	
SKYDROL 7000	NR at 70°F	AB to 70°F	NR at 70°F	C/NR at 70°F		
SLAKED LIME						
SLUDGE ACID						
SOAP LIQUOR						
SOAP OIL	AB to 70°F					
SOAP SOLUTIONS	+4.6% vol 3 days 70°F NR at 70°F	AB to 70°F	A to 70°F	A to 70°F	A to 70°F	A to 70°F
SOCONY MOBIL TYPE A						
SOCONY PD959B VACUUM	A to 70°F					
SOCONY VACUUM AMV AC781 GREASE						
SODA						
SODA ALUM						
SODA ASH	NR at 70°F	AB to 70°F	A to 70°F	AB to 70°F		A to 70°F
SODA, BAKING	C at 70°F		A to 70°F	AB to 70°F		A to 70°F

	ETHYLENE ACRYLIC EA	POLYALLOMER LINEAR COPOLYMER	NYLON 11 POLYAMIDE	NYLON 12 POLYAMIDE	ETHYLENE VINYL ACETATE EVA	POLYVINYLCHLORIDE FLEXIBLE PVC
SHELL DD						
SHELL DIALA						
SHELL IRUS 905			A to 70°F 902 & 905			
SHELL LO HYDRAX 27 & 29						
SHELL MACOME 72						
SHELL TELLUS 27 (PETRO BASE)						
SHELL TELLUS 33						
SHELL TURBINE OIL 307						
SHELL UMF (5% AROMATIC)						
SHELL 3XF MINE FLUID						
SHELLAC		A to 70°F	A to 70°F		AB to 70°F	NR at 70°F
SHERWOOD OIL						
SHORTENING						B at 70°F C at 150°F
SIEMANS GAS						
SILANE						
SILENT SPIRIT						
SILICATE ESTERS			AB to 70°F		NR at 70°F	NR at 70°F
SILICATE OF SODA						
SILICIC ACID						
SILICOFLUORIC ACID						
SILICONE GREASES	A to 158°F		A to 70°F			A to 70°F
SILICONE OILS	A to 70°F		A to 70°F	AB to 70°F	B at 70°F	AB to 70°F
SILVER ACETATE		A to 122°F				B to 122°F
SILVER CYANIDE						*AB to 140°F
SILVER NITRATE		A to 70°F B at 122°F	A to 70°F		AB to 70°F *AB to 140°F	AB to 70°F B at 122°F
SILVER SALTS				AB to 70°F		
SILVER SULFATE						
SINCLAIR OPALINE CX-EP LUBE						
SPELLY SOLVENT B, C & E						
SKYDROL HYDRAULIC FLUID	NR at 70°F		A to 70°F (all) AB to 200°F			NR at 70°F
SKYDROL 500	NR at 70°F		+18% vol 180 days 70°F AB to 250°F			NR at 70°F
SKYDROL 7000	NR at 70°F		A to 70°F AB to 200°F			NR at 70°F
SLAKED LIME	A to 70°F		A to 194°F		AB to 140°F	A to 70°F *AB to 130°F
SLUDGE ACID						
SOAP LIQUOR					A to 70°F	B at 70°F
SOAP OIL			A to 70°F			
SOAP SOLUTIONS	A to 70°F	A to 70°F	A to 70°F	AB 10% conc to 70°F	AB to 70°F *AB to 140°F	A to 70°F *AB to 130°F
SOCONY MOBIL TYPE A	A to 70°F					
SOCONY PD959B VACUUM						
SOCONY VACUUM AMV AC781 GREASE						
SODA				AB 25% to 70°F		A to 70°F NR at 150°F
SODA ALUM						
SODA ASH	A to 70°F	A to 70°F	AB to 140°F NR at 194°F	AB 10% to 70°F	A to 70°F AB to 140°F	AB to 70°F *AB to 130°F
SODA, BAKING	A to 70°F	A to 70°F	A to 70°F AB to 140°F	AB to 70°F aqueous	A to 70°F AB to 140°F	A to 70°F *AB to 130°F

	POLYETHYLENE LOW DENSITY LDPE	®TEFLON FEP	®KALREZ PERFLUORINATED ELASTOMER	®FLUORAZ FLUORINATED COPOLYMER	®AFLAS FLUORINATED COPOLYMER	®NORPRENE COPOLYMER TPO
SHELL DD		A to 70°F				
SHELL DIALA				A to 70°F		
SHELL IRUS 905		A to 70°F 902 & 905		A to 70°F		
SHELL LO HYDRAX 27 & 29				A to 70°F		
SHELL MACOME 72				A to 70°F		
SHELL TELLUS 27 (PETRO BASE)				A to 70°F		
SHELL TELLUS 33				A to 70°F		
SHELL TURBINE OIL 307						
SHELL UMF (5% AROMATIC)				A to 70°F		
SHELL 3XF MINE FLUID				A to 70°F		
SHELLAC	A to 70°F	A to 70°F bleached or orange	A to 212°F	NR at 70°F		
SHERWOOD OIL		A to 70°F				
SHORTENING						
SIEMANS GAS		A to 70°F				
SILANE	AB to 70°F	AB to 70°F	A to 212°F			
SILENT SPIRIT		A to 70°F				
SILICATE ESTERS	C/NR at 70°F	A to 70°F	A to 212°F	A to 70°F		
SILICATE OF SODA	A to 70°F	A to 300°F	A to 212°F	A to 70°F		
SILICIC ACID	A to 70°F AB to 140°F	A to 70°F				
SILICOFLUORIC ACID		A to 70°F				
SILICONE GREASES		A to 300°F	A to 212°F	A to 70°F	A to 70°F	TEST
SILICONE OILS	A to 70°F	A to 300°F	A to 212°F	A to 70°F	A to 70°F	TEST
SILVER ACETATE	A to 122°F	A to 122°F				
SILVER CYANIDE		A to 300°F	A to 212°F			
SILVER NITRATE	AB to 140°F	A to 300°F	A to 212°F	A to 100% to 70°F		A to 70°F
SILVER SALTS	A to 70°F AB to 140°F	A to 70°F	A to 212°F			A to 70°F
SILVER SULFATE		A to 70°F	A to 212°F			
SINCLAIR OPALINE CX-EP LUBE				A to 70°F		
SPELLY SOLVENT B, C & E		A to 70°F		A to 70°F		
SKYDROL HYDRAULIC FLUID	NR at 70°F	A to 200°F	AB to 400°F	A to 70°F	A to 70°F AB to 212°F	BC to 212°F
SKYDROL 500	NR at 70°F	A to 200°F	A to 212°F AB to 400°F	A to 70°F	+14% vol 3 days 212°F AB to 212°F	BC to 212°F
SKYDROL 7000	NR at 70°F	A to 200°F	A to 212°F	A to 70°F		
SLAKED LIME	A to 122°F AB to 140°F	A to 122°F	A to 70°F	A to 70°F		
SLUDGE ACID		A to 70°F				
SOAP LIQUOR	A to 70°F			A to 70°F		
SOAP OIL		A to 70°F				
SOAP SOLUTIONS	AC to 70°F	A to 300°F	A to 212°F	A to 70°F		A to 70°F
SOCONY MOBIL TYPE A				A to 70°F		
SOCONY PD959B VACUUM				A to 70°F		
SOCONY VACUUM AMV AC781 GREASE				A to 70°F		
SODA		A to 70°F				
SODA ALUM		A to 70°F				
SODA ASH	AB to 100% to 140°F	A to 100% to 300°F	A to 212°F	A to 70°F		A to 70°F
SODA, BAKING	A to 70°F AB to 140°F	A to 300°F	A to 212°F	A aqueous to 70°F		A to 70°F

C 331

	RECOMMENDATIONS	NITRILE NBR	ETHYLENE PROPYLENE EP, EPDM	FLUOROCARBON FKM	CHLOROPRENE CR	HYDROGENATED NITRILE HNBR
SODA, CAUSTIC	EP,IIR,NBR,CR	NR 80-100% at 70°F A to 50% to 176°F	A to 100% to 70°F A to 50% to 176°F	C/NR 20-50% at 70°F NR 20% at 212°F	A to conc to 70°F A to 50% to 140°F	B at 70°F
SODA LIME		B at 70°F	A to 70°F	B at 70°F	B to 100% to 70°F	
SODA NITER	EP,IIR,NBR,CR,NR	AB any conc to 150°F	A to 200°F	A to 212°F	B to 200°F	
SODA SALTPETER		AB any conc to 150°F	A to 70°F	A to 70°F	AB any conc to 200°F	
SODA, WASHING		A to 70°F	A to 70°F	A to 70°F	A to 70°F	
SODAN		AB to 70°F			AB to 70°F	
SODIUM	FKM	A to 140°F B at 176°F	A to 176°F	A to 212°F		
SODIUM ACETATE	EP,IIR,NBR	AB to 140°F B at 176°F	A to conc to 176°F A sol'n to 200°F	NR any conc at 70°F	AB to conc to 140°F	B at 70°F
SODIUM ACID CARBONATE		A to 70°F AB any conc to 200°F	A to 70°F	A to 70°F	A to 70°F AB any conc to 200°F	
SODIUM ACID SULFATE		AB any conc to 200°F	A to 70°F	A to 70°F	A to 70°F AB any conc to 200°F	
SODIUM ACID SULFITE		AB any conc to 160°F		AB to 70°F	AB any conc to 200°F	
SODIUM ALUM		A to 140°F B at 176°F	A to 176°F	A to 212°F	A to 70°F	
SODIUM ALUMINATE		A to 140°F	A to 140°F	A to 140°F	A to 140°F	
SODIUM ALUMINUM SULFATE		A to 70°F AB any conc to 200°F	A to 70°F	A to 70°F	A to 70°F AB any conc to 200°F	
SODIUM ARSENATE		AB to 70°F		AB to 70°F	A/NR at 70°F	
SODIUM BENZOATE		AB to 70°F	A to 70°F	A to 70°F	A to 70°F	
SODIUM BICARBONATE	NBR,EP,FKM,CR,IIR	A to 140°F AB any conc to 200°F	A to 100% to 176°F B at 212°F	A to 100% to 212°F	A to 100% to 140°F AB any conc to 200°F	A to 70°F
SODIUM BICHROMATE		A to 140°F B at 176°F	A to 176°F	A to 212°F	AB to 70°F	
SODIUM BISULFATE	NBR,EP,FKM,CR,IIR	AB any conc to 160°F C at 212°F	A to 176°F AB to 200°F	A to 212°F	A to 140°F AB any conc to 200°F	A to 70°F
SODIUM BISULFITE	NBR,EP,FKM,CR,IIR	A any conc to 160°F B at 212°F	A any conc to 176°F B at 212°F	A any conc to 212°F	A any conc to 140°F AB any conc to 200°F	A to 70°F
SODIUM BILSULPHATE	NBR,EP,FKM,CR,IIR	AB any conc to 160°F C at 212°F	A to 176°F AB to 200°F	A to 212°F	A to 140°F AB any conc to 200°F	A to 70°F
SODIUM BORATE	NBR,EP,FKM,CR,IIR	A to 70°F AB any conc to 180°F	A to 140°F	A to 176°F	A to 70°F AB any conc to 200°F	A to 70°F
SODIUM BROMIDES			A to 70°F	A to 100% to 70°F	A to 70°F	
SODIUM CARBONATE	NBR,EP,FKM,CR,IIR	A to 100% to 160°F AB to 100% to 200°F	A to 100% to 176°F B at 212°F	A to 100% to 212°F	A to 100% to 176°F AB 100% to 200°F	A to 70°F
SODIUM CHLORATE		A to 70°F AB any conc to 200°F	A to 130°F AB to 200°F	A to 140°F AB to 200°F	A to 140°F AB any conc to 200°F	
SODIUM CHLORIDE	NBR,EP,FKM,CR,IIR	-0.8% vol 3 days 212°F A to 160°F	A to 100% to 176°F AB to 200°F	A to 100% to 212°F	+0.4% vol 3 days 212°F A to 150°F	A to 70°F
SODIUM CHLORITE		NR 25% at 70°F	NR 25-100% at 70°F	NR 25-100% at 70°F	NR at 70°F	
SODIUM CHROMATE		A to 140°F		A to 140°F	A to 140°F	
SODIUM CYANIDE	NBR,EP,CR,SBR,IIR	A to 100% to 140°F AB to 200°F	A to 100% to 176°F B at 212°F	A to 100% to 176°F B at 212°F	A to 100% to 140°F B 176-200°F	
SODIUM DICHROMATE	NBR,EP,IIR	A to 70°F	A 20-100% to 70°F	A 20-100% to 70°F	BC 100% at 70°F B 20% at 70°F	
SODIUM DIMETABORATE		B at 70°F	A to 70°F	A to 70°F	A to 70°F	
SODIUM DIOXIDE		AB any conc to 200°F	B at 70°F	A to 70°F	AB any conc to 200°F	
SODIUM DIPHOSPHATE		A to 70°F		AB to 70°F	B at 70°F	
SODIUM DISULFITE	NBR,EP,FKM,CR,IIR	A to 70°F AB any conc to 160°F	A to 176°F AB to 200°F	A to 212°F	A to 70°F AB any conc to 160°F	
SODIUM ETHOXIDE		AB to 70°F		AB to 70°F	AB to 70°F	
SODIUM FERRICYANIDE		A to 70°F	A to 140°F	A to 140°F	A to 70°F	
SODIUM FERROCYANIDE		A to 70°F	A to 140°F	A to 140°F	A to 70°F	
SODIUM FLUOALUMINATE		A to 70°F	A 10% to 70°F	A to 70°F	A to 70°F	
SODIUM FLUOBORATE		AB to 70°F		A to 70°F	A to 70°F	
SODIUM FLUORIDE	NBR,CR,EP	A to 100% to 70°F C at 140°F	A to 100% to 140°F	A to 140°F	A to 140°F	
SODIUM HEXAMETAPHOSPHATE	NBR,EP	A to 200°F	A to 70°F	A to 70°F	A any conc to 150°F	
SODIUM HYDRATE		AB any conc to 150°F	A to 70°F	B at 70°F	AB any conc to 200°F	
SODIUM HYDROCHLORITE		C at 70°F	B at 70°F	A to 70°F	C at 70°F	
SODIUM HYDROGEN CARBONATE		A to 70°F	A to 70°F	A to 70°F	A to 70°F	

	STYRENE BUTADIENE SBR	POLYACRYLATE ACM	POLYURETHANE AU, EU	ISOBUTYLENE ISOPRENE IIR	POLYBUTADIENE BR	® AEROQUIP AQP
SODA, CAUSTIC	-1.1% vol 3 days 70°F A to 70°F	+0.3% vol 3 days 70°F A/NR at 70°F	B to 100% at 70°F NR 70% at 140°F	A to 100% to 70°F A to 80% to 175°F	A to 100% to 70°F	C 50% at 70°F
SODA LIME			C at 70°F	A to 70°F		
SODA NITER	B at 70°F	B at 70°F	AB to 185°F	A to 70°F AB any conc to 185°F	A to 70°F	AB to 70°F
SODA SALTPETER				A to 70°F AB any conc to 185°F		
SODA, WASHING				A to 70°F		
SODAN				B at 70°F		
SODIUM						
SODIUM ACETATE	NR at 70°F C sol'n at 70°F	NR any conc at 70°F	C/NR at 70°F	A to 70°F	C/NR at 70°F	
SODIUM ACID CARBONATE	AB to 70°F			A to 70°F AB any conc to 185°F		
SODIUM ACID SULFATE	AB to 70°F		A to 70°F	A to 70°F AB any conc to 185°F		
SODIUM ACID SULFITE	AB to 70°F			AB any conc to 185°F		
SODIUM ALUM	AB to 70°F		A to 70°F	A to 70°F		
SODIUM ALUMINATE	AB to 70°F			A to 70°F		
SODIUM ALUMINUM SULFATE	AB to 70°F		A to 70°F	A to 70°F AB any conc to 185°F		
SODIUM ARSENATE	AB to 70°F			A to 70°F		
SODIUM BENZOATE	AB to 70°F			A to 70°F		
SODIUM BICARBONATE	A to 100% to 70°F	NR sol'n at 70°F	A to 70°F	A to 70°F AB any conc to 185°F	A to 70°F	
SODIUM BICHROMATE	AB to 70°F		A to 70°F	A to 70°F		
SODIUM BISULFATE	B any conc at 70°F	NR any conc at 70°F	A to 100% to 70°F	A to 70°F AB any conc to 185°F	B at 70°F	NR at 70°F
SODIUM BISULFITE	B any conc at 70°F	NR any conc at 70°F	A to 100% to 70°F	A to 70°F AB any conc to 185°F	B at 70°F	
SODIUM BILSULPHATE	B any conc at 70°F	NR any conc at 70°F	A to 100% to 70°F	A to 70°F AB any conc to 185°F	B at 70°F	NR at 70°F
SODIUM BORATE	A to 70°F		AB to 70°F	A to 70°F AB any conc to 185°F	A to 70°F	
SODIUM BROMIDES	AB to 70°F			A to 70°F		
SODIUM CARBONATE	A to 100% to 70°F	NR at 70°F	AB to 100% to 70°F	A to 70°F AB any conc to 185°F	A to 70°F	AB to 70°F
SODIUM CHLORATE	AB to 70°F		AB to 70°F	B at 70°F		
SODIUM CHLORIDE	+0.1% vol 3 days 212°F A to 70°F	+2% vol 3 days 212°F A to 70°F	A to 100% to 70°F	-0.1% vol 3 days 212°F AB any conc to 185°F	A to 70°F	AB to 70°F
SODIUM CHLORITE	AB to 70°F			AB to 70°F		
SODIUM CHROMATE	AB to 70°F			B at 70°F		
SODIUM CYANIDE	A to 100% to 70°F		AB to 70°F	A to 70°F AB any conc to 150°F	A to 70°F	AB to 70°F
SODIUM DICHROMATE	AB to 70°F		AB to 70°F discolors	A to 70°F		
SODIUM DIMETABORATE	AB to 70°F		A to 70°F	A to 70°F		
SODIUM DIOXIDE	AB to 70°F		NR at 70°F	A to 70°F AB any conc to 185°F		
SODIUM DIPHOSPHATE	AB to 70°F			A to 70°F		
SODIUM DISULFITE	B any conc at 70°F	NR any conc at 70°F	A to 100% to 70°F	A to 70°F AB any conc to 185°F	B at 70°F	
SODIUM ETHOXIDE	AB to 70°F			AB to 70°F		
SODIUM FERRICYANIDE	AB to 70°F			AB to 70°F		
SODIUM FERROCYANIDE	AB to 70°F		AB to 70°F	AB to 70°F		
SODIUM FLUOALUMINATE	AB to 70°F			A to 70°F		
SODIUM FLUOBORATE	AB to 70°F			AB to 70°F		
SODIUM FLUORIDE	AB to 70°F		AB to 70°F	A to 70°F		
SODIUM HEXAMETAPHOSPHATE	AB to 70°F	NR at 70°F	NR at 70°F	A to 70°F AB any conc to 185°F		
SODIUM HYDRATE	AB to 70°F		C at 70°F	A to 70°F AB any conc to 185°F		
SODIUM HYDROCHLORITE	AB to 70°F		B/NR at 70°F	B at 70°F		
SODIUM HYDROGEN CARBONATE	AB to 70°F			A to 70°F		

	SYNTHETIC ISOPRENE IR	NATURAL ISOPRENE NR	CHLOROSULFONATED POLYETHYLENE CSM	FLUOROSILICONE FVMQ	SILICONE VMQ	CHEMRAZ FFKM
SODA, CAUSTIC	A to 100% to 70°F	AB any conc to 150°F NR 20% at 160°F	A to 100% to 70°F A to 73% to 280°F	B to 100% at 70°F	A to conc to 70°F A 20% to 212°F	AB to 70°F C at 70°F white
SODA LIME	A to 70°F	A to 70°F	B at 70°F		C at 70°F	
SODA NITER	AB to 70°F	B any conc to 150°F	A to 70°F AB to 200°F		C/NR at 70°F NR sol'n at 70°F	A to 70°F
SODA SALTPETER	B at 70°F	AB any conc to 150°F	A to 70°F		C at 70°F	
SODA, WASHING	A to 70°F	A to 70°F	A to 70°F	A to 70°F	A to 70°F	
SODAN	B at 70°F	B at 70°F				
SODIUM						NR at 70°F
SODIUM ACETATE	A to 70°F	A to 70°F	B/NR to 200°F	NR at 70°F	NR at 70°F	A to 70°F
SODIUM ACID CARBONATE	A to 70°F	A to 70°F AB any conc to 150°F	A to 70°F		A to 70°F	
SODIUM ACID SULFATE	A to 70°F	A to 70°F AB any conc to 150°F	AB to 70°F			
SODIUM ACID SULFITE		AB any conc to 150°F	AB to 70°F			
SODIUM ALUM	A to 70°F	A to 70°F	A to 70°F		A to 70°F	
SODIUM ALUMINATE	A to 70°F	AB to 70°F	A to 70°F			A to 70°F
SODIUM ALUMINUM SULFATE	A to 70°F	A to 70°F AB any conc to 150°F	A to 70°F		A to 70°F	A to 70°F
SODIUM ARSENATE	A to 70°F	A to 70°F	AB to 70°F			
SODIUM BENZOATE	A to 70°F	A to 70°F	AB to 70°F			
SODIUM BICARBONATE	A to 70°F	A to 70°F AB to 185°F	A to 70°F AB to 200°F	A to 100% to 70°F	A to 100% to 70°F	A to 70°F
SODIUM BICHROMATE	NR at 70°F	NR at 70°F	B at 70°F			
SODIUM BISULFATE	AB to 70°F	A to 70°F AB to 185°F	A to 70°F AB to 200°F	A to 100% to 70°F	AC 100% at 70°F A sol'n to 70°F	A to 70°F
SODIUM BISULFITE	AB to 70°F	A to 70°F AB any conc to 150°F	A to 70°F	A to 100% to 70°F	A to 100% to 70°F	A to 70°F
SODIUM BILSULPHATE	AB to 70°F	A to 70°F AB to 185°F	A to 70°F AB to 200°F	A to 100% to 70°F	AC 100% at 70°F A sol'n to 70°F	A to 70°F
SODIUM BORATE	A to 70°F	A to 70°F AB any conc to 150°F	A to 70°F	A to 70°F	A to 70°F	A to 70°F
SODIUM BROMIDES	A to 70°F	A to 70°F	AB to 70°F			
SODIUM CARBONATE	A to 100% to 70°F	A to 100% to 70°F AB any conc to 185°F	A to 100% to 70°F AB to 200°F	A to 100% to 70°F	A to 100% to 70°F	A to 70°F
SODIUM CHLORATE	A to 70°F	A to 70°F	A to 70°F AB to 200°F		C at 70°F	
SODIUM CHLORIDE	A to 70°F	A any conc to 150°F	A to 70°F AB to 200°F	A to 70°F	no swell solution 3 days 212°F	A to 70°F
SODIUM CHLORITE		AB to 70°F	AB to 70°F			
SODIUM CHROMATE		AB to 70°F	C at 70°F			
SODIUM CYANIDE	A to 70°F	A to 70°F AB to 185°F	A to 70°F AB to 200°F	A to 70°F	A to 100% to 70°F	A to 70°F
SODIUM DICHROMATE	NR at 70°F	NR at 70°F	AB to 200°F	C at 70°F	AB to 70°F	
SODIUM DIMETABORATE	A to 70°F	A to 70°F	A to 70°F		B at 70°F	
SODIUM DIOXIDE	B at 70°F	AB any conc to 150°F	A to 70°F		C at 70°F	
SODIUM DIPHOSPHATE	A to 70°F	A to 70°F	AB to 70°F			A to 70°F
SODIUM DISULFITE	AB to 70°F	A to 70°F AB any conc to 150°F	A to 70°F	A to 100% to 70°F	A to 100% to 70°F	A to 70°F
SODIUM ETHOXIDE		AB to 70°F	AB to 70°F			
SODIUM FERRICYANIDE		AB to 70°F	AB to 70°F			A to 70°F
SODIUM FERROCYANIDE		AB to 70°F	AB to 70°F			A to 70°F
SODIUM FLUOALUMINATE	A 10% to 70°F	A to 70°F	A to 70°F			
SODIUM FLUOBORATE		AB to 70°F	AB to 70°F			
SODIUM FLUORIDE	A to 70°F	A/NR at 70°F	AB to 70°F			A to 70°F
SODIUM HEXAMETAPHOSPHATE	A to 70°F	A to 70°F AB any conc to 150°F	AB to 70°F	A to 70°F	A to 70°F	
SODIUM HYDRATE	A to 70°F	A to 70°F AB any conc to 150°F	B at 70°F		C at 70°F	
SODIUM HYDROCHLORITE	C at 70°F	C at 70°F	A to 70°F		B at 70°F	
SODIUM HYDROGEN CARBONATE	A to 70°F	A to 70°F	A to 70°F		A to 70°F	

C 334

	POLYSULFIDE T	CHLORINATED POLYETHYLENE CM	EPICHLOROHYDRIN CO, ECO	® HYTREL COPOLYESTER TPE	® SANTOPRENE COPOLYMER TPO	® C-FLEX STYRENIC TPE
SODA, CAUSTIC	+0.2% vol 3 days 70°F AB 20-50% to 70°F	C at 70°F	B at 70°F	NR 73% at 70°F BC 46-50% at 70°F	A to 50% to 70°F	A to conc to 70°F
SODA LIME						
SODA NITER	AB to 70°F	AB to 70°F	A to 70°F	AB to 70°F		A to 70°F
SODA SALTPETER						
SODA, WASHING						
SODAN						
SODIUM						
SODIUM ACETATE	NR at 70°F					
SODIUM ACID CARBONATE	AB to 70°F					
SODIUM ACID SULFATE	AB to 70°F					
SODIUM ACID SULFITE	AB to 70°F					
SODIUM ALUM	AB to 70°F					
SODIUM ALUMINATE	AB to 70°F					
SODIUM ALUMINUM SULFATE	AB to 70°F					
SODIUM ARSENATE	AB to 70°F					
SODIUM BENZOATE	AB to 70°F					
SODIUM BICARBONATE	C at 70°F		A to 70°F	AB to 70°F		A to 70°F
SODIUM BICHROMATE	AB to 70°F					
SODIUM BISULFATE	C at 70°F	NR at 70°F	A to 70°F	B/NR at 70°F		A to 70°F
SODIUM BISULFITE	C at 70°F		A to 70°F	AB to 70°F		A to 70°F
SODIUM BILSULPHATE	C at 70°F	NR at 70°F	A to 70°F	B/NR at 70°F		A to 70°F
SODIUM BORATE	A to 70°F		A to 70°F	AB to 70°F		A to 70°F
SODIUM BROMIDES	AB to 70°F					
SODIUM CARBONATE	NR at 70°F	AB to 70°F	A to 70°F	AB to 70°F		A to 70°F
SODIUM CHLORATE	AB to 70°F					A to 70°F
SODIUM CHLORIDE	-0.8% vol 3 days 212°F C conc at 70°F	AB to 70°F	A to 70°F	A to conc to 70°F	no swell 15% conc 7 days 70°F	A to 70°F
SODIUM CHLORITE	AB to 70°F					
SODIUM CHROMATE	AB to 70°F					
SODIUM CYANIDE	A to 70°F	AB to 70°F	A to 70°F	AB to 70°F		
SODIUM DICHROMATE	AB to 70°F			*AB 30% to 70°F		
SODIUM DIMETABORATE	AB to 70°F					
SODIUM DIOXIDE	AB to 70°F					
SODIUM DIPHOSPHATE	AB to 70°F					
SODIUM DISULFITE	C at 70°F		A to 70°F	AB to 70°F		A to 70°F
SODIUM ETHOXIDE	AB to 70°F					
SODIUM FERRICYANIDE	AB to 70°F					
SODIUM FERROCYANIDE	AB to 70°F					A to 70°F
SODIUM FLUOALUMINATE	AB to 70°F					
SODIUM FLUOBORATE	AB to 70°F					
SODIUM FLUORIDE	AB to 70°F					
SODIUM HEXAMETAPHOSPHATE	AB to 70°F					
SODIUM HYDRATE	AB to 70°F					
SODIUM HYDROCHLORITE	AB to 70°F					
SODIUM HYDROGEN CARBONATE	AB to 70°F					

	ETHYLENE ACRYLIC EA	POLYALLOMER LINEAR COPOLYMER	NYLON 11 POLYAMIDE	NYLON 12 POLYAMIDE	ETHYLENE VINYL ACETATE EVA	POLYVINYLCHLORIDE FLEXIBLE PVC
SODA, CAUSTIC	NR 73% at 70°F A 20-46% to 70°F	A to conc to 70°F	AB to 100% to 70°F NR 50% at 140°F	AB 40% to 70°F	A to 70°F AB to 140°F	A to conc to 70°F B conc at 122°F
SODA LIME						
SODA NITER		A to 70°F	A to 70°F	AB 10% to 70°F	A to 70°F	A to 70°F
SODA SALTPETER						
SODA, WASHING						
SODAN						
SODIUM						
SODIUM ACETATE		A to 122°F	AB to 70°F			B at 70°F C at 122°F
SODIUM ACID CARBONATE			AB to 70°F			AB to 70°F
SODIUM ACID SULFATE			AB to 70°F			AB to 70°F
SODIUM ACID SULFITE			AB to 70°F			AB to 70°F
SODIUM ALUM			AB to 70°F			AB to 70°F
SODIUM ALUMINATE			AB to 70°F			AB to 70°F
SODIUM ALUMINUM SULFATE			AB to 70°F			AB to 70°F
SODIUM ARSENATE			AB to 70°F			AB to 70°F
SODIUM BENZOATE			AB to 70°F			*AB to 70°F *NR at 140°F
SODIUM BICARBONATE	A to 70°F	A to 70°F	A to 70°F AB to 140°F	AB to 70°F aqueous	A to 70°F AB to 140°F	A to 70°F *AB to 140°F
SODIUM BICHROMATE			AB to 70°F			AB to 70°F
SODIUM BISULFATE	A to 70°F	A to 70°F	A/NR at 70°F		A to 70°F	A to 70°F
SODIUM BISULFITE	A to 70°F	A to 70°F	A to 70°F	AB 10% to 70°F	A to 70°F	A to 70°F
SODIUM BILSULPHATE	A to 70°F	A to 70°F	A/NR at 70°F		A to 70°F	A to 70°F
SODIUM BORATE	A to 70°F	A to 70°F	A to 70°F		A to 70°F	A to 70°F
SODIUM BROMIDES			AB to 70°F	AB 10% to 70°F		*AB to 140°F
SODIUM CARBONATE	A to 70°F	A to 70°F	AB to 140°F NR at 194°F	AB 10% to 70°F	AB to 70°F	AB to 70°F *AB to 130°F
SODIUM CHLORATE		A to 70°F	AC to 70°F		B/NR at 70°F	AB to 70°F *AB to 140°F
SODIUM CHLORIDE	A to 70°F	A to 70°F	AB to 100% to 194°F	AB to 70°F aqueous		AB to 70°F
SODIUM CHLORITE			AB to 70°F	NR 5% at 70°F		AB to 70°F
SODIUM CHROMATE			AB to 70°F			AB to 70°F
SODIUM CYANIDE		A to 70°F	A to 70°F		A to 70°F	A to 150°F
SODIUM DICHROMATE			AB to 70°F			AB to 70°F
SODIUM DIMETABORATE			AB to 70°F			AB to 70°F
SODIUM DIOXIDE			AB to 70°F			AB to 70°F
SODIUM DIPHOSPHATE			AB to 70°F			AB to 70°F
SODIUM DISULFITE	A to 70°F	A to 70°F	A to 70°F	AB 10% to 70°F	A to 70°F	A to 70°F
SODIUM ETHOXIDE			AB to 70°F			AB to 70°F
SODIUM FERRICYANIDE			AB to 70°F			*AB to 140°F
SODIUM FERROCYANIDE			AB to 70°F			A to 70°F *AB to 140°F
SODIUM FLUOALUMINATE			AB to 70°F			AB to 70°F
SODIUM FLUOBORATE			AB to 70°F			AB to 70°F
SODIUM FLUORIDE			AB to 70°F			NR at 70°F
SODIUM HEXAMETAPHOSPHATE			AB to 70°F			AB to 70°F
SODIUM HYDRATE			AB to 70°F			AB to 70°F
SODIUM HYDROCHLORITE			AB to 70°F		B at 70°F	B at 70°F
SODIUM HYDROGEN CARBONATE			AB to 70°F			AB to 70°F

	POLYETHYLENE LOW DENSITY LDPE	®TEFLON FEP	® KALREZ PERFLUORINATED ELASTOMER	® FLUORAZ FLUORINATED COPOLYMER	® AFLAS FLUORINATED COPOLYMER	®NORPRENE COPOLYMER TPO
SODA, CAUSTIC	AB conc to 140°F C 50-80% at 70°F	A to 100% to 300°F	A to 100% to 212°F A to 50% to 302°F	A to 100% to 250°F	A to 50% to 212°F	AB 50% to conc to 70°F BC 46% at 70°F
SODA LIME		A to 70°F				
SODA NITER	A to 70°F	A to 300°F	A to 70°F			AB to 70°F
SODA SALTPETER		A to 300°F				
SODA, WASHING		A to 70°F				
SODAN		A to 70°F				
SODIUM		A to 70°F NR molten	NR molten			
SODIUM ACETATE	B at 70°F	A to 300°F	A to 212°F	NR at 70°F anhydrous		
SODIUM ACID CARBONATE	AB to 70°F	A to 70°F				
SODIUM ACID SULFATE	A to 122°F	A to 122°F	A to 212°F			
SODIUM ACID SULFITE	AB to 70°F					
SODIUM ALUM	AB to 70°F	A to 70°F	A to 212°F			
SODIUM ALUMINATE	AB to 70°F	A to 70°F	A to 212°F	A to 70°F		
SODIUM ALUMINUM SULFATE	AB to 70°F	A to 70°F	A to 212°F			
SODIUM ARSENATE	AB to 70°F	A to 70°F	A to 212°F			
SODIUM BENZOATE	AB to 70°F	A to 300°F	A to 212°F			
SODIUM BICARBONATE	A to 70°F AB to 140°F	A to 300°F	A to 212°F	A to 70°F		A to 70°F
SODIUM BICHROMATE	AB to 70°F	A to 300°F	A to 212°F			
SODIUM BISULFATE	AB to 70°F	A to 300°F	A to 212°F	A to 70°F		A to 70°F
SODIUM BISULFITE	A to 70°F	A to 200°F	A to 212°F	A to 70°F	+1% vol 7 days 212°F A to 212°F	A to 70°F
SODIUM BILSULPHATE	AB to 70°F	A to 300°F	A to 212°F	A to 70°F		A to 70°F
SODIUM BORATE	A to 70°F AB to 140°F	A to 300°F	A to 212°F	A to 70°F	+0% vol 7 days 212°F A to 212°F	A to 70°F
SODIUM BROMIDES	A to 70°F	A to 200°F	A to 212°F			
SODIUM CARBONATE	AB to 100% to 140°F	A to 100% to 300°F	A to 212°F	A to 70°F	+0% vol 7 days 212°F A to 212°F	A to 70°F
SODIUM CHLORATE	AB to 140°F	A to 200°F	A to 212°F		+0% vol 3 days 48% conc 176°F	A to 70°F
SODIUM CHLORIDE	AB to 140°F	A to 100% to 300°F	A to 212°F	A to 100% to 70°F	A 35% to 212°F	A to 70°F
SODIUM CHLORITE	AB to 70°F	B at 70°F	A to 212°F		+22% vol 10% conc 7 days 212°F	
SODIUM CHROMATE	AB to 70°F	A to 70°F	A to 212°F			
SODIUM CYANIDE	A to 70°F	A to 300°F	A to 212°F	A to 70°F		
SODIUM DICHROMATE	A to 70°F	A to 70°F	A to 212°F			
SODIUM DIMETABORATE	AB to 70°F	A to 70°F				
SODIUM DIOXIDE	AB to 70°F	A to 70°F				
SODIUM DIPHOSPHATE	AB to 70°F	A to 70°F	A to 212°F			
SODIUM DISULFITE	A to 70°F	A to 200°F	A to 212°F	A to 70°F		A to 70°F
SODIUM ETHOXIDE	AB to 70°F	A to 70°F				
SODIUM FERRICYANIDE	A to 70°F	A to 300°F	A to 212°F			
SODIUM FERROCYANIDE	A to 70°F	A to 70°F	A to 212°F			A to 70°F
SODIUM FLUOALUMINATE	AB to 70°F	A 10% to 70°F				
SODIUM FLUOBORATE	AB to 70°F					
SODIUM FLUORIDE	AC to 70°F	A to 300°F	A to 212°F			
SODIUM HEXAMETAPHOSPHATE	NR at 70°F	A to 70°F				
SODIUM HYDRATE	AB to 70°F	A to 70°F				
SODIUM HYDROCHLORITE	A to 70°F	A to 70°F				
SODIUM HYDROGEN CARBONATE	AB to 70°F	A to 70°F				

	RECOMMENDATIONS	NITRILE NBR	ETHYLENE PROPYLENE EP, EPDM	FLUOROCARBON FKM	CHLOROPRENE CR	HYDROGENATED NITRILE HNBR
SODIUM HYDROGEN SULFATE		A to 70°F	A to 70°F	A to 70°F	A to 70°F	
SODIUM HYDROSULFITE	NBR	A/NR at 70°F A sol'n to 160°F	B at 70°F A sol'n to 70°F	AB to 70°F A sol'n to 70°F	AB to 70°F AB sol'n to 70°F	
SODIUM HYDROXIDE	EP,IIR,NBR,CR	+1% vol 100% conc 3 days 70°F	no swell 100% conc 3 days 70°F	no swell 100% conc 28 days 70°F	+1% vol 100% conc 3 days 70°F	B at 70°F
SODIUM HYDROXIDE		−5% vol 20% conc 3 days 212°F	−1.2% vol 20% conc 3 days 212°F	+46% vol 20% conc 3 days 212°F	−1% vol 20% conc 3 days 212°F	
SODIUM HYDROXIDE		NR 80-100% at 70°F B/NR 50-100% at 175°F	A to 100% to 70°F AB 100% to 125°F	B 100% at 70°F NR 100% at 104°F	A to conc to 70°F A to 50% to 140°F	
SODIUM HYDROXIDE		A to 50% to 176°F C/NR 30% at 212°F	A to 50% to 176°F B 30-50% at 212°F	B 80% at 70°F C/NR 40-80% at 175°F	AB to 50% to 200°F NR 50% at 212°F	
SODIUM HYDROXIDE		A to 20% to 212°F NR molten	AB 20% at 200°F B 20% at 212°F	AC 15-50% to 140°F NR 15-30% at 150°F		
SODIUM HYDROXIDE			B 15% at 176°F A sol'n to 250°F	A sol'n to 70°F		
SODIUM HYPOCHLORIDE		NR at 70°F	C at 70°F	A to 70°F	C at 70°F	
SODIUM HYPOCHLORITE	EP,CR,IIR,NBR,FKM	B/NR 100% at 70°F NR 100% at 130°F	AB 20-100% to 130°F AB 20% at 200°F	+24% vol 20% conc 28 days 158°F	B/NR 100% at 70°F BC 20% at 70°F	B at 70°F
SODIUM HYPOCHLORITE		NR 50% at 70°F B/NR to 30% at 70°F	A sol'n to 160°F	A to conc to 130°F BC 20% at 158°F	B to 50% at 70°F C 5% at 104°F	
SODIUM HYPOCHLORITE		B sol'n at 70°F		NR 5% at 140°F	NR 5% at 140°F	
SODIUM HYPOSULFITE	NBR,FKM,CR,EP	A to 70°F AB any conc to 200°F	A to 104°F	A to 140°F	A to 140°F AB any conc to 200°F	A to 70°F
SODIUM IODINE		C at 70°F	B at 70°F	A to 70°F	AB to 70°F	
SODIUM, LIQUID				A to 350°F		
SODIUM METABORATE PEROXY-HYDRATE		A to 70°F AB any conc to 200°F		AB to 70°F	A to 70°F AB any conc to 200°F	
SODIUM METAPHOSPHATE	NBR,EP,FKM,CR,IIR SBR	A to 100% to 140°F AB any conc to 200°F	A to 100% to 212°F	A to 100% to 140°F	B to 100% to 200°F	A to 70°F
SODIUM METASILICATE		A to 176°F	A to 176°F B at 212°F	A to 212°F	A to 70°F	
SODIUM NITRATE	EP,IIR,NBR,CR,NR	AB any conc to 150°F B at 176°F	A to 100% to 200°F	A to 212°F	B to 100% to 200°F	
SODIUM NITRITE		A to 140°F B at 176°F	A to 176°F	A to 212°F	A to 70°F	
SODIUM ORTHOPHENYLPHENATE	CR,NR	NR at 70°F		AB to 70°F	A to 70°F	
SODIUM PERBORATE	EP,NBR,IIR,FKM,CR	B to 100% to 200°F	A to 100% to 140°F A sol'n to 200°F	A to 140°F A sol'n to 70°F	B any conc to 200°F	B at 70°F
SODIUM PEROXIDE	EP,NBR,IIR,FKM,CR	B to 100% to 200°F	A to 100% to 140°F	A to 100% to boiling	B to 100% at 70°F NR conc at 140°F	B at 70°F
SODIUM PHOSPHATE DIBASIC	NBR,EP,FKM,SBR	A to 70°F B at 140°F	A to 100% to 140°F	A to 100% to 140°F	AB to 100% to 70°F NR at 140°F	A to 70°F
SODIUM PHOSPHATE MONOBASIC	NBR,EP,FKM,SBR	A to 70°F B at 140°F	A to 100% to 140°F	A to 100% to 140°F	BC to 100% at 70°F NR at 140°F	A to 70°F
SODIUM PHOSPHATE TRIBASIC	NBR,EP,FKM,SBR	A to 70°F B at 140°F	A to 100% to 140°F	A to 100% to 140°F B to 300°F	BC to 100% at 70°F NR at 140°F	A to 70°F
SODIUM PHOSPHATES	NBR,EP,IIR,CR,FKM	A to 200°F	A to 176°F B at 212°F	A to 176°F B at 212°F	A to 140°F AB any conc to 200°F	A to 70°F
SODIUM POLYPHOSPHATE		A to 70°F	A to 70°F	A to 70°F	A/NR at 70°F	
SODIUM PLUMBITE		AB to 70°F		AB to 70°F		
SODIUM PYROBORATE		B any conc to 180°F	A to 70°F	A to 70°F	A to 70°F AB any conc to 200°F	
SODIUM SACCHARIN		AB to 70°F		AB to 70°F		
SODIUM SALT	NBR,EP,FKM,SBR	A to 70°F	AB to 70°F	A to 70°F	AB to 70°F	A to 70°F
SODIUM SESQUICARBONATE		A to 70°F AB any conc to 200°F		AB to 70°F	A to 70°F AB any conc to 200°F	
SODIUM SILICATE	NBR,EP,IIR,CR	A to 100% to 140°F A sol'n to 200°F	A to 100% to 176°F AB to 100% to 200°F	A to 100% to 212°F	A to 100% to 140°F AB to 100% to 200°F	A to 70°F
SODIUM SILICATE SULFATE		A to 70°F AB any conc to 200°F		AB to 70°F	A to 70°F AB any conc to 200°F	
SODIUM STANNIC CHLORIDE		B any conc to 150°F		AB to 70°F	NR at 70°F	
SODIUM SUBSULFITE		A to 70°F AB any conc to 200°F	A to 70°F	A to 70°F	A to 70°F AB any conc to 200°F	
SODIUM SULFATE DECAHYDRATE		AB any conc to 200°F		AB to 70°F	AB any conc to 200°F	
SODIUM SULFATES	NBR,EP,FKM,CR,IIR	A to 100% to 200°F B at 212°F	A to 100% to 176°F AB to 200°F	A to 100% to 212°F	A to 100% to 140°F AB to 100% to 200°F	NR at 70°F
SODIUM SULFHYDRATE	EP	A sol'n to 70°F	A sol'n to 200°F	A sol'n to 70°F	A sol'n to 70°F	
SODIUM SULFIDE	NBR,EP,FKM,IIR,CR	A to 100% to 140°F AB to 200°F	A to 100% to 176°F B at 212°F	A to 100% to 176°F B at 212°F	A to 100% to 140°F AB to 200°F	
SODIUM SULFITE	NBR,EP,FKM,IIR,CR	A to 100% to 70°F AB to 100% to 200°F	A to 100% to 140°F AB to 100% to 200°F	A to 100% to 140°F B at 176°F	A to 100% to 140°F AB to 100% to 200°F	A to 70°F
SODIUM SUPEROXIDE		B any conc to 200°F	B at 70°F	A to 70°F	B any conc to 200°F	
SODIUM TETRABORATE	NBR,CR,EP	A to 140°F AB any conc to 180°F	A to 100% to 70°F A sol'n to 200°F	A to 100% to 140°F	AB any conc to 200°F	A to 70°F

	STYRENE BUTADIENE SBR	POLYACRYLATE ACM	POLYURETHANE AU, EU	ISOBUTYLENE ISOPRENE IIR	POLYBUTADIENE BR	® AEROQUIP AQP
SODIUM HYDROGEN SULFATE	AB to 70°F			A to 70°F		
SODIUM HYDROSULFITE	B sol'n at 70°F	NR sol'n at 70°F	B at 70°F	B at 70°F		
SODIUM HYDROXIDE	-2% vol 3 days 70°F A conc to 70°F	no swell 3 days 70°F A/NR conc at 70°F	+2% vol 3 days 70°F B to 100% at 70°F	+0.6% vol 100% conc 3 days 70°F	A to 100% to 70°F	C 50% at 70°F
SODIUM HYDROXIDE	B sol'n at 70°F B 3 molar at 70°F	NR sol'n at 70°F NR 3 molar at 70°F	NR 70% at 140°F AB 50% fumes to 70°F	-2.7% vol 20% conc 3 days 212°F		
SODIUM HYDROXIDE				A to 100% at 70°F A to 80% to 175°F		
SODIUM HYDROXIDE				AB any conc to 185°F A 20% to 212°F		
SODIUM HYDROXIDE						
SODIUM HYDROXIDE						
SODIUM HYPOCHLORIDE	AB to 70°F		NR at 70°F *AB dilute to 70°F	B at 70°F		
SODIUM HYPOCHLORITE	BC 100% at 70°F B to 20% at 70°F	NR any conc at 70°F	NR 5-100% at 70°F AB 1% to 70°F	B 100% at 70°F B to 20% to 80°F	BC 100% at 70°F B 20% at 70°F	
SODIUM HYPOCHLORITE				AB 5% to 120°F		
SODIUM HYPOCHLORITE						
SODIUM HYPOSULFITE	AB to 70°F		A to 70°F	A to 70°F AB any conc to 185°F		
SODIUM IODINE	AB to 70°F			B at 70°F		
SODIUM, LIQUID			NR at 70°F			
SODIUM METABORATE PEROXY-HYDRATE	AB to 70°F			A to 70°F AB any conc to 185°F		
SODIUM METAPHOSPHATE	A to 100% to 70°F		B at 70°F	AB to 70°F		
SODIUM METASILICATE	AB to 70°F		B at 70°F	A to 70°F AB any conc to 185°F	A to 70°F	
SODIUM NITRATE	B to 100% at 70°F		B at 70°F	A to 70°F AB any conc to 185°F	A to 70°F	AB to 70°F
SODIUM NITRITE	AB to 70°F			AB to 70°F		
SODIUM ORTHOPHENYLPHENATE	AB to 70°F			AB to 70°F		
SODIUM PERBORATE	B to 100% at 70°F		B at 70°F	A to 70°F AB any conc to 185°F	B at 70°F	NR at 70°F
SODIUM PEROXIDE	B to 100% at 70°F	NR any conc at 70°F	NR any conc at 70°F	A to boiling	B at 70°F	NR at 70°F
SODIUM PHOSPHATE DIBASIC	A to 100% to 70°F	A to 70°F	A to 70°F	A to 70°F	A to 70°F	NR at 70°F
SODIUM PHOSPHATE MONOBASIC	A to 100% to 70°F	A to 70°F	A to 70°F	A to 70°F AB to 185°F	A to 70°F	NR at 70°F
SODIUM PHOSPHATE TRIBASIC	A to 100% to 70°F	A to 70°F	A to 70°F	A to 70°F	A to 70°F	NR at 70°F
SODIUM PHOSPHATES	A to 70°F	A to 70°F	A to 70°F	A to 70°F AB any conc to 185°F	A to 70°F	NR at 70°F
SODIUM POLYPHOSPHATE	AB to 70°F			AB to 70°F		
SODIUM PLUMBITE	AB to 70°F		B at 70°F	AB to 70°F		
SODIUM PYROBORATE	AB to 70°F		A to 70°F	A to 70°F AB any conc to 185°F		
SODIUM SACCHARIN	AB to 70°F		B at 70°F	AB to 70°F		
SODIUM SALT	A to 70°F	A to 70°F	A to 70°F	A to 70°F	A to 70°F	
SODIUM SESQUICARBONATE	AB to 70°F			A to 70°F AB any conc to 185°F		
SODIUM SILICATE	A to 100% to 70°F		AB to 70°F	A to 70°F	A to 70°F	AB to 70°F
SODIUM SILICATE SULFATE	AB to 70°F			A to 70°F AB any conc to 185°F		
SODIUM STANNIC CHLORIDE	AB to 70°F			A to 70°F AB any conc to 150°F		
SODIUM SUBSULFITE	AB to 70°F		A to 70°F	A to 70°F AB any conc to 185°F		
SODIUM SULFATE DECAHYDRATE	AB to 70°F			AB any conc to 185°F		
SODIUM SULFATES	B to 100% at 70°F	NR any conc at 70°F	A to 100% to 70°F	A to 70°F AB any conc to 150°F	B/NR at 70°F	
SODIUM SULFHYDRATE	B sol'n at 70°F	NR sol'n at 70°F		AB to 70°F		
SODIUM SULFIDE	B to 100% at 70°F	NR any conc at 70°F	A to 100% to 70°F	A to 70°F	B at 70°F	AB to 70°F
SODIUM SULFITE	B to 100% at 70°F	NR any conc at 70°F	A to 100% to 70°F A to 10% to 150°F	A to 70°F AB any conc to 185°F	B at 70°F	AB to 70°F
SODIUM SUPEROXIDE	AB to 70°F		NR at 70°F	A to 70°F AB any conc to 185°F		
SODIUM TETRABORATE	B sol'n at 70°F	NR sol'n at 70°F	*AB to 70°F	A to 70°F AB any conc to 185°F		

C 339

	SYNTHETIC ISOPRENE IR	NATURAL ISOPRENE NR	CHLOROSULFONATED POLYETHYLENE CSM	FLUOROSILICONE FVMQ	SILICONE VMQ	CHEMRAZ FFKM
SODIUM HYDROGEN SULFATE	A to 70°F	A to 70°F	A to 70°F		A to 70°F	A to 70°F
SODIUM HYDROSULFITE		AC to 70°F	B at 70°F	A sol'n to 70°F	AC to 70°F A sol'n to 70°F	A to 70°F
SODIUM HYDROXIDE	A conc to 70°F A 3 molar to 70°F	A to conc to 70°F AB any conc to 150°F	-2% vol 20% conc 3 days 212°F	no swell 50% conc 7 days 75°F	no swell 100% conc 3 days 70°F	AB to 100% to 70°F C at 70°F white
SODIUM HYDROXIDE		NR 20% at 160°F	A to 100% to 70°F A to 73% to 280°F	B to 100% to 70°F	+9% vol 50% conc 7 days 70°F	
SODIUM HYDROXIDE					-2% vol 20% conc 7 days 70°F	
SODIUM HYDROXIDE					+1.2% vol 20% conc 3 days 212°F	
SODIUM HYDROXIDE					A to conc to 70°F A 20% to 212°F	
SODIUM HYDROXIDE						
SODIUM HYPOCHLORIDE	C at 70°F	C at 70°F	A to 70°F			
SODIUM HYPOCHLORITE	BC conc at 70°F B/NR 20% at 70°F	C/NR 5-100% at 70°F	AB 30-100% to 200°F A to 20% to 158°F	B to 100% to 70°F	B to 100% to 70°F	A 20-100% to 70°F
SODIUM HYPOCHLORITE	C 5% at 70°F					
SODIUM HYPOCHLORITE						
SODIUM HYPOSULFITE	A to 70°F	A to 70°F AB any conc to 150°F	A to 70°F	A to 70°F	AB to 70°F	A to 70°F
SODIUM IODINE		C at 70°F	A to 70°F		NR at 70°F	
SODIUM, LIQUID						NR at 70°F
SODIUM METABORATE PEROXY-HYDRATE	A to 70°F	A to 70°F AB any conc to 150°F	AB to 70°F			
SODIUM METAPHOSPHATE	A to 70°F	A to 70°F AB any conc to 150°F	B at 70°F	A to 100% to 70°F	A to 70°F	A to 70°F
SODIUM METASILICATE		AB to 70°F	AB to 70°F			A to 70°F
SODIUM NITRATE	AB to 70°F	B any conc to 150°F	A to 70°F AB to 200°F		C/NR conc at 70°F NR sol'n at 70°F	A to 70°F
SODIUM NITRITE		AB to 70°F	A to 70°F		AB to 70°F	A to 70°F
SODIUM ORTHOPHENYLPHENATE		A to 70°F	AB to 70°F			
SODIUM PERBORATE	B at 70°F	B any conc to 150°F	B at 70°F	A to 100% to 70°F	B to 100% to 70°F	A to 70°F
SODIUM PEROXIDE	B at 70°F	B any conc to 150°F	B at 70°F A sol'n to 200°F	A to 100% to 70°F	C/NR at 70°F NR sol'n at 70°F	A to 70°F
SODIUM PHOSPHATE DIBASIC	A to 70°F	A to 70°F	A to 70°F		NR any conc at 70°F	A to 70°F
SODIUM PHOSPHATE MONOBASIC	A to 70°F	A to 70°F	A to 70°F		NR any conc at 70°F	A to 70°F
SODIUM PHOSPHATE TRIBASIC	A to 70°F	A to 70°F	A to 70°F	B at 70°F	A to 70°F NR sol'n at 70°F	A to 70°F
SODIUM PHOSPHATES	A to 70°F	A to 70°F AB any conc to 150°F	A to 70°F		C/NR at 70°F	A to 70°F
SODIUM POLYPHOSPHATE		AC to 70°F	AB to 70°F		NR at 70°F	
SODIUM PLUMBITE		AB to 70°F	AB to 70°F			
SODIUM PYROBORATE	A to 70°F	A to 70°F AB any conc to 150°F	A to 70°F		B at 70°F	
SODIUM SACCHARIN		AB to 70°F	AB to 70°F			
SODIUM SALT	A to 70°F	A to 70°F	A to 70°F	A to 70°F	A to 70°F	A to 70°F
SODIUM SESQUICARBONATE	A to 70°F	A to 70°F AB any conc to 150°F	AB to 70°F			A to 70°F
SODIUM SILICATE	A to 70°F	A to 70°F AB any conc to 185°F	A to 70°F AB to 200°F		A to 70°F	A to 70°F
SODIUM SILICATE SULFATE	A to 70°F	A to 70°F AB any conc to 150°F	AB to 70°F			
SODIUM STANNIC CHLORIDE	A to 70°F	A to 70°F	AB to 70°F			
SODIUM SUBSULFITE	A to 70°F	A to 70°F	A to 70°F		A to 70°F	
SODIUM SULFATE DECAHYDRATE		AB any conc to 150°F	AB to 70°F			
SODIUM SULFATES	AB to 70°F	AB any conc to 185°F	A to 70°F AB to 200°F	A to 100% to 70°F	A to 100% to 70°F	A to 70°F
SODIUM SULFHYDRATE		AB to 70°F	AB to 70°F	A sol'n to 70°F	A sol'n to 70°F	
SODIUM SULFIDE	AB to 70°F	AB to 185°F	A to 70°F AB to 200°F	A to 100% to 70°F	A to 100% to 70°F	A to 70°F
SODIUM SULFITE	AB to 70°F	AB any conc to 150°F	A to 70°F	A to 100% to 70°F	A to 100% to 70°F	A to 70°F
SODIUM SUPEROXIDE	B at 70°F	AB any conc to 150°F	A to 70°F		C at 70°F	
SODIUM TETRABORATE	A to 70°F	A to 70°F AB any conc to 150°F	A to 70°F	A to 70°F	A to 70°F	A to 70°F

	POLYSULFIDE T	CHLORINATED POLYETHYLENE CM	EPICHLOROHYDRIN CO, ECO	® HYTREL COPOLYESTER TPE	® SANTOPRENE COPOLYMER TPO	® C-FLEX STYRENIC TPE
SODIUM HYDROGEN SULFATE	AB to 70°F					
SODIUM HYDROSULFITE	AB to 70°F					A to 70°F
SODIUM HYDROXIDE	– 0.2% vol 50% conc 3 days 70°F	C at 70°F	B conc at 70°F B 3 molar at 70°F	+1% vol 20% conc 3 days 70°F	no swell 15-20% conc 7 days 70°F	A to conc to 70°F
SODIUM HYDROXIDE	NR conc at 70°F NR 3 molar at 70°F			NR 73% at 70°F BC 46-50% at 70°F	A to 50% to 70°F	
SODIUM HYDROXIDE	AB 20-50% to 70°F			AC to 20% to 70°F		
SODIUM HYDROXIDE						
SODIUM HYDROXIDE						
SODIUM HYDROXIDE						
SODIUM HYPOCHLORIDE	AB to 70°F					
SODIUM HYPOCHLORITE	NR 20-100% at 70°F	NR at 70°F	A 20-100% to 70°F	BC 100% at 70°F AB to 20% to 70°F	A to 70°F aqueous	A to 100% to 70°F
SODIUM HYPOCHLORITE				A to 5% to 70°F		
SODIUM HYPOCHLORITE						
SODIUM HYPOSULFITE	AB to 70°F					
SODIUM IODINE	AB to 70°F					
SODIUM, LIQUID						
SODIUM METABORATE PEROXY-HYDRATE	AB to 70°F					
SODIUM METAPHOSPHATE	AB to 70°F					
SODIUM METASILICATE	AB to 70°F					
SODIUM NITRATE	AB to 70°F	AB to 70°F	A to 70°F	AB to 70°F		A to 70°F
SODIUM NITRITE	AB to 70°F					
SODIUM ORTHOPHENYLPHENATE	AB to 70°F					
SODIUM PERBORATE	B at 70°F	NR at 70°F		AB to 70°F		
SODIUM PEROXIDE	AB to 70°F	NR at 70°F		AB to 70°F		
SODIUM PHOSPHATE DIBASIC	AB to 70°F			AB to 70°F		
SODIUM PHOSPHATE MONOBASIC	AB to 70°F			AB to 70°F		
SODIUM PHOSPHATE TRIBASIC	AB to 70°F		NR at 70°F	AB to 70°F		
SODIUM PHOSPHATES	AB to 70°F	NR at 70°F		AC to 70°F		
SODIUM POLYPHOSPHATE	AB to 70°F					
SODIUM PLUMBITE	AB to 70°F					
SODIUM PYROBORATE	AB to 70°F					
SODIUM SACCHARIN	AB to 70°F					
SODIUM SALT	A to 70°F					
SODIUM SESQUICARBONATE	AB to 70°F					
SODIUM SILICATE	AB to 70°F	AB to 70°F		AB to 70°F		A to 70°F
SODIUM SILICATE SULFATE	AB to 70°F					
SODIUM STANNIC CHLORIDE	AB to 70°F					
SODIUM SUBSULFITE	AB to 70°F					
SODIUM SULFATE DECAHYDRATE	AB to 70°F					
SODIUM SULFATES	B at 70°F	AB to 70°F	A to 70°F	AB to 70°F		
SODIUM SULFHYDRATE	AB to 70°F					
SODIUM SULFIDE	B at 70°F	AB to 70°F		AB to 70°F		A to 70°F
SODIUM SULFITE	B at 70°F			*AB to 70°F		A to 70°F
SODIUM SUPEROXIDE	AB to 70°F					
SODIUM TETRABORATE	AB to 70°F			AB to 70°F *AB to 140°F		

	ETHYLENE ACRYLIC EA	POLYALLOMER LINEAR COPOLYMER	NYLON 11 POLYAMIDE	NYLON 12 POLYAMIDE	ETHYLENE VINYL ACETATE EVA	POLYVINYLCHLORIDE FLEXIBLE PVC
SODIUM HYDROGEN SULFATE			AB to 70°F			AB to 70°F
SODIUM HYDROSULFITE			AB to 70°F		B at 70°F	AB to 70°F
SODIUM HYDROXIDE	NR 73% at 70°F A 20-46% to 70°F	A to conc to 122°F	+2.5% vol 0-10% conc 180 days 70°F	AB 50% to 70°F	A to 70°F AB to 140°F	A to conc to 70°F B conc at 122°F
SODIUM HYDROXIDE			AB to 100% to 70°F NR 50% at 140°F			NR conc at 140°F B/NR 40-80% at 130°F
SODIUM HYDROXIDE						BC 10% at 150°F A 1% to 122°F
SODIUM HYDROXIDE						
SODIUM HYDROXIDE						
SODIUM HYDROXIDE						
SODIUM HYPOCHLORIDE			AB to 70°F			AB to 70°F
SODIUM HYPOCHLORITE	A to 20% to 70°F	A to 100% to 70°F A to 15% to 122°F	B/NR conc at 70°F C dilute at 70°F	C 5% at 70°F	AB conc to 70°F *AB 15% to 140°F	AB 5-100% to 70°F C 20-100% at 140°F
SODIUM HYPOCHLORITE						
SODIUM HYPOCHLORITE						
SODIUM HYPOSULFITE			AB to 70°F			AB to 70°F
SODIUM IODINE			AB to 70°F			B at 70°F
SODIUM, LIQUID			AB to 70°F		NR at 70°F	NR at 70°F
SODIUM METABORATE PEROXY-HYDRATE						
SODIUM METAPHOSPHATE		A to 70°F	A to 70°F		AB to 70°F	B at 70°F *B to 140°F
SODIUM METASILICATE			AB to 70°F		B at 70°F	B at 70°F
SODIUM NITRATE		A to 70°F	A to 70°F *AB to 140°F	AB 10% to 70°F	A to 70°F AB to 140°F	A 30-100% to 70°F *AB 30-100% at 140°F
SODIUM NITRITE			AB to 70°F	C 5% at 70°F		AB to 70°F
SODIUM ORTHOPHENYLPHENATE			AB to 70°F			AB to 70°F
SODIUM PERBORATE		A to 70°F	A to 70°F	AB 5% to 70°F	A to 70°F	A to 70°F
SODIUM PEROXIDE			NR at 70°F		NR at 70°F	C/NR at 70°F
SODIUM PHOSPHATE DIBASIC			A/NR at 70°F			AB to 70°F
SODIUM PHOSPHATE MONOBASIC			A/NR at 70°F			AB to 70°F
SODIUM PHOSPHATE TRIBASIC	B at 70°F		AB to 140°F C at 194°F		*NR at 70°F	B/NR at 70°F
SODIUM PHOSPHATES		A to 70°F	A/NR at 70°F	AB 10% to 70°F	A to 70°F	A to 70°F
SODIUM POLYPHOSPHATE			AB to 70°F			AB to 70°F
SODIUM PLUMBITE			AB to 70°F		A to 70°F	A to 70°F
SODIUM PYROBORATE			AB to 70°F			AB to 70°F
SODIUM SACCHARIN			AB to 70°F		A to 70°F	B at 70°F
SODIUM SALT			AB to 70°F			A to 70°F
SODIUM SESQUICARBONATE			AB to 70°F			AB to 70°F
SODIUM SILICATE		A to 70°F	A to 70°F AB to 194°F	AB to 70°F	A to 70°F	A to 70°F
SODIUM SILICATE SULFATE			AB to 70°F			AB to 70°F
SODIUM STANNIC CHLORIDE			AB to 70°F			AB to 70°F
SODIUM SUBSULFITE			AB to 70°F			AB to 70°F
SODIUM SULFATE DECAHYDRATE			AB to 70°F			AB to 70°F
SODIUM SULFATES	A sol'n to 70°F	A to 70°F	A to 70°F	AB to 15% to 70°F	AB to 70°F	A to 70°F *AB to 140°F
SODIUM SULFHYDRATE			AB to 70°F			AB to 70°F
SODIUM SULFIDE		A to 70°F	A to 70°F C 105-140°F	AB 10% to 70°F	A to 70°F *AB to 100% to 140°F	A to 70°F AB to 100% to 140°F
SODIUM SULFITE		A to 70°F	A to 70°F *C at 140°F	AB 10% to 70°F	AB to 70°F *AB to 140°F	A to 70°F
SODIUM SUPEROXIDE			AB to 70°F			AB to 70°F
SODIUM TETRABORATE			*AB to 140°F		AB to 140°F	*AB to 130°F

	POLYETHYLENE LOW DENSITY LDPE	® TEFLON FEP	® KALREZ PERFLUORINATED ELASTOMER	® FLUORAZ FLUORINATED COPOLYMER	® AFLAS FLUORINATED COPOLYMER	® NORPRENE COPOLYMER TPO
SODIUM HYDROGEN SULFATE	AB to 70°F	A to 70°F	A to 212°F			
SODIUM HYDROSULFITE	A to 70°F	A to 70°F	A to 212°F	A to 70°F		AB to 70°F
SODIUM HYDROXIDE	AB conc to 140°F C 50-80% at 70°F	no swell to 50% conc 3 days 212°F	A to 100% to 212°F A to 50% to 302°F	A to 100% to 250°F	+2% vol to 50% conc 3 days 212°F	AB 50% to conc to 70°F BC 46% at 70°F
SODIUM HYDROXIDE	AB 5-20% to 140°F	A to 100% to 300°F			A to 50% to 212°F	A to 40% to 70°F
SODIUM HYDROXIDE		NR molten				
SODIUM HYDROXIDE						
SODIUM HYDROXIDE						
SODIUM HYDROXIDE						
SODIUM HYPOCHLORIDE	AB to 15% to 140°F	A to 70°F				
SODIUM HYPOCHLORITE	AB to conc to 140°F	A to 100% to 300°F	A to 212°F	A to 100% to 70°F	+1% vol 10% conc 7 days 212°F	A to conc to 70°F
SODIUM HYPOCHLORITE					A to 10% to 212°F	
SODIUM HYPOCHLORITE						
SODIUM HYPOSULFITE	A to 104°F B at 140°F	A to 300°F	A to 212°F	AC to 70°F		
SODIUM IODINE	AB to 70°F	A to 70°F				
SODIUM, LIQUID	NR	NR hot	NR molten			
SODIUM METABORATE PEROXY-HYDRATE	AB to 70°F	A to 70°F				
SODIUM METAPHOSPHATE	A to 70°F	A to 300°F	A to 212°F	A to 70°F		
SODIUM METASILICATE	A to 70°F	A to 70°F	A to 212°F	C at 70°F		
SODIUM NITRATE	A to 70°F AB to 140°F	A to 300°F NR molten	A to 212°F	A/NR at 70°F	A sat'd to 70°F	AB to 70°F
SODIUM NITRITE	AB to 70°F	A to 300°F				
SODIUM ORTHOPHENYLPHENATE	AB to 70°F					
SODIUM PERBORATE	A to 70°F	A to 300°F	A to 212°F	A to 70°F		
SODIUM PEROXIDE	C/NR at 70°F	A to 300°F	A to 212°F	A to 100% to 70°F		
SODIUM PHOSPHATE DIBASIC	AB to 70°F	A to 300°F		A to 70°F		
SODIUM PHOSPHATE MONOBASIC	AB to 70°F	A to 300°F		A to 70°F		
SODIUM PHOSPHATE TRIBASIC	AB to 70°F	A to 300°F	A to 70°F	A to 70°F		
SODIUM PHOSPHATES	AB to 70°F	A to 300°F	A to 212°F	A to 70°F	+0% vol 7 days 212°F A to 212°F	
SODIUM POLYPHOSPHATE	AB to 70°F	A to 70°F				
SODIUM PLUMBITE	A to 70°F	A to 70°F	A to 212°F	A to 70°F		
SODIUM PYROBORATE	AB to 70°F	A to 70°F				
SODIUM SACCHARIN	A to 70°F					
SODIUM SALT	AB to 140°F	A to 70°F	A to 212°F			A to 70°F
SODIUM SESQUICARBONATE	AB to 70°F	A to 70°F				
SODIUM SILICATE	A to 70°F	A to 300°F	A to 212°F	A to 70°F	+0% vol 7 days 212°F A to 212°F	
SODIUM SILICATE SULFATE	AB to 70°F	A to 70°F				
SODIUM STANNIC CHLORIDE	AB to 70°F	A to 70°F				
SODIUM SUBSULFITE	AB to 70°F	A to 70°F				
SODIUM SULFATE DECAHYDRATE	AB to 70°F					
SODIUM SULFATES	A to 70°F	A to 300°F	A to 212°F	A to 70°F	A sat'd to 70°F	
SODIUM SULFHYDRATE	AB to 70°F					
SODIUM SULFIDE	A to 70°F AB to 140°F	A to 300°F	A to 212°F	A to 70°F		
SODIUM SULFITE	A to 70°F AB to 140°F	A to 300°F	A to 212°F	A to 70°F		
SODIUM SUPEROXIDE	AB to 70°F	A to 70°F				
SODIUM TETRABORATE	A to 70°F AB to 140°F	A to 300°F	A to 212°F	A to 70°F		A to 70°F

C 343

	RECOMMENDATIONS	NITRILE NBR	ETHYLENE PROPYLENE EP, EPDM	FLUOROCARBON FKM	CHLOROPRENE CR	HYDROGENATED NITRILE HNBR
SODIUM THIOCYANATE	NBR, EP, FKM	A to 104°F	A to 140°F	A to 176°F		
SODIUM THIOPHOSPHATES	EP, NBR, FKM, CR, IIR AU	AB any conc to 200°F	A to 100% to 176°F B at 212°F	A to 100% to 212°F	A to 100% to 140°F NR at 212°F	
SODIUM TRIPHOSPHATES		A to 70°F AB any conc to 200°F	AB to 70°F	AB to 70°F	A to 70°F AB any conc to 200°F	
SODIUM TRIPOLYPHOSPHATE		A to 70°F AB any conc to 200°F	AB to 70°F	AB to 70°F	A to 70°F AB any conc to 200°F	
SOLENE		A to 70°F	NR at 70°F	A to 70°F	B at 70°F	
SOLUBLE GLASS	NBR, EP, IIR, CR	A to 100% to 140°F A sol'n to 200°F	A to 100% to 176°F AB to 100% to 200°F	A to 100% to 212°F	A to 100% to 140°F AB to 100% to 200°F	A to 70°F
SOLVASOL 1, 2 & 3	NBR, FKM	A to 70°F	NR at 70°F	A to 70°F	B at 70°F	A to 70°F
SOLVASOL 73	FKM	B at 70°F	NR at 70°F	A to 70°F	B at 70°F	B at 70°F
SOLVASOL 74	FKM	B at 70°F	NR at 70°F	A to 70°F	B at 70°F	B at 70°F
SOLVENTS, GENERAL		A/NR at 70°F	NR at 70°F	AB to 70°F	C/NR at 70°F	
SORGHUM		A to 140°F		A to 140°F	A to 140°F	
SOY SAUCE		A to 140°F		A to 140°F	A to 140°F	
SOYA OIL (SOY OIL)	NBR	A to 250°F	B/NR at 70°F	A to 70°F	AB to 80°F	
SOYBEAN OIL (SOYABEAN OIL)	NBR, FKM, CR, ACM	A to 250°F	NR at 70°F	A to 250°F	A to 80°F	A to 70°F
SPERM OIL	NBR	A to 250°F	NR at 70°F	A to 70°F	C/NR at 70°F	
SPERRY OIL						
SPIRIT		A to 70°F	A to 70°F	B at 70°F	A to 70°F	
SPIRITS OF TURPENTINE		A to 70°F AB to 150°F	NR at 70°F	A to 70°F	C/NR at 70°F	
SPIRITS OF VINEGAR		C at 70°F	A to 70°F	C at 70°F	B at 70°F	
SPIRITS OF WINE		A to 70°F AB to 150°F	A to 70°F	B at 70°F	A to 70°F AB to 150°F	
SPRY		A to 70°F	B at 70°F	A to 70°F	B at 70°F	A to 70°F
SR6 FUEL	FKM	B at 70°F	NR at 70°F	A to 70°F	NR at 70°F	B at 70°F
SR10 FUEL	NBR, FKM	A to 70°F	NR at 70°F	A to 70°F	NR at 70°F	A to 70°F
STANDARD OIL MULTILUBE GX90-EP	NBR, FKM, ACM, AU	A to 70°F	NR at 70°F	A to 70°F	B at 70°F	B at 70°F
STANNIC CHLORIDE	NBR, EP, FKM, IIR	A to 100% to 140°F A sol'n to 200°F	A to 100% to 140°F	A to 100% to 140°F AB to 200°F	BC conc at 70°F NR 50% at 70°F	A to 70°F
STANNIC FLUOBORATE		A to 140°F		A to 140°F	A to 140°F	
STANNIC SALTS		A to 70°F AB any conc to 150°F	AB to 70°F	A to 70°F	A/NR at 70°F	
STANNIC SULFIDE		A to 70°F				
STANNOUS CHLORIDE	NBR, EP, FKM, CR, IIR	A to 100% to 70°F AB any conc to 150°F	BC to 100% to 104°F	A to 100% to 70°F	A to 100% to 140°F B at 176°F	A to 70°F
STANNOUS SULFIDE		A to 70°F				
STARCH		A to 140°F AB to 176°F	A to 176°F	A to 212°F	A to 70°F	
STARCH SYRUP		A to 100% to 140°F	A to 70°F	A to 140°F	A to 100% to 140°F	
STAUFFER 7700	FKM	B at 70°F	NR at 70°F	+19% vol 7 days 392°F A to 70°F	NR at 70°F	B at 70°F
STEAM	EP	NR at 212°F	+1% vol 7 days 320°F +2% vol 7 days 550°F	+13% vol 7 days 320°F +20% vol 3 days 392°F	C at 212°F C/NR to 450°F	
STEAM			A to 350°F B to 450°F	A to 212°F NR at 225°F		
STEAM & HOT WATER	EP	NR at 225°F	A to 350°F	BC at 350°F NR at 400°F	A/NR to 275°F	
STEARIC ACID	NBR, EP, IIR	A to 100% to 70°F B to 100% to 250°F	B to 140°F	A to 140°F C at 176°F	B to 140°F BC at 158°F	B at 70°F
STEARIN			A to 176°F			
STODDARDS SOLVENT	NBR, FKM, FVMQ, ACM AU	A to 250°F	NR at 70°F	A to 158°F	BC at 70°F NR at 140°F	A to 70°F
STRIPPER S.A.					NR at 70°F	
STRONTIEM NITRATE						
STYRENE MONOMER	FKM, FVMQ	NR at 70°F	NR at 70°F	+11.4% vol 28 days 75°F +31% vol 7 days 122°F	NR at 70°F	NR at 70°F
STYRENE MONOMER				AB to 70°F		
STYRENE POLYMER	FKM, FVMQ	NR at 70°F	NR at 70°F	A to 120°F	NR at 70°F	

	STYRENE BUTADIENE SBR	POLYACRYLATE ACM	POLYURETHANE AU, EU	ISOBUTYLENE ISOPRENE IIR	POLYBUTADIENE BR	®AEROQUIP AQP
SODIUM THIOCYANATE	AB to 70°F			AB to 70°F		
SODIUM THIOPHOSPHATES	B to 100% at 70°F	NR any conc at 70°F	A to 100% to 70°F	A to 70°F AB any conc to 185°F	B at 70°F	AB to 70°F
SODIUM TRIPHOSPHATES	AB to 70°F			A to 70°F AB any conc to 185°F		
SODIUM TRIPOLYPHOSPHATE	AB to 70°F			A to 70°F AB any conc to 185°F		
SOLENE			B at 70°F	NR at 70°F		
SOLUBLE GLASS	A to 100% to 70°F		AB to 70°F	A to 70°F	A to 70°F	AB to 70°F
SOLVASOL 1, 2 & 3	NR at 70°F	B at 70°F	B at 70°F	NR at 70°F	NR at 70°F	
SOLVASOL 73	NR at 70°F	B at 70°F	B at 70°F	NR at 70°F	NR at 70°F	
SOLVASOL 74	NR at 70°F	B at 70°F	B at 70°F	NR at 70°F	NR at 70°F	
SOLVENTS, GENERAL			B at 70°F			
SORGHUM						
SOY SAUCE						
SOYA OIL (SOY OIL)	NR at 70°F	A to 70°F	AB to 70°F	A/NR at 70°F		
SOYBEAN OIL (SOYABEAN OIL)	NR at 70°F	A to 70°F	B at 70°F	AC to 120°F	NR at 70°F	AB to 70°F
SPERM OIL	NR at 70°F	A to 70°F				
SPERRY OIL			AB to 70°F			
SPIRIT			NR at 70°F	A to 70°F		
SPIRITS OF TURPENTINE			NR at 70°F	NR at 70°F		
SPIRITS OF VINEGAR			B at 70°F	B at 70°F		
SPIRITS OF WINE			NR at 70°F	A to 70°F AB to 120°F		
SPRY	NR at 70°F	A to 70°F	A to 70°F	B at 70°F	NR at 70°F	
SR6 FUEL	NR at 70°F	B at 70°F	B at 70°F	NR at 70°F	NR at 70°F	
SR10 FUEL	NR at 70°F	B at 70°F	B at 70°F	NR at 70°F	NR at 70°F	
STANDARD OIL MULTILUBE GX90-EP	NR at 70°F	A to 70°F	A to 70°F	NR at 70°F	NR at 70°F	
STANNIC CHLORIDE	A to 100% to 70°F		AB to 70°F	AB any conc to 150°F	A to 100% to 70°F	
STANNIC FLUOBORATE						
STANNIC SALTS			B at 70°F	AB any conc to 150°F		
STANNIC SULFIDE				A to 70°F		
STANNOUS CHLORIDE	A to 100% to 70°F		BC at 70°F	AB any conc to 150°F	A to 100% to 70°F	
STANNOUS SULFIDE				A to 70°F		
STARCH			A to 70°F			
STARCH SYRUP			A to 70°F	A to 70°F AB any conc to 185°F		
STAUFFER 7700	NR at 70°F	B at 70°F		NR at 70°F	NR at 70°F	
STEAM	B to 225°F NR at 300°F	NR at 225°F	NR at 225°F	+2.1% vol 7 days 320°F +7.8% vol 3 days 392°F	NR at 225°F	TEST
STEAM				B to 350°F C/NR 350–450°F		
STEAM & HOT WATER	NR at 275°F	NR at 275°F	NR at 275°F			
STEARIC ACID	B at 70°F	NR at 70°F	A to 70°F	B any conc to 75°F	B/NR at 70°F	AB to 70°F
STEARIN			AB to 70°F			
STODDARDS SOLVENT	NR at 70°F	A to 70°F	AB to 110°F C at 140°F	NR at 70°F	NR at 70°F	AB to 70°F
STRIPPER S.A.						
STRONTIEM NITRATE			C/NR at 70°F			
STYRENE MONOMER	NR at 70°F	NR at 70°F	BC at 70°F	NR at 70°F	NR at 70°F	
STYRENE MONOMER						
STYRENE POLYMER	NR at 70°F	NR at 70°F	A/NR at 70°F	NR at 70°F	NR at 70°F	

C 345

	SYNTHETIC ISOPRENE IR	NATURAL ISOPRENE NR	CHLOROSULFONATED POLYETHYLENE CSM	FLUOROSILICONE FVMQ	SILICONE VMQ	CHEMRAZ FFKM
SODIUM THIOCYANATE		AB to 70°F	AB to 70°F			A to 70°F
SODIUM THIOPHOSPHATES	AB to 70°F	AB any conc to 150°F	A to 70°F AB to 125°F	A to 100% to 70°F	A to 100% to 70°F AB to 125°F	A to 70°F
SODIUM TRIPHOSPHATES	A to 70°F	A to 70°F AB any conc to 150°F	AB to 70°F			A to 70°F
SODIUM TRIPOLYPHOSPHATE	A to 70°F	A to 70°F AB any conc to 150°F	AB to 70°F			
SOLENE	NR at 70°F	NR at 70°F	NR at 70°F		C at 70°F	
SOLUBLE GLASS	A to 70°F	A to 70°F AB any conc to 150°F	A to 70°F AB to 200°F		A to 70°F	A to 70°F
SOLVASOL 1, 2 & 3	NR at 70°F	NR at 70°F	B at 70°F	A to 70°F	NR at 70°F	A to 70°F
SOLVASOL 73	NR at 70°F	NR at 70°F	B at 70°F	A to 70°F	NR at 70°F	A to 70°F
SOLVASOL 74	NR at 70°F	NR at 70°F	B at 70°F	A to 70°F	NR at 70°F	A to 70°F
SOLVENTS, GENERAL						
SORGHUM		A to 70°F				
SOY SAUCE		A/NR at 70°F				
SOYA OIL (SOY OIL)	NR at 70°F	NR at 70°F	A to 70°F	A to 70°F	NR at 70°F	
SOYBEAN OIL (SOYABEAN OIL)	NR at 70°F	NR at 70°F	A to 70°F	A to 70°F	A to 70°F	A to 70°F
SPERM OIL		NR at 70°F		A to 70°F	NR at 70°F	
SPERRY OIL						
SPIRIT	A to 70°F	A to 70°F	A to 70°F		A to 70°F	
SPIRITS OF TURPENTINE	NR at 70°F	NR at 70°F	NR at 70°F			
SPIRITS OF VINEGAR	B at 70°F	B at 70°F	B at 70°F		A to 70°F	
SPIRITS OF WINE	A to 70°F	A to 70°F AB to 150°F	A to 70°F		A to 70°F	
SPRY	NR at 70°F	NR at 70°F	NR at 70°F	A to 70°F	A to 70°F	A to 70°F
SR6 FUEL	NR at 70°F	NR at 70°F	NR at 70°F	A to 70°F	NR at 70°F	A to 70°F
SR10 FUEL	NR at 70°F	NR at 70°F	NR at 70°F	A to 70°F	NR at 70°F	A to 70°F
STANDARD OIL MULTILUBE GX90-EP	NR at 70°F	NR at 70°F	B at 70°F	A to 70°F	NR at 70°F	A to 70°F
STANNIC CHLORIDE	A 50-100% to 70°F	A 50-100% to 70°F AB to 185°F	AC conc to 70°F NR 50% at 70°F	A to 100% to 70°F	B 50-100% at 70°F	AB to 70°F
STANNIC FLUOBORATE						
STANNIC SALTS	A to 70°F	A to 70°F AB any conc to 150°F	A to 70°F		B at 70°F	
STANNIC SULFIDE	A to 70°F	A to 70°F	A to 70°F			
STANNOUS CHLORIDE	A to 100% to 70°F	A to 100% to 70°F AB to 100% to 150°F	A to 100% to 70°F A to 15% to 158°F	A to 100% to 70°F	B 15-100% at 70°F	A 15-100% to 70°F
STANNOUS SULFIDE	A to 70°F	A to 70°F	A to 70°F			
STARCH		A to 70°F	A to 70°F			
STARCH SYRUP	B at 70°F	B any conc to 120°F	A to 70°F		A to 70°F	
STAUFFER 7700	NR at 70°F	NR at 70°F	NR at 70°F	B at 70°F	NR at 70°F	A to 70°F
STEAM	C at 225°F NR at 300°F	C at 225°F NR at 300°F	B to 225°F C/NR at 300°F	no swell 1 day 225°F NR at 225°F	B/NR at 225°F C/NR at 300°F	A to 350°F
STEAM			NR at 350°F		NR at 350°F disintegrated	
STEAM & HOT WATER			A to 275°F	NR at 275°F	NR at 275°F	
STEARIC ACID	B/NR at 70°F	B/NR at 70°F	BC to 158°F	BC at 70°F	B at 70°F	A to 70°F
STEARIN						
STODDARDS SOLVENT	NR at 70°F	NR at 70°F	C/NR at 70°F	A to 70°F	NR at 70°F	A to 70°F
STRIPPER S.A.	NR at 70°F	NR at 70°F				
STRONTIEM NITRATE						A aqueous sol'n to 70°F
STYRENE MONOMER	NR at 70°F	NR at 70°F	NR at 70°F	C at 70°F	NR at 70°F	A to 70°F
STYRENE MONOMER						
STYRENE POLYMER	NR at 70°F	NR at 70°F	NR at 70°F	C at 70°F	NR at 70°F	A to 70°F

	POLYSULFIDE T	CHLORINATED POLYETHYLENE CM	EPICHLOROHYDRIN CO, ECO	® HYTREL COPOLYESTER TPE	® SANTOPRENE COPOLYMER TPO	® C-FLEX STYRENIC TPE
SODIUM THIOCYANATE	AB to 70°F					
SODIUM THIOPHOSPHATES	B at 70°F	AB to 70°F		AB to 70°F		
SODIUM TRIPHOSPHATES	AB to 70°F					
SODIUM TRIPOLYPHOSPHATE	AB to 70°F					
SOLENE						
SOLUBLE GLASS	AB to 70°F	AB to 70°F		AB to 70°F		A to 70°F
SOLVASOL 1, 2 & 3	B at 70°F					
SOLVASOL 73	B at 70°F					
SOLVASOL 74	B at 70°F					
SOLVENTS, GENERAL			A to 70°F kerosene base			
SORGHUM						
SOY SAUCE						
SOYA OIL (SOY OIL)						
SOYBEAN OIL (SOYABEAN OIL)	NR at 70°F	AB to 70°F	A to 70°F	AB to 70°F		
SPERM OIL						
SPERRY OIL						
SPIRIT						
SPIRITS OF TURPENTINE						
SPIRITS OF VINEGAR						
SPIRITS OF WINE						
SPRY	NR at 70°F					
SR6 FUEL	A to 70°F					
SR10 FUEL	A to 70°F					
STANDARD OIL MULTILUBE GX90-EP	B at 70°F					
STANNIC CHLORIDE	AB to 70°F	NR at 70°F		AB to 70°F		
STANNIC FLUOBORATE						
STANNIC SALTS						A to 70°F
STANNIC SULFIDE						
STANNOUS CHLORIDE	AB to 70°F			AB 15-100% to 70°F		
STANNOUS SULFIDE						
STARCH				*AB to 70°F		
STARCH SYRUP						
STAUFFER 7700	A to 70°F					
STEAM	NR at 350°F	C at 450°F	NR at 300°F	NR at 225°F		TEST
STEAM						
STEAM & HOT WATER				BC at 212°F NR at 230°F		
STEARIC ACID	AB to 70°F	AB to 70°F	B at 70°F	AB to 70°F	A to 70°F	A to 70°F
STEARIN				AB to 70°F		
STODDARDS SOLVENT	B at 70°F	AB to 70°F	A to 70°F	A to 70°F	NR at 70°F	NR at 70°F
STRIPPER S.A.						
STRONTIEM NITRATE						
STYRENE MONOMER	NR at 70°F			*NR at 70°F		*NR at 70°F
STYRENE MONOMER						
STYRENE POLYMER	NR at 70°F			C at 70°F		NR at 70°F

C 347

	ETHYLENE ACRYLIC EA	POLYALLOMER LINEAR COPOLYMER	NYLON 11 POLYAMIDE	NYLON 12 POLYAMIDE	ETHYLENE VINYL ACETATE EVA	POLYVINYLCHLORIDE FLEXIBLE PVC
SODIUM THIOCYANATE			AB to 70°F			AB to 70°F
SODIUM THIOPHOSPHATES		A to 70°F	A to 70°F	AB to 70°F	A to 70°F	A to 70°F AB to 140°F
SODIUM TRIPHOSPHATES			AB to 70°F			AB to 70°F
SODIUM TRIPOLYPHOSPHATE			AB to 70°F			AB to 70°F
SOLENE						
SOLUBLE GLASS		A to 70°F	A to 70°F AB to 194°F		A to 70°F	A to 70°F
SOLVASOL 1, 2 & 3						
SOLVASOL 73						
SOLVASOL 74						
SOLVENTS, GENERAL				AB to 70°F naphtha base	A to 70°F	NR at 70°F AB to 70°F kerosene
SORGHUM						
SOY SAUCE						
SOYA OIL (SOY OIL)				AB to 70°F		A to 70°F B at 150°F
SOYBEAN OIL (SOYABEAN OIL)	*AB to 70°F		A to 70°F		NR at 70°F	AB to 70°F C at 150°F
SPERM OIL						
SPERRY OIL					NR at 70°F	B at 70°F
SPIRIT						
SPIRITS OF TURPENTINE						
SPIRITS OF VINEGAR						
SPIRITS OF WINE						NR at 70°F
SPRY						
SR6 FUEL						
SR10 FUEL						
STANDARD OIL MULTILUBE GX90-EP						
STANNIC CHLORIDE	*AB to 70°F		A/NR at 70°F		B at 70°F	AB to 140°F
STANNIC FLUOBORATE						
STANNIC SALTS						A to 70°F
STANNIC SULFIDE						
STANNOUS CHLORIDE	*AB 15% to 70°F		C at 70°F		B at 70°F	AB to 140°F
STANNOUS SULFIDE						
STARCH			*AB to 140°F	AB to 70°F aqueous	AB to 140°F	A to 70°F *AB to 140°F
STARCH SYRUP						
STAUFFER 7700						
STEAM			C at 212°F NR at 225°F		NR at 275°F	NR at 225°F
STEAM						
STEAM & HOT WATER	A to 212°F				NR at 275°F	NR at 225°F
STEARIC ACID		A to 70°F A crystals to 122°F	A to 70°F C at 194°F		BC at 70°F	A to 70°F B crystals at 122°F
STEARIN			AB to 140°F			
STODDARDS SOLVENT	BC at 70°F	B at 70°F C at 122°F	A to 70°F swells at 140°F		NR at 70°F	AC to 70°F A to 122°F (special
STRIPPER S.A.						
STRONTIEM NITRATE					B at 70°F	A to 70°F
STYRENE MONOMER	*NR at 70°F			AB to 70°F		NR at 70°F
STYRENE MONOMER						
STYRENE POLYMER	NR at 70°F		A to 70°F AB to 105°F swells	AB to 70°F		NR at 70°F

C 348

	POLYETHYLENE LOW DENSITY LDPE	®TEFLON FEP	®KALREZ PERFLUORINATED ELASTOMER	®FLUORAZ FLUORINATED COPOLYMER	®AFLAS FLUORINATED COPOLYMER	®NORPRENE COPOLYMER TPO
SODIUM THIOCYANATE	AB to 70°F	A to 70°F	A to 212°F			
SODIUM THIOPHOSPHATES	A to 70°F	A to 300°F	A to 70°F	A to 70°F		
SODIUM TRIPHOSPHATES	AB to 70°F	A to 70°F	A to 212°F			
SODIUM TRIPOLYPHOSPHATE	AB to 70°F					
SOLENE		A to 70°F				
SOLUBLE GLASS	A to 70°F	A to 300°F	A to 212°F	A to 70°F	A to 212°F	
SOLVASOL 1, 2 & 3		A to 70°F		NR at 70°F		
SOLVASOL 73		A to 70°F		NR at 70°F		
SOLVASOL 74		A to 70°F		NR at 70°F		
SOLVENTS, GENERAL	A to 70°F	A to boiling aliphatic & aromatic				
SORGHUM						
SOY SAUCE						
SOYA OIL (SOY OIL)		A to 70°F		A to 70°F		
SOYBEAN OIL (SOYABEAN OIL)	A to 70°F	A to 200°F	A to 212°F	A to 70°F		
SPERM OIL				A to 70°F		
SPERRY OIL	A to 70°F					
SPIRIT		A to 70°F				
SPIRITS OF TURPENTINE		A to 70°F				
SPIRITS OF VINEGAR		A to 70°F				
SPIRITS OF WINE		A to 70°F				
SPRY				A to 70°F		
SR6 FUEL				A to 70°F		
SR10 FUEL				A to 70°F		
STANDARD OIL MULTILUBE GX90-EP				A to 70°F		
STANNIC CHLORIDE	A to 70°F AB to 140°F	A to 300°F	A to 70°F *A aqueous to 212°F	A 15-100% to 70°F		
STANNIC FLUOBORATE						
STANNIC SALTS	A to 70°F	A to 70°F				AB to 70°F
STANNIC SULFIDE	AB to 140°F	A to 70°F				
STANNOUS CHLORIDE	A to 70°F AB to 140°F	A to 300°F	A to 212°F	A to 70°F		
STANNOUS SULFIDE	AB to 140°F	A to 70°F				
STARCH	A to 140°F	A to 70°F		A to 70°F		
STARCH SYRUP		A to 70°F				
STAUFFER 7700			A to 212°F	B at 70°F	+22% vol 3 days 350°F +26% vol 3 days 400°F	
STEAM	NR at 225°F	A to 450°F	A to 540°F 1018 compound	A to 350°F	+4.6% vol 7 days 320°F +4.9% vol 3 days 392°F	TEST
STEAM					+1.6% vol 4 days 550°F 90 shore compound	
STEAM & HOT WATER			A to 355°F 1018 compound		+1.1% vol 3 days 212°F +18% vol 3 days 350°F	
STEARIC ACID	AC to 70°F A crystals to 122°F	A to 300°F A crystals to 122°F	A to 212°F	AC at 70°F	A to 70°F	A to 70°F
STEARIN						
STODDARDS SOLVENT	C at 70°F NR at 122°F	A to 300°F	A to 212°F *A to 400°F	AC to 70°F	+5% vol 7 days 70°F A to 70°F	NR at 70°F
STRIPPER S.A.		A to 70°F				
STRONTIEM NITRATE	A to 70°F		A aqueous to 212°F	NR at 70°F aqueous		
STYRENE MONOMER		A to 70°F	*A to 212°F	NR at 70°F	+27% vol 7 days 77°F BC at 70°F	*NR at 70°F
STYRENE MONOMER						
STYRENE POLYMER		A to 70°F	A to 75°F *A to 212°F			NR at 70°F

C 349

	RECOMMENDATIONS	NITRILE NBR	ETHYLENE PROPYLENE EP, EPDM	FLUOROCARBON FKM	CHLOROPRENE CR	HYDROGENATED NITRILE HNBR
SUBLIMED WHITE LEAD		AB any conc to 120°F			A to 70°F AB any conc to 180°F	
SUCCINIC ACID		A to 70°F	A to 70°F	A to 70°F		
SUCROSE SOLUTION	NBR,EP,FKM,SBR	A to 100% to 140°F	A to 250°F	A to 140°F	A to 70°F AB any conc to 120°F	B at 70°F
SUGAR OF LEAD	EP,IIR,NBR	AB any conc to 140°F B at 176°F	A to 200°F	A to 140°F B at 176°F	AB any conc to 80°F	
SUGAR LIQUORS, CANE, BEET & MAPLE		A to 70°F	A to 70°F	A to 70°F	A to 70°F	
SUGAR SOLUTIONS	NBR,EP,FKM,SBR	A to 100% to 140°F	A to 250°F	A to 140°F	AB any conc to 120°F	
SULFAMIC ACID		B/NR at 70°F	NR at 70°F	A to 70°F	AB any conc to 150°F	
SULFATE LIQUORS (GREEN & BLACK)	EP,NBR,FKM,IIR,CR	AB to 150°F B/NR 212–250°F	A to 200°F AB to 250°F	A to 176°F B at 212°F	AB any conc to 150°F	
SULFITE CELLULOSE LIQUORS		A to 70°F	B at 70°F	A to 70°F	A to 70°F	
SULFITE LIQUORS	NBR,CR,EP	AB to 70°F	B at 70°F A 6% to 140°F	A to 70°F A 6% to 140°F	AB to 70°F	
SULFITIC LIQUORS		AB to 70°F	B at 70°F	A to 70°F	A to 70°F	
SULFONIC ACID		NR at 70°F		NR at 70°F		
SULFUR (MOLTEN 250° F)	EP,CR,FKM,IIR	NR at 70°F	A to 250°F	A to 266°F	A to 70°F NR at 250°F	NR at 70°F
SULFUR CHLORIDE	FKM,FVMQ	C/NR at 70°F NR at 140°F	NR at 70°F	A to 140°F	C/NR at 70°F NR at 140°F	NR at 70°F
SULFUR DICHLORIDE		NR at 70°F	NR at 70°F	A to 70°F		
SULFUR DIOXIDE GAS DRY	EP	NR at 70°F	A to 140°F NR at 212°F	A to 250°F	NR at 70°F	NR at 70°F
SULFUR DIOXIDE GAS WET	EP,IIR	NR at 70°F	A to 140°F NR at 212°F	A to 140°F B/NR at 176°F	B to 104°F NR at 176°F	NR at 70°F
SULFUR DIOXIDE LIQUID	EP	NR at 70°F anhydrous	A to 70°F anhydrous	AB to 70°F NR at 70°F anhydrous	A to 70°F NR at 70°F anhydrous	
SULFUR DIOXIDE, PRESSURIZED	EP	NR at 70°F	A to 70°F	NR at 70°F	NR at 70°F	NR at 70°F
SULFUR HEXAFLUORIDE	CR,EP,IIR,VMQ	B at 70°F	A to 120°F	A/NR at 70°F	A to 70°F	B at 70°F
SULFUR LIQUORS	FKM	B at 70°F	B at 70°F	A to 70°F	B at 70°F	
SULFUR MONOCHLORIDE	FKM,FVMQ	C/NR at 70°F NR at 140°F	NR at 70°F	A to 140°F	C/NR at 70°F NR at 140°F	
SULFUR SALTS		C/NR at 70°F	NR at 70°F	A to 140°F	C/NR at 70°F	
SULFUR SUBCHLORIDE		C/NR at 70°F	NR at 70°F	A to 70°F	C/NR at 70°F	
SULFUR TRIOXIDE	FKM,EP,IIR	NR at 70°F	BC at 70°F B/NR at 140°F	A to 140°F	NR at 70°F	
SULFUR TRIOXIDE, DRY	FKM,EP,IIR	NR at 70°F	AB to 120°F	A to 70°F	NR at 70°F	NR at 70°F
SULFURATED LIME		A to 70°F AB any conc to 200°F	A to 70°F	A to 70°F	AB any conc to 150°F	
SULFURETTED HYDROGEN		NR at 70°F	A to 70°F	B at 70°F	C/NR at 70°F	
SULFURIC ACID	FKM,EP,IIR	dissolved 100% conc 3 days 70°F	+7% vol 100% conc 3 days 158°F	+3% vol 100% conc 3 days 158°F	dissolved 100% conc 3 days 70°F	
SULFURIC ACID		disintegrated 93% conc 3 days 212°F	disintegrated 95% conc 3 days 212°F	+20% vol 98% conc 28 days 158°F	no swell 50% conc 3 days 158°F	
SULFURIC ACID		+1% vol 50% conc 3 days 70°F	-1% vol 50% conc 3 days 158°F	+28% vol 98% conc 14 days 212°F	NR 75–100% at 70°F BC 60% to 104°F	
SULFURIC ACID		NR 93–100% at 70°F C 90% to 176°F	NR 95–100% at 70°F B 93% to 104°F	+5% vol 95% conc 28 days 158°F	C/NR 60% at 140°F NR 60% at 176°F	
SULFURIC ACID		B 80% to 104°F C 70–80% 140–176°F	NR 90–93% at 176°F A to 90% to 70°F	no swell 60% conc 28 days 158°F	A to 50% to 158°F B to 30% at 176°F	
SULFURIC ACID		NR 30–80% at 212°F B 70% to 140°F	B 90% at 104°F C 90% at 140°F	A to 100% to 158°F NR 95% at 212°F	NR 10–30% at 212°F AB 10% at 180°F	
SULFURIC ACID		A 60% to 140°F B 60% at 176°F	A to 80% to 140°F C 80% at 176°F	B 80–90% 176–212°F NR 80–90% at 248°F		
SULFURIC ACID		A 50% to 70°F B 50% to 104°F	B 25–70% 140–176°F C 70% at 212°F	A to 70% to 176°F B 60–70% at 212°F		
SULFURIC ACID		C 50% at 176°F A to 30% to 140°F	NR 10–60% at 212°F A 10% to 176°F	C 60–70% at 248°F A to 50% to 212°F		
SULFURIC ACID		B to 30% at 176°F	A dilute to 70°F	B 50% at 248°F A to 30% at 248°F		
SULFURIC ACID FUMING	FKM	NR at 70°F	NR at 70°F	+23% vol 7 days 70°F B to 140°F	NR at 70°F disintegrated	
SULFURIC ACID (20-25% OLEUM)	FKM	NR at 70°F	NR at 70°F	+5–12% vol 7 days 75°F +51% vol 5 days 158°F	NR at 70°F	B at 70°F
SULFURIC ETHER		AB any conc to 120°F	NR at 70°F	NR at 70°F	NR at 70°F	
SULFUROUS ACID	FKM,EP,IIR	B/NR 100% at 70°F C/NR 100% at 140°F	AB to 176°F B at 212°F	A to 140°F B to 200°F	NR 75–100% at 70°F C 30% at 70°F	B at 70°F
SULFUROUS ACID		A/NR 10–75% at 70°F A 5% to 70°F	A to 75% to 70°F	C/NR at 212°F A to 75% to 70°F	AB 10% to 70°F	
SULFUROUS ANHYDRIDE		C/NR at 70°F	C at 70°F	A to 70°F	BC at 70°F NR at 140°F	

	STYRENE BUTADIENE SBR	POLYACRYLATE ACM	POLYURETHANE AU, EU	ISOBUTYLENE ISOPRENE IIR	POLYBUTADIENE BR	® AEROQUIP AQP
SUBLIMED WHITE LEAD				A to 70°F AB any conc to 185°F		
SUCCINIC ACID						
SUCROSE SOLUTION	A to 70°F	NR at 70°F	A/NR at 70°F	A to 70°F AB any conc to 120°F	A to 70°F	
SUGAR OF LEAD	NR at 70°F	NR at 70°F	B/NR at 70°F	A to 70°F AB any conc to 120°F	NR at 70°F	
SUGAR LIQUORS, CANE, BEET & MAPLE				A to 70°F		
SUGAR SOLUTIONS	A to 70°F	NR at 70°F	A/NR at 70°F	A to 70°F AB any conc to 120°F	A to 70°F	
SULFAMIC ACID			NR at 70°F	A to 70°F AB any conc to 150°F		
SULFATE LIQUORS (GREEN & BLACK)	B at 70°F	NR at 70°F	A/NR at 70°F	A to 70°F AB any conc to 150°F	B at 70°F	NR at 70°F
SULFITE CELLULOSE LIQUORS				A to 70°F		
SULFITE LIQUORS	B at 70°F	NR at 70°F	C/NR at 70°F	B at 70°F	B at 70°F	
SULFITIC LIQUORS				A to 70°F AB any conc to 120°F		
SULFONIC ACID				NR at 70°F		
SULFUR (MOLTEN 250° F)	NR at 70°F	NR at 70°F	B at 70°F NR at 250°F	A to 70°F C at 250°F	NR at 70°F	AB to 70°F
SULFUR CHLORIDE	NR at 70°F	NR at 70°F	C at 70°F	NR at 70°F	NR at 70°F	AB to 70°F
SULFUR DICHLORIDE						
SULFUR DIOXIDE GAS DRY	BC at 70°F	NR at 70°F	B/NR at 70°F	B any conc to 185°F	BC at 70°F	NR at 70°F
SULFUR DIOXIDE GAS WET	B/NR at 70°F	NR at 70°F	*C at 70°F	A to 70°F	NR at 70°F	
SULFUR DIOXIDE LIQUID	NR at 70°F anhydrous	NR at 70°F anhydrous	B at 70°F	B at 70°F		
SULFUR DIOXIDE, PRESSURIZED	NR at 70°F	NR at 70°F		B at 70°F	NR at 70°F	
SULFUR HEXAFLUORIDE	NR at 70°F	NR at 70°F	AB to 70°F	A any conc to 80°F	NR at 70°F	
SULFUR LIQUORS	B at 70°F	NR at 70°F		B at 70°F	B at 70°F	
SULFUR MONOCHLORIDE	NR at 70°F	NR at 70°F	C at 70°F	NR at 70°F	NR at 70°F	AB to 70°F
SULFUR SALTS						
SULFUR SUBCHLORIDE						
SULFUR TRIOXIDE	C/NR at 70°F	NR at 70°F	C/NR at 70°F	B at 70°F	NR at 70°F	NR at 70°F
SULFUR TRIOXIDE, DRY	NR at 70°F	NR at 70°F	B at 70°F	BC at 70°F	B at 70°F	
SULFURATED LIME			A to 70°F	A to 70°F AB any conc to 185°F		
SULFURETTED HYDROGEN				A to 70°F AB any conc to 185°F		
SULFURIC ACID	dissolved 100% conc 3 days 70°F	dissolved 100% conc 3 days 70°F	dissolved 100% conc 3 days 158°F	dissolved 100% conc 3 days 70°F	NR 100% at 70°F C dilute at 70°F	
SULFURIC ACID	+3% vol 50% conc 3 days 70°F	+3% vol 50% conc 3 days 70°F	+36% vol 50% conc 3 days 158°F	dissolved 95% conc 3 days 212°F		
SULFURIC ACID	NR 100% at 70°F B 10-50% at 70°F	NR 50-100% at 70°F NR dilute at 70°F	NR 60-100% at 70°F B/NR 10-50% at 70°F	+0.6% vol 50% conc 3 days 70°F		
SULFURIC ACID	C dilute at 70°F		AB 5-10% to 70°F B/NR dilute at 70°F	NR 95-100% at 70°F B 75% to 100°F		
SULFURIC ACID				NR 75% at 160°F B 50% to 150°F		
SULFURIC ACID				B 25-30% to 200°F A to 10% to 150°F		
SULFURIC ACID				B 10% to 185°F A dilute to 70°F		
SULFURIC ACID						
SULFURIC ACID						
SULFURIC ACID						
SULFURIC ACID FUMING	NR at 70°F	NR at 70°F	NR at 70°F	NR at 70°F disintegrated	NR at 70°F	NR at 70°F
SULFURIC ACID (20-25% OLEUM)	NR at 70°F	NR at 70°F	NR at 70°F	NR at 70°F	NR at 70°F	NR at 70°F
SULFURIC ETHER			C at 70°F	C/NR at 70°F		
SULFUROUS ACID	B at 70°F	NR at 70°F	NR 30-100% at 70°F	AB 100% to 100°F AB 75% to 125°F	B at 75°F	NR at 70°F
SULFUROUS ACID				A to 75% to 70°F AB 50% to 175°F		
SULFUROUS ANHYDRIDE				B/NR at 70°F		

	SYNTHETIC ISOPRENE IR	NATURAL ISOPRENE NR	CHLOROSULFONATED POLYETHYLENE CSM	FLUOROSILICONE FVMQ	SILICONE VMQ	CHEMRAZ FFKM
SUBLIMED WHITE LEAD	B at 70°F	AB any conc to 120°F				
SUCCINIC ACID						A to 70°F
SUCROSE SOLUTION	A to 70°F	A to 70°F AB any conc to 120°F	AB to 70°F	A to 70°F	A to 70°F	A to 70°F
SUGAR OF LEAD	AB to 70°F	A any conc to 80°F AB to 185°F	A/NR at 70°F	NR at 70°F	NR at 70°F	A to 70°F
SUGAR LIQUORS, CANE, BEET & MAPLE	A to 70°F	A to 70°F	A to 70°F		A to 70°F	
SUGAR SOLUTIONS	A to 70°F	A to 70°F AB any conc to 120°F	AB to 70°F	A to 70°F	A to 70°F	
SULFAMIC ACID	B at 70°F	AB any conc to 150°F	A to 70°F			A to 70°F
SULFATE LIQUORS (GREEN & BLACK)	AB to 70°F	AB any conc to 75°F	AB to 70°F	B at 70°F	AB to 70°F	A to 70°F
SULFITE CELLULOSE LIQUORS	B at 70°F	B at 70°F	A to 70°F		NR at 70°F	
SULFITE LIQUORS	B at 70°F	B any conc to 120°F	B at 70°F	B at 70°F	NR at 70°F	
SULFITIC LIQUORS	B at 70°F	B at 70°F	A to 70°F		NR at 70°F	
SULFONIC ACID	NR at 70°F	NR at 70°F	BC at 70°F			A to 70°F
SULFUR (MOLTEN 250° F)	C/NR at 70°F NR at 250°F	C/NR at 70°F NR at 250°F	A to 250°F	A to 70°F C at 250°F	A to 70°F C at 250°F	A to 250°F
SULFUR CHLORIDE	NR at 70°F	NR at 70°F	A/NR at 70°F	A to 70°F	C at 70°F	A to 70°F
SULFUR DICHLORIDE						
SULFUR DIOXIDE GAS DRY	BC at 70°F	BC at 70°F	A/NR conc at 70°F A 1% to 100°F	B at 70°F	AB to 70°F	A to 70°F
SULFUR DIOXIDE GAS WET	NR at 70°F	B/NR at 70°F	AC to 70°F	B at 70°F	B at 70°F	A to 70°F
SULFUR DIOXIDE LIQUID	B at 70°F	B at 70°F	A to 70°F	B at 70°F anhydrous	B at 70°F anhydrous	A to 70°F
SULFUR DIOXIDE, PRESSURIZED	NR at 70°F	NR at 70°F	NR at 70°F	B at 70°F	B at 70°F	A to 70°F
SULFUR HEXAFLUORIDE	NR at 70°F	NR at 70°F	AB to 70°F	B at 70°F	AB to 70°F	*B at 70°F
SULFUR LIQUORS	B at 70°F	B at 70°F	B at 70°F	B at 70°F	NR at 70°F	*B at 70°F
SULFUR MONOCHLORIDE	NR at 70°F	NR at 70°F	A/NR at 70°F	A to 70°F	C at 70°F	A to 70°F
SULFUR SALTS						
SULFUR SUBCHLORIDE	NR at 70°F	NR at 70°F	A to 70°F		C at 70°F	
SULFUR TRIOXIDE	B at 70°F	BC at 70°F	NR at 70°F	B at 70°F	B at 70°F	AB to 70°F A to 70°F white
SULFUR TRIOXIDE, DRY	B at 70°F	B/NR at 70°F	B/NR at 70°F	B at 70°F	B at 70°F	AB to 70°F A to 70°F white
SULFURATED LIME	A to 70°F	A to 70°F AB any conc to 150°F	A to 70°F		B at 70°F	
SULFURETTED HYDROGEN	C at 70°F	C/NR at 70°F	C at 70°F		C at 70°F	
SULFURIC ACID	NR 75-100% at 70°F C 60% at 70°F	NR 75-100% at 70°F C/NR 60% at 70°F	disintegrated 100%conc 3 days 212°F	disintegrated 100%conc 3 days 70°F	dissolved 100% conc 3 days 70°F	A 3M to conc to 70°
SULFURIC ACID	B 25-50% at 70°F A 10% to 70°F	B 50% to 80°F AB 37% to 100°F	C/NR 100% at 70°F BC 95% at 70°F	dissolved 95% conc 3 days 212°F	dissolved 95% conc 3 days 212°F	
SULFURIC ACID	A dilute to 70°F	AB to 30% to 150°F A dilute to 70°F	NR 95% at 212°F AB 60-75% to 158°F	no swell 50% conc 3 days 158°F	no swell 50% conc 3 days 158°F	
SULFURIC ACID			AB 25-50% to 250°F A 10% to 250°F	no swell 10% conc 7 days 70°F	+5% vol 10% conc 7 days 70°F	
SULFURIC ACID			A dilute 70°F	NR 95-100% at 70°F A to 50% to 70°F	NR 25-100% at 70°F B/NR 10% at 70°F	
SULFURIC ACID				C dilute at 70°F	B/NR dilute at 70°F	
SULFURIC ACID						
SULFURIC ACID						
SULFURIC ACID						
SULFURIC ACID						
SULFURIC ACID FUMING	NR at 70°F	NR at 70°F	NR at 70°F disintegrated	NR at 70°F	NR at 70°F disintegrated	AB to 70°F NR at 70°F white
SULFURIC ACID (20-25% OLEUM)	NR at 70°F	NR at 70°F	NR at 70°F	NR at 70°F	NR at 70°F	AB to 70°F NR at 70°F white
SULFURIC ETHER	NR at 70°F	NR at 70°F	B at 70°F		C at 70°F	
SULFUROUS ACID	B 10-100% at 70°F	B/NR 100% at 70°F B 75% at 70°F	A to 100% to 158°F		NR at 70°F	A to 70°F
SULFUROUS ACID		C 75% at 100°F AB 10-25% to 70°F				
SULFUROUS ANHYDRIDE	C at 70°F	C/NR at 70°F	A to 70°F		A to 70°F	

	POLYSULFIDE T	CHLORINATED POLYETHYLENE CM	EPICHLOROHYDRIN CO, ECO	®HYTREL COPOLYESTER TPE	®SANTOPRENE COPOLYMER TPO	®C-FLEX STYRENIC TPE
SUBLIMED WHITE LEAD						
SUCCINIC ACID						
SUCROSE SOLUTION	NR at 70°F			*AB to 140°F		
SUGAR OF LEAD	NR at 70°F		B at 70°F			
SUGAR LIQUORS, CANE, BEET & MAPLE						
SUGAR SOLUTIONS	NR at 70°F			*AB to 140°F		
SULFAMIC ACID						
SULFATE LIQUORS (GREEN & BLACK)	NR at 70°F	C at 70°F	A to 70°F	A/NR at 70°F		
SULFITE CELLULOSE LIQUORS						
SULFITE LIQUORS	NR at 70°F		B at 70°F			
SULFITIC LIQUORS						
SULFONIC ACID						
SULFUR (MOLTEN 250° F)	NR at 70°F	AB to 70°F	C at 70°F	AB to 70°F *AB to 226°F		
SULFUR CHLORIDE	NR at 70°F	AB to 70°F		C at 70°F		A to 70°F
SULFUR DICHLORIDE						
SULFUR DIOXIDE GAS DRY	NR at 70°F	NR at 70°F		NR at 70°F		A to 70°F
SULFUR DIOXIDE GAS WET	NR at 70°F			NR at 70°F		
SULFUR DIOXIDE LIQUID						
SULFUR DIOXIDE, PRESSURIZED	NR at 70°F					
SULFUR HEXAFLUORIDE	C at 70°F		A to 70°F	AB to 70°F		A to 70°F
SULFUR LIQUORS	NR at 70°F					
SULFUR MONOCHLORIDE	NR at 70°F	AB to 70°F		C at 70°F		A to 70°F
SULFUR SALTS						
SULFUR SUBCHLORIDE						
SULFUR TRIOXIDE	NR at 70°F	AB to 70°F		NR at 70°F		A to 70°F
SULFUR TRIOXIDE, DRY	NR at 70°F			NR at 70°F		
SULFURATED LIME						
SULFURETTED HYDROGEN						
SULFURIC ACID	dissolved 50-100% conc 3 days 70°F	NR any conc at 70°F	NR 100% at 70°F B 3 molar at 70°F	no swell 25% conc 7 days 70°F	no swell 10-25% conc 7 days 70°F	A to 100% to 70°F
SULFURIC ACID	NR any conc at 70°F		B dilute at 70°F	NR 60-100% at 70°F A 5-50% to 70°F	A to 25% to 70°F	
SULFURIC ACID				AB 25-50% at 140°F A dilute to 70°F		
SULFURIC ACID						
SULFURIC ACID						
SULFURIC ACID						
SULFURIC ACID						
SULFURIC ACID						
SULFURIC ACID						
SULFURIC ACID FUMING	NR at 70°F	NR at 70°F	NR at 70°F	NR at 70°F		
SULFURIC ACID (20-25% OLEUM)	NR at 70°F	NR at 70°F	NR at 70°F	NR at 70°F		
SULFURIC ETHER						
SULFUROUS ACID	NR at 70°F	NR at 70°F		BC at 70°F		A to 70°F
SULFUROUS ACID						
SULFUROUS ANHYDRIDE						

	ETHYLENE ACRYLIC EA	POLYALLOMER LINEAR COPOLYMER	NYLON 11 POLYAMIDE	NYLON 12 POLYAMIDE	ETHYLENE VINYL ACETATE EVA	POLYVINYLCHLORIDE FLEXIBLE PVC
SUBLIMED WHITE LEAD						
SUCCINIC ACID		A to 70°F	A to 70°F		A to 70°F	
SUCROSE SOLUTION			*AB to 140°F	AB to 70°F	A to 70°F *AB to 140°F	AB to 70°F *AB to 140°F
SUGAR OF LEAD		A to 70°F	A to 70°F		A to 70°F AB to 140°F	A to 70°F *AB to 140°F
SUGAR LIQUORS, CANE, BEET & MAPLE						
SUGAR SOLUTIONS			*AB to 140°F	AB to 70°F	A to 70°F *AB to 140°F	AB to 70°F *AB to 140°F
SULFAMIC ACID						
SULFATE LIQUORS (GREEN & BLACK)		A to 70°F	B/NR at 70°F		A to 70°F	A to 150°F
SULFITE CELLULOSE LIQUORS						
SULFITE LIQUORS					C/NR at 70°F	A to 70°F
SULFITIC LIQUORS						
SULFONIC ACID						
SULFUR (MOLTEN 250° F)	*AB to 250°F	C at 70°F	A to 70°F AB to 105°F	AB to 70°F	AB to 70°F *AB to 140°F	AB to 70°F
SULFUR CHLORIDE		C at 70°F	B/NR at 70°F		B at 70°F	C at 70°F
SULFUR DICHLORIDE						
SULFUR DIOXIDE GAS DRY	*AB to 70°F	A to 122°F	C/NR at 70°F C 10% at 70°F	C to 5% at 70°F	A to 70°F *AB to 140°F	A to 70°F B at 122°F
SULFUR DIOXIDE GAS WET	*AB to 70°F	A to 122°F	NR at 70°F		*AB to 70°F *NR at 140°F	A to 70°F B at 122°F
SULFUR DIOXIDE LIQUID	*AB to 70°F					C at 70°F NR at 122°F
SULFUR DIOXIDE, PRESSURIZED		NR at 70°F at 46 PSI stress				C at 70°F at 46 PSI NR at 122°F at 46 PSI
SULFUR HEXAFLUORIDE			AB to 70°F			AB to 70°F
SULFUR LIQUORS						
SULFUR MONOCHLORIDE		C at 70°F	B/NR at 70°F		B at 70°F	C at 70°F
SULFUR SALTS		C at 70°F NR at 122°F				NR at 70°F
SULFUR SUBCHLORIDE						
SULFUR TRIOXIDE			C at 70°F NR at 105°F		B at 70°F	BC at 70°F
SULFUR TRIOXIDE, DRY						
SULFURATED LIME						
SULFURETTED HYDROGEN						
SULFURIC ACID	NR 95-100% at 70°F *AC 60-75% at 70°F	C 98% at 70°F NR 98% at 122°F	NR 20-100% at 70°F C/NR 10% at 70°F	NR 20-100% at 70°F C 2-10% at 70°F	C/NR 50-100% at 70°F BC 10-50% at 70°F	no swell to 30% conc 14 days 70°F
SULFURIC ACID	AC 20-50% at 70°F A to 10% to 70°F	A to 98% to 70°F B 20-60% at 122°F	AB 1% to 70°F C 1% 105-140°F		A dilute to 70°F *AB dilute to 140°F	NR 95-100% at 70°F BC 75% at 70°F
SULFURIC ACID		A to 6% to 122°F	NR 1% at 194°F			B 60% to 122°F C 60% at 140°F
SULFURIC ACID						A to 50% to 70°F BC 50% at 150°F
SULFURIC ACID						AB to 40% to 140°F
SULFURIC ACID						
SULFURIC ACID						
SULFURIC ACID						
SULFURIC ACID						
SULFURIC ACID						
SULFURIC ACID FUMING	NR at 70°F		NR at 70°F	NR at 70°F	NR at 70°F	NR at 70°F
SULFURIC ACID (20-25% OLEUM)	NR at 70°F		NR at 70°F	NR at 70°F	NR at 70°F	NR at 70°F
SULFURIC ETHER						
SULFUROUS ACID		A to 70°F	NR 100% at 70°F C 30% at 70°F		C at 70°F	AB to 70°F C at 150°F
SULFUROUS ACID						NR 75% at 70°F AB 30% to 70°F
SULFUROUS ANHYDRIDE			AB to 70°F			

	POLYETHYLENE LOW DENSITY LDPE	® TEFLON FEP	® KALREZ PERFLUORINATED ELASTOMER	® FLUORAZ FLUORINATED COPOLYMER	® AFLAS FLUORINATED COPOLYMER	® NORPRENE COPOLYMER TPO
SUBLIMED WHITE LEAD		A to 70°F	A to 212°F			
SUCCINIC ACID		A to 70°F	A to 212°F			
SUCROSE SOLUTION	A to 140°F	A to 200°F	A to 212°F	A to 70°F		
SUGAR OF LEAD	A to 70°F AB to 140°F	A to 300°F	A to 212°F	NR at 70°F		A to 70°F
SUGAR LIQUORS, CANE, BEET & MAPLE		A to 70°F	A to 212°F			
SUGAR SOLUTIONS	A to 140°F	A to 200°F	A to 212°F	A to 70°F		
SULFAMIC ACID		A to 70°F	A to 212°F			
SULFATE LIQUORS (GREEN & BLACK)	A to 70°F	A to 300°F	A to 212°F	AC to 70°F		
SULFITE CELLULOSE LIQUORS		A to 70°F				
SULFITE LIQUORS	A to 70°F	A to 70°F	A to 212°F			
SULFITIC LIQUORS		A to 70°F				
SULFONIC ACID		A to 70°F	A to 212°F	C dilute at 140°F		
SULFUR (MOLTEN 250° F)	AB to 100% to 70°F B/NR at 140°F	A to 100% to boiling	A to 212°F	A to 70°F C at 250°F		
SULFUR CHLORIDE	AB to 70°F NR at 122°F	A to 300°F	A to 212°F	A/NR at 70°F		TEST
SULFUR DICHLORIDE						
SULFUR DIOXIDE GAS DRY	A to 140°F	A to conc to 300°F	A to 212°F	A to 70°F A 1% to 100°F	+7.8% vol 5% conc 2 days 104°F	A to 70°F
SULFUR DIOXIDE GAS WET	A to 70°F A/NR at 140°F	A to conc to 300°F	A to 212°F	A to 70°F		
SULFUR DIOXIDE LIQUID	B/NR at 70°F	A to 122°F	A to 212°F	A to 70°F		
SULFUR DIOXIDE, PRESSURIZED	NR at 70°F	A to 122°F		B at 70°F		
SULFUR HEXAFLUORIDE	AB to 70°F	A to 70°F	B at 70°F	C at 70°F		AB to 70°F
SULFUR LIQUORS				B at 70°F		
SULFUR MONOCHLORIDE	AB to 70°F NR at 122°F	A to 300°F	A to 212°F	NR at 70°F		TEST
SULFUR SALTS	C at 70°F NR at 122°F	A to 122°F				
SULFUR SUBCHLORIDE		A to 70°F				
SULFUR TRIOXIDE	A/NR at 70°F	A to 300°F	A to 212°F	A to 70°F		BC at 70°F
SULFUR TRIOXIDE, DRY	B/NR at 70°F	A to 70°F		A to 70°F		
SULFURATED LIME		A to 70°F				
SULFURETTED HYDROGEN		A to 70°F				
SULFURIC ACID	NR 100% at 70°F C/NR 90-96% at 70°F	no swell to 30% conc 3 days 158°F	A to conc to 212°F *A to 60% to 250°F	A conc to 70°F	+2.3% vol 95% conc 180 days 73°F	TEST 100% BC 90% at 70°F
SULFURIC ACID	NR 96% at 140°F AB 80% to 104°F	A to 100% to 400°F			+4.4% vol 95% conc 3 days 212°F	A to 50% to 70°F
SULFURIC ACID	C 80% at 140°F BC 75% at 70°F				no swell 60% conc 7 days 73°F	
SULFURIC ACID	NR 75% at 140°F A to 60% to 140°F				+0.4% 25-60% conc 3 days 212°F	
SULFURIC ACID					A to 100% to 70°F	
SULFURIC ACID						
SULFURIC ACID						
SULFURIC ACID						
SULFURIC ACID						
SULFURIC ACID						
SULFURIC ACID FUMING	NR at 70°F	A to 400°F	A to 212°F	B at 70°F	+7% vol 180 days 70°F A to 70°F	NR at 70°F
SULFURIC ACID (20-25% OLEUM)	NR at 70°F	A to 400°F	A to 212°F	A to 70°F		TEST
SULFURIC ETHER		A to 70°F				
SULFUROUS ACID	AB to 100% to 140°F	A to 100% to 300°F	A to 212°F	A to 100% to 70°F		A to 70°F
SULFUROUS ACID						
SULFUROUS ANHYDRIDE		A to 70°F				

C 355

	RECOMMENDATIONS	NITRILE NBR	ETHYLENE PROPYLENE EP, EPDM	FLUOROCARBON FKM	CHLOROPRENE CR	HYDROGENATED NITRILE HNBR
SULFUROUS OXYCHLORIDE		NR at 70°F			NR at 70°F	
SULPHITE LIQUORS	NBR, CR, EP	AB to 70°F	B at 70°F / A 6% to 140°F	A to 70°F / A 6% to 140°F	AB to 70°F	
SULPHITE PULP	FKM	B at 70°F	B at 70°F	A to 160°F	B at 70°F	
SUMMER OIL		A to 70°F		A to 70°F	A to 70°F	
SUNOCO ALL PURPOSE GREASE	NBR, FKM, ACM, AU	A to 70°F	NR at 70°F	A to 70°F	B at 70°F	
SUNOCO SAE 10	NBR, FKM, ACM, AU	A to 70°F	NR at 70°F	A to 70°F	B at 70°F	A to 70°F
SUNOCO XS-820 (EP LUBRICANT)			NR at 70°F	A to 300°F		
SUNOCO 3661	NBR, FKM, ACM, AU	A to 70°F	NR at 70°F	A to 70°F	B at 70°F	A to 70°F
SUNLIGHT			A to 70°F		B at 70°F	
SUNSAFE (FIRE RESISTANT FLUID)	NBR, FKM	A to 70°F	NR at 70°F	A to 70°F	B at 70°F	A to 70°F
SUPERPHOSPHORIC ACID (100-115%)		NR at 70°F			NR at 70°F	
SUPERSHELL GASOLINE	NBR, FKM	A to 70°F	NR at 70°F	+2.3% vol 28 days 75°F / A to 70°F	B at 70°F	A to 70°F
SWANFINCH EP LUBRICANT	NBR, FKM, ACM, AU	A to 70°F	NR at 70°F	A to 70°F	B/NR at 70°F	
SWANFINCH HYPOID 90	NBR, FKM, ACM, AU	A to 70°F	NR at 70°F	A to 70°F	B at 70°F	
SWEET BIRCH OIL		NR at 70°F	C at 70°F	B at 70°F	NR at 70°F	
SWEET OIL		AB to 120°F	A to 70°F	A to 70°F	B to 80°F	
SYRUP		A to 70°F	A to 70°F	A to 70°F	AB to 70°F	
TABLE SALT	NBR, EP, FKM, CR, IIR	A to 160°F	A to 176°F / AB to 200°F	A to 212°F	A to 150°F	A to 70°F
TALL OIL	NBR	A to 200°F	NR at 70°F	A to 158°F	B/NR at 70°F	
TALLOL	NBR	A to 200°F	NR at 70°F	A to 158°F	B/NR at 70°F	
TALLOW	NBR, ACM, FKM, AU	A to 200°F	A to 140°F	A to molten	B at 70°F / C at 140°F	
TAN		C at 70°F	C at 70°F	A to 70°F	A to 70°F	
TANNIC ACID (TANNIN)	NBR, EP, FKM, IIR, CR	A to 100% to 200°F	A to 100% to 140°F / AB to 200°F	A to 100% to 140°F / AB to 200°F	A to 100% to 70°F / AB to 200°F	A to 70°F
TANNING LIQUORS	NBR	AB to 140°F / AC to 200°F	AB to 140°F	A to 70°F / AB to 140°F	A to 70°F	
TANNING OILS	NBR, FKM	A to 70°F		A to 70°F	NR at 70°F	
TANNING SOLUTIONS		A to 70°F / AB to 160°F		AB to 70°F extracts	A to 70°F / AB to 160°F	
TAR, BITUMINOUS	FKM, NBR	B at 70°F	NR at 70°F	A to 300°F	C at 70°F	B at 70°F
TAR, CAMPHOR	FKM, FVMQ, T	NR at 70°F	NR at 70°F	A to 176°F	NR at 70°F	NR at 70°F
TAR, HOT	FKM	B/NR	NR	A to 300°F	C	
TAR & AMMONIA						
TAR OIL						
TARTARIC ACID (DIOXYSUCCINIC ACID)	NBR, FKM, FVMQ, CR	A to 200°F	BC at 70°F	A to 140°F / AB to 200°F	A to 158°F / NR at 212°F	A to 70°F
TAXAPHENE		B 12% at 70°F			B 12% at 70°F	
T.C.A.		NR at 70°F			NR at 70°F	
T.C.P. (TRICRESYL PHOSPHATE)	EP, IIR, FKM, SBR	NR at 70°F	A to 100% to 212°F	A to 140°F / B to 300°F	BC to 100% at 70°F / NR at 140°F	NR at 70°F
T.E.A.	EP, IIR, NBR	B to 100°F / AB to 80% to 120°F	A to 160°F	NR at 70°F	NR at 70°F	C at 70°F
TELLUS (SHELL)		A to 70°F	NR at 70°F		A to 70°F	
TERPENE		C at 70°F	NR at 70°F	A to 70°F	NR at 70°F	
TERPINEOL	FKM, NBR, CR	B at 70°F	C at 70°F	A to 70°F	NR at 70°F	B at 70°F
TERTIARY BUTYL ALCOHOL	FKM, NBR, EP	AB to 70°F	B at 70°F	A to 70°F	B at 70°F	B at 70°F
TERTIARY BUTYL CALTECHOL	FKM, EP, IIR	NR at 70°F	B at 70°F	A to 70°F	B any conc to 80°F	
P-TERTIARY BUTYL CALTECHOL	FKM, EP, CR	NR at 70°F	B at 70°F	A to 70°F	B at 70°F	
TERTIARY BUTYL MERCAPTAN	FKM	NR at 70°F	NR at 70°F	A to 70°F	NR at 70°F	NR at 70°F
TETRABROMOETHANE	FKM, FVMQ	NR at 70°F	NR at 70°F	A to 70°F	NR at 70°F	NR at 70°F

	STYRENE BUTADIENE SBR	POLYACRYLATE ACM	POLYURETHANE AU, EU	ISOBUTYLENE ISOPRENE IIR	POLYBUTADIENE BR	®AEROQUIP AQP
SULFUROUS OXYCHLORIDE				NR at 70°F		
SULPHITE LIQUORS				A to 70°F		
SULPHITE PULP	B at 70°F	NR at 70°F				
SUMMER OIL						
SUNOCO ALL PURPOSE GREASE	NR at 70°F	A to 70°F	A to 70°F	NR at 70°F	NR at 70°F	
SUNOCO SAE 10	NR at 70°F	A to 70°F	A to 70°F	NR at 70°F	NR at 70°F	
SUNOCO XS-820 (EP LUBRICANT)			A to 70°F			
SUNOCO 3661	NR at 70°F	A to 70°F	A to 70°F	NR at 70°F	NR at 70°F	
SUNLIGHT			C/NR at 70°F			
SUNSAFE (FIRE RESISTANT FLUID)	NR at 70°F	NR at 70°F	NR at 70°F	NR at 70°F	NR at 70°F	AB to 200°F
SUPERPHOSPHORIC ACID (100-115%)			NR at 70°F	A to 70°F AB 106% to 160°F	C/NR at 70°F	AB to 70°F
SUPERSHELL GASOLINE	NR at 70°F	B at 70°F	B at 70°F	NR at 70°F	NR at 70°F	
SWANFINCH EP LUBRICANT	NR at 70°F	A to 70°F	A to 70°F	NR at 70°F	NR at 70°F	
SWANFINCH HYPOID 90	NR at 70°F	A to 70°F	A to 70°F	NR at 70°F	NR at 70°F	
SWEET BIRCH OIL				B to 80°F		
SWEET OIL			B at 70°F	A to 100°F		
SYRUP						
TABLE SALT	A to 70°F	A to 70°F	A to 70°F	A to 70°F AB to 185°F	A to 70°F	
TALL OIL	NR at 70°F	A to 70°F	A to 70°F	NR at 70°F		
TALLOL				NR at 70°F		
TALLOW	NR at 150°F	A to 150°F	A to 140°F	B at 70°F		
TAN			A to 70°F	C at 70°F		
TANNIC ACID (TANNIN)	B to 100% at 70°F	NR 10-100% at 70°F	A to 70°F	A to 100% to 70°F AB to 100% to 185°F	B 100% at 70°F A 10% to 70°F	NR at 70°F
TANNING LIQUORS	B at 70°F	NR at 70°F	C/NR at 70°F	NR at 70°F		
TANNING OILS						
TANNING SOLUTIONS				A to 70°F AB to 150°F		
TAR, BITUMINOUS	NR at 70°F	NR at 70°F	B at 70°F	NR at 70°F	NR at 70°F	AB to 70°F
TAR, CAMPHOR	NR at 70°F		B at 70°F	NR at 70°F	NR at 70°F	AB to 70°F
TAR, HOT	NR	NR		NR	NR	
TAR & AMMONIA			C/NR at 70°F			
TAR OIL			AB to 70°F			
TARTARIC ACID (DIOXYSUCCINIC ACID)	B at 70°F		A to 70°F	B to 185°F	B at 70°F	NR at 70°F
TAXAPHENE				NR 12% at 70°F		
T.C.A.				NR at 70°F		
T.C.P. (TRICRESYL PHOSPHATE)	NR at 70°F	NR at 70°F	C/NR at 70°F	A to 70°F AB to 185°F	NR at 70°F	NR at 70°F
T.E.A.				A to 70°F AB to 185°F		
TELLUS (SHELL)						
TERPENE				NR at 70°F		
TERPINEOL	NR at 70°F		B at 70°F	C/NR at 70°F	NR at 70°F	
TERTIARY BUTYL ALCOHOL	B at 70°F	NR at 70°F	NR at 70°F	B at 70°F	B at 70°F	
TERTIARY BUTYL CALTECHOL	C at 70°F	NR at 70°F	NR at 70°F	B any conc to 80°F	C at 70°F	
P-TERTIARY BUTYL CALTECHOL	B at 70°F	NR at 70°F		B at 70°F	B at 70°F	
TERTIARY BUTYL MERCAPTAN	NR at 70°F	NR at 70°F	NR at 70°F	NR at 70°F	NR at 70°F	
TETRABROMOETHANE	NR at 70°F	NR at 70°F		NR at 70°F	NR at 70°F	

	SYNTHETIC ISOPRENE IR	NATURAL ISOPRENE NR	CHLOROSULFONATED POLYETHYLENE CSM	FLUOROSILICONE FVMQ	SILICONE VMQ	CHEMRAZ FFKM
SULFUROUS OXYCHLORIDE	NR at 70°F	NR at 70°F				
SULPHITE LIQUORS	B at 70°F	B at 70°F	A to 70°F	B at 70°F	NR at 70°F	
SULPHITE PULP				B at 70°F	NR at 70°F	
SUMMER OIL	A to 70°F	A to 70°F	AB to 70°F			
SUNOCO ALL PURPOSE GREASE	NR at 70°F	NR at 70°F	B at 70°F	A to 70°F	NR at 70°F	
SUNOCO SAE 10	NR at 70°F	NR at 70°F	B at 70°F	A to 70°F	NR at 70°F	A to 70°F
SUNOCO XS-820 (EP LUBRICANT)						
SUNOCO 3661	NR at 70°F	NR at 70°F	B at 70°F	A to 70°F	NR at 70°F	A to 70°F
SUNLIGHT			A to 70°F			
SUNSAFE (FIRE RESISTANT FLUID)	NR at 70°F	NR at 70°F	B at 70°F	A to 70°F		A to 70°F
SUPERPHOSPHORIC ACID (100-115%)	NR at 70°F	NR at 70°F			C/NR at 70°F	
SUPERSHELL GASOLINE	NR at 70°F	NR at 70°F	NR at 70°F	B at 70°F	NR at 70°F	A to 70°F
SWANFINCH EP LUBRICANT	NR at 70°F	NR at 70°F	NR at 70°F	A to 70°F	NR at 70°F	A to 70°F
SWANFINCH HYPOID 90	NR at 70°F	NR at 70°F	NR at 70°F	A to 70°F	NR at 70°F	A to 70°F
SWEET BIRCH OIL	NR at 70°F	NR at 70°F				
SWEET OIL	NR at 70°F	NR at 70°F	B at 70°F		NR at 70°F	
SYRUP		A to 70°F				
TABLE SALT	A to 70°F	A to 150°F	A to 70°F AB to 200°F	A to 70°F	A to 212°F	A to 70°F
TALL OIL	NR at 70°F	NR at 70°F	C at 70°F	A to 70°F	NR at 70°F	
TALLOL	NR at 70°F	NR at 70°F	C at 70°F	A to 70°F	NR at 70°F	
TALLOW	NR at 70°F	B/NR at 70°F	BC at 70°F	A to 140°F	B at 70°F NR at 140°F	A to 70°F
TAN	A to 70°F	A to 70°F	A to 70°F		B at 70°F	
TANNIC ACID (TANNIN)	A to 100% to 70°F	A to 100% to 100°F AB to 185°F	A to 100% to 140°F AB to 200°F	A 10% to 70°F	B to 100% at 70°F	A 10-100% to 70°
TANNING LIQUORS		C at 70°F	B at 70°F		B at 70°F	
TANNING OILS						
TANNING SOLUTIONS	B at 70°F	B at 70°F				
TAR, BITUMINOUS	B/NR at 70°F	C/NR at 70°F	C/NR at 70°F	A to 140°F	B at 140°F	A to 70°F
TAR, CAMPHOR	NR at 70°F	NR at 70°F	NR at 70°F	A to 70°F	NR at 70°F	A to 70°F
TAR, HOT				A	B	
TAR & AMMONIA						
TAR OIL						
TARTARIC ACID (DIOXYSUCCINIC ACID)	A to 70°F	A to 70°F AB any conc to 150°F	A to 200°F	A to 70°F	A to 70°F	A to 70°F
TAXAPHENE	NR 12% at 70°F	NR 12% at 70°F	NR 12% at 70°F			
T.C.A.	NR at 70°F	NR at 70°F				
T.C.P. (TRICRESYL PHOSPHATE)	C/NR at 70°F	BC to 80°F	C/NR at 70°F	B at 70°F	C at 70°F	A to 70°F
T.E.A.	B at 70°F	B any conc to 80°F	A to 158°F	NR at 70°F	NR at 70°F	B at 70°F
TELLUS (SHELL)						A to 70°F
TERPENE	NR at 70°F	NR at 70°F	NR at 70°F			
TERPINEOL	NR at 70°F	NR at 70°F	NR at 70°F	A to 70°F		A to 70°F
TERTIARY BUTYL ALCOHOL	B at 70°F	B at 70°F	B at 70°F	B at 70°F	B at 70°F	A to 70°F
TERTIARY BUTYL CALTECHOL	NR at 70°F	NR at 70°F	B at 70°F	A to 70°F		A to 70°F
P-TERTIARY BUTYL CALTECHOL	NR at 70°F	NR at 70°F	B at 70°F	A to 70°F		A to 70°F
TERTIARY BUTYL MERCAPTAN	NR at 70°F	NR at 70°F	NR at 70°F		NR at 70°F	A to 70°F
TETRABROMOETHANE	NR at 70°F	NR at 70°F	NR at 70°F	B at 70°F	NR at 70°F	A to 70°F

	POLYSULFIDE T	CHLORINATED POLYETHYLENE CM	EPICHLOROHYDRIN CO, ECO	® HYTREL COPOLYESTER TPE	® SANTOPRENE COPOLYMER TPO	® C-FLEX STYRENIC TPE
SULFUROUS OXYCHLORIDE						
SULPHITE LIQUORS						
SULPHITE PULP						
SUMMER OIL						
SUNOCO ALL PURPOSE GREASE	A to 70°F					
SUNOCO SAE 10	A to 70°F					
SUNOCO XS-820 (EP LUBRICANT)				*AB to 70°F		
SUNOCO 3661	A to 70°F					
SUNLIGHT						
SUNSAFE (FIRE RESISTANT FLUID)				AB to 150°F		
SUPERPHOSPHORIC ACID (100-115%)	NR at 70°F	NR at 70°F		NR at 70°F		
SUPERSHELL GASOLINE	B at 70°F					
SWANFINCH EP LUBRICANT	A to 70°F					
SWANFINCH HYPOID 90	A to 70°F					
SWEET BIRCH OIL						
SWEET OIL						
SYRUP						
TABLE SALT	C at 70°F	AB to 70°F	A to 70°F	A to 70°F	A 15% to 70°F	A to 70°F
TALL OIL						
TALLOL						
TALLOW						
TAN						
TANNIC ACID (TANNIN)	A to 70°F	NR at 70°F	B 10% at 70°F	AB 100% at 70°F A 10% to 70°F	A to 70°F	A to 70°F
TANNING LIQUORS						
TANNING OILS						
TANNING SOLUTIONS						A to 70°F
TAR, BITUMINOUS	AB to 70°F	AB to 70°F	B at 70°F	AB to 70°F		
TAR, CAMPHOR	B at 70°F	C at 70°F		A to 70°F hard BC at 70°F soft	B/NR at 70°F	
TAR, HOT						
TAR & AMMONIA						
TAR OIL				AB to 70°F		
TARTARIC ACID (DIOXYSUCCINIC ACID)	NR at 70°F	NR at 70°F	B at 70°F	AB to 70°F	A to 70°F	A to 70°F
TAXAPHENE						
T.C.A.						
T.C.P. (TRICRESYL PHOSPHATE)	AB to 70°F		NR at 70°F	AC to 70°F		
T.E.A.						
TELLUS (SHELL)						
TERPENE						
TERPINEOL	A to 70°F					
TERTIARY BUTYL ALCOHOL	B at 70°F					
TERTIARY BUTYL CALTECHOL	NR at 70°F					
P-TERTIARY BUTYL CALTECHOL	NR at 70°F					
TERTIARY BUTYL MERCAPTAN	NR at 70°F					
TETRABROMOETHANE						

	ETHYLENE ACRYLIC EA	POLYALLOMER LINEAR COPOLYMER	NYLON 11 POLYAMIDE	NYLON 12 POLYAMIDE	ETHYLENE VINYL ACETATE EVA	POLYVINYLCHLORIDE FLEXIBLE PVC
SULFUROUS OXYCHLORIDE						NR at 70°F
SULPHITE LIQUORS						
SULPHITE PULP						
SUMMER OIL						
SUNOCO ALL PURPOSE GREASE						
SUNOCO SAE 10						
SUNOCO XS-820 (EP LUBRICANT)	*AB to 70°F					
SUNOCO 3661						
SUNLIGHT					B at 70°F	NR at 70°F
SUNSAFE (FIRE RESISTANT FLUID)			A to 70°F AB to 180°F			
SUPERPHOSPHORIC ACID (100-115%)			NR at 70°F	NR at 70°F	NR at 70°F	
SUPERSHELL GASOLINE						
SWANFINCH EP LUBRICANT						
SWANFINCH HYPOID 90						
SWEET BIRCH OIL						NR at 70°F
SWEET OIL						
SYRUP					A to 70°F	A to 70°F
TABLE SALT	A to 70°F	A to 70°F	AB to 194°F	AB to 70°F aqueous		AB to 70°F
TALL OIL						NR at 70°F
TALLOL						
TALLOW				AB to 70°F	B at 70°F	B at 70°F
TAN						
TANNIC ACID (TANNIN)	*AB 10% to 70°F	A to 70°F	AB to 70°F		BC at 70°F	A to 150°F
TANNING LIQUORS					B at 70°F	B at 70°F
TANNING OILS						
TANNING SOLUTIONS			AB to 70°F extracts			A to 70°F AB to 70°F extracts
TAR, BITUMINOUS		A to 70°F	AB to 70°F	AB to 70°F	C/NR at 70°F	B at 70°F
TAR, CAMPHOR			AB to 140°F C at 194°F		*NR at 70°F	C/NR at 70°F NR at 150°F
TAR, HOT						
TAR & AMMONIA					NR at 70°F	B at 70°F
TAR OIL			AB to 70°F			A to 70°F
TARTARIC ACID (DIOXYSUCCINIC ACID)	*AB to 70°F	A to 122°F	A to 70°F C at 194°F	AB 10% to 70°F	AB to 70°F *AB 10% to 140°F	A to 70°F B at 122°F
TAXAPHENE						
T.C.A.						NR at 70°F
T.C.P. (TRICRESYL PHOSPHATE)	B at 70°F		AB to 140°F C at 194°F		*NR at 70°F	NR at 70°F
T.E.A.						
TELLUS (SHELL)			A to 70°F			
TERPENE						
TERPINEOL						
TERTIARY BUTYL ALCOHOL						
TERTIARY BUTYL CALTECHOL						NR at 70°F
P-TERTIARY BUTYL CALTECHOL						
TERTIARY BUTYL MERCAPTAN						
TETRABROMOETHANE						NR at 70°F

	POLYETHYLENE LOW DENSITY LDPE	®TEFLON FEP	®KALREZ PERFLUORINATED ELASTOMER	®FLUORAZ FLUORINATED COPOLYMER	®AFLAS FLUORINATED COPOLYMER	®NORPRENE COPOLYMER TPO
SULFUROUS OXYCHLORIDE		A to 70°F				
SULPHITE LIQUORS	A to 70°F	A to 70°F	A to 212°F			
SULPHITE PULP						
SUMMER OIL		A to 70°F				
SUNOCO ALL PURPOSE GREASE						
SUNOCO SAE 10				A to 70°F		
SUNOCO XS-820 (EP LUBRICANT)						
SUNOCO 3661				A to 70°F		
SUNLIGHT	C/NR at 70°F					
SUNSAFE (FIRE RESISTANT FLUID)		A to 70°F AB to 250°F		A to 70°F		
SUPERPHOSPHORIC ACID (100-115%)		A to 70°F	A to 70°F	A to 70°F		A to 70°F
SUPERSHELL GASOLINE				C at 70°F		
SWANFINCH EP LUBRICANT				B at 70°F		
SWANFINCH HYPOID 90				A to 70°F		
SWEET BIRCH OIL		A to 70°F				
SWEET OIL		A to 70°F				
SYRUP	C/NR at 70°F			A to 70°F		
TABLE SALT	AB to 140°F	A to 300°F	A to 212°F	A to 70°F	A to 35% to 212°F	A to 75°F
TALL OIL		A to 300°F		A to 200°F		
TALLOL		A to 70°F		A to 200°F		
TALLOW	AC to 70°F NR at 140°F	A to 300°F	A to 212°F	A to 100% to 70°F		
TAN		A to 70°F				
TANNIC ACID (TANNIN)	AB 100% to 70°F A 10% to 70°F	A to 100% to 300°F	A to 212°F	A to 150°F		A to 70°F
TANNING LIQUORS	A to 70°F	A to 70°F		A to 70°F		
TANNING OILS		A to 70°F				
TANNING SOLUTIONS	AB to 70°F extracts	A to 300°F			A to 70°F	A to 70°F
TAR, BITUMINOUS	B/NR at 70°F	A to 300°F	A to 212°F	A to 70°F		
TAR, CAMPHOR	NR at 70°F	A to 424°F	A to 70°F	C/NR at 70°F		
TAR, HOT		*A to 200°F				
TAR & AMMONIA	BC at 70°F			A to 70°F		
TAR OIL	AB to 70°F	*A to 200°F				
TARTARIC ACID (DIOXYSUCCINIC ACID)	A to 122°F AB to 100% at 140°F	A to 300°F	A to 212°F	A to 140°F aqueous		TEST
TAXAPHENE		A 12% to 70°F				
T.C.A.		A to 70°F				
T.C.P. (TRICRESYL PHOSPHATE)	NR at 70°F	A to boiling	A to 212°F	A to 70°F	+8.5% vol 7 days 302°F AB to 302°F	
T.E.A.		A to 300°F	A to 70°F *A to 212°F	A to 70°F	A to 70°F	
TELLUS (SHELL)		A to 70°F				
TERPENE		A to 70°F				
TERPINEOL		A to 70°F	A to 212°F			
TERTIARY BUTYL ALCOHOL		A to 70°F	A to 212°F	A to 70°F		
TERTIARY BUTYL CALTECHOL		A to 70°F	A to 212°F	A to 70°F		
P-TERTIARY BUTYL CALTECHOL		A to 70°F		A to 70°F		
TERTIARY BUTYL MERCAPTAN		A to 70°F	A to 212°F	A to 70°F		
TETRABROMOETHANE		A to boiling	A to 212°F	C at 70°F		

C 361

	RECOMMENDATIONS	NITRILE NBR	ETHYLENE PROPYLENE EP, EPDM	FLUOROCARBON FKM	CHLOROPRENE CR	HYDROGENATED NITRILE HNBR
TETRABROMOMETHANE	FKM, FVMQ	NR at 70°F	NR at 70°F	A to 70°F	NR at 70°F	
TETRABUTYL TITANATE	EP, FKM	B at 70°F	A to 70°F	A to 70°F	AB to 70°F	B at 70°F
TETRACHLORIDE OF TIN	NBR, EP, FKM, IIR	A to 100% to 140°F AB any conc to 150°F	A to 100% to 140°F	A to 100% to 140°F AB to 200°F	BC at 70°F NR 50% at 70°F	A to 70°F
TETRACHLOROBENZENE		NR at 70°F		B at 70°F		
TETRACHLORODIFLUOROETHANE (F112)	CR, NBR, T, FKM	AB to 250°F	NR at 70°F	A to 70°F	B at 70°F	B at 70°F
TETRACHLORODIFLUOROMETHANE		NR at 70°F			NR at 70°F	
TETRACHLOROETHANE	FKM, FVMQ	NR at 70°F	NR at 70°F	+2.7% vol 21 days 75°F A to 200°F	NR at 70°F	NR at 70°F
TETRACHLOROETHYLENE	FKM, FVMQ	B/NR at 70°F	NR at 70°F	A to 212°F	NR at 70°F	B/NR at 70°F
TETRACHLOROMETHANE	FKM, FVMQ	NR at 70°F	NR at 70°F	A to 158°F	NR at 70°F	B at 70°F
TETRACHLORONAPHTHALENE		NR at 70°F		B at 70°F		
TETRADECANOL		A to 70°F		B at 70°F		
TETRAETHYL LEAD	FKM, FVMQ, NBR	B to 120°F	NR at 70°F	A to 120°F	BC at 70°F	B at 70°F
TETRAETHYL LEAD BLEND		B at 70°F	NR at 70°F	A to 70°F	NR at 70°F	B at 70°F
TETRAETHYL ORTHOSILICATE		A to 70°F AB to 180°F			A to 70°F AB to 180°F	
TETRAETHYLENE GLYCOL		A to 70°F		A to 70°F		
TETRAFLUOROMETHANE	CR, NBR, EP, FKM	A to 250°F	A to 70°F	A to 70°F	A to 250°F	
TETRAHYDROFURAN	T, IIR	NR at 70°F	NR at 70°F	+200% vol 7 days 70°F NR at 70°F	NR at 70°F	NR at 70°F
TETRAHYDRONAPHTHALENE	FKM, FVMQ	NR at 70°F	NR at 70°F	A to 70°F	NR at 70°F	NR at 70°F
TETRALIN (SULFOLANE)	FKM, FVMQ	NR at 70°F	NR at 70°F	A to 70°F	NR at 70°F	NR at 70°F
TETRANE		A to 70°F	C at 70°F	A to 70°F	B at 70°F	
TETRANITROMETHANE						
TETROL		NR at 70°F	NR at 70°F	C at 70°F	NR at 70°F	
TEXACO CAPELLA A & AA	NBR, FKM, ACM	A to 70°F	NR at 70°F	A to 70°F	B at 70°F	A to 70°F
TEXACO MEROPA 220 NO. 3 (NO LEAD)	NBR, FKM, ACM	A to 70°F	NR at 70°F	A to 70°F	B at 70°F	A to 70°F
TEXACO REGAL B	NBR, FKM, ACM, AU	A to 70°F	NR at 70°F	A to 70°F	NR at 70°F	A to 70°F
TEXACO UNI-TEMP GREASE	NBR, FKM, ACM, AU	A to 70°F	NR at 70°F	A to 70°F	B at 70°F	A to 70°F
TEXACO 3450 GEAR OIL	NBR, FKM, ACM, AU	A to 70°F	NR at 70°F	+26% vol 21 days 300°F A to 70°F	NR at 70°F	NR at 70°F
TEXAMATIC "A" TRANSMISSION OIL	NBR, FKM, ACM	A to 70°F	NR at 70°F	A to 70°F	B at 70°F	
TEXAMATIC 1581 FLUID	NBR, FKM, ACM	A to 70°F	NR at 70°F	+3% vol 7 days 350°F A to 70°F	B at 70°F	
TEXAMATIC 3401 FLUID	NBR, FKM, ACM	A to 70°F	NR at 70°F	+1.5% vol 7 days 350°F A to 70°F	B at 70°F	A to 70°F
TEXAMATIC 3525 FLUID	NBR, FKM, ACM	A to 70°F	NR at 70°F	+2% vol 7 days 350°F A to 70°F	B at 70°F	A to 70°F
TEXAMATIC 3528 FLUID	NBR, FKM, ACM	A to 70°F	NR at 70°F	+2% vol 7 days 350°F A to 70°F	B at 70°F	A to 70°F
TEXAS 1500 OIL	NBR, FKM, ACM, AU	A to 70°F	NR at 70°F	A to 70°F	B at 70°F	A to 70°F
THENARDITE		A to 70°F AB any conc to 200°F	A to 70°F	A to 70°F	A to 70°F AB any conc to 200°F	
THERMINOL VP1, 60 & 66	FKM	NR at 70°F	NR at 70°F	A to 70°F	NR at 70°F	
THERMINOL 44	FKM	NR at 70°F	NR at 70°F	A to 70°F	NR at 70°F	
THERMINOL 45	FKM	B at 70°F	NR at 70°F	A to 70°F	NR at 70°F	
THETA OCTADECANOIC ACID		NR at 70°F			B at 70°F	
T.H.F.	T, IIR	NR at 70°F	NR at 70°F	+200% vol 7 days 70°F NR at 70°F	NR at 70°F	NR at 70°F
THIOETHYL ALCOHOL		NR at 70°F	NR at 70°F	B at 70°F	NR at 70°F	
THIONYL CHLORIDE	FKM	NR at 70°F	NR at 70°F	AB to 70°F	NR at 70°F	
THIOKOL TP-90B	EP, FKM, IIR	NR at 70°F	A to 70°F	A to 70°F	B at 70°F	NR at 70°F
THIOKOL TP-95	EP, FKM, IIR	NR at 70°F	A to 70°F	A to 70°F	B at 70°F	NR at 70°F
THIOPHENE (THIOPEN)		NR at 70°F	NR at 70°F	C at 70°F	NR at 70°F	

	STYRENE BUTADIENE SBR	POLYACRYLATE ACM	POLYURETHANE AU, EU	ISOBUTYLENE ISOPRENE IIR	POLYBUTADIENE BR	® AEROQUIP AQP
TETRABROMOMETHANE	NR at 70°F			NR at 70°F	NR at 70°F	
TETRABUTYL TITANATE	B at 70°F			B at 70°F	B at 70°F	
TETRACHLORIDE OF TIN			B at 70°F	A to 70°F AB any conc to 150°F		
TETRACHLOROBENZENE				NR at 70°F		
TETRACHLORODIFLUOROETHANE (F112)	NR at 70°F		B at 70°F	NR at 70°F	NR at 70°F	
TETRACHLORODIFLUOROMETHANE				NR at 70°F		
TETRACHLOROETHANE	NR at 70°F	NR at 70°F	NR at 70°F	NR at 70°F	NR at 70°F	
TETRACHLOROETHYLENE	NR at 70°F	NR at 70°F	NR at 70°F	NR at 70°F	NR at 70°F	
TETRACHLOROMETHANE			C at 70°F	NR at 70°F		
TETRACHLORONAPHTHALENE				NR at 70°F		
TETRADECANOL				A to 70°F		
TETRAETHYL LEAD	NR at 70°F		AB to 70°F	NR at 70°F	NR at 70°F	
TETRAETHYL LEAD BLEND	NR at 70°F			NR at 70°F	NR at 70°F	
TETRAETHYL ORTHOSILICATE				A to 70°F AB to 180°F		
TETRAETHYLENE GLYCOL				A to 70°F		
TETRAFLUOROMETHANE				B /NR at 70°F		
TETRAHYDROFURAN	NR at 70°F	NR at 70°F	C /NR at 70°F	B at 70°F	NR at 70°F	
TETRAHYDRONAPHTHALENE	NR at 70°F	NR at 70°F	NR at 70°F	NR at 70°F	NR at 70°F	
TETRALIN (SULFOLANE)	NR at 70°F	NR at 70°F	NR at 70°F	NR at 70°F	NR at 70°F	
TETRANE			A to 70°F	NR at 70°F		
TETRANITROMETHANE			A to 70°F			
TETROL				NR at 70°F		
TEXACO CAPELLA A & AA	NR at 70°F	A to 70°F	B at 70°F	NR at 70°F	NR at 70°F	
TEXACO MEROPA 220 NO. 3 (NO LEAD)	NR at 70°F	A to 70°F	B at 70°F	NR at 70°F	NR at 70°F	
TEXACO REGAL B	NR at 70°F	A to 70°F	A to 70°F	NR at 70°F	NR at 70°F	
TEXACO UNI-TEMP GREASE	NR at 70°F	A to 70°F	A to 70°F	NR at 70°F	NR at 70°F	
TEXACO 3450 GEAR OIL	NR at 70°F	A to 70°F	A to 70°F	NR at 70°F	NR at 70°F	
TEXAMATIC "A" TRANSMISSION OIL	NR at 70°F	A to 70°F	B at 70°F	NR at 70°F	NR at 70°F	
TEXAMATIC 1581 FLUID	NR at 70°F	A to 70°F	B at 70°F	NR at 70°F	NR at 70°F	
TEXAMATIC 3401 FLUID	NR at 70°F	A to 70°F	B at 70°F	NR at 70°F	NR at 70°F	
TEXAMATIC 3525 FLUID	NR at 70°F	A to 70°F	B at 70°F	NR at 70°F	NR at 70°F	
TEXAMATIC 3528 FLUID	NR at 70°F	A to 70°F	B at 70°F	NR at 70°F	NR at 70°F	
TEXAS 1500 OIL	NR at 70°F	A to 70°F	A to 70°F	NR at 70°F	NR at 70°F	
THENARDITE			A to 70°F	A to 70°F AB any conc to 185°F		
THERMINOL VP1, 60 & 66		NR at 70°F		NR at 70°F		
THERMINOL 44		NR at 70°F		NR at 70°F		
THERMINOL 45		B at 70°F		NR at 70°F		
THETA OCTADECANOIC ACID				NR at 70°F		
T.H.F.	NR at 70°F	NR at 70°F	C /NR at 70°F	B at 70°F	NR at 70°F	
THIOETHYL ALCOHOL				NR at 70°F		
THIONYL CHLORIDE				NR at 70°F	NR at 70°F	
THIOKOL TP-90B	NR at 70°F			A to 70°F		
THIOKOL TP-95	NR at 70°F			A to 70°F		
THIOPHENE (THIOPEN)				C at 70°F		

	SYNTHETIC ISOPRENE IR	NATURAL ISOPRENE NR	CHLOROSULFONATED POLYETHYLENE CSM	FLUOROSILICONE FVMQ	SILICONE VMQ	CHEMRAZ FFKM
TETRABROMOMETHANE	NR at 70°F	NR at 70°F	NR at 70°F	B at 70°F	NR at 70°F	
TETRABUTYL TITANATE	B at 70°F	B at 70°F	A to 70°F	A to 70°F		A to 70°F
TETRACHLORIDE OF TIN	A 50-100% to 70°F	A 50-100% to 70°F AB any conc to 185°F	AC to 70°F NR 50% at 70°F	A to 100% to 70°F	B 50-100% at 70°F	A to 70°F
TETRACHLOROBENZENE	NR at 70°F	NR at 70°F	NR at 70°F			
TETRACHLORODIFLUOROETHANE (F112)	NR at 70°F	NR at 70°F	B at 70°F		NR at 70°F	A to 70°F
TETRACHLORODIFLUOROMETHANE	NR at 70°F	NR at 70°F				*B at 70°F
TETRACHLOROETHANE	NR at 70°F	NR at 70°F	NR at 70°F	B at 70°F	C/NR at 70°F	A to 70°F
TETRACHLOROETHYLENE	NR at 70°F	NR at 70°F	NR at 70°F	B at 70°F	B/NR at 70°F	A to 70°F
TETRACHLOROMETHANE	NR at 70°F	NR at 70°F	NR at 70°F	BC at 70°F	NR at 70°F	B at 70°F
TETRACHLORONAPHTHALENE	NR at 70°F	NR at 70°F	NR at 70°F			
TETRADECANOL	A to 70°F	A to 70°F	A to 70°F			
TETRAETHYL LEAD	NR at 70°F	NR at 70°F	NR at 70°F	B/NR at 70°F		A to 70°F
TETRAETHYL LEAD BLEND	NR at 70°F	NR at 70°F	NR at 70°F			A to 70°F
TETRAETHYL ORTHOSILICATE	NR at 70°F	NR at 70°F				
TETRAETHYLENE GLYCOL	A to 70°F	A to 70°F	A to 70°F			
TETRAFLUOROMETHANE	NR at 70°F	A/NR at 70°F	A/NR at 70°F		NR at 70°F	*B at 70°F
TETRAHYDROFURAN	NR at 70°F	NR at 70°F	NR at 70°F	NR at 70°F	NR at 70°F	A to 70°F
TETRAHYDRONAPHTHALENE	NR at 70°F	NR at 70°F	NR at 70°F	A to 70°F	NR at 70°F	A to 70°F
TETRALIN (SULFOLANE)	NR at 70°F	NR at 70°F	NR at 70°F	A to 70°F	NR at 70°F	A to 70°F
TETRANE	NR at 70°F	NR at 70°F	A to 70°F			
TETRANITROMETHANE						
TETROL	NR at 70°F	NR at 70°F	NR at 70°F			
TEXACO CAPELLA A & AA	NR at 70°F	NR at 70°F	NR at 70°F	A to 70°F	NR at 70°F	A to 70°F
TEXACO MEROPA 220 NO. 3 (NO LEAD)	NR at 70°F	NR at 70°F	NR at 70°F	A to 70°F	NR at 70°F	A to 70°F
TEXACO REGAL B	NR at 70°F	NR at 70°F	NR at 70°F	NR at 70°F	NR at 70°F	A to 70°F
TEXACO UNI-TEMP GREASE	NR at 70°F	NR at 70°F	NR at 70°F	A to 70°F	B at 70°F	A to 70°F
TEXACO 3450 GEAR OIL	NR at 70°F	NR at 70°F	NR at 70°F	A to 70°F	NR at 70°F	A to 70°F
TEXAMATIC "A" TRANSMISSION OIL	NR at 70°F	NR at 70°F	NR at 70°F	B at 70°F	NR at 70°F	A to 70°F
TEXAMATIC 1581 FLUID	NR at 70°F	NR at 70°F	NR at 70°F	B at 70°F	NR at 70°F	A to 70°F
TEXAMATIC 3401 FLUID	NR at 70°F	NR at 70°F	NR at 70°F	B at 70°F	NR at 70°F	A to 70°F
TEXAMATIC 3525 FLUID	NR at 70°F	NR at 70°F	NR at 70°F	B at 70°F	NR at 70°F	A to 70°F
TEXAMATIC 3528 FLUID	NR at 70°F	NR at 70°F	NR at 70°F	B at 70°F	NR at 70°F	A to 70°F
TEXAS 1500 OIL	NR at 70°F	NR at 70°F	NR at 70°F	A to 70°F	B at 70°F	A to 70°F
THENARDITE	A to 70°F	A to 70°F AB any conc to 150°F	A to 70°F		A to 70°F	
THERMINOL VP1, 60 & 66					B at 70°F	
THERMINOL 44					NR at 70°F	
THERMINOL 45					NR at 70°F	
THETA OCTADECANOIC ACID	NR at 70°F	NR at 70°F				
T.H.F.	NR at 70°F	NR at 70°F	NR at 70°F	NR at 70°F	NR at 70°F	A to 70°F
THIOETHYL ALCOHOL	NR at 70°F	NR at 70°F	NR at 70°F			A to 70°F
THIONYL CHLORIDE	NR at 70°F	NR at 70°F			NR at 70°F	A to 70°F
THIOKOL TP-90B			B at 70°F	B at 70°F		A to 70°F
THIOKOL TP-95			B at 70°F	B at 70°F		A to 70°F
THIOPHENE (THIOPEN)	NR at 70°F	NR at 70°F				A to 70°F

	POLYSULFIDE T	CHLORINATED POLYETHYLENE CM	EPICHLOROHYDRIN CO, ECO	® HYTREL COPOLYESTER TPE	® SANTOPRENE COPOLYMER TPO	® C-FLEX STYRENIC TPE
TETRABROMOMETHANE						
TETRABUTYL TITANATE						
TETRACHLORIDE OF TIN						
TETRACHLOROBENZENE						
TETRACHLORODIFLUOROETHANE (F112)	A to 70°F					
TETRACHLORODIFLUOROMETHANE						
TETRACHLOROETHANE	NR at 70°F				B/NR at 70°F	
TETRACHLOROETHYLENE	NR at 70°F					
TETRACHLOROMETHANE						
TETRACHLORONAPHTHALENE						
TETRADECANOL						
TETRAETHYL LEAD						
TETRAETHYL LEAD BLEND						
TETRAETHYL ORTHOSILICATE						
TETRAETHYLENE GLYCOL						
TETRAFLUOROMETHANE						
TETRAHYDROFURAN	A to 70°F			AC to 70°F	NR at 70°F	
TETRAHYDRONAPHTHALENE	NR at 70°F					
TETRALIN (SULFOLANE)	NR at 70°F					
TETRANE						
TETRANITROMETHANE						
TETROL						
TEXACO CAPELLA A & AA						
TEXACO MEROPA 220 NO. 3 (NO LEAD)						
TEXACO REGAL B	A to 70°F					
TEXACO UNI-TEMP GREASE	A to 70°F					
TEXACO 3450 GEAR OIL	A to 70°F					
TEXAMATIC "A" TRANSMISSION OIL						
TEXAMATIC 1581 FLUID						
TEXAMATIC 3401 FLUID						
TEXAMATIC 3525 FLUID						
TEXAMATIC 3528 FLUID						
TEXAS 1500 OIL	A to 70°F					
THENARDITE						
THERMINOL VP1, 60 & 66						
THERMINOL 44						
THERMINOL 45						
THETA OCTADECANOIC ACID						
T.H.F.	A to 70°F			AC to 70°F	NR at 70°F	
THIOETHYL ALCOHOL						
THIONYL CHLORIDE						
THIOKOL TP-90B						
THIOKOL TP-95						
THIOPHENE (THIOPEN)						

	ETHYLENE ACRYLIC EA	POLYALLOMER LINEAR COPOLYMER	NYLON 11 POLYAMIDE	NYLON 12 POLYAMIDE	ETHYLENE VINYL ACETATE EVA	POLYVINYLCHLORIDE FLEXIBLE PVC
TETRABROMOMETHANE						
TETRABUTYL TITANATE						
TETRACHLORIDE OF TIN					NR at 70°F	NR at 70°F
TETRACHLOROBENZENE						
TETRACHLORODIFLUOROETHANE (F112)	BC at 70°F		NR at 70°F			NR at 70°F
TETRACHLORODIFLUOROMETHANE						NR at 70°F
TETRACHLOROETHANE						NR at 70°F
TETRACHLOROETHYLENE			C at 70°F			
TETRACHLOROMETHANE						
TETRACHLORONAPHTHALENE						
TETRADECANOL						
TETRAETHYL LEAD			AB to 70°F		NR at 70°F	NR at 70°F
TETRAETHYL LEAD BLEND						
TETRAETHYL ORTHOSILICATE						
TETRAETHYLENE GLYCOL						
TETRAFLUOROMETHANE						NR at 70°F
TETRAHYDROFURAN	NR at 70°F	AB to 70°F C at 122°F	AB to 70°F	AB to 70°F	NR at 70°F	NR at 70°F
TETRAHYDRONAPHTHALENE		NR at 70°F	A to 70°F	AB to 70°F	NR at 70°F	
TETRALIN (SULFOLANE)		NR at 70°F	A to 70°F	AB to 70°F	NR at 70°F	
TETRANE						
TETRANITROMETHANE					NR at 70°F	NR at 70°F
TETROL						
TEXACO CAPELLA A & AA						
TEXACO MEROPA 220 NO. 3 (NO LEAD)						
TEXACO REGAL B						
TEXACO UNI-TEMP GREASE	A to 70°F					
TEXACO 3450 GEAR OIL	A to 70°F					
TEXAMATIC "A" TRANSMISSION OIL	A to 70°F					
TEXAMATIC 1581 FLUID	A to 70°F					
TEXAMATIC 3401 FLUID	A to 70°F					
TEXAMATIC 3525 FLUID	A to 70°F					
TEXAMATIC 3528 FLUID	A to 70°F					
TEXAS 1500 OIL						
THENARDITE						
THERMINOL VP1, 60 & 66						
THERMINOL 44						
THERMINOL 45						
THETA OCTADECANOIC ACID						
T.H.F.	NR at 70°F	AB to 70°F C at 122°F	AB to 70°F	AB to 70°F	NR at 70°F	NR at 70°F
THIOETHYL ALCOHOL						
THIONYL CHLORIDE		NR at 70°F		NR at 70°F		NR at 70°F
THIOKOL TP-90B						
THIOKOL TP-95						
THIOPHENE (THIOPEN)		B at 70°F	A to 70°F		NR at 70°F	

	POLYETHYLENE LOW DENSITY LDPE	®TEFLON FEP	®KALREZ PERFLUORINATED ELASTOMER	®FLUORAZ FLUORINATED COPOLYMER	®AFLAS FLUORINATED COPOLYMER	®NORPRENE COPOLYMER TPO
TETRABROMOMETHANE		A to 70°F	A to 212°F			
TETRABUTYL TITANATE		A to 70°F	A to 212°F	A to 70°F		
TETRACHLORIDE OF TIN	A to 70°F AB to 140°F	A to 300°F	A to 212°F	A 15-100% to 70°F		
TETRACHLOROBENZENE		A to 70°F				
TETRACHLORODIFLUOROETHANE (F112)		A to 70°F	BC at 70°F	NR at 70°F		
TETRACHLORODIFLUOROMETHANE		A to 70°F		NR at 70°F		
TETRACHLOROETHANE	B at 70°F	A to 125°F	*A to 70°F	NR at 70°F		
TETRACHLOROETHYLENE	NR at 70°F	A to 250°F	A to 212°F	NR at 70°F	+51% 1 day 70°F NR at 70°F	
TETRACHLOROMETHANE	C/NR at 70°F NR at 122°F	A to 200°F	AB to 70°F B at 113°F	NR 10-100% at 70°F	NR at 70°F	NR at 70°F
TETRACHLORONAPHTHALENE		A to 70°F				
TETRADECANOL		A to 70°F				
TETRAETHYL LEAD	AC to 70°F NR at 140°F	A to 300°F	A to 212°F	C/NR at 70°F		
TETRAETHYL LEAD BLEND				C at 70°F		
TETRAETHYL ORTHOSILICATE		A to 70°F				
TETRAETHYLENE GLYCOL		A to 70°F				
TETRAFLUOROMETHANE		A to 70°F	B at 70°F	NR at 70°F		
TETRAHYDROFURAN	NR at 70°F	A to 300°F	+1% vol 7 days 70°F A to 212°F	NR at 70°F		NR at 70°F
TETRAHYDRONAPHTHALENE	NR at 70°F	A to 70°F	A to 212°F	NR at 70°F		NR at 70°F
TETRALIN (SULFOLANE)	NR at 70°F	A to 70°F	A to 212°F	NR at 70°F		NR at 70°F
TETRANE		A to 70°F				
TETRANITROMETHANE	A to 70°F					
TETROL		A to 70°F				
TEXACO CAPELLA A & AA				A to 70°F		
TEXACO MEROPA 220 NO. 3 (NO LEAD)				A to 70°F		
TEXACO REGAL B				A to 70°F		
TEXACO UNI-TEMP GREASE				A to 70°F		
TEXACO 3450 GEAR OIL				A to 70°F		
TEXAMATIC "A" TRANSMISSION OIL				A to 70°F		
TEXAMATIC 1581 FLUID				A to 70°F		
TEXAMATIC 3401 FLUID				A to 70°F		
TEXAMATIC 3525 FLUID				A to 70°F		
TEXAMATIC 3528 FLUID				A to 70°F		
TEXAS 1500 OIL				A to 70°F		
THENARDITE		A to 70°F				
THERMINOL VP1, 60 & 66			A to 212°F			
THERMINOL 44				A to 70°F		
THERMINOL 45						
THETA OCTADECANOIC ACID		A to 70°F				
T.H.F.	NR at 70°F	A to 300°F	+1% vol 7 days 70°F A to 212°F	NR at 70°F		NR at 70°F
THIOETHYL ALCOHOL		A to 70°F	A to 212°F			
THIONYL CHLORIDE	NR at 70°F	A to 300°F	A to 212°F			
THIOKOL TP-90B				A to 70°F		
THIOKOL TP-95				A to 70°F		
THIOPHENE (THIOPEN)	NR at 70°F	A to 70°F	A to 212°F			

	RECOMMENDATIONS	NITRILE NBR	ETHYLENE PROPYLENE EP, EPDM	FLUOROCARBON FKM	CHLOROPRENE CR	HYDROGENATED NITRILE HNBR
TIDEWATER MULTIGEAR 140, EP LUBE	NBR,FKM,ACM,AU	A to 70°F	NR at 70°F	A to 70°F	B at 70°F	A to 70°F
TIDEWATER OIL, BEEDOL	NBR,FKM,ACM,AU	A to 70°F	NR at 70°F	A to 70°F	B at 70°F	A to 70°F
TILE GLAZE SLURRY		NR at 70°F	C at 70°F	C at 70°F	A to 70°F	
TIN BICHLORIDE		A to 70°F AB any conc to 150°F	B at 70°F	A to 70°F	A/NR at 70°F	A to 70°F
TIN CHLORIDES	NBR,EP,FKM,CR	A to 100% to 140°F AB any conc to 150°F	A to 100% to 140°F	A to 100% to 140°F AB to 200°F	BC at 70°F NR 50% at 70°F	A to 70°F
TIN CRYSTALS		A to 70°F AB to 150°F	B at 70°F	A to 70°F	A/NR at 70°F	
TIN DICHLORIDE	NBR,EP,FKM,CR	A to 70°F AB any conc to 150°F	B at 70°F	A to 70°F	A/NR at 70°F	A to 70°F
TIN PROTOCHLORIDE		A to 70°F AB any conc to 150°F	B at 70°F	A to 70°F	A/NR at 70°F	
TIN SALTS		A to 70°F AB any conc to 150°F	AB to 70°F	A to 70°F	A/NR at 70°F	
TIN TETRACHLORIDE	NBR,EP,FKM,IIR	A to 100% to 140°F AB any conc to 150°F	A to 100% to 140°F	A to 100% to 140°F AB to 200°F	BC at 70°F NR 50% at 70°F	A to 70°F
TINCAL		AB any conc to 180°F	A to 70°F	A to 70°F	A to 70°F AB any conc to 200°F	
TITANIUM SALTS		B/NR at 70°F	NR at 70°F	A to 70°F		
TITANIUM TETRACHLORIDE	FKM,FVMQ	B/NR at 70°F	NR at 70°F	A to 160°F	NR at 160°F	B at 70°F
T.N.T.	FKM,CR,FVMQ	NR at 70°F	NR at 70°F	B at 70°F	A to 70°F	NR at 70°F
TOILET VENEGAR		C at 70°F	A to 70°F	A to 70°F	B at 70°F	
TOLUENE	FKM	+141% vol 3 days 70°F NR 30–100% at 70°F	+132% vol 3 days 70°F NR 30–100% at 70°F	+15% vol 28 days 70°F A to 100°F	+159% vol 3 days 70°F NR 30–100% at 70°F	NR at 70°F
TOLUENE		+31% vol 30% conc 3 days 70°F	+128% vol 30% conc 3 days 70°F	BC to 200°F	+65% vol 30% conc 3 days 70°F	
TOLUENE DIISOCYANATE (HYLENE)	EP	NR at 70°F	AB to 70°F	B/NR at 70°F	NR at 70°F	NR at 70°F
TOLUIDINE		NR at 70°F		B at 70°F		
TOLUNE TRICHLORIDE		NR at 70°F				
TOLUOL	FKM	NR 30–100% at 70°F	NR 30–100% at 70°F	A to 100°F BC to 200°F	NR 30–100% at 70°F	NR at 70°F
TOMATO		A to 140°F juice & pulp	A to 70°F juice	A to 140°F juice & pulp	A to 140°F juice & pulp	
TOXAPHENE		AB 12% at 70°F			AB 12% to 70°F	
TRANSFORMER OIL	NBR,FKM,FVMQ,AU	A to 150°F	NR at 70°F	A to 300°F	B to 120°F	A to 70°F
TRANSFORMER PYRANOL		NR at 70°F		A to 70°F	NR at 70°F	
TRANSMISSION FLUID TYPE A	NBR,FKM,ACM,AU	A to 70°F AB to 100°F	NR at 70°F	no swell 7 days 212°F A to 350°F ATF 1,2,3,4,5	B at 70°F	A to 70°F
TRANSMISSION OIL		AB to 70°F	NR at 70°F		AB to 70°F	
TRIACETIN	EP,IIR,NBR,CR	B any conc to 80°F	A to 70°F	NR at 70°F	B any conc to 80°F	B at 70°F
TRIAMMONIUM PHOSPHATE		A to 70°F	A to 70°F	A to 70°F	A to 70°F	
TRIARYL PHOSPHATE	EP,FKM,IIR	NR at 70°F	A to 70°F	A to 70°F	NR at 70°F	NR at 70°F
TRIBUTOXY ETHYL PHOSPHATE	EP,FKM,IIR	NR at 70°F	A to 70°F	A to 70°F	NR at 70°F	NR at 70°F
TRIBUTYL AMINE		B at 70°F				
TRIBUTYL CITRATE						
TRIBUTYL MERCAPTAN	FKM,EP,IIR	NR at 70°F	NR at 70°F	A to 70°F	NR at 70°F	
TRIBUTYL PHOSPHATE	EP,IIR,T	+135% vol 3 days 212°F NR at 70°F	+15% vol 3 days 212°F A to 250°F	+422% vol 28 days 212°F NR at 70°F	+138% vol 3 days 212°F NR at 70°F	NR at 70°F
TRICHLOROACETIC ACID	EP,NBR,IIR	B/NR at 70°F	B at 70°F	+55% vol 28 days 158°F C/NR at 70°F	NR at 70°F	B at 70°F
TRICHLOROBENZENE	FKM	NR at 70°F		AB to 160°F	NR at 70°F	
TRICHLOROFLUOROMETHANE (F11)	NBR,T,FKM	AB to 250°F	NR at 70°F	BC to 140°F	NR at 70°F	B at 70°F
TRICHLOROETHANE	FKM,FVMQ	NR at 70°F	NR at 70°F	A to 140°F AB to 200°F	NR at 70°F	NR at 70°F
TRICHLOROETHYLENE (TRIAD)	FKM,FVMQ	+180% vol 7 days 70°F NR at 70°F	+163% vol 3 days 70°F NR at 70°F	+10% vol 7 days 75°F +15% vol 28 days 158°F	+280% vol 7 days 70°F NR at 70°F	C at 70°F
TRICHLOROMETHANE	FKM,FVMQ	NR at 70°F	NR at 70°F	A to 70°F AB to 200°F	NR at 70°F	NR at 70°F
TRICHLOROMONOFLUOROMETHANE		NR at 70°F			NR at 70°F	
TRICHLOROPANE		NR at 70°F		A to 70°F	NR at 70°F	
TRICHLOROPROPANE		A to 140°F		A to 140°F	A to 140°F	

	STYRENE BUTADIENE SBR	POLYACRYLATE ACM	POLYURETHANE AU, EU	ISOBUTYLENE ISOPRENE IIR	POLYBUTADIENE BR	® AEROQUIP AQP
TIDEWATER MULTIGEAR 140, EP LUBE	NR at 70°F	A to 70°F	A to 70°F	NR at 70°F	NR at 70°F	
TIDEWATER OIL, BEEDOL	NR at 70°F	A to 70°F	A to 70°F	NR at 70°F	NR at 70°F	
TILE GLAZE SLURRY						
TIN BICHLORIDE			B at 70°F	AB any conc to 150°F		
TIN CHLORIDES			B at 70°F	AB any conc to 150°F		
TIN CRYSTALS			B at 70°F	AB any conc to 150°F		
TIN DICHLORIDE			B at 70°F	AB any conc to 150°F		
TIN PROTOCHLORIDE			B at 70°F	AB any conc to 150°F		
TIN SALTS			B at 70°F	AB any conc to 150°F		
TIN TETRACHLORIDE			B at 70°F	AB any conc to 150°F		
TINCAL			A to 70°F	A to 70°F AB any conc to 150°F		
TITANIUM SALTS			B at 70°F			
TITANIUM TETRACHLORIDE	NR at 70°F	NR at 70°F	NR at 70°F	NR at 70°F	NR at 70°F	
T.N.T.				NR at 70°F		
TOILET VENEGAR			B at 70°F	A to 70°F		
TOLUENE	+202% vol 3 days 70°F NR 30-100% at 70°F	+179% vol 3 days 70°F NR 30-100% at 70°F	+99% vol 3 days 70°F NR 100% at 70°F	NR at 70°F	NR at 70°F	TEST
TOLUENE	+112% vol 30% conc 3 days 70°F	+39% vol 30% conc 3 days 70°F	+12% vol 30% conc 3 days 70°F			
TOLUENE DIISOCYANATE (HYLENE)	NR at 70°F	NR at 70°F		AB to 70°F	C/NR at 70°F	
TOLUIDINE				NR at 70°F		
TOLUNE TRICHLORIDE						
TOLUOL	NR 30-100% at 70°F	NR 30-100% at 70°F	NR 100% at 70°F BC 30% at 70°F	NR at 70°F	NR at 70°F	TEST
TOMATO			A to 70°F			
TOXAPHENE						
TRANSFORMER OIL	NR at 70°F	B at 70°F	A to 70°F C/NR at 140°F	NR at 70°F	NR at 70°F	
TRANSFORMER PYRANOL			C at 70°F			
TRANSMISSION FLUID TYPE A	NR at 70°F	A to 70°F	A to 70°F	NR at 70°F	NR at 70°F	
TRANSMISSION OIL						
TRIACETIN	C at 70°F	NR at 70°F	NR at 70°F	A any conc to 80°F	BC at 70°F	
TRIAMMONIUM PHOSPHATE				A to 70°F		
TRIARYL PHOSPHATE	NR at 70°F	NR at 70°F	NR at 70°F	A to 70°F	NR at 70°F	
TRIBUTOXY ETHYL PHOSPHATE	B at 70°F	NR at 70°F	NR at 70°F	A to 70°F	B at 70°F	
TRIBUTYL AMINE				A to 70°F		
TRIBUTYL CITRATE						
TRIBUTYL MERCAPTAN	NR at 70°F	NR at 70°F		NR at 70°F	NR at 70°F	
TRIBUTYL PHOSPHATE	+72% vol 3 days 212°F NR at 70°F	+158% vol 3 days 212°F NR at 70°F	+234% vol 3 days 212°F NR at 70°F	+18% vol 3 days 212°F B at 70°F	NR at 70°F	
TRICHLOROACETIC ACID	NR at 70°F	NR at 70°F	NR at 70°F	B at 70°F	B at 70°F	
TRICHLOROBENZENE	NR at 70°F	NR at 70°F	NR at 70°F	NR at 70°F		
TRICHLOROFLUOROMETHANE (F11)	NR at 70°F		B/NR at 70°F	NR at 70°F	B/NR at 70°F	
TRICHLOROETHANE	NR at 70°F	NR at 70°F	NR at 70°F	NR at 70°F	NR at 70°F	
TRICHLOROETHYLENE (TRIAD)	+212% vol 3 days 70°F NR at 70°F	+248% vol 3 days 70°F NR at 70°F	+170% vol 7 days 70°F NR at 70°F	+240% vol 7 days 70°F NR at 70°F	NR at 70°F	TEST
TRICHLOROMETHANE	NR at 70°F	NR at 70°F	NR at 70°F	NR at 70°F	NR at 70°F	NR at 70°F
TRICHLOROMONOFLUOROMETHANE				NR at 70°F		
TRICHLOROPANE						
TRICHLOROPROPANE			A to 70°F	NR at 70°F		

	SYNTHETIC ISOPRENE IR	NATURAL ISOPRENE NR	CHLOROSULFONATED POLYETHYLENE CSM	FLUOROSILICONE FVMQ	SILICONE VMQ	CHEMRAZ FFKM
TIDEWATER MULTIGEAR 140, EP LUBE	NR at 70°F	NR at 70°F	B at 70°F	A to 70°F	NR at 70°F	A to 70°F
TIDEWATER OIL, BEEDOL	NR at 70°F	NR at 70°F	NR at 70°F	A to 70°F	B at 70°F	A to 70°F
TILE GLAZE SLURRY		A to 70°F	C at 70°F			
TIN BICHLORIDE	A to 70°F	A to 70°F AB any conc to 150°F	A to 70°F		B at 70°F	A 15-100% to 70°F
TIN CHLORIDES	A to 70°F	A to 70°F AB any conc to 150°F	A to 70°F	A to 100% to 70°F	B 15-100% at 70°F	A to 70°F
TIN CRYSTALS	A to 70°F	A to 70°F AB any conc to 150°F	A to 70°F		B at 70°F	
TIN DICHLORIDE	A to 100% to 70°F	A to 100% to 70°F AB any conc to 150°F	A to 100% to 70°F A to 15% to 150°F	A to 100% to 70°F	B 15-100% to 70°F	A 15-100% to 70°F
TIN PROTOCHLORIDE	A to 70°F	A to 70°F AB any conc to 150°F	A to 70°F		B at 70°F	
TIN SALTS	A to 70°F	A to 70°F AB any conc to 150°F	A to 70°F		B at 70°F	
TIN TETRACHLORIDE	A 50-100% to 70°F	A 50-100% to 70°F AB any conc to 185°F	AC at 70°F NR 50% at 70°F	A to 100% to 70°F	B 50-100% at 70°F	A to 70°F
TINCAL	A to 70°F	A to 70°F AB any conc to 150°F	A to 70°F		B at 70°F	
TITANIUM SALTS						
TITANIUM TETRACHLORIDE	NR at 70°F	NR at 70°F	NR at 70°F	B at 70°F	NR at 70°F	*B at 70°F
T.N.T.	NR at 70°F	NR at 70°F	B at 70°F	B at 70°F	C at 70°F	A to 70°F
TOILET VENEGAR	B at 70°F	B at 70°F	B at 70°F		A to 70°F	
TOLUENE	NR at 70°F	NR at 70°F	NR at 70°F	+23% vol 3 days 70°F B at 70°F	+104% vol 3 days 70°F NR 30-100% at 70°F	A to 70°F
TOLUENE				+24% vol 30% conc 3 days 70°F	+109% vol 30% conc 3 days 70°F	
TOLUENE DIISOCYANATE (HYLENE)	C/NR at 70°F	C/NR at 70°F	NR at 70°F	NR at 70°F	NR at 70°F	A to 70°F
TOLUIDINE	NR at 70°F	NR at 70°F	NR at 70°F			A to 70°F
TOLUNE TRICHLORIDE						
TOLUOL	NR at 70°F	NR at 70°F	NR at 70°F	B at 70°F	NR 30-100% at 70°F	A to 70°F
TOMATO						
TOXAPHENE						
TRANSFORMER OIL	NR at 70°F	NR at 70°F	B/NR at 70°F	A to 70°F	B at 70°F	A to 70°F
TRANSFORMER PYRANOL						
TRANSMISSION FLUID TYPE A	NR at 70°F	NR at 70°F	B at 70°F	A to 70°F	B at 70°F	A to 70°F
TRANSMISSION OIL						
TRIACETIN	B at 70°F	B at 70°F	B at 70°F	NR at 70°F		A to 70°F
TRIAMMONIUM PHOSPHATE	A to 70°F	A to 70°F	A to 70°F		A to 70°F	
TRIARYL PHOSPHATE	NR at 70°F	NR at 70°F	C/NR at 70°F	B at 70°F	C at 70°F	A to 70°F
TRIBUTOXY ETHYL PHOSPHATE	B/NR at 70°F	B at 70°F	NR at 70°F	B at 70°F		A to 70°F
TRIBUTYL AMINE	B at 70°F	B at 70°F	C at 70°F			A to 70°F
TRIBUTYL CITRATE						A to 70°F
TRIBUTYL MERCAPTAN	NR at 70°F	NR at 70°F	NR at 70°F	C at 70°F	NR at 70°F	A to 70°F
TRIBUTYL PHOSPHATE	B at 70°F	B/NR at 70°F	NR at 70°F	dissolved 3 days 212°F NR at 70°F	+36% vol 3 days 212°F C/NR at 70°F	A to 70°F
TRICHLOROACETIC ACID	BC at 70°F	BC at 70°F	B/NR at 70°F	NR at 70°F	NR at 70°F	A to 70°F
TRICHLOROBENZENE	NR at 70°F	NR at 70°F	NR at 70°F	NR at 70°F	NR at 70°F	A to 70°F
TRICHLOROFLUOROMETHANE (F11)	NR at 70°F	NR at 70°F	A to 70°F *AB to 140°F	B at 70°F	NR at 70°F	B at 70°F
TRICHLOROETHANE	NR at 70°F	NR at 70°F	NR at 70°F	B/NR at 70°F	NR at 70°F	A to 70°F
TRICHLOROETHYLENE (TRIAD)	NR at 70°F	NR at 70°F	+500% vol 7 days 70°F NR at 70°F	+26% vol 3 days 70°F BC at 70°F	+240% vol 7 days 70°F NR at 70°F	A to 70°F
TRICHLOROMETHANE	NR at 70°F	NR at 70°F	NR at 70°F	B/NR at 70°F	NR at 70°F	A to 70°F
TRICHLOROMONOFLUOROMETHANE	NR at 70°F	NR at 70°F				
TRICHLOROPANE		NR at 70°F				
TRICHLOROPROPANE	NR at 70°F	NR at 70°F	NR at 70°F			A to 70°F

C 370

	POLYSULFIDE T	CHLORINATED POLYETHYLENE CM	EPICHLOROHYDRIN CO, ECO	® HYTREL COPOLYESTER TPE	® SANTOPRENE COPOLYMER TPO	® C-FLEX STYRENIC TPE
TIDEWATER MULTIGEAR 140, EP LUBE	A to 70°F					
TIDEWATER OIL, BEEDOL	A to 70°F					
TILE GLAZE SLURRY						
TIN BICHLORIDE						
TIN CHLORIDES						
TIN CRYSTALS						
TIN DICHLORIDE						
TIN PROTOCHLORIDE						
TIN SALTS						A to 70°F
TIN TETRACHLORIDE						
TINCAL						
TITANIUM SALTS						A to 70°F
TITANIUM TETRACHLORIDE	C at 70°F					
T.N.T.						
TOILET VENEGAR						
TOLUENE	B/NR at 70°F	C at 70°F	NR at 70°F	BC at 70°F NR at 140°F	+59% vol 7 days 70°F NR at 70°F	NR at 70°F
TOLUENE						
TOLUENE DIISOCYANATE (HYLENE)				AB to 70°F		
TOLUIDINE	AB to 70°F					
TOLUNE TRICHLORIDE						
TOLUOL	B/NR at 70°F	C at 70°F	NR at 70°F	BC at 70°F NR at 140°F	+59% vol 7 days 70°F NR at 70°F	NR at 70°F
TOMATO						
TOXAPHENE	AB 12% to 70°F					
TRANSFORMER OIL	A to 70°F					
TRANSFORMER PYRANOL						
TRANSMISSION FLUID TYPE A	A to 70°F		A to 70°F	AB to 70°F		
TRANSMISSION OIL						
TRIACETIN	B at 70°F					
TRIAMMONIUM PHOSPHATE						
TRIARYL PHOSPHATE	B at 70°F					
TRIBUTOXY ETHYL PHOSPHATE	A to 70°F					
TRIBUTYL AMINE						
TRIBUTYL CITRATE						
TRIBUTYL MERCAPTAN	NR at 70°F					
TRIBUTYL PHOSPHATE	+78% vol 3 days 212°F A to 70°F			C at 70°F		
TRICHLOROACETIC ACID				NR at 70°F		A to 70°F
TRICHLOROBENZENE						
TRICHLOROFLUOROMETHANE (F11)	A to 70°F			A to 70°F	*AB to 70°F	BC at 70°F
TRICHLOROETHANE	NR at 70°F			*NR at 70°F	*NR at 70°F	
TRICHLOROETHYLENE (TRIAD)	+150% vol 3 day 70°F NR at 70°F	C at 70°F	NR at 70°F	NR at 70°F	+179% vol 7 days 70°F NR at 70°F	NR at 70°F
TRICHLOROMETHANE	NR at 70°F	NR at 70°F		NR at 70°F	NR at 70°F	NR at 70°F
TRICHLOROMONOFLUOROMETHANE						
TRICHLOROPANE						
TRICHLOROPROPANE						

C 371

	ETHYLENE ACRYLIC EA	POLYALLOMER LINEAR COPOLYMER	NYLON 11 POLYAMIDE	NYLON 12 POLYAMIDE	ETHYLENE VINYL ACETATE EVA	POLYVINYLCHLORIDE FLEXIBLE PVC
TIDEWATER MULTIGEAR 140, EP LUBE	A to 70°F					
TIDEWATER OIL, BEEDOL						
TILE GLAZE SLURRY						
TIN BICHLORIDE						
TIN CHLORIDES						
TIN CRYSTALS						
TIN DICHLORIDE						
TIN PROTOCHLORIDE						
TIN SALTS						A to 70°F
TIN TETRACHLORIDE						
TINCAL						
TITANIUM SALTS			AB to 70°F			A to 70°F
TITANIUM TETRACHLORIDE						
T.N.T.						
TOILET VENEGAR						
TOLUENE	NR at 70°F	B at 70°F NR at 122°F	+1% vol 180 days 70°F A to 70°F	AB to 70°F	NR at 70°F	-35% vol 140 days 70° NR at 70°F
TOLUENE			BC at 105°F C at 194°F			
TOLUENE DIISOCYANATE (HYLENE)						
TOLUIDINE						
TOLUNE TRICHLORIDE						
TOLUOL	NR at 70°F	B at 70°F NR at 122°F	A to 70°F C at 194°F	AB to 70°F	NR at 70°F	-3% vol 7 days 70°F NR at 70°F
TOMATO		A to 70°F	A to 70°F		A to 70°F	A to 150°F
TOXAPHENE						
TRANSFORMER OIL	A to 70°F		*AB to 70°F	AB to 70°F	NR at 70°F	AC to 70°F NR at 140°F
TRANSFORMER PYRANOL						
TRANSMISSION FLUID TYPE A	A to 70°F		AB to 70°F			NR at 70°F
TRANSMISSION OIL			AB to 70°F			
TRIACETIN						
TRIAMMONIUM PHOSPHATE						
TRIARYL PHOSPHATE						
TRIBUTOXY ETHYL PHOSPHATE						NR at 70°F
TRIBUTYL AMINE						
TRIBUTYL CITRATE		B at 70°F C at 122°F				C at 70°F NR at 122°F
TRIBUTYL MERCAPTAN						
TRIBUTYL PHOSPHATE	NR at 70°F		AB to 140°F C at 194°F		NR at 70°F	NR at 70°F
TRICHLOROACETIC ACID			NR at 70°F			C/NR at 70°F
TRICHLOROBENZENE						
TRICHLOROFLUOROMETHANE (F11)	BC at 70°F					NR at 70°F
TRICHLOROETHANE	NR at 70°F	NR at 70°F	C at 70°F NR at 140°F		NR at 70°F	NR at 70°F
TRICHLOROETHYLENE (TRIAD)	NR at 70°F	B/NR at 70°F	AC to 70°F NR at 105°F	BC at 70°F	NR at 70°F	C/NR at 70°F
TRICHLOROMETHANE	NR at 70°F	B at 70°F C at 122°F	B/NR at 70°F		NR at 70°F	NR at 70°F
TRICHLOROMONOFLUOROMETHANE						NR at 70°F
TRICHLOROPANE						
TRICHLOROPROPANE						

	POLYETHYLENE LOW DENSITY LDPE	®TEFLON FEP	®KALREZ PERFLUORINATED ELASTOMER	®FLUORAZ FLUORINATED COPOLYMER	®AFLAS FLUORINATED COPOLYMER	®NORPRENE COPOLYMER TPO
TIDEWATER MULTIGEAR 140, EP LUBE				A to 70°F		
TIDEWATER OIL, BEEDOL				A to 70°F		
TILE GLAZE SLURRY						
TIN BICHLORIDE	A to 70°F AB to 140°F	A to 70°F				
TIN CHLORIDES	AB to 140°F	A to 300°F	A to 212°F	A to 100% to 70°F		
TIN CRYSTALS		A to 70°F				
TIN DICHLORIDE	A to 70°F AB to 140°F	A to 300°F				
TIN PROTOCHLORIDE		A to 70°F	A to 70°F	A to 70°F		
TIN SALTS	A to 70°F	A to 70°F				AB to 70°F
TIN TETRACHLORIDE	A to 70°F AB to 140°F	A to 300°F	A to 212°F	A 15-100% to 70°F		
TINCAL		A to 70°F				
TITANIUM SALTS	AB to 70°F					AB to 70°F
TITANIUM TETRACHLORIDE		A to 300°F	AB to 75°F	B/NR at 70°F		
T.N.T.		A to 70°F	A to 212°F	B at 70°F		
TOILET VENEGAR		A to 70°F				
TOLUENE	B/NR at 70°F NR at 122°F	+0.6% vol 14 days 158°F A to 300°F	+1% vol 7 days 75°F A to 212°F	NR at 70°F	+41% vol 7 days 70°F NR at 70°F	NR at 70°F
TOLUENE						
TOLUENE DIISOCYANATE (HYLENE)		A to 70°F	A to 212°F	NR at 70°F		
TOLUIDINE		A to 70°F	A to 257°F			
TOLUNE TRICHLORIDE		A to 70°F				
TOLUOL	B/NR at 70°F NR at 122°F	+0.6% vol 14 days 158°F A to 300°F	+1% vol 7 days 75°F A to 212°F	NR at 70°F	+41% vol 7 days 70°F NR at 70°F	NR at 70°F
TOMATO	A to 70°F	A to 300°F		A to 70°F		
TOXAPHENE		A 12% to 70°F		C at 70°F		
TRANSFORMER OIL	C at 70°F NR at 140°F	A to 300°F	A to 212°F	A to 70°F		
TRANSFORMER PYRANOL						
TRANSMISSION FLUID TYPE A		A to 70°F	A to 212°F	A to 70°F		
TRANSMISSION OIL		A to 70°F				
TRIACETIN		A to 70°F	A to 212°F	NR at 70°F		
TRIAMMONIUM PHOSPHATE		A to 70°F				
TRIARYL PHOSPHATE		A to 70°F	A to 212°F	A to 70°F		
TRIBUTOXY ETHYL PHOSPHATE		A to 70°F	A to 212°F	A to 70°F		
TRIBUTYL AMINE		A to 70°F	*A to 212°F			
TRIBUTYL CITRATE	B at 70°F C at 122°F	A to 125°F	A to 212°F			
TRIBUTYL MERCAPTAN		A to 70°F	A to 212°F	A to 70°F		
TRIBUTYL PHOSPHATE	A/NR at 70°F NR at 140°F	A to 392°F	A 30-100% to 212°F	B at 70°F	+36% vol 7 days 212°F A to 70°F	
TRICHLOROACETIC ACID	BC at 70°F	A to 384°F	*A to 100% to 212°F	C at 70°F		NR at 70°F
TRICHLOROBENZENE		A to 70°F	A to 212°F	NR at 70°F	B at 70°F	
TRICHLOROFLUOROMETHANE (F11)	BC at 70°F	A to 70°F	BC at 70°F	NR at 70°F		
TRICHLOROETHANE	NR at 70°F	A to 200°F	A to 70°F *A to 212°F	NR at 70°F wet or dry	NR at 70°F	NR at 70°F
TRICHLOROETHYLENE (TRIAD)	NR at 70°F	A to 300°F	A to 212°F	NR at 70°F wet or dry	+95% vol 7 days 70°F NR at 70°F	NR at 70°F
TRICHLOROMETHANE	NR at 70°F	A to 300°F	A to 212°F	NR at 70°F	NR at 70°F	NR at 70°F
TRICHLOROMONOFLUOROMETHANE		A to 70°F				
TRICHLOROPANE						
TRICHLOROPROPANE		A to 70°F	A to 212°F			

	RECOMMENDATIONS	NITRILE NBR	ETHYLENE PROPYLENE EP, EPDM	FLUOROCARBON FKM	CHLOROPRENE CR	HYDROGENATED NITRILE HNBR
TRICHLOROTRIFLUOROETHANE (F113)	NBR,CR,AU,T	A to 250°F	NR at 70°F	AB to 140°F	A to 130°F NR at 140°F	A to 70°F
TRICRESYL PHOSPHATE	EP,IIR,FKM,SBR	NR at 70°F	A to 100% to 212°F	A to 140°F B to 300°F	BC to 100% at 70°F NR at 140°F	NR at 70°F
TRIDECANOL (TRIDECYL ALCOHOL)		A to 70°F		B at 70°F		
TRIETHANOLAMINE	EP,IIR,NBR	BC 100% at 100°F AB to 80% to 120°F	A to 160°F	NR at 70°F	NR at 70°F	C at 70°F
TRIETHYL ALUMINUM	FKM	NR at 70°F		+30% vol 30 days 160°F B at 70°F	NR at 70°F	
TRIETHYLAMINE	NBR	A to 140°F	A to 160°F	A to 140°F	AB to 70°F	
TRIETHYL BORANE	FKM	NR at 70°F		+5% vol 30 days 160°F A to 70°F	NR at 70°F	
TRIETHYLENE GLYCOL		A to 70°F		A to 70°F		
TRIFLUORETHANE	FKM	NR at 70°F	NR at 70°F	A to 70°F	NR at 70°F	NR at 70°F
TRIFLUOROVINYL CHLORIDE		NR at 70°F		A to 70°F		
TRIHYDROXYBENZOIC ACID	FKM,FVMQ,EP	AB to 70°F	AB to 70°F	A to 120°F	B at 70°F	B at 70°F
TRIHYDROXYETHYLAMINE		B any conc to 120°F			B/NR at 70°F	
TRIETHYLAMINE				NR at 70°F	A/NR at 70°F	
TRIMETHYLMETHANE		A to 70°F		A to 70°F		
TRIMETHYLPENTANE	NBR,FKM,ACM,FVMQ	+2% vol 3 days 70°F A to 250°F	+103% vol 3 days 70°F NR at 70°F	no swell 3 days 70°F A to 140°F	+14% vol 3 days 70°F B/NR to 80°F	A to 70°F
TRIMETHYLENE GLYCOL		A to 70°F	A to 70°F	A to 70°F		
TRINIDAD PITCH		A to 70°F	NR at 70°F	A to 70°F	B at 70°F	
TRINITROPHENOL	NBR,EP,FKM,CR,IIR	NR conc at 70°F AB to 10% to 160°F	BC conc at 70°F AB 10% at 200°F	A to conc to 70°F A to 10% to 140°F	BC conc at 70°F A 10% to 70°F	
TRINITROTOLUENE	FKM,CR,FVMQ	NR at 70°F	NR at 70°F	B at 70°F	A to 70°F	NR at 70°F
TRI-NORMAL-BUTYL PHOSPHATE	EP,IIR,T	NR at 70°F	A to 250°F	NR at 70°F	NR at 70°F	NR at 70°F
TRIOCTYL PHOSPHATE	EP,IIR	NR at 70°F	A to 70°F	B at 70°F	NR at 70°F	
TRIOLEIN		B to 120°F			BC to 80°F	
TRIPHENYL PHOSPHATE		NR at 70°F		C at 70°F		
TRIPOLYPHOSPHATE	EP,IIR	NR at 70°F	A to 70°F	B at 70°F	B/NR at 70°F	NR at 70°F
TRISODIUM PHOSPHATE (TSP)	NBR,EP,IIR	A to 100% to 70°F AB to 100% to 200°F	A to 100% to 70°F	A to 100% to 70°F	A to 100% to 70°F AB to 100% to 200°F	
TRITOYL		NR at 70°F	NR at 70°F	C at 70°F	A to 70°F	
TRITOYL PHOSPHATE	EP,IIR	NR at 70°F	A to 212°F	A to 140°F B to 300°F	C/NR at 70°F	
TT-I-735b		A to 70°F	A to 70°F	A to 70°F	B at 70°F	
TT-N-95a		A to 70°F	NR at 70°F	A to 70°F	C at 70°F	
TT-N-97b		B at 70°F	NR at 70°F	A to 70°F	C at 70°F	
TT-S-735 TYPE I		A to 70°F	NR at 70°F	A to 70°F	B at 70°F	
TT-S-735 TYPE II		A to 70°F	NR at 70°F	A to 70°F	C at 70°F	
TT-S-735 TYPE III		A to 70°F	NR at 70°F	A to 70°F	C at 70°F	
TT-S-735 TYPE IV		A to 70°F	NR at 70°F	A to 70°F	A to 70°F	
TT-S-735 TYPE V		A to 70°F	NR at 70°F	A to 70°F	B at 70°F	
TT-S-735 TYPE VI		A to 70°F	NR at 70°F	A to 70°F	B at 70°F	
TT-S-735 TYPE VII		A to 70°F	NR at 70°F	A to 70°F	C at 70°F	
TT-T-656B		NR at 70°F	A to 70°F	NR at 70°F	NR at 70°F	
TUNG OIL	NBR,FKM,FVMQ	A to 250°F	NR at 70°F	A to 70°F	A to 70°F AB to 120°F	A to 70°F
TURBINE OIL	FKM,NBR	AB to 70°F	NR at 70°F	A to 140°F	B to 70°F NR at 140°F	A to 70°F
TURBINE OIL NO. 15 (MIL-L-7808A)	FKM	AB to 70°F	NR at 70°F	A to 350°F B at 400°F	NR at 70°F	B at 70°F
TURBO OIL NO. 35	NBR,FKM,ACM,AU	A to 70°F	NR at 70°F	A to 70°F	B at 70°F	A to 70°F
TURPENTINE	NBR,FKM,FVMQ	+9% vol 3 days 70°F A to 200°F	+163% vol 3 days 70°F NR at 70°F	no swell 3 days 70°F A to 158°F	+60% vol 3 days 70°F NR at 70°F	A to 70°F
TURPENTINE SUBSTITUTE		NR at 70°F			NR at 70°F	

	STYRENE BUTADIENE SBR	POLYACRYLATE ACM	POLYURETHANE AU, EU	ISOBUTYLENE ISOPRENE IIR	POLYBUTADIENE BR	® AEROQUIP AQP
TRICHLOROTRIFLUOROETHANE (F113)	B at 70°F		AB to 70°F	NR at 70°F	B at 70°F	
TRICRESYL PHOSPHATE	NR at 70°F	NR at 70°F	C /NR at 70°F	A to 70°F AB to 185°F	NR at 70°F	NR at 70°F
TRIDECANOL (TRIDECYL ALCOHOL)				A to 70°F		
TRIETHANOLAMINE	B at 70°F	NR at 70°F	NR at 70°F	AB any conc to 150°F	B at 70°F	
TRIETHYL ALUMINUM						
TRIETHYLAMINE	NR at 70°F	NR at 70°F	NR at 70°F	C at 70°F		
TRIETHYL BORANE						
TRIETHYLENE GLYCOL				A to 70°F		
TRIFLUORETHANE	NR at 70°F	NR at 70°F	NR at 70°F	NR at 70°F	NR at 70°F	
TRIFLUOROVINYL CHLORIDE				NR at 70°F		
TRIHYDROXYBENZOIC ACID	B at 70°F	NR at 70°F	NR at 70°F	B any conc to 150°F	B at 70°F	
TRIHYDROXYETHYLAMINE				A to 70°F AB any conc to 150°F		
TRIETHYLAMINE				AB to 70°F		
TRIMETHYLMETHANE				NR at 70°F		
TRIMETHYLPENTANE	+52% vol 3 days 70°F NR at 70°F	−2% vol 3 days 70°F A to 70°F	no swell 3 days 70°F B to 158°F	+111% vol 3 days 70°F NR at 70°F	NR at 70°F	
TRIMETHYLENE GLYCOL				A to 70°F		
TRINIDAD PITCH			B at 70°F	NR at 70°F		
TRINITROPHENOL	B sol'n at 70°F B molten		B to conc at 70°F B molten	C conc at 70°F AB sol'n to 70°F	B sol'n at 70°F B molten	NR sol'n at 70°F NR molten
TRINITROTOLUENE	NR at 70°F	NR at 70°F		NR at 70°F	NR at 70°F	
TRI-NORMAL-BUTYL PHOSPHATE	NR at 70°F	NR at 70°F	NR at 70°F			
TRIOCTYL PHOSPHATE	NR at 70°F	NR at 70°F	NR at 70°F	A to 70°F	NR at 70°F	
TRIOLEIN				B to 100°F		
TRIPHENYL PHOSPHATE				A to 70°F		
TRIPOLYPHOSPHATE	NR at 70°F	NR at 70°F	NR at 70°F	A to 70°F	NR at 70°F	
TRISODIUM PHOSPHATE (TSP)	A sol'n to 70°F	NR sol'n at 70°F	AB to 100% to 70°F *C /NR at 140°F	A to 70°F AB any conc to 185°F		
TRITOYL				NR at 70°F		
TRITOYL PHOSPHATE	NR at 70°F	NR at 70°F	C /NR at 70°F	A to 70°F AB to 185°F	NR at 70°F	
TT-I-735b	A to 70°F		B at 70°F	A to 70°F	A to 70°F	
TT-N-95a	NR at 70°F		C at 70°F	NR at 70°F	NR at 70°F	
TT-N-97b	NR at 70°F	C at 70°F	C at 70°F	NR at 70°F	NR at 70°F	
TT-S-735 TYPE I	NR at 70°F	B at 70°F	B at 70°F	NR at 70°F	NR at 70°F	
TT-S-735 TYPE II	NR at 70°F		C at 70°F	NR at 70°F	NR at 70°F	
TT-S-735 TYPE III	NR at 70°F		C at 70°F	NR at 70°F	NR at 70°F	
TT-S-735 TYPE IV	NR at 70°F	A to 70°F	A to 70°F	NR at 70°F	NR at 70°F	
TT-S-735 TYPE V	NR at 70°F	A to 70°F	B at 70°F	NR at 70°F	NR at 70°F	
TT-S-735 TYPE VI	NR at 70°F	A to 70°F	B at 70°F	NR at 70°F	NR at 70°F	
TT-S-735 TYPE VII	NR at 70°F		C at 70°F	NR at 70°F	NR at 70°F	
TT-T-656B	NR at 70°F	NR at 70°F	NR at 70°F	A to 70°F	NR at 70°F	
TUNG OIL	NR at 70°F	A to 70°F	BC at 70°F	B /NR at 70°F	NR at 70°F	AB to 70°F
TURBINE OIL	NR at 70°F	AB to 70°F	A to 70°F	NR at 70°F	NR at 70°F	
TURBINE OIL NO. 15 (MIL-L-7808A)	NR at 70°F	B at 70°F	NR at 70°F	NR at 70°F	NR at 70°F	
TURBO OIL NO. 35	NR at 70°F	A to 70°F	A to 70°F	NR at 70°F	NR at 70°F	
TURPENTINE	+153% vol 3 days 70°F NR at 70°F	+22% vol 3 days 70°F B at 70°F	+21% vol 3 days 70°F NR at 70°F	+182% vol 3 days 70°F NR at 70°F	NR at 70°F	AB to 70°F
TURPENTINE SUBSTITUTE			AB to 70°F C at 140°F	NR at 70°F		

	SYNTHETIC ISOPRENE IR	NATURAL ISOPRENE NR	CHLOROSULFONATED POLYETHYLENE CSM	FLUOROSILICONE FVMQ	SILICONE VMQ	CHEMRAZ FFKM
TRICHLOROTRIFLUOROETHANE (F113)	C/NR at 70°F	C/NR at 70°F	A to 130°F	NR at 70°F	NR at 70°F	C at 70°F
TRICRESYL PHOSPHATE	C/NR at 70°F	BC to 80°F	C/NR at 70°F	B at 70°F	C at 70°F	A to 70°F
TRIDECANOL (TRIDECYL ALCOHOL)	A to 70°F	A to 70°F	A to 70°F			
TRIETHANOLAMINE	B at 70°F	B any conc to 80°F	A to 158°F	NR at 70°F	NR at 70°F	B at 70°F
TRIETHYL ALUMINUM	NR at 70°F	NR at 70°F				
TRIETHYLAMINE	B at 70°F	B at 70°F			NR at 70°F	A to 70°F
TRIETHYL BORANE	NR at 70°F	NR at 70°F				
TRIETHYLENE GLYCOL	A to 70°F	A to 70°F	A to 70°F			A to 70°F
TRIFLUORETHANE	NR at 70°F	NR at 70°F	NR at 70°F	B at 70°F	NR at 70°F	A to 70°F
TRIFLUOROVINYL CHLORIDE	NR at 70°F	NR at 70°F				
TRIHYDROXYBENZOIC ACID	A to 70°F	A to 70°F AB any conc to 150°F	B at 70°F	A to 70°F		A to 70°F
TRIHYDROXYETHYLAMINE	C at 70°F	BC at 70°F				
TRIETHYLAMINE						
TRIMETHYLMETHANE	NR at 70°F	NR at 70°F	NR at 70°F			
TRIMETHYLPENTANE	NR at 70°F	NR at 70°F	B/NR at 70°F	+3% vol 3 days 70°F +15% vol 7 days 70°F	+150% vol 7 days 70°F NR at 70°F	A to 70°F
TRIMETHYLENE GLYCOL	A to 70°F	A to 70°F	A to 70°F			
TRINIDAD PITCH	NR at 70°F	NR at 70°F	B at 70°F		NR at 70°F	
TRINITROPHENOL	C conc at 70°F AB sol'n to 70°F	NR conc at 70°F AB sol'n to 70°F	A to conc to 70°F AB molten	B to conc at 70°F B molten	NR any conc at 70°F NR molten	A to 100% to molten
TRINITROTOLUENE	NR at 70°F	NR at 70°F	B at 70°F	B at 70°F	C at 70°F	A to 70°F
TRI-NORMAL-BUTYL PHOSPHATE		B/NR at 70°F	NR at 70°F	NR at 70°F	C/NR at 70°F	A to 70°F
TRIOCTYL PHOSPHATE	NR at 70°F	NR at 70°F	NR at 70°F	B at 70°F	C at 70°F	A to 70°F
TRIOLEIN	NR at 70°F	NR at 70°F				
TRIPHENYL PHOSPHATE	NR at 70°F	NR at 70°F	C at 70°F			
TRIPOLYPHOSPHATE	NR at 70°F	NR at 70°F	NR at 70°F	AB to 70°F	C at 70°F	A to 70°F
TRISODIUM PHOSPHATE (TSP)	A to 70°F	A to 70°F AB any conc to 150°F	A to 100% to 70°F	NR 100% at 70°F A sol'n to 70°F	A sol'n to 70°F	A to 70°F
TRITOYL	NR at 70°F	NR at 70°F	B at 70°F			
TRITOYL PHOSPHATE	C/NR at 70°F	B/NR at 70°F	C/NR at 70°F	B at 70°F	C at 70°F	
TT-I-735b	A to 70°F	A to 70°F	B at 70°F	A to 70°F	A to 70°F	
TT-N-95a	NR at 70°F	NR at 70°F	C at 70°F	A to 70°F	NR at 70°F	
TT-N-97b	NR at 70°F	NR at 70°F	C at 70°F	B at 70°F	NR at 70°F	
TT-S-735 TYPE I	NR at 70°F	NR at 70°F	C at 70°F	A to 70°F	NR at 70°F	
TT-S-735 TYPE II	NR at 70°F	NR at 70°F	C at 70°F	A to 70°F	NR at 70°F	
TT-S-735 TYPE III	NR at 70°F	NR at 70°F	C at 70°F	A to 70°F	NR at 70°F	
TT-S-735 TYPE IV	NR at 70°F	NR at 70°F	A to 70°F	A to 70°F	C at 70°F	
TT-S-735 TYPE V	NR at 70°F	NR at 70°F	B at 70°F	A to 70°F	C at 70°F	
TT-S-735 TYPE VI	NR at 70°F	NR at 70°F	B at 70°F	A to 70°F	C at 70°F	
TT-S-735 TYPE VII	NR at 70°F	NR at 70°F	C at 70°F	A to 70°F	NR at 70°F	
TT-T-656B	NR at 70°F	NR at 70°F	NR at 70°F	C at 70°F	NR at 70°F	
TUNG OIL	NR at 70°F	NR at 70°F	A to 70°F	B at 70°F	NR at 70°F	A to 70°F
TURBINE OIL	NR at 70°F	NR at 70°F	B/NR at 70°F	AB to 70°F	NR at 70°F	A to 70°F
TURBINE OIL NO. 15 (MIL-L-7808A)	NR at 70°F	NR at 70°F	NR at 70°F	+10% vol 3 days 300°F B at 70°F	+10-30% vol 3 days 300°F A to 300°F	A to 70°F
TURBO OIL NO. 35	NR at 70°F	NR at 70°F	NR at 70°F	A to 70°F	NR at 70°F	A to 70°F
TURPENTINE	NR at 70°F	NR at 70°F	NR at 70°F	+13% vol 3 days 70°F B at 70°F	+98% vol 3 days 70°F NR at 70°F	A to 70°F
TURPENTINE SUBSTITUTE	NR at 70°F	NR at 70°F				

C 376

	POLYSULFIDE T	CHLORINATED POLYETHYLENE CM	EPICHLOROHYDRIN CO, ECO	® HYTREL COPOLYESTER TPE	® SANTOPRENE COPOLYMER TPO	® C-FLEX STYRENIC TPE
TRICHLOROTRIFLUOROETHANE (F113)	A to 70°F		A to 70°F	A to 130°F	NR at 70°F	
TRICRESYL PHOSPHATE	AB to 70°F		NR at 70°F	AC at 70°F		
TRIDECANOL (TRIDECYL ALCOHOL)						
TRIETHANOLAMINE	NR at 70°F			NR at 70°F		
TRIETHYL ALUMINUM						
TRIETHYLAMINE						
TRIETHYL BORANE						
TRIETHYLENE GLYCOL						
TRIFLUORETHANE	NR at 70°F					
TRIFLUOROVINYL CHLORIDE						
TRIHYDROXYBENZOIC ACID				NR at 70°F		
TRIHYDROXYETHYLAMINE						
TRIETHYLAMINE						
TRIMETHYLMETHANE						
TRIMETHYLPENTANE	+0.2% vol 3 days 70°F A to 70°F	A to 70°F	A to 70°F	+1–8% vol 7 days 70°F A to 158°F	NR at 70°F	
TRIMETHYLENE GLYCOL						
TRINIDAD PITCH						
TRINITROPHENOL	NR sol'n at 70°F NR molten	NR sol'n at 70°F NR molten		NR sol'n at 70°F NR molten	NR sol'n at 70°F	
TRINITROTOLUENE	B at 70°F					
TRI-NORMAL-BUTYL PHOSPHATE						
TRIOCTYL PHOSPHATE	B at 70°F					
TRIOLEIN						
TRIPHENYL PHOSPHATE						
TRIPOLYPHOSPHATE	B at 70°F					
TRISODIUM PHOSPHATE (TSP)				A to 100% to 70°F		A to 70°F
TRITOYL						
TRITOYL PHOSPHATE	B at 70°F		NR at 70°F	C at 70°F		
TT-I-735b	A to 70°F		B at 70°F			
TT-N-95a	A to 70°F		A to 70°F			
TT-N-97b						
TT-S-735 TYPE I	A to 70°F		A to 70°F			
TT-S-735 TYPE II	A to 70°F		A to 70°F			
TT-S-735 TYPE III	A to 70°F		A to 70°F			
TT-S-735 TYPE IV	A to 70°F		A to 70°F			
TT-S-735 TYPE V	A to 70°F		A to 70°F			
TT-S-735 TYPE VI	A to 70°F		A to 70°F			
TT-S-735 TYPE VII	A to 70°F		A to 70°F			
TT-T-656B	NR at 70°F		NR at 70°F			
TUNG OIL	BC at 70°F	C at 70°F		AB to 70°F		
TURBINE OIL	A to 70°F		A to 70°F			
TURBINE OIL NO. 15 (MIL-L-7808A)	A to 70°F					
TURBO OIL NO. 35	A to 70°F					
TURPENTINE	-3.7% vol 3 days 70°F B at 70°F	AB to 70°F	A to 70°F	AB to 70°F	NR at 70°F	NR at 70°F
TURPENTINE SUBSTITUTE				AB to 70°F		

	ETHYLENE ACRYLIC EA	POLYALLOMER LINEAR COPOLYMER	NYLON 11 POLYAMIDE	NYLON 12 POLYAMIDE	ETHYLENE VINYL ACETATE EVA	POLYVINYLCHLORIDE FLEXIBLE PVC
TRICHLOROTRIFLUOROETHANE (F113)	BC at 70°F		NR at 70°F			NR at 70°F
TRICRESYL PHOSPHATE	B at 70°F		AB to 140°F C at 194°F		*NR at 70°F	NR at 70°F
TRIDECANOL (TRIDECYL ALCOHOL)						A to 150°F
TRIETHANOLAMINE	A to 158°F					
TRIETHYL ALUMINUM						
TRIETHYLAMINE						
TRIETHYL BORANE						
TRIETHYLENE GLYCOL		A to 122°F				B at 70°F C at 122°F
TRIFLUORETHANE						
TRIFLUOROVINYL CHLORIDE						NR at 70°F
TRIHYDROXYBENZOIC ACID			AB to 70°F			A to 150°F
TRIHYDROXYETHYLAMINE						
TRIETHYLAMINE						
TRIMETHYLMETHANE						
TRIMETHYLPENTANE	A to 70°F		A to 70°F	AB to 70°F	NR at 70°F	AB to 70°F C at 150°F
TRIMETHYLENE GLYCOL						
TRINIDAD PITCH		A sol'n to 70°F	C sol'n at 70°F NR sol'n at 105°F		C/NR sol'n at 70°F C 1% to 140°F	C/NR sol'n at 70°F AB 1% to 130°F
TRINITROPHENOL						NR at 70°F
TRINITROTOLUENE						
TRI-NORMAL-BUTYL PHOSPHATE						
TRIOCTYL PHOSPHATE						
TRIOLEIN	NR at 70°F					
TRIPHENYL PHOSPHATE						
TRIPOLYPHOSPHATE						
TRISODIUM PHOSPHATE (TSP)	A sol'n to 70°F		A to 70°F AB to 194°F		*AB to 140°F	A to 70°F AB to 140°F
TRITOYL						
TRITOYL PHOSPHATE	B at 70°F		AB to 140°F C at 194°F		*NR at 70°F	NR at 70°F
TT-I-735b						
TT-N-95a						
TT-N-97b						
TT-S-735 TYPE I						
TT-S-735 TYPE II						
TT-S-735 TYPE III						
TT-S-735 TYPE IV						
TT-S-735 TYPE V						
TT-S-735 TYPE VI						
TT-S-735 TYPE VII						
TT-T-656B						
TUNG OIL	*AB to 70°F		AB to 70°F			
TURBINE OIL						
TURBINE OIL NO. 15 (MIL-L-7808A)						
TURBO OIL NO. 35						
TURPENTINE	B/NR at 70°F	AB to 70°F C at 122°F	AB to 105°F BC at 140°F	AB to 70°F	NR at 70°F	AB to 70°F NR at 130°F
TURPENTINE SUBSTITUTE			AB to 70°F BC at 140°F		*C at 70°F *NR at 140°F	NR at 70°F

	POLYETHYLENE LOW DENSITY LDPE	® TEFLON FEP	® KALREZ PERFLUORINATED ELASTOMER	® FLUORAZ FLUORINATED COPOLYMER	® AFLAS FLUORINATED COPOLYMER	® NORPRENE COPOLYMER TPO
TRICHLOROTRIFLUOROETHANE (F113)	NR at 70°F	A to 200°F	BC at 70°F	NR at 70°F	+249% vol 7 days 70°F NR at 70°F	
TRICRESYL PHOSPHATE	NR at 70°F	A to boiling	A to 212°F	A to 70°F	+8% vol 7 days 302°F A to 302°F	
TRIDECANOL (TRIDECYL ALCOHOL)		A to 70°F				
TRIETHANOLAMINE		A to 300°F	A to 70°F *A to 212°F	A to 70°F	A to 70°F	
TRIETHYL ALUMINUM		A to 70°F	A to 212°F			
TRIETHYLAMINE		A to 70°F	*A to 212°F	A to 70°F		
TRIETHYL BORANE		A to 70°F	A to 212°F			
TRIETHYLENE GLYCOL	A to 122°F	A to 125°F	A to 212°F			
TRIFLUORETHANE				B at 70°F		
TRIFLUOROVINYL CHLORIDE		A to 70°F	A to 212°F	NR at 70°F		
TRIHYDROXYBENZOIC ACID	B /NR at 70°F	A to 200°F	A to 70°F	A to 70°F		
TRIHYDROXYETHYLAMINE		A to 70°F				
TRIETHYLAMINE	NR at 70°F	A to 300°F	*A to 212°F			
TRIMETHYLMETHANE		A to 70°F				
TRIMETHYLPENTANE	B at 70°F	+1% wt gain 3 days 212°F A to 300°F	A to 212°F	BC at 70°F	+19% vol 7 days 70°F BC at 70°F	
TRIMETHYLENE GLYCOL		A to 70°F				
TRINIDAD PITCH		A to 70°F				
TRINITROPHENOL	B /NR sol'n at 70°F	A to 300°F	A sol'n to 70°F	A sol'n to 70°F A molten		
TRINITROTOLUENE		A to 70°F	A to 212°F	B at 70°F		
TRI-NORMAL-BUTYL PHOSPHATE	A /NR at 70°F NR at 140°F	A to 392°F	A 30-100% to 212°F	B at 70°F		
TRIOCTYL PHOSPHATE		A to 70°F	A to 212°F	A to 70°F		
TRIOLEIN		A to 70°F				
TRIPHENYL PHOSPHATE		A to 200°F	A to 212°F		+7% vol 7 days 302°F A to 302°F	
TRIPOLYPHOSPHATE		A to 70°F		A to 70°F		
TRISODIUM PHOSPHATE (TSP)	A to 70°F AB to 140°F	A to 300°F	A to 212°F	A to 70°F		A to 70°F
TRITOYL		A to 70°F				
TRITOYL PHOSPHATE	NR at 70°F	A to boiling	A to 70°F	A to 70°F		
TT-I-735b						
TT-N-95a						
TT-N-97b						
TT-S-735 TYPE I						
TT-S-735 TYPE II						
TT-S-735 TYPE III						
TT-S-735 TYPE IV						
TT-S-735 TYPE V						
TT-S-735 TYPE VI						
TT-S-735 TYPE VII						
TT-T-656B						
TUNG OIL		A to 70°F	A to 212°F	A to 70°F		
TURBINE OIL		A to 70°F	A to 212°F	A to 70°F	AB to 70°F	
TURBINE OIL NO. 15 (MIL-L-7808A)		A to 70°F	A to 212°F	A to 70°F	+6% vol 3 days 212°F AB to 212°F	
TURBO OIL NO. 35			A to 212°F	A to 70°F		
TURPENTINE	NR at 70°F	A to 300°F	A to 212°F	A to 70°F	C at 70°F	BC at 70°F
TURPENTINE SUBSTITUTE	*C at 70°F *NR at 140°F	A to 70°F				

C 379

	RECOMMENDATIONS	NITRILE NBR	ETHYLENE PROPYLENE EP, EPDM	FLUOROCARBON FKM	CHLOROPRENE CR	HYDROGENATED NITRILE HNBR
TURPS		A to 70°F AB to 150°F	NR at 70°F	A to 70°F	NR at 70°F	
TWO FOUR D WITH 10% FUEL OIL		A to 70°F	NR at 70°F	A to 70°F	A to 70°F	
TYPE I FUEL (MIL-S-3136, ASTM FUEL A)	NBR, FKM, ACM, AU FVMQ	A to 250°F	NR at 70°F	A to 70°F	B to 80°F	A to 70°F
TYPE II FUEL (MIL-S-3136)	FKM	B/NR at 70°F	NR at 70°F	A to 70°F	NR at 70°F	
TYPE III FUEL (MIL-S-3136, ASTM FUEL B)	FKM	B/NR at 70°F	NR at 70°F	A to 70°F	NR at 70°F	A to 70°F
UCON HYDROLUBE J-4	NBR, EP, FKM, SBR	A to 70°F	A to 70°F	A to 70°F	B at 70°F	B at 70°F
UCON HYDROLUBE OILS	NBR	A to 200°F		AB to 70°F	AB to 70°F	
UCON LUBRICANT LB65	NBR, EP, FKM	A to 70°F	A to 70°F	A to 70°F	A to 70°F	B at 70°F
UCON LUBRICANT 50-HB55, 100, 260, 660 & 5100	NBR, EP, FKM, CR	A to 70°F	A to 70°F	A to 70°F	A to 70°F	B at 70°F
UCON LUBRICANT 135, 285, 300X, 625 & 1145	NBR, EP, FKM, CR	A to 70°F	A to 70°F	A to 70°F	A to 70°F	
UCON OIL LB-385 & 400X	NBR, EP, FKM, CR	A to 70°F AB to 250°F	A to 70°F	A to 70°F AB to 400°F	A to 70°F	B at 70°F
UCON OIL 50-HB-280X	NBR, EP, FKM, CR	A to 70°F AB to 250°F	A to 70°F	A to 70°F AB to 400°F	A to 70°F	B at 70°F
UDMH	EP, IIR	B at 70°F	A to 70°F	dissolved 1 day 70°F NR at 70°F	B at 70°F	B at 70°F
UNDECANOL (UNDECYL ALCOHOL)		A to 70°F		B at 70°F		
UNIVIS J-43	NBR, FKM, FVMQ	A to 275°F	BC at 70°F NR at 140°F	A to 212°F B at 248°F	AB to 70°F NR at 140°F	A to 70°F
UNIVIS 40 HYDRAULIC FLUID	NBR, FKM, ACM, AU	A to 70°F	NR at 70°F	A to 70°F	B at 70°F	A to 70°F
UNIVOLT NO. 35 (MINERAL OIL)	NBR, FKM, ACM, AU	A to 70°F	NR at 70°F	A to 70°F	B at 70°F	A to 70°F
UNSLAKED LIME		A to 70°F	A to 70°F		A to 70°F	
UNSYMETRICAL DIMETHYL HYDRAZINE	EP, IIR	B at 70°F	A to 70°F	dissolved 1 day 70°F NR at 70°F	B at 70°F	B at 70°F
URAN		B at 70°F			B at 70°F	
UREA		AB any conc to 150°F	A to 70°F AB to 200°F	A to 70°F AB to 200°F	AB any conc to 150°F	
UREA FORMALDEHYDE		A to 70°F			C at 70°F	
URIC ACID						
URINE		A to 140°F	A to 140°F	A to 140°F	NR at 70°F	
VALERIC ACID		NR at 70°F	A to 70°F		NR at 70°F	
VALERONE		NR at 70°F	B at 70°F	NR at 70°F	NR at 70°F	
VANILLA EXTRACT		A to 70°F		TEST	A to 70°F	
VARNISH	FKM, T, FVMQ	B at 70°F NR at 150°F	NR at 70°F	A to 70°F	NR at 70°F	B at 70°F
VASOLINE	NBR, FKM, FVMQ, ACM AU	A to 140°F	NR at 70°F	A to 140°F	AB to 140°F	
VEGETABLE JUICES	NBR	A to 250°F	A to 70°F	A to 140°F	B/NR at 70°F NR at 140°F	
VEGETABLE OILS	NBR, FKM, ACM	A to 200°F	AC to 200°F	A to 200°F	B/NR at 70°F	A to 70°F
VERSILUBE F44, F50 & F55	NBR, EP, FKM, CR	A to 70°F AB to 150°F	A to 70°F (F44, F50) NR at 70°F (F55)	+3% vol 28 days 350°F A to 500°F	A to 70°F	A to 70°F
VINEGAR	NBR, EP, FKM, IIR, CR	B at 70°F B/NR at 140°F	A to 140°F AB to 200°F	A to 200°F	B to 200°F AB 8% to 150°F	
VINEGAR NAPHTHA		NR at 70°F	B at 70°F	NR at 70°F	NR at 70°F	
VINEGAR SALTS		AB any conc to 120°F	A to 70°F	NR at 70°F	B at 70°F	
VINYL ACETATE	EP	NR at 70°F	B to 160°F	A to 70°F	NR at 70°F	NR at 70°F
VINYL BENZENE	FKM, FVMQ	NR at 70°F	NR at 70°F	AB to 70°F	NR at 70°F	NR at 70°F
VINYL BROMIDE					AB to 70°F	
VINYL CHLORIDE GAS			B at 70°F			
VINYL CHLORIDE LIQUID	FKM	NR at 70°F	BC at 70°F	A to 70°F	NR at 70°F	
VINYL CYANIDE	NR, CR	NR at 70°F	NR at 70°F	C/NR at 70°F	B/NR at 70°F	NR at 70°F
VINYL FLUORIDE				A to 70°F	AB to 70°F	
VINYL ETHER		B at 70°F		NR at 70°F		
VINYL OXIDE		B at 70°F		NR at 70°F		

	STYRENE BUTADIENE SBR	POLYACRYLATE ACM	POLYURETHANE AU, EU	ISOBUTYLENE ISOPRENE IIR	POLYBUTADIENE BR	® AEROQUIP AQP
TURPS			AB to 70°F C at 140°F	NR at 70°F		
TWO FOUR D WITH 10% FUEL OIL						
TYPE I FUEL (MIL-S-3136, ASTM FUEL A)	NR at 70°F	B /NR at 70°F	B to 158°F	NR at 70°F	NR at 70°F	
TYPE II FUEL (MIL-S-3136)	NR at 70°F	B at 70°F	B at 70°F	NR at 70°F	NR at 70°F	
TYPE III FUEL (MIL-S-3136, ASTM FUEL B)	NR at 70°F	B /NR at 70°F	B at 70°F	NR at 70°F	NR at 70°F	
UCON HYDROLUBE J-4	A to 70°F	NR at 70°F	NR at 70°F	A to 70°F	B at 70°F	
UCON HYDROLUBE OILS		NR at 70°F	C /NR at 70°F	A to 70°F		
UCON LUBRICANT LB65	B at 70°F		NR at 70°F	A to 70°F	B at 70°F	
UCON LUBRICANT 50-HB55, 100, 260, 660 & 5100	A to 70°F			A to 70°F	A to 70°F	
UCON LUBRICANT 135, 285, 300X, 625 & 1145	A to 70°F			A to 70°F	A to 70°F	
UCON OIL LB-385 & 400X	A to 70°F		AB to 70°F	A to 70°F	A to 70°F	
UCON OIL 50-HB-280X	A to 70°F		AB to 70°F	A to 70°F	A to 70°F	
UDMH	B /NR at 70°F	NR at 70°F	NR at 70°F	A to 70°F	A to 70°F	
UNDECANOL (UNDECYL ALCOHOL)				A to 70°F		
UNIVIS J-43	NR at 70°F	A to 70°F	A to 70°F			
UNIVIS 40 HYDRAULIC FLUID	NR at 70°F	A to 70°F	A to 70°F	NR at 70°F	NR at 70°F	
UNIVOLT NO. 35 (MINERAL OIL)	NR at 70°F	A to 70°F	A to 70°F	NR at 70°F	NR at 70°F	
UNSLAKED LIME				A to 70°F		
UNSYMETRICAL DIMETHYL HYDRAZINE	B /NR at 70°F	NR at 70°F	NR at 70°F	A to 70°F	A to 70°F	
URAN				B at 70°F		
UREA			AB to 70°F	A to 70°F AB any conc to 150°F		
UREA FORMALDEHYDE						
URIC ACID			NR at 70°F			
URINE						
VALERIC ACID				A to 70°F		
VALERONE				C at 70°F		
VANILLA EXTRACT						
VARNISH	NR at 70°F	NR at 70°F	B /NR at 70°F	NR at 70°F	NR at 70°F	TEST
VASOLINE	NR at 70°F	A to 70°F	A /NR at 70°F	NR at 70°F	NR at 70°F	
VEGETABLE JUICES	B at 70°F	NR at 70°F	B /NR at 70°F			
VEGETABLE OILS	NR at 70°F	A to 70°F	A to 70°F	AC to 100°F	NR at 70°F	
VERSILUBE F44, F50 & F55	A to 70°F	A to 70°F	A to 70°F	A to 70°F	A to 70°F	
VINEGAR	B at 70°F	NR at 70°F	B /NR at 70°F	A to 70°F A 8% to 150°F	B at 70°F	
VINEGAR NAPHTHA			NR at 70°F	B any conc to 100°F		
VINEGAR SALTS				A to 70°F AB any conc to 180°F		
VINYL ACETATE	NR at 70°F	NR at 70°F	NR at 70°F	A to 70°F		
VINYL BENZENE	NR at 70°F	NR at 70°F	BC at 70°F	NR at 70°F	NR at 70°F	
VINYL BROMIDE				AB to 70°F		
VINYL CHLORIDE GAS				NR at 70°F		
VINYL CHLORIDE LIQUID				B at 70°F		
VINYL CYANIDE	C at 70°F	NR at 70°F	NR at 70°F	NR at 70°F	C at 70°F	
VINYL FLUORIDE				AB to 70°F		
VINYL ETHER				NR at 70°F		
VINYL OXIDE				NR at 70°F		

C 381

	SYNTHETIC ISOPRENE IR	NATURAL ISOPRENE NR	CHLOROSULFONATED POLYETHYLENE CSM	FLUOROSILICONE FVMQ	SILICONE VMQ	CHEMRAZ FFKM
TURPS	NR at 70°F	NR at 70°F	NR at 70°F		NR at 70°F	
TWO FOUR D WITH 10% FUEL OIL	NR at 70°F	NR at 70°F				
TYPE I FUEL (MIL-S-3136, ASTM FUEL A)	NR at 70°F	NR at 70°F	AB to 70°F	AB to 70°F	NR at 70°F	A to 70°F
TYPE II FUEL (MIL-S-3136)	NR at 70°F	NR at 70°F	NR at 70°F	B at 70°F	NR at 70°F	
TYPE III FUEL (MIL-S-3136, ASTM FUEL B)	NR at 70°F	NR at 70°F	NR at 70°F	B to 150°F	NR at 70°F	A to 70°F
UCON HYDROLUBE J-4				B at 70°F	A to 70°F	A to 70°F
UCON HYDROLUBE OILS	NR at 70°F	NR at 70°F				A to 70°F
UCON LUBRICANT LB65	B at 70°F	B at 70°F	B at 70°F	A to 70°F	A to 70°F	A to 70°F
UCON LUBRICANT 50-HB55, 100, 260, 660 & 5100	A to 70°F	A to 70°F	A to 70°F	A to 70°F	A to 70°F	
UCON LUBRICANT 135, 285, 300X, 625 & 1145	A to 70°F	A to 70°F	A to 70°F	A to 70°F	A to 70°F	A to 70°F
UCON OIL LB-385 & 400X	A to 70°F	A to 70°F	A to 70°F	A to 70°F	A to 70°F	A to 70°F
UCON OIL 50-HB-280X	A to 70°F	A to 70°F	A to 70°F	A to 70°F	A to 70°F	A to 70°F
UDMH	A to 70°F	A to 70°F	A to 70°F	NR at 70°F	NR at 70°F	AB to 70°F
UNDECANOL (UNDECYL ALCOHOL)	A to 70°F	A to 70°F	A to 70°F			
UNIVIS J-43	NR at 70°F	NR at 70°F	B at 70°F	+10% vol 3 days 300°F A to 400°F	+150% vol 3 days 300°F NR at 70°F	
UNIVIS 40 HYDRAULIC FLUID	NR at 70°F	NR at 70°F	B at 70°F	A to 70°F	NR at 70°F	A to 70°F
UNIVOLT NO. 35 (MINERAL OIL)	NR at 70°F	NR at 70°F	NR at 70°F	A to 70°F	NR at 70°F	A to 70°F
UNSLAKED LIME	A to 70°F	A to 70°F	A to 70°F			
UNSYMETRICAL DIMETHYL HYDRAZINE	A to 70°F	A to 70°F	A to 70°F	NR at 70°F	NR at 70°F	AB to 70°F
URAN	B at 70°F	B at 70°F	AB to 70°F			
UREA	A to 70°F	A to 70°F AB any conc to 150°F	A to 70°F AB to 200°F		AB to 70°F	
UREA FORMALDEHYDE						
URIC ACID						A to 70°F
URINE		NR at 70°F	AB to 70°F			
VALERIC ACID	A to 70°F	A to 70°F				A to 70°F
VALERONE	NR at 70°F	NR at 70°F	NR at 70°F			
VANILLA EXTRACT						
VARNISH	NR at 70°F	NR at 70°F	NR at 70°F	B at 70°F	NR at 70°F	A to 70°F
VASOLINE	NR at 70°F	C/NR at 70°F	B at 70°F	A to 70°F	NR at 70°F	A to 70°F
VEGETABLE JUICES		C/NR at 70°F		A to 70°F	AB to 70°F	
VEGETABLE OILS	NR at 70°F	NR at 70°F	B at 200°F	A to 70°F	A to 70°F	A to 70°F
VERSILUBE F44, F50 & F55	A to 70°F	A to 70°F	A to 70°F	A to 70°F	C at 70°F	A to 70°F
VINEGAR	B at 70°F	B at 70°F AB 8% to 150°F	A to 70°F AB to 200°F	C at 70°F	A to 70°F	
VINEGAR NAPHTHA	NR at 70°F	NR at 70°F	NR at 70°F		B at 70°F	
VINEGAR SALTS	B at 70°F	B any conc to 80°F	A to 70°F			
VINYL ACETATE	NR at 70°F	NR at 70°F	C at 70°F		NR at 70°F	A to 70°F
VINYL BENZENE	NR at 70°F	NR at 70°F	NR at 70°F	C at 70°F	NR at 70°F	A to 70°F
VINYL BROMIDE						
VINYL CHLORIDE GAS		AB to 70°F				A to 70°F
VINYL CHLORIDE LIQUID	C at 70°F	BC at 70°F	NR at 70°F			A to 70°F
VINYL CYANIDE	BC at 70°F	BC any conc to 80°F	BC at 70°F	NR at 70°F	NR at 70°F	A to 70°F
VINYL FLUORIDE						A to 70°F
VINYL ETHER	NR at 70°F	NR at 70°F	B at 70°F			
VINYL OXIDE	NR at 70°F	NR at 70°F	B at 70°F			

	POLYSULFIDE T	CHLORINATED POLYETHYLENE CM	EPICHLOROHYDRIN CO, ECO	® HYTREL COPOLYESTER TPE	® SANTOPRENE COPOLYMER TPO	® C-FLEX STYRENIC TPE
TURPS				AB to 70°F		
TWO FOUR D WITH 10% FUEL OIL	AB to 70°F					
TYPE I FUEL (MIL-S-3136, ASTM FUEL A)	A to 70°F	A to 70°F	A to 70°F	A to 158°F	NR at 70°F	
TYPE II FUEL (MIL-S-3136)	A to 70°F		A to 70°F			
TYPE III FUEL (MIL-S-3136, ASTM FUEL B)	A to 70°F	B at 70°F	A to 70°F	A to 158°F	NR at 70°F	
UCON HYDROLUBE J-4				AB to 70°F		
UCON HYDROLUBE OILS				BC at 70°F		
UCON LUBRICANT LB65						
UCON LUBRICANT 50-HB55, 100, 260, 660 & 5100	A to 70°F				A to 70°F B at 257°F	
UCON LUBRICANT 135, 285, 300X, 625 & 1145						
UCON OIL LB-385 & 400X						
UCON OIL 50-HB-280X						
UDMH	NR at 70°F					
UNDECANOL (UNDECYL ALCOHOL)						
UNIVIS J-43						
UNIVIS 40 HYDRAULIC FLUID	A to 70°F					
UNIVOLT NO. 35 (MINERAL OIL)	A to 70°F					
UNSLAKED LIME						
UNSYMETRICAL DIMETHYL HYDRAZINE	NR at 70°F					
URAN	AB to 70°F					
UREA	AB to 70°F			AB to 100% to 70°F		A to 70°F
UREA FORMALDEHYDE				*AB to 70°F		
URIC ACID				NR at 70°F *AB to 140°F dilute	A to 70°F	A to 70°F
URINE	AB to 70°F					
VALERIC ACID						
VALERONE						
VANILLA EXTRACT						
VARNISH	A to 70°F	C at 70°F				
VASOLINE	B at 70°F			AB to 70°F		
VEGETABLE JUICES						
VEGETABLE OILS	NR at 70°F		A to 70°F			TEST
VERSILUBE F44, F50 & F55	B at 70°F		A to 70°F			
VINEGAR	B at 70°F			AC to 70°F		
VINEGAR NAPHTHA						
VINEGAR SALTS						
VINYL ACETATE						
VINYL BENZENE	NR at 70°F			*NR at 70°F		*NR at 70°F
VINYL BROMIDE						
VINYL CHLORIDE GAS						
VINYL CHLORIDE LIQUID						
VINYL CYANIDE	NR at 70°F	A to 70°F			*NR at 70°F	
VINYL FLUORIDE						
VINYL ETHER						
VINYL OXIDE						

C 383

	ETHYLENE ACRYLIC EA	POLYALLOMER LINEAR COPOLYMER	NYLON 11 POLYAMIDE	NYLON 12 POLYAMIDE	ETHYLENE VINYL ACETATE EVA	POLYVINYLCHLORIDE FLEXIBLE PVC
TURPS			AB to 70°F BC at 140°F		*C at 70°F *NR at 140°F	NR at 70°F
TWO FOUR D WITH 10% FUEL OIL						
TYPE I FUEL (MIL-S-3136, ASTM FUEL A)	A to 70°F		A to 70°F	AB to 70°F	NR at 70°F	AB to 70°F C at 150°F
TYPE II FUEL (MIL-S-3136)						
TYPE III FUEL (MIL-S-3136, ASTM FUEL B)	B/NR at 70°F		A to 70°F			NR at 70°F
UCON HYDROLUBE J-4			A to 70°F			AB to 70°F
UCON HYDROLUBE OILS			A to 70°F			AB to 70°F
UCON LUBRICANT LB65						
UCON LUBRICANT 50-HB55, 100, 260, 660 & 5100						
UCON LUBRICANT 135, 285, 300X, 625 & 1145						
UCON OIL LB-385 & 400X			A to 70°F			AB to 70°F
UCON OIL 50-HB-280X			A to 70°F			AB to 70°F
UDMH						
UNDECANOL (UNDECYL ALCOHOL)		A to 70°F B at 122°F				A to 70°F C at 122°F
UNIVIS J-43						
UNIVIS 40 HYDRAULIC FLUID						
UNIVOLT NO. 35 (MINERAL OIL)						
UNSLAKED LIME						
UNSYMETRICAL DIMETHYL HYDRAZINE						
URAN						
UREA	A to 70°F	A to 122°F	A to 70°F C 140–194°F	AB 20% to 70°F	A to 100% to 70°F *A sol'n to 140°F	AB to 70°F NR at 122°F
UREA FORMALDEHYDE			*AB to 70°F			NR at 70°F
URIC ACID			AB to 140°F C at 194°F		*AB to 140°F dilute	A to 70°F
URINE				AB to 70°F		
VALERIC ACID						
VALERONE						
VANILLA EXTRACT						
VARNISH	A to 70°F	A to 70°F	A to 70°F		BC at 70°F	NR at 70°F
VASOLINE	A to 70°F			AB to 70°F		B at 70°F
VEGETABLE JUICES					A to 70°F	A to 70°F
VEGETABLE OILS	A to 70°F		AB to 70°F	AB to 70°F	NR at 70°F	C/NR at 70°F
VERSILUBE F44, F50 & F55			A to 70°F			
VINEGAR	AB to 70°F	A to 70°F	A to 70°F AB to 140°F	AB to 70°F	AC to 70°F	A to 70°F AB to 150°F
VINEGAR NAPHTHA						NR at 70°F
VINEGAR SALTS						
VINYL ACETATE						
VINYL BENZENE	*NR at 70°F					NR at 70°F
VINYL BROMIDE						AB to 70°F
VINYL CHLORIDE GAS						
VINYL CHLORIDE LIQUID						NR at 70°F
VINYL CYANIDE		C at 70°F NR at 122°F	A to 70°F			NR at 70°F
VINYL FLUORIDE						AB to 70°F
VINYL ETHER						
VINYL OXIDE						

C 384

	POLYETHYLENE LOW DENSITY LDPE	®TEFLON FEP	®KALREZ PERFLUORINATED ELASTOMER	®FLUORAZ FLUORINATED COPOLYMER	®AFLAS FLUORINATED COPOLYMER	®NORPRENE COPOLYMER TPO
TURPS		A to 70°F				
TWO FOUR D WITH 10% FUEL OIL		A to 70°F				
TYPE I FUEL (MIL-S-3136, ASTM FUEL A)	B at 70°F	A to 300°F	A to 212°F	BC at 70°F	BC at 70°F	
TYPE II FUEL (MIL-S-3136)						
TYPE III FUEL (MIL-S-3136, ASTM FUEL B)		A to 70°F	A to 212°F	NR at 70°F		
UCON HYDROLUBE J-4		A to 70°F	A to 212°F	A to 70°F		
UCON HYDROLUBE OILS	AB to 70°F	A to 70°F	A to 212°F			
UCON LUBRICANT LB65			A to 212°F	A to 70°F		
UCON LUBRICANT 50-HB55, 100, 260, 660 & 5100			A to 212°F			
UCON LUBRICANT 135, 285, 300X, 625 & 1145			A to 212°F			
UCON OIL LB-385 & 400X	AB to 70°F		A to 212°F	A to 70°F		
UCON OIL 50-HB-280X	AB to 70°F		A to 212°F	A to 70°F		
UDMH		A to 70°F	A to 70°F	C at 70°F		
UNDECANOL (UNDECYL ALCOHOL)	A to 70°F C at 122°F	A to 125°F				
UNIVIS J-43						
UNIVIS 40 HYDRAULIC FLUID				A to 70°F		
UNIVOLT NO. 35 (MINERAL OIL)				A to 70°F		
UNSLAKED LIME		A to 70°F				
UNSYMETRICAL DIMETHYL HYDRAZINE		A to 70°F	A to 70°F *A to 212°F	C at 70°F		
URAN		A to 70°F				
UREA	AB to 100% to 140°F	A to 300°F		C at 70°F wet or dry		A to 70°F
UREA FORMALDEHYDE						
URIC ACID	AB to 100% to 70°F AB to 140°F dilute	*A to 200°F	A to 212°F			A to 70°F
URINE	AB to 140°F	A to 300°F		C at 70°F		
VALERIC ACID		A to 70°F	A to 212°F			
VALERONE		A to 70°F				
VANILLA EXTRACT	C/NR at 70°F	A to 70°F	*A to 212°F			
VARNISH	A to 70°F	A to 70°F	A to 212°F	C at 70°F aromatic A to 70°F non aromatic		
VASOLINE		A to 70°F		A to 70°F		
VEGETABLE JUICES	A to 70°F			A to 70°F		
VEGETABLE OILS	A to 70°F	A to 70°F	A to 212°F	A to 70°F		
VERSILUBE F44, F50 & F55		A to 70°F	A to 212°F	A to 70°F		
VINEGAR	A to 70°F AB to 140°F	A to 300°F	A to 212°F	A to 70°F		
VINEGAR NAPHTHA		A to 70°F				
VINEGAR SALTS		A to 70°F				
VINYL ACETATE		A to 300°F	A to 212°F	NR at 70°F	NR at 70°F	
VINYL BENZENE		A to 70°F	A to 212°F	NR at 70°F		*NR at 70°F
VINYL BROMIDE	AB to 70°F	A to 70°F				
VINYL CHLORIDE GAS		A to 300°F			AB to 70°F	
VINYL CHLORIDE LIQUID	AB to 70°F	A to 300°F	A to 302°F	A to 70°F		
VINYL CYANIDE		A to 200°F	A to 100°F	BC at 70°F		
VINYL FLUORIDE	AB to 70°F	A to 70°F	A to 212°F			
VINYL ETHER		A to 70°F				
VINYL OXIDE		A to 70°F				

	RECOMMENDATIONS	NITRILE NBR	ETHYLENE PROPYLENE EP, EPDM	FLUOROCARBON FKM	CHLOROPRENE CR	HYDROGENATED NITRILE HNBR
VINYL TOLUENE		NR at 70°F		A to 70°F		
VINYL TRICHLORIDE		NR at 70°F		A to 70°F	NR at 70°F	
VINYLIDENE CHLORIDE	FKM	NR at 70°F	B/NR at 70°F	B to 160°F	NR at 70°F	
VITRIOL, OIL OF	FKM	NR at 70°F	C at 70°F	A to 158°F	NR at 70°F	NR at 70°F
V.M. & P. NAPHTHA		A/NR at 70°F	NR at 70°F	A to 70°F	B/NR at 70°F	
VV-B-680		B at 70°F	A to 70°F	A to 70°F	B at 70°F	
VV-G-632		A to 70°F	NR at 70°F	A to 70°F	A to 70°F	
VV-G-671c		A to 70°F	NR at 70°F	A to 70°F	A to 70°F	
VV-H-910	EP, FKM	C at 70°F	A to 70°F	A to 70°F	B at 70°F	C at 70°F
VV-I-530a		A to 70°F	NR at 70°F	A to 70°F	B at 70°F	
VV-K-211d		A to 70°F	NR at 70°F	A to 70°F	C at 70°F	
VV-K-220a		A to 70°F	NR at 70°F	A to 70°F	B at 70°F	
VV-L-751b		B at 70°F	NR at 70°F	A to 70°F	B at 70°F	
VV-L-800		A to 70°F	NR at 70°F	A to 70°F	B at 70°F	
VV-L-820b		A to 70°F	NR at 70°F	A to 70°F	B at 70°F	
VV-L-825a TYPE I		A to 70°F	NR at 70°F	A to 70°F	A to 70°F	
VV-L-825a TYPE II		A to 70°F	NR at 70°F	A to 70°F	A to 70°F	
VV-L-825a TYPE III		B at 70°F	NR at 70°F	A to 70°F	B at 70°F	
VV-O-526		A to 70°F	NR at 70°F	A to 70°F	A to 70°F	
VV-P-216a		A to 70°F	NR at 70°F	A to 70°F	B at 70°F	
VV-P-236		B at 70°F	NR at 70°F	A to 70°F	B at 70°F	
WAGNER 21B BRAKE FLUID	EP, SBR	+47% vol 3 days 212°F C at 70°F	no swell 3 days 212°F A to 70°F	+46% vol 7 days 75°F NR at 70°F	+13% vol 3 days 212°F AB to 70°F	C at 70°F
WALNUT OIL		A to 70°F		A to 70°F	B at 70°F	
WASHING SODA	NBR, EP, FKM, CR	A to 100% to 160°F AB to 100% to 200°F	A to 100% to 176°F B at 212°F	A to 100% to 212°F	A to 100% to 176°F AB 100% to 200°F	A to 70°F
WATER, ACID MINE	FKM, EP, IIR	A to 140°F	A to 70°F	A to 180°F	C to 140°F	
WATER, BRINE (PROCESS BEVERAGE)	NBR, CR, EP	A to 70°F	A to 70°F	A to 70°F	A to 70°F	
WATER, DEIONIZED	NBR, EP, CR, IIR	A to 70°F AB to 200°F	A to 70°F AB to 200°F	A to 70°F AB to 200°F	A to 70°F AB to 150°F	
WATER, DETERGENT SOLUTION	EP	B at 180°F	A to 180°F	B at 180°F	C at 180°F	
WATER, DISTILLED		A to 140°F	A to 140°F	A to 140°F	AC to 140°F	
WATER, FRESH	NBR, EP, SBR, IIR	+5% vol 3 days 212°F A to 180°F	+2% vol 3 days 212°F A to 275°F	+7% vol 21 days 212°F B at 275°F	+12% vol 3 days 212°F AB to 212°F	A to 70°F
WATER, GLYCOL MIXTURE			B at 70°F	B at 70°F	C at 70°F	
WATER, PROOFING SALT		B any conc to 120°F			B any conc to 120°F	
WATER, SALT	EP, NBR, FVMQ, SBR	A to 140°F AB to 200°F	A to 176°F AB to 200°F	A to 176°F AB to 200°F	A to 140°F AB to 200°F	A to 70°F
WATER, SEA	EP, NBR, SBR	A to 140°F B to 200°F	A to 180°F AB to 200°F	+4.5% vol 30 days 212°F A to 212°F	AB to 140°F BC to 200°F	A to 70°F
WATER, SOAP SOLUTION	EP	B at 180°F	A to 180°F	B at 180°F	C at 180°F	
WATER, SOLUBLE OIL SOLUTION	NBR	A to 180°F	NR at 70°F	A to 70°F B at 180°F	A to 70°F B at 180°F	
WATERGLASS	NBR, EP, IIR, CR	A to 100% to 140°F A sol'n to 200°F	A to 100% to 176°F AB to 100% to 200°F	A to 100% to 212°F	A to 100% to 140°F AB to 100% to 200°F	A to 70°F
WAX		A to 70°F	A/NR at 70°F	A to 70°F	A to 70°F	
WEMCO C	NBR, FKM, ACM, AU	A to 70°F	NR at 70°F	A to 70°F	B at 70°F	A to 70°F
WHEAT GERM OIL		AB to 70°F	NR at 70°F	AB to 70°F	C at 70°F	
WHISKEY	NBR, EP, FKM, CR	A to 200°F NR boiling	A to 200°F	A to boiling	A to 140°F C at 200°F	A to 70°F
WHISKEY & ALCOHOLIC BEVERAGES	NBR	A to 200°F	A to 200°F	A to 200°F	C at 200°F	
WHITE CAUSTIC		AB any conc to 150°F	A to 70°F	B at 70°F	AB any conc to 200°F	
WHITE COPPERAS		A to 70°F AB any conc to 150°F	A to 70°F	A to 70°F	A to 70°F AB any conc to 150°F	

	STYRENE BUTADIENE SBR	POLYACRYLATE ACM	POLYURETHANE AU, EU	ISOBUTYLENE ISOPRENE IIR	POLYBUTADIENE BR	® AEROQUIP AQP
VINYL TOLUENE				NR at 70°F		
VINYL TRICHLORIDE				NR at 70°F		
VINYLIDENE CHLORIDE	NR at 70°F	NR at 70°F	NR at 70°F	AC to 70°F		
VITRIOL, OIL OF			A to 70°F	NR at 70°F		
V.M. & P. NAPHTHA				NR at 70°F		
VV-B-680	A to 70°F	B at 70°F	C at 70°F	B at 70°F	A to 70°F	
VV-G-632	NR at 70°F	A to 70°F	A to 70°F	NR at 70°F	NR at 70°F	
VV-G-671c	NR at 70°F	A to 70°F	A to 70°F	NR at 70°F	NR at 70°F	
VV-H-910	A to 70°F	B /NR at 70°F	NR at 70°F	B /NR at 70°F	AB to 70°F	
VV-I-530a	NR at 70°F	A to 70°F	B at 70°F	NR at 70°F	NR at 70°F	
VV-K-211d	NR at 70°F		C at 70°F	NR at 70°F	NR at 70°F	
VV-K-220a	NR at 70°F	B at 70°F	B at 70°F	NR at 70°F	NR at 70°F	
VV-L-751b	NR at 70°F	B at 70°F	C at 70°F	NR at 70°F	NR at 70°F	
VV-L-800	NR at 70°F	A to 70°F	B at 70°F	NR at 70°F	NR at 70°F	
VV-L-820b	NR at 70°F	A to 70°F	B at 70°F	NR at 70°F	NR at 70°F	
VV-L-825a TYPE I	NR at 70°F	A to 70°F	A to 70°F	NR at 70°F	NR at 70°F	
VV-L-825a TYPE II	NR at 70°F	A to 70°F	A to 70°F	NR at 70°F	NR at 70°F	
VV-L-825a TYPE III	NR at 70°F	B at 70°F	C at 70°F	NR at 70°F	NR at 70°F	
VV-O-526	NR at 70°F	A to 70°F	A to 70°F	NR at 70°F	NR at 70°F	
VV-P-216a	NR at 70°F	A to 70°F	B at 70°F	NR at 70°F	NR at 70°F	
VV-P-236	NR at 70°F	B at 70°F	C at 70°F	NR at 70°F	NR at 70°F	
WAGNER 21B BRAKE FLUID	+10% vol 3 days 212°F A to 70°F	+166% vol 3 days 212°F NR at 70°F	dissolved 3 days 212°F NR at 70°F	+3% vol 3 days 212°F B at 70°F	A to 70°F	
WALNUT OIL						
WASHING SODA				A to 70°F		
WATER, ACID MINE	NR at 70°F	NR at 70°F	C /NR at 70°F	A to 70°F		
WATER, BRINE (PROCESS BEVERAGE)			NR at 70°F	AB to 185°F		
WATER, DEIONIZED				A to 70°F AB to 150°F		
WATER, DETERGENT SOLUTION	B at 180°F	NR at 180°F	NR at 180°F			
WATER, DISTILLED			A to 70°F	A to 70°F		
WATER, FRESH	+10% vol 3 days 212°F AB to 180°F	+166% vol 3 days 212°F NR at 70°F	+15% vol 3 days 212°F A to 122°F	no swell 3 days 212°F A to 70°F	A to 70°F	AB to 150°F
WATER, GLYCOL MIXTURE			C at 70°F	B at 70°F		
WATER, PROOFING SALT				AB any conc to 120°F		
WATER, SALT	A to 70°F	A /NR at 70°F	A /NR at 70°F	A to 70°F AB to 185°F	A to 70°F	
WATER, SEA	A to 70°F	A /NR at 70°F	A /NR at 70°F	A to 70°F AB to 120°F	A to 70°F	
WATER, SOAP SOLUTION	B at 180°F	NR at 180°F	NR at 180°F			
WATER, SOLUBLE OIL SOLUTION	NR at 180°F	NR at 180°F	C to 150°F NR at 180°F			
WATERGLASS	A to 100% to 70°F		AB to 70°F	A to 70°F	A to 70°F	AB to 70°F
WAX			A to 70°F	NR at 70°F		
WEMCO C	NR at 70°F	A to 70°F	A to 70°F	NR at 70°F	NR at 70°F	
WHEAT GERM OIL	NR at 70°F	AB to 70°F	AB to 70°F			
WHISKEY	A to 70°F	NR at 70°F	NR at 70°F	A to 70°F AB to 150°F	A to 70°F	
WHISKEY & ALCOHOLIC BEVERAGES	A to 70°F	NR at 70°F	NR at 70°F			
WHITE CAUSTIC			C at 70°F	A to 70°F AB any conc to 180°F		
WHITE COPPERAS				A to 70°F AB any conc to 150°F		

C 387

	SYNTHETIC ISOPRENE IR	NATURAL ISOPRENE NR	CHLOROSULFONATED POLYETHYLENE CSM	FLUOROSILICONE FVMQ	SILICONE VMQ	CHEMRAZ FFKM
VINYL TOLUENE	NR at 70°F	NR at 70°F	NR at 70°F			
VINYL TRICHLORIDE	NR at 70°F	NR at 70°F	NR at 70°F			
VINYLIDENE CHLORIDE	NR at 70°F	NR at 70°F	NR at 70°F	NR at 70°F	C/NR at 70°F	A to 70°F
VITRIOL, OIL OF	NR at 70°F	NR at 70°F	NR at 70°F	NR at 70°F	NR at 70°F	A to 70°F
V.M. & P. NAPHTHA	NR at 70°F	NR at 70°F	NR at 70°F	+1% vol 120 days 70°F A to 70°F	+4% vol 120 days 70°F A to 70°F	
VV-B-680	B at 70°F	B at 70°F	B at 70°F	B at 70°F	NR at 70°F	
VV-G-632	NR at 70°F	NR at 70°F	A to 70°F	A to 70°F	C at 70°F	
VV-G-671c	NR at 70°F	NR at 70°F	A to 70°F	A to 70°F	C at 70°F	
VV-H-910	B at 70°F	B at 70°F	B at 70°F	B at 70°F	B/NR at 70°F	A to 70°F
VV-I-530a	NR at 70°F	NR at 70°F	B at 70°F	A to 70°F	C at 70°F	
VV-K-211d	NR at 70°F	NR at 70°F	C at 70°F	A to 70°F	NR at 70°F	
VV-K-220a	NR at 70°F	NR at 70°F	C at 70°F	A to 70°F	NR at 70°F	
VV-L-751b	NR at 70°F	NR at 70°F	C at 70°F	A to 70°F	NR at 70°F	
VV-L-800	NR at 70°F	NR at 70°F	B at 70°F	A to 70°F	C at 70°F	
VV-L-820b	NR at 70°F	NR at 70°F	B at 70°F	A to 70°F	C at 70°F	
VV-L-825a TYPE I	NR at 70°F	NR at 70°F	A to 70°F	A to 70°F	C at 70°F	
VV-L-825a TYPE II	NR at 70°F	NR at 70°F	A to 70°F	A to 70°F	C at 70°F	
VV-L-825a TYPE III	NR at 70°F	NR at 70°F	C at 70°F	A to 70°F	NR at 70°F	
VV-O-526	NR at 70°F	NR at 70°F	A to 70°F	A to 70°F	C at 70°F	
VV-P-216a	NR at 70°F	NR at 70°F	B at 70°F	A to 70°F	C at 70°F	
VV-P-236	NR at 70°F	NR at 70°F	C at 70°F	A to 70°F	NR at 70°F	
WAGNER 21B BRAKE FLUID			B at 70°F	dissolved 3 days 212°F NR at 70°F	+8% vol 3 days 212°F C at 70°F	A to 70°F
WALNUT OIL	NR at 70°F	NR at 70°F				
WASHING SODA	A to 100% to 70°F	A to 100% to 70°F AB any conc to 185°F	A to 100% to 70°F AB to 200°F	A to 100% to 70°F	A to 100% to 70°F	A to 70°F
WATER, ACID MINE	A to 70°F	AB to 70°F	A to 70°F		AB to 70°F	
WATER, BRINE (PROCESS BEVERAGE)					A to 70°F	
WATER, DEIONIZED	A to 70°F	A to 70°F AB to 150°F	A to 70°F AB to 200°F			A to 70°F NR at 70°F white
WATER, DETERGENT SOLUTION				A to 180°F	B at 180°F	
WATER, DISTILLED	A to 70°F	A to 70°F	A to 70°F	A to 70°F	C at 70°F	
WATER, FRESH	A to 70°F	A to 70°F AB to 180°F	+4% vol 3 days 212°F A to 212°F	no swell 3 days 212°F A to 275°F	+3% vol 3 days 212°F B at 180°F	A to 70°F B at 70°F white
WATER, GLYCOL MIXTURE		B at 70°F	NR at 70°F		B at 70°F	
WATER, PROOFING SALT	A to 70°F	A to 70°F AB any conc to 150°F				
WATER, SALT	A to 70°F	A to 70°F NR at 200°F	A to 70°F AB to 200°F	A to 70°F	A to 70°F	A to 70°F B at 70°F white
WATER, SEA	A to 70°F	A to 70°F NR at 200°F	A to 70°F AB to 200°F	A to 70°F	A to 70°F B at 180°F	A to 70°F B at 70°F white
WATER, SOAP SOLUTION				A to 180°F	NR at 70°F	
WATER, SOLUBLE OIL SOLUTION				B at 180°F		
WATERGLASS	A to 70°F	A to 70°F AB any conc to 185°F	A to 70°F AB to 200°F		A to 70°F	A to 70°F
WAX	NR at 70°F	NR at 70°F	A to 70°F			
WEMCO C	NR at 70°F	NR at 70°F	NR at 70°F	A to 70°F	NR at 70°F	A to 70°F
WHEAT GERM OIL					AB to 70°F	
WHISKEY	A to 70°F	A to 70°F AB to 150°F	A to 70°F	A to 70°F	A to 70°F	A to 70°F B at 70°F white
WHISKEY & ALCOHOLIC BEVERAGES				A to 70°F	A to 70°F	
WHITE CAUSTIC	A to 70°F	A to 70°F AB any conc to 150°F	B at 70°F		C at 70°F	
WHITE COPPERAS	A to 70°F	A to 70°F AB any conc to 150°F	A to 70°F		A to 70°F	

	POLYSULFIDE T	CHLORINATED POLYETHYLENE CM	EPICHLOROHYDRIN CO, ECO	® HYTREL COPOLYESTER TPE	® SANTOPRENE COPOLYMER TPO	® C-FLEX STYRENIC TPE
VINYL TOLUENE						
VINYL TRICHLORIDE						
VINYLIDENE CHLORIDE						
VITRIOL, OIL OF						
V.M. & P. NAPHTHA						
VV-B-680	C at 70°F		B at 70°F			
VV-G-632	A to 70°F		A to 70°F			
VV-G-671c	A to 70°F		A to 70°F			
VV-H-910	NR at 70°F		B at 70°F			
VV-I-530a	A to 70°F		A to 70°F			
VV-K-211d	A to 70°F		A to 70°F			
VV-K-220a	A to 70°F		A to 70°F			
VV-L-751b	C at 70°F		B at 70°F			
VV-L-800	A to 70°F		A to 70°F			
VV-L-820b	A to 70°F		A to 70°F			
VV-L-825a TYPE I	A to 70°F		A to 70°F			
VV-L-825a TYPE II	A to 70°F		A to 70°F			
VV-L-825a TYPE III	C at 70°F		B at 70°F			
VV-O-526	A to 70°F		A to 70°F			
VV-P-216a	A to 70°F		A to 70°F			
VV-P-236	C at 70°F		B at 70°F			
WAGNER 21B BRAKE FLUID	+23% vol 3 days 212°F NR at 70°F					
WALNUT OIL						
WASHING SODA						
WATER, ACID MINE						
WATER, BRINE (PROCESS BEVERAGE)						A to 70°F
WATER, DEIONIZED						
WATER, DETERGENT SOLUTION						
WATER, DISTILLED					A to 70°F	
WATER, FRESH	NR at 70°F	AB to 70°F	B at 70°F	A to 70°F AC at 212°F	+2-5% vol 7 days 212°F A to boiling	A to 70°F
WATER, GLYCOL MIXTURE				C at 70°F		
WATER, PROOFING SALT						
WATER, SALT	C/NR at 70°F			A to 70°F	no swell 15% conc 7 days 70°F	
WATER, SEA	NR at 70°F			A to 70°F	A to 70°F	
WATER, SOAP SOLUTION						
WATER, SOLUBLE OIL SOLUTION				C to 150°F NR at 180°F		
WATERGLASS	AB to 70°F	AB to 70°F		AB to 70°F		A to 70°F
WAX						
WEMCO C	A to 70°F					
WHEAT GERM OIL						
WHISKEY	NR at 70°F			AB to 70°F		
WHISKEY & ALCOHOLIC BEVERAGES						
WHITE CAUSTIC						
WHITE COPPERAS						

	ETHYLENE ACRYLIC EA	POLYALLOMER LINEAR COPOLYMER	NYLON 11 POLYAMIDE	NYLON 12 POLYAMIDE	ETHYLENE VINYL ACETATE EVA	POLYVINYLCHLORIDE FLEXIBLE PVC
VINYL TOLUENE						
VINYL TRICHLORIDE						NR at 70°F
VINYLIDENE CHLORIDE		NR at 70°F				NR at 70°F
VITRIOL, OIL OF					A to 70°F	A to 70°F
V.M. & P. NAPHTHA						NR at 70°F
VV-B-680						
VV-G-632						
VV-G-671c						
VV-H-910						
VV-I-530a						
VV-K-211d						
VV-K-220a						
VV-L-751b						
VV-L-800						
VV-L-820b						
VV-L-825a TYPE I						
VV-L-825a TYPE II						
VV-L-825a TYPE III						
VV-O-526						
VV-P-216a						
VV-P-236						
WAGNER 21B BRAKE FLUID	NR at 70°F					
WALNUT OIL						
WASHING SODA						
WATER, ACID MINE					C at 70°F	B/NR at 70°F
WATER, BRINE (PROCESS BEVERAGE)						A to 70°F
WATER, DEIONIZED						
WATER, DETERGENT SOLUTION						
WATER, DISTILLED						B at 70°F
WATER, FRESH	A to 212°F	A to 70°F	AB to 194°F NR at 212°F	AB to 70°F	AB to 140°F NR at 212°F	AB to 150°F NR at 170°F
WATER, GLYCOL MIXTURE			AB to 70°F			NR at 70°F
WATER, PROOFING SALT						
WATER, SALT		A to 70°F	AB to 70°F		A to 70°F	AB to 70°F *AB to 130°F
WATER, SEA	A to 70°F		A to 194°F		A to 70°F AB to 140°F	AB to 70°F *AB to 130°F
WATER, SOAP SOLUTION			AB to 70°F			
WATER, SOLUBLE OIL SOLUTION			AB to 150°F		C/NR at 70°F	AB to 150°F C at 170°F
WATERGLASS		A to 70°F	A to 70°F AB to 194°F	AB to 70°F	A to 70°F	A to 70°F
WAX				AB to 70°F		
WEMCO C						
WHEAT GERM OIL						
WHISKEY	A to 70°F	A to 70°F	A to 70°F		AB to 70°F	A to 70°F B at 150°F
WHISKEY & ALCOHOLIC BEVERAGES						
WHITE CAUSTIC						
WHITE COPPERAS						

	POLYETHYLENE LOW DENSITY LDPE	®TEFLON FEP	® KALREZ PERFLUORINATED ELASTOMER	® FLUORAZ FLUORINATED COPOLYMER	® AFLAS FLUORINATED COPOLYMER	® NORPRENE COPOLYMER TPO
VINYL TOLUENE		A to 70°F				
VINYL TRICHLORIDE		A to 70°F				
VINYLIDENE CHLORIDE	NR at 70°F	A to 200°F	A to 212°F	A to 70°F		
VITRIOL, OIL OF	NR at 70°F	A to 400°F	A to 133°F	AB to 70°F	A to 70°F	TEST
V.M. & P. NAPHTHA		A to 70°F				
VV-B-680						
VV-G-632						
VV-G-671c						
VV-H-910				A to 70°F		
VV-I-530a						
VV-K-211d						
VV-K-220a						
VV-L-751b						
VV-L-800						
VV-L-820b						
VV-L-825a TYPE I						
VV-L-825a TYPE II						
VV-L-825a TYPE III						
VV-O-526						
VV-P-216a						
VV-P-236						
WAGNER 21B BRAKE FLUID		A to 70°F	A to 212°F	A to 70°F	+9% vol 3 days 300°F A to 70°F	
WALNUT OIL		A to 70°F				
WASHING SODA	AB to 100% to 140°F	A to 100% to 300°F	A to 70°F	A to 70°F		A to 70°F
WATER, ACID MINE	A to 70°F	A to 300°F		A to 70°F		
WATER, BRINE (PROCESS BEVERAGE)					A to 70°F	A to 70°F
WATER, DEIONIZED	A to 70°F	A to 212°F				
WATER, DETERGENT SOLUTION						
WATER, DISTILLED		A to 300°F		A to 70°F		
WATER, FRESH	A to 140°F NR at 212°F	A to boiling	A to 194°F	A to 70°F	+1-9% vol 3 days 300°F A to 212°F	A to 212°F
WATER, GLYCOL MIXTURE						
WATER, PROOFING SALT		A to 70°F				
WATER, SALT	A to 122°F AB to 140°F	A to 300°F		A to 70°F		
WATER, SEA	A to 70°F AB to 140°F	A to 300°F	A to 70°F	A to 70°F		
WATER, SOAP SOLUTION						
WATER, SOLUBLE OIL SOLUTION	B at 70°F			A to 70°F		
WATERGLASS	A to 70°F	A to 300°F	A to 70°F	A to 70°F		
WAX		A to 70°F				
WEMCO C				A to 70°F		
WHEAT GERM OIL						
WHISKEY	AB to 70°F NR at 140°F	A to 300°F	A to 70°F	A to 70°F		
WHISKEY & ALCOHOLIC BEVERAGES		A to 300°F				
WHITE CAUSTIC		A to 70°F				
WHITE COPPERAS		A to 70°F				

C 391

	RECOMMENDATIONS	NITRILE NBR	ETHYLENE PROPYLENE EP, EPDM	FLUOROCARBON FKM	CHLOROPRENE CR	HYDROGENATED NITRILE HNBR
WHITE LEAD SULFATE		AB any conc to 120°F		A to 70°F	A to 70°F AB any conc to 180°F	
WHITE LIQUOR (PULP MILL)		A to 130°F	A to 176°F	A to 140°F	A to 140°F	A to 70°F
WHITE OIL	NBR	A to 100% to 70°F	NR at 70°F	A to 70°F	B at 70°F AB 10% to 70°F	A to 70°F
WHITE PINE TAR	NBR,FKM,ACM,FVMQ AU	B at 70°F	NR at 70°F	A to 70°F	NR at 70°F	B at 70°F
WHITE SPIRIT	NBR,FKM,FVMQ,ACM AU	A to 250°F	NR at 70°F	A to 158°F	BC at 70°F	A to 70°F
WHITE VITRIOL	NBR,EP,FKM,CR	A to 100% to 140°F AB any conc to 150°F	A to 100% to 176°F B at 212°F	A to 100% to boiling	A to 100% to 140°F AB any conc to 150°F	A to 70°F
WHITING	NBR,CR	A to 70°F	A to 70°F	A to 70°F	A to 70°F	
WINES	NBR,CR	A to 200°F all colors	A to 200°F all colors	A to 200°F all colors	A to 140°F all colors C at 200°F all colors	A to 70°F
WINTERGREEN OIL		NR at 70°F	C at 70°F	B at 70°F	NR at 70°F	
WOLMAR SALTS	NBR,EP,FKM,SBR	A to 70°F	A to 70°F	A to 70°F	B at 70°F	A to 70°F
WOOD ALCOHOL	NBR,EP,SBR,CR	A to 70°F AB any conc to 150°F	A to 160°F AB to 176°F	BC at 70°F C/NR at 140°F	A to 100% to 140°F NR at 212°F	A to 70°F
WOOD NAPHTHA		A to 70°F	B at 70°F	C at 70°F	A to 70°F	
WOOD OIL	NBR,FKM	A to 70°F AB to 120°F	NR at 70°F	A to 70°F	B to 120°F	
WOOD SPIRIT		A to 70°F	B at 70°F	C at 70°F	A to 70°F	
WOOD TAR		B/NR at 70°F	NR at 70°F	A to 70°F	NR at 70°F	
WOOD VINEGAR		C/NR at 70°F	B at 70°F	A to 70°F	C/NR at 70°F	
WOOL OIL		A to 70°F			A to 70°F	
WORT	NBR	A to 250°F	A to 70°F	A to 70°F	AB to 70°F	
XENON	NBR,EP,FKM,CR	A to 70°F	A to 70°F	A to 70°F	A to 70°F	A to 70°F
XYLENE (XYLOL)	FKM,FVMQ,T	NR at 70°F	NR at 70°F	+18% vol 28 days 158°F A to 140°F	NR at 70°F	NR at 70°F
XYLIDINES		C/NR at 70°F	C/NR at 70°F	C/NR at 70°F	NR at 70°F	
XYLIDINES, MIXED		C/NR at 70°F	NR at 70°F	NR at 70°F	NR at 70°F	C at 70°F
XYLOL & STODDARDS SOLVENT		NR at 70°F	NR at 70°F	A to 70°F	NR at 70°F	
YEAST		A to 70°F	A to 70°F	A to 70°F	AB to 70°F	
ZALA		B at 70°F	A to 70°F	A to 70°F	A to 70°F	
ZEOLITE (ZEOLITIC)		A/NR at 70°F	A to 70°F	A to 70°F	A/NR at 70°F	
ZEOLITES	NBR,EP,FKM,CR	A to 70°F	A to 70°F	A to 70°F	A to 70°F	A to 70°F
ZINC ACETATE	EP,NBR,IIR	AB to 140°F B at 176°F	A to 140°F	A/NR to 176°F	B at 70°F	B at 70°F
ZINC BUTTER		B at 70°F	A to 70°F	A to 70°F	B at 70°F	
ZINC CARBONATE		A to 70°F	A to 70°F	A to 70°F	A to 70°F	
ZINC CHLORIDE	NBR,EP,FKM,CR,IIR	A to 140°F AB any conc to 200°F	A to 100% to 176°F B at 212°F	A to 100% to 212°F	A to 100% to 140°F B 176-200°F	A to 70°F
ZINC CHROMATE						
ZINC HYDRATE						
ZINC HYDROSULFITE		A to 140°F	A to 140°F		A to 140°F	
ZINC SALTS	NBR,EP,FKM,CR	A to 70°F	A to 70°F	A to 70°F	A to 70°F	A to 70°F
ZINC STEARATE						
ZINC SULFATE	NBR,EP,FKM,CR,IIR	A to 100% to 140°F B at 176°F	A to 100% to 176°F B at 212°F	A to 100% to boiling	A to 100% to 140°F B 176-200°F	A to 70°F
ZINC SULPHATE	NBR,EP,FKM,CR,IIR	A to 100% to 140°F B at 176°F	A to 100% to 176°F B at 212°F	A to 100% to boiling	A to 100% to 140°F B 176-200°F	
ZINC VITRIOL		A to 70°F AB any conc to 150°F	A to 70°F	A to 70°F	A to 70°F AB any conc to 150°F	
ZIRLITE		B at 70°F	A to 70°F	C at 70°F	A to 70°F	
51-F-23		A to 70°F	NR at 70°F	A to 70°F	B at 70°F	
ASTM METHOD D-471.1		A	NR	A	A	
ASTM METHOD D-471.2		A	NR	A	B	
ASTM METHOD D-471.3		A	NR	A	B	

	STYRENE BUTADIENE SBR	POLYACRYLATE ACM	POLYURETHANE AU, EU	ISOBUTYLENE ISOPRENE IIR	POLYBUTADIENE BR	® AEROQUIP AQP
WHITE LEAD SULFATE				A to 70°F AB any conc to 185°F		
WHITE LIQUOR (PULP MILL)			C/NR at 70°F			
WHITE OIL	NR at 70°F	A to 70°F	A to 70°F	NR at 70°F	NR at 70°F	
WHITE PINE TAR	NR at 70°F			NR at 70°F	NR at 70°F	
WHITE SPIRIT	NR at 70°F	A to 70°F	AB to 110°F C at 140°F	NR at 70°F	NR at 70°F	AB to 70°F
WHITE VITRIOL			A to 70°F	A to 70°F AB any conc to 150°F		
WHITING				A to 70°F		
WINES	A to 70°F	NR at 70°F	NR at 70°F	A to 70°F AB to 150°F	A to 70°F	
WINTERGREEN OIL				B any conc to 80°F		
WOLMAR SALTS	A to 70°F	B at 70°F	A to 70°F	A to 70°F	A to 70°F	
WOOD ALCOHOL	A to 70°F	NR at 70°F	NR at 70°F	A to 70°F AB any conc to 185°F	A to 70°F	
WOOD NAPHTHA			NR at 70°F	A to 70°F		
WOOD OIL	NR at 70°F	A to 70°F	C at 70°F	C/NR at 70°F	NR at 70°F	
WOOD SPIRIT			NR at 70°F	A to 70°F		
WOOD TAR			NR at 70°F	NR at 70°F		
WOOD VINEGAR			B at 70°F	A to 70°F AB any conc to 100°F		
WOOL OIL			AB to 70°F			
WORT	A to 70°F	NR at 70°F	B at 70°F			
XENON	A to 70°F	A to 70°F	A to 70°F	A to 70°F	A to 70°F	
XYLENE (XYLOL)	NR at 70°F	NR at 70°F	C/NR at 70°F	NR at 70°F	NR at 70°F	TEST
XYLIDINES	NR at 70°F			B any conc to 100°F	NR at 70°F	
XYLIDINES, MIXED	NR at 70°F	NR at 70°F	NR at 70°F	NR at 70°F	NR at 70°F	
XYLOL & STODDARDS SOLVENT				NR at 70°F		
YEAST			C at 70°F	AB to 70°F		
ZALA			A to 70°F	A to 70°F		
ZEOLITE (ZEOLITIC)				A to 70°F		
ZEOLITES	A to 70°F			A to 70°F	A to 70°F	
ZINC ACETATE	NR at 70°F	NR at 70°F	NR at 70°F	A to 70°F AB any conc to 150°F	C/NR at 70°F	
ZINC BUTTER			A to 70°F	A to 70°F		
ZINC CARBONATE			A to 70°F	A to 70°F		
ZINC CHLORIDE	A conc to 70°F B sol'n at 70°F	NR any conc at 70°F	AB to 100% to 70°F	A to 70°F AB any conc to 185°F	A to 70°F	
ZINC CHROMATE				A to 70°F		
ZINC HYDRATE			C at 70°F			
ZINC HYDROSULFITE						
ZINC SALTS	A to 70°F	NR at 70°F	A to 70°F	A to 70°F	A to 70°F	
ZINC STEARATE						
ZINC SULFATE	B at 70°F	NR at 70°F	AB to 70°F	A to 70°F AB any conc to 150°F	B at 70°F	NR at 70°F
ZINC SULPHATE	B at 70°F	NR at 70°F	AB to 70°F	A to 70°F AB any conc to 150°F	B at 70°F	NR at 70°F
ZINC VITRIOL				A to 70°F AB any conc to 150°F		
ZIRLITE				A to 70°F		
51-F-23	NR at 70°F	A to 70°F	B at 70°F	NR at 70°F	NR at 70°F	
ASTM METHOD D-471.1	NR	A	A	NR	NR	
ASTM METHOD D-471.2	NR	A	B	NR	NR	
ASTM METHOD D-471.3	NR	A	B	NR	NR	

	SYNTHETIC ISOPRENE IR	NATURAL ISOPRENE NR	CHLOROSULFONATED POLYETHYLENE CSM	FLUOROSILICONE FVMQ	SILICONE VMQ	CHEMRAZ FFKM
WHITE LEAD SULFATE	A to 70°F	A to 70°F AB any conc to 120°F	A to 70°F			
WHITE LIQUOR (PULP MILL)	A to 70°F	A to 70°F	A to 70°F			AB to 70°F NR at 70°F white
WHITE OIL	NR at 70°F	NR at 70°F AB 10% to 70°F	NR at 70°F AB 10% to 70°F	A to 70°F	NR at 70°F	A to 70°F
WHITE PINE TAR	NR at 70°F	NR at 70°F	NR at 70°F	A to 70°F	NR at 70°F	A to 70°F
WHITE SPIRIT	NR at 70°F	NR at 70°F	C/NR at 70°F	A to 70°F	NR at 70°F	A to 70°F
WHITE VITRIOL	AB to 70°F	A to 70°F AB any conc to 185°F	A to 70°F AB to 200°F	A to 158°F	A to 70°F	A to 70°F
WHITING	A to 70°F	A to 70°F	A to 70°F			
WINES	A to 70°F	A to 70°F AB to 150°F	A to 70°F	A to 70°F	A to 70°F	A to 70°F B at 70°F white
WINTERGREEN OIL	NR at 70°F	NR at 70°F				
WOLMAR SALTS	A to 70°F	A to 70°F	A to 70°F	A to 70°F	A to 70°F	A to 70°F
WOOD ALCOHOL	A to 70°F	A to 100% to 70°F AB any conc to 100°F	A to 100% to 70°F A 100% to 158°F	A to 158°F	A to 100% to 70°F A 100% to 158°F	A to 70°F
WOOD NAPHTHA	A to 70°F	A to 70°F	A to 70°F		A to 70°F	
WOOD OIL	NR at 70°F	NR at 70°F	BC at 70°F	B at 70°F	NR at 70°F	A to 70°F
WOOD SPIRIT	A to 70°F	A to 70°F	A to 70°F		A to 70°F	
WOOD TAR	NR at 70°F	NR at 70°F	NR at 70°F		C at 70°F	
WOOD VINEGAR	A to 70°F	A to 70°F	A to 70°F		A to 70°F	
WOOL OIL	B at 70°F	B at 70°F	AB at 70°F			
WORT				A to 70°F	AB to 70°F	
XENON	A to 70°F	A to 70°F	A to 70°F	A to 70°F	A to 70°F	A to 70°F
XYLENE (XYLOL)	NR at 70°F	NR at 70°F	NR at 70°F	+20% vol 7 days 70°F AC to 70°F	+150% vol 7 days 70°F NR at 70°F	A to 70°F
XYLIDINES	NR at 70°F	NR at 70°F	NR at 70°F	NR at 70°F	NR at 70°F	
XYLIDINES, MIXED	NR at 70°F	NR at 70°F	NR at 70°F	NR at 70°F	NR at 70°F	A to 70°F
XYLOL & STODDARDS SOLVENT	NR at 70°F	NR at 70°F	NR at 70°F			
YEAST					AB to 70°F	
ZALA	A to 70°F	A to 70°F	A to 70°F		B at 70°F	
ZEOLITE (ZEOLITIC)	B at 70°F	AB any conc to 80°F	A to 70°F			
ZEOLITES	A to 70°F	A to 70°F	A to 70°F	A to 70°F		A to 70°F
ZINC ACETATE	A to 70°F	A to 70°F AB any conc to 150°F	B/NR at 70°F	NR at 70°F	NR at 70°F	A to 70°F
ZINC BUTTER	B at 70°F	B at 70°F	A to 70°F		A to 70°F	
ZINC CARBONATE	A to 70°F	A to 70°F	A to 70°F			A to 70°F
ZINC CHLORIDE	A to 70°F	A to 70°F AB any conc to 185°F	A to 100% to 200°F	A to 70°F B sol'n at 70°F	AB to 100% to 70°F	A to 70°F
ZINC CHROMATE			C at 70°F			A to 70°F
ZINC HYDRATE						
ZINC HYDROSULFITE						A to 70°F
ZINC SALTS	A to 70°F	A to 70°F	A to 70°F	A to 70°F	A to 70°F	A to 70°F
ZINC STEARATE						A to 70°F
ZINC SULFATE	AB to 70°F	A to 70°F AB any conc to 185°F	A to 70°F AB to 200°F	A to 70°F	A to 70°F	A to 70°F
ZINC SULPHATE	AB to 70°F	AB any conc to 185°F	A to 70°F AB to 200°F	A to 70°F	A to 70°F	
ZINC VITRIOL	A to 70°F	A to 70°F AB any conc to 150°F	A to 70°F		A to 70°F	
ZIRLITE	A to 70°F	A to 70°F	B at 70°F			
51-F-23	NR at 70°F	NR at 70°F	B at 70°F	A to 70°F	C at 70°F	
ASTM METHOD D-471.1	NR	NR	A	A	C	
ASTM METHOD D-471.2	NR	NR	B	A	C	
ASTM METHOD D-471.3	NR	NR	B	A	C	

	POLYSULFIDE T	CHLORINATED POLYETHYLENE CM	EPICHLOROHYDRIN CO, ECO	® HYTREL COPOLYESTER TPE	® SANTOPRENE COPOLYMER TPO	® C-FLEX STYRENIC TPE
WHITE LEAD SULFATE						
WHITE LIQUOR (PULP MILL)						
WHITE OIL	A to 70°F					
WHITE PINE TAR	B at 70°F					
WHITE SPIRIT	B at 70°F	AB to 70°F	A to 70°F	A to 70°F	NR at 70°F	NR at 70°F
WHITE VITRIOL						
WHITING						
WINES	NR at 70°F			AB to 70°F		
WINTERGREEN OIL						
WOLMAR SALTS	A to 70°F					
WOOD ALCOHOL	B at 70°F					A to 70°F
WOOD NAPHTHA						
WOOD OIL	B at 70°F			AB to 70°F		
WOOD SPIRIT						
WOOD TAR						
WOOD VINEGAR						
WOOL OIL	AB to 70°F			AB to 70°F		
WORT						
XENON	A to 70°F					
XYLENE (XYLOL)	B at 70°F	C at 70°F	NR at 70°F	BC at 70°F *C /NR at 140°F	NR at 70°F	NR at 70°F
XYLIDINES	NR at 70°F					
XYLIDINES, MIXED	B /NR at 70°F					
XYLOL & STODDARDS SOLVENT						
YEAST						
ZALA						
ZEOLITE (ZEOLITIC)						
ZEOLITES						
ZINC ACETATE	NR at 70°F					
ZINC BUTTER						
ZINC CARBONATE						
ZINC CHLORIDE	C at 70°F	NR at 70°F		A to 100% to 70°F *AB to 140°F	+1% vol 10% conc 7 days 70°F	A to 70°F
ZINC CHROMATE						
ZINC HYDRATE				NR at 70°F		
ZINC HYDROSULFITE						
ZINC SALTS	NR at 70°F					
ZINC STEARATE						
ZINC SULFATE	NR at 70°F	NR at 70°F		B /NR at 70°F		
ZINC SULPHATE	NR at 70°F	NR at 70°F		B /NR at 70°F		
ZINC VITRIOL						
ZIRLITE						
51-F-23	A to 70°F		A to 70°F			
ASTM METHOD D-471.1	A		A			
ASTM METHOD D-471.2	A		A			
ASTM METHOD D-471.3	A		A			

	ETHYLENE ACRYLIC EA	POLYALLOMER LINEAR COPOLYMER	NYLON 11 POLYAMIDE	NYLON 12 POLYAMIDE	ETHYLENE VINYL ACETATE EVA	POLYVINYLCHLORIDE FLEXIBLE PVC
WHITE LEAD SULFATE						
WHITE LIQUOR (PULP MILL)					C/NR at 70°F	A to 150°F
WHITE OIL						
WHITE PINE TAR						
WHITE SPIRIT	BC at 70°F	B at 70°F C at 122°F	A to 70°F swells at 140°F		NR at 70°F	AC to 70°F A to 122°F (special)
WHITE VITRIOL					A to 70°F	A to 70°F
WHITING						
WINES	A to 70°F	A to 70°F	A to 70°F	AB to 70°F	AB to 70°F	A to 70°F B at 150°F
WINTERGREEN OIL						NR at 70°F
WOLMAR SALTS						
WOOD ALCOHOL	A to 70°F				B at 70°F	B at 70°F
WOOD NAPHTHA						
WOOD OIL			AB to 70°F			A to 70°F
WOOD SPIRIT						
WOOD TAR					.	
WOOD VINEGAR					A to 70°F	A to 70°F
WOOL OIL			AB to 70°F			AB to 70°F
WORT				AB to 70°F	AB to 70°F	AB to 70°F
XENON						AB to 70°F
XYLENE (XYLOL)	NR at 70°F	C at 70°F NR at 122°F	AB to 105°F C 140–194°F	AB to 70°F	NR at 70°F	NR at 70°F
XYLIDINES						NR at 70°F
XYLIDINES, MIXED						
XYLOL & STODDARDS SOLVENT						NR at 70°F
YEAST					C/NR at 70°F	B/NR at 70°F NR at 150°F
ZALA						
ZEOLITE (ZEOLITIC)						
ZEOLITES						AB to 70°F
ZINC ACETATE						
ZINC BUTTER						
ZINC CARBONATE						
ZINC CHLORIDE	AB to 70°F	A to 70°F	+1.2% vol 180 days 70°F *AB to 140°F	AB to 50% to 70°F	A to 70°F *AB to 140°F	A to 150°F
ZINC CHROMATE						
ZINC HYDRATE			NR at 70°F			A to 70°F
ZINC HYDROSULFITE						
ZINC SALTS						AB to 70°F
ZINC STEARATE		AB to 122°F				A to 70°F B at 122°F
ZINC SULFATE		A to 70°F	NR at 70°F		A to 70°F	A to 150°F
ZINC SULPHATE		A to 70°F	NR at 70°F		A to 70°F	A to 150°F
ZINC VITRIOL						
ZIRLITE						
51-F-23						
ASTM METHOD D-471.1						
ASTM METHOD D-471.2						
ASTM METHOD D-471.3						

	POLYETHYLENE LOW DENSITY LDPE	®TEFLON FEP	®KALREZ PERFLUORINATED ELASTOMER	®FLUORAZ FLUORINATED COPOLYMER	®AFLAS FLUORINATED COPOLYMER	®NORPRENE COPOLYMER TPO
WHITE LEAD SULFATE		A to 70°F				
WHITE LIQUOR (PULP MILL)	A to 70°F	A to 300°F		A to 70°F		
WHITE OIL		A to 100% to 70°F	A to 212°F	A to 70°F		
WHITE PINE TAR		A to 70°F	A to 70°F	A to 70°F		
WHITE SPIRIT	C at 70°F NR at 122°F	A to conc to 300°F	A to 113°F *A to 400°F	AC to 70°F	A to 70°F	NR at 70°F
WHITE VITRIOL	AB to 70°F	A to 300°F	A to 212°F	A to 70°F		
WHITING		A to 70°F				
WINES	A to 70°F all colors AB to 140°F all colors	A to 300°F	A to 70°F	A to 70°F		
WINTERGREEN OIL		A to 70°F				
WOLMAR SALTS		A to 70°F		A to 70°F		
WOOD ALCOHOL	AB 100% to 122°F B/NR at 140°F	A to 300°F	A 50-100% to 212°F	A to 70°F	A to 70°F	A to 70°F
WOOD NAPHTHA		A to 70°F				
WOOD OIL		A to 70°F	A to 212°F	A to 70°F		
WOOD SPIRIT		A to 70°F				
WOOD TAR		A to 70°F				
WOOD VINEGAR	A to 70°F	A to 70°F		A to 70°F		
WOOL OIL	AB to 70°F	A to 70°F				
WORT	AB to 70°F			A to 70°F		
XENON	AB to 70°F	A to 70°F	A to 212°F	A to 70°F		
XYLENE (XYLOL)	B/NR at 70°F NR at 122°F	A to 300°F	A to 450°F	NR at 70°F	+30% vol 3 days 73°F C/NR at 70°F	NR at 70°F
XYLIDINES		A to 70°F	A to 212°F			
XYLIDINES, MIXED		A to 70°F		A to 70°F		
XYLOL & STODDARDS SOLVENT		A to 70°F				
YEAST	A to 70°F AB to 140°F	A to 70°F		A to 70°F		
ZALA		A to 70°F				
ZEOLITE (ZEOLITIC)		A to 70°F		A to 70°F in treated water		
ZEOLITES			A to 212°F	A to 70°F		
ZINC ACETATE		A to 70°F	A to 212°F	C at 70°F		
ZINC BUTTER		A to 70°F				
ZINC CARBONATE		A to 70°F				
ZINC CHLORIDE	A to 70°F AB to 140°F	A to 300°F	A to 212°F	A to 70°F wet or dry	A sat'd to 70°F	AB to 70°F
ZINC CHROMATE		A to 70°F	A to 212°F			
ZINC HYDRATE	AB to 70°F					
ZINC HYDROSULFITE			A to 212°F			
ZINC SALTS	A to 70°F AB to 140°F	A to 70°F	A to 212°F	A to 70°F		A to 70°F
ZINC STEARATE	A to 122°F	A to 125°F	A to 212°F			
ZINC SULFATE	AB to 70°F	A to 300°F	A to 212°F	A to 70°F		
ZINC SULPHATE	AB to 70°F	A to 300°F	A to 212°F	A to 70°F	A sat'd to 70°F	
ZINC VITRIOL		A to 70°F				
ZIRLITE		A to 70°F				
51-F-23						
ASTM METHOD D-471.1						
ASTM METHOD D-471.2						
ASTM METHOD D-471.3						

	RECOMMENDATIONS	NITRILE NBR	ETHYLENE PROPYLENE EP, EPDM	FLUOROCARBON FKM	CHLOROPRENE CR	HYDROGENATED NITRILE HNBR
MIL-L-644B		A to 70°F	C at 70°F		C at 70°F	
MIL-L-2104	NBR, FKM, ACM, IIR	A to 70°F	NR at 70°F		A to 70°F	
MIL-L-2104B	NBR	A to 70°F	NR at 70°F	A to 70°F	AB to 70°F	
MIL-L-2105B		A to 70°F	NR at 70°F	A to 300°F	A to 70°F	
MIL-G-2108		A to 70°F	NR at 70°F	A to 70°F	A to 70°F	
MIL-S-3136B TYPE I (ASTM FUEL A)	NBR, FKM, ACM, AU FVMQ	A to 250°F	NR at 70°F	A to 70°F	B at 80°F	A to 70°F
MIL-S-3136B TYPE II	FKM	B /NR at 70°F	NR at 70°F	A to 70°F	NR at 70°F	
MIL-S-3136B TYPE III (ASTM FUEL B)	FKM	B /NR at 70°F	NR at 70°F	A to 70°F	NR at 70°F	A to 70°F
MIL-S-3136B TYPE IV, LOW SWELL OIL	NBR, FKM, CR, ACM	A to 70°F	NR at 70°F	A to 70°F	A to 70°F	
MIL-S-3136B TYPE V, MEDIUM SWELL OIL	NBR, FKM, ACM, AU	A to 70°F	NR at 70°F	A to 70°F	B at 70°F	
MIL-S-3136B TYPE VI, HIGH SWELL OIL	NBR, FKM, ACM, AU	A to 70°F	NR at 70°F	A to 70°F	NR at 70°F	
MIL-S-3136B TYPE VII		A to 70°F	NR at 70°F	A to 70°F	C at 70°F	
MIL-L-3150A	NBR, FKM	A to 70°F	NR at 70°F	A to 70°F	B at 70°F	
MIL-G-3278	FKM, ACM	B at 70°F	NR at 70°F	A to 70°F	NR at 70°F	
MIL-L-3503	NBR, FKM, AU	A to 70°F	NR at 70°F	A to 70°F	B at 70°F	
MIL-L-3545B	NBR, FKM, ACM, AU	AB to 70°F	NR at 70°F	A to 70°F	B at 70°F	
MIL-L-4339C	NBR, FKM	A to 70°F	NR at 70°F	A to 70°F	NR at 70°F	
MIL-G-4343B	FKM	AB to 70°F	AC to 70°F	A to 70°F	AB to 70°F	
MIL-L-5020A		A to 70°F	NR at 70°F	A to 70°F	B at 70°F	
MIL-J-5161F	FKM, ACM	B at 70°F	NR at 70°F	A to 70°F	NR at 70°F	
MIL-C-5545A		B at 70°F	NR at 70°F	A to 70°F	B at 70°F	
MIL-M-5559A		A to 70°F	A to 70°F	B at 70°F	B at 70°F	
MIL-F-5566	EP, FKM, AU	B at 70°F	A to 70°F	A to 70°F	B at 70°F	
MIL-G-5572	NBR, FKM	A to 70°F	NR at 70°F	A to 70°F	NR at 70°F	
MIL-F-5602		A to 70°F	NR at 70°F	A to 70°F	B at 70°F	
MIL-O-5606	NBR, FKM	A to 70°F		+1.8% vol 28 days 158°F A to 300°F		
MIL-H-5606B	NBR, FKM, FVMQ, CR	A to 275°F	BC at 70°F NR at 140°F	A to 212°F B at 248°F	B at 70°F NR at 140°F	A to 70°F
MIL-J-5624G, JP3	NBR, FKM, FVMQ	A to 70°F B at 140°F	NR at 70°F	A to 140°F	NR at 70°F	A to 70°F
MIL-J-5624G, JP4	NBR, FKM, FVMQ	A to 200°F ground or airborn	NR at 70°F	+12% vol 3 days 400°F A to 400°F	NR at 70°F	A to 70°F
MIL-J-5624G, JP5	NBR, FKM, FVMQ	A to 200°F ground or airborn	NR at 70°F	+4% vol 3 days 400°F NR at 500°F	NR at 70°F	A to 70°F
MIL-O-6081C	NBR, FKM, ACM, AU	A to 70°F	NR at 70°F	A to 70°F	B at 70°F	
MIL-L-6082C	NBR, FKM, ACM, AU	A to 70°F	NR at 70°F	A to 70°F	B at 70°F	
MIL-H-6083C	NBR, FKM, CR, ACM	A to 70°F	NR at 70°F	A to 140°F	AB to 70°F	
MIL-L-6085A	FKM, AU	B at 70°F	NR at 70°F	A to 70°F	NR at 70°F	B at 70°F
MIL-L-6086B		A to 70°F	NR at 70°F	A to 70°F	A to 70°F	
MIL-A-6091	EP, FKM, CR, SBR	B at 70°F	A to 70°F	A to 70°F	A to 70°F	
MIL-L-6387A	FKM, AU	B at 70°F	NR at 70°F	A to 70°F	NR at 70°F	
MIL-C-6529C		B at 70°F	NR at 70°F	A to 70°F	B at 70°F	
MIL-F-7024A	NBR, FKM, AU	A to 70°F	NR at 70°F	A to 70°F	NR at 70°F	
MIL-H-7083A	NBR, EP, IIR	A to 70°F	A to 70°F	B at 70°F	B at 70°F	
MIL-G-7118A	FKM, NBR	B at 70°F	NR at 70°F	A to 70°F	BC at 70°F	
MIL-G-7187	NBR, FKM, ACM, AU	A to 70°F	NR at 70°F	A to 70°F	NR at 70°F	
MIL-G-7421A	FKM	B at 70°F	NR at 70°F	A to 70°F	BC at 70°F	
MIL-H-7644		B at 70°F	A to 70°F	A to 70°F	B at 70°F	

	STYRENE BUTADIENE SBR	POLYACRYLATE ACM	POLYURETHANE AU, EU	ISOBUTYLENE ISOPRENE IIR	POLYBUTADIENE BR	EPICHLOROHYDRIN CO, ECO
MIL-L-644B	C at 70°F	B at 70°F	C at 70°F	C at 70°F	C at 70°F	A to 70°F
MIL-L-2104						
MIL-L-2104B	NR at 70°F	A to 70°F	A to 70°F	NR at 70°F	NR at 70°F	A to 70°F
MIL-L-2105B	NR at 70°F	A to 70°F	A to 70°F	NR at 70°F	NR at 70°F	A to 70°F
MIL-G-2108	NR at 70°F	A to 70°F	A to 70°F	NR at 70°F	NR at 70°F	A to 70°F
MIL-S-3136B TYPE I (ASTM FUEL A)	NR at 70°F	B /NR at 70°F	B to 158°F	NR at 70°F	NR at 70°F	A to 70°F
MIL-S-3136B TYPE II	NR at 70°F	B at 70°F	B at 70°F	NR at 70°F	NR at 70°F	A to 70°F
MIL-S-3136B TYPE III (ASTM FUEL B)	NR at 70°F	B /NR at 70°F	B at 70°F	NR at 70°F	NR at 70°F	A to 70°F
MIL-S-3136B TYPE IV, LOW SWELL OIL	NR at 70°F	A to 70°F	A to 70°F	NR at 70°F	NR at 70°F	A to 70°F
MIL-S-3136B TYPE V, MEDIUM SWELL OIL	NR at 70°F	A to 70°F	AB to 70°F	NR at 70°F	NR at 70°F	A to 70°F
MIL-S-3136B TYPE VI, HIGH SWELL OIL	NR at 70°F	A to 70°F	AB to 70°F	NR at 70°F	NR at 70°F	A to 70°F
MIL-S-3136B TYPE VII	NR at 70°F	A to 70°F	C at 70°F	NR at 70°F	NR at 70°F	A to 70°F
MIL-L-3150A	NR at 70°F	AB to 70°F	B at 70°F	NR at 70°F	NR at 70°F	A to 70°F
MIL-G-3278	NR at 70°F	A to 70°F	B at 70°F	NR at 70°F	NR at 70°F	
MIL-L-3503	NR at 70°F	AB to 70°F	AB to 70°F	NR at 70°F	NR at 70°F	A to 70°F
MIL-L-3545B	NR at 70°F	AB to 70°F	AC to 70°F	NR at 70°F	NR at 70°F	B at 70°F
MIL-L-4339C	NR at 70°F	A /NR at 70°F	NR at 70°F	NR at 70°F	NR at 70°F	A to 70°F
MIL-G-4343B	A /NR at 70°F	A to 70°F	A to 70°F	AC to 70°F	A to 70°F	
MIL-L-5020A	NR at 70°F	B at 70°F	B at 70°F	NR at 70°F	NR at 70°F	A to 70°F
MIL-J-5161F	NR at 70°F	A to 70°F	BC at 70°F	NR at 70°F	NR at 70°F	A to 70°F
MIL-C-5545A	NR at 70°F	B at 70°F	C at 70°F	NR at 70°F	NR at 70°F	B at 70°F
MIL-M-5559A	A to 70°F	C at 70°F	C at 70°F	A to 70°F	A to 70°F	B at 70°F
MIL-F-5566	B at 70°F	NR at 70°F	NR at 70°F	A to 70°F	AB to 70°F	B at 70°F
MIL-G-5572	NR at 70°F	B at 70°F	B at 70°F	NR at 70°F	NR at 70°F	
MIL-F-5602	NR at 70°F	A to 70°F	B at 70°F	NR at 70°F	NR at 70°F	A to 70°F
MIL-O-5606			C at 70°F			
MIL-H-5606B	NR at 70°F	AB to 70°F	B at 70°F AB to 70°F fumes	NR at 70°F	NR at 70°F	A to 70°F
MIL-J-5624G, JP3	NR at 70°F	B at 70°F	BC at 70°F	NR at 70°F	NR at 70°F	A to 70°F
MIL-J-5624G, JP4	NR at 70°F	AB to 70°F ground NR at 70°F airborn	BC at 70°F	NR at 70°F	NR at 70°F	A to 70°F
MIL-J-5624G, JP5	NR at 70°F	AB to 70°F ground NR at 70°F airborn	B at 70°F	NR at 70°F	NR at 70°F	A to 70°F
MIL-O-6081C	NR at 70°F	A to 70°F	AB to 70°F	NR at 70°F	NR at 70°F	A to 70°F
MIL-L-6082C	NR at 70°F	A to 70°F	A to 70°F	NR at 70°F	NR at 70°F	A to 70°F
MIL-H-6083C	NR at 70°F	A to 70°F	AB to 70°F	NR at 70°F	NR at 70°F	A to 70°F
MIL-L-6085A	NR at 70°F	BC at 70°F	AC to 70°F	NR at 70°F	NR at 70°F	B at 70°F
MIL-L-6086B	NR at 70°F	A to 70°F	A to 70°F	NR at 70°F	NR at 70°F	A to 70°F
MIL-A-6091	A to 70°F	NR at 70°F	NR at 70°F	A to 70°F	A to 70°F	
MIL-L-6387A	NR at 70°F	B at 70°F	AB to 70°F	NR at 70°F	NR at 70°F	B at 70°F
MIL-C-6529C	NR at 70°F	B at 70°F	C at 70°F	NR at 70°F	NR at 70°F	B at 70°F
MIL-F-7024A	NR at 70°F	B at 70°F	AB to 70°F	NR at 70°F	NR at 70°F	AB to 70°F
MIL-H-7083A	B at 70°F	C /NR at 70°F	NR at 70°F	A to 70°F	AC to 70°F	B at 70°F
MIL-G-7118A	NR at 70°F	AC to 70°F	AC to 70°F	NR at 70°F	NR at 70°F	B at 70°F
MIL-G-7187	NR at 70°F	A to 70°F	A to 70°F	NR at 70°F	NR at 70°F	A to 70°F
MIL-G-7421A	NR at 70°F	NR at 70°F	B at 70°F	NR at 70°F	NR at 70°F	B at 70°F
MIL-H-7644	A to 70°F	B at 70°F	C at 70°F	B at 70°F	A to 70°F	B at 70°F

	SYNTHETIC ISOPRENE IR	NATURAL ISOPRENE NR	CHLOROSULFONATED POLYETHYLENE CSM	FLUOROSILICONE FVMQ	SILICONE VMQ	POLYSULFIDE T
MIL-L-644B	C at 70°F	C at 70°F	C at 70°F		C at 70°F	
MIL-L-2104						
MIL-L-2104B	NR at 70°F	NR at 70°F	C at 70°F	A to 70°F	NR at 70°F	A to 70°F
MIL-L-2105B	NR at 70°F	NR at 70°F	A to 70°F	A to 70°F	C at 70°F	A to 70°F
MIL-G-2108	NR at 70°F	NR at 70°F	A to 70°F	A to 70°F	C at 70°F	A to 70°F
MIL-S-3136B TYPE I (ASTM FUEL A)	NR at 70°F	NR at 70°F	AB to 70°F	AB to 70°F	NR at 70°F	A to 70°F
MIL-S-3136B TYPE II	NR at 70°F	NR at 70°F	NR at 70°F	B at 70°F	NR at 70°F	A to 70°F
MIL-S-3136B TYPE III (ASTM FUEL B)	NR at 70°F	NR at 70°F	NR at 70°F	B at 70°F	NR at 70°F	A to 70°F
MIL-S-3136B TYPE IV, LOW SWELL OIL	NR at 70°F	NR at 70°F	A to 70°F	A to 70°F	AC to 70°F	A to 70°F
MIL-S-3136B TYPE V, MEDIUM SWELL OIL	NR at 70°F	NR at 70°F	B at 70°F	A to 70°F	BC to 70°F	A to 70°F
MIL-S-3136B TYPE VI, HIGH SWELL OIL	NR at 70°F	NR at 70°F	NR at 70°F	A to 70°F	BC to 70°F	A to 70°F
MIL-S-3136B TYPE VII	NR at 70°F	NR at 70°F	C at 70°F	A to 70°F	NR at 70°F	A to 70°F
MIL-L-3150A	NR at 70°F	NR at 70°F	B at 70°F	A to 70°F	NR at 70°F	A to 70°F
MIL-G-3278	NR at 70°F	NR at 70°F	NR at 70°F	B at 70°F	NR at 70°F	B at 70°F
MIL-L-3503	NR at 70°F	NR at 70°F	B at 70°F	A to 70°F	NR at 70°F	A to 70°F
MIL-L-3545B	NR at 70°F	NR at 70°F	BC at 70°F	A to 70°F	NR at 70°F	AC to 70°F
MIL-L-4339C	NR at 70°F	NR at 70°F	NR at 70°F	A to 70°F	C at 70°F	A to 70°F
MIL-G-4343B	A/NR at 70°F	A/NR at 70°F	AB to 70°F	AB to 70°F	C/NR at 70°F	A to 70°F
MIL-L-5020A	NR at 70°F	NR at 70°F	C at 70°F	A to 70°F	NR at 70°F	A to 70°F
MIL-J-5161F	NR at 70°F	NR at 70°F	NR at 70°F	A to 70°F	NR at 70°F	A to 70°F
MIL-C-5545A	NR at 70°F	NR at 70°F	C at 70°F	A to 70°F	NR at 70°F	C at 70°F
MIL-M-5559A	B at 70°F	B at 70°F	B at 70°F	B at 70°F	B at 70°F	C at 70°F
MIL-F-5566	A to 70°F	A to 70°F	AB to 70°F	A to 70°F	A to 70°F	A to 70°F
MIL-G-5572	NR at 70°F	NR at 70°F	NR at 70°F	A to 70°F	NR at 70°F	A to 70°F
MIL-F-5602	NR at 70°F	NR at 70°F	B at 70°F	A to 70°F	C at 70°F	A to 70°F
MIL-O-5606				+10% vol 3 days 300°F A to 70°F	+150% vol 3 days 300°F NR at 300°F	
MIL-H-5606B	NR at 70°F	NR at 70°F	B at 70°F	A to 400°F	NR at 70°F	A to 70°F
MIL-J-5624G, JP3	NR at 70°F	NR at 70°F	NR at 70°F	AB to 70°F	NR at 70°F	B at 70°F
MIL-J-5624G, JP4	NR at 70°F	NR at 70°F	NR at 70°F	+10% vol 7 days 75°F B at 70°F	+150% vol 7 days 70°F NR at 70°F	B at 70°F
MIL-J-5624G, JP5	NR at 70°F	NR at 70°F	NR at 70°F	B at 70°F	NR at 70°F	B at 70°F
MIL-O-6081C	NR at 70°F	NR at 70°F	B at 70°F	A to 70°F	NR at 70°F	A to 70°F
MIL-L-6082C	NR at 70°F	NR at 70°F	B at 70°F	A to 70°F	AC to 70°F	A to 70°F
MIL-H-6083C	NR at 70°F	B/NR at 70°F	B at 70°F	A to 70°F	NR at 70°F	A to 70°F
MIL-L-6085A	NR at 70°F	NR at 70°F	NR at 70°F	B at 70°F	NR at 70°F	B at 70°F
MIL-L-6086B	NR at 70°F	NR at 70°F	A to 70°F	A to 70°F	C at 70°F	A to 70°F
MIL-A-6091	A to 70°F	A to 70°F	A to 70°F	A to 70°F	A to 70°F	A to 70°F
MIL-L-6387A	NR at 70°F	NR at 70°F	NR at 70°F	B at 70°F	NR at 70°F	B at 70°F
MIL-C-6529C	NR at 70°F	NR at 70°F	C at 70°F	A to 70°F	NR at 70°F	C at 70°F
MIL-F-7024A	NR at 70°F	NR at 70°F	NR at 70°F	A to 70°F	NR at 70°F	A to 70°F
MIL-H-7083A	BC at 70°F	B at 70°F	B at 70°F	AB to 70°F	AB to 70°F	AC to 70°F
MIL-G-7118A	NR at 70°F	NR at 70°F	BC at 70°F	A to 70°F	NR at 70°F	AB to 70°F
MIL-G-7187	NR at 70°F	NR at 70°F	NR at 70°F	A to 70°F	NR at 70°F	A to 70°F
MIL-G-7421A	NR at 70°F	NR at 70°F	BC at 70°F	B at 70°F	NR at 70°F	AB to 70°F
MIL-H-7644	B at 70°F	B at 70°F	B at 70°F	B at 70°F	NR at 70°F	C at 70°F

	RECOMMENDATIONS	NITRILE NBR	ETHYLENE PROPYLENE EP, EPDM	FLUOROCARBON FKM	CHLOROPRENE CR	HYDROGENATED NITRILE HNBR
MIL-L-7645		B at 70°F	NR at 70°F	A to 70°F	B at 70°F	
MIL-G-7711A	NBR, FKM, AU	A to 70°F	NR at 70°F	A to 70°F	NR at 70°F	
MIL-O-7808	FKM, FVMQ, NBR, ACM	AB to 70°F	NR at 70°F	A to 350°F B at 400°F	NR at 70°F	B at 70°F
MIL-L-7808C		+27% vol 3 days 300°F BC at 300°F	+81% vol 3 days 300°F NR at 300°F	+3% vol 3 days 300°F A to 350°F	+87% vol 3 days 300°F NR at 70°F	
MIL-L-7808F	FKM	A to 275°F	NR at 70°F	A to 70°F	NR at 70°F	
MIL-L-7808H		A to 70°F	NR at 70°F		A to 70°F	
MIL-L-7870A	NBR	A to 70°F	NR at 70°F	A to 70°F	B at 70°F	
MIL-C-8188C	FKM	B at 70°F	NR at 70°F	B at 70°F	NR at 70°F	
MIL-A-8243B		A to 70°F	A to 70°F	B at 70°F	B at 70°F	
MIL-L-8383B		A to 70°F	NR at 70°F	A to 70°F	A to 70°F	
MIL-H-8446B	FKM, CR, AU	B at 70°F	NR at 70°F	+4.9% vol 14 days 350°F A to 70°F	AB to 70°F	B at 70°F
MIL-I-8660B		A to 70°F	A to 70°F	A to 70°F	A to 70°F	
MIL-L-9000F	NBR, FKM, ACM	A to 70°F	NR at 70°F	A to 70°F	B at 70°F	
MIL-T-9188B		NR at 70°F	A to 70°F	NR at 70°F	NR at 70°F	
MIL-L-9236B	FKM	B at 70°F	NR at 70°F	A to 70°F BC at 400°F	NR at 70°F	
MIL-E-9500	EP, NBR, FKM, CR	A to 70°F	A to 70°F	A to 70°F AB to 400°F	A to 70°F	
MIL-L-10295A		A to 70°F	NR at 70°F	A to 70°F	B at 70°F	
MIL-L-10324A		A to 70°F	NR at 70°F	A to 70°F	B at 70°F	
MIL-G-10924B	NBR, FKM, AU	A to 70°F	NR at 70°F	A to 70°F	NR at 70°F	
MIL-L-11734B		A to 70°F	NR at 70°F	A to 70°F	C at 70°F	
MIL-O-11773		A to 70°F	NR at 70°F	A to 70°F	C at 70°F	
MIL-P-12098		B at 70°F	A to 70°F	A to 70°F	B at 70°F	
MIL-H-13862		A to 70°F	NR at 70°F	A to 70°F	B at 70°F	
MIL-H-13866A		A to 70°F	NR at 70°F	A to 70°F	B at 70°F	
MIL-H-13910B	NBR, EP, FKM, CR	AB to 70°F	A to 70°F	A to 70°F	AB to 70°F	
MIL-H-13919A		A to 70°F	NR at 70°F	A to 70°F	B at 70°F	
MIL-L-14107B		C at 70°F	NR at 70°F	A to 70°F	A to 70°F	
MIL-L-15016	NBR, FKM, ACM, AU	A to 70°F	NR at 70°F	A to 70°F	B at 70°F	
MIL-L-15017	NBR, FKM, ACM, AU	A to 70°F	NR at 70°F	A to 70°F	B at 70°F	
MIL-L-15018B		A to 70°F	NR at 70°F	A to 70°F	A to 70°F	
MIL-L-15019A		A to 70°F	NR at 70°F	A to 70°F	A to 70°F	
MIL-L-15719A		B at 70°F	B at 70°F	A to 70°F	B at 70°F	
MIL-G-15793	NBR, FKM, ACM, AU	A to 70°F	NR at 70°F	A to 70°F	BC at 70°F	
MIL-F-16884	NBR, FKM, ACM	A to 70°F	NR at 70°F	A to 70°F	C at 70°F	
MIL-F-16929A		A to 70°F	NR at 70°F	A to 70°F	C at 70°F	
MIL-L-16958A		A to 70°F	NR at 70°F	A to 70°F	B at 70°F	
MIL-F-17111	NBR, FKM, ACM	A to 70°F	NR at 70°F	A to 70°F	B at 70°F	
MIL-L-17331D	NBR, FKM, ACM	A to 70°F	NR at 70°F	A to 70°F	B at 70°F	
MIL-L-17353A		A to 70°F	NR at 70°F	A to 70°F	C at 70°F	
MIL-L-17672B		A to 70°F	NR at 70°F	A to 70°F	A to 70°F	
MIL-L-18486A		A to 70°F	NR at 70°F	A to 70°F	A to 70°F	
MIL-G-18709A		A to 70°F	NR at 70°F	A to 70°F	A to 70°F	
MIL-H-19457B	EP, IIR	NR at 70°F	A to 70°F	B/NR to 158°F	NR at 70°F	
MIL-F-19605		A to 70°F	NR at 70°F	A to 70°F	C at 70°F	

	STYRENE BUTADIENE SBR	POLYACRYLATE ACM	POLYURETHANE AU, EU	ISOBUTYLENE ISOPRENE IIR	POLYBUTADIENE BR	EPICHLOROHYDRIN CO, ECO
MIL-L-7645	NR at 70°F	B at 70°F	C at 70°F	NR at 70°F	NR at 70°F	B at 70°F
MIL-G-7711A	NR at 70°F	AB to 70°F	A to 70°F	NR at 70°F	NR at 70°F	A to 70°F
MIL-O-7808	NR at 70°F	B at 70°F	NR at 70°F	NR at 70°F	NR at 70°F	
MIL-L-7808C	dissolved 3 days 300°F NR at 300°F	+36% vol 3 days 300°F C/NR at 300°F	softened 3 days 300°F	NR at 70°F	NR at 70°F	
MIL-L-7808F	NR at 70°F	BC at 70°F	NR at 70°F	NR at 70°F	NR at 70°F	B at 70°F
MIL-L-7808H			AB to 70°F			
MIL-L-7870A	NR at 70°F	A to 70°F	NR at 70°F	NR at 70°F	NR at 70°F	A to 70°F
MIL-C-8188C	NR at 70°F	C at 70°F	NR at 70°F	NR at 70°F	NR at 70°F	B at 70°F
MIL-A-8243B	A to 70°F	C at 70°F	C at 70°F	A to 70°F	A to 70°F	B at 70°F
MIL-L-8383B	NR at 70°F	A to 70°F	A to 70°F	NR at 70°F	NR at 70°F	A to 70°F
MIL-H-8446B	NR at 70°F	C at 70°F	A/NR at 70°F	NR at 70°F	NR at 70°F	C at 70°F
MIL-I-8660B	A to 70°F		A to 70°F	A to 70°F	A to 70°F	A to 70°F
MIL-L-9000F	NR at 70°F	AB to 70°F	AC to 70°F	NR at 70°F	NR at 70°F	A to 70°F
MIL-T-9188B	NR at 70°F	NR at 70°F	NR at 70°F	A to 70°F	NR at 70°F	NR at 70°F
MIL-L-9236B	NR at 70°F	BC at 70°F	NR at 70°F	NR at 70°F	C/NR at 70°F	B at 70°F
MIL-E-9500	A to 70°F	NR at 70°F	NR at 70°F	A to 70°F	A to 70°F	
MIL-L-10295A	NR at 70°F	A to 70°F	B at 70°F	NR at 70°F	NR at 70°F	A to 70°F
MIL-L-10324A	NR at 70°F	A to 70°F	B at 70°F	NR at 70°F	NR at 70°F	A to 70°F
MIL-G-10924B	NR at 70°F	AB to 70°F	AB to 70°F	NR at 70°F	NR at 70°F	A to 70°F
MIL-L-11734B	NR at 70°F	C at 70°F	C at 70°F	C at 70°F	NR at 70°F	B at 70°F
MIL-O-11773	NR at 70°F	C at 70°F	C at 70°F	C at 70°F	NR at 70°F	B at 70°F
MIL-P-12098	A to 70°F	B at 70°F	C at 70°F	B at 70°F	A to 70°F	B at 70°F
MIL-H-13862	NR at 70°F	A to 70°F	B at 70°F	NR at 70°F	NR at 70°F	A to 70°F
MIL-H-13866A	NR at 70°F	A to 70°F	B at 70°F	NR at 70°F	NR at 70°F	A to 70°F
MIL-H-13910B	A to 70°F	B/NR at 70°F	NR at 70°F	AB to 70°F	A to 70°F	B at 70°F
MIL-H-13919A	NR at 70°F	A to 70°F	B at 70°F	NR at 70°F	NR at 70°F	A to 70°F
MIL-L-14107B	NR at 70°F			NR at 70°F	NR at 70°F	
MIL-L-15016	NR at 70°F	A to 70°F	A to 70°F	NR at 70°F	NR at 70°F	
MIL-L-15017	NR at 70°F	A to 70°F	A to 70°F	NR at 70°F	NR at 70°F	A to 70°F
MIL-L-15018B	NR at 70°F	A to 70°F	A to 70°F	NR at 70°F	NR at 70°F	A to 70°F
MIL-L-15019A	NR at 70°F	A to 70°F	A to 70°F	NR at 70°F	NR at 70°F	A to 70°F
MIL-L-15719A	B at 70°F	B at 70°F	NR at 70°F	B at 70°F	B at 70°F	B at 70°F
MIL-G-15793	NR at 70°F	AC to 70°F	AC to 70°F	NR at 70°F	NR at 70°F	B at 70°F
MIL-F-16884	NR at 70°F	A to 70°F	C at 70°F	NR at 70°F	NR at 70°F	
MIL-F-16929A	NR at 70°F	C at 70°F	C at 70°F	C at 70°F	NR at 70°F	B at 70°F
MIL-L-16958A	NR at 70°F	A to 70°F	B at 70°F	NR at 70°F	NR at 70°F	A to 70°F
MIL-F-17111	NR at 70°F	A to 70°F	C at 70°F	NR at 70°F	NR at 70°F	A to 70°F
MIL-L-17331D	NR at 70°F	A to 70°F	A to 70°F	NR at 70°F	NR at 70°F	A to 70°F
MIL-L-17353A	NR at 70°F		B at 70°F	NR at 70°F	NR at 70°F	B at 70°F
MIL-L-17672B	NR at 70°F	A to 70°F	A to 70°F	NR at 70°F	NR at 70°F	A to 70°F
MIL-L-18486A	NR at 70°F	A to 70°F	A to 70°F	NR at 70°F	NR at 70°F	A to 70°F
MIL-G-18709A	NR at 70°F	A to 70°F	A to 70°F	NR at 70°F	NR at 70°F	A to 70°F
MIL-H-19457B	NR at 70°F	NR at 70°F	NR at 70°F	A to 70°F	NR at 70°F	NR at 70°F
MIL-F-19605	NR at 70°F		C at 70°F	NR at 70°F	NR at 70°F	A to 70°F

	SYNTHETIC ISOPRENE IR	NATURAL ISOPRENE NR	CHLOROSULFONATED POLYETHYLENE CSM	FLUOROSILICONE FVMQ	SILICONE VMQ	POLYSULFIDE T
MIL-L-7645	NR at 70°F	NR at 70°F	C at 70°F	A to 70°F	NR at 70°F	C at 70°F
MIL-G-7711A	NR at 70°F	NR at 70°F	NR at 70°F	A to 70°F	BC at 70°F	A to 70°F
MIL-O-7808	NR at 70°F	NR at 70°F	NR at 70°F	+10% vol 3 days 300°F A to 300°F	+10-30% vol 3 days 300°F NR at 70°F	B at 70°F
MIL-L-7808C	NR at 70°F	NR at 70°F	NR at 70°F	+8% vol 3 days 300°F A to 300°F	+16% vol 3 days 300°F C at 300°F	B at 70°F
MIL-L-7808F	NR at 70°F	NR at 70°F	NR at 70°F	B at 70°F	NR at 70°F	AB to 70°F
MIL-L-7808H						
MIL-L-7870A	NR at 70°F	NR at 70°F	NR at 70°F	A to 70°F	NR at 70°F	A to 70°F
MIL-C-8188C	NR at 70°F	NR at 70°F	NR at 70°F	B at 70°F	NR at 70°F	B at 70°F
MIL-A-8243B	B at 70°F	B at 70°F	B at 70°F	B at 70°F	B at 70°F	C at 70°F
MIL-L-8383B	NR at 70°F	NR at 70°F	A to 70°F	A to 70°F	C at 70°F	A to 70°F
MIL-H-8446B	NR at 70°F	NR at 70°F	C at 70°F	A to 70°F	NR at 70°F	BC at 70°F
MIL-I-8660B	A to 70°F	A to 70°F	A to 70°F	A to 70°F	NR at 70°F	A to 70°F
MIL-L-9000F	NR at 70°F	NR at 70°F	BC at 70°F	B at 70°F	NR at 70°F	AB to 70°F
MIL-T-9188B	NR at 70°F	NR at 70°F	NR at 70°F	C at 70°F	NR at 70°F	NR at 70°F
MIL-L-9236B	C/NR at 70°F	NR at 70°F	NR at 70°F	B at 70°F	NR at 70°F	AB to 70°F
MIL-E-9500	A to 70°F	A to 70°F	A to 70°F	A to 70°F	A to 70°F	A to 70°F
MIL-L-10295A	NR at 70°F	NR at 70°F	B at 70°F	A to 70°F	C at 70°F	A to 70°F
MIL-L-10324A	NR at 70°F	NR at 70°F	B at 70°F	A to 70°F	C at 70°F	A to 70°F
MIL-G-10924B	NR at 70°F	NR at 70°F	B at 70°F	A to 70°F	NR at 70°F	A to 70°F
MIL-L-11734B	NR at 70°F	NR at 70°F	C at 70°F	A to 70°F	C at 70°F	B at 70°F
MIL-O-11773	NR at 70°F	NR at 70°F	C at 70°F	A to 70°F	C at 70°F	B at 70°F
MIL-P-12098	B at 70°F	B at 70°F	B at 70°F	B at 70°F	NR at 70°F	C at 70°F
MIL-H-13862	NR at 70°F	NR at 70°F	B at 70°F	A to 70°F	C at 70°F	A to 70°F
MIL-H-13866A	NR at 70°F	NR at 70°F	B at 70°F	A to 70°F	C at 70°F	A to 70°F
MIL-H-13910B	AB to 70°F	AB to 70°F	AB to 70°F	AB to 70°F	A/NR at 70°F	AC to 70°F
MIL-H-13919A	NR at 70°F	NR at 70°F	B at 70°F	A to 70°F	C at 70°F	A to 70°F
MIL-L-14107B	NR at 70°F	NR at 70°F		A to 70°F	NR at 70°F	NR at 70°F
MIL-L-15016	NR at 70°F	NR at 70°F	B at 70°F	B at 70°F	NR at 70°F	A to 70°F
MIL-L-15017	NR at 70°F	NR at 70°F	B at 70°F	B at 70°F	NR at 70°F	C at 70°F
MIL-L-15018B	NR at 70°F	NR at 70°F	A to 70°F	A to 70°F	C at 70°F	A to 70°F
MIL-L-15019A	NR at 70°F	NR at 70°F	A to 70°F	A to 70°F	C at 70°F	A to 70°F
MIL-L-15719A	C at 70°F	C at 70°F	B at 70°F	B at 70°F	NR at 70°F	NR at 70°F
MIL-G-15793	NR at 70°F	NR at 70°F	BC at 70°F	B at 70°F	NR at 70°F	AB to 70°F
MIL-F-16884	NR at 70°F	NR at 70°F	C at 70°F	A to 70°F	NR at 70°F	A to 70°F
MIL-F-16929A	NR at 70°F	NR at 70°F	C at 70°F	A to 70°F	C at 70°F	B at 70°F
MIL-L-16958A	NR at 70°F	NR at 70°F	B at 70°F	A to 70°F	C at 70°F	A to 70°F
MIL-F-17111	NR at 70°F	NR at 70°F	NR at 70°F	B at 70°F	NR at 70°F	A to 70°F
MIL-L-17331D	NR at 70°F	NR at 70°F	B at 70°F	A to 70°F	NR at 70°F	A to 70°F
MIL-L-17353A	NR at 70°F	NR at 70°F	C at 70°F	A to 70°F	C at 70°F	B at 70°F
MIL-L-17672B	NR at 70°F	NR at 70°F	A to 70°F	A to 70°F	C at 70°F	A to 70°F
MIL-L-18486A	NR at 70°F	NR at 70°F	A to 70°F	A to 70°F	C at 70°F	A to 70°F
MIL-G-18709A	NR at 70°F	NR at 70°F	A to 70°F	A to 70°F	C at 70°F	A to 70°F
MIL-H-19457B	NR at 70°F	NR at 70°F	NR at 70°F	NR at 70°F	C/NR at 70°F	NR at 70°F
MIL-F-19605	NR at 70°F	NR at 70°F	C at 70°F	A to 70°F	NR at 70°F	A to 70°F

	RECOMMENDATIONS	NITRILE NBR	ETHYLENE PROPYLENE EP, EPDM	FLUOROCARBON FKM	CHLOROPRENE CR	HYDROGENATED NITRILE HNBR
MIL-L-19701		A to 70°F	NR at 70°F	A to 70°F	C at 70°F	
MIL-L-21260	NBR,FKM,ACM,AU	A to 70°F	NR at 70°F	A to 70°F	B at 70°F	
MIL-S-21568A	EP,NBR,FKM,CR	A to 70°F	A to 70°F	A to 70°F	A to 70°F	
MIL-H-22072		A to 70°F	A to 70°F	B at 70°F	B at 70°F	
MIL-H-22251	EP,IIR	B at 70°F	A to 70°F	A to 70°F	B at 70°F	
MIL-L-22396		A to 70°F	NR at 70°F	A to 70°F	A to 70°F	
MIL-L-23699A	FKM	B at 70°F	NR at 70°F	A to 70°F	C at 70°F	
MIL-G-23827A		A to 70°F	NR at 70°F	A to 70°F	C at 70°F	
MIL-G-25013D	FKM,NBR,EP	A to 70°F	NR at 70°F	A to 70°F	B at 70°F	
MIL-F-25172		A to 70°F	NR at 70°F	A to 70°F	C at 70°F	
MIL-L-25336B		A to 70°F	NR at 70°F	A to 70°F	C at 70°F	
MIL-F-25524A		A to 70°F	NR at 70°F	A to 70°F	C at 70°F	
MIL-G-25537A	NBR,FKM,AU	A to 70°F	NR at 70°F	A to 70°F	B at 70°F	
MIL-F-25558B	NBR,FKM,ACM,AU	A to 70°F	NR at 70°F	−1% vol 28 days 75°F A to 70°F	B at 70°F	A to 70°F
MIL-F-25576C	NBR,FKM,ACM,AU	A to 70°F	NR at 70°F	+1% vol 28 days 75°F A to 70°F	BC at 70°F	A to 70°F
MIL-H-25598		A to 70°F	NR at 70°F	A to 70°F	B at 70°F	
MIL-F-25656B	NBR,FKM,FVMQ	A to 70°F AB to 120°F	NR at 70°F	A to 100°F NR at 550°F	NR at 70°F	A to 70°F
MIL-L-25681C	FKM,EP	B at 70°F	A to 70°F	A to 70°F	B at 70°F	
MIL-G-25760A	FKM	B at 70°F	NR at 70°F	A to 70°F	BC at 70°F	
MIL-L-25968		A to 70°F	NR at 70°F	A to 70°F	C at 70°F	
MIL-L-26087A		A to 70°F	NR at 70°F	A to 70°F	A to 70°F	
MIL-G-27343		A to 70°F	A to 70°F	A to 70°F	A to 70°F	
MIL-P-27402	EP,IIR	B at 70°F	A to 70°F		B at 70°F	
MIL-H-27601A	NBR,FKM,ACM	AB to 70°F	NR at 70°F	A to 70°F	B at 70°F	
MIL-G-27617		NR at 70°F	A to 70°F	A to 70°F		
MIL-I-27686D		A to 70°F	A to 70°F	B at 70°F	B at 70°F	
MIL-L-27694A		A to 70°F	A to 70°F	A to 70°F	A to 70°F	
MIL-L-46000A		A to 70°F	NR at 70°F	A to 70°F	C at 70°F	
MIL-H-46001A		A to 70°F	NR at 70°F	A to 70°F	A to 70°F	
MIL-L-46002		A to 70°F	NR at 70°F	A to 70°F	A to 70°F	
MIL-H-46004		A to 70°F	NR at 70°F	A to 70°F	B at 70°F	
MIL-P-46046A		B at 70°F	A to 70°F	A to 70°F	B at 70°F	
MIL-H-81019B		A to 70°F	NR at 70°F	A to 70°F	B at 70°F	
MIL-S-81087	EP,NBR,FKM,CR	A to 70°F	A to 70°F	A to 70°F	A to 70°F	
MIL-H-83282	NBR,FKM	A to 275°F	NR at 70°F	A to 400°F	B at 70°F	

	STYRENE BUTADIENE SBR	POLYACRYLATE ACM	POLYURETHANE AU, EU	ISOBUTYLENE ISOPRENE IIR	POLYBUTADIENE BR	EPICHLOROHYDRIN CO, ECO
MIL-L-19701	NR at 70°F	C at 70°F	C at 70°F	C at 70°F	NR at 70°F	B at 70°F
MIL-L-21260	NR at 70°F	A to 70°F	A to 70°F	NR at 70°F	NR at 70°F	A to 70°F
MIL-S-21568A	A to 70°F	A to 70°F	A to 70°F	A to 70°F	A to 70°F	A to 70°F
MIL-H-22072	A to 70°F	C at 70°F	C at 70°F	A to 70°F	A to 70°F	B at 70°F
MIL-H-22251	B at 70°F			A to 70°F		
MIL-L-22396	NR at 70°F	A to 70°F	A to 70°F	NR at 70°F	NR at 70°F	A to 70°F
MIL-L-23699A	NR at 70°F	C at 70°F	C at 70°F	NR at 70°F	NR at 70°F	B at 70°F
MIL-G-23827A	NR at 70°F	C at 70°F	C at 70°F	C at 70°F	NR at 70°F	B at 70°F
MIL-G-25013D	NR at 70°F	AB to 70°F	AC to 70°F	NR at 70°F	A/NR at 70°F	A to 70°F
MIL-F-25172	NR at 70°F		C at 70°F	NR at 70°F	NR at 70°F	A to 70°F
MIL-L-25336B	NR at 70°F	C at 70°F	C at 70°F	C at 70°F	NR at 70°F	B at 70°F
MIL-F-25524A	NR at 70°F		C at 70°F	NR at 70°F	NR at 70°F	A to 70°F
MIL-G-25537A	NR at 70°F	AB to 70°F	AB to 70°F	NR at 70°F	NR at 70°F	A to 70°F
MIL-F-25558B	NR at 70°F	A to 70°F	AB to 70°F	NR at 70°F	NR at 70°F	A to 70°F
MIL-F-25576C	NR at 70°F	A to 70°F	AC to 70°F	NR at 70°F	NR at 70°F	A to 70°F
MIL-H-25598	NR at 70°F	A to 70°F	B at 70°F	NR at 70°F	NR at 70°F	A to 70°F
MIL-F-25656B	NR at 70°F	B at 70°F	BC at 70°F	NR at 70°F	NR at 70°F	A to 70°F
MIL-L-25681C	B at 70°F	B at 70°F	C at 70°F	A to 70°F	AB to 70°F	A to 70°F
MIL-G-25760A	NR at 70°F	BC at 70°F	B at 70°F	NR at 70°F	C/NR at 70°F	B at 70°F
MIL-L-25968	NR at 70°F	C at 70°F	C at 70°F	C at 70°F	NR at 70°F	B at 70°F
MIL-L-26087A	NR at 70°F	A to 70°F	A to 70°F	NR at 70°F	NR at 70°F	A to 70°F
MIL-G-27343	A to 70°F		A to 70°F	A to 70°F	A to 70°F	A to 70°F
MIL-P-27402	B at 70°F			A to 70°F		
MIL-H-27601A	NR at 70°F	AB to 70°F	C at 70°F	NR at 70°F	NR at 70°F	B at 70°F
MIL-G-27617	B at 70°F			A to 70°F	B at 70°F	
MIL-I-27686D	A to 70°F	C at 70°F	C at 70°F	A to 70°F	A to 70°F	B at 70°F
MIL-L-27694A	A to 70°F		A to 70°F	A to 70°F	A to 70°F	A to 70°F
MIL-L-46000A	NR at 70°F	C at 70°F	C at 70°F	C at 70°F	NR at 70°F	B at 70°F
MIL-H-46001A	NR at 70°F	A to 70°F	A to 70°F	NR at 70°F	NR at 70°F	A to 70°F
MIL-L-46002	NR at 70°F		A to 70°F	NR at 70°F	NR at 70°F	A to 70°F
MIL-H-46004	NR at 70°F	A to 70°F	B at 70°F	NR at 70°F	NR at 70°F	A to 70°F
MIL-P-46046A	A to 70°F	B at 70°F	C at 70°F	B at 70°F	A to 70°F	B at 70°F
MIL-H-81019B	NR at 70°F	A to 70°F	B at 70°F	NR at 70°F	NR at 70°F	A to 70°F
MIL-S-81087	A to 70°F	A to 70°F	A to 70°F	A to 70°F	A to 70°F	A to 70°F
MIL-H-83282	NR at 70°F	B at 70°F	B at 70°F	NR at 70°F	NR at 70°F	

	SYNTHETIC ISOPRENE IR	NATURAL ISOPRENE NR	CHLOROSULFONATED POLYETHYLENE CSM	FLUOROSILICONE FVMQ	SILICONE VMQ	POLYSULFIDE T
MIL-L-19701	NR at 70°F	NR at 70°F	C at 70°F	A to 70°F	C at 70°F	B at 70°F
MIL-L-21260	NR at 70°F	NR at 70°F	B at 70°F	A to 70°F	NR at 70°F	A to 70°F
MIL-S-21568A	AB to 70°F	AB to 70°F	A to 70°F	AB to 70°F	NR at 70°F	A to 70°F
MIL-H-22072	B at 70°F	B at 70°F	B at 70°F	B at 70°F	B at 70°F	C at 70°F
MIL-H-22251			B at 70°F		NR at 70°F	NR at 70°F
MIL-L-22396	NR at 70°F	NR at 70°F	A to 70°F	A to 70°F	C at 70°F	A to 70°F
MIL-L-23699A	NR at 70°F	NR at 70°F	C at 70°F	B at 70°F	NR at 70°F	B at 70°F
MIL-G-23827A	NR at 70°F	NR at 70°F	C at 70°F	A to 70°F	C at 70°F	B at 70°F
MIL-G-25013D	NR at 70°F	NR at 70°F	B at 70°F	AB to 70°F	NR at 70°F	AB to 70°F
MIL-F-25172	NR at 70°F	NR at 70°F	C at 70°F	A to 70°F	NR at 70°F	A to 70°F
MIL-L-25336B	NR at 70°F	NR at 70°F	C at 70°F	A to 70°F	C at 70°F	B at 70°F
MIL-F-25524A	NR at 70°F	NR at 70°F	C at 70°F	A to 70°F	NR at 70°F	A to 70°F
MIL-G-25537A	NR at 70°F	NR at 70°F	B at 70°F	A to 70°F	NR at 70°F	A to 70°F
MIL-F-25558B	NR at 70°F	NR at 70°F	B at 70°F	A to 70°F	NR at 70°F	A to 70°F
MIL-F-25576C	NR at 70°F	NR at 70°F	BC at 70°F	A to 70°F	NR at 70°F	A to 70°F
MIL-H-25598	NR at 70°F	NR at 70°F	B at 70°F	A to 70°F	C at 70°F	A to 70°F
MIL-F-25656B	NR at 70°F	NR at 70°F	NR at 70°F	B at 70°F	NR at 70°F	B at 70°F
MIL-L-25681C	B at 70°F	B at 70°F	B at 70°F	B at 70°F	NR at 70°F	B at 70°F
MIL-G-25760A	C/NR at 70°F	NR at 70°F	BC at 70°F	B at 70°F	NR at 70°F	B at 70°F
MIL-L-25968	NR at 70°F	NR at 70°F	C at 70°F	A to 70°F	C at 70°F	B at 70°F
MIL-L-26087A	NR at 70°F	NR at 70°F	A to 70°F	A to 70°F	C at 70°F	A to 70°F
MIL-G-27343	A to 70°F	A to 70°F	A to 70°F	A to 70°F	NR at 70°F	A to 70°F
MIL-P-27402			B at 70°F		NR at 70°F	NR at 70°F
MIL-H-27601A	NR at 70°F	NR at 70°F	C at 70°F	B at 70°F	C/NR at 70°F	C at 70°F
MIL-G-27617				A to 70°F	NR at 70°F	
MIL-I-27686D	B at 70°F	B at 70°F	B at 70°F	B at 70°F	B at 70°F	C at 70°F
MIL-L-27694A	A to 70°F	A to 70°F	A to 70°F	A to 70°F	NR at 70°F	A to 70°F
MIL-L-46000A	NR at 70°F	NR at 70°F	C at 70°F	A to 70°F	C at 70°F	B at 70°F
MIL-H-46001A	NR at 70°F	NR at 70°F	A to 70°F	A to 70°F	C at 70°F	A to 70°F
MIL-L-46002	NR at 70°F	NR at 70°F	A to 70°F	A to 70°F	C at 70°F	A to 70°F
MIL-H-46004	NR at 70°F	NR at 70°F	B at 70°F	A to 70°F	C at 70°F	A to 70°F
MIL-P-46046A	B at 70°F	B at 70°F	B at 70°F	B at 70°F	NR at 70°F	C at 70°F
MIL-H-81019B	NR at 70°F	NR at 70°F	B at 70°F	A to 70°F	C at 70°F	A to 70°F
MIL-S-81087	A to 70°F	A to 70°F	A to 70°F	AB to 70°F	C/NR at 70°F	B at 70°F
MIL-H-83282	NR at 70°F	NR at 70°F	B at 70°F	A to 70°F	NR at 70°F	A to 70°F

ABBREVIATIONS APPEARING IN THIS PUBLICATION

A.P. = All Purpose

BTU = British Thermal Units

Conc = Concentrated or Concentration

°F = Degrees Fahrenheit

FDA = Federal Drug Administration

F.R. = Fire Resistant

Ft = Feet

Ga Gy = Gamma Grey

gn = Gain

H.P. = High Pressure

Lb = Pound

Liq = Liquid

Lube = Lubricant

M = Molar

Max = Maximum

Neut = Neutral

PLI = Pounds Per Linear Inch

PPM = Parts Per Million

PSI = Pounds Per Square Inch

R.H. = Relative Humidity

Sat'd = Saturated

Sec = Second

Temp = Temperature

T.S. = Tensile Strength

Ult = Ultimate

Vol = Volume

vs = Versus

Wt = Weight

Yld = Yield

	HIGH RANGE / LOW RANGE	NITRILE	ETHYLENE PROPYLENE	FLUOROCARBON	CHLOROPRENE	HYDROGENATED NITRILE
		NBR	EP, EPDM	FKM	CR	HNBR
COMMON NAME		Buna-N, Nitrile	Ethylene Propylene	Fluorinated Hydrocarbon	Chloroprene	Hydrogenated Nitrile
TRADE NAMES		Oilace, Mincar Chemigum, Hycar	Nordel, Epcar, Vistalon Epsyn, Coyaene	Viton, Fluorel, Vi-Chem	Neoprene, Matchless Mirprene	Therban Tornac, Zetpol
CHEMICAL DEFINITION		Butadiene Acrylonitrile	Ethylene Propylene Diene Co & Terpolymer	Vinylidene Fluoride Hexafluoropropylene	Polychloroprene	Hydrogenated Butadiene Acrylonitrile
ASTM D-2000 & SAE J 200		BF BG BK CH	CA BA AA DH	HK	BC BE	DH
ASTM D-735, SAE J-14 MIL-R-3065, (MIL-STD-417)		SB SA	R(N)	TB	SC	
ASTM (D 1418)		NBR	EPM EPDM	FKM	CR	HNBR
DENSITY (gm/cm3)	FVMQ VMQ FKM CO EP EVA IR NR IIR	0.98-1.00	0.86	1.40-1.95	1.23-1.25	1.1-1.3
MELTING TEMPERATURE (°F)	VMQ FFKM AU EU					
MINIMUM CONTINUOUS USE TEMPERATURE (°F)	VMQ FVMQ BR PVA	−40 to −70	−75	−20 to −70	−80	−40
MAXIMUM CONTINUOUS USE TEMPERATURE (°F)	FVMQ VMQ FKM PVA IR PVC	+250 to +300	+400	+550 to +650	+300	+350
STIFFENING TEMPERATURE (°F)	VMQ FVMQ BR EVA NBR ACM FKM FFKM	−40 to +30	−20 to −50	−10 to +10	−50 to +10	
BRITTLE TEMPERATURE (°F)	FVMQ AU EU EVA ACM CO PVC FKM	−40 to −85	−90	−35 to −60	−85	
HOT TENSILE STRENGTH DECREASE (%)	FKKM ACM NBR IIR VMQ T NR PVC	−55 at 212°F −76 at 350°F	−49 at 212°F −78 at 350°F	−72 at 212°F −87 at 350°F	−50 at 212°F −74 at 350°F	
HOT ELONGATION CHANGE (%)	CSM ACM CO ECO NR FVMQ IR BR IIR	−35 at 212°F −60 at 350°F	−21 at 212°F −48 at 350°F	−33 at 212°F −63 at 350°F	−35 at 212°F −45 at 350°F	
STRAIN RELAXATION AT ROOM TEMPERATURE (CREEP)	NR IR SBR BR T CSM PVC	Good	Fair-Good	Good	Good	
STRAIN RELAXATION AT 212°F	VMQ FKM FVMQ AU EU T	Good	Fair-Good	Good-Excellent	Good	
GLASS TRANSITION TEMP. Tg (°F)	BR IR NR IIR FVMQ CSM FKM EA	−13 to −21		−5	−40 to −45	
HEAT AGING AT 212°F	FKM VMQ FVMQ ACM PVC	Good	Good-Outstanding	Outstanding	Good-Excellent	Excellent
AGING IN OXYGEN OZONE & WEATHER	VMQ ACM AU NBR SBR BR	Poor-Fair	Very good	Very good-Excellent	Good	Good
STEAM RESISTANCE	EP IIR FFKM PVA ACM AU EU CM	Fair-Good	Excellent	Poor-Good	Good	Excellent
FLAME RESISTANCE	FKM CR FVMQ CSM NBR EP PVA IR NR	Poor	Poor	Very good-Excellent	Very good-Excellent	Poor
LIMITING OXYGEN INDEX	FKM CR EA CSM CM EP AU EU NBR IIR	17-20	10-20	50-100	38-45	
THERMAL CONDUCTIVITY Btu/hr/sq ft/°F/ft	EP NBR VMQ SBR FKM, CSM IIR PVC	0.143	0.15	0.06-0.13	0.11	
CUBICAL COEFFICIENT OF THERMAL EXPANSION in/in/°F X 10-5	VMQ FVMQ T PVC PVC AU EU CR	23-39	26-32	25-33	18.6-36	
LINEAR COEFFICIENT OF THERMAL EXPANSION in/in/°F X 10-5	VMQ FVMQ T PVC PVC AU EU CR	7.6-13.0	8.8-10.7	8.3-11.0	6.2-12.0	
SHELF LIFE, COOL DRY, NO RADIATION (YEARS)	PVA FKM VMQ VMQ NBR IR NR SBR BR	2-5	5-10	5-20	5-10	
WEATHER RESISTANCE	CSM EA EP FKM VMQ NBR IR NR	Poor-Fair	Excellent	Excellent	Poor-Good	Good-Excellent
SUNLIGHT RESISTANCE	EP CSM CM EA PVA NBR IR NR SBR BR	Poor	Outstanding	Good Outstanding	Good-Very good	
OZONE RESISTANCE	FKM CSM VMQ EA AU IR NR SBR BR	Poor-Fair 1 Hour 150 ppm	Good-Excellent 8 Hhors 150 ppm	Outstanding 2 Weeks 150 ppm	Very good-Excellent 24 Hours 150 ppm	Good-Excellent
RADIATION RESISTANCE	EP AU EU PVC SBR BR CO ECO	Good 1 X10⁵Ga Gy Fair 6 X10⁵Ga Gy	Good 5 X10⁵Ga Gy Fair 2 X10⁶Ga Gy	Good 7 X10⁴Ga Gy Fair 2 X10⁵Ga Gy	Good 1 X10⁵Ga Gy Fair 9 X10⁵Ga Gy	Good
OXIDIZATION RESISTANCE	FKM CSM EA EP CR	Good	Excellent	Outstanding	Excellent	Excellent
AIR	CSM VMQ EVA EP FKM NBR SBR	Fair	Very good	Very good	Good	
FUNGUS RESISTANCE AVAILABLE		Yes	Yes	Yes	Yes	
FDA APPROVED GRADES AVAILABLE		Yes	Yes	Yes	Yes	
STERILIZABLE	VMQ FVMQ AU EU					
WATER RESISTANCE	EP FKM IR NR VMQ PVA AU ACM	Good-Excellent	Excellent	Excellent	Fair-Good	Excellent
WATER ABSORBTION	EVA VMQ EA EP SBR PVA T ACM	Fair-Good	Very good-Excellent	Very good	Good	
GAS PERMEABILITY CO₂ cc·cm X 10-8/cm²·sec·atm	FKM T IIR PVC NBR VMQ FVMQ NR	5.76-14.1 at 25°C 22.4 at 50°C		0.04 at 25°C	13.9-25.8 at 25°C 47.6 at 50°C	
GAS PERMEABILITY H₂ cc·cm X 10-8/cm²·sec·atm	IIR NBR AU EU VMQ CR	5.42-8.97 at 25°C 17-23 at 50°C	29-111 at 38°C	160 at 93°C	180 at 38°C	
GAS PERMEABILITY He cc·cm X 10-8/cm²·sec·atm	CR T FKM AU EU VMQ FVMQ	6.2 at 25°C 14.2-19.6 at 50°C	19.7 at 25°C	0.9-1.27 at 25°C 13.1 at 80°C	0.6-6.5 at 25°C 26 at 50°C	
GAS PERMEABILITY N₂ cc·cm X 10-8/cm²·sec·atm	IIR NBR FKM ECO VMQ FVMQ	0.18-0.50 at 25°C 1.08-6.9 at 50°C	6.4 at 25°C	0.23-0.50 at 25°C 0.98 at 50°C	0.01-2.0 at 25°C 4.3 at 50°C	
GAS PERMEABILITY O₂ cc·cm X 10-8/cm²·sec·atm	AU EU IIR PVC CR VMQ FVMQ	0.73-6.5 at 25°C 3.5-18.9 at 50°C		1.7 at 25°C	1.1-4.0 at 25°C 4.73 at 50°C	
GAS PERMEABILITY RATING	FKM AU EU CO EA VMQ FVMQ	Fair-Good	Fair-Good	Good-Excellent	Fair-Good	Good
ODOR	NR AU EU EVA VMQ T	Good	Good	Good	Fair-Good	

	STYRENE BUTADIENE SBR	POLYACRYLATE ACM	POLYURETHANE AU, EU	ISOBUTYLENE ISOPRENE IIR	POLYBUTADIENE BR	® AEROQUIP AQP
COMMON NAME	Buna-S	Polyacrylate	Polyurethane	Butyl	Butadiene	
TRADE NAMES	Ameripol Pliofles, Stereon	ACM Cyanacryl Hycar Krynac	Adiprene Cyanaprene Formrex Vibrathane	Exxon Butyl, K-Resin Polysar Butyl Ironside		Aeroquip
CHEMICAL DEFINITION	Styrene Butadiene	Copolymer Ethyl & Butyl Acrylate	AU = Polyester Urethane EU = Polyether Urethane	Isobutene Isoprene	Polybutadiene	Modified Polyethylene Elastomer
ASTM D-2000 & SAE J 200	AA BA	DF DH	BG	AA BA	AA	
ASTM D-735, SAE J-14 MIL-R-3065, (MIL-STD-417)	R(S)	TB	SB	R(S)	R(S)	
ASTM (D 1418)	SBR	ACM ANM	AU EU	IIR	BR	
DENSITY (gm/cm^3)	0.94	1.09-1.10	AU = 1.2-1.25 EU = 1.04-1.07	0.92	0.91-0.94	1.3-1.6
MELTING TEMPERATURE (°F)			275-375			
MINIMUM CONTINUOUS USE TEMPERATURE (°F)	-75	-40	-65	-75	-150	-56
MAXIMUM CONTINUOUS USE TEMPERATURE (°F)	+250	+400	+250	+300	+200	+300
STIFFENING TEMPERATURE (°F)	-1 to -50	+10 to +35	-10 to -35	-10 to -40	-30 to -60	
BRITTLE TEMPERATURE (°F)	-89	-14 to -40	-60 to -200	-80	-100	
HOT TENSILE STRENGTH DECREASE (%)	-44 at 212°F -72 at 350°F	-67 at 212°F -82 at 350°F	-56 at 212°F -83 at 350°F	-57 at 212°F -87 at 350°F	-45 at 212°F -62 at 350°F	
HOT ELONGATION CHANGE (%)	-25 at 212°F -47 at 350°F	-45 at 212°F -61 at 350°F	-32 at 212°F -54 at 350°F	-10 at 212°F -32 at 350°F	-7 at 212°F -50 at 350°F	
STRAIN RELAXATION AT ROOM TEMPERATURE (CREEP)	Good-Excellent	Fair	Fair-Excellent	Fair-Good	Good-Excellent	
STRAIN RELAXATION AT 212°F	Good	Fair	Poor	Fair	Good	
GLASS TRANSITION TEMP. Tg (°F)	-80		-87	-98	-151	
HEAT AGING AT 212°F	Good	Excellent	Good	Good-Excellent	Fair-Good	
AGING IN OXYGEN OZONE & WEATHER	Poor	Excellent	Excellent	Very good	Poor	Good to 250°F
STEAM RESISTANCE	Fair-Good	Poor	Poor	Good-Excellent	Fair-Good	Fair
FLAME RESISTANCE	Poor	Poor	Poor-Good	Poor	Poor	
LIMITING OXYGEN INDEX			15-20 Standard 28-32 Flame ret. grades	18-19		
THERMAL CONDUCTIVITY Btu/hr/sq ft/°F/ft	0.143	0.092	0.09-0.10	0.053		
CUBICAL COEFFICIENT OF THERMAL EXPANSION in/in/°F X 10-5	27-37	27-36	17-39	28-32	37.5	
LINEAR COEFFICIENT OF THERMAL EXPANSION in/in/°F X 10-5	9.0-12.3	9.0-12.0	5.6-12.9	9.3-10.7	12.5	
SHELF LIFE, COOL DRY, NO RADIATION (YEARS)	2-5	20	2-15	5-10	2-5	Indefinite
WEATHER RESISTANCE	Fair-Good	Excellent	Excellent	Excellent	Poor-Good	
SUNLIGHT RESISTANCE	Poor	Good-Excellent	Very good-Excellent	Excellent	Poor	
OZONE RESISTANCE	Poor Fails quickly 150 ppm	Good-Excellent 1 Hour 150 ppm	Excellent 1-21 Days 150 ppm	Excellent 7 Days 150 ppm	Poor Fails quickly 150 ppm	
RADIATION RESISTANCE	Good 3 X10^5 Ga Gy Poor 2 X10^6 Ga Gy	Good 1.5 X10^4 Ga Gy Poor 6 X10^5 Ga Gy	Good-Excellent 50 X10^{-7} Rads Gamma	Good 1 X10^4 Ga Gy Poor 3 X10^4 Ga Gy	Poor	
OXIDIZATION RESISTANCE	Fair-Excellent	Excellent	Good-Excellent	Excellent	Good-Excellent	Outstanding
AIR	Fair	Very good	Good	Good		Good
FUNGUS RESISTANCE AVAILABLE	Yes	Poor, supports fungus	Yes, Special Supports fungus, Std.	Yes		
FDA APPROVED GRADES AVAILABLE			Yes			
STERILIZABLE			Autoclave briefly Ethylene oxide, Gamma			
WATER RESISTANCE	Good-Excellent	Poor-Fair	AU = Poor EU = Good	Good-Excellent	Excellent	
WATER ABSORBTION	Very good	Poor	Good-Very good to 70°F Poor at 212°F	Very good		
GAS PERMEABLITY CO$_2$ cc·cm X 10-8/cm²·sec·atm	92.8 at 25°C		5-30 at 25°C	3.92-5.18 at 25°C 14.3 at 50°C	35-105 at 25°C 197-200 at 50°C	
GAS PERMEABILITY H$_2$ cc·cm X 10-8/cm²·sec·atm	30.1 at 25°C 46 at 38°C	49.6 at 38°C	5.0 at 38°C	5.5 at 25°C 17.2 at 50°C	32 at 25°C 77 at 50°C	
GAS PERMEABILITY He cc·cm X 10-8/cm²·sec·atm	17.3 at 25°C	16.3 at 25°C	3.6 at 25°C 33.5 at 80°C	6.4 at 25°C 17.3 at 50°C	11.8 at 25°C	
GAS PERMEABILITY N$_2$ cc·cm X 10-8/cm²·sec·atm	4.7 at 25°C	0.9-1.9 at 25°C	1.3 at 25°C	0.24 at 25°C 1.27 at 50°C	3.0-4.9 at 25°C 14.5 at 50°C	
GAS PERMEABILITY O$_2$ cc·cm X 10-8/cm²·sec·atm	12.8 at 25°C		0.8 at 25°C 1.3-4.0 at 32°C	0.99-4.0 at 25°C 4.0 at 50°C	8.4-14.5 at 25°C 36.0 at 50°C	
GAS PERMEABILITY RATING	Fair	Good-Excellent	Good-Excellent	Good	Good	
ODOR	Good	Fair-Good	Excellent (none)	Good	Good	

	SYNTHETIC ISOPRENE IR	NATURAL ISOPRENE NR	CHLOROSULFONATED POLYETHYLENE CSM	FLUOROSILICONE FVMQ	SILICONE VMQ	CHEMRAZ FFKM
COMMON NAME	Synthetic Natural Rubber	Natural Rubber	Chlorosulphonated Polyethylene	Fluorosilicone	Silicone	
TRADE NAMES	Latex	Qualatex Ametax Duro Micro	Hypalon Chlorosol CSM	Fluorosilicone LS	Thermoflex Silastic	Chemraz
CHEMICAL DEFINITION	Polyisoprene	Polyisoprene	Chlorosulphonated Polyethylene	Fluorovinyl Methyl siloxane	Polydimethylsiloxane	Perfluoroelastomer
ASTM D-2000 & SAE J 200	AA	AA	CE DE	FK	FC FE GE	
ASTM D-735, SAE J-14 MIL-R-3065, (MIL-STD-417)	R(S)	R(N)	SC	TA TB	TA	
ASTM (D 1418)	IR	NR	CSM	FVMQ FVS	MQ VMQ PMQ PVMQ PSi PVSi Si VSi	FFKM
DENSITY (gm/cm³)	0.91-0.93	0.92-0.95	1.08-1.28	1.13-2.20	1.07-2.05	2.0
MELTING TEMPERATURE (°F)						
MINIMUM CONTINUOUS USE TEMPERATURE (°F)	-60	-76	-65	-90 to -112	-178	-20
MAXIMUM CONTINUOUS USE TEMPERATURE (°F)	+180	+250	+275	+450	+600	+450
STIFFENING TEMPERATURE (°F)	-20 to -50	-20 to -65	-30 to -50	-70 to -75	-60 to -180	
BRITTLE TEMPERATURE (°F)	-80	-80	-70	-85 to -90	-90 to -180	
HOT TENSILE STRENGTH DECREASE (%)	-44 at 212°F -90 at 350°F	-32 at 212°F -84 at 350°F	-57 at 212°F -82 at 350°F	-56 at 212°F -60 at 350°F	-12 at 212°F -41 at 350°F	+9 at 450°F
HOT ELONGATION CHANGE (%)	+6 at 212°F -26 at 350°F	+17 at 212°F -33 at 350°F	-49 at 212°F -60 at 350°F	+9 at 212°F -17 at 350°F	-29 at 212°F -48 at 350°F	+40 at 450°F
STRAIN RELAXATION AT ROOM TEMPERATURE (CREEP)	Good-Excellent	Excellent	Fair	Good	Fair-Excellent	
STRAIN RELAXATION AT 212°F	Good	Good	Fair	Good-Excellent	Excellent	
GLASS TRANSITION TEMP. Tg (°F)	-99	-99	+1	-90		2
HEAT AGING AT 212°F	Fair-Good	Fair-Good	Good-Excellent	Excellent-Outstanding	Excellent-Outstanding	Excellent
AGING IN OXYGEN OZONE & WEATHER	Fair	Fair	Very good	Excellent	Excellent	Good-Excellent
STEAM RESISTANCE	Fair-Good	Fair-Good	Poor-Good	Fair-Good	Fair-Good	Good-Excellent
FLAME RESISTANCE	Poor	Poor	Good-Excellent	Excellent	Fair-Excellent	Excellent
LIMITING OXYGEN INDEX			30-36		20-30	95+
THERMAL CONDUCTIVITY Btu/hr/sq ft/°F/ft	0.082	0.082	0.065	0.104-0.130	0.09-0.18	0.19
CUBICAL COEFFICIENT OF THERMAL EXPANSION in/in/°F X 10⁻⁵	37	37	27-38.7	26.1-45.0	30.0-56.4	22.5
LINEAR COEFFICIENT OF THERMAL EXPANSION in/in/°F X 10⁻⁵	12.0	12.0	12.9	8.7-14.0	10.0-18.8	7.5
SHELF LIFE, COOL DRY, NO RADIATION (YEARS)	2-5	2-5	5-10	20	20	
WEATHER RESISTANCE	Poor-Fair	Poor-Fair	Excellent-Outstanding	Excellent	Excellent	Excellent
SUNLIGHT RESISTANCE	Poor	Poor	Excellent-Outstanding	Excellent	Excellent	
OZONE RESISTANCE	Poor Fails quickly 150 ppm	Poor Fails quickly 150 ppm	Excellent-Outstanding 2 Weeks 150 ppm	Excellent	Excellent-Outstanding 2 Weeks 150 ppm	Excellent
RADIATION RESISTANCE	Fair-Good	Fair-Good	Good 1 X10⁵ Ga Gy Poor 7 X10⁵ Ga Gy	Fair-Excellent	Good 7 X10⁴ Ga Gy Poor 2 X10⁵ Ga Gy	Very Good
OXIDIZATION RESISTANCE	Good	Good	Excellent-Outstanding	Excellent	Excellent	Excellent
AIR	Good	Good	Excellent	Very good	Excellent	Excellent
FUNGUS RESISTANCE AVAILABLE				Yes	Yes	
FDA APPROVED GRADES AVAILABLE					Yes	
STERILIZABLE				Steam, Dry heat, Ethylene oxide	Autoclave, Boiling, E.O. Dry heat, Steam, Gamma	
WATER RESISTANCE	Excellent	Excellent	Good	Excellent	Excellent	Good-Excellent
WATER ABSORBTION		Very good	Good-Very good		Good 0.01% 24 Hours	
GAS PERMEABILITY CO₂ cc·cm X 10⁻⁸/cm²·sec·atm		99.6-153 at 25°C 218-221 at 50°C		514 at 25°C	1028-1530 at 25°C 1043-1538 at 50°C	
GAS PERMEABILITY H₂ cc·cm X 10⁻⁸/cm²·sec·atm	37.9 at 25°C 89 at 50°C	37.4 at 25°C 90.8 at 50°C			188-500 at 25°C 1010 at 50°C	53 at 25°C
GAS PERMEABILITY He cc·cm X 10⁻⁸/cm²·sec·atm		17-32 at 25°C 52.3 at 50°C		143 at 25°C 461 at 80°C	238-263 at 25°C 560 at 80°C	108 at 25°C
GAS PERMEABILITY N₂ cc·cm X 10⁻⁸/cm²·sec·atm	5.4-6.1 at 25°C 17.0 at 50°C	6.1-9.0 at 25°C 19.4 at 50°C	0.7-0.9 at 25°C	33-40 at 25°C	75-210 at 25°C 240 at 50°C	6.7 at 25°C
GAS PERMEABILITY O₂ cc·cm X 10⁻⁸/cm²·sec·atm		13.0-23.3 at 25°C 46-47 at 50°C		78-82 at 25°C	195-605 at 25°C 257 at 50°C	16 at 25°C
GAS PERMEABILITY RATING	Good	Fair-Good	Good-Excellent	Poor-Good	Poor-Fair	Good
ODOR	Good	Good-Excellent	Good	Good	Good	

	POLYSULFIDE T	CHLORINATED POLYETHYLENE CM	EPICHLOROHYDRIN CO, ECO	® HYTREL COPOLYESTER TPE	® SANTOPRENE COPOLYMER TPO	® C-FLEX STYRENIC TPE
COMMON NAME	Polysulfide	Chlorinated Polyethylene	Epichlorohydrin	Co Polyester	Copolymer TPO	Block Copolymer
TRADE NAMES	Thiokol ST FA	CPE	Hydrin Herchlor	Hytrel	Santoprene	C-Flex
CHEMICAL DEFINITION	Polysulfide	Chlorinated Polyethylene	Epichlorohydrin		EPDM-Polypropylene T-plastic Vulcanate	Styrene + Butadiene Butylene or Isoprene
ASTM D-2000 & SAE J 200	AK BK	BC BE CE	CH DJ DK		AA BA BC BE BF	
ASTM D-735, SAE J-14 MIL-R-3065, (MIL-STD-417)	SA	SB SC	SA SB			
ASTM (D 1418)	T PTR	CM CPE	CO ECO			
DENSITY (gm/cm³)	1.34	1.16-1.32	CO = 1.36-1.49 ECO = 1.27	1.17-1.25	0.94-0.98	0.90-1.20
MELTING TEMPERATURE (°F)				410-465	380-400	
MINIMUM CONTINUOUS USE TEMPERATURE (°F)	-50 to -75	-60	-15 to -80	-90	-40 to -81	-50
MAXIMUM CONTINUOUS USE TEMPERATURE (°F)	+250	+300	+225 to +325	+250 to +300	+275 to +300	+300
STIFFENING TEMPERATURE (°F)	-10 to -45		-15 to -40	-40 to -65		
BRITTLE TEMPERATURE (°F)	-65 to -68	-40	-10 to -50	-95	-29 to -81	-100
HOT TENSILE STRENGTH DECREASE (%)	-26 at 250°F -44 at 350°F		-45 at 212°F -67 at 350°F		-1 to -8 at 257°F -5 to -22 at 350°F	
HOT ELONGATION CHANGE (%)	-34 at 250°F -38 at 350°F		-45 at 212°F -60 at 350°F		-2 to -14 at 257°F -2 to -22 at 350°F	
STRAIN RELAXATION AT ROOM TEMPERATURE (CREEP)	Poor		Good	Good		
STRAIN RELAXATION AT 212°F	Poor		Fair-Good			
GLASS TRANSITION TEMP. Tg (°F)	-75		-50			
HEAT AGING AT 212°F	Fair-Good	Good-Very good	Good-Excellent	Good	Excellent	Good
AGING IN OXYGEN OZONE & WEATHER	Excellent		Very good	Good, with additives		Excellent
STEAM RESISTANCE	Poor	Poor	Fair-Good	Poor-Fair		TEST
FLAME RESISTANCE	Poor	Good	Poor-Good	Good, with additives	Poor-Good	
LIMITING OXYGEN INDEX		30-35	25-33	26-32		
THERMAL CONDUCTIVITY Btu/hr/sq ft/°F/ft				0.089-0.109		0.0873
CUBICAL COEFFICIENT OF THERMAL EXPANSION in/in/°F X 10-5	42		36	15-31.5		11.1-22.4
LINEAR COEFFICIENT OF THERMAL EXPANSION in/in/°F X 10-5	14		12	5-10.5		3.7-7.78
SHELF LIFE, COOL DRY, NO RADIATION (YEARS)	20		5-10		Several	Excellent
WEATHER RESISTANCE	Excellent	Excellent	Good-Excellent	Good-Excellent	Good-Excellent	
SUNLIGHT RESISTANCE	Good	Outstanding	Good	Very good, with additives	Good-Excellent	
OZONE RESISTANCE	Good-Excellent 8 Hours 150 ppm	Excellent	Good-Excellent	Excellent	Excellent 70 Hours 50 ppm	
RADIATION RESISTANCE	Fair	Excellent	Poor			
OXIDIZATION RESISTANCE	Good-Excellent	Excellent	Good-Excellent	Excellent		
AIR	Good		Good	Good	Excellent	Good
FUNGUS RESISTANCE AVAILABLE	Yes			Yes		
FDA APPROVED GRADES AVAILABLE				Yes	Yes	Yes
STERILIZABLE				Ethylene oxide		Steam autoclave, ethylene oxide, gamma
WATER RESISTANCE	Good	Good-Excellent	Good	Good	Good	Good
WATER ABSORBTION	Poor	Very good	Good	Very good to 212°F 0.3-1.6% 24 Hrs .125 in.		Very good 0.01-0.4% 24 Hrs .125 in.
GAS PERMEABILITY CO₂ cc·cm X 10-8/cm²·sec·atm	2.37 at 25°C					
GAS PERMEABILITY H₂ cc·cm X 10-8/cm²·sec·atm						
GAS PERMEABILITY He cc·cm X 10-8/cm²·sec·atm	1.2 at 25°C					
GAS PERMEABILITY N₂ cc·cm X 10-8/cm²·sec·atm			CO = 0.17 at 25°C ECO = 0.66 at 25°C			
GAS PERMEABILITY O₂ cc·cm X 10-8/cm²·sec·atm	0.2-6.0 at 25°C					
GAS PERMEABILITY RATING	Good	Good-Excellent	Excellent	Fair-Good	Fair	Fair
ODOR	Poor		Good			

	ETHYLENE ACRYLIC EA	POLYALLOMER LINEAR COPOLYMER	NYLON 11 POLYAMIDE	NYLON 12 POLYAMIDE	ETHYLENE VINYL ACETATE EVA	POLYVINYLCHLOR FLEXIBLE PVC
COMMON NAME	Ethylene Acrylic	Polyallomer	Nylon 11	Nylon 12	Ethylene Vinyl Acetate	PVC, Plasticized Vi
TRADE NAMES	Vamac	Impolene	Grilamide Rilsan Vecton	Grilamide Rilsan	EVA	Geon Tygon
CHEMICAL DEFINITION	Ethylene/Acrylic	Ethylene/Propylene Copolymer	Polyamide	Polyamide	Copolymer 5-50% Vinyl Acetate & Ethylene	Polyvinylchloride
ASTM D-2000 & SAE J 200	CH EF					
ASTM D-735, SAE J-14 MIL-R-3065, (MIL-STD-417)						
ASTM (D 1418)	EA					
DENSITY (gm/cm³)	1.03-1.12	0.89-0.90	1.04-1.11	1.04-1.06	0.922-0.943	1.16-1.40
MELTING TEMPERATURE (°F)		301	320-396	320-390	300 Standard 350 Special	290-450
MINIMUM CONTINUOUS USE TEMPERATURE (°F)	-30 to -55	-40	-65		-50 to -60	-30 to -40
MAXIMUM CONTINUOUS USE TEMPERATURE (°F)	+350 to +400	+250 to +266	+190 to +230	+212	+180 Standard +350 Crosslinked	+185 to +221
STIFFENING TEMPERATURE (°F)	-41 to -49		-65		-29 to -130	-40
BRITTLE TEMPERATURE (°F)	-65	-40	-94	-76	-148 to -159	-12 to -50
HOT TENSILE STRENGTH DECREASE (%)						
HOT ELONGATION CHANGE (%)						-35 at 266°F
STRAIN RELAXATION AT ROOM TEMPERATURE (CREEP)	Good		Good			Poor
STRAIN RELAXATION AT 212°F	Good			Excellent		
GLASS TRANSITION TEMP. Tg (°F)	-10					
HEAT AGING AT 212°F	Excellent		Good	Good to 212°F with additives	Execllent	Poor
AGING IN OXYGEN OZONE & WEATHER	Excellent		Excellent	Good	Excellent	Excellent
STEAM RESISTANCE	Poor		Poor	Poor	Fair	Poor (30 min. max
FLAME RESISTANCE	Poor		Poor Burns to self ext.	Poor Burns HB to V2	Poor Slow burn	Poor Burns to self ext
LIMITING OXYGEN INDEX	48			26		23-26
THERMAL CONDUCTIVITY Btu/hr/sq ft/°F/ft		0.0485-0.097	0.167-0.192	0.126-0.177		0.073-0.097
CUBICAL COEFFICIENT OF THERMAL EXPANSION in/in/°F X 10-5			15.6-20.1	18.3-21.6	28.8-36.0	11.7-41.7
LINEAR COEFFICIENT OF THERMAL EXPANSION in/in/°F X 10-5			5.2-6.7	6.1-7.2	9.6-12.0	3.9-13.9
SHELF LIFE, COOL DRY, NO RADIATION (YEARS)			Several		Excellent	5-10
WEATHER RESISTANCE	Excellent-Outstanding		Excellent	Good	Excellent	Fair-Good
SUNLIGHT RESISTANCE	Outstanding		Fair with additives	Fair stabilized	Good-Excellent	Fair
OZONE RESISTANCE	Excellent-Outstanding	Good-Excellent	Fair at 70°F Poor at 100°F	Fair, 0.5 ppm	Excellent	Fair-Good
RADIATION RESISTANCE	Good	Good 1 X10⁵ Ga Gy Poor 4 X10⁵ Ga Gy	Good 1 X10⁵ Ga Gy Poor 9 X10⁵ Ga Gy	Good 1 X10⁵ Ga Gy Poor 9 X10⁵ Ga Gy		Fair 9 X10⁵ Ga Gy Poor 6 X10⁶ Ga Gy
OXIDIZATION RESISTANCE	Excellent-Outstanding	Fair	Poor	Poor	Excellent	Excellent
AIR			Excellent	Excellent	Excellent	Good
FUNGUS RESISTANCE AVAILABLE			Yes			Yes
FDA APPROVED GRADES AVAILABLE			Yes	Yes	Yes	Yes
STERILIZABLE		Autoclave, Gas, Ethylene oxide briefly	Autoclave, Ethylene oxide	Gamma, Steam, Ethylene oxide	Autoclave (special) Dry steam heat to 350°F	Autoclave *15 ps *Ethylene oxide *Ga
WATER RESISTANCE	Good-Excellent	Good	Good cold Slow attack hot	Good to boiling	Good-Excellent	Excellent
WATER ABSORBTION	Very good to 212°F	0.015% 24 Hr .125 in.	1.8-1.9% Saturated 2.0% 100 Hr 100% R.H.	0.15-0.90% Saturated 0.25-0.30% 24 Hr	0.005-0.14% 24 Hr	0.13-0.50% 48 Hr Negligible 24 Hr
GAS PERMEABILITY CO₂ cc·cm X 10-8/cm²·sec·atm		2.8 at 30°C	0.69-1.51 at 25°C	0.69-1.51 at 25°C	27.1 at 25°C	0.09-2.7 at 25°C
GAS PERMEABILITY H₂ cc·cm X 10-8/cm²·sec·atm			1.45 at 25°C	1.45 at 25°C		
GAS PERMEABILITY He cc·cm X 10-8/cm²·sec·atm						
GAS PERMEABILITY N₂ cc·cm X 10-8/cm²·sec·atm		0.17 at 30°C	0.015-0.081 at 25°C	0.015-0.081 at 25°C	1.81 at 25°C	
GAS PERMEABILITY O₂ cc·cm X 10-8/cm²·sec·atm		0.84 at 30°C	0.15-0.41 at 25°C	0.15-0.41 at 25°C	3.79 at 25°C	0.01-2.7 at 25°C
GAS PERMEABILITY RATING	Excellent	Good	Excellent	Excellent	Poor-Fair	Fair-Good
ODOR		Excellent (none)	Excellent		Excellent (none)	Good-Excellent (none to slight)

* Some Grades

	POLYETHYLENE LOW DENSITY LDPE	® TEFLON FEP	® KALREZ PERFLUORINATED ELASTOMER	® FLUORAZ FLUORINATED COPOLYMER	® AFLAS FLUORINATED COPOLYMER	® NORPRENE COPOLYMER TPO
COMMON NAME	Polyethylene	FEP		TFE/Propylene	TFE/Propylene	
TRADE NAMES	LDPE Type I	Teflon	Kalrez	Fluoraz	Aflas	Norprene
CHEMICAL DEFINITION	Low Density Polyethylene	Fluorinated Ethylene Propylene	Perfluoroelastomer	Polytetrafluoroethylene Propylene copolymer	Polytetrafluoroethylene Propylene Copolymer	Copolymer Thermoplastic Olefin
ASTM D-2000 & SAE J 200					HK	AA BA BC BE CA CE
ASTM D-735, SAE J-14 MIL-R-3065, (MIL-STD-417)						
ASTM (D 1418)			FFKM	FXM	FKM	
DENSITY (gm/cm³)	0.91-0.92	2.12-2.17	2.01-2.02	1.61-1.89	1.55-1.86	0.98
MELTING TEMPERATURE (°F)	250-400	525-563				
MINIMUM CONTINUOUS USE TEMPERATURE (°F)	-40 to -60	-148	-36	-50	-20	-60
MAXIMUM CONTINUOUS USE TEMPERATURE (°F)	+150 to +176	+500	+554 to +600	+450 to +500	+400 to +446	+275
STIFFENING TEMPERATURE (°F)	-29		+5	-10 to +32		
BRITTLE TEMPERATURE (°F)	-80 to -148	-454	-29 to -42	-35 to -42	-45 to -55	-75
HOT TENSILE STRENGTH DECREASE (%)			-8 at 450°F	-9.6 at 212°F -14.0 at 300°F		
HOT ELONGATION CHANGE (%)			+83 at 450°F	-6.3 at 212°F -11 at 300°F		
STRAIN RELAXATION AT ROOM TEMPERATURE (CREEP)	Poor	Poor	Good			
STRAIN RELAXATION AT 212°F	Poor					
GLASS TRANSITION TEMP. Tg (°F)						
HEAT AGING AT 212°F	Poor	Outstanding	Excellent	Excellent	Excellent	Good
AGING IN OXYGEN OZONE & WEATHER	Excellent, Dark	Outstanding	Excellent	Excellent	Excellent	Good
STEAM RESISTANCE	Poor-Fair	Excellent	Excellent	Good	Good-Excellent	Good-Excellent
FLAME RESISTANCE	Poor	Outstanding		Excellent		
LIMITING OXYGEN INDEX	17.4 Standard 25 Flame Ret. Grades	95+				
THERMAL CONDUCTIVITY Btu/hr/sq ft/°F/ft	0.194	0.116				
CUBICAL COEFFICIENT OF THERMAL EXPANSION in/in/°F X 10⁻⁵	16.8-36.6	13.8-17.4	39.0			
LINEAR COEFFICIENT OF THERMAL EXPANSION in/in/°F X 10⁻⁵	5.6-12.2	4.6-5.8	13.0			
SHELF LIFE, COOL DRY, NO RADIATION (YEARS)		Indefinite		20	Excellent	
WEATHER RESISTANCE		Excellent		Excellent	Excellent	
SUNLIGHT RESISTANCE	Fair, with carbon black	Outstanding	Good			
OZONE RESISTANCE	Good-Excellent	Outstanding	Excellent	Excellent	Excellent	Good
RADIATION RESISTANCE		Good 1 X10³ Ga Gy Poor 2 X10³ Ga Gy	Good 1 X10⁵ Ga Gy Poor 6 X10⁵ Ga Gy		Good 50M Rads Gamma Fair 200M Rads Gamma	
OXIDIZATION RESISTANCE	Fair	Outstanding	Excellent		Good-Excellent	
AIR	Good	Excellent	Excellent	Excellent	Excellent	Good-Excellent
FUNGUS RESISTANCE AVAILABLE	Yes	Inherent				
FDA APPROVED GRADES AVAILABLE	Yes	Yes				Yes
STERILIZABLE	Ethylene oxide, Briefly	Autoclave, Hot air, Steam, Ethylene oxide	Ethylene oxide	Steam	Steam	Steam
WATER RESISTANCE	Good	Outstanding	Excellent	Good-Excellent	Excellent	Good-Excellent
WATER ABSORBTION	Good 0.001% 24 Hr .125 in.	Good 0.001% 24 Hr .125 in.				
GAS PERMEABILITY CO₂ cc·cm X 10⁻⁸/cm²·sec·atm	0.84 at 30°C	7.51 at 25°C				
GAS PERMEABILITY H₂ cc·cm X 10⁻⁸/cm²·sec·atm		9.93 at 25°C 24.7 at 50°C				
GAS PERMEABILITY He cc·cm X 10⁻⁸/cm²·sec·atm		30 at 25°C 58-128 at 50°C				
GAS PERMEABILITY N₂ cc·cm X 10⁻⁸/cm²·sec·atm	0.007 at 30°C 0.22 at 50°C	1.44 at 25°C 4.4 at 50°C				
GAS PERMEABILITY O₂ cc·cm X 10⁻⁸/cm²·sec·atm	0.27 at 30°C	3.37 at 25°C 9.22 at 50°C				
GAS PERMEABILITY RATING	Excellent	Excellent thick Moderate thin		Good	Good	
ODOR	Excellent (none)	Excellent (none)			Excellent (none)	

C413

	HIGH RANGE / LOW RANGE	NITRILE NBR	ETHYLENE PROPYLENE EP, EPDM	FLUOROCARBON FKM	CHLOROPRENE CR	HYDROGENATED NITRILE HNBR
COST ($/pound)	SBR BR IIR NR FVMQ FKM VMQ	1.25-2.07	0.75-1.00	16.50-19.50	1.15-1.32	
TASTE	EP VMQ FVMQ CO T	Fair-Good	Good	Fair-Good	Fair-Good	
ACIDS, DILUTE	EP CR FKM CSM PVC PVA	Good	Excellent	Good-Excellent	Excellent	Excellent
ACIDS, CONCENTRATED	EP CSM CR AU EU ACM EA	Fair-Good	Excellent	Good-Excellent	Poor	Fair-Good
ACID, INORGANIC	FKM CR CSM IIR PVA ACM AU EU	Good Dilute Fair Conc.	Good Conc.	Excellent Dilute Excellent Conc.	Excellent Dilute Good Conc.	Good Dilute Fair-Good Conc.
ACID, ORGANIC	CSM T EA PVC IIR ACM AU EU	Good Dilute Poor Conc.	Excellent Dilute Fair-Good Conc.	Poor-Good Conc.	Excellent Dilute Poor-Good Conc.	Good Dilute Fair-Good Conc.
ACID, OXIDIZING	FKM PVC IR NR NBR CR SBR	Poor Dilute Poor Conc.	Good Dilute Poor Conc.	Excellent Dilute Good Conc.	Fair Dilute Poor Conc.	
ALCOHOLS	CR EP CSM NR IIR ACM	Fair-Good	Good-Excellent	Fair-Excellent	Excellent	Excellent
ALDEHYDES	EP IR NR VMQ IIR BR FKM FVMQ ACM AU EU	Poor-Fair	Very good	Poor	Poor-Fair	Fair-Good
ALKALIES, DILUTE	EP FVMQ CM EA PVC VMQ	Good	Excellent	Good	Good	Excellent
ALKALIES, CONCENTRATED	EP CSM IIR CM EVA FKM CR AU EU EA	Poor-Good	Excellent	Poor	Poor	Poor-Good
AMINES	VMQ EP FVMQ NBR EP IR CSM	Poor	Fair-Very good	Poor	Poor-Very good	Good
ANIMAL & VEGETABLE OILS	FKM FVMQ T CO ECO CM EVA	Good-Excellent	Good	Excellent	Good	Excellent
BRAKE FLUID, NON PETRO BASED	FVMQ VMQ EP NBR ACM AU EU T	Poor	Good-Excellent	Fair	Fair	Fair
CHLORINATED HYDROCARBONS	FKM FVMQ EP CR IR NR SBR ACM	Fair-Good	Poor	Good-Excellent	Poor	Poor
DIESTER OILS	FKM FVMQ EP CR IR NR CSM SBR	Fair-Good	Poor	Good-Excellent	Poor	Good
ESTERS, ALKYL PHOSPHATE	EP VMQ IIR NBR FKM CR IR NR	Poor	Excellent	Poor	Poor	Poor
ESTERS, ARYL PHOSPHATE	IIR EP FKM FVMQ IR NR SBR ACM AU	Poor-Fair	Excellent	Excellent	Poor-Fair	Poor-Fair
ETHERS	T NBR FKM CR IR NR	Poor	Fair	Poor	Poor	Poor-Fair
HALOGENATED SOLVENTS	FKM FVMQ NBR EP CR IR NR	Poor	Poor	Good-Excellent	Poor	Poor-Fair
HYDROCARBONS, HALOGENATED	FKM PVA FVMQ EP CR IR NR VMQ	Poor-Fair	Poor	Excellent	Poor	Poor
HYDROCARBONS, ALIPHATIC	NBR FKM PVA FVMQ EP IR NR SBR IIR BR	Good-Excellent	Poor	Excellent	Poor-Good	Good-Excellent
HYDROCARBONS, AROMATIC	FKM PVA FVMQ CO EP IR NR VMQ SBR	Fair-Good	Poor	Excellent	Poor-Fair	Poor-Fair
HYDROCARBON OIL, HIGH ANILINE	NBR FKM ACM AU CO EP SBR IIR BR	Excellent	Poor	Excellent	Good	Excellent
HYDROCARBON OIL, LOW ANILINE	FKM ACM AU CO T EP SBR IIR BR	Very good	Poor	Excellent	Poor-Fair	Good-Excellent
KETONES	EP T NBR FKM ACM AU EA	Poor	Good-Excellent	Poor	Poor-Fair	Poor
LACQUER SOLVENTS	T EVA EP FKM CR SBR ACM	Fair	Poor	Poor	Poor	Fair
L.P. GASSES & FUEL OILS	NBR FKM T CO CM EP SBR IIR BR EA	Excellent	Poor	Excellent	Good	Excellent
MINERAL OIL, LOW ANILINE	NBR FKM PVA ACM EP SBR IIR BR	Excellent	Poor	Excellent	Fair-Good	
MINERAL OIL, HIGH ANILINE	NBR FKM ACM AU CO EP PVA SBR IIR BR	Excellent	Poor	Excellent	Good-Excellent	
PETROLEUM AROMATIC	FKM PVA T EP SBR IIR BR CO	Good	Poor	Excellent	Good	Good-Excellent
PETROLEUM NON AROMATIC	NBR FKM PVA EP SBR IIR BR CO	Excellent	Poor	Excellent	Good	
PHOSPHATE ESTERS	EP PVA IIR FKM NBR CR ACM AU CO	Poor	Excellent	Good	Poor	
REFRIGERANT AMMONIA	CR EVA NBR EP SBR FKM AU EU CO	Good	Good	Poor	Excellent	Good Cold Poor Hot
REFRIGERANT HALOFLUOROCARBONS	CR ACM T NBR EP	R-11, 12, 13	R-12, 13, 22	R-11, 12, 13	R-11, 12, 13, 21, 22	R-11, 12, 112, 113
REFRIGERANT HALOFLUOROCARBONS WITH OIL	ACM T CR FKM CR EP SBR IIR BR EA	R-11, 12	Poor	R-11, 12	R-11, 12, 22	R-12, 112
REFRIGERANT METHYL CHLORIDE	FKM EA NBR EP CR SBR ACM	Poor	Poor	Good	Poor	Poor
SILICATE ESTERS	FKM NBR EP CO SBR ACM CM PVC EVA	Good	Good-Excellent	Excellent	Poor-Good	
SILICONE OILS	EP FKM ACM AU T CO SBR BR IIR	Good	Excellent	Excellent	Fair-Excellent	Good-Excellent
ADHESION TO FABRICS	CR PVA FKM ACM AU T	Good	Good	Good-Excellent	Excellent	
ADHESION TO METALS	NBR CR PVA SBR AU CM	Excellent	Good-Excellent	Good-Excellent	Excellent	
ADHESION TO RIGID MATERIALS	NBR CR SBR BR IR	Good-Excellent	Fair-Good	Fair-Good	Good-Excellent	
COLORABILITY	NBR AU EU PVC CM CR T	Excellent	Good-Excellent	Very good	Fair	Good
VULCANIZING PROPERTIES	NBR CR SBR AU IR T	Excellent	Very good-Excellent	Good-Excellent	Excellent	

C414

	STYRENE BUTADIENE SBR	POLYACRYLATE ACM	POLYURETHANE AU, EU	ISOBUTYLENE ISOPRENE IIR	POLYBUTADIENE BR	® AEROQUIP AQP
COST ($/pound)	0.36-0.68	1.26-2.78	AU = 2.02-2.58 EU = 1.50-3.75	0.75-0.90	0.38-0.57	
TASTE	Fair-Good	Fair-Good	Good (none)	Fair-Good	Fair-Good	
ACIDS, DILUTE	Fair-Good	Fair	Fair-Good	Good-Excellent	Fair-Good	Poor
ACIDS, CONCENTRATED	Poor-Good	Poor-Fair	Poor	Fair-Excellent	Fair-Good	Poor
ACID, INORGANIC	Good Dilute Fair-Good Conc.	Fair Dilute Poor Conc.	Fair Dilute Poor Conc.	Good Dilute Good Conc.	Good Dilute Good Conc.	Poor Dilute Poor Conc.
ACID, ORGANIC	Good Dilute Poor-Good Conc.	Poor Dilute Poor Conc.	Fair Dilute Poor Conc.	Good Dilute Fair-Good Conc.	Good Dilute Poor Conc.	Poor-Good Conc.
ACID, OXIDIZING	Poor Dilute Poor Conc.	Poor Dilute Poor Conc.	Poor Dilute Poor Conc.	Poor Dilute Poor Conc.	Poor Dilute Poor Conc.	Poor Dilute Poor Conc.
ALCOHOLS	Good	Poor	Good	Good-Excellent	Fair-Good	Good
ALDEHYDES	Poor-Fair	Poor	Poor	Good	Good	Good
ALKALIES, DILUTE	Fair-Good	Fair	Poor-Excellent	Good-Excellent	Fair-Good	Fair 50%
ALKALIES, CONCENTRATED	Fair-Good	Fair	Poor	Good-Excellent	Fair-Good	Poor
AMINES	Poor-Good	Poor	Poor-Fair	Good	Poor-Good	
ANIMAL & VEGETABLE OILS	Poor-Good	Good	Fair-Excellent	Good-Excellent	Poor-Good	
BRAKE FLUID, NON PETRO BASED	Poor-Good	Poor	Poor	Good	Poor-Good	
CHLORINATED HYDROCARBONS	Poor	Poor	Poor-Good	Poor	Poor	Varies
DIESTER OILS	Poor	Good	Poor-Good	Poor-Good	Poor	
ESTERS, ALKYL PHOSPHATE	Poor	Poor	Poor	Very good	Poor	Good
ESTERS, ARYL PHOSPHATE	Poor	Poor	Poor	Excellent	Poor	Good
ETHERS	Poor	Poor-Fair	Fair	Poor-Fair	Poor	Good
HALOGENATED SOLVENTS	Poor	Poor-Good	Poor-Good	Poor	Poor	Poor
HYDROCARBONS, HALOGENATED	Poor	Poor-Good	Fair-Good	Poor	Poor	Poor
HYDROCARBONS, ALIPHATIC	Poor	Excellent	Good-Excellent	Poor	Poor	Good
HYDROCARBONS, AROMATIC	Poor	Poor-Good	Poor-Fair	Poor	Poor	Varies
HYDROCARBON OIL, HIGH ANILINE	Poor	Excellent	Excellent	Poor	Poor	Good
HYDROCARBON OIL, LOW ANILINE	Poor	Excellent to 300°F	Good-Excellent	Poor	Poor	Good
KETONES	Poor-Good	Poor	Poor	Poor-Excellent	Good	Fair-Good
LACQUER SOLVENTS	Poor	Poor	Poor	Fair-Good	Poor	Fair
L.P. GASSES & FUEL OILS	Poor	Good	Fair-Good	Poor	Poor	Good
MINERAL OIL, LOW ANILINE	Poor	Excellent	Good-Excellent	Poor	Poor	Good
MINERAL OIL, HIGH ANILINE	Poor	Excellent	Good-Excellent	Poor	Poor	Good
PETROLEUM AROMATIC	Poor	Fair	Good	Poor	Poor	Good
PETROLEUM NON AROMATIC	Poor	Poor	Good	Poor	Poor	Fair-Good
PHOSPHATE ESTERS	Poor-Good	Poor	Poor	Excellent to 250°F	Good	Good
REFRIGERANT AMMONIA	Good	Fair	Poor	Good	Good	Poor-Fair
REFRIGERANT HALOFLUOROCARBONS	R-12, 13, 22	R-11, 12, 13, 22	R-12	R-12, 13, 22	R-12, 13, 22	Poor
REFRIGERANT HALOFLUOROCARBONS WITH OIL	Poor	R-11, 12, 13, 22	AU R-12 EU Poor	Poor	Poor	R-11
REFRIGERANT METHYL CHLORIDE	Poor	Poor	Poor	Fair	Poor	Poor-Fair
SILICATE ESTERS	Poor	Poor	Poor-Good	Poor-Excellent	Poor-Excellent	
SILICONE OILS	Poor	Excellent	Excellent	Poor	Poor	Good
ADHESION TO FABRICS	Good	Very good	Very good	Good	Good	
ADHESION TO METALS	Excellent	Good	Excellent	Good	Excellent	
ADHESION TO RIGID MATERIALS	Excellent	Good	Good	Fair-Excellent	Excellent	
COLORABILITY	Good	Good	Good-Excellent	Good		
VULCANIZING PROPERTIES	Excellent	Very good-Excellent	Excellent	Good		

	SYNTHETIC ISOPRENE IR	NATURAL ISOPRENE NR	CHLOROSULFONATED POLYETHYLENE CSM	FLUOROSILICONE FVMQ	SILICONE VMQ	CHEMRAZ FFKM
COST ($/pound)	0.75-0.93	0.69-0.93	1.17-1.35	18.00-58.00	3.00-12.00	
TASTE	Fair-Good	Fair-Good	Fair-Good	Good	Good	
ACIDS, DILUTE	Fair-Good	Good-Excellent	Excellent	Excellent	Fair-Good	Excellent
ACIDS, CONCENTRATED	Poor-Good	Fair-Good	Good-Excellent	Good	Poor-Fair	Excellent
ACID, INORGANIC	Good Dilute Good Conc.	Good Dilute Fair Conc.	Excellent Dilute Good-Excellent Conc.	Excellent Dilute Fair-Good Conc.	Excellent Dilute Fair-Good Conc.	Excellent Dilute Excellent Conc.
ACID, ORGANIC	Good Dilute Fair-Good Conc.	Good Dilute Fair-Good Conc.	Excellent Dilute Good Conc.	Good Dilute Fair Conc.	Good Dilute Fair Conc.	Excellent Dilute Excellent Conc.
ACID, OXIDIZING	Poor Dilute Poor Conc.	Poor Dilute Poor Conc.	Good Dilute Poor Conc.	Fair Dilute Fair Conc.	Fair Dilute Fair Conc.	Excellent Dilute Excellent Conc.
ALCOHOLS	Good	Good-Excellent	Excellent	Fair-Good	Fair-Good	Excellent
ALDEHYDES	Good	Good	Fair-Good	Poor	Good	Good
ALKALIES, DILUTE	Fair-Good	Good-Excellent	Good-Excellent	Excellent	Poor-Fair	Excellent
ALKALIES, CONCENTRATED	Fair-Good	Fair-Good	Good-Excellent	Good	Poor-Excellent	Good
AMINES	Poor	Fair-Poor	Poor	Poor	Good	Good-Excellent
ANIMAL & VEGETABLE OILS	Poor-Good	Poor-Good	Good	Excellent	Good-Excellent	Excellent
BRAKE FLUID, NON PETRO BASED	Good	Good	Fair	Excellent	Excellent	
CHLORINATED HYDROCARBONS	Poor	Poor	Poor-Fair	Good-Excellent	Poor-Good	Excellent
DIESTER OILS	Poor	Poor	Poor	Good-Excellent	Poor-Fair	
ESTERS, ALKYL PHOSPHATE	Poor	Poor	Poor	Poor-Fair	Good	
ESTERS, ARYL PHOSPHATE	Poor	Poor	Fair	Very good	Good	Good
ETHERS	Poor	Poor	Poor	Fair	Poor	Good
HALOGENATED SOLVENTS	Poor	Poor	Poor	Very good	Poor	
HYDROCARBONS, HALOGENATED	Poor	Poor	Poor-Fair	Good-Very good	Poor	Excellent
HYDROCARBONS, ALIPHATIC	Poor	Poor	Fair-Good	Excellent	Poor-Fair	Excellent
HYDROCARBONS, AROMATIC	Poor	Poor	Fair	Good-Excellent	Poor	Excellent
HYDROCARBON OIL, HIGH ANILINE	Poor	Poor	Excellent	Excellent	Very good	
HYDROCARBON OIL, LOW ANILINE	Poor	Poor	Very good	Very good	Fair-Very good	
KETONES	Good	Fair-Good	Poor	Poor	Poor	Excellent
LACQUER SOLVENTS	Poor	Poor	Poor	Poor	Poor	
L.P. GASSES & FUEL OILS	Poor	Poor	Good	Excellent	Fair	
MINERAL OIL, LOW ANILINE	Poor	Poor-Fair	Good	Good-Excellent	Poor	
MINERAL OIL, HIGH ANILINE	Poor	Poor-Fair	Good-Excellent	Good-Excellent	Poor-Good	
PETROLEUM AROMATIC	Poor	Poor	Poor	Good	Poor	
PETROLEUM NON AROMATIC	Poor	Poor	Fair	Good	Good	
PHOSPHATE ESTERS	Poor-Good	Poor-Good	Poor-Fair	Poor-Good	Poor-Good	Excellent
REFRIGERANT AMMONIA	Good	Good	Good	Excellent	Excellent	Good
REFRIGERANT HALOFLUOROCARBONS	R-12, 13, 22	R-12, 13, 22	R-11, 12, 13, 22	R-11, 12	Poor	Fair-Good
REFRIGERANT HALOFLUOROCARBONS WITH OIL	Poor	Poor	R-11, 12, 22	R-11, 12	Poor	Fair-Good
REFRIGERANT METHYL CHLORIDE	Poor	Poor	Poor	Poor	Poor	Good-Excellent
SILICATE ESTERS	Poor-Excellent	Poor-Excellent	Poor-Good	Good-Excellent	Poor	Excellent
SILICONE OILS	Good	Good	Excellent	Excellent	Poor Oils Fair Greases	Excellent
ADHESION TO FABRICS	Excellent	Excellent	Good	Excellent	Excellent	
ADHESION TO METALS	Excellent	Excellent	Excellent	Good-Excellent	Good-Excellent	
ADHESION TO RIGID MATERIALS	Excellent	Excellent	Excellent	Fair-Good	Fair-Excellent	
COLORABILITY		Natural & Black	Excellent	Very good	Excellent	
VULCANIZING PROPERTIES	Excellent	Excellent	Excellent	Excellent	Good-Excellent	

	POLYSULFIDE T	CHLORINATED POLYETHYLENE CM	EPICHLOROHYDRIN CO, ECO	® HYTREL COPOLYESTER TPE	® SANTOPRENE COPOLYMER TPO	® C-FLEX STYRENIC TPE
COST ($/pound)	1.50-5.00	0.73	CO = 2.26 ECO = 2.11	2.35-3.50	1.38-1.85	3.00-3.50
TASTE	Poor-Fair		Good			
ACIDS, DILUTE	Poor-Fair	Excellent	Good	Good	Excellent	Good
ACIDS, CONCENTRATED	Poor	Good-Excellent	Poor-Fair	Poor	Poor-Excellent	Good
ACID, INORGANIC	Excellent Dilute Poor Conc.	Excellent Dilute Good Conc.	Good Dilute Fair Conc.	Good Dilute Poor Conc.	Good-Excellent Dilute Poor-Good Conc.	Good Dilute Good Conc
ACID, ORGANIC	Good Dilute Good Conc.	Poor-Good Dilute Poor-Good Conc.	Fair Dilute Poor Conc.	Good Dilute Poor-Good Conc.	Poor-Excellent Dilute Poor-Excellent Conc.	Good Dilute Good Conc.
ACID, OXIDIZING	Poor Dilute Poor Conc.	Good Dilute Poor Conc.	Poor Dilute Poor Conc.	Poor Conc.	Poor-Good Dilute Poor-Good Conc.	Good Dilute Good Conc.
ALCOHOLS	Good-Excellent	Excellent	Fair-Good	Good	Good	Poor-Good
ALDEHYDES	Fair-Good		Poor	Good		Good
ALKALIES, DILUTE	Poor-Good	Excellent	Fair-Good	Fair-Good	Excellent	Good
ALKALIES, CONCENTRATED	Poor-Good	Excellent	Poor-Fair	Poor	Excellent	Good
AMINES	Poor-Good		Poor-Good	Fair-Good		Good
ANIMAL & VEGETABLE OILS	Excellent	Fair-Good	Excellent	Excellent	Excellent	Good
BRAKE FLUID, NON PETRO BASED	Poor		Poor	Fair-Good		
CHLORINATED HYDROCARBONS	Fair-Excellent	Poor	Poor-Excellent	Poor	Poor-Good	Poor-Good
DIESTER OILS	Poor-Good		Poor-Excellent	Fair		
ESTERS, ALKYL PHOSPHATE	Poor	Good	Poor	Fair-Good	Excellent	
ESTERS, ARYL PHOSPHATE	Poor-Good	Good	Poor	Poor-Good	Good to 212°F	Good
ETHERS	Excellent	Good	Good	Fair		Poor
HALOGENATED SOLVENTS	Poor-Fair	Poor	Poor	Poor	Poor	Poor-Good
HYDROCARBONS, HALOGENATED	Fair-Good	Poor	Excellent	Good	Poor-Good	Poor-Good
HYDROCARBONS, ALIPHATIC	Excellent	Fair-Good	Good-Excellent	Excellent	Good	Poor
HYDROCARBONS, AROMATIC	Good	Poor-Fair	Good-Excellent	Good	Poor	Poor
HYDROCARBON OIL, HIGH ANILINE	Very good-Excellent	Excellent	Excellent	Good-Excellent	Fair-Good	
HYDROCARBON OIL, LOW ANILINE	Good-Excellent	Good	Excellent	Good-Excellent	Poor-Fair at 212°F	
KETONES	Good	Fair-Good	Fair	Fair-Good	Excellent	Good (water soluble)
LACQUER SOLVENTS	Good	Fair	Fair	Fair-Good		Poor
L.P. GASSES & FUEL OILS	Excellent	Good-Excellent	Excellent	Good	Poor	Poor
MINERAL OIL, LOW ANILINE	Good-Excellent	Good-Excellent	Excellent	Good	Poor	Good
MINERAL OIL, HIGH ANILINE	Good-Excellent	Good-Excellent	Excellent	Good	Poor	Good
PETROLEUM AROMATIC	Good	Good	Good-Excellent	Good-Excellent	Poor	Poor
PETROLEUM NON AROMATIC	Excellent	Fair-Good	Poor	Fair-Good	Poor	Poor
PHOSPHATE ESTERS	Fair-Excellent	Good	Poor	Poor-Good	Excellent	Good
REFRIGERANT AMMONIA	Good Gas Poor Liquid	Poor Gas Good Liquid	Poor	Poor		Good
REFRIGERANT HALOFLUOROCARBONS	R-11,12,12,22	Poor-Good	CO = R-12,22 ECO = R-12	Good	Poor-Good	Poor-Good
REFRIGERANT HALOFLUOROCARBONS WITH OIL	R-11,12,12,22		Good-Excellent			
REFRIGERANT METHYL CHLORIDE	Fair-Good	Fair	Poor	Poor	Poor	Poor-Good
SILICATE ESTERS	Fair-Good	Poor	Good	Fair-Good		
SILICONE OILS	Good-Excellent		Good-Excellent	Good-Excellent		Good
ADHESION TO FABRICS	Fair	Fair	Fair-Good	Good		
ADHESION TO METALS	Excellent	Fair-Good	Fair-Good	Good		
ADHESION TO RIGID MATERIALS	Fair-Good	Fair-Good	Fair-Excellent			Good-Excellent
COLORABILITY	Fair	Excellent	Good	Excellent	Excellent	Good
VULCANIZING PROPERTIES	Fair		Good	Not applicable	Not applicable	Not applicable

	ETHYLENE ACRYLIC EA	POLYALLOMER LINEAR COPOLYMER	NYLON 11 POLYAMIDE	NYLON 12 POLYAMIDE	ETHYLENE VINYL ACETATE EVA	POLYVINYLCHLORIDE FLEXIBLE PVC
COST ($/pound)	2.02-2.25	9.00-12.80 Processed parts	2.40-3.70	2.40-3.70	0.70-1.32	6.00-19.00 Tubing
TASTE		Excellent (none)	Excellent		Excellent (none)	Good-Excellent (none to slight)
ACIDS, DILUTE	Good	Excellent	Good pH-5 or more	Good pH-5 or more	Good-Excellent	Excellent
ACIDS, CONCENTRATED	Poor-Fair	Poor-Excellent	Poor	Poor	Poor-Fair	Good
ACID, INORGANIC	Good-Excellent Dilute Poor Conc.	Excellent Dilute Poor-Excellent Conc.	Poor-Good Dilute Poor Conc.	Good Dilute Poor Conc.	Good Dilute Fair-Excellent Conc.	Good-Excellent Dilute Poor-Excellent Conc.
ACID, ORGANIC	Good-Excellent Dilute Poor-Excellent Conc.	Excellent Dilute Excellent Conc.	Poor-Good Dilute Poor-Good Conc.	Good Dilute Poor Conc.	Good Dilute Poor-Good Conc.	Good-Excellent Dilute Poor-Excellent Conc.
ACID, OXIDIZING	Good Dilute Poor Conc.	Excellent Dilute Poor Conc.	Poor Conc.	Poor-Good Dilute Poor Conc.	Poor-Excellent Conc.	Good-Excellent Dilute Poor-Excellent Conc.
ALCOHOLS	Good-Excellent	Excellent	Good	Good (swells)	Excellent (polyhydric)	Fair-Good
ALDEHYDES	Fair-Good	Fair-Good	Good	Fair	Poor	Poor
ALKALIES, DILUTE	Good-Excellent	Excellent	Good to pH-11	Good to pH-11	Fair-Excellent	Excellent
ALKALIES, CONCENTRATED	Poor	Excellent	Poor-Good	Good	Good-Excellent	Good
AMINES			Good			Poor-Good
ANIMAL & VEGETABLE OILS	Good		Good	Good	Poor-Fair	Good
BRAKE FLUID, NON PETRO BASED	Poor		Good		Poor	Good
CHLORINATED HYDROCARBONS	Poor	Poor-Fair	Poor-Good	Good (swells)	Poor	Poor
DIESTER OILS			Good			Poor
ESTERS, ALKYL PHOSPHATE	Poor		Good		Poor	Poor
ESTERS, ARYL PHOSPHATE	Poor	Good	Good		Poor	Poor
ETHERS		Poor-Excellent	Good-Excellent	Good-Excellent	Poor	Fair
HALOGENATED SOLVENTS	Poor-Good	Fair-Good	Good	Good	Fair	Poor-Fair
HYDROCARBONS, HALOGENATED	Poor-Good	Fair-Good	Good	Good	Fair	Poor-Good
HYDROCARBONS, ALIPHATIC	Good	Good	Good-Excellent	Good-Excellent	Fair	Fair-Excellent
HYDROCARBONS, AROMATIC	Poor-Fair	Poor-Fair	Good-Excellent	Excellent (swells)	Poor	Poor-Excellent
HYDROCARBON OIL, HIGH ANILINE	Good		Good		Fair-Good	Good
HYDROCARBON OIL, LOW ANILINE	Good		Good		Fair-Good	Excellent
KETONES	Poor	Good	Good-Excellent	Good-Excellent	Poor	Poor-Fair
LACQUER SOLVENTS	Poor	Good-Excellent	Good		Good	Poor
L.P. GASSES & FUEL OILS	Poor	Poor-Excellent	Good	Excellent	Poor	Poor-Good
MINERAL OIL, LOW ANILINE	Good-Excellent	Excellent	Good	Good	Poor	Good
MINERAL OIL, HIGH ANILINE	Good-Excellent	Excellent	Good	Good	Poor	Good
PETROLEUM AROMATIC	Poor	Fair-Good	Good	Good	Poor	Poor-Fair
PETROLEUM NON AROMATIC	Poor	Fair-Good	Good	Good	Poor	Poor-Good
PHOSPHATE ESTERS	Poor	Good	Good		Poor	Good
REFRIGERANT AMMONIA	Poor Gas Poor Liquid	Excellent Gas Excellent Liquid	Good Gas Good Liquid	Good Gas Good 10% Liquid	Good Gas Good Liquid	Good Gas Poor Liquid
REFRIGERANT HALOFLUOROCARBONS	Poor-Good	Fair-Good	Good-Excellent	Excellent	Poor-Good	Poor
REFRIGERANT HALOFLUOROCARBONS WITH OIL	Poor					
REFRIGERANT METHYL CHLORIDE	Good	Poor-Fair	Good	Good	Poor	Poor-Fair
SILICATE ESTERS			Good		Poor	Poor
SILICONE OILS	Good-Excellent		Good-Excellent	Good	Good	Good
ADHESION TO FABRICS	Good					
ADHESION TO METALS	Good		Excellent (primed)		Good	
ADHESION TO RIGID MATERIALS			Excellent		Good	Poor
COLORABILITY			Excellent	Excellent	Excellent	Good
VULCANIZING PROPERTIES	Good	Not applicable	Not applicable	Not applicable	Not applicable	Not applicable

	POLYETHYLENE LOW DENSITY LDPE	®TEFLON FEP	®KALREZ PERFLUORINATED ELASTOMER	®FLUORAZ FLUORINATED COPOLYMER	®AFLAS FLUORINATED COPOLYMER	®NORPRENE COPOLYMER TPO
COST ($/pound)	0.62	32.00–40.00 Tubing		70.00 Minimum Processed parts	554.00 Minimum Processed parts	9.00 Minimum Tubing
TASTE	Excellent (none)	Excellent (none)				
ACIDS, DILUTE	Excellent	Excellent	Excellent	Good–Excellent	Excellent	Good
ACIDS, CONCENTRATED	Good–Excellent	Excellent	Excellent	Poor–Excellent	Poor–Excellent	Fair–Good
ACID, INORGANIC	Excellent Dilute Fair-Excellent Conc.	Outstanding Dilute Outstanding Conc.	Excellent Dilute Excellent Conc.	Good Dilute Good Conc.	Good-Excellent Dilute Poor-Excellent Conc.	Good-Excellent Dilute Poor Conc.
ACID, ORGANIC	Excellent Dilute Good-Excellent Conc.	Outstanding Dilute Outstanding Conc.	Excellent Dilute Excellent Conc.	Good Dilute Good Conc.	Excellent Dilute Good-Excellent Conc.	Excellent Dilute Excellent Conc.
ACID, OXIDIZING	Excellent Dilute Fair-Excellent Conc.	Excellent Dilute Excellenr Conc.	Excellent Dilute Excellent Conc.	Good Dilute Good Conc.	Excellent Dilute Good Conc.	Poor-Excellent Conc.
ALCOHOLS	Excellent	Excellent	Excellent	Good–Excellent	Good–Excellent	
ALDEHYDES	Fair–Good	Excellent	Fair–Good	Poor–Good	Good–Excellent	Good–Excellent
ALKALIES, DILUTE	Excellent	Excellent	Excellent	Good–Excellent	Excellent	Good
ALKALIES, CONCENTRATED	Good	Excellent	Excellent	Good	Excellent	
AMINES	Good–Excellent	Excellent	Good–Excellent	Poor–Good	Good–Excellent	
ANIMAL & VEGETABLE OILS	Poor–Good	Excellent	Excellent	Good	Excellent	Good–Excellent
BRAKE FLUID, NON PETRO BASED	Poor	Excellent			Fair–Excellent	Good–Excellent
CHLORINATED HYDROCARBONS	Poor	Good	Good	Poor	Poor	Poor
DIESTER OILS	Poor	Excellent	Excellent	Good–Excellent	Good–Excellent	
ESTERS, ALKYL PHOSPHATE	Poor	Excellent	Excellent	Good	Excellent	Fair at 212°F
ESTERS, ARYL PHOSPHATE	Good	Excellent	Excellent	Poor	Fair	
ETHERS	Excellent	Excellent–Outstanding	Excellent	Poor	Poor	Fair–Good
HALOGENATED SOLVENTS	Poor	Excellent	Poor	Poor	Poor	Poor
HYDROCARBONS, HALOGENATED	Poor	Excellent	Poor	Poor	Poor	Poor
HYDROCARBONS, ALIPHATIC	Fair Cold Poor Hot	Outstanding	Excellent	Fair	Poor–Fair	Fair–Good
HYDROCARBONS, AROMATIC	Poor–Fair	Outstanding	Excellent	Poor	Poor–Good	Poor
HYDROCARBON OIL, HIGH ANILINE	Fair	Excellent		Good–Excellent	Excellent	Good to 242°F
HYDROCARBON OIL, LOW ANILINE		Excellent		Good–Excellent	Excellent	
KETONES	Good	Outstanding	Excellent	Poor	Poor–Fair	Poor
LACQUER SOLVENTS	Good		Excellent	Poor	Poor	Poor
L.P. GASSES & FUEL OILS	Poor	Outstanding	Excellent	Good	Good–Excellent	
MINERAL OIL, LOW ANILINE	Poor	Excellent	Excellent	Good–Excellent	Good	Good–Excellent
MINERAL OIL, HIGH ANILINE	Poor	Excellent	Excellent	Good–Excellent	Good	Good–Excellent
PETROLEUM AROMATIC	Poor	Outstanding	Excellent	Fair	Excellent	Poor
PETROLEUM NON AROMATIC	Poor	Outstanding	Excellent	Poor	Poor	
PHOSPHATE ESTERS	Poor		Excellent	Good–Excellent	Fair–Excellent	Good
REFRIGERANT AMMONIA	Excellent	Excellent	Excellent	Good Gas Poor–Fair Liquid	Excellent	Good–Excellent
REFRIGERANT HALOFLUOROCARBONS	Fair	Excellent	Poor–Good	Poor	Poor	Poor
REFRIGERANT HALOFLUOROCARBONS WITH OIL		Excellent	Fair–Good	Poor		
REFRIGERANT METHYL CHLORIDE	Poor–Fair	Excellent	Excellent	Poor		
SILICATE ESTERS	Poor–Fair	Excellent	Excellent	Poor		
SILICONE OILS	Excellent	Excellent	Excellent	Good–Excellent	Good–Excellent	
ADHESION TO FABRICS		Poor			Good	
ADHESION TO METALS		Poor				
ADHESION TO RIGID MATERIALS		Poor Untreated	Good			
COLORABILITY	Excellent	Good				Good
VULCANIZING PROPERTIES	Not applicable	Not applicable	Not available			Not applicable

C419

	HIGH RANGE LOW RANGE	NITRILE NBR	ETHYLENE PROPYLENE EP, EPDM	FLUOROCARBON FKM	CHLOROPRENE CR	HYDROGENATED NITRILE HNBR
TENSILE STRENGTH (X 10³ psi)	AU NR IR AU VMQ EP	1.0-4.0	0.3-3.5	0.5-3.0	0.5-4.0	1.5-6.0
TENSILE MODULUS AT 100% ELONGATION (psi)	AU EP CR AU VMQ	490-550	100-3000	200-2000	100-3000	300-2900
FLEXURAL MODULUS (psi)	Nylon					
HARDNESS DUROMETER	AU NR IR CSM AU IIR CR	20A-100A	30A-90A	50A-95A	15A-95A	55A-77A
HARDNESS ROCKWELL						
HARDNESS SHORE	AU NR IR CSM AU IIR CR	12D-58D	12D-39D	12D-46D	12D-46D	60D
ELONGATION (%)	AU IIR NR VMQ PVA EP FKM CR	400-650	100-700	100-500	100-800	150-550
COMPRESSION SET RATING	NBR FKM SBR IR NR T PVC	Good-Excellent	Poor-Good	Good-Excellent	Poor-Good	Good-Excellent
COMPRESSION SET ASTM D395, METHOD B (%)		5-20	20-60 70 Hrs at 212°F	9-16 70 Hrs 75°F 10-30 70 Hrs 392°F	20-60 70 Hrs at 212°F	10
RESILIENCE OR REBOUND RATING	IR NR VMQ FVMQ IIR PVC T	Good Cold Good Hot	Fair-Very good Cold Fair-Very good Hot	Fair-Good Cold Good-Excellent Hot	Fair-Very good Cold Very good Hot	
RESILIENCE, YERZLEY (ASTM 945) (%)			40-75	40-70	50-80	
RESILIENCE, REBOUND (BASHORE) (%)			40-75	40-70	50-80	
HYSTERESIS RESISTANCE	IR NR CR		Good	Good	Very good	Fair-Good
FLEX CRACKING RESISTANCE	AU EU IIR EP CR T	Good	Good-Very good	Good	Good-Very good	Good
TEAR RESISTANCE	AU EU EA NBR IIR T	Good-Excellent	90 PLI at 77°F 182 PLI at 212°F	Fair-Very good 110 PLI at 70°F	110 PLI at 77°F 230 PLI at 212°F	Fair-Good
ABRASION RESISTANCE	AU EU SBR BR IR NR T FVMQ	Good-Excellent	Good-Excellent	Good	Very good-Excellent	Good-Excellent
IMPACT RESISTANCE	AU EU IR NR SBR ACM VMQ	Fair-Good	Very good	Good	Good-Excellent	Excellent
STAINING	SBR IR NR CSM FVMQ	Poor-Good	Good	Fair-Good	Good-Excellent	
COEFFICIENT OF FRICTION, DYNAMIC	AU EU NBR AU EU	0.38-0.87 vs steel		0.24-0.42 vs steel		
COEFFICIENT OF FRICTION STATIC	EP VMQ PVC	0.50-0.59 vs steel	0.27 vs steel	0.57-0.77 vs steel	0.89-0.95 vs steel	
VOLUME RESISTIVITY, ohm-cm	EP VMQ IIR EVA PVA CR NBR	3.5×10^{10}	2×10^{16} to 1×10^{17}	2.0×10^{13} at 50% R.H. 1.5×10^{13} at 90% R.H.	1.0×10^{11} to 2.0×10^{13}	
DIELECTRIC STRENGTH, volts/mil	EP IIR SBR NBR AU EU FVMQ	250	Good-Excellent 500-1000	Good-Excellent 400-700	Good-Excellent 400-700	Fair
DIELECTRIC CONSTANT AT 60 Hz	CR AU EU CSM FVMQ EP SBR		2.25-3.00		8.0	
DIELECTRIC CONSTANT AT 1 MHz	AU EU CR T FVMQ EP SBR IR		2.20-2.85		6.7	
VACUUM WEIGHT LOSS 14 DAYS AT 1 X 10⁻⁶ TORR (%)	NBR AU EU EP FKM VMQ	1.1-3.5	0.39-0.93	0.07-0.09		

	STYRENE BUTADIENE SBR	POLYACRYLATE ACM	POLYURETHANE AU, EU	ISOBUTYLENE ISOPRENE IIR	POLYBUTADIENE BR	® AEROQUIP AQP
TENSILE STRENGTH (X 10³ psi)	0.5-3.5	1.25-2.50	1-10	2.0-3.0	2.5-3.0	2-3
TENSILE MODULUS AT 100% ELONGATION (psi)	300-1500	100-1500	25-5000	50-500	300-1500	
FLEXURAL MODULUS (psi)			75 X10³			
HARDNESS DUROMETER	30A-100A	40A-90A	10A-100A	10A-100A	45A-80A	70A-90A
HARDNESS ROCKWELL						
HARDNESS SHORE	12D-58D	12D-39D	12D-80D	12D-58D	12D-29D	22D-39D
ELONGATION (%)	450-600	100-450	250-900	300-850	450-650	100-300
COMPRESSION SET RATING	Good-Excellent	Poor-Good	Poor-Good	Poor-Good	Good	
COMPRESSION SET ASTM D395, METHOD B (%)	5-30	10-60 70 Hrs 302°F	10-45 22 Hrs 158°F 25 22 Hrs 250°F	25 70 Hrs 257°F	10-30	
RESILIENCE OR REBOUND RATING	Poor-Excellent	Fair-Good Cold Fair-Good Hot	Poor-Very good	Poor-Fair	Fair-Excellent	
RESILIENCE, YERZLEY (ASTM 945) (%)	20-90		5-75	30	50-90	
RESILIENCE, REBOUND (BASHORE) (%)	10-60		20-65			
HYSTERESIS RESISTANCE	Fair-Good		Fair-Good		Good	
FLEX CRACKING RESISTANCE	Good	Fair-Good	Good-Excellent	Good-Excellent	Fair-Excellent	
TEAR RESISTANCE	Fair-Excellent	Poor-Good	Excellent 500-650 PLI	Good	Good-Excellent	
ABRASION RESISTANCE	Excellent	Fair-Good	Excellent-Outstanding	Fair-Good	Excellent	
IMPACT RESISTANCE	Excellent	Poor	Excellent-Outstanding No break Izod	Good	Good	
STAINING	Poor-Fair	Good	Good	Good	Good	
COEFFICIENT OF FRICTION, DYNAMIC			0.22-0.58			
COEFFICIENT OF FRICTION STATIC				0.37 vs steel		
VOLUME RESISTIVITY, ohm-cm	5.0-8.4×10^{13}	7×10^{12}	Good 0.3×10^{10} to 5.0×10^{13}	2×10^{16}		
DIELECTRIC STRENGTH, volts/mil	Excellent 600-800	Fair-Excellent 800	Fair-Good 330-700	Excellent 600-900	Fair-Excellent 400-600	
DIELECTRIC CONSTANT AT 60 Hz	2.31		4.7-9.5 at 60 Hz 5.2-7.5 at 1 kHz			
DIELECTRIC CONSTANT AT 1 MHz	2.25		5.5-8.5		3.3	
VACUUM WEIGHT LOSS 14 DAYS AT 1 X 10⁻⁶ TORR (%)			1.29	0.18		

	SYNTHETIC ISOPRENE IR	NATURAL ISOPRENE NR	CHLOROSULFONATED POLYETHYLENE CSM	FLUOROSILICONE FVMQ	SILICONE VMQ	CHEMRAZ FFKM
TENSILE STRENGTH (X 10³ psi)	2.5–4.5	0.5–5.0		0.5–1.4	0.2–1.5	1.1–3.15
TENSILE MODULUS AT 100% ELONGATION (psi)		70–120 at 100% 1050–1770 at 300%		450–500	900	665–2450
FLEXURAL MODULUS (psi)						
HARDNESS DUROMETER	30A–100A	20A–100A	40A–100A	35A–80A	20A–90A	60A–90A
HARDNESS ROCKWELL						
HARDNESS SHORE	12D–58D	12D–58D	12D–58D	12D–29D	12D–39D	
ELONGATION (%)	300–750	500–900	100–700	100–480	100–900	60–155
COMPRESSION SET RATING	Excellent	Excellent	Poor-Fair	Fair-Good	Good-Excellent	Good
COMPRESSION SET ASTM D395, METHOD B (%)			35–80 22 Hrs 212°F	17–25 22 Hrs 300°F	4–10 22 Hrs 77°F 20 70 Hrs 250°F	14–21 Hrs 77°F 34–65 Hrs 400°F
RESILIENCE OR REBOUND RATING	Excellent	Good-Excellent Cold Excellent Hot	Fair-Good Cold Good Hot	Excellent Cold Excellent Hot	Good-Excellent Cold Excellent Hot	
RESILIENCE, YERZLEY (ASTM 945) (%)		80	30–70		30–60	12
RESILIENCE, REBOUND (BASHORE) (%)		40–48	30–70			
HYSTERESIS RESISTANCE	Excellent	Excellent	Fair-Good	Good	Fair-Good	
FLEX CRACKING RESISTANCE	Excellent	Excellent	Fair-Very good	Poor-Good	Poor-Excellent	
TEAR RESISTANCE	Good-Excellent	Good-Excellent 118–436 PLI	Fair-Good 100–213 PLI	Poor-Excellent 40–265 PLI	Poor-Good 30–180 PLI	Poor-Fair
ABRASION RESISTANCE	Excellent	Excellent	Very good-Excellent	Poor	Poor-Good	Poor
IMPACT RESISTANCE	Excellent	Excellent	Good-Very good	Poor-Good	Poor-Fair	
STAINING	Excellent	Excellent	Excellent	Excellent	Excellent	
COEFFICIENT OF FRICTION, DYNAMIC						
COEFFICIENT OF FRICTION STATIC					1.03–2.36 vs steel	
VOLUME RESISTIVITY, ohm-cm			Good 1 X 10¹⁴	1.1×10^{12} to 1.6×10^{14}	8×10^{13} to 1×10^{16}	1.4×10^{17}
DIELECTRIC STRENGTH, volts/mil	Excellent 600	Good-Excellent 400–600	Good-Excellent 400–700	Fair-Good 340–380	Good-Excellent 400–700	Excellent
DIELECTRIC CONSTANT AT 60 Hz			7.0	6.1–7.4 at 100 Hz	3–4	
DIELECTRIC CONSTANT AT 1 MHz		2.9	6.0 Estimate	5.7–6.2	3–4	2.4
VACUUM WEIGHT LOSS 14 DAYS AT 1 X 10⁻⁶ TORR (%)				0.25–0.28	0.03–0.31	

	POLYSULFIDE T	CHLORINATED POLYETHYLENE CM	EPICHLOROHYDRIN CO, ECO	® HYTREL COPOLYESTER TPE	® SANTOPRENE COPOLYMER TPO	® C-FLEX STYRENIC TPE
TENSILE STRENGTH (X 10³ psi)	0.5–1.5	0.9–3.0		1.35–3.90 yield 4.6–7.6 ultimate	0.64–4.00	0.99–2.08
TENSILE MODULUS AT 100% ELONGATION (psi)				$1.1–75.0 \times 10^3$	$0.29–1.45 \times 10^3$	100–1010 at 100% 180–1170 at 300%
FLEXURAL MODULUS (psi)		700–2200		$7–91 \times 10^3$	$3–50 \times 10^3$	
HARDNESS DUROMETER	20A–85A	40A–95A	30A–95A	90A–100A	55A–97A	30A–90A
HARDNESS ROCKWELL					50R max.	
HARDNESS SHORE	12D–33D	12D–46D	12D–46D	46D–72D	14D–50D	12D–39D
ELONGATION (%)	210–450	100–700	200–800	170–800	330–600	20–1350
COMPRESSION SET RATING	Poor–Fair	Good–Excellent	Good–Excellent	Poor–Excellent	Fair–Good	Fair–Good
COMPRESSION SET ASTM D395, METHOD B (%)	29–38	5–30	20 70 Hrs 212°F	2–27 22 Hrs 158°F	20–47 at 77°F 25–81 at 212°F	10–55 at 70°F 39–100 at 158°F
RESILIENCE OR REBOUND RATING	Poor–Fair	Poor–Fair Cold Good Hot	Good–Very good Cold Good Hot	Fair–Good Cold Good–Excellent Hot	Excellent	Excellent
RESILIENCE, YERZLEY (ASTM 945) (%)			50–80	40		
RESILIENCE, REBOUND (BASHORE) (%)		15–40	45–75	43–62		
HYSTERESIS RESISTANCE		Good	Good	Excellent		
FLEX CRACKING RESISTANCE	Poor–Good		Very good	Excellent	Outstanding	
TEAR RESISTANCE	Poor–Fair	Fair–Good	Fair–Excellent	Excellent–Outstanding 400–1355 PLI	108–514 PLI at 77°F 58–364 PLI at 212°F	Good–Excellent 190–360 PLI Un–notched
ABRASION RESISTANCE	Poor–Fair	Good–Excellent	Fair–Good	Outstanding	Good	
IMPACT RESISTANCE	Poor–Fair	Excellent	Fair–Excellent	Excellent No break Izod notched		Excellent No break Izod notched
STAINING	Poor–Good		Good			
COEFFICIENT OF FRICTION, DYNAMIC						
COEFFICIENT OF FRICTION STATIC					0.31–0.33 50 Shore 0.99 25 Shore	0.81
VOLUME RESISTIVITY, ohm-cm	Fair 50×10^{13}			1.1×10^{12} to 1.8×10^{14}		$0.8–3.0 \times 10^{16}$
DIELECTRIC STRENGTH, volts/mil	Good	Fair–Good	Good	Fair–Excellent 400–900 1/8 in.		Fair–Excellent 300–800 1/8 in.
DIELECTRIC CONSTANT AT 60 Hz						2.27–2.33
DIELECTRIC CONSTANT AT 1 MHz	7.3 at 1 kHz 6.8 at 1 MHz			3.9–5.1 at 1 kHz		2.15–2.31
VACUUM WEIGHT LOSS 14 DAYS AT 1 X 10-6 TORR (%)						

	ETHYLENE ACRYLIC EA	POLYALLOMER LINEAR COPOLYMER	NYLON 11 POLYAMIDE	NYLON 12 POLYAMIDE	ETHYLENE VINYL ACETATE EVA	POLYVINYLCHLORIDE FLEXIBLE PVC
TENSILE STRENGTH (X 10^3 psi)	1.0–3.0	3.0–3.4 yield 3.0–4.3 ultimate	3.8–12.0	3.6–6.1 yield 5.1–9.0 ultimate	1.0–6.0	1.4–3.5
TENSILE MODULUS AT 100% ELONGATION (psi)	100–1500		15–180 X 10^3	180 X 10^3	7–29 X 10^3	850–1200 at 100% 1540–2240 at 300%
FLEXURAL MODULUS (psi)		70–130 X 10^3	17–150 X 10^3	33–270 X 10^3	7.7 X 10^3	
HARDNESS DUROMETER	35A–95A		100A	95A	62A–94A	50A–100A
HARDNESS ROCKWELL		50R–85R	42R–108R	105R–109R		
HARDNESS SHORE	12D–46D		63D–75D	56D–91D	17D–45D	12D–58D
ELONGATION (%)	200–650	300–500	100–300	250–350	300–800	100–500
COMPRESSION SET RATING	Poor–Good				Good	Poor–Fair
COMPRESSION SET ASTM D395, METHOD B (%)	10–60					
RESILIENCE OR REBOUND RATING	Poor–Fair Cold Fair Hot					Poor
RESILIENCE, YERZLEY (ASTM 945) (%)	40					
RESILIENCE, REBOUND (BASHORE) (%)	20–25					
HYSTERESIS RESISTANCE	Good					
FLEX CRACKING RESISTANCE	Good	Excellent	Excellent	Good	Good	
TEAR RESISTANCE	Good–Excellent		Excellent 1000–1100 PLI	Excellent 1000–1100 PLI	Fair–Excellent 80–500 PLI	Fair–Good 100–400 PLI
ABRASION RESISTANCE	Good–Excellent	Good–Excellent	Excellent	Outstanding	Poor	Fair–Good
IMPACT RESISTANCE	Good–Very good	1.5–3.8 ft.lb./in. Izod 1/8 in.	1.8–3.3 ft. lb./in. Izod 1/8 in.	No break to 1.2 ft.lb./in. Izod 1/8 in.	Excellent No break Izod 1/8 in.	Excellent
STAINING			Good		Excellent except iodine at 140°F	Fair
COEFFICIENT OF FRICTION, DYNAMIC			0.12–0.18 vs steel			
COEFFICIENT OF FRICTION STATIC			0.20–0.32 vs steel			1.32 vs steel (std.) 2.5–3.5 vs steel (fuel)
VOLUME RESISTIVITY, ohm-cm	1.9 X 10^{12} to 1.0 X 10^{13}	1 X 10^{16}+	1 X 10^{11} to 1 X 10^{16}	1 X 10^{13} to 3 X 10^{16}	1 X 10^9 to 1 X 10^{16}	1 X 10^{11} to 1 X 10^{15}
DIELECTRIC STRENGTH, volts/mil	Good–Excellent 700–730	Good–Excellent 500–800 1/8 in.	1500–1800 1 mill 425 1/8 in.	1500–1800 1 mill 450 1/8 in.	5000 1 mill 620–760 1/8 in.	1500–1800 1 mill 300–400 1/8 in.
DIELECTRIC CONSTANT AT 60 Hz	7.0	2.3	4.0	3.5 50–100 Hz	3.16	
DIELECTRIC CONSTANT AT 1 MHz			3.3–3.8 at 1 kHz 3.1 at 1 MHz	3.5–3.8 at 1 kHz 3.1–3.3 at 1 MHz	2.7–2.9	4.0–8.0 at 1 kHz 3.3–4.5 at 1 MHz
VACUUM WEIGHT LOSS 14 DAYS AT 1 X 10^{-6} TORR (%)			Excellent	Excellent		Excellent Vacuum grades

	POLYETHYLENE LOW DENSITY LDPE	® TEFLON FEP	® KALREZ PERFLUORINATED ELASTOMER	® FLUORAZ FLUORINATED COPOLYMER	® AFLAS FLUORINATED COPOLYMER	® NORPRENE COPOLYMER TPO
TENSILE STRENGTH (X 10^3 psi)	0.6-2.3	2.5-3.1	1.90-2.66	3.00-3.81	2.0-3.2	1.0-1.5
TENSILE MODULUS AT 100% ELONGATION (psi)	14-38 X 10^3	50-250 X 10^3	0.9-1.9 X 10^3	1.24-3.17 X 10^3	600-2500 at 100% 200-3100 at 50%	410-540 at 100% 800-975 at 300%
FLEXURAL MODULUS (psi)	8-60 X 10^3	80-95 X 10^3				
HARDNESS DUROMETER		98A-100A	65A-95A	75A-100A	60A-100A	61A-73A
HARDNESS ROCKWELL	10R					
HARDNESS SHORE		55D-65D	19D-46D	25D-60D	16D-70D	16D-24D
ELONGATION (%)	90-800	250-330	60-170	117-198	50-400	375-425
COMPRESSION SET RATING	Poor	Poor	Fair-Good at 70°F Good at 500°F	Good at 70°F Good at 400°F	Good at low temp. Poor at high temp.	Good
COMPRESSION SET ASTM D395, METHOD B (%)			20-40 70 Hrs 70°F 32-71 70 Hrs 400°F	16-19 70 Hrs 70°F 21-27 22 Hrs 392°F	25 70 Hrs 200°F 20-50 70 Hrs 392°F	27-33 22 Hrs 212°F
RESILIENCE OR REBOUND RATING	Poor	Poor	Fair			
RESILIENCE, YERZLEY (ASTM 945) (%)						
RESILIENCE, REBOUND (BASHORE) (%)						
HYSTERESIS RESISTANCE						
FLEX CRACKING RESISTANCE	Good	Excellent to 1400 psi stress	Fair-Good			
TEAR RESISTANCE	Fair-Excellent 65-575 PLI	Excellent 600 PLI	Fair-Good	Fair	Fair-Good 120-300 PLI	Fair 120-150 PLI
ABRASION RESISTANCE	Good Excellent	Poor-Good	Fair-Good	Good	Good-Excellent	
IMPACT RESISTANCE	No break Izod notch	No break Izod notch	Good			
STAINING		Excellent				
COEFFICIENT OF FRICTION, DYNAMIC		0.06-0.09 vs steel				
COEFFICIENT OF FRICTION STATIC	0.22 vs steel	0.06-0.09 vs steel	0.23 vs steel		0.23-0.37 vs steel	0.81 vs steel (std.) 0.37 vs steel (food)
VOLUME RESISTIVITY, ohm-cm	1 X 10^{15} to 1 X 10^{16}	2 X 10^{18} to 1 X 10^{19}	5 X 10^{17}		3 X 10^{16} at 70°F 1.7 X 10^{13} at 400°F	
DIELECTRIC STRENGTH, volts/mil	5000 1 mill 460-700 1/8 in.	6500 1 mill 600 1/8 in.	Good 450+	Good	580 at 70°F 200 at 400°F	
DIELECTRIC CONSTANT AT 60 Hz	2.30-2.35	2.1			2.5 at 70°F 3.0 at 300°F	
DIELECTRIC CONSTANT AT 1 MHz	2.2-2.3 at 1 kHz 2.2 to 1 GHz	2.1 at 1 MHz 2.1 at 1 GHz	4.1 at 1 kHz		2.6 at 70°F 2.8 at 300°F	
VACUUM WEIGHT LOSS 14 DAYS AT 1 X 10^{-6} TORR (%)			Excellent			Excellent

SYNONYM	LISTED UNDER	SEE PAGES
Absolut Alcohol	Vodka	C2-C7
Acetaldehyde	Ethanal	C2-C7 & C134-C139
Acetamide	Acryimide	C2-C7
Acetic Acid Amide	Acetamide	C2-C7
Acetic Acid, Crude	Pyroligineous Acid	C314-C319
Acetic Acid Ethenyl Ester	Vinyl Acetate	C380-C385
Acetic Acid + Methanol	Pyroligineous Acid	C314-C319
Acetic Acid 2-Methyl-2 Butyl Ester	Amyl Acetate	C20-C25
Acetic Acid Methyl Ester	Methyl Acetate	C230-C235
Acetic Aldehyde	Acetaldehyde	C2-C7
Acetic Anhydride	Ethanoic Anhydride	C2-C7
Acetic Ester	Ethyl Acetate	C2-C7 & C140-C145
Acetol	Diacetone Alcohol	C110-C115
Acetone	2-Propanone	C2-C7 & C302-C307
Acetonitrile	Methyl Cyanide	C2-C7 & C235-C241
Acetyl Benzene	Acetophenone	C2-C7
Acetylene Dichloride	Dichloroethylene	C2-C7 & C116-C121
Acetylene Tetrachloride	Tetrachloroethane	C2-C7 & C362-C367
Acryimide	Acetamide	C2-C7
Acrylonitrile	Propanoic Acid Nitrile	C8-C13
Acrylonitrile	Vinyl Cyanide	C8-C13 & C380-C385
Alkazene	Alkasene	C8-C13
Alcohol, 2-Aminoethanol	Ethanolamine	C140-C145
Alum Potash	Potassium Hydroxide	C296-C301
Aluminum Hydrate	Aluminum Hydroxide	C14-C19
Amino Benzene	Aniline	C14-C19 & C26-C31
Amino Ethanol	Ethanolamine	C14-C19 & C140-C145
Aminomethane	Methylamine	C230-C235
2-Amino 2-Methyl Propanol	Butylamine	C62-C67
Ammonium Chloride	Sal Ammoniac	C20-C25 & C320-C325
Ammonium Hydroxide + Water	Ammonia Liquors	C14-C19
Amyl Alcohol	Methyl Butanol	C20-C25 & C230-C235
Amyl Hydride	Pentane	C26-C31 & C278-C283
Amylum	Starch	C344-C349
Anethole	Anethol	C26-C31
Aniline	Phenyl Amine	C26-C31 & C284-C289
Animal Fat	Lard	C26-C31 & C212-C217
Animal Oil	Lard Oil	C26-C31
Ant Oil	Furfural	C26-C31 & C164-C169
Apple Acid	Malic Acid	C32-C37 & C224-C229
Aqua Regia	3 Parts Hydrochloric Acid + 1 Part Nitric Acid	C32-C37
Aqueous Sodium Nitrate	Soda Niter	C332-C337
ASTM Reference Fuel A	Isooctane	C32-C37 & C206-C211
ASTM Reference Fuel B	70% Isooctane + 30% Toluene	C32-C37 & C380-C385
ASTM Reference Fuel C	50% Isooctane + 50% Toluene	C32-C37
Azine	Pyridine	C308-C313
Azotic Acid	Nitric Acid	C254-C259
Baking Soda	Sodium Bicarbonate	C38-C43 & C332-C337
Barium Hydroxide	Caustic Baryta	C38-C43 & C74-C79
O-Benzedicarbolic Acid	Phthalic Acid	C290-C295
Benzene	Benzol	C44-C49
Benzene Carbonal	Benzaldehyde	C38-C43
Benzene Carboxylic Acid	Benzoic Acid	C38-C43 & C44-C49
Benzine	Gasoline	C43-C49 & C170-C175
Benzine	Petroleum Ether	C44-C49 & C278-C283

SYNONYM	LISTED UNDER	SEE PAGES
Benzine	Petroleum Ether + Petroleum Naphtha	C44-C49
Benzoic Acid	Benzene Carboxylic Acid	C38-C43 & C44-C49
Biphenyl	Diphenyl	C44-C49 & C128-C133
Bischofite	Magnesium Chloride	C244-C229
Blank Fixe	Barium Sulfate	C38-C43 & C44-C49
Bleaching Powder Solutions	Calcium Hypochlorite	C44-C49 & C68-C73
Blue Copperas	Copper Sulfate	C50-C55 & C98-C103
Blue Verdigris	Copper Acetate	C92-C97
Blue Vitriol	Copper Sulfate	C50-C55 & C98-C103
Boracic Acid	Boric Acid	C50-C55
Borax	Sodium Borate	C50-C55 & C332-C337
Borax	Sodium Tetraborate	C50-C55 & C338-C343
Boron Fluids	High Energy Fuels (HEF)	C50-C55 & C176-C181
Bran Oil	Furfural	C164-C169
Bromomethane	Methyl Bromide	C56-C61 & C230-C235
Brucite	Magnesium Hydrozide	C56-C61 & C224-C229
Butanal	Butal	C56-C61
Butanethiol	Butyl Mercaptan	C62-C67
Butanoic Acid	Butyric Acid	C56-C61 & C62-C67
Butanol	Butyl Alcohol	C56-C61
2-Butanone	Methyl Ethyl Ketone	C56-C61 & C236-C241
1-Butene	Croton	C56-C61
2-Butene	Butylene	C61-C67
Butenedioic Acid	Maleic Acid	C224-C229
2-Butoxyethanol	Butyl Cellosolve	C56-C61 & C62-C67
Butter	Animal Fat	C26-C31 & C56-C61
Butter of Antimony	Antimony Trichloride	C56-C61 & C32-C37
Butyl Benzene	Phenyl Butane	C61-C67
Butyl Butanoate	N-Butyl Butyrate	C61-C67
Butyl Cellosolve	Ethylene Glycol Monobutyl Ether	C61-C67 & C146-C151
Butyl Cellosolve	Butoxyethanol	C56-C61 & C62-C67
Butyl Chloride	Chlorobutane	C61-C67 & C86-C91
Butyl Citrate	n Hexene-1	C182-C187
Bultyl Phthalate	Tributyl Phthalate	C61-C67
Calcium Hydroxide	Slaked Lime	C68-C73 & C326-C331
Calcium Hypochlorite	Chloride of Lime	C68-C73 & C80-C85
Calcium Oxide	Lime	C68-C73 & C218-C223
Caliche Liquors	Salt Peter	C68-C73 & C320-C325
Caprylic Acid	Octanoic Acid	C68-C73
Carbamide	Urea	C380-C385
Carbinol	Methanol	C230-C235
Carbitol	2-Ethoxy Ethoxy Ethanol	C74-C79
Carbolic Acid	Phenol	C74-C79 & C284-C289
Carbon Bisulfide	Carbon Disulfide	C74-C79
Carbon Tetrachloride	Tetrachloromethane	C74-C79 & C362-C367
Carbonyl Chloride	Phosgene	C284-C289
Carbonyl Diamide	Urea	C380-C385
Casein	Casymen	C74-C79
Caustic Lime	Calcium Hydroxide	C68-C73 & C74-C79
Caustic Potash	Potassium Hydroxide	C74-C79 & C296-C301
Caustic Soda	Sodium Hydroxide	C74-C79 & C338-C343
Cellosolve Acetate	Ethylene Glycol Monoethyl Ether Acetate	C74-C79 & C146-C151
Cellosolve Acetate	2 Ethoxyethyl Acetate	C74-C79
Cetane	Hexadecane	C80-C85
Chinawood Oil	Tung Oil	C80-C85 & C374-C379

SYNONYM	LISTED UNDER	SEE PAGES
Chile Nitrate	Sodium Nitrate	C80-C85 & C338-C343
Chile Saltpeter	Sodium Nitrate	C80-C85 & C338-C343
Chlorate of Lime	Chloride of Lime	C80-C85
Chlorinated Biphenyl	Chlorinated Diphenyl	C80-C85
Chloro-1-Hydroxybenzene	Chlorophenol	C86-C91
Chloro, 2, 3 Epoxy Propane	Epichlorohydrin	C134-C139
Chloroacetic Acid	Ethyl Chloroacetate	C80-C85
Chloroazotic Acid	Aqua Regia	C32-C37 & C80-C85
Chlorobenzene	Phenyl Chloride	C86-C91 & C284-C289
Chlorobenzol	Chlorobenzene	C86-C91
Chlorobutanol	Chlorobutane	C86-C91
Chlorodifluoromethane	Freon 22	C-158-C169
Chloroethane	Ethyl Chloride	C86-C91 & C140-C145
Chloroethanoic Acid	Chloroacetic Acid	C80-C85 & C86-C91
1- Chloro, 2, 3 Epoxypropane	Epichlorhydrin	C134-C139
Chloroethene	Vinyl Chloride	C380-C385
2- Chloroethanol	Ethylene Chlorohydrin	C86-C91 & C146-C151
Chloroethylene	Vinyl Chloride	C380-C385
Chloromethane	Methyl Chloride	C86-C91 & C230-V235
Chloromethyl	Chloromethane	C86-C91
Chloropentafluoroethane	Freon F-115	C164-C169
Chloropentane	Amyl Chloride	C26-C31 & C86-C91
Chioropicrin	Larvacide	C212-C217
Choroprene	Chlorobutadiene	C86-C91
3- Chloropropene	Allyl Chloride	C8-C13 & C86-C91
Chlorothene	Methyl Chloroform	C86-C91 & C230-C235
Chlorothene	Trichloroethane	C86-C91 & C368-C373
Chlorotoluene	Benzyl Chloride	C44-C49 & C86-C91
Chrome Alum	Chromium Potassium Sulfate	C86-C91
Coal Tar	Bituminious	C92-C97
Coal Tar	Creosote	C92-C97 & C98-C103
Copper Acetate	Blue Verdgris	C92-C97
Copper Sulfate	Blue Vitriol	C50-C55 & C98-C103
Copperas	Ferrous Sulfate	C98-C103 & C152-C157
Corn Oil	Maize Oil	C98-C103 & C225-C229
Corrosive Sublimate	Mercury Chloride	C98-C103 & C230-C235
Creosote	Coal Tar	C92-C97 & C98-C103
Creosote	Wood Tar	C98-C103 & C392-C397
m Cresol	m Methyl Phenol	C98-C103 & C236-C241
o Cresol + m Cresol + p Cresol Mixture	Cresylic Acid	C98-C103
Croton	1-Butene	C56-C61
Cyanic Acid	Isocyanic Acid	C104-C109
Cyclic Esters	Lactams	C212-C217
Cyclohexane	Hexahydrobenzene	C104-C109 & C182-C187
Cyclohexanone	Pimlic Ketone	C104-C109 & C290-C295
DBP	Dibutyl Phthalate	C104-C109 & C110-C115
DC200, 510, 710	Silicone Oil	C104-C109 & C326-C331
DDT	Dichloro Diphenyl Trichloroethane	C104-C109
Decalin (Deklin)	Decahydronaphthalene	C104-C109
Decanol	Decyl Alcohol	C104-C109
Developing Fluids	Sodium Thiosulfate	C110-C115
Dextrin	Starch Gum	C110-C115
Dextrose	Glucose	C110-C115 & C170-C175
Diacetone Alcohol	Acetol	C110-C115
Diamine	Diamidogen	C110-C115

SYNONYM	LISTED UNDER	SEE PAGES
Diaminoethane	Ethylene Diamine	C110-C115 & C146-C151
Diamylamine	Dipentylamine	C110-C115
Dibenzofuran	Diphenyl Oxide	C128-C133
1,2, Dibromoethane	Ethylene Bromide	C110-C115 & C146-C151
1,2, Dibromoethane	Ethylene Dibromide	C110-C115 & C146-C151
Dibromotetrafluoroethane	Freon 114B2	C163-C169
Dibutyl Ether	Butyl Ether	C62-C67 & C110-C115
Dibutyl Phthalate	Butyl Phthalate	C61-C67 & C110-C115
Dichloro Difluoro Methane	F-12 Refrigerant	C110-C115 & C158-C163
Dichloro Diphenyl Trichloroethane	DDT	C104-C109
Dichloroethane	Ethylene Chloride	C116-C121 & C146-C151
Dichloroethane	Ethylene Dichloride	C116-C121 & C146-C151
1-1 Dichloroethanol	Dicloroethylene	C116-C121
1-1 Dichloroethanol	Vinylidine Chloride	C386-C391
Dichloromethane	Methylene Chloride	C116-C121 & C236-C241
Dichloropropane	Propylene Dichloride	C116-C121 & C308-C313
Dichlorotetrafluoroethane	F-114 Refrigerant	C116-C121 & C164-C169
Diester Lubricant	Mil-L-7808	C116-C121 & C401-C403
Diethyl Carbonate	Ethyl Carbonate	C116-C121
Diethyl Ether	Ether	C116-C121 & C140-C145
Di-2-Ethylhexylphthalate	Dioctyl Phthalate	C122-C127
Diethylene Dioxide	P-Dioxane	C128-C133
Diethylenimide Oxide	Morpholine	C248-C253
m Digallic Acid	Tannic Acid	C122-C127 & C356-C361
Diisopropylidene Acetone	Phorone	C284-C289
Dihydrogen Dioxide	Hydrogen Peroxide	C194-C199
2,2, Dihydroxyamine	Diethanolamine	C116-C121
1,3, Dihydroxybenzene	Resorcinol	C314-C319
1,2,3, Dihydroxybutanedioic Acid	Tartaric Acid	C356-C361
2,2, Dihydroxydiethyl Ether	Diethylene Glycol	C116-C121 & C122-C127
Dihydroxyethane	Glycol, Ethylene	C176-C181
Dihydroxyethane	Ethylene Glycol	C146-C151
Diisopropyle Ether	Isopropyle Ether	C122-C127 & C206-C211
Dimethyle	Ethane	C134-C139
Dimethyl Benzene	Xylene	C122-C127 & C392-C397
2,6, Dimethyl 4 Heptane	Diisobutyl Ketone	C122-C127
Dimethyl Ether	Methyl Ether	C122-C127 & C236-C241
Dimethyl Ketone	Acetone	C2-C7 & C122-C127
Dimethyl Methane	Propane	C122-C127 & C302-C307
Dimethyl Phthalate	DMP	C122-C127 & C128-C133
Dimethyl Polysiloxane	Silicone Oil	C326-C331
Dinitrochlorobenzene	DNCB	C122-C127
Dioctyl Phthalate	DOP	C122-C127
p Dioxan	Dioxane	C128-C133
Dioxysuccinic Acid	Tartaric Acid	C356-C361
Dipentylamine	Diamylamine	C110-C115
Diphenyl	26% diphenyl (Dowtherm A)	C128-C133 & C134-C139
Diphenyl	Phenyl Benzene	C128-C133 & C284-C289
Dipropyl	Hexane	C182-C187
DMF	Dimethyl Formamide	C122-C127
DMP	Dimethyl Phthalate	C122-C127 & C128-C133
DMT	Dimethyl Terephthalate	C122-C127 & C128-C133
Dodecyl Alcohol	Dodecanol	C128-C133
DOP	Dioctyl Phthalate	C122-C127
EMK	Ethyl Methyl Ketone	C134-C139 & C146-C151
1,2, Epoxyethane	Ethylene Oxide	C146-C151
1,2, Epoxypropane	Propylene Oxide	C308-C313
Epsom Salts	Magnesium Sulfate	C224-C229

C429

SYNONYM	LISTED UNDER	SEE PAGE
1,2,Ethanediol	Ethylene Glycol	C146-C151
Ethanedionic Acid	Oxalic Acid	C134-C139 & C266-C271
Ethanethiol	Ethyl Mercaptan	C134-C139 & C146-C151
Ethanoic Acid	Acetic Acid	C2-C7 & C134-C139
Ethanoic Anhydride	Acetic Anhydride	C2-C7
Ethanol	Ethyl Alcohol	C134-C139 & C140-C145
Ethanolamine	Monethanolamine	C140-C145 & C248-C253
Ethanonitrile	Acetonitrile	C2-C7
Ethanoyl Chloride	Acetyl Chloride	C2-C7 & C140-C145
Ethene	Ethylene	C146-C151
Ethenyl Benzene	Styrene	C344-C349
Ethoxy Ethane	Ethyl Ether	C140-C145
2, Ethoxy Ethanol	Glycol Monoethyl Ether	C140-C145 & C176-C181
2, Ethoxy Ethanol	Ethyl Cellosolve	C140-C145 & C230-C235
2-2 Ethoxy Ethoxy Ethanol	Carbitol	C74-C79
Ethoxyethyl Acetate	Cellosolve Acetate	C74-C79
Ethyl Benzene	Phenyl Ethane	C140-C145 & C284-C289
Ethyl Butanoate	Ethyl Butyrate	C140-C145
Ethyl Carbonate	Diethyl Carbonate	C116-C121
Ethyl Chloride	Chloroethane	C86-C91 & C140-C145
Ethyl Chloroacetate	Chloroacetic Acid	C80-C85
Ethyl Ethanoate	Ethyl Acetate	C140-C145
Ethyl Hexyl Alcohol	Ethyl Hexanol	C146-C151
Ethyl Propyl Carbinol	Hexanol	C146-C151
Ethyl Sulfate	Diethyl Sulfate	C116-C121 & C146-C151
Ethylene	Ethene	C146-C151
Ethylene Bromide	Ethylene Dibromide	C146-C151
Ethylene Dichloride	Ethylene Chloride	C146-C151
Ethylene Dichloride	Dichloroethane	C116-C121 & C146-C151
Ethylic Acid	Acetic Acid	C2-C7 & C152-C157
Ethylidine Chloride	1,2 Dichloroethane	C116-C121
Ethyle	Acetylene	C2-C7
Ethylene Glycol	Glycol, Ethylene	C146-C151 & C176-C181
Ethyrene	Butadiene	C56-C61
Fluosilicic Acid	Hydrofluorosilicic	C158-C163 & C188-C193
Formaldehyde	Formalin	C158-C163
Formaldehyde	Methanal	C158-C163 & C188-C193
Formamide	Formylamine	C158-C163
Formylic Acid	Formic Acid	C158-C163
Freon BF	Freon 112	C158-C163 & C164-C169
Freon MF	Freon 11	C158-C163 & C164-C169
Freon TF	Freon 113	C164-C169
Freon 11	Trichlorofluoromethane	C158-C163 & C368-C373
Freon 12	Dichlorodifluoromethane	C110-C115 & C158-C163
Freon 13	Monochlorotrifluoromethane	C158-C163 & C248-C253
Freon 13B1	Monobromotrifluoromethane	C158-C163 & C242-C247
Freon 14	Tetrafluoromethane	C158-C163 & C362 & C367
Freon 22	Monochlorodifluoromethane	C158-C163 & C242-C247
Freon 112	Tetrachlorodifluoroethane	C164-C169 & C362-C367
Freon 113	Trichlorotrifluoroethane	C164-C169 & C374-C379
Freon 114	Dichlorotetrafluoroethane	C116-C121 & C164-C169
Freon 14B2	Dibromotetrafluoroethane	C164-C169
Freon 115	Monochloropentafluoroethane	C164-C169
Freon C-318	Octafluorotetraethylene	C164-C169
Freon 502	Freon 22 + Freon 316	C164-C169
Fructose	Sugar	C164-C169
Fumaric Acid	Boletic Acid	C50-C55 & C164-C169
Fuming Sulfuric Acid	20-25% Oleum	C164-C169 & C350-C355

SYNONYM	LISTED UNDER	SEE PAGES
Furan	Furfural	C 264-C 169
Furan	Furane	C 164-C 169
Furfural	Furfurol	C 164-C 169
Furfural	Ant Oil	C 26-C 31
Furfural	Furancarbonal	C 26-C 31
Furfuraldehyde	Furfural	C 164-C 169
Furfuran	Furan	C 164-C 169 & C 170-C 175
2-Furyl Methanol	Furfural Alcohol	C 170-C 176
Galotannic Acid	Tannic Acid	C 170-C 175 & C 356-C 361
Gelatin	Aqueous Sodium Sulfate	C 170-C 175
Glaubers Salt	Sodium Sulfate	C 170-C 175 & C 338-C 343
Glucose	Sugar	C 170-C 175
Glycerin	Glycerine	C 170-C 175
Glycerin	Glycerol	C 170-C 175
Glycerine	1, 2, 3 Propanetriol	C 302-C 307
Glycol	Ethylene Glycol	C 146-C 151 & C 176-C 181
Green Copperas	Vitriol	C 176-C 181
HCL	Hydrochloric Acid	C 188-C 193
HEF-2	High Energy Fuel 2	C 176-C 181
HEF-2	Trialkylpentaborane	C 176-C 181
HEF-3	High Energy Fuel 3	C 176-C 181
1 Hendecanol	Undecanol	C 380-C 385
Heptaldehyde	Heptanal	C 176-C 181
Heptalene	Heptane	C 176-C 181 & C 272-C 277
3-Heptanone	Ethyl Butyl Ketone	C 140-C 145
Hexadecane	Cetane	C 80-C 85
Hexadecanoic Acid	Palmitic Acid	C 182-C 187
Hexanedioic Acid	Adipic Acid	C 8-C 13
Hexanol	Hexyl Alcohol	C 182-C 187
2-Hexanone	Methyl Butyl Ketone	C 230-C 235
n-Hexene 1	Butyl Ethylene	C 182-C 187
Hexone	Hexon	C 182-C 187
Hexone	Methyl Isobutyl Ketone	C 182-C 187 & C 236-C 241
Hydrochloric Acid	Muriatic Acid	C 188-C 193 & C 248-C 253
Hydrocyanic Acid	Prussic Acid	C 188-C 193 & C 308-C 313
Hydrofluosilicic Acid	Sand Acid	C 188-C 193 & C 320-C 325
Hydrofluosilicic Acid	Fluosilicic Acid	C 158-C 163 & C 188-C 193
Hydrogen Chloride	Hydrochloric Acid	C 188-C 193
Hydrogen Cyanide	Hydrocyanic Acid	C 188-C 193 & C 194-C 199
Hydrogen Fluoride	Hydrofluoric Acid	C 188-C 193 & C 194-C 199
Hydrogen Oxide	Water	C 194-C 199 & C 386-C 391
Hydrogen Peroxide	Dihydrogen Dioxide	C 194-C 199
Hydroxyacetic Acid	Glycolic Acid	C 176-C 181 & C 194-C 199
2-Hydroxy-Benzaldehyde	Salicyaldehyde	C 320-C 325
2-Hydroxy-Benzene Sulfonic Acid	Phenol Sulfonic Acid	C 284-C 289
2-Hydroxybenzoic Acid	Salicylic Acid	C 320-C 325
4-Hydroxy - 4 Methyl -2 Pentone	Diacetone Alcohol	C 110-C 115
2-Hydroxy -1, 2, 3-Propane Tricarboxylic Acid	Citric Acid	C 92-C 97
Hydroxysuccinic Acid	Malic Acid	C 194-C 199 & C 224-C 229
Hydroxy Toluene	Benzyl Alcohol	C 44-C 49
Hylene	Toluene Diisocyanate (TDI)	C 368-C 373
Hypo Photographic Solution	Sodium Thiosulphate	C 194-C 199
Ice Spar	Icestone	C 194-C 199
Ice Spar	Stone	C 194-C 199
Iodine	Iodum	C 200-C 205
IPA	Isopropyl Alcohol	C 200-C 205 & C 206-C 211
Isobutanol	Isobutyl Alcohol	C 200-C 205
Isocyanic Acid	Cyanic Acid	C 104-C 109

SYNONYM	LISTED UNDER	SEE PAGES
Isooctane	ASTM Fuel A	C32-C37 & C206-C211
Isooctane	2,2,4 Trimethylpentane	C206-C211 & C374-C379
Isopropanol	Isopropyl Alcohol	C206-C211
Isopropyl Acetone	Isobutyl Methyl Ketone	C206-C211
Isopropyl Benzene	Cumene	C98-C103 & C206-C211
JP3	MIL-J-5624 Jet Fuel	C206-C211 & C398-C403
JP4	MIL-J-5624 Jet Fuel	C206-C211 & C398-C403
JP5	MIL-J-5624 Jet Fuel	C206-C211 & C398-C403
JP6	MIL-J-25656 Jet Fuel	C206-C211
JPX	MIL-F-25604 Jet Fuel	C206-C211
Jet Fuel Type B	Gasoline + Kerosene	C206-C211
Kerosene	Kerosine	C206-C211
Ketohexamethylene	Cyclohexanone	C104-C109 & C206-C211
Ketone	Isophorone	C206-C211
Ketone	Metsityl Oxide	C230-C235
Lactams	Amino Acids	C212-C217
Lactones	Cyclic Esters	C212-C217
Lard	Animal Fat	C26-C31 & C212-C217
Lard Oil	Animal Oil	C26-C31
Larvacide	Trichloronitromethane	C212-C217
Larvacide	Chloropicrin	C212-C217
Latex	Synthetic Natural Rubber	C212-C217
Laughing Gas	Nitrous Oxide	C212-C217 & C260-C265
Lead Acetate	Sugar of Lead	C212-C217 & C350-C355
Ligroin	Ligroine	C218-C223
Ligroin	Petroleum Ether	C218-C233 & C278-C283
Lime	Calcium Oxide	C68-C73 & C218-C223
Lime + Water	Calcium Hydroxide	C68-C73 & C218-C223
Limewater	Milk of Lime	C218-C223
Lime Sulfur	Calcium Sulfide	C68-C73 & C218-C223
Lindol	Tricresyl Phosphate	C218-C223 & C374-C379
Linoleic Acid	Linolenic Acid	C218-C223
Liquid Rosin	Retinol	C218-C223 & C314-C319
Lithic Acid	Uric Acid	C380-C385
LOX	Liquid Oxygen	C218-C223
LPG	Liquified Petroleum Gas	C218-C223
LPG	Butane	C56-C61
LPG	Propane	C302-C307
Lye	Sodium Hydroxide	C218-C223 & C338-C343
Magnesium Chloride	Bischofite	C224-C229
Magnesium Hydroxide	Brucite	C56-C61 & C224-C229
Magnesium Hydroxide	Milk of Magnesia	C224-C229 & C242-C247
Magnesium Sulfate	Epsom Salts	C224-C229
Maize Oil	Corn Oil	C98-C103 & C224-C229
Malic Acid	Apple Acid	C32-C37 & C224-C229
Marsh Gas	Methane	C224-C229 & C230-C235
MEA	Monoethanolamine	C224-C229 & C248-C253
MEK	Methyl Ethyl Ketone	C224-C229 & C236-C241
MEKP	Methyl Ethyl Ketone Peroxide	C236-C241
Melamine Resins	Triazane	C224-C229
Mesityl Oxide	Ketone	C230-C235
Methanal	Formaldehyde	C158-C163 & C230-C235
Methane	Marsh Gas	C224-C229 & C230-C235
Methanoic Acid	Formic Acid	C158-C163
Methanol	Methyl Alcohol	C230-C235
2-Methoxy-4 Methyl Phenol	Cresols	C98-C103
Methyl Acetone	Methyl Acetate, Methyl Alcohol, Acetone mixture	C230-C235
Methyl Amyl Alcohol	Methyl Isobutyl Carbinol	C230-C235 & C236-C241

SYNONYM	LISTED UNDER	SEE PAGES
Methyl Benzene	Toluene	C230–C235 & C368–C373
Methyl Bromide	Bromomethane	C56–C61 & C230–C235
Methyl Butanol	Amyl Alcohol	C20–C25 & C230–C235
Methyl Chloride	Chloromethane	C86–C91 & C230–C235
Methyl Chloroform	Trichloroethane	C230–C235 & C368–C373
Methyl Ether	Dimethyl Ether	C122–C127 & C236–C241
Methyl Ethyl Ketone	MEK	C224–C229 & C236–C241
Methyl Ethyl Ketone Peroxide	MEKP	C236–C241
Methyl Hexyl Ketone	2-Octanone	C236–C241 & C260–C265
Methyl Isobutyl Ketone	MIBK	C236–C241 & C242–C247
4-Methyl-2 Pentanone	Methyl Isobutyl Ketone	C236–C241
m Methyl Phenol	m-Cresol	C98–C103 & C236–C241
m Methyl Phenyl Ketone	Acetophenone	C2–C7
Methyl Polysiloxanes	Silicone Oils	C236–C241 & C326–C331
Methyl Salicylate	Wintergreen Oil	C392–C397
Methyl Sulfoxide	Dimethyl Sulfoxide	C122–C127
Methylene Chloride	Dichloromethane	C116–C121 & C236–C241
MIBK	Methyl Isobutyl Ketone	C236–C241 & C242–C247
MIL-H-5606	Red Oil	C314–C319 & C398–C403
MIL-H-7808A	Turbine Oil No. 15	C116–C121 & C374–C379
MIL-F-25555B	RJ-1	C314–C319 & C404–C406
MIL-R-25576	RP-1	C314–C319 & C404–C406
Milk of Lime	Lime Water	C218–C223
Milk of Magnesia	Magnesium Hydroxide	C224–C229 & C242–C247
Monobromotrifluoromethane	F-13B1	C158–C163 & C242–C247
Monochlorobenzene	Chlorobenzene	C86–C91 & C242–C247
Monochlorodifluoromethane	F-22	C158–C163 & C242–C247
Monochlorotrifluoromethane	F-13	C158–C163 & C242–C247
Monochloropentafluoroethane	F-115	C164–C169
Mosiac Gold	Tin Sulfide	C248–C253
Muriatic Acid	Hydrochloric Acid	C188–C193 & C248–C253
Naphtha, Crude	Petroleum	C278–C283
Napthalene	Tar Camphor	C248–C253 & C356–C361
Natural Gas	75-99% Methane, 0.3-18% Nitrogen, 0.2-14% Ethane	C248–C253
	0-26% Carbon Dioxide, 0.1-12% Higher Hydrocarbon	
	0.1-0.6% Other	
Neetsfoot Oil	Neatsfoot Oil	C248–C253
Neville Acid	1-Naphthol-4 Sulfonic Acid Sodium Salt	C248–C253
Niacin	Nicotinic Acid	C254–C259
Niter	Potassium Nitrate	C254–C259 & C302–C307
Niter	Sodium Nitrate	C254–C259 & C338–C343
Nitrobenzene	Nitrobenzol	C254–C259
Nitrobenzene	Oil of Mirbane	C254–C259 & C266–C271
Nitrobenzine	Ligroin	C218–C223 & C254–C259
Nitrogen Monoxide	Nitrous Oxide	C260–C265
Nitrohydrochloroic Acid	Aqua Regia	C32–C37 & C260–C265
Nitromuriatic Acid	Aqua Regia	C32–C37 & C260–C265
Octadecadienoic Acid	Linoeic Acid	C218–C223
Octadecanoic Acid	Stearic Acid	C260–C265 & C344–C349
9-Octadecanoic Acid	Oleic Acid	C266–C271
Octafluorotetraethylene	Freon C-318	C164–C169
Octanoic Acid	Caprylic Acid	C68–C73
Octanol	Octyl Alcohol	C260–C265 & C266–C271
Octoic Acid	Caprylic Acid	C68–C73 & C260–C265
Oil of Mirbane	Nitrobenzene	C254–C259 & C266–C271
Oil of Turpentine	Turpentine	C266–C271 & C374–C379
Oil of Vitriol	Concentrated Sulfuric Acid	C266–C271 & C350–C355
Oleic Acid	Red Oil	C266–C271 & C314–C319

C433

SYNONYM	LISTED UNDER	SEE PAGES
Oleum	Fuming Sulfuric Acid	C266-C271 & C350-C355
OS45 Type III & IV	Silicate Ester Oil	C266-C271
Palmetic Acid	Hexadecanoic Acid	C182-C187 & C272-C277
Pearl Ash	Potassium Carbonate	C272-C277 & C296-C301
Pentane	Amyl Hydride	C26-C31 & C278-C283
2-Pentanone	Methyl Propyl Ketone	C236-C241 & C278-C283
Perchloroethylene	Tetrachloroethylene	C278-C283 & C362 & C367
Petroleum Ether	Ligroin	C218-C223 & C278-C283
Petroleum Ether + Petroleum Naphtha	Benzine	C44-C49
Petroleum Jelly	Petrolatum	C278-C283
Phenetole	Phenyl Ethyl Ether	C284-C289
Phenol	Carbolic Acid	C74-C79 & C284-C289
Phenyl Amine	Aniline	C26-C31 & C284-C289
Phenyl Benzene	Diphenyl	C128-C133 & C284-C289
Phenyl Butane	Butyl Benzene	C62-C67
Phenyl Chloride	Chlorobenzene	C86-C91 & C284-C289
Phenyl Ethane	Ethyl Benzene	C140-C145 ¢ C284-C289
Phenyl Ethylene	Styrene	C344-C349
Phenyl Sulfonic Acid	Benzene Sulfonic Acid	C44-C49
Phosphoric Acid Triphenyl Ester	Tricresyl Phosphate	C374-C379
Phthalic Acid	Terephthalic Acid	C290-C295
Phthalic Acid Anhydride	Phthalic Anhydride	C290-C295
Pimilic Ketone	Cyclohexanone	C104-C109 & C290-C295
Pinacol	Pinacolin	C290-C295
Potash	Potassium Carbonate	C296-C301
Potash Caustic	Potassium Hydroxide	C296-C301
Potassium Bichromate	Potassium Dichromate	C296-C301
Potassium Nitrate	Niter	C254-C259 & C302-C307
Propane	LPG	C302-C307
Propane	Dimethyl Methane	C122-C127 & C302-C307
1,2 Propanediol	Propylene Glycol	C302-C307 & C308-C313
1,2,3 Propanetriol	Glycerine	C302-C307
Propanoic Acid, 2 Hydroxy	Lactic Acid	C212-C217
Propanoic Acid Nitrile	Acrylonitrile	C8-C13
2-Propanone	Acetone	C2-C7 & C302-C307
Propanol	Propyl Alcohol	C302-C307 & C308-C313
Propene	Propylene	C302-C307 & C308-C313
Propenyl Alcohol	Allyl Alcohol	C8-C13
Propylene Glycol	1,2 Propanediol	C302-C307 & C308-C313
Propyne	Methyl Acetylene	C230-C235
Prussic Acid	Hydrocyanic Acid	C188-C193 & C308-C313
2,6,8 Purinetrione	Uric Acid	C380-C385
Pyrogallic Acid	Pyrogallol	C308-C313
Red Oil	Oleic Acid	C266-C271 & C314-C319
Red Oil	MIL-H-5606	C314-C319 & C398-C400
Retinol	Liquid Rosin	C218-C223 & C314-C319
Ricinus Oil	Ricinine	C314-C319
RJ-1	MIL-F-25555B	C314-C319 & C404-C406
RP-1	MIL-R-25576	C314-C319 & C404-C406
Saccharose	Sucrose	C320-C325 & C350-C355
Sand Acid	Hydrofluosilicic Acid	C188-C193 & C320-C325
Sal Ammonia	Ammonium Chloride	C20-C25 & C320-C325
Sal Ammoniac	Ammonium Chloride	C20-C25 & C320-C325
Sal Soda	Sodium Carbonate	C320-C325 & C332-C337
Sewerage	Sludge Acid	C320-C325 & C326 & C331
Silicate of Soda	Sodium Silicate	C326-C331 & C338-C343
Silicone Oils	Methyl Polysiloxanes	C236-C241 & C326-C331
Slaked Lime	Lime + Water	C218-C223 & C326-C331

SYNONYM	LISTED UNDER	SEE PAGES
Sludge Acid	Sewerage	C320-C325 & C326-C331
Soda Ash	Sodium Carbonate	C326-C331 & C332-C337
Soda, Baking	Bicarbonate of Soda	C44-C49 & C326-C331
Soda Caustic	Sodium Hydroxide	C332-C337 & C338-C343
Soda Niter	Aqueous Sodium Nitrate	C332-C337 & C338-C343
Sodium Bicarbonate	Baking Soda	C38-C43 & C332-C337
Sodium Bisulfate	Hypo	C194-C199 & C332-C337
Sodium Carbonate	Sal Soda	C320-C325 & C332-C337
Sodium Carbonate	Soda Ash	C320-C325 & C332-C337
Sodium Chloride	Salt	C320-C325 & C332-C337
Sodium Hydroxide	Caustic Soda	C74-C79 & C338-C343
Sodium Hypochlorite	Bleach	C44-C49 & C338-C343
Sodium Niter	Sodium Nitrate	C258-C263 & C338-C343
Sodium Phosphate Tribasic	Trisodium Phosphate	C338-C343 & C374-C379
Sodium Silicate	Water Glass	C338-C343 & C386-C391
Sodium Thiosulphate	Hypo	C194-C199
Sodium Thiosulphate	Developing Fluids	C110-C115
Soluble Glass	Sodium Silicate	C334-C349 & C338-C343
Soya Oil	Soy Oil	C344-C349
Soybean Oil	Soya Bean Oil	C344-C349
Stannic Chloride	Tin Chloride	C344-C349 & C368-C373
Stannic Salts	Tin Salts	C344-C349 & C368-C373
Stannic Sulfide	Mosiac Gold	C248-C253 & C344-C349
Starch	Amylum	C344-C349
Starch Gum	Dextrin	C110-C115
Starch Syrup	Sugar	C344-C349
Stearic Acid	Octadecanoic Acid	C260-C265 & C344-C349
Stoddards Solvent	Kerosene	C206-C211 & C344-C349
Stoddards Solvent	White Spirit	C344-C349 & C392-C397
Styrene Monomer	Vinyl Benzene	C344-C349 & C380-C385
Sucrose	Saccharose	C320-C325 & C350-C355
Sugar Dextrose Syrup	Corn Syrup	C98-C103
Sugar Juice	Saccharose	C320-C325 & C350-C355
Sugar of Lead	Lead Acetate	C212-C217 & C350-C355
Sulfolane	Tetralin	C362-C367
Sulfuric Acid Diethyl Ester	Ethyl Sulfate	C146-C152
Sulfuric Acid, Fuming	Oleum	C266-C271 & C350-C355
Sulfuric Chlorohydrin	Chlorosulfonic Acid	C86-C91
Super Phosphoric Acid	100-115% Phosphoric Acid	C356-C361
Synthetic Natural Rubber	Latex	C212-C217
Table Salt	Sodium Chloride	C332-C337 & C365-C361
Tall Oil	Tallol	C356-C361
Tallow	Animal Fat	C26-C31 & C356-C361
Tannin	Tannic Acid	C356-C361
Tar Camphor	Naphthalene	C258-C253 & C356-C361
Tartaric Acid	Dioxysuccinic Acid	C356-C361
TCP	Tricresyl Phosphate	C356-C361 & C374-C379
TEA	Triethanolamine	C356-C361 & C374-C379
Terephthalic Acid	Phthalic Acid	C290-C295
Tetrachlorodifluoroethane	F-112	C164-C169 & C362-C367
Tetrachloroethane	Acetylene Tetrachloride	C2-C7
Tetrachloroethylene	Perchloroethylene	C278-C283 & C362-C367
Tetrachloromethane	Carbon Tetrachloride	C74-C79 & C362-C367
Tetrafluoromethane	F-14	C158-C163
Tetrahydrofuran	THF	C362-C367
Tetrahydronaphthelene	Tetralin	C362-C367
Tetrahydrothiophene 1, 1-Dioxide	Tetralin	C362-C367
Tetramethylene Oxide	Tetrahydrofuran	C362-C367

SYNONYM	LISTED UNDER	SEE PAGES
THF	Tetrahydrofuran	C362-C367
Thiofuran	Thiophene	C362-C367
Tin Chloride	Stannous Chloride	C344-C349 & C368-C373
Tin Dichloride	Stannous Chloride	C344-C349 & C368-C373
Tin Sulfide	Mosiac Gold	C248-C253 & C344-C349
Tin Tetrachloride	Stannic Chloride	C344-C349 & C368-C373
Titanic Chloride	Titanium Tetrachloride	C368-C373
TNT	Trinitrotoluene	C368-C373 & C374-C379
Toluene	Toluol	C368-C373
Toluene	Methyl Benzene	C230-C235 & C368-C373
Toluene Diisocyanate	Hylene	C368-C373
Town Gas	Coal Gas	C92-C97
Trialkylpentaborane	High Energy Fuel 2	C176-C181
Triazane	Melamine Resins	C224-C229
TriButyl Phthalate	Butyl Phthalate	C62-C67
Trichloroethylene	Triad	C368-C373
Trichlorofluoromethane	F-11	C158-C163 & C374-C379
Trichloromethane	Chloroform	C86-C91 & C368-C373
Trichloronitromethane	Larvacide	C212-C217
Trichlorotrifluoroethane	F-113	C164-C169 & C374-C379
Tricresyl Phosphate	TCP	C356-C361 & C374-C379
1,2,3-Trihydroxy Benzene	Pyrogallic Acid	C308-C313
3,4,5-Trihydroxy Benzoic Acid	Gallic Acid	C170-C175 & C374-C379
2,2,2-Trihydroxyamine	Triethanolamine	C374-C379
2,2,4-Trimethyl Pentane	Diisobutylene	C122-C127 & C374-C379
2,2,4-Trimethyl Pentane	Isooctane	C206-C211 & C374-C379
2,4,6-Trinitrophenol	Picric Acid	C290-C295 & C374-C379
Trisodium Phosphate	TSP	C374-C379
Tung Oil	China Wood Oil	C80-C85 & C374-C379
Turbine Oil	MIL-L-7808A	C374-C379
Type 1 Fuel (MIL-S-3136)	ASTM Reference Fuel A	C32-C37 & C380-C385
Type III Fuel (MIL-S-3136)	ASTM Reference Fuel B	C32-C37 & C380-C385
UDMH	Unsymetrical Dimethyl Hydrazine	C380-C385
Undecanol	Undecyl Alcohol	C380-C385
Vasoline	Petrolatum	C278-C283 & C380-C385
Vinyl Benzene	Styrene	C344-C349 & C380-C385
Vitriol	Green Coperas	C176-C181
Vitriol, Blue	Copper Sulfate	C50-C55 & C98-C103
Vitriol, Oil of	Concentrated Sulfuric Acid	C350-C355 & C386-C391
Vodka	Absolut Alcohol	C2-C7
Washing Soda	Sodium Carbonate	C322-C337 & C386-C391
Water Brine	Process Beverage	C386-C391
Water Glass	Sodium Silicate	C338-C343 & C386-C391
White Lead	Lead Carbonate, Sublimed	C350-C355
White Spirit	Kerosene	C206-C211 & C392-C397
White Vitriol	Zinc Sulfate	C392-C397
Wintergreen Oil	Methyl Salicylate	C392-C397
Wood Alcohol	Methanol	C230-C235 & C392-C397
Xylene	Xylol	C392-C397
Xylene	Dimethyl Benzene	C122-C127 & C392-C397
Xylidines	Amino Dimethyl Benzene	C14-C19 & C392-C397
Xylidines	Mixed Aromatic Amines	C392-C397
Zeolite	Zeolitic	C392-C397

NAME	MANUFACTURER	PRODUCT
Airshow W		Deicing Fluid
Aero Lubriplate	Fish Brothers	Lubricant
Aerosafe 2300 & 2300W	Stauffer	
Aeroshell 1AC & 750	Shell Oil Co.	
Aeroshell 7A & 17	Shell Oil Co.	Grease
Ambrex 33 & 830	Mobil Oil Co.	Lubricating Oil
Anderol L-774, 826 & 829	Tenneco Chemicals	Diester Base Oil
ANG-25	Texaco	Diester Base & Glycerol Ester
Ansul Ether 161 & 181	Fire Engineers Inc.	
AN-0-3 Grade M	GAF Corp.	E.P. Grease
AN-0-6	GAF Corp.	Oil No. 6
AN-0-366	GAF Corp.	Oil
Arochlor 1248, 1254 & 1260	Monsanto	F.R. Chlorinated Hydraulic Fluid
Aro-Tox		Spray
Askarel	Monsanto	Chlorinated Transformer Oil
Astral Oil	D.A. Stewart Oil Co.	Lubricating Oil
Atlantic Utro Gear		E.P. Lubricant
Atlantic Dominion F		
Aurex 903R	Mobil Oil	
Bardol B	Bardahl Manufacturing Co.	Oil Additives
Bayol D & 35		
Blackpoint 77		
Bray GG-130	Bray Oil Co.	
Brayco 719-R, 885 & 910	Bray Oil Co.	
Bret 710		
Brom-113 & 114		
Bunker C		6000 Second Fluid Oil
Calgon		Sodium Hexametaphosphate
Cellosolves	Union Carbide	Alcohols
Celluguard		
Cellulube A60	Celanese Corp.	Triaryl Phosphate Ester
Cellutherm 2505A	Celanese Corp.	Trimethylol Propane Ester
Chlorextol	Allis-Chalmers	Transformer Oil
Chlordane		
Chlorox	Clorox Co.	Sodium Hypochlorite
Chlorowax	Diamond Shamrock	Liquid Chlorinated Parafin
Coco-Cola	Coca-Cola Co.	Beverage
Circo Light Processing Oil	Sunmark Industries	
City Service Kool Motor		A.P. Gear Oil
City Service Pacemaker No. 2		Glycol FR15, 20 & 25
City Service 65, 120 & 250		
Convelex 10		
Coolanol 25 & 45	Monsanto Co.	Dielectric Heat Transfer Fluid
Cryolite		
Cryscoat F.H. Rinse, L.T. & S.W.	Oakite Products	Phosphate Coatings
Cryscoat 42, 87, 89 & 89M	Oakite Products	Phosphate Coatings
DC-200, 500 & 710	Dow Corning	Silicone Fluids
Delco Brake Fluid	General Motors	Hydraulic Fluids
Dextron	General Motors	Automatic Transmission Fluid
Diazon	Insecticide	
Dow Chemical 50-4, ET588 & ET378	Dow Chemical	
Dow Corning Oil 3, 4 & 11	Dow Corning Corp.	Silicone Fluids
Dow Corning 5, 33, 44, 55, 220, 510, 550, 704 & 705	Dow Corning Corp.	Silicone Fluids
Dow Corning 1265	Dow Corning Corp.	Fluorosilicone Fluid
Dow Corning 1208, 4050, 6620, F-60 & XF-60	Dow Corning Corp.	Chlorinated Silicone Fluid
Dow Corning F-61	Dow Corning Corp.	
Dow General Weed Killer	Dow Chemical	Phenol & Water Base Weed Killer
Dow Gage Fluid R-200		Pressure Gage Fluid
Dow Per	Dow Chemical	Dry Cleaning Fluid
Dow Purifloc C-31		

NAME	MANUFACTURER	PRODUCT
Dowanols	Dow Chemical Co.	Glycol Ethers
Dowtherm A	Dow Chemical Co.	Heat Transfer Fluid 26.5% Diphenyl, 73.5% Diphenyl Oxide
Dowtherm E & 209	Dow Chemical Co.	Heat Transfer Fluid
Dowtherm S.R.1	Dow Chemical Co.	Heat Transfer Fluid
Drinox	Morton Chemical Co.	
DTE Light Oil	Mobil Oil	Lubricating Oil
Duco Paint Thinner	Dupont Co.	Paint Thinner
Elco 28	Detrex Chemical Industries	E.P. Lubricant
Esso Fuel 208	Exxon Corp.	
Esso Golden Gasoline	Exxon Corp.	Gasoline
Esso Transmission Fluid	Exxon Corp.	Automatic Transmission Fluid
Esso WS2812	Exxon Corp.	MIL-L-7808A
Esso XP90	Exxon Corp.	E.P. Lubricant
Esso Turbo Oil	Exxon Corp.	Turbine Oil
Esstic 42 & 43		
Exxon 2380 Turbo Oil	Exxon Corp.	Turbine Oil
F-60 & 61	Dow Corning	
FC-43		Heptacosoflourotributylamine
FC75	3M Co.	Fluorocarbon
Fluorolube	Hooker Chemical Co.	
Freon	E.I. du Pont Co.	Fluorocarbon Refrigerants
Fyrquel 90, 100, 150, 220, 300, 500, 550 & A60	Stauffer Chemical	F.R. Hydraulic Fluids
Girling Brake Fluid	Lucas Service	Brake Fluid
Gulf Endurance Oils	Gulf Refining	Lubricating Oil
Gulf FR Fluids	Gulf Refining	F.R. Fluids
Gulf FR G-Fluids G100, 150, 200 & 250	Gulf Refining	F.R. Fluids
Gulf FR P-Fluids P37, 40, 43, 45 & 47	Gulf Refining	F.R. Fluids
Gulf Harmony Oils	Gulf Refining	Oils
Gulf High Temperature Grease	Gulf Refining	H.T. Grease
Gulf Legion Oils	Gulf Refining	Oils
Gulf Paramont Oils	Gulf Refining	Oils
Gulf Security Oils	Gulf Refining	Oils
Gulfcrown Grease	Gulf Refining	Grease
Hannifin Lube A	Parker Hannifin	Lubricant
Hi-Lo MS No. 1		
Houghto-Safe 271 & 600	E.F. Houghton & Co.	MIL-H-27072 Water & Glycol Base
Houghto-Safe 416 & 500	E.F. Houghton & Co.	
Houghto-Safe 1010, 1055 & 1120	E.F. Houghton & Co.	MIL-H-19547 Phosphate Ester Base
Houghto-Safe 5040	E.F. Houghton & Co.	Water & Oil Emulsion
Hydro-Drive MIH50 & MIH10	E.F. Houghton & Co.	Petroleum Base
Hydrolube		Water & Ethylene Glycol
Hypoid Lubes		Hypoid Gear E.P. Lubes
Hy Kil No. 6		
Hyjet	Chevron	Phosphate Ester
Irus 902		
Isopar G	Exxon Corp.	Solvent
Karo	CPC International	Syrup
Kel-F Liquids	3M Co.	Fluorocarbon Liquids
Keystone No. 87HX	United Refining Co.	Grease
Klenzade		Sanitizer
Lehigh X1169 & X1170	Lehigh	
Lindoil		Phosphate Ester Hydraulic Fluid
Lindol	Stauffer Chemical Co.	F.R. Plasticizer
Liqui-Moly	Lockney Co.	Oils & Greases
Lysol	National Laboratories	Cleanser
Master Kill		
MCS 312, 352 & 463	Monsanto	Jet Lubricant

NAME	MANUFACTURER	PRODUCT
MIL-L-2104B		Motor Oil SAE 10W
MIL-S-3136B Type I		ASTM Reference Fuel A
MIL-S-3136B Type III		ASTM Reference Fuel B
MIL-H-5606		High Energy Fuel A & J-43
MIL-J-5624F		JP4 & JP5 Jet Fuel
MIL-L-7808C		Dibasic Ester
MIL-L-7808E	Braco 880D	
MIL-L-7808E	Stauffer Jet I	
MIL-L-7808G		Ester Blend Oil
MIL-C-8188C		Diester Fluid
MIL-H-8446B		MLO-8515 Silicone Fluid
MIL-L-14107B		Silicate Ester Oil
MIL-L-17672B		Turbine Oil
MIL-L-19457B		F.R. Fluid, Phosphate Ester
MIL-L-23699		Neopentyl Ester Oil
MIL-G-25013D		Silicone Grease
MIL-F-25558		RJ1 Ram Jet Fuel Petroleum Based
MIL-F-25558B		RJ1 Ram Jet Fuel Petroleum Based
MIL-F-25576C		RP1 Rocket Fuel Petroleum Based
MIL-F-25656		JP6 Jet Fuel
MIL-L-46000A		Diester Oil
MIL-S-81087A		Chlorinated Phenyl Silicone Fluid
Mineguard FR		F.R. Hydraulic Fluid
MLO-8515		MIL-H-8446B Silicone Fluid
Mobil Delvac 1100, 1110, 1120 & 1130	Mobil Co.	
Mobil HF	Mobil Co.	
Mobil Nyvac 20 & 30	Mobil Co.	
Mobil Therm 600	Mobil Co.	
Mobil Velocite C	Mobil Co.	
Mobil XRM 206A	Mobil Co.	
Mobil 24DTE	Mobil Co.	
Mobilgas WA200 ATF	Mobil Co.	Automatic Transmission Fluid
Mobiloil SAE20	Mobil Co.	SAE 20W Oil
Mobilux		
Mopar	Chrysler Corp.	Hydraulic Fluid
Navee		Deicing Fluid
Nitrana 2 & 3		
Noryl	General Electric Co.	Thermoplastic
Oakite Solutions	Oakite Product	Fluids
Oronite 8200	Chevron Chemical	Disiloxane Hydraulic Fluid
Oronite 8515	Chevron Chemical	Hydraulic Fluid 85% Disiloxane 15% Diester
OS45 & OS70	Monsanto	Silicate Ester Base
Par-Al-Ketone (Paralketone)	Black Bear Co. & Emco Chemical Co.	
Paraplex G62	Rohm & Haas Co.	
Parapoid 10-C	Exxon Chemical Co.	Hypoid Gear Oil
Parker O-Lube	Parker Hannifin	Grease
Parker Super O-Lube	Parker Hannifin	Silicone Grease
Penda Oil		Dibasic Ester Oil
Polyol Ester		Hydraulic Fluid
Prestone	Union Carbide	Antifreeze
PRL	Rohm & Haas Co.	High Temperature Hydraulic Oil
Purina	Ralston Purina	
Pydraul F9 & 150	Monsanto	Aryl Phosphate Ester Hydraulic Fluid
Pyranol	General Electric Co.	Chlorinated Transformer Oil
Pyrex	Corning Glass Works	High Temp. & Low Expansion Glass
Pyroguard	Mobil Oil	F.R. Oils
Pyrolube	Kano Laboratories	High Temperature Lubes
Red Line 100	Union Oil Co. Of California	Oil
Richfield A & B	Richfield Oil Co.	Weed Killer
RJ-1 (MIL-F-25558B)		Ram Jet Fuel Petroleum Based

TRADE NAMES APPEARING IN THIS PUBLICATION

NAME	MANUFACTURER	PRODUCT
RP-1 (MIL-F-25576A)		Rocket Fuel Petroleum Based
Sanitizer 160	Monsanto	F.R Additive
Santosafe 300	Monsanto	
Separan NP-10	Dow Chemical	Flocculant
SF 96, SF 1147, SF 1153, SF 1154	General Electric	Silicone Fluid
Shell Alvania Grease No. 2	Shell Oil Co.	Grease
Shell Carnea 19 & 29	Shell Oil Co.	
Shell DD	Shell Oil Co.	
Shell Diala	Shell Oil Co.	
Shell Irus 905	Shell Oil Co.	F.R Hydraulic Fluid
Shell LO Hydrax 27 & 29	Shell Oil Co.	
Shell Macome 72	Shell Oil Co.	
Shell Tellus 22 & 23	Shell Oil Co.	F.R Hydraulic Fluid Petro. Based
Shell Turbine Oil 307	Shell Oil Co.	Turbine Oil
Shell UMF	Shell Oil Co.	
Shell 3XF Mine Fluid	Shell Oil Co.	F.R Hydraulic Fluid
Skelly Solvent B, C & E	Getty Refining	Solvent
Skydrol 500	Monsanto	Isooctyl Diphenyl Phosphate Hyd. Fluid
Skydrol 7000	Monsanto	Hydraulic Fluid
Socony Mobil Type A	Mobil Oil Co.	Transmission Fluid
Socony Vacuum AMV AC781	Mobil Oil	Grease
Socony PD959B	Mobil Oil	
Solvasol 1, 2, 3, 73 & 74		
Standard Oil Mobilube GX90-EP	Mobil Oil	Gear Lube
Stauffer 7700	Stauffer Chemical Co.	
Sunoco All Purpose Grease	Sunmark Industries	A.P. Grease
Sunoco SAE 10	Sunmark Industries	
Sunoco 3661	Sunmark Industries	
Sunoco XS-820	Sunmark Industries	E.P. Lubricant
Sunsafe	Sun Refining	F.R. Hydraulic Fluid
Supershell Gas	Shell Oil Co.	Gasoline
Swan Finch E.P. Lube	Swan Finch	E.P. Lubricant
Swan Finch Hypoid 90	Swan Finch	Hypoid Gear E.P. Lubricant
Tellus	Shell Oil Co.	F.R. Hydraulic Fluid
Texaco Capell A & AA	Texaco	
Texaco Meropa 220 No Lead	Texaco	Gasoline
Texaco Regal B	Texaco	
Texaco Uni-Temp Grease	Texaco	Grease
Texaco 3450	Texaco	Rear Axel Oil
Texamatic A	Texaco	Automatic Transmission Fluid
Texamatic 1581, 3401, 3525 & 3528	Texaco	Transmission Fluid
Texas 1500 Oil		
Therminol 44, 45, 60, 66 & VP-1		H.D. Concentrate
Thiokol TP-90B & TP-95		
Tidewater Multigear 140		E.P. Lubricant
Tidewater Oil, Beedol		Oil
Turbine Oil No. 15	Exxon Corp.	Turbo Oil 15
Ucon Hydrolube J-4	Union Carbide	
Ucon Lubricant 135, 285, 300X, 625 & 1145	Union Carbide	
Ucon Lubricant 50-HB55, HB100, HB260, HB660	Union Carbide	
Ucon Oil LB65, LB385 & 400X	Union Carbide	
Ucon Oil 50-HB280X	Union Carbide	Polyacrylon Glycol Derivative
Univis 40 Hydraulic Fluid	Exxon Corp.	Hydraulic Fluid
Univis J43	Exxon Corp.	MIL-H-5606 F.R. Oil
Univolt No. 35	Exxon Corp.	Mineral Oil, Transformer Oil
Vasoline	Chesebrough-Ponds, Inc.	Petroleum Jelly
Versilube F44, F50 & F55	General Electric	Silicone Hydraulic Fluid
VV-H-910		Glycol Brake Fluid
Wagner 21B	Wagner Div. Mc Graw Edison	Brake Fluid
Wemco C		Transformer Oil

Compass Publications

P.O. Box 1275
La Jolla, Ca 92038-1275 USA
Tel (619) 551-9240
Fax (619) 551-9340

Chemical Resistance Guide for Elastomers II

20% NEW DATA!

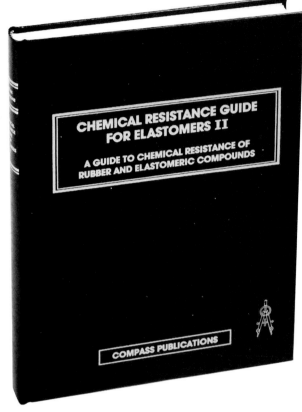

Data for more than 48,800 combinations of corrodents vs. rubber and other elastometric compounds and representing 20% new data.

Now you can avoid having your sales contacts look elsewhere while you consider the chemical compatibility of your product.

RUBBER AND ELASTOMERS INCLUDED

Chlorinated Polyethylene	CM	
Chlorosulfonated Polyethylene	CSM	® Hypalon
Copolyester TPE		® Hytrel
Copolymer TPO		® Norprene
Copolymer TPO		® Santoprene
Epichlorohydrin	CO, ECO	® Hydrin
Ethylene Acrylic	EA	® Vamac
Ethylene Propylene	EP, EPDM	
Ethylene Vinyl Acetate	EVA	
Fluorinated Copolymer	FKM	® Aflas
Fluorinated Copolymer	FXM	® Fluoraz
Fluorinated Ethylene Propylene	FEP	® Teflon
Fluorocarbon	FKM	® Viton
Fluorosilicone	FVMQ	
Hydrogenated Nitrile	HNBR	
Isobutylene Isoprene	IIR	Butyl
Linear Copolymer		Polyallomer
Low Density Polyethylene	LDPE	
Modified Polyethylene	AQP	® Aeroquip
Nitrile	NBR	Buna N
Perfluorinated Elastomer	FFKM	® Chemraz
Perfluorinated Elastomer	FFKM	® Kalrez
Polyacrylate	ACM	
Polyamide		Nylon 11
Polyamide		Nylon 12
Polybutadiene	BR	Butadiene
Polychloroprene	CR	Neoprene
Polyisoprene, Natural	NR	Natural Rubber
Polyisoprene, Synthetic	IR	SyntheticRubber
Polysulfide	T	
Polyvinyl Chloride, Flexible	PVC	
Polyurethane	AU, EU	
Silicone	VMQ	
Styrene Butadiene	SBR	Buna S
Styrenic TPE		® C-Flex

Features

• 3000+ liquid or dry chemicals, gases, lubricants, household fluids, foods and other environments are covered.

• 500+ chemical trade names are covered.
• 1000+ synonyms of covered chemicals, gases etc. indexed to page numbers.

• Corrodents are listed in alphabetical order.

• Data are presented in symbolic format (A, B, C, NR) with some data at specific rates of deterioration, time and temperature. A sample page is shown for clarity.

• Where known chemical resistance varies with concentration and temperature, data is presented in decending order of concentration and temperature.

• A quick reference summary of chemical resistance for each elastomer compound is provided.

• Mechanical, physical and electrical properties data for each elastomer is provided.

• Printed on semigloss 50 pound plastic coated bond paper to last several years of use.

• Case bound hardcover.

• 444 pages, 8 1/2 x 11 page sizes.

	STYRENE BUTADIENE SBR	POLYACRYLATE ACM	POLYURETHANE AU, EU	ISOBUTYLENE ISOPRENE IIR	POLYBUTADIENE BR	®AEROQUIP AQP
POTASSIUM MURIATE	AB to 70°F		A to 70°F	A to 70°F		
POTASSIUM NITRATE	A to 70°F	A to 70°F	A to 100% to 70°F	A to 70°F AB any conc to 185°F	A to 70°F	
POTASSIUM NITRITE	AB to 70°F		A to 70°F	A to 70°F		
POTASSIUM OXIDE	AB to 70°F			A to 70°F AB any conc to 185°F		
POTASSIUM PERBORATE	AB to 70°F			AB to 70°F		
POTASSIUM PERCHLORATE	AB to 70°F			AB to 70°F		
POTASSIUM PERFLUOROACETATE	AB to 70°F			AB to 70°F		
POTASSIUM PERMANGANATE	AB to 70°F		B 100% at 70°F NR 5% at 70°F	A to 70°F AB any conc to 130°F		
POTASSIUM PERSULFATE	AB to 70°F		AB to 70°F	AB to 70°F		
POTASSIUM PHOSPHATE	A to 70°F	NR at 70°F	B/NR at 70°F	AB to 70°F		
POTASSIUM SALTS	A to 70°F	A to 70°F	A to 70°F	A to 70°F	A to 70°F	
POTASSIUM SILICATE	A to 70°F	A to 70°F	A to 70°F	A to 70°F		
POTASSIUM SULFATE	B at 70°F	NR at 70°F	A to 100% to 70°F	A to 70°F AB any conc to 185°F	B at 70°F	AB to 70°F
POTASSIUM SULFIDE	AB to 70°F		A to 70°F			
POTASSIUM SULFITE	B at 70°F	NR at 70°F	A to 70°F	A to 70°F AB any conc to 180°F	A to 70°F	
POTASSIUM SULPHATE	B at 70°F	NR at 70°F	A to 70°F	A to 70°F	A to 70°F	
POTASSIUM THIOSULPHATE	AB to 70°F		AB to 70°F	AB to 70°F		
POTASSIUM TRIPHOSPHATE	A to 70°F	NR at 70°F	NR at 70°F	AB to 70°F		
POTATO OIL			C at 70°F	A to 70°F AB any conc to 180°F		
POTATO SPIRIT			C at 70°F	A to 70°F AB any conc to 180°F		
PRESTONE ANTIFREEZE	A to 70°F	NR at 70°F	NR at 70°F	A to 70°F	A to 70°F	AB to 70°F
PRL-HIGH TEMP. HYDRAULIC OIL	NR at 70°F	A to 70°F	B at 70°F	NR at 70°F	NR at 70°F	
PRODUCER GAS	NR at 70°F	B at 70°F	A to 70°F	NR at 70°F	NR at 70°F	
PROPANE	NR at 70°F	A/NR at 70°F	AC to 70°F	NR at 70°F	NR at 70°F	
PROPANE PROPIONITRILE	NR at 70°F	A to 70°F	NR at 70°F	NR at 70°F	NR at 70°F	
PROPANEDIAMINE				A to 70°F		
PROPANEDIOL	A to 70°F	NR at 70°F	AB to 70°F	AC to 70°F		
PROPANETRIOL	A to 70°F	NR at 70°F	A/NR at 70°F	A to 70°F AB any conc to 150°F	A to 70°F	AB to 70°F
PROPANOL	A to 250°F	NR at 70°F	NR at 70°F	A to 70°F AB any conc to 120°F	A to 70°F	
2-PROPANONE	BC at 70°F	NR at 70°F	NR at 70°F	A to 70°F AB any conc to 150°F	NR at 70°F	AB to 70°F
PROPELLER OIL						
PROPENOL				A to 70°F		
PROPENE	NR at 70°F	NR at 70°F	NR at 70°F	NR at 70°F	NR at 70°F	
PROPENE NITRILE				NR at 70°F		
PROPENE OXIDE				B at 70°F		
PROPENEOL				A to 70°F		
PROPENYL HYDRATE			A to 70°F	A to 70°F		
PROPENYL ANISOLE				NR at 70°F		
PROPION ALDEHYDE	NR at 70°F	NR at 70°F	NR at 70°F	A to 70°F		
PROPIONIC ACID	NR at 70°F	NR at 70°F	NR at 70°F	A to 70°F		
PROPIONITRILE				A to 70°F		
PROPYL ACETATE	NR at 70°F	NR at 70°F	NR at 70°F	AB to 80°F	NR at 70°F	
n-PROPYL ACETATE	NR at 70°F	NR at 70°F	NR at 70°F	AB to 70°F	NR at 70°F	
PROPYL ACETONE				A to 70°F AB any conc to 150°F		

C 303

Chemical Resistance Guide for Elastomers II

Chemical Resistance Guide for Metals & Alloys

features & specifications

• 963 liquid or dry chemicals, gases, lubricants, household fluids, foods, atmospheres, and other environments are covered.

• 70 chemical trade names covered.

• 500 synonyms of covered chemicals, gases etc. indexed to page numbers.

• Corrodents listed in alphabetical order.

• Data are presented in symbolic format (A, B, C, NR). A sample page is shown for clarity (figure 1).

• Where known chemical resistance varies with concentration and temperature, data is presented in decending order of concentration and temperature.

• Mechanical, physical and electrical properties data for each metal is provided.

• A flex thumb index is provided at the right hand margin of the right hand pages to facilitate rapid access to the desired data.

• An electromotive or galvanic series list covering 120 metals alloys and carbon is included.

• Printed on semigloss, 70 pound, plastic coated bond paper to last several years.

• Machinability ratings for most metals, including some specific S.F.M. rates.

• Creep or stress relaxation rates at various levels of stress, temperature and time are included.

• Case bound hardcover.

• 436 pages, 8 1/2 x 11 page size.

what's new!

Spotlighting METALS & ALLOYS only.

72 METALS & ALLOYS compared to 24 in the previous Compass Corrosion Guide II, 800% NEW DATA.

963 chemical or harsh environments – up from 560.

More categories covered in the mechanical & physical properties section.

Twice as many synonyms listed to help you find covered combinations.

Expanded galvanic series now covers 120 metals.

data for 29,000 combinations of corrodents vs. metals, metal alloys and carbon.

Figure 1. Sample Page

Corrodent	STAINLESS STEEL 18-8 301 302	STAINLESS STEEL 303	STAINLESS STEEL 304	STAINLESS STEEL 305	STAINLESS STEEL 309 310 312	STELLITE 1 6 12
ACETALDEHYDE	A to 70°F		A to 100% to 200°F / A to 75% to 239°F / A to 35% to 410°F	A to 70°F		A to 70°F
ACETAMIDE	B at 70°F	B at 70°F	B at 70°F	B at 70°F		
ACETATE ACID, CRUDE						
ACETATE ACID, PURE	B 100% at 70°F / AB 20% to 70°F	B 100% at 70°F / AB 20% to 70°F	B 100% at 70°F / AB 20% to 70°F	B 100% at 70°F / AB 20% to 70°F		
ACETATE SOLVENTS, CRUDE	A to 70°F	A to 70°F	A to 70°F	A to 70°F		
ACETATE SOLVENTS, PURE	C at 70°F		B to 70°F			
ACETIC ACID, CRUDE						
ACETIC ACID, AIR FREE	AB glacial at 70°F / A to 100% to 70°F / B 99.7% at 248°F / NR 70% at 242°F / NR 50-70% boiling / NR 47% at 572°F / AB 20% boiling / NR 10% boiling / C 2.2% boiling / A 1% to boiling	A 20 & 100% to 180°F / AB to 99% to 120°F / AB to 15% to 165°F / NR 5-100% boiling / BC 50% at 180°F	AB 95-100% to 230°F / A/NR 90% 250-390°F / NR 95-100% 300-370°F / C 80-98% 255-400°F / C 80-98% 255°F / A to 95% to 210°F / A to 70% to 210°F / B 40-70% 255-390°F / B 40-50% 255-390°F / B 30-40% 210-345°F / C 20-28% 285-345°F / A 10% to 345°F	AB to 100% to 70°F / AB 100% to 180°F / NR 80% to 400°F / NR 100% at 180°F / BC 50% at 180°F / A 6% to 70°F	C 100% at 70°F / NR 50-100% 300°F / A 10% at 200°F	
ACETYL CHLORIDE	BC 70°F to boiling					
ACETYLENE	A to 70°F	A to 400°F	AC to 70°F	A to 70°F		
ACETYLENE TETRACHLORIDE	A to 70°F	A to 80°F	A 90-100% to 70°F / AB 100% to 600°F	A to 70°F		
ACRYLIC ACID			A crude to 120°F pits / B vapors 185-220°F			
ACRYLONITRILE	A to 190°F / NR at 25°F / NR not vapor	A to 190°F / NR at 25°F / NR not vapor	A 90% to 77°F / A 45% to 150°F	A to 190°F / NR at 25°F		

Legend:
A < .002 ipy (< .05 mm/y)
B < .020 ipy (< .50 mm/y)
C < .050 ipy (< 1.27 mm/y)
NR > .050 ipy, per year or explosive; not recommended

72 Metals & Alloys Covered

Metal & Alloy	Chemical Entries
Aluminum, 3003	(838)
Aluminum, 5052	(125)
Aluminum, B356	(483)
Aluminum Bronze, C613, C614, C952, C953	(475)
Aluminum Nickel Bronze, C630, C955, C958	(177)
Beryllium Copper, C170-C176	(103)
Beryllium Copper-Nickel, C717, C966	(73)
Brass, Admiralty, C443-C445	(29)
Brass, Cartridge, C240-C260	(224)
Brass, Leaded, C345-C377	(458)
Brass, Leaded Red, C836	(609)
Brass, Muntz Metal/Naval/Yellow, C268, C280, C462	(459)
Brass, Undefined	(189)
Bronze, Phosphor, C505, C510, C521, C524	(430)
Bronze, Silicon, C651, C655, C973	(566)
Bronze, Undefined	(155)
Carbon, Nickel Impregnated	(738)
Carbon, Resin Impregnated	(515)
Cast Iron, Ductile, 60-40-18, 60-45-12	(380)
Cast Iron, Ductile, Nickel Plated, A-536/B-320	(489)
Cast Iron, Grey	(300)
Cast Iron, High Nickel, NI-RESIST®	(258)
Cast Iron, High Silicon, DURIRON®	(206)
Cast Iron, 3% Nickel	(721)
Cast Iron, Undefined Grey	(190)
Columbium (Niobium)	(735)
Copper, C102-C110	(131)
Copper Nickel, 55-18, C770/WAUKESHA	(429)
Copper Nickel, 70-30, C715, C964	(413)
Copper Nickel, 90-10, C706, C962	(209)
Galvinized Iron & Steel (Zinc Coated)	(531)
HASTELLOY® B & B-2, UNS #N10665	(671)
HASTELLOY® C & C-276, UNS #N10276	(241)
HASTELLOY® C-22, UNS #N06022	(62)
HASTELLOY® D	(121)
ILLIUM® G	(339)
INCOLO® 800	(416)
INCOLOY® 825	(465)
INCONEL® 600	(222)
INCONEL® X-750/WAUKESHA® 88	(346)
Lead, (Plumbum)	(206)
Magnesium	(101)
Molybdenum	(663)
Monel®	(473)
Nickel	(88)
Nickel Silver 20%, C976/WAUKESHA 118	(352)
Silver, UNS #P07001	(707)
Stainless Steel, 18-8, 301, 302	(681)
Stainless Steel, 303	(842)
Stainless Steel, 304	(402)
Stainless Steel, 305	(70)
Stainless Steel, 309, 310, 312	(936)
Stainless Steel, 316, 317	(192)
Stainless Steel, 321	(338)
Stainless Steel, 347, 348	(729)
Stainless Steel, 403, 410	(688)
Stainless Steel, 416, 420	(506)
Stainless Steel, 430	(57)
Stainless Steel, 431, 436	(626)
Stainless Steel, 440A, 440B, 440C	(225)
Stainless Steel, 446	(410)
Stainless Steel, 17-4PH	(232)
Stainless Steel, CF-8M Cast	(729)
Stainless Steel, CARPENTER® 20Cb-3	(829)
Steel, Low Carbon, 1010, 1045	(247)
Steel, Free Machining, 12L14	(555)
Steel, Hardenable, 1075, 1095	(183)
STELLITE® 1, 6, 12	(483)
Tantalum	(96)
Titanium	(209)
Zinc	
Zirconium	